"As challenges of inequality, unsustainability, and insecurity ramp up across the world, the need for hard-hitting gender analysis and action has never been greater. Bringing together top scholars and practitioners from across the world, this collection provides an extraordinarily comprehensive overview of the fast-changing gender and development field past and present, and a vital set of signposts to a more gender-equal future as development moves into a post-2015 era."

Professor Melissa Leach, Director of the Institute of Development
Studies, University of Sussex, UK

"This Handbook is an invaluable and accessibly written resource for students, scholars, and practitioners engaged with international development and its gendered effects. The multi-national contributors draw on first-hand experiences in diverse environmental, economic, and political settings. They address scales from the body and the local to the national and international, highlighting the challenges of survival and approaches that can foster change and enhance women's well-being."

Professor Janice Monk, Research Professor of Geography and
Development, University of Arizona, USA

"*The Routledge Handbook of Gender and Development* is timely, thoughtful, and historically grounded. Bringing together an extraordinary group of academics, practitioners, and professionals, it revisits past development approaches and contemplates futuristic perspectives for the study of gender and development globally. A must-read."

Amal Hassan Fadlalla, Associate Professor in the Departments of
Women's Studies, Anthropology, and Afroamerican and African
Studies, University of Michigan, USA

THE ROUTLEDGE HANDBOOK OF GENDER AND DEVELOPMENT

The Routledge Handbook of Gender and Development provides a comprehensive statement and reference point for gender and development policymaking and practice in an international and multidisciplinary context. Specifically, it provides critical reviews and appraisals of the current state of gender and development and considers future trends. It includes theoretical and practical approaches as well as empirical studies. The international reach and scope of the Handbook and the contributors' experiences allow engagement with and reflection upon these bridging and linking themes, as well as an examination of the politics and policy of how we think about and practice gender and development.

Organized into eight inter-related sections, the *Handbook* contains more than 50 contributions from leading scholars, looking at conceptual and theoretical approaches, environmental resources, poverty and families, women and health-related services, migration and mobility, the effect of civil and international conflict, and international economies and development. This Handbook provides a wealth of interdisciplinary information and will appeal to students and practitioners in Geography, Development Studies, Gender Studies, and related disciplines.

Anne Coles is a Research Associate at the International Gender Studies Centre, Oxford University, UK. She was previously a senior social development adviser in Britain's Department for International Development and has chaired two development NGOs. Her research interests include migration, people's responses to harsh environments, and public health. Recent publications include *Gender, Water and Development*, *Gender and Family among Transnational Professionals* (as co-editor and contributor), and *Windtower* (2007, reprinted 2009).

Leslie Gray is a geographer and Executive Director of the Environmental Studies Institute at Santa Clara University, USA. Her current research considers agrarian and environmental change in Burkina Faso and food justice in California. She has published articles on environmental policy, land degradation, and women's access to resources in Burkina Faso and Sudan. This research has been funded by the National Science Foundation, Fulbright/IIE and the Social Science Research Council.

Janet Momsen has taught at universities in the UK, Canada, Brazil, Switzerland, the Netherlands, Singapore, and the USA. She is Emerita Professor of Geography at the University of California, Davis, USA and was a Board member of AWID. She is currently a Senior Research Associate in the Oxford University Centre for the Environment, a Research Associate in the International Gender Studies Centre at Oxford University, and a Trustee of the development NGO, INTRAC. She has published over 170 articles in refereed journals and chapters in books and authored or edited 18 books.

THE ROUTLEDGE HANDBOOK OF GENDER AND DEVELOPMENT

Edited by
Anne Coles, Leslie Gray,
and Janet Momsen

Routledge
Taylor & Francis Group

LONDON AND NEW YORK

First published 2015
by Routledge
2 Park Square, Milton Park, Abingdon, Oxon OX14 4RN

and by Routledge
52 Vanderbilt Avenue, New York, NY 10017, USA

First issued in paperback 2020

Routledge is an imprint of the Taylor & Francis Group, an informa business

British Library Cataloguing in Publication Data
A catalogue record for this book is available from the British Library

Library of Congress Cataloging-in-Publication Data
The Routledge handbook of gender and development / edited by
Anne Coles, Leslie Gray, and Janet Momsen.
pages cm
Includes bibliographical references and index.
1. Women in development. I. Coles, Anne. II. Gray, Leslie.
III. Momsen, Janet Henshall.
HQ1240.R68 2015
305.42—dc23
2014030172

ISBN 13: 978-0-367-58185-5 (pbk)
ISBN 13: 978-0-415-82908-3 (hbk)

Typeset in Bembo and Minion Pro
by Florence Production Ltd, Stoodleigh, Devon, UK

CONTENTS

Contents

Contents

Contents

Contents

ACKNOWLEDGEMENTS

We first wish to thank the contributors to this handbook. They have patiently put up with our repeated requests for revisions and the lengthy gestation period of the project. Some people were defeated by the precise requirements laid down by the publisher for the Handbook. Others were unable to complete their chapters at the last minute because of other commitments or illness. We thank everyone for their interest in our project.

We are also grateful to our editor Andrew Mould for persuading us to undertake the Handbook, and for later relaxing the format for contributors. Sarah Gilkes, his editorial assistant, has been very supportive and helpful.

The three editors have worked together very smoothly and efficiently despite the eight-hour time difference between England and California. Anne and Janet wish to thank their colleagues at the International NGO Training and Research Centre (INTRAC) and at International Gender Studies (IGS) at Lady Margaret Hall, Oxford for their valuable cross-fertilization of ideas. Leslie would like to thank Anne and Janet for this incredible opportunity to work with so many gender and development specialists.

We all wish to thank our collective families: spouses, children and grandchildren for supporting us and providing us with joy when this project took over our lives.

CONTRIBUTORS

Seela Aladuwaka is an Assistant Professor of Geography at the Department of Criminal Justice and Social Sciences at Alabama State University, Montgomery, Alabama. Her PhD is from West Virginia University. Her publications mainly focus on microfinance, disaster management and development in Sri Lanka.

Salma Abbasi, a technologist and development practitioner, started her career in Silicon Valley in 1980 and has a PhD in Geography. She works in the field of security and development, focusing on sustainable economic, social and political empowerment for women and youth.

Christine Barrow is Professor Emerita, University of the West Indies (Barbados), where she served as Professor of Social Development, Head of Department and Deputy Principal. Publications include 12 books/monographs and articles on family and gender, child rights and development, HIV/AIDS and stigma, gender-based violence, sexual and reproductive rights.

Jo Boyden is based in the Department of International Development at the University of Oxford. She is an anthropologist and Director of Young Lives, a comparative, longitudinal study of childhood poverty. Her research has mainly focused on children and political violence, childhood poverty and child labor.

Sylvia Chant, FRSA, is Professor of Development Geography at the LSE. A specialist in Gender and Development, with particular interests in gender and poverty and female-headed households, Sylvia has conducted extensive primary research in Mexico, Costa Rica, Philippines and the Gambia.

Yi'En Cheng is a DPhil candidate at the School of Geography and the Environment, University of Oxford, studying youth transitions and higher education in Singapore. His research interest in the geographies of identities and change are organized around contemporary issues on education, youth and transnational migration.

Gina Crivello is an anthropologist working with Young Lives. Her research explores young people's everyday experiences of poverty and risk, mobility and migration, and gender and generational relations in transitions to adulthood. In Young Lives, she leads on a longitudinal qualitative study of childhood poverty in four developing countries.

Patricia Daley is an associate professor of human geography in the School of Geography and the Environment at the University of Oxford. She is the author of *Gender and Genocide in Burundi: The Search for Spaces of Peace in Central Africa*, published by James Currey.

Rebecca Elmhirst is a human geographer based at the University of Brighton, in the United Kingdom. Her research interests lie in feminist political ecology with a regional focus on Southeast Asia. She is co-editor of *Gender and Natural Resource Management: Livelihoods, Mobility and Interventions* (Earthscan, 2008).

Christine Eriksen is a social geographer at the Australian Centre for Cultural Environmental Research, University of Wollongong. Her research examines the role of local environmental knowledge in building resilience to natural disasters. She has a book on *Gender and Wildfire* (Routledge 2014).

Rosalind Eyben is a former development practitioner and Professorial Fellow at the Institute of Development Studies. Publications include *Relationships for Aid* (2006), *Feminists in Development Organizations* (2013) and *International Aid and the Making of a Better World: Reflexive Practice* (2014).

Anne-Meike Fechter is Senior Lecturer in Anthropology, University of Sussex. Her current research focuses on aid workers as mobile professionals; her books include *Inside the Everyday Lives of Development Workers* (2011) and *The Personal and the Professional in Aid Work* (2013).

Elena Fiddian-Qasmiyeh is Senior Research Officer at the Oxford Department of International Development and Research Fellow at Lady Margaret Hall, both at the University of Oxford. Her research explores the intersections between gender, religion and forced migration in the Middle East, North Africa and the Caribbean.

Bettina Fredrich is head of the social policy desk at the NGO Caritas Switzerland. She has a PhD in Human Geography and Gender Studies from the University of Bern (Switzerland) in which she focused on Swiss peace politics from a gender perspective.

Clara Greed is Professor Emerita of Inclusive Urban Planning in the Department of Planning and Architecture at the University of the West of England, Bristol, UK. *Inclusive Urban Design: Public Toilets* (Architectural Press. 2004) epitomizes her ongoing interests.

Don L. Hankins is an Associate Professor in the Department of Geography and Planning at California State University, Chico. Don is of Native American descent and is currently engaged in fire and land stewardship research involving indigenous communities in California and Australia.

Colette Harris is Reader in Gender and Development, Deptartment of Development Studies and a member of the Centre for Gender Studies, SOAS, University of London. Her research has focused on gender and conflict in Tajikistan and in Africa.

Patricia Holden is currently an independent social development and social policy consultant. From 1988 to 2008 she was Senior Social Development Adviser and Senior Gender and Rights Adviser at the UK Department for International Development (DFID). From 1997 to 2001 she was Senior Gender Adviser to the UK Mission to the UN in New York, and from 2001 to 2003, Senior Poverty Adviser at the International Labour Organization.

Alice J. Hovorka is Professor in the Department of Geography, Queen's University, Canada. Her research focuses on gender, environment and development issues with a particular emphasis on human–animal relations in Botswana.

Patricia L. Howard is Professor of Gender and Agriculture at Wageningen University in the Netherlands, and Honorary Professor in Biocultural Diversity and Ethnobotany at the University of Kent in the UK. She worked for a decade as project officer for FAO's Women and Rural Development Unit in Rome, and in Central America as a gender expert for the United Nations system.

Shirlena Huang is Associate Professor at the Department of Geography, National University of Singapore. Her research focuses mainly on issues at the intersection of migration and gender (with a focus on labor migration and transnational families within the Asia-Pacific region), as well as urbanization and heritage conservation (particularly in Singapore).

Shahnaz Huq-Hussain, PhD (London), is a Professor of Geography at the University of Dhaka, Bangladesh. Her research interests are gender, GIS and cartography and she published the *Gender Atlas of Bangladeshi* (with Amanat Khan and Janet Momsen) in 2006.

F. Munira Ismail is currently an Adjunct in the Economics-Government-History Department at Union County College, Cranford, NJ, USA. She holds a PhD and MA in Geography from the University of California-Davis. Her research interests include gender and the use of technology in the social sciences.

Susie Jacobs is a Reader in Comparative Sociology at Manchester Metropolitan University, UK. Recent publications include *Gender and Agrarian Reforms* (Routledge, 2010) and with C. Klesse, a Special Issue on 'Gender, Sexuality and Political Economy', *International Journal of Politics, Culture and Society*, vol. 2, 2014.

Elaine Jones is the Director of the WIEGO Global Trade Programme and an independent advisor in the fields of ethical and fair trade. Together with Carol Wills, she coordinated a multi-country series of case studies on women's economic empowerment with a focus on women producers engaged with global fair trade markets from 2009 to 2011.

Maria Jaschok is Director of the International Gender Studies Centre at Lady Margaret Hall, Oxford and a Research Fellow there. A recent publication with Shi Jingjun is *Women, Religion, and Space in China: Islamic Mosques & Daoist Temples, Catholic Convents and Chinese Virgins* (Routledge, 2011).

Michael Kevane is Associate Professor of Economics at Santa Clara University and Co-Director of Friends of African Village Libraries. Recent research evaluates impacts of libraries and reading programs. His books include *Women and Development in Africa: How Gender Works* (2014).

Contributors

Eleonore Kofman is Professor of Gender, Migration and Citizenship and Co-Director of the Social Policy Research Centre, Middlesex University, London. Her principal research interests include family migration, gender and skilled migration and the different sites and sectors of gendered migration and social reproduction.

Isabelle Kunze is doing her PhD at the University of Hanover, Germany on agrobiodiversity, and gender equity in South India. Isabelle holds a Master of Social Sciences from the University of Waikato, New Zealand in Feminist Geography.

Kuntala Lahiri-Dutt is a Senior Fellow at the Resource, Environment & Development group of Crawford School of Public Policy in the College of Asia and the Pacific in the Australian National University. She researches the formal and informal extractive industries through feminist and political economic lenses.

Nina Laurie is Professor of Development and Environment at Newcastle University. Her work explores how culture and identity intersect in diverse development settings, in knowledge co-production and professionalization. With Robert Andolina and Sarah Radcliffe she is co-author of *Indigenous Development*.

Margareta Amy Lelea, has a PhD in Geography from the University of California, Davis and is currently carrying out research in Africa for the German Institute for Tropical and Subtropical Agriculture, Faculty of Organic Agriculture, University of Kassel.

Li Sun obtained her doctorate from the University of Bielefeld in Germany and is currently a Postdoctoral Fellow at Delft University of Technology in the Netherlands. Her research interests focus on the roles of rural women in China.

Ragnhild Lund is Professor of Geography, specializing in Development Studies, at the Norwegian University of Science and Technology (NTNU). Her research interests are in theories of development and geography, gender and place, development-induced displacement, transnational feminism and women's activism.

Faranak Miraftab is Professor of Urban and Regional Planning at the University of Illinois, Urbana-Champaign. Her work crosses planning, geography and transnational studies empirically based in cities of Latin America, Africa and North America. Her forthcoming book concerns global production and social reproduction of migrant labor and how this makes for local development in the heartland US.

Virginia Morrow is Deputy Director of Young Lives. She is a sociologist and her research focuses on children's work in developed and developing countries, sociological approaches to the study of childhood and children's rights, and the ethics of social research with children. She leads on research ethics in Young Lives, and works closely with the qualitative research teams.

Ayesha Nibbe got her PhD at the University of California, Davis. She is an Assistant Professor of Anthropology at Hawaii Pacific University. Her current book project focuses on the sociopolitical effects of the international humanitarian aid intervention in the conflict zone of northern Uganda.

Barbara Parfitt CBE, PhD, ALBC, RGN. RM, FNP is Founding Principal of the Grameen Caledonian College of Nursing, Dhaka, and Director at Global Health Development Glasgow Caledonian University.

Jane L. Parpart is Research Professor in the Department of Conflict Resolution, Human Security and Global Governance at the University of Massachusetts, Boston. She writes on gender mainstreaming; urban life in Southern Africa; and conflict situations. She has just published 'Exploring the Transformative Potential of Gender Mainstreaming in International Development Institutions,' *Journal of International Development*, 2014.

Adriana Parra-Fox has degrees from the University of California-Davis in Community Development and Geography and currently teaches at the Universidad Tecnológica del Chocó where she also provides advice on research. She continues to work with microfinance and carries out participatory action research on youth empowerment and community health.

Meena Poudel works with the IOM in Kathmandu as a Nepalese social researcher-practitioner, experienced in development work and feminist activism with women vulnerable to or having experienced trafficking. She is the author of *Dealing with Hidden Issues: Social Rejection Experienced by Trafficked Women in Nepal*.

Sarah A. Radcliffe is Professor of Latin American Geography at the University of Cambridge. Her research focuses on the contradictions of and contestations over social hierarchy and development in the Andes. Recent publications include *Indigenous Development in the Andes: Culture, Power and Transnationalism* (Duke University Press, 2009).

Parvati Raghuram is Reader in Geography at the Open University. She has published widely on gender, migration and development. She has also been exploring the use of 'care' as a concept in social policy, postcolonial theory and feminist ethics.

Diane Richardson is Professor of Sociology at Newcastle University, with a particular focus on sexual and gender minorities' rights. Recent books include: *Intersections Between Feminist and Queer Theory*, co-edited with Mark Casey and Janice McLaughlin; *Sexuality, Equality and Diversity*, co-authored with Surya Monro.

Vidyamali Samarasinghe, PhD, is Professor in the International Development Program at American University, Washington, DC. Her research interests include poverty and rural areas in Sri Lanka. Her book *Female Sex Trafficking in Asia* (Routledge, 2008) examines human trafficking in Nepal, Cambodia and the Philippines.

Carolin Schurr is a 'Society in Science – Branco Weiss Fellow' in Economic Geography at the University of Zürich researching the transnational economies of assisted reproduction in Mexico and Central America. Carolin holds a PhD in Human Geography and Gender Studies from the University of Bern (Switzerland).

Sally Smith is an independent research consultant, previously of the Institute of Development Studies at Sussex University. Her work is focused on ethical and fair trade, gender and poverty, and covers multiple sectors and geographical regions. She is a member of WIEGO and various advisory bodies for non-profit organizations.

Ines Smyth is Senior Gender Advisor at Oxfam GB. Her interests include gender, disasters and DRR, as well as the organizational change necessary to promote women rights in development. She is a feminist activist.

Anita Spring is Professor Emeritus in the Department of Anthropology and African Studies, University of Florida, and Director of the Sub-Saharan Africa Business Environment Report Project, She has been conducting a multi-country pilot study of African entrepreneurs, of which 22 percent are women.

Farhana Sultana is Associate Professor of Geography in the Maxwell School of Citizenship and Public Affairs of Syracuse University. Her research broadly engages issues of gender, environment and development, and particularly the politics of water governance in South Asia.

Margaret B. Swain, an anthropologist at University of California, Davis, researched tourism development among the Kuna of San Blas, Panama, and the Sani of Shilin, Yunnan. She co-edited *Gender/Tourism/Fun* (2002), and currently writes on feminist analysis of cosmopolitanisms in tourism.

Caroline Sweetman is editor of the journal *Gender & Development* published by Oxfam. She has a Master's in Gender Analysis in Development Studies from the University of East Anglia, and a PhD on Poverty and Women's Empowerment in Ethiopia from the University of Leeds.

Elizabeth Thomas-Hope is Professor Emerita of the University of the West Indies, where she was the James Seivright Moss-Solomon Professor of Environmental Management. Among her publications are *Solid Waste Management: Critical Issues for Developing Countries* (UWI Press, 1998) and *Environmental Management in the Caribbean: Policy and Practice* (UWI Press, 2013).

Rebecca Maria Torres is an Associate Professor in the Department of Geography and the Environment at the University of Texas at Austin. Her research interests include transnational migration, rural development, agrarian transformations and gender. She is co-editor of *Tourism and Agriculture* (Routledge 2011).

Janet G. Townsend is a Visiting Fellow at Newcastle University. She researches gender with non-governmental organizations in poorer countries. She is co-author of *Women and Power: Fighting Patriarchies and Poverty* (Zed Books, 1999).

Ann Varley is Professor of Human Geography and convenes the Gender Studies program at UCL (University College London). Her research has focused on housing and land in urban Mexico and she edits the *Bulletin of Latin American Research*.

Hannelore Verbrugge is a PhD student at the Institute for Anthropological Research in Africa, Faculty of Social Sciences, KU Leuven, Belgium. At the time of writing, she was conducting her field research on gold-mining and gender in southwest Tanzania.

Vu Minh Hai is the Senior Technical Adviser for Building Resilience for Oxfam in Vietnam. She is involved with approaches to gender integration into disaster risk reduction, climate change adaptation and resilience-building at a local level and advocacy at the national level.

Meryl J. Williams, PhD, is a fishery and environmental scientist with experience in leading national and international research institutes. For almost 20 years, through research, conferences, encouraging key institutions and social media, she has been working to achieve greater awareness and action on gender in aquaculture and fisheries.

Carol Wills is a WIEGO consultant, supporting Elaine Jones in the coordination of Global Trade Programme women's empowerment projects. She is an Honorary Member of the World Fair Trade Organization and a non-executive Director of the Divine Chocolate company and Twin.

Steven Van Wolputte is Associate Professor of Social and Cultural Anthropology at KU Leuven, Belgium. His research interest is in political anthropology and in material/popular culture. Among his recent publications is a volume on beer in Africa, which he co-edited with Mattia Fumanti.

Brenda S.A. Yeoh is Professor (Provost's Chair), Department of Geography, National University of Singapore, as well as Research Leader of the Asian Migration Cluster at the Asia Research Institute, NUS. Her research interests focus on the geographies of migration and include key themes such as cosmopolitanism and mobility; gender, social reproduction and care migration; and cultural politics, family dynamics and international marriage migrants.

FIGURES

TABLES

BOXES

1

INTRODUCTION TO
THE HANDBOOK OF GENDER
AND DEVELOPMENT

In 2015 we shall commemorate the twentieth anniversary of the Beijing Platform for Action drawn up at the Fourth World Conference on Women held in Beijing in 1995. The Executive Director of UN Women, Phumzile Mlambo-Ngcuka, has said that, although the Beijing Platform has now been adopted by 189 governments, it is as yet unfulfilled: "Our goal is straightforward: renewed commitment, strengthened action and increased resources to realize gender equality, women's empowerment and the human rights of women and girls" (press release, May 22, 2014). At the same time we are moving towards achieving the Millennium Development Goals (MDGs) by 2015 and defining a new global development framework. Another benchmark is the 1979 Convention on the Elimination of All Forms of Discrimination against Women (CEDAW), now ratified by 188 states that have committed to advancing gender equality by confronting "any distinction, exclusion, or restriction made on the basis of sex which [impairs] the enjoyment or exercise by women . . . of human rights and fundamental freedoms." However, not all signature countries have fully implemented CEDAW and not all countries (most notably the United States) have ratified the Convention. Unfortunately ratification does not lead to an immediate reduction in gender discrimination but it does enforce regular reporting on progress.

A third benchmark relevant to gender and development is the twentieth anniversary of the 1993 World Conference of Human Rights held in Vienna. At this conference, after much lobbying from women's movements, women's rights were recognized as human rights and violence against women was seen as a human rights violation. Hina Julani, advocate of the Supreme Court of Pakistan, said that "any challenges to the affirmation that 'the human rights of women and of the girl child are an inalienable, integral and indivisible part of universal human rights' have been effectively marginalized" (Tolmay 2014). As recent events have shown, Pakistan still has far to go in enforcing these rights. At the individual level, full participation in rights such as sexual and reproductive health, freedom from gender-based violence, and equal access and control for men and women to land, requires agency and voice, sometimes also called empowerment (Klugman et al. 2014). Poverty is linked to multiple deprivations including a lack of personal agency. Consequently girls from poor families are more likely to marry before the age of 18 and to experience domestic violence. Ethnic minority status can further increase deprivation with three-quarters of the girls currently out of school globally belonging to ethnic minorities in their countries (Klugman et al. 2014). However, women are now living longer than men in all parts of the world and the participation of women in the paid labor force has been gradually increasing.

In the past two decades, global poverty has declined and the world reached the MDG of halving world poverty between 1990 and 2015 five years early. China accounts for most of the decline but the number of poor that had been rising in other world regions has fallen since 2000 in Latin America, Eastern Europe, and Central Asia. In Africa the number of poor people almost doubled between 1981 and 2005 but then declined so that for the first time in 2008 less than half of Africans were living below the poverty line (*Economist* 2012). Progress has also been made generally, with many people living longer and healthier lives, becoming educated, and having greater access to goods and services. Yet today in the world 1.2 billion people still live in extreme poverty, 774 million adults are illiterate, 783 million have no access to clean water, and 2.5 billion people lack adequate sanitation (Klugman et al. 2014).

The World Development Report for 2012 (World Bank 2011) sees gender equality as a core development objective that is also "smart economics." It goes on to say that greater gender equality can enhance productivity, raise development outcomes for the next generation, and, by making women's voices heard, can improve governance. The Food and Agriculture Organization (FAO) estimates that equalizing access to productive resources such as land, labor, fertilizer, and improved seeds for female and male farmers could raise output of food and fiber in developing countries by 2.5 to 4 percent. Evidence from several countries shows that increasing the share of income controlled by women, either through their own earnings or cash transfers, changes spending in ways that benefit children (World Bank 2011: 5). In Ghana female landownership is linked with higher food expenditure; in Brazil women's own labor income has a positive impact on the height of their daughters and in China it is linked to a higher survival rate for girls and more schooling for both boys and girls (World Bank 2011). Globalization has encouraged the spread of cheaper information and communication technologies (ICTs), helping to reshape attitudes among women and men about gender relations. The current acceptance of such economic benefits to gender equality may help to explain why in 2013, developing countries received $134.8 billion in aid, the highest amount ever (*Economist* 2014), while at the same time insisting on gender equality in aid programs. Donor aid had fallen in rich countries in 2011 and 2012 as these countries dealt with financial recession, but in 2013 Britain and four other countries met the United Nations target of 0.7 percent for aid as a share of gross national income. The USA remains the largest donor but gives less than 0.2 percent of its gross national income in foreign aid.

Development paradigms that influence the global development framework change over time and are influenced by aid agency policies, donor fashions, think tanks, private sector norms, and, occasionally, academics. The earliest development approaches saw women as mothers and focused on reducing fertility and improving mother and child health, assuming that economic growth would trickle down to women within the family. Boserup (1970), in her groundbreaking book, challenged this view showing that women did not always benefit as the male house-hold head's income increased and that women were increasingly being seen as backward and traditionalist. The rise of the women's movement in Western Europe and North America, and the 1975 UN International Year for Women and the International Women's Decade (1976–1985) led to the establishment of women's ministries in many countries and the institu-tionalization of Women in Development (WID) policies in governments, donor agencies, and NGOs. The aim of WID was to integrate women into economic development mainly by focusing on income generation projects for women. Many of these projects were only marginally successful as they assumed that women of the Global South had spare time to undertake new projects, and failed to take into account diversity among women. Gender and development

(GAD) approaches criticized WID for treating women as a homogeneous category. Instead GAD approaches saw women as agents of change and considered how development reshaped gender power relations. Proponents distinguished between "practical" gender interests, that is changes that would improve women's lives within their existing roles, and "strategic" gender interests that help to increase women's ability to take on new roles and to empower them (Molyneux 1985; Moser 1993). In the 1980s came the approach of gender and the environment (GED), which was at first based on ecofeminist views, especially those of Vandana Shiva (1989) and Carolyn Merchant (1992), who identified an essentialist link between women and the environment. This led to the belief that women were responsible for managing the environment and so gender and development programs were seen as also benefiting the environment. This approach was later critiqued by feminist political ecologists who looked at issues of power and politics in natural resource management through a gendered lens. This more materialist ecological feminism contextualized the relationship between environment and development in terms of access and control over natural resources, explicitly highlighting struggles over knowledge, power, and practice (see Elmhirst, Chapter 7 in this volume).

Millennium Goals

The Millennium Declaration signed at the United Nations Millennium Summit in 2000 set out the United Nations' goals for the next 15 years. These goals were based on the resolutions of various international conferences, most of which were organized by the United Nations during the 1990s. These goals were seen as targets that could be monitored. There were eight main goals, among which were: halve the proportion of people living in extreme poverty between 1990 and 2015; enroll all children in primary school by 2015; reduce infant and child mortality rates by two-thirds between 1990 and 2015; implement national strategies for sustainable development by 2005 and develop a global partnership for development.

Goal three was the main one to focus on gender issues aiming to empower women and eliminate gender disparities in primary and secondary education by 2005 and in all levels of education by 2015. Indicators measuring success in achieving this goal were the ratio of literate females to males aged 15 to 24 years of age; the share of women in paid employment in the non-agricultural sector; and the proportion of seats held by women in national parliaments. Other goals particularly relevant to women were MDG 5—to reduce maternal mortality rates by three-quarters between 1990 and 2015—and MDG 6—to provide access to all who need reproductive health services by 2015. The 2010 Millennium Development Goal Summit recognized the intrinsic importance of gender equality more clearly than in the earlier proposals of 2000. The meeting in 2010 concluded with the adoption of a global action plan to achieve the eight goals by 2015. At the Summit, a resolution was passed calling for gender parity in education and health, economic opportunities, and decision-making through gender mainstreaming in development policymaking. It was noted that MDG 3 and MDG 5 not only were development objectives in their own right but also served as "critical channels for achieving the other MDGs and reducing income and non-income poverty"(World Bank 2011: 4). However, the MDGs have been criticized for linking developing countries into a top down neoliberal market-led development approach. As Cornwall points out, the MDGs "have focused more attention on formal politics than on the myriad of other spaces that exist for decision making and influence . . . such as through [women's] movement building to . . . influence public opinion" (Cornwall and Edwards 2014: 14).

In the past decade, several stumbling blocks have emerged to a focus on gender and development. One arose from the intensified debate following the Paris Declaration (2005) and the Accra Agenda for Action (2008) about the effectiveness of development cooperation and the concentration on quantified measurable results (Sancar et al. 2012). Budget support to governments was promoted in the Paris Declaration and concepts of accountability and transparency were prevalent. Management principles of timeliness, efficiency, and profitability have encouraged large-scale global projects. Small-scale, grassroots qualitative studies that reveal gender relations, the living conditions of women, and the side-effects and long-term consequences of change are rarely funded.

Structure of the *Handbook*

Our *Handbook* attempts to address a wide range of gender issues from a variety of viewpoints. Some chapters focus on theoretical perspectives in the development of different subfields in the broader field of gender and development; others demonstrate the experiences of practitioners, consultants, and policymakers. Many chapter authors have track records as development professionals. Some are academics who have bridged the academic/practitioner divide with occasional consultancies for development organizations. All of our authors have long years of field experience that has influenced their thinking about gender and development issues. The bringing together of academics and professional practitioners is relatively rare and is therefore, we believe, a major strength of this *Handbook*.

The topics covered vary widely. Our chapters are split into eight broad parts that address the making of the field of gender and development, environmental resources, population, health and services, mobilities, conflict and post-conflict, economics, and development organizations. Our goal in the introductions to each of these eight parts is to frame the topics covered in the chapters. Many of the chapters are broad reviews of topics, while others are case studies of specific issues. In many cases, we have a general broad piece and a parallel more detailed case study. We combine short (3,000 words) and long (6,000 words) pieces. Although we have split this volume into different parts, there is significant overlap between many topics; the editors have attempted to cross-reference where appropriate. Our objective in this handbook is to tackle foundational issues as well as newer emerging topics in gender and development. This diversity of approaches will, we hope, give a nuanced and accessible view that informs both students and practitioners of gender and development. While the hope of this volume is to be as extensive as possible, we make no claim to be completely comprehensive.

Our goal has been to represent a wide range of geographic areas—while many of the chapters are set in regions that are generally understood to be "developing" regions, in places such as Africa, Asia, Latin America, and the Caribbean, we have chapters that consider indigenous peoples in developed countries and gender in post-communist countries. Many of the chapter authors are from the regions that they are writing about, working both as practitioners and academics.

This volume also has a wide-ranging environmental resources section. While many gender and development volumes do not address this element of gender, the editors—all geographers—believe that rights to environmental resources are extremely important to women across the world, but particularly in developing countries, where livelihoods often depend on direct access to and control over environmental resources, as agriculturalists, pastoralists, workers, nurturers, and consumers.

Methodologies

The writers of many of these chapters also make clear their personal positions, recognizing the importance of that well-worn catch phrase that "the personal is political" and therefore honesty may require that the author's perceptions are known to readers. Some are explicitly personal. Autobiography comes into its own most clearly in Lund's chapter, while Holden provides an overtly 'insider' perspective. Others simply make use of their lifetime of experience (Eyben). Townsend recognizes her contribution and limitations as an outsider. Ethnography, in the form of careful long-term fieldwork, is well represented by Li, Nibbe, Verbrugge and Van Wolputte, and Spring.

Most contributors have used mixed methods, both quantitative and qualitative, in field research. Qualitative field methods of structured and unstructured interviews, focus groups (Aladuwaka, Kunze and Momsen), participant observation (Aladuwaka, Ismail), participant appraisal workshops (Torres), and gender activity profiles (Aladuwaka) were undertaken. Life histories were found useful in migration studies (Lelea, Torres). Many interviewed key informants such as community leaders (Torres, Aladuwaka), bankers (Aladuwaka), NGO officials (Townsend et al.), and even brothel madams (Samarasinghe) as well as participants in the particular project under study. Lelea used snowball techniques to find suitable interviewees by getting additional participants by asking the first people interviewed for contacts. Ismail utilized geospatial techniques tracking the movements of participants using GPS.

Most of the case studies described are of small projects but some are bigger and undertaken over a longer period. The chapter by Boyden et al. describes part of the Young Lives project, which is a 15-year study of childhood poverty being carried out with 12,000 children in Ethiopia, India, Peru, and Vietnam. Spring describes her long-term project on African entrepreneurs in several sub-Saharan countries and, through interviews, is able to identify transnational networks of women entrepreneurs. Abbasi carried out both quantitative and qualitative research in Pakistan involving 768 questionnaires, 127 interviews, and 99 focus groups. Additionally she interviewed 39 elite women, of whom 19 were from Pakistan and 20 from seven other Muslim countries enabling both geographical and class implications to be drawn. She used her focus groups to look at variation in attitudes to ICTs between rural and urban individuals and those at different education levels. Many contributors including Aladuwaka, Huq-Hussain, Sultana, Ismail, Parra-Fox, Li, and Vu wrote chapters based in their home countries. Others such as Yeoh et al. and Samarasinghe wrote about the broader regions including their home countries.

The combination of methods used in connection with humanitarian aid and development projects are each rather different from those used in a purely research setting. In an emergency (whether manmade as in conflict or the result of a natural disaster) the quantitative tends to predominate. Answers to "How many?" and "Where?" are in the immediate, short term considered more important than answers to "Why?", "Who?", and "How?" In contrast, in development programs and projects, while quantitative data is needed to ensure coverage and representativeness, qualitative information throws light on people's norms, opinions, and behavior in relation to the project's intended aims. To return to the positionality mentioned earlier, Daley, in particular, but also Harris, questions the preconceptions and motives of aid players collecting information in emergency situations, raising issues of selectivity and objectivity.

Not all contributors describe their field methodologies explicitly, but backgrounds in empirical research ground the more theoretical chapters. They navigate multiple scales, both conceptually and empirically, in order to challenge conventional boundaries that situate systems and processes within local, regional, and global situations. They follow cross-border and cross-scalar connections with gendered perspectives to throw a new light on gender and development.

Contributors

We have 66 contributors who have written 47 original chapters. Two of the editors have also contributed to chapters. Our contributors come from 20 countries and are currently working and living in 17 different countries. They are a very mobile group with only 21 residing in their natal countries. Almost two-thirds live and work in the United Kingdom. They are based at universities ranging from Glasgow to Newcastle, Manchester Metropolitan, the Open University, Oxford and Cambridge, London, Brighton, and Sussex with two employed by Oxfam in Oxford. Some are on secondment overseas, some are retired from the UK Department for International Development, and one is an overseas postgraduate student. The second largest group work and live in the United States (12) but only three of these were born there and one of them is a Native American. We have European contributors from Germany, Switzerland, Norway, Denmark, and the Netherlands. From South America and the Caribbean we have contributions from Colombia, Barbados, and Jamaica. We have two Australian contributors and many from Asia.

The combination of academic training and development work found among our contributors has enhanced this *Handbook*. In terms of experience, our contributors range from graduate students in the final stages of their doctoral research and postdoctoral students to some who are retired but still active researchers. Disciplines include geography, anthropology, sociology, development studies, economics, marine biology, nursing, planning, and technology.

Editors

Anne Coles has combined work as a development professional with university teaching and research. She has lived and worked in Nigeria, Sudan, the United Arab Emirates, Egypt, and Jordan and has taught at universities in Sudan, Britain, Canada, and Australia. Her long-term research interest is on how people cope with uncertain and difficult environments. She has studied food and public health issues and migration (ranging from nomadism to expatriatism). Recently she has written on development organizations. Anne has undertaken more than 20 consultancies for international agencies and banks, national governments, international and local NGOs. She was five years with DFID, leading on gender as a senior social development adviser after Beijing (1995) and representing the UK in the OECD/DAC expert group on Women in Development. She has been a trustee of three development NGOs, chairing two of them. Apart from articles and book chapters, her publications include *Windtower*, and the co-edited collections *Gender and Family among Transnational Professionals* and *Gender, Water and Development*.

Janet Momsen is Emerita Professor of Geography at the University of California, Davis, USA. She is also a Senior Research Fellow at the Oxford University Centre for the Environment and a Research Fellow of International Gender Studies at Lady Margaret Hall, Oxford. She was a Board Member of the Association for Women's Rights in Development (AWID) and is currently a Board Member of the International NGO Training and Research Centre (INTRAC). She has taught at universities in the UK, USA, Canada, the Netherlands, Switzerland, and Brazil and has carried out field research in the Caribbean, Costa Rica, Mexico, Brazil, Bangladesh, and Hungary. She is author or editor of 18 books and has published many journal articles and chapters in books.

Leslie Gray is an Associate Professor in the Department of Environmental Studies and Sciences at Santa Clara University, USA. She is a geographer who teaches classes that emphasize global environment, development, and population issues. Her current research considers agrarian change issues in Burkina Faso and food systems in the United States. She has published numerous articles on environmental policy, land degradation, and women's access to resources in Africa. This research has been funded by the National Science Foundation, Fulbright/IIE and the Social Science Research Council. She has also done work for several international organizations, including CARE, Catholic Relief Services, UNDP, ILO, and the World Bank.

References

Boserup, E. (1970) *Women's Role in Economic Development*, New York: St Martin's Press.

Cornwall, A. and Edwards, J. (eds.) (2014) *Feminisms, Empowerment and Development: Changing Women's Lives*, London and New York: Zed Press.

Economist (2012) "Global poverty: a fall to cheer," March 3: 75–76.

—— (2014) "Aid," April 12: 97.

Klugman, J., Hammer, L., Twigg, S., Hasan, T., McCleary-Sills, J., and Santamaria Bonilla, A.J. (2014) *Voice and Agency: Empowering Women and Girls for Shared Prosperity*, Washington, DC: World Bank.

Merchant, C. (1992) *Radical Ecology: Global Issues and Local Experiences*, London and New York: Routledge.

Molyneux, M. (1985) "Mobilisation without emancipation? Women's interests, the state and revolution in Nicaragua," *Feminist Studies* 11(2): 227–254.

Moser, C.O.N.(1993) *Gender, Planning and Development: Theory, Practice and Training*, London and New York: Routledge.

Sancar, A., Bieri, S., Fankhauser, L., and Stolz, N. (eds.) (2012) *Added Value: Contributions to Gender Equitable Economic Development*, Bern: Swiss Agency for Development and Cooperation and Caritas.

Shiva, V. (1989) *Staying Alive: Women, Ecology and Development*, London: Zed Books.

Tolmay, S. (2014) *Vienna + 20: Some Advances and Setbacks for Women's Human Rights in Asia*, www.awid.org/eng/layout/set/print/News-Analysis/Friday-Files/Vienna-20-Some-Advances-and-Setbacks-for Women's-Human-Rights-in Asia, 07/02/2014 (accessed February 8, 2014).

World Bank (2011) *World Development Report 2012: Gender Equality and Development*, Washington, DC: World Bank.

PART I

The making of the field: concepts and case studies

2

INTRODUCTION
TO PART I

Although gender and development has been recognized as a separate field of study for less than half a century, it has developed several new strands since the last United Nations Conference on Women held in Beijing in 1995. The Beijing Plan for Action provided the initiatives for work on gender mainstreaming and masculinities, but global political changes in the past two decades have stimulated interest in postcolonialism, gender and religion, and feminist political ecology. In this section we look at five of these innovative directions and end with an autobiographical essay that shows how these changing trends have influenced one individual over her working life.

As Cornwall et al. (2007: 1) indicate "there are feminisms, not 'feminism', and 'development' covers a multitude of theoretical and political stances and a wide diversity of practices." This section elaborates on several of these relatively recent theoretical approaches. It includes chapters by academic geographers, a historian, and an INGO worker. All of them have done extensive fieldwork in many parts of the world and draw on these experiences to illustrate theoretical aspects of gender and development. Many of these conceptual analyses are picked up in other chapters later in the book.

Debates on masculinities began in the 1980s and highlighted concerns with hegemonic masculinity and its link with women's movements. The 1994 Cairo Conference on Population and Development was the first international gathering to challenge men to play their part in the fight for gender equality. Jane L. Parpart's chapter sees the concept's recent growth as rooted in the post-Beijing era after the Beijing Plan of Action of 1995 had stated the principle of men and women as partners in the development agenda. It called on men to support women and argued that women's concerns could only be addressed through shared responsibility (Cornwall et al. 2011). At first men were seen principally as impediments to women's progress because of their role as a source of violence against women and in the spread of HIV/AIDS. There has been less interest in men's role in the public sphere and how being a man culturally and socially privileges and empowers men, or in men's various sexualities. More recently a growing awareness of the often inferior educational achievements of boys when compared to girls has led to a policy response aimed at understanding male school-based deviance. Yet, as Parpart points out, this is mainly a concern for middle and upper income countries, while in poor countries boys are more likely to be able to access education than girls.

Caroline Sweetman's contribution is on gender mainstreaming. This became a central focus of the field following the 1995 United Nations Fourth World Conference on Women held in Beijing. The difficulties of this policy soon became clear when the DAC Working Party on Gender Equality reported that for its members "implementing a gender mainstreaming strategy to support gender equality goals is a longer and more difficult process than some had anticipated" (OECD/DAC 1996: 7). A study in South Africa found that, despite the post-apartheid gender equal constitution, gender mainstreaming was widely paid lip-service to but rarely implemented (Reppenaar-Joseph 2009). Many of the problems identified by Sweetman and academic colleagues are also consistently found in recent external evaluations commissioned by donor agencies. Sweetman's plea for donors to provide more support to women's grassroots organizations is timely and important but many donors, bilateral ones in particular, may be hesitant to engage with movements that may be looked at askance by the national governments concerned.

The next three chapters consider current global changes and their influence on development issues in relation to women: first, postcoloniality; second, religion; and third, environmental degradation, conservation, and climate change. Sarah Radcliffe argues that critical postcolonial and gender analyses provide insights into the nature of development and destabilize the dominant discourses of Northern activists. Feminism's interest in situated knowledges complements postcolonialism's focus on sources of power. She shows that postcolonialism highlights the diversity of gender, race, class, and location relations in the South. This chapter indicates that postcolonial feminists welcome a development policy premised on women's empowerment, yet this power is often embedded within colonialist assumptions.

Another major current issue is that of religion on gender and development with fundamentalist Islam increasingly limiting women's access to education and rights to work in many countries. Maria Jaschok indicates that we are now in an era of post-secular feminism and looks particularly at the role of Islam in legitimizing beliefs and policies that disempower women through the instrumentalization of religion. She stresses the vital importance of the specificities of local contexts and indigenous knowledge, indicating that religion does not shape women's condition in the same way everywhere.

Rebecca Elmhirst considers the place of feminist political ecology (FPE) as an influential subfield of gender and development. She sees its foundation in the book edited by Rocheleau et al. (1996), which treated gender as a critical variable in shaping resource access and control. Since then there have been new environmental challenges to the role of women in development. Increases in rural–urban and transnational linkages, greater and more widespread mobility and accelerating processes of environmental degradation (deforestation, desertification, climate change, and urbanization) have undermined traditional livelihoods. Women are often dependent for firewood, grazing rights, and gathering of plants on common lands that are becoming increasingly privatized (see the chapter in Part II by Kevane). Women's land rights are increasingly threatened reducing the productivity of female farmers (see Part II chapters by Kevane and Jacobs). Elmhirst sees these problems as reinvigorating FPE around four foci: analyses of gendered resource access and control; gendered subjectivity and power; relationships between human and non-human nature; and a new feminist ethics of environmental care.

The final essay in Part I is an autobiographical chapter by a Norwegian academic and development expert. This provides a strong counterpoint to the other chapters in this section, which are written by Anglo-American authors. The Norwegian Agency for Development Assistance (NORAD) and the Scandinavian aid programs in general were regarded as taking women/gender into consideration earlier than many other bilateral agencies. The impact of seminal works such as Boserup's *Women's Role in Economic Development* in 1970 and the concepts of practical and strategic interests inspired this early interest in the field. Ragnhild Lund explores the problems of combining academic research and working for an agency where one's loyalties are torn between the people at the grassroots and the organization providing one's salary. Her field research has long focused on Asia. The development of new ideas in gender and development and changes in Asian economies have led her to link globalization and related mobilities (Lund et al. 2014). Many decades of field research enable her to call for revisiting sites of earlier research and increasingly working with researchers from the Global South and thus contributing to reducing the Western lens and building transnational knowledge. Thus from her early experiences of working with the DAWN Network, which grew out of the rejection by Southern women of Northern feminist views at the 1975 UN Women's World Conference in Mexico City, to her recent focus on running long-term multi-nation research projects across Asia (Lund et al. 2014), Ragnhild Lund offers a personal view of the long trajectory of the field of gender and development.

References

Boserup, E. (1970) *Women's Role in Economic Development*, London: Allen and Unwin.

Cornwall, A., Harrison, E., and Whitehead, A. (eds.) (2007) *Feminisms in Development: Contradictions, Contestations and Challenges*, London: Zed Books.

Cornwall, A., Edstrom, J. and Greig, E. (eds.) (2011) *Men and Development: Politicizing Masculinities*, London: Zed Books.

Lund, R., Kusakabe, K., Mishra Panda, S., and Wang, Y. (2014) *Gender, Mobilities, and Livelihood Transformations: Comparing Indigenous People in China, India and Laos*, London: Routledge.

OECD/DAC (1996) *Progress Towards Gender Equality in the Perspective of Beijing +5: Beijing and the DAC Statement on Gender Equality*, Paris: Organisation for Economic Co-operation and Development.

Reppenaar-Joseph, T. (2009) *Mainstreaming Women in Development? A Gender Analysis of the United Nations Development Programme in South Africa*, unpublished PhD thesis, University of Stellenbosch, South Africa.

Rocheleau, D., Thomas-Slayter, B., and Wangari, E. (eds.) (1996) *Feminist Political Ecology: Global Issues and Local Experiences*, London and New York: Routledge.

3

MEN, MASCULINITIES, AND DEVELOPMENT

Jane L. Parpart[1]

Introduction

Men and masculinities only moved on to the development agenda in the 1990s. Initial efforts to enhance economic development in colonial and postcolonial nations and territories paid little attention either to women or inequality between the sexes. Development was regarded as an economic issue that, when done properly, would benefit all citizens. This orthodoxy only began to be challenged in the 1970s, when a few key studies discovered that women had very different experiences of development projects than men. Women gradually became a development issue, women in development (WID) programs emerged, and projects addressing women's issues slowly expanded. Men were generally ignored or regarded as impediments to these efforts.

However, as the limitations of WID programs became more apparent in the late 1980s, the focus began to shift from women to gender, with its emphasis on the social construction of norms and practices around masculinity and femininity and their impact on gendered opportunities and relations. Gender and development (GAD) programs began to explore the way attitudes and practices of men affected women's prosperity and positions. The Fourth International Meeting on the Status of Women held in Beijing in 1995 reinforced this growing concern, and the resulting *Plan of Action* placed men as well as women on the development agenda, calling for mainstreaming gender into all policies, programs, and institutions (see chapter by Sweetman in this volume). Men began to surface in development programs, but more often as impediments to women's advancement than as a factor in the gendered distribution of power and authority. Early projects focused on the role of men in the spread of HIV/AIDS, violence against women, and opposition to women's empowerment. Since the turn of the century, increasing attention has been paid to the developmental problems of men and boys, particularly their underperformance in schools, high levels of youth unemployment, involvement in crime, and the spread of HIV/AIDS. There has been much talk about a crisis of masculinity, of the need to change toxic masculine practices and to create sensitized men who will be allies in the struggle for gender equality. Less has been said about the material and social consequences of patriarchal privilege and its consequences for men and women around the world.

The chapter documents the introduction of men, boys, and masculinities into development policies and programs, first as development problems requiring new attitudes and practices, and later also as potential allies in efforts to achieve gender equality. It seeks to evaluate the transformative potential of these efforts, paying particular attention to the silences about gendered hierarchies of power, their implications for deep-seated resistance to gender equality, and the limitations of current approaches for achieving gender equality in an increasingly complex, unequal and gendered world.

Bringing men and boys into development

Attempts to organize and promote economic development in colonial and later postcolonial nations and territories generally assumed that women and men would benefit equally from efforts to improve the economies of the "Third World." The resurgence of feminism in the 1960s raised questions about women's inequality and inspired investigations into the impact of development programs and policies on women. Studies revealed very different and unequal consequences for women and men, and spurred the emergence of programs to address these inequalities. In the 1970s, WID programs began to emerge in key development institutions, and programs to address the developmental problems facing women became increasingly common. The first UN-sponsored World Conference on Women took place in Mexico in 1975, where the Decade of Women was declared (1976–1985) and the organization UNIFEM was established (later evolving into UN Women). In 1979, the United Nations passed the Convention on the Elimination of All Forms of Discrimination against Women (CEDAW). The focus centered on women and the role of men was rarely addressed.

The 1980 World Conference on Women in Copenhagen concluded that men should be more involved in efforts to improve women's position in societies around the world, but little emerged from these discussions. For the next 15 years, the role of men in promoting (or inhibiting) women's rights received little attention. During that time, the limitations of women-oriented development projects made for depressing reading. Women had entered employment in increasing numbers, yet rarely led important economic institutions. They were more active in politics, but still largely absent from positions of political power. Thus development programs and projects had done little to break male dominance over social, economic, and political power. The critiques of WID, particularly its failure to pay attention to the complex realities of poverty, race, ethnicity, and class for women living in the Global South, inspired a gradual shift to gender and development (GAD), with its emphasis on the power relations shaping the lives of poor women and men in the Global South (and North). The concept of gender focused on the socially constructed attitudes and practices associated with women, femininities, men, and masculinities and their impact on relations between men and women as well as the gendered distribution of power and influence in particular societies. Men thus entered into development discussions, albeit largely as impediments to women's empowerment rather than as solutions to gender inequality (Sweetman 2013; White 1997).

The 1995 Beijing conference reinforced this shift and the resulting *Plan of Action* declared that men and women should share power and responsibility and work in partnership towards gender equality, especially in the areas of education, socialization of children, childcare and housework, sexual health, gender-based violence, and the establishment of more equitable work–life balance for both sexes. The *Plan of Action* introduced gender mainstreaming (GM) as the primary mechanism for achieving gender equality in all institutions, programs, and policies in governments and non-governmental organizations (NGOs). GM became the vehicle of choice for ensuring the empowerment of women and gender equality around the world. The *Plan of Action* thus shifted the discussion of men's roles in gender and development to more concrete issues. Yet, while men were presented as essential to this undertaking, in practice the emphasis continued to focus on women's empowerment and the need to pay attention to differences among women, particularly around race, ethnicity, class, and age. GAD projects aimed to transfer resources and support to women in the Global South in order to enable them "to put their own agendas and priorities into action and challenge the current top-down, male-biased model of global development" (Sweetman 2013: 3). Programs for men generally focused on their role in this process rather than on the deeper issues facing men and boys, particularly the masculine norms associated with manly "success" and the difficulties of achieving these goals in an increasingly global, neoliberal world (Connell 2005; Cornwall et al. 2012).

The focus on women soon foundered as projects designed to empower women often challenged power relations between women and men, inadvertently fueling male hostility and raising tensions between the sexes that sometimes flared into violence at home and in communities. Male hostility to women's advancement troubled advocates of gender equality and women's empowerment, raising questions about men and masculinity. Leaving men out of efforts to improve women's lives and achieve gender equality became increasingly difficult to defend. Drawing on the expanding research on men and masculinities, particularly by Raewyn Connell (2005), some gender and development scholars and practitioners began to explore the impact of masculine norms and practices on the behavior of men, particularly hostility towards women-centered empowerment projects. As Connell points out, only a small minority of men are able to fulfill the expectations associated with hegemonic masculinity. The norms and practices associated with and expected of male elites, particularly the emphasis on masculine power, superior strength and intelligence, as well as the ability to protect and control women, children, and subordinate males, are out of reach for many men. Thus, while patriarchal ideals based on male power over females are regarded as a social ideal in most societies, this ideal has become increasingly difficult for many men to achieve. As Sweetman (2013: 4) points out, "for the majority of men, there is a level of anxiety around living up to ideals of masculinity." This anxiety has fueled a backlash against development projects that are seen as threatening masculine privilege, at any level, and raised concern about the need to understand and change the patriarchal attitudes underwriting this behavior.

This concern has intensified in the twenty-first century, which has witnessed an expansion of educational and employment opportunities for women, along with a global economic crisis that has undermined many traditionally male-dominated sources of employment such as construction, manufacturing, and investment. Young men in particular have found it difficult to live up to expected roles as breadwinners. Many (along with some women) have turned to illegal activities, including the drug trade and smuggling of persons and goods, in order to survive (see chapters by Townsend et al. and Samarasinghe in this volume). Armed conflicts have drawn young men (and some women) into battle, inuring them to violence and highlighting militarized masculine values (see chapter by Harris in this volume). The growing contradiction between masculine ideals as breadwinners and leaders and the realities on the ground are fueling resentment against more "successful" males and females, often leading to violence against the few people "failed" men can control, particularly the women and children in their households and communities (Catala et al. 2012).

In such an atmosphere, it is not surprising that many GAD projects aiming to empower women have encountered resistance from many males and undermined cooperation between women and men. The expansion of gender-based violence, even in a society such as South Africa with its exceptional gender-sensitive constitution and its high rate of women parliamentarians, highlights the difficulties facing efforts to rein in this behavior. The continuing spread of HIV/AIDS through unprotected sex, even among married couples, has raised questions about patriarchal privilege and the sexual behavior of men (Cornwall et al. 2012). The underperformance of males in education has raised concerns about male underachievement and the long-term consequences for young men and society as a whole (Reddock 2004). All of these development crises highlight the limits of development projects that focus mainly on empowering women, and reaffirm the need to pay attention to the consequences of male frustration over their inability to live up to gender norms, particularly the expectation that they should protect and dominate women, children, and subordinate males.

Organizations such as the UN Commission on the Status of Women and the United Nations Population Fund (UNFPA) have called for greater attention to the role of men and boys in the search for women's empowerment and gender equality. The emphasis has been on identifying the socialization processes that maintain unequal gender roles and power relations, and designing programs that challenge these practices and encourage the adoption of more progressive gender norms built around ideals of non-violence, and "a sense of male pride and dignity based on progressive, gender based ideals" (Sweetman 2013: 5). The attempts to foster progressive gender relations that support women's rights and gender equality in ways that benefit men, women, and the social worlds they live in have often focused on HIV/AIDS, education, and gender-based violence. Case studies on these flash points provide an entry point for evaluating both the strengths and weaknesses of these undertakings.

Men and development projects: case studies on HIV/AIDS, education, and violence

HIV/AIDS

The spread of HIV/AIDS in the 1980s and after raised questions about male sexuality, sexual relations and practices and their impact on women's lives, empowerment, and gender equality. The disease surfaced as a public health issue in North America, particularly among homosexuals, prostitutes, heroin addicts, and Haitians. The initial fight against HIV/AIDS centered on the risk-taking of these "social deviants," often associated with people of African descent. However, as HIV/AIDS spread around the world, particularly to Africa, the focus shifted from risk-taking behavior to heterosexual couples, particularly women's vulnerability. The statistics in Africa reported ten million young men and women aged 15–24 living with HIV/AIDS in 2003, with more than 75 percent being female. These statistics reflected the worldwide feminization of the epidemic and help to explain the shift from risks to the vulnerabilities causing this dramatic imbalance, particularly women's disempowerment and the sexual dominance of heterosexual men over their wives, girlfriends, and contractual sexual partners (Barker and Ricardo 2006).

Male sexual behavior became the central explanation for women's vulnerability to HIV/AIDS. Older men were regarded as particularly problematic as they often had higher rates of infection as well as the resources to convince younger women to exchange sex for much-needed money and support. More broadly, the unequal balance of social power between males and females as well as patterns of risk behavior among men—such as the prevalence of multiple partners, the widely held assumption that men had a right to unprotected sex with wives, girlfriends, and especially in contracted sex, the distrust towards condoms and a widespread preference for dry sex—became central explanations for women's vulnerability to the epidemic. Initial attempts to educate men (and women) about avoiding the disease had little success. Men continued to be seen as the problem, but development projects primarily focused on women's vulnerabilities and ways to address them.

A growing awareness of the importance of men and masculinities in the struggle to contain and manage the HIV/AIDS crisis encouraged new approaches to the disease in the late 1990s.

The tendency to ignore or demonize men began to shift towards a more systematic incorporation of men and masculinities into HIV/AIDS projects. The language of gender mainstreaming in Beijing, followed by increasing attention to sexual and reproductive health and rights as well as the World AIDS Day campaign of 2000, identified men as potential allies in the struggle against the epidemic. Projects to deal with the epidemic flourished. Stepping Stones, originally developed in Uganda at the turn of the twenty-first century, aimed to improve sexual health by building stronger, more equitable and interactive relationships between the sexes that challenge destructive gender practices, particularly among young men. This approach has been adopted in more than 40 countries (Cornwall et al. 2012). Engender Health, an international non-governmental organization (INGO) focusing on gender equality and HIV/AIDS through the Men as Partners Project, has mobilized men to take an active stand for gender equality and against gender-based violence that encourages unequal gender relations and the vulnerability of women and some men (engenderhealth.org). These are just a sample of the many projects engaging men and boys in the fight against HIV/AIDS by encouraging changes in male attitudes and behavior towards women, sexuality, and family life (Bannon and Correia 2006; Sweetman 2013).

Men, masculinity/ies and gender-based violence

Gender-based violence and the attitudes that underwrite it have been a matter of concern for development projects seeking to contain and manage HIV/AIDS. The increasing attention to the attitudes and behavior of men and boys in HIV/AIDS projects in the 1990s coincided with a dramatic expansion of conflicts in the post-Cold War world. The increasing violence of men against women, as well as between men, in these struggles became a matter of grave concern for policymakers, academics, and development practitioners. While the violence between males fit easily with long-held notions of men, war, and conflict, the attacks on women—especially in Bosnia and Rwanda—raised serious questions about gender-based violence, especially in the developing world. Relations between men and women in the home provided a basis for thinking about gender-based violence, particularly the role of norms of male superiority over all females and subordinate males, as well as cultural practices that sanctioned male violence against women who challenged the standards expected of "good women." The development community began to focus on these destructive patterns, setting up projects to challenge masculine practices that legitimated violence against women and urging men to change their behavior and take personal responsibility for confronting these practices (Bannon and Correia 2006; Cleaver 2002; Cornwall et al. 2012).

The use of rape as a weapon of war in the post-Cold War conflicts also dramatized the need to address sexual violence in conflicts as well in the home and community. In 2008 the United Nations passed Security Resolution 1820, which officially recognized rape as a weapon of war (see Section 6). Civil society organizations such as the UN Women's project PeaceWomen.org and the Global Network for Women Peacebuilders began to develop indicators to track and report evidence of rapes in order to highlight the issue and motivate states to take it seriously both during and after conflicts. Women Under Siege, a journalism project working to document sexualized violence as a weapon in global conflict in the twentieth and twenty-first centuries, has gone to great lengths to gather women's stories and map incidences of rape worldwide. For example, the degree to which rape has been used as a weapon of war has been thoroughly documented in Syria through personal storytelling and ethnographic mapping.

Yet for the most part, rape as a weapon of war continues to be regarded as a discussion of rape *of women* by men. The emphasis has continued to focus on changing the attitudes and behavior of men that encourage and participate in such actions. This framing neglects the reality that gender-based violence can provide an outlet for various power struggles, particularly in post-conflict environments and developing countries. The rape of men by men is rarely discussed, let alone confronted. The current conversations about gender-based violence also underplay the rape of LGBT and gender non-conforming individuals as well as the role of homophobia in this process, both locally and internationally. This bias towards rape as male violence against women has shaped development agencies' response to rape in war and conflict-ridden societies, encouraging an assumption that all men are inclined to violence, that all men need to change their attitudes and practices towards women, and that heterosexual violence is the main issue that needs to be addressed (Sweetman 2013).

Some NGOs and development agencies have begun to recognize the role socialization plays in shaping and responding to traditional patriarchal masculine ideals and expectations. Small, local workshops where men are able to openly discuss and negotiate their masculinity with other men have provided an alternative vision of how men might explore their understanding and acting out of masculinity, as practiced in the CBC (Centre Bartolmé de las Casas) Masculinities Programme initiative in El Salvador (Bird et al. 2007). In 2000, UNICEF produced a series of educational films on masculinities designed to help young men explore the role of masculinity and how it relates to gender-based violence in South Asia. A UNFPA project worked with boys aged between 10 and 15 years to reflect on machismo culture and its impact on sexual violence. These are just a few examples of the many development projects confronting gender-based violence by addressing the role of masculinity as it impacts and shapes men on an individual psychological level, but broader structural practices also need to be considered, particularly those values being reinforced by patriarchal heteronormative state governments. Finally, to address the full spectrum of those impacted by gender-based violence, development policy and practice needs to develop and implement indicators that take into account those experiencing sexual violence because of their LGBT orientation. Attention to men raped by men would aid in closing the current gap in thinking about gender-based violence and sexuality.

Men, masculinity and education

Development discourse on women, gender, and education initially focused on unequal access to education, particularly for women and minorities. Applied research on gender and education in high income economies focused on the issues raised by diversity in the classroom. This challenge was compounded by the need to consider the concept of gender equality[2] in the classroom in the late 1970s–early 1980s. The issue of equality based on gender as well as other factors such as race and class gained relevance as legal access to education for all became a reality for much of the world's population. Researchers and development specialists began to explore the impact of gender in educational practices. A gender analysis of classroom reading material and textbooks highlighted their often negative impact on the relationship between subject choices and career options for girls. This research inspired development projects aimed at improving girls' access to education, particularly if it enhanced their skills as future mothers and as contributing members of the family economy.

As the issue of male underachievement in education and its possible links to violence and the HIV/AIDS epidemic became matters of growing public concern in regions such as the Caribbean, policymakers increasingly shifted their focus from gender equality in the classroom to the implications of learnt masculinities for law and order. Much of this early research explored the relationship between the teacher and pupils and the perception of teachers about the disruptive behavior of some boys. The research focused on how this behavior translated into male performance on the street for many 10–16-year-olds, ultimately contributing to the loss of capacity of young males to fulfill the masculine ideal of the male breadwinner, with its accompanying norms of masculine authority over women and subordinate males and its legitimation of patriarchal power and gendered hierarchies (Reddock 2004).

Concern with the declining capacity of young male students to fulfill the expectations of masculine roles has preoccupied many researchers and development practitioners concerned with gender and education, particularly in regions with high levels of crime and violence. The lawless, aggressive, anti-education culture of masculinity found in many parts of the world has become the focus of much of the work on masculinity and education. This work has interpreted male students' disruption in the classroom as a precursor to the anti-social, often criminal behavior found on the streets around the world, particularly in low-income areas. The shift to the street has been seen as leading young men to choose criminality over work activities, thereby making the goals of education irrelevant and inconsistent with the expectations of adult manhood. This research has raised concerns among policymakers and development practitioners in areas such as the Caribbean, where male underachievement in school has been widely publicized and debated since the 1980s. It has been driven by a belief in the nexus between less than ideal male behavior in the classroom, criminality, and—by extension—a compromised ability of many males to assume their traditional roles as breadwinners and leaders in the home and society (Reddock 2004).

Although some of this research discussed the social construction of men, boys, and masculinity, most of it focused on developing a remedial policy response to masculine subjects in education. Boys were to be treated as a distinct category from girls. Teacher–pupil and classroom interaction was seen as a medium through which issues such as sexual harassment in schools, rape, sexual abuse of children, pornography, and deviant male sexuality could be approached and corrected. The research shifted from the behavior of masculine subjects in various schools and systems of education in the 1990s to a focus on the construction of masculinity and the recognition that schools are crucial socializing agents for young men (and women). This emerging area of inquiry explored the way institutions of learning fostered specific types of masculinities, while policing others. Instead of trying to investigate the behavior of various local students, the discourse around education and masculinity deepened to explore the relationship between the school and the student, the complex meanings around masculinities that were created by relationships between individuals and the schools they attend, and the diverse cultures of masculinities created as products of the sociology of schooling. Much of this work produced a policy response that endeavored to understand male school-based deviance by exploring the relationship between male anti-education behavior, schooling, and criminality (Reddock 2004).

However, male underachievement in schools remains largely a concern for middle and high income countries, where women and girls are increasingly outperforming men and boys. Consequently, most of the research, policy and programs addressing this issue occur in these regions. For example, the Caribbean Development Bank between 2004 and 2008 supported a research project exploring gender differentials between female and male enrolment and performance in secondary and tertiary education in the region. Government policies and development programs have drawn on such research to improve lagging male enrollment and performance at these levels. Yet male underachievement has little traction in the poorer regions of the world or among most development agencies, where the concern remains overwhelmingly centered on gender parity in access to schooling. The focus on equal access for women and girls influenced the efforts of the Millennium Development Goals (MDGs)[3] to ensure gender parity in education by 2015 around the world. Indeed, the sexed differences around access to education continue to frame development policy responses to gender and education, especially in the poorest regions of the world.

Conclusion

The gradual shift from women to gender has provided an entry point for bringing men and masculinities into development discourse and practice. The assumption that women's position could be improved solely through women-centered development projects proved illusory, doing little to unseat masculine privilege and masculinist definitions of power and authority. The move to consider gender opened the possibility for dialoguing with men in order to explore the way masculine attitudes and behavior have affected the drive for gender equality and women's empowerment in particular contexts. Despite a tendency to revert to the more comfortable focus on women and girls, development practitioners and researchers concerned with men and masculinity have sought to understand and address the beliefs and practices among men and boys (and some females) that legitimate male privilege and oppose efforts to produce a more gender equitable world. Challenging these toxic beliefs and practices and encouraging new more gender-sensitive ways of thinking and behaving have become important strategies for encouraging support for gender equality and reducing gender-based violence, HIV/AIDS, and gender disparities in education and employment.

At the same time, the effort to challenge and alter masculine attitudes and behavior that oppose gender equality takes place in a world where gendered assumptions still largely define which sexed bodies, performances, sexuality, and positionalities are seen as deserving the material and social rewards offered in particular societies. Trying to change long-held assumptions about what kinds of men (and women) should wield power is extremely difficult, as those who benefit from such systems are rarely willing to step aside for a more gender-equitable system. Convincing the powerful to give up their privileges is never easy, and it is even more difficult if many men (and some women) who benefit more indirectly from masculinist privilege still believe even their minimal power and authority will be threatened if the gendered structures of power are altered. Power has too often been seen by development agencies as "something that can be bestowed or acquired rather than a structural relation that is in itself gendered. And targeted 'investment' has come to displace any consideration of the broader social changes that need to take place if the persistent inequalities associated with gender difference are to be eradicated" (Cornwall et al. 2012: 1). These broader social changes threaten the status quo of many and thus will rarely be supported by policymakers and even large sections of the public. Indeed, it is not surprising that the World Bank—despite its 2012 *World Development Report* committed to enhancing gender equality—is increasingly focusing on women and girls as the "solution" to gender equality (Chant 2012).

At the same time, the twenty-first century has produced conditions that are beginning to shift perceptions about gender relations, men, and masculinity/ies, particularly their role in causing or challenging gender-based violence. Some development agencies now see gender-based violence not only as a threat to women's empowerment, but also as a systemic reaction of men grappling with their own struggles with (dis)empowerment on shifting economic and political sands. As most of the world continues to suffer from growing inequality, high levels of unemployment—especially among young men—and competition from increasingly skilled women entering the workforce are giving rise to new social tensions that sometimes have dangerous outcomes (Catala et al. 2012). Recognizing the larger psychological implications of these social changes for men is essential to understanding and responding to gender-based violence as well as gender equality in general. Shifting power dynamics driven by women moving into the workforce are challenging traditional patriarchal structures. The same can be said for women taking on new roles in government. A recognition of the complexities for men in adjusting to a society where women are gaining access to new forms of power can inform an understanding of how to respond to these changes and challenges in ways that potentially bode well for the prospects of both men and women.

Responses to recent crises have too often focused on the inherent violence of males and the need to convince men to change their attitudes and behavior. Certainly masculinist privilege continues to be a crucial factor reinforcing gender inequality around the world. However, the changing international political economy privileges hegemonic masculine authority, but the reach of that privilege is smaller, leaving many men as well as women struggling to survive in an increasingly hostile world (Catala et al. 2012; Connell 2005). Growing attention given to the problem of men and violence demonstrates the importance of environment and socialization, culturally dominant ideas and beliefs regarding acceptable masculine behavior, and images and stereotypes present in men's lives. The current focus on the responsibility of individual males is not enough. A deeper examination of the problematic nature of continued masculine authority and violence of the state needs to be incorporated into development policies and practices (Cornwall et al. 2012). Support for gender equality and a more gender equitable world will require collaboration with both men and women, a profound understanding of the patriarchal structures that underwrite and support gender bias in societies around the world and a set of policies and programs designed to undertake long-term change in the complex, highly rooted gender systems around the world. The current economic and political crises provide an opening, but also a challenge, as patriarchy could regroup and become more entrenched—witness much of what has happened in the Middle Eastern "Arab Spring." The need for broad social change has never been greater, but it will require determination and a commitment for women and men to work together to create a more gender equitable and fairer world.

Notes

1   I would like to thank my research assistants from the PhD program on Global Governance and Human Security at the University of Massachusetts, Boston, for their assistance with this article. They are: Deborah McFee, Jamie Hogan, and Polliann Hardeo.
2   Gender equality needs to be understood not as nuanced analysis of the differences between groups of women and men but, in this research, more apt to focus on differences in the classroom between an unproblematized sexed category of female and male students.
3   The MDGs were orchestrated by the United Nations in 2000 and widely supported.

References

Bannon, I. and Correia, M. (eds.) (2006) *The Other Half of Gender: Men's Issues in Development*, Washington, DC: World Bank.

Barker, G. and Ricardo, C. (2006) "Young men and the construction of masculinity in sub-Saharan Africa," in I. Bannon and M. Correia (eds.), *The Other Half of Gender: Men's Issues in Development*, Washington, DC: World Bank.

Bird, S., Delgado, R., Madrigal, L., Ochoa, J.B., and Tejeda, W. (2007) "Constructing an alternative masculine identity: the experience of the Centro Bartolomé de las Casas and Oxfam American in El Salvador," *Gender and Development* 15(1): 112–121.

Catala, V., Colom, S., Santamaria, L., and Casajust, A. (2012) "Male hegemony in decline? Reflections on the Spanish case," *Men and Masculinities* 15(4): 406–423.

Chant, S. (2012) "The disappearing of 'smart economics'?" The *World Development Report* 2012 on gender equality," *Global Social Policy* 12(2): 198–218.

Cleaver, F. (ed.) (2002) *Masculinities Matter: Men, Gender and Development*, London: Zed Books.

Connell, R. (2005) "Change among the gatekeepers: men, masculinities, and gender equality in the global arena," *SIGNS* 30(3): 1801–1825.

Cornwall, A., Edstrom, J., and Greig, A. (eds.) (2012) *Men and Development: Politicizing Masculinities*, London: Zed Books.

Reddock, R. (ed.) (2004) *Interrogating Caribbean Masculinities*, Kingston, Jamaica: University of the West Indies Press, Mona.

Sweetman, C. (2013) "Introduction: working with men on gender equality," *Gender and Development* 21(1): 1–13.

White, S. (1997) "Men, masculinities, and the politics of development," *Gender and Development* 5(2): 14–22.

4

GENDER MAINSTREAMING

Changing the course of development?

Caroline Sweetman

Introduction

In 2011–2012, the journal *Gender & Development*[1] instigated a discussion between women's rights activists and gender and development policymakers and practitioners on the progress of gender mainstreaming. Gender mainstreaming is an approach to women's rights and gender issues, widely adopted by governments and development institutions of all sizes, shapes and hues after it acquired currency at the UN Fourth World Conference on Women, held in Beijing, China, in 1995.

Today, nearly 20 years on from Beijing, 'gender' is spoken of all the time. It is supposed to be 'mainstreamed' into humanitarian responses, and in long-term development programmes; at different stages of the project cycle; and in all types of interventions – from advocacy initiatives to community-level development. But what has this meant for women and men, girls and boys, in the communities and nations that have signed up to work on challenging and changing gender inequality? In the *Gender & Development* Beyond Gender Mainstreaming Learning Project, a total of around 100 participants was involved to discuss these and other concerns.[2] Participants came from markedly different personal, professional and institutional locations. We pooled our thinking and expertise; challenged each other's perspectives on the contribution gender mainstreaming has made to the pursuit of gender equality and women's rights worldwide; and considered the future of gender mainstreaming as an approach to the feminist goals of gender equality and women's rights.

Some of the insights gained are offered here. Gender mainstreaming has not had uniform effects, nor should we expect it to have had; yet we can make some general points to help our diagnosis of success and failure, and this chapter attempts to raise some of the most important among these. The first section gives a brief account of the coining of the concept of gender mainstreaming, recalling that this was an outcome of a process lasting 20 years, punctuated by formal UN-convened encounters between the global women's rights movements[3] and the international development community. The chapter then goes on to discuss some of the ways in which gender mainstreaming has played out in practice. In such a discussion, it is crucial to consider how tensions around the underlying vision of development inform gender mainstreaming in international institutions, donor governments, governments in the Global South and NGOs of different types.

The origins of gender mainstreaming

To the international women's movements in 1995, gender mainstreaming represented the most encouraging and largest step forward yet, in a struggle that had been marked by the series of UN Conferences on Women, which had run from the first one in Mexico City, in 1975, to Beijing in 1995, and had taken in the UN Decade for Women 1976–1985. Women's rights activists from all over the world had engaged with states and international development organisations over this time. They had debated differences in dialogue with each other, honing a shared vision and priorities, and built a powerful coalition that aimed to transform the course of global development itself.

Gender mainstreaming was a response to the failure of women in development (WID) programming of the 1970s and 1980s. Evidence had proliferated of 'project misbehaviour' (Buvinic 1986), as women responded to ill-conceived development initiatives foisted on them by pursuing strategies to mitigate their negative effects and capture any benefits to be had. Despite varying political and economic rationales for undertaking work with women (Moser 1989), the vast majority of WID programs began with promoting women's income-generating activities as an entry point. Feminist and women's rights organisations from the Global South argued that WID projects designed to harness women's productive potential to unsustainable national growth strategies needed to be jettisoned. Informed by a broadly socialist analysis of poverty as rooted in inequality, they argued for a strategy of empowerment and transformation. In this, the role of Northern-based development organisations was to support Southern women's rights organisations to use as they saw fit in pursuit of home-grown solutions to political, economic and social marginalisation of women. Ultimately, this would change the course of development.

Perhaps the most famous articulation of an alternative vision of development informed fully by the perspectives of women in the global South is that of the DAWN (Development Alternatives with Women for a New era) network:

> For many women, problems of nationality, class, and race are inextricably linked to their specific oppression as women. Defining feminism to include the struggle against all forms of oppression is both legitimate and necessary. In many instances gender equality must be accompanied by changes on these other fronts, at the same time, the struggle against gender subordination cannot be compromised during the struggle against other forms of oppression, or be relegated to a future when they may be wiped out . . . we strongly affirm that feminism strives for the broadest and deepest development of society and human beings, free of all systems of domination.
>
> *(Sen and Grown 1987: 19)*

Feminist activism at the UN Beijing Conference 1995

It is clear from the last section that feminist activists saw (and still see) dramatic change in the ways development addressed the concerns of gender equality and women's rights as necessary not only to women, but to the whole of humanity. Development organisations needed to change their sense of what their purpose was, and who they were intended to serve, in addition to their ways of working, in order to be capable of delivering the transformative vision of development set out by international women's movements over the two decades leading up to the UN Fourth World Conference on Women at Beijing in 1995.

At Beijing, an enormous NGO forum (a gathering of civil society organisations) took place parallel to the conference proper. The vibrancy – and the diversity – of the international women's movements were evident. Differences in identity, location and hence in priorities did not hinder powerful lobbying and advocacy to achieve a positive outcome at the conference, resulting in the UN commitment to gender mainstreaming in the Beijing Platform for Action. Yet – as is so often the case when passionate activism and radical political thinking attempts to change the course of history – the commitment given by governments at Beijing was a compromise between widely varying views from radicals and conservatives, often relying for the establishment of consensus on the use of words that can be widely interpreted according to the motive and context of the policymaker.

This was the context for the UN's eventual definition of gender mainstreaming, with its rather mechanistic understanding of gender analysis as a means to transform development. Interpretation and implementation has been left to individual development organizations to figure out for themselves:

> Mainstreaming a gender perspective is the process of assessing the implications for women and men of any planned action, including legislation, policies or programmes, in all areas and at all levels. It is a strategy for making women's as well as men's concerns and experiences an integral dimension of the design, implementation, monitoring and evaluation of policies and programmes in all political, economic and societal spheres so that women and men benefit equally and inequality is not perpetuated. The ultimate goal is to achieve gender equality.
>
> *(United Nations 1997: 1)*

We are now almost two decades on from the Beijing Conference and the start of widespread commitments to gender mainstreaming, and 'almost all international development organizations and governments have adopted gender mainstreaming in some form' (Derbyshire 2012: 406).

In the next section, I draw on discussions from the Beyond Gender Mainstreaming Learning Project to examine perhaps the key concern in assessing the 'fit' between the radical vision of change of feminists at Beijing, and the progress made thus far.

Gender mainstreaming, policy and organisational approaches to development

It is a challenge for development and humanitarian organisations to assess the impact of their activities on the lives of women and girls, men and boys 'on the ground', whose lives are altered profoundly and continuously by all sorts of factors. In a review of gender mainstreaming undertaken in the early 2000s, Caroline Moser and Annalise Moser (2005) pointed out that many analyses of the impact of gender mainstreaming have tended to focus on the impact of gender mainstreaming on development organisations themselves, rather than looking at the impact on supposed 'beneficiaries' (and others). This focus on 'who we are' rather than 'what we do' has of course a great deal to do with the focus of feminist enquiry in the 1990s into the key role of social institutions, including the market, state and NGOs, in perpetuating or challenging gender inequality. Institutions reflect the norms and values of surrounding society – an aspect of which is 'male bias' (Elson 1991). Getting development institutions 'right for women' (Goetz 1995) has thus been a priority in the era of gender mainstreaming. Ramya Subrahmanian (2007: 113) has observed that 'a key criticism about gender mainstreaming has been the "narrowness" of the strategy despite the complexity of gender relations and the contextual variations in the processes and outcomes related to gender inequalities'.

In her research into the initial impact of gender mainstreaming, Rounaq Jahan (1995) distinguished between integrationist and agenda-setting mainstreaming. Integrationist mainstreaming is a process whereby aspects of the Beijing agenda are taken up without organisations essentially altering their vision of development, policies or ways of working. It is seen as a technical concern – as Jahan states, work that integrates gender analysis of women's and men's concerns throughout policies and projects. In contrast, agenda-setting mainstreaming changes the course of the organisation's work – in the way gender mainstreaming was intended by the women's movement – and, as such, requires understanding of the deeply rooted normative beliefs that inform an organisation's culture and therefore underpin all its activities, keeping existing power relations in place and unchallenged.

For many working in governments and international organisations disbursing development funding to states and NGOs in developing countries, 'gender mainstreaming' took a decidedly integrationist form. At worst, it has been reduced into a magic bullet solution to development, in which the North-to-South transfer of resources to women entrepreneurs is often in the form of repayable loans. The hope is that this will result in wealth creation for women, their families and society. These small injections of capital are often linked in the minds of the donors to socially desirable outcomes associated with the empowerment of women. These range from fertility reduction to investment in daughters' education. Giving resources to women is seen as 'smart economics' – policymakers and practitioners in the international financial institutions in particular argue for funding for programs with gender equality aims on the basis of broader economic and social impact (Chant and Sweetman 2012).

Yet, while it focuses on women and channels resources to them, programing undertaken from a 'smart economics' perspective fails to challenge the complex structural inequalities shaping the realities of individual women living in poverty in the Global South. While it may benefit some women in a limited way, smart economics depoliticises the notion of gender mainstreaming by using the language of radical change, empowerment and women's rights, while abdicating responsibility that politicians and economists should take to remove gender, race and class-based inequalities in social institutions, which shape the realities of women and constrain their ability to realise real, sustained and systemic change.

In contrast to this focus on economic prosperity as the key to development *and* gender inequality, 'alternative' development organisations – that is, those whose vision of development is founded on left-wing political ideals of equality, rights and justice – are more likely to see the need for political voice to be built for women as a constituency, through work addressing political and social dimensions of powerlessness head-on. However, experience of gender mainstreaming shows progress has been patchy. Participants in the Beyond Gender Mainstreaming project confirmed that there is still a tendency not to look beyond the economic aspects of development to focus on inequality holistically. Programmes may end up defaulting to look very much like smart economics approaches, focusing on 'economic empowerment' of women and leaving the dirty, dangerous work of dismantling political and social barriers to women's rights to women themselves – just as the WID projects discussed above did in the past.

In addition, some participants highlighted the fact that – even after all this time, and the incredibly high-profile activism of feminists from the Global South – gender mainstreaming in development is still dogged by assertions that women's human rights agendas and feminist change are cultural impositions on developing countries. There is, of course, a vast range of evidence of the thriving multiplicity of women's rights movements in the Global South, which may or may not use the term 'feminist', and may or may not be wholly 'home-grown' (we live in an age of globalised ideas and political movements, after all), but all of which represent constituencies of women in developing countries fighting for their rights.

Research focusing on the impact of gender mainstreaming on women and girls 'on the ground' tends to suggest that development policymakers and practitioners need to base our mainstreaming in a more conscious analysis of how change happens, and gain a clearer – and more realistic? – awareness of the role to be played by development organisations. Gender mainstreaming strategies that prioritise the top-down North-to-South transfer of resources and see gender inequality as remedied by women earning income fail to root their analysis of change in the empowerment of women living in poverty in the Global South; yet this was the central vision of mainstreaming as originally articulated by the global women's movements.

Institutionalising gender mainstreaming

What strategies have feminists inside and outside development organisations employed to change development organisations internally, so they are more able to deliver real change for women in the Global South? The institutional aspects of gender mainstreaming are critical to dismantle bias informed by gender, race and class-based inequalities, development institutions (from states to international financial institutions to NGOs of all types and sizes) need to be 'right for women' (Goetz 1995). Gender mainstreaming has thus involved attention to policies, personnel, organisational 'architecture' and structures, human resource recruitment and management procedures, as well as the analytical tools employed in programme planning and implementation, to ensure these support gender equality and women's rights. Gender policies have been drafted, redrafted and ratified. Debates have been had about adopting the terminology of 'gender': gender analysis, gender tools and frameworks, gender equality – or equity, or justice. Decisions have been taken on how to develop organisational infrastructure to support gender mainstreaming. Methods for programme planning, implementation and evaluation have been developed, rolled out, discarded, forgotten and reinvented. Senior management comes and goes, and gender issues are sometimes championed, and occasionally seemingly almost forgotten. Consultants are asked to evaluate progress on gender issues periodically, and their findings acted on, or placed on a (literal or virtual) shelf to gather dust.

The trajectories followed by different organisations have been usefully typologised by Helen Derbyshire (2012), building on her participation in the Beyond Gender Mainstreaming Project; some of the key elements are discussed here.

The need for twin-track programming

Some success has been won by arguing for a 'twin-track' approach to mainstreaming (van Eerdewijk and Dubel 2012), in which the organisation integrates gender analysis into planning and implements a programme that has 'stand-alone' women's rights-related aims alongside primary economic goals concerned with livelihoods, the provision of basic services or humanitarian response. Stand-alone work can take the form of advocacy or programme work addressing key concerns for women, which aims to improve their status rather than simply meeting their needs within the existing gender division of labour (Molyneux 1985). Examples are campaigning against gender-based violence, or promoting women's reproductive and sexual rights. This kind of work has an explicit primary aim of furthering women's rights and gender equality, and complements the work happening in other sectors, which may meet women's practical needs rather than challenging power relations. Stand-alone work potentially acts as a catalyst to radicalise programmes, so that comparatively politically conservative projects can end up having a real impact on gender power relations. Examples of this happening were discussed in the Beyond Gender Mainstreaming project (Holt Zachariassen 2012).

The role of the gender specialist in mainstreaming

Research has shown that staff in development institutions often feel poorly equipped and informed on gender issues (see Mehra and Rao Gupta 2006), and are unsure of the right methods to employ to gain understanding of the perspectives and perceived needs of potential beneficiaries of development programmes. Yet if the emphasis in gender mainstreaming is on ensuring that *all* the activities of development organisations focus on gender equality, then everyone has a role to play in working towards gender equality (see, for example, van Eerdewijk and Dubel 2012).

Planning and programming in ways that support gender equality and women's rights requires staff with the right knowledge, skills and attitudes. A key element of gender mainstreaming is the use of conceptual frameworks and 'gender analysis tools' that enable development and humanitarian workers to ask the right questions in their baseline research, build up a picture of women's and men's roles and of gender power relations in a particular context, and plan and implement a programme that responds fully to these varying realities. At the other end of the project cycle is impact assessment, so that we can see what we have actually done to improve the lives of women and men, girls and boys, from a gender perspective. All development organisations need structures and accounting procedures that enable us to see the impact of institutional gender mainstreaming, via 'gender budgeting' – part of mainstreaming in the case of some national governments, and equally important in development organisations.

Having staff with a specific gender job description in a department or organisation can create a 'corporate home to guide the incorporation of gender equality considerations' (Murison 2002), thus supporting the entire gender mainstreaming project. Such a presence can offer a 'location' to which other equality advocates can refer to forge important relationships and find support of different types (Wakefield 2012, in the context of Oxfam International). Gender experts within institutions, whether governments or large INGOs, can be (at their best) 'entre-preneurial actors' who take advantage of internal opportunities (for example, an organisational focus on social justice) and external opportunities (for example, global conferences such as Beijing) to facilitate the flow of information and stronger bonds between themselves and the women's movement – between gender advisors in INGOs and women's rights organisations, and between themselves and other similar institutions.

Once the battle is won to get organisations to accept the need for gender specialists to work in an empowering way with colleagues and project participants, the next task is to consider their role and scope. Rarely, if ever, is the presence of gender expertise an explicit part of the theory of change that underlies a programme, project or initiative, yet 'the work [gender advocates] are involved in plays a critical direct (as well as behind the scene) role in shaping the development decisions that ultimately affect women's and girls' lives' (Derbyshire 2012: 407).

It will have been obvious that different terms have been used so far: gender experts, gender advocates, gender advisor and gender focal points. The difference in terminology reflects concrete differences in the types of roles that individuals can take within development bodies, as well as existing negative or positive perceptions of such roles.

There are a range of other factors associated with the personal characteristics of gender specialists, which affect their ability to drive effective mainstreaming. They not only need the right knowledge and skills, and credibility (Wong 2012), but their substantive post needs to give them a mandate to do this work – which is not the case for most 'focal points' (a notoriously weak formulation). Also, evidence suggests that they need to identify to some extent as feminist (Derbyshire 2012).

However, employing gender specialists and cultivating gender expertise among other staff, while necessary, will not deliver positive outcomes in the absence of other important institutional, policy and operational steps that change structures, processes, hearts and minds. The Beyond Gender Mainstreaming debates focused, for example, on the critical point of committed leadership from senior management as integral for effective gender mainstreaming. That leadership also needs to be informed, and focused accompaniment by a gender specialist might be a strategy to follow to ensure leaders' understanding of the very complex gender issues in different contexts is approached in the spirit of seeing transformative potential wherever this lies. While there is no denying that the words and actions of a CEO or head of department in relation to women's rights is crucial, they can create new dilemmas, as their interpretation of issues and strategies may not necessarily match what gender advocates would promote, while at the same time carrying more weight within the organisation.

Gender mainstreaming and collaboration with women's movements

Another key issue for transformative gender mainstreaming is working with women's rights movements, feminist organisations and networks, as key and essential allies in main-streaming.

In the wake of gender mainstreaming being embraced by the UN and states as a priority, what has happened to the relationship between feminists inside and outside development organisations, and to the wider relationship between the development organisations and the women's rights movements, which hoped for so much from gender mainstreaming?

Working with feminist movements, organisations and networks is perhaps, above all others, the key to ensure gender mainstreaming is truly transformative. Alliances with such organisations promote effective gender mainstreaming for a number of reasons. At an immediate level, such organisations are intimately connected on a daily basis with the gender equality and women's rights issues that form just one part of the work of development organisations. They are critical allies for feminists working within development organisations and, at best, this relationship results in powerful, nuanced and politically 'savvy' activism that advances women's rights either nationally or internationally, and results in community development work that actually furthers the perceived interests of women on the ground, as it comes from a locally informed perspective. Feminist movements can exchange different forms of support with development practitioners as well as pool information and strategies. Organisations and movements working on women's rights are frequently directly linked to women's and girls' lives (O'Connell 2012), and thus help organisations with whom they collaborate to increase their legitimacy and relevance. This throws up a very significant challenge to development organisations, of course. The ultimate challenge for these organisations is to let go of the agenda and permit Southern women to determine the agenda themselves – which may or may not resemble organisations' ideas of the changes to be achieved in the name of gender mainstreaming.

By definition, women's rights organisations and movements bring a much more transform-ative element to gender mainstreaming because of the approaches and methodology they use. They also contribute to a holistic, well-rounded approach; mainstreaming gender considerations in sectors or in specialised bodies is liable to bring with it the kind of fragmented view that can easily lead to simplistic (or even mistaken) approaches; for example, those based on the notion that working towards better income for women is *the* way to gender equality and women's empowerment, rather than part of a more comprehensive and complex understanding of both.

Having stressed the importance of such alliances and collaborations, it is necessary to remind ourselves that there are – in the women's rights community – serious concerns about the consequences of women's rights organisations and networks working with governments and other actors in the pursuit of gender justice. Srilatha Batliwala (2012) sees the dangers of an 'NGO-isation' process that derails feminist activism in the Global South, distorting the agenda of local feminist and women's organisations as international donors begin to shift priorities to a focus on anti-poverty service delivery. This line of argument is that UN agencies and international donors are turning women's organisations into NGOs and, in the process, despoiling them of the characteristics that make them 'social movements' capable of representing and furthering the social changes desired by local populations. Another view is that the imposition from Northern universities and NGOs of 'gender' is to be blamed for work on gender issues not 'translating into real change, either within NGOs or within communities' (Wendoh and Wallace 2006).

'Women's rights organizations and movements are a vital catalyst for the realisation of women's rights' and their partnership with donors and international NGOs has huge benefits for women's needs and interests (Esplen 2013). While recognising that problems and pitfalls exist against which we must all guard, national governments and other international organisations have an obligation to 'support' women's organisations and networks, both because it has long been established that collective action is the most powerful avenue to promoting women's rights, and because of the constant struggle such networks have in securing sufficient financial resources. Indeed, in a recent path-breaking study on the impact of local feminist movements on violence against women in 70 countries over 40 years, Mala Htun and Laurel Weldon confirm that the existence of a strong vibrant local movement is the single most important factor in successful work to address violence against women, in part due to their role in holding governments to account on their international commitments on women's rights and ending violence against women (Htun and Weldon 2013).

What, then, has been the impact of mainstreaming?

This chapter set out to examine a few of the most important issues that the experience of gender mainstreaming has thrown up. Twenty years after Beijing, women are still waiting for the transformation to global and national governance that they require in order to participate and demand accountability, and that would logically result in a changed world order reflecting the values and priorities of women living in poverty in the Global South.

We face complex crises globally, presenting enormous challenges to the whole of humanity. Conflicts, social turmoil owing to changes in economic and political power between regions and countries, the impact of climate change, food crises, and the continuing impact of the 2007–2008 economic crisis that began in Europe and North America, are among the factors threatening many in the Global South. Other factors challenging women's rights include continuing struggles for citizenship in the context of globalization (see chapter by Townsend et al. in this volume); struggles for democracy in regions including the Middle East in the Arab Spring; ongoing conflicts; and failed states. Women are also facing the threat of growing social and political conservatisms in many societies, including fundamentalisms and religious extremism, in both the North and South, which particularly threatens women's human rights, including their sexual and reproductive rights. The wider picture is terrifying, but in many countries feminists – in particular, young feminists – are organising and taking action, often using new technologies to create new forms of activism.

Gender mainstreaming in most international development organisations, including the IFIs and the UN system, has fallen very far short of visions of transformative gender mainstreaming. Inequality is becoming the development issue of our times, but understandings of it tend to focus on inter-household inequality and not on the intra-household issues that are gendered and intrinsically linked to class and race inequalities throughout societies. The leadership of key development organisations that shape our world is still overwhelmingly male, white and elite. Women's priorities and perspectives are missing from key decision-making bodies. Gender mainstreaming is a discredited concept for many feminist activists. The radical visions of Beijing have not resulted in a transformed global order; and there is widespread 'sobering recognition of the enormous gap between feminists' aspirations for social transformation and the limited, though important, gains that have been made' (Cornwall et al. 2007: 1).

Nearly 20 years after Beijing and the inception of gender mainstreaming, what is its future? Perhaps the time has come to move 'Beyond Gender Mainstreaming' entirely. Do we need perhaps to move away from gender mainstreaming? Have we been there, done that and got the T-shirt? Is it now an outmoded and tired concept that has lost its currency? We may acknowledge the progress made in development organisations in taking on a concern for gender equality, but we also need to respond to the current backlash against these advances. Some argue that what is needed now is a shift away from gender mainstreaming, towards strategies that aim to end gender discrimination, focusing on social justice, transformation and change (Sandler and Rao 2012). Yet for others, adoption of gender mainstreaming language by states, development and humanitarian organisations is seen as a critical 'foot in the door'. 'Gender' is now on the agenda of all organisations involved in development and humanitarian work. Can integrationist approaches be used as entry points for transformation and a genuine redistribution of power in development?

In the past year, there have been discussions in different quarters about the possibility of a UN Fifth Conference on Women, the need to lobby for the voices of women and girls to inform the successor to the Millennium Development Goals, and an assessment and refocusing of the UN Commission for the Status of Women, among other high-level changes to the gender equality and women's rights agenda. Rather than adding women to recipes for poisonous, unsustainable development based on exploitation – in particular, of women in poverty in the Global South, or portraying these women as virtuous victims while failing to support their activism and demands for justice – development organisations need to take on a commitment to transformative gender mainstreaming. Development requires the perspectives and solutions of women, and women require equality, human rights and justice.

Notes

1   *Gender & Development* is an international journal published by Routledge for Oxfam. It aims to support transformative gender mainstreaming by providing a forum for the exchange of experience and analysis of development and humanitarian work. Its readers include development policymakers and practitioners, researchers, students and academics, and feminist activists. For more information, visit www.genderand development.org.
2   The *Gender & Development* Learning Project involved consultations with women's organisations and researchers on gender issues, including a forum in Beirut, Lebanon, a two-day international online discussion hosted on the Eldis Communities website of the Institute of Development Studies (IDS), University of Sussex; and a face-to-face learning event in London in February 2012. A special issue of *Gender & Development* was published in November 2012.

3   I use the terms 'women's rights' and 'feminist' interchangeably to refer to organisations and movements focusing on transformative change in gender power relations. Not every women's organisation is concerned with such change (Molyneux 1998).

References

Batliwala, S. (2012), *Changing Their World: Concepts and Practices of Women's Movements*, 2nd edition, Toronto, Mexico City, Cape Town: Association for Women's Rights in Development.

Boserup, E. (1970) *Women's Role in Economic Development*, London: Allen and Unwin.

Buvinic, M. (1986) 'Projects for women in the Third World: explaining their misbehavior', *World Development* 14(5): 653–664.

Chant, S. and Sweetman, C. (2012) 'Fixing women or fixing the world? Smart economics, efficiency approaches, and gender equality', *Gender & Development* 20(3): 517–530.

Cornwall, A., Harrison, E. and Whitehead, A. (2007) 'Introduction: feminisms in development: contradictions, contestations and challenges', in A. Cornwall, E. Harrison and A. Whitehead (eds.), *Feminisms in Development: Contradictions, Contestations and Challenges*, London: Zed Books, pp. 1–17.

Derbyshire, H, (2012) 'Gender mainstreaming: recognising and building on progress. Views from the UK Gender and Development Network', *Gender & Development* 20(3): 405–422.

Elson, E. (1991) *Male Bias In The Development Process*, Manchester: University of Manchester Press.

Esplen, E. (2013) *Leaders for Change: Why Support Women's Rights Organisations?* London: Womankind Worldwide, www.womankind.org.uk/wp-content/uploads/downloads/2013/03/LeadersForChange-FINAL.pdf.

Goetz, A.M. (1995) 'Institutionalising women's interests and gender-sensitive accountability in development', *IDS Bulletin, Getting Institutions Right for Women in Development* 26(3): 1–10.

Holt Zachariassen, H. (2012) 'From the bottom up: lessons about gender mainstreaming in the Andes from Digni's Women's Empowerment and Gender Equality (WEGE) Programme', *Gender & Development* 20(3): 481–490.

Htun, M. and Weldon, S.L. (2013) 'Feminist mobilisation and progressive policy change: why governments take action to combat violence against women', *Gender & Development* 21(2): 231–248.

Jahan, R. (1995) *The Elusive Agenda: Mainstreaming Women in Development*, London: Zed Books

Kabeer, N. (1999) 'Resources, agency, achievements: reflections on the measurement of women's empowerment', *Development and Change* 30: 435–464.

Mehra, R. and Rao Gupta, G. (2006) *Gender Mainstreaming: Making It Happen*, Ottawa: International Centre for Research on Women (ICRW).

Molyneux, M. (1985) 'Mobilisation without emancipation? Women's interests, the state and revolution in Nicaragua', *Feminist Studies* 11(2): 227–254.

—— (1998) 'Analysing women's movements', *Development and Change* 29: 219–245.

Moser, C. (1989) 'Gender planning in the Third World: meeting practical and strategic gender needs', *World Development* 17(11): 1799–1825.

Moser, C. and Moser, A. (2005) 'Gender mainstreaming since Beijing: a review of success and limitations in international institutions', *Gender & Development* 13(2): 11–22.

Murison, S. (2002) *Oxfam GB Gender Review: Institutional Arrangements Assessment*, internal unpublished document, Oxford: Oxfam.

O'Connell, H, (2012), *What Added-Value do Organisations that are Led and Managed by Women and Girls Bring to Work Addressing the Rights, Needs and Priorities of Women and Girls?* Comic Relief Review Paper, London: Comic Relief.

Sandler, J. and Rao, A. (2012) 'The elephant in the room and the dragons at the gate: strategising for gender equality in the 21st century', *Gender & Development* 20(3): 547–562.

Sen, G, and C. Grown (1987) *Development Crises and Alternative Visions: Third World Women's Perspectives*, New York: Monthly Review Press.

Subrahmanian, R. (2007) 'Making sense of gender in shifting institutional contexts: some reflections on gender mainstreaming', in A. Cornwall, E. Harrison and A. Whitehead (eds.), *Feminisms: Contradictions, Contestations and Challenges in Development,* London: Zed Books, pp. 112–121.

Smyth, I. (2007) 'Talking of gender: words and meanings in development', in A. Cornwall and D. Eade (eds.), *Deconstructing Development Discourse: Buzzwords and Fuzzwords*, Rugby: Practical Action Publishing and Oxfam.

United Nations (1997) *Report of the Economic and Social Council for 1997*, A/52/3, 18 September.

van Eerdewijk, A. and Dubel, I. (2012) 'Substantive gender mainstreaming and the missing middle: a view from Dutch development agencies', *Gender & Development* 20(3): 491–504.

Wakefield, S, (2012) 'Better than the sum of our parts? Reflections on gender mainstreaming in a confederation', *Gender & Development* 20(3): 585–598.

Wendoh, S. and Wallace, T. (2006) 'Re-thinking gender mainstreaming in African NGOs and communities', *Gender & Development* 13(2): 70–79.

Wong, F. (2012) 'The micro-politics of gender mainstreaming: the administration of policy in humanitarian work in Cambodia', *Gender & Development* 20(3): 467–480.

5

GENDER AND POSTCOLONIALISM

Sarah A. Radcliffe

One consequence of "development" thus far has been the (not uncontested) marginalization and misrepresentation of women in the South. What Alexander and Mohanty (1997) refer to as "historic and newly emergent forms of colonization" continue to play their role in keeping women of the "South" in their place as the other, while simultaneously stabilizing gendered, racialized and classed identities in the "North."

(Peake and Trotz 2002: 335)

This chapter examines the impact of majority world of the Global South women's activism and postcolonial analyses in relation to development. Postcolonial approaches destabilize the North Atlantic's dominant discourses and knowledges, challenging the representation of ordinary people's lives in the majority world. To replace Northern narratives, postcolonialism highlights the diversity of the world's most marginalized populations. Key to this diversity is postcolonial and Global South feminists' focus on the complexity of gender-race-class-location relations and power dynamics after formal decolonization. The North–South divide is in part organized through and dependent upon a gendered axis. Chandra Mohanty (1988) introduced us to the "discursive colonialism" exerted by development and Western feminists over majority-world women. For these majority-world feminists, postcolonialism offers a means to explore the intersecting power-laden divisions of class, gender, race-ethnicity, migrant status, location, age, and sexuality in diverse contexts, while making analytical connections and networks of resistance between distant places.

After an introduction, the chapter addresses how to conceptualize the relationship between coloniality, male–female relations and development; the continued postcolonial complexities of race, gender and culture; the political economies of development; and the security and development field. The conclusion summarizes the chapter's argument, and notes several upcoming issues.

Introduction

The chapter discusses the ways in which critical postcolonial and gender analyses continue to provide powerful critical insights into the nature of development, insights that—due to ongoing

South–North power imbalances—are generally difficult to incorporate into development practice and mainstream thinking, although some aspects are in the MDG goals. The chapter's argument is that analyses of the power relations expressed through and around female–male difference,[1] and critiques of the material, discursive, and embodied legacies of colonialism have had profound effects on the ways in which we understand questions of development, and the problems facing low-income groups in the majority world. Postcolonial societies are often riven by class-race-location-religious differences that contribute to the marginalization of low-income women; hence addressing wellbeing becomes entangled with complex articulations of power. For women in the Global South, postcolonial perspectives are strengthened by openness (by whom?) to non-metropolitan understandings of female–male power dynamics.

In some respects, development takes on board key insights from postcolonial feminist writers. One mainstream text notes "different goals and practices that drive the postcolonial focus on equity (rather than [liberal feminism's] equality) must be integrated into policy and action-oriented efforts to improve women's lives" (Jacquette and Summerfield 2006: 5). Rather than homogenizing and dismissing Third World women as they did in the past, today's interventions now place "Third World women" at the center of development projects, with a specific Millennium Development Goal. Current development lauds low-income women's potential as micro-entrepreneurs, their responsible behavior regarding children's wellbeing, and their wisdom concerning nutrition and environments. Yet as this chapter argues, these new positionalities for Third World women need to be carefully scrutinized from women's perspective, informed by their experiences and diversity. Serious questions need to be asked about how majority-world women are positioned in unequal global economies and the material consequences of new development models. Low-income women and men in postcolonial contexts face numerous constraints on the leverage for change, not least because of pervasive market-oriented frameworks, and the growing impacts of securitization, resource extraction, and worsening income inequalities.

Political and ethical issues lie at the very heart of postcolonial gender questions, as development policy and practice comprise in-between sites of (dis)encounter (from the Spanish *desencuentro*). Development continues to operate as a site for the reproduction (unwitting or otherwise) of coloniality's power, and hence becomes the source of rightful anger from subalterns, especially women who are often least able to challenge (neo-)colonial misrepresentation. In light of these dangers, postcolonial writers carry out research—and rethink politics—examining critically the effects of positionality. Postcolonial feminism questions the boundaries established between theory and practice, positing development practice as often ignoring the impact of community on development, and failing to achieve a real dialogue with the grassroots sites. As "polyvalent theorists and practitioners," postcolonial feminists advocate instead focusing on "the ways in which theory and practice interact . . . the real life solutions that women craft often navigate them between theory and practice, the local and the global" (Perry and Schenck 2001: 2). Considerable methodological challenges hence remain to translating postcolonial analyses into programs for change, not least because postcolonialism implies a rupture in development's configurations of power and knowledge. Postcolonialism and feminism both take a critical approach to questions of knowledge and difference. Feminism's interest in situated knowledges and partial knowledges complements postcolonialism's interest in whose knowledge is used to wield power over subaltern Others and how that knowledge is codified and normalized in ways that underpin imperialism and coloniality (Marchand 2009: 923). Pursuing this agenda, Suárez and Hernández (2008) point to the need to decolonize knowledge, i.e., the recognition of the specificities of feminist and women's struggles, jettisoning once and for all the tendency for (Western, liberal) feminism to speak on behalf of all women.

Development intervenes in societies that share features of the power hierarchies coming out of colonialism, features such as multicultural and multiracial societies, overdeveloped militarization, authoritarian features, and dependency upon economic ties with former colonies. Yet majority-world contexts of postcolonial existence are highly varied, depending upon the country and period of colonial rule, pre-colonial institutional and economic structures, and the outcomes of political struggles to define after-colonial societies and economies. Whereas development focused on modernization tended to treat Third World societies and women as homogeneous, careful ethnographic, historical, and feminist research has largely exploded that representation. Moreover, postcolonial scholars such as Homi Bhabha document the enduringly ambivalent relations between colonizer and colonized, relations often mediated by colonized elites. In postcolonial settings, one subpopulation may be trained by colonizers as "mimic men," local adopters of colonial priorities and attitudes who perpetuate Western forms of governance, identity, and goals. Gender is central to these processes as ethno-cultural and national differences between women become a symbol of national independence, as well as setting standards for femininities, femininities that become the basis of development interventions and policy priorities (Anthias and Yuval-Davis 1993).

Yet gender and development policy approaches originating in the North often view (traditional, patriarchal) culture as the prime reason for Third World women's impoverishment and marginalization by limiting their empowerment. A postcolonial perspective counters this assumption in a number of ways. First they point to the messy, globalized, and locally made quality of culture that does not exist in neatly compartmentalized national "containers" (McEwan 2009). Diverse forms of gender- and culture-based subjectivities exist where there is a deeply felt tension between tradition and modernity (Chua et al. 2000). Second, postcolonial feminist accounts document extensively how grassroots women in the South may strategically draw upon global human rights discourses for local and national struggles, highlighting the slippery and contested boundary between local cultures and global frameworks (Bhavnani et al. 2003: 12). Third, postcolonial feminist writers document the process by which female policy makers in Third World states (femocrats) extend (neo-)colonial forms of power against minority women—becoming "mimic women" in an amendment of Homi Bhabha's term. Fourth and finally, postcolonial feminist writers highlight how women in the Global South are not subject merely to development's internal or external interventions but also to a number of diverse economic, religious, legal, sociocultural "scattered hegemonies."

These diverse factors help explain the response of women's groups in the Global South as at the Johannesburg World Summit on Sustainable Development in 2002, to top-down policy initiatives such as the Millennium Development Goals (Momsen 2010: 110). "Although gender mainstreaming is a major concern of the UN, seven of the eight [MDG] goals were not formulated in gender-sensitive terms" (Marchand 2009: 932). Critics highlight how the MDG program marginalizes ethnic minority women within nation-states committed to gendered and racialized solutions. Others point to the Western cultural assumptions built into the MDGs (Olowu 2012), while others highlight the ways that international policy—despite MDGs— tends to minimize the agency of grassroots women, rolling back on policy commitments since the Beijing International Conference. Kabeer (2005: 22) moreover reminds us that "The visions and values of women's groups and organizations across the world have been translated into a series of technical goals, to be implemented mainly by the very actors and institutions that have blocked their implementation in the past."

Gender and development (GAD) and postcolonialism: core issues

Postcolonial scholar Gayatri Spivak (1993) places the new international division of labor at the heart of her account of inequalities in the postcolonial dynamics between women and men of different nations, races. She characterizes the current global situation as producing diverse human subjects, who are constituted with widely divergent interests, knowledges, and ideologies. Within these institutionalized power relations, whiteness and First World privilege work to exclude Third World subjects. Yet nationalism, modernity, and development work through and across women's bodies. In this sense "the protection of woman (today's 'third world woman') becomes a signifier of the establishment of a *good* society" (Spivak 1993: 94, emphasis in original). Women in the Global South are represented as both not-yet-modern and to-be-modernized, "caught between tradition and modernization" (Spivak 1993: 102). Hence, the "Third World woman" has become a privileged site for development to carry forward agendas for modernization (e.g., microcredit).

Yet Spivak goes onto famously argue that—given this power-driven set of agendas set by powerful Northern nations, by capitalist agendas, through masculinist states and institutions —subaltern women consequently cannot speak, their agendas being over-determined by the multiscale and interlocking power relations within which they find themselves. Such insights are critical to understanding the dynamics of postcolonialism and gender in development. Arguably too, women's movements have been erased or weakened with the establishment of state-centered technical feminisms. With the explosion of women's movements in the majority world over 20 years, subaltern women increasingly raise their voice in protest, contesting their silencing, and speaking to an array of agendas which often lie outside the parameters of mainstream development (see chapter by Parra-Fox in this volume).

Postcolonial accounts of development documented in detail how colonial administration mutated into after-colonial development, both institutionally and even in terms of the personnel involved (Kothari 2001). According to Spivak's analysis of British colonial policy in India, "white men are saving brown women from brown men" (Spivak 1993: 92). Spivak understands these transformations as part of the multifaceted processes by which imperialist subjects are produced, ensuring white Northern men retain their unmarked privileged status. Relatedly the global politics around after-colonialism has gendered dimensions—the figure of "woman in need" comprises an integral part of international geopolitics and development policies. As feminist postcolonial scholars make clear, these claims are far from innocent. They position Northern development workers as holders of liberating knowledge that they can apply on behalf of Others. In the Global South, these claims and the gender policies that follow are often contested by subaltern women.

The GAD development policy is premised on women's empowerment. Postcolonial feminist accounts welcome such an agenda, yet a close attention to this policy's normalization of power demonstrates their embeddedness within colonialist assumptions. On the one hand, empowerment is offered as a means of modernizing "Third World women." Moreover, First World policy often pictures majority world women as primarily domestic and private, thereby justifying expansion of "public roles" as a means of empowerment. Such representations oversimplify the complex and blurred connections between "public" and "private" in many societies of the South. GAD also applies a standard model over a diverse set of agendas of low-income women, equating domesticity with powerlessness and public roles with power. Examining GAD projects on the ground reveals these oversimplifications. In an empowerment project in Rajasthan, India, female project staff and female beneficiaries negotiated over the types of empowerment constructed. Fieldworkers subverted domestic/public, modern/tradition dichotomies by noting women's power at home, while low-caste women demanded pay for public "community" work (O'Reilly 2006).

A recent GAD policy shift regarding empowerment comprises "girl effect" development programs that argue that young women are the catalysts for change in families, communities, and countries. While rightly disaggregating the category of Third World women, "girl effect" programs remain profoundly influenced by colonialist assumptions. A young women's sports and development program in Eastern Uganda used the imagery of the North "saving" distant Others, as it reproduced the hierarchical donor/recipient relations. In another case, a Nike video promoted a "girl effect" message, positioning viewers as active "investors" in the (passive) lives of 600 million young women, who are positioned as self-helping subjects (Wilson 2011). The "Third World girl" is thereby distanced discursively from material embeddedness in crisscrossing institutions and arenas for action such as the household, community, social movements, citizenship, and transnational networks.

Another key point from postcolonial analyses of gender in development is that women in the South do not necessarily consider their agendas to be opposed to, or separate from, those of low-income men in similar circumstances. This representation works to justify and legitimate Northern interventions—feminist and otherwise—in female–male relations of the South. Many grassroots women in the majority world connect with and share political agendas with men in households, communities, and "mixed" organizations. In this sense they "recognize that growing insecurities among young men and difficult negotiations over salaries and control of money in homes where men do not have access to stable employment or income cannot be separated from the increasing instances of alcoholism, gambling and violence in the poorest rural communities" (Sangtin Writers and Nagar 2006: 142–143). Yet development often works from the notion that Third World women are subject to patriarchal cultures; this colonialist representation belies the disempowerment of Third World men relative to the North (see chapter by Parpart in this volume).

Building on these insights, studies explore the consequences of these representations. Development literature often reinforces gender binaries in Third World subjects, differentiating the "good" Third World woman from the "bad" men. Mainstream frameworks authorize development for the heroic or victimized "Third World woman," yet "posits 'third world men' as absent or irredeemable" and "recalcitrant" (Wood 2001: 439). Third World men can also be racialized as lazy, irresponsible, and preoccupied with sex, thereby reiterating nineteenth-century distinctions between the deserving and undeserving poor (Wilson 2011: 318). Despite the "masculinities in crisis" literature on men in the wealthy world, "Third World men" remain firmly fixed in binaries that normalize an employed, competent, rational, and racially unmarked masculinity. The widespread development model of the conditional cash transfer (CCT), as in Mexico and Brazil, adopts such gendered premises, targeting funds at women-as-mothers, while making invisible men's role in social reproduction and childcare, a fact largely unremarked-upon in the policy literature. Although these CCT programs have had noticeable positive effects on the health and educational levels of children from poor families.

In summary, the NIDL and international political economy remain key shapers of the lives of Third World women, despite empowerment discourses and models. Moreover images of Third World women as amenable to empowerment rework colonialist representations in ways that link closely to the prevailing neoliberal context for development. Women of the Global South continue to be ambivalently positioned in international development geopolitics, as subjects who are "traditional-to-become-modern" by means of development.

The fraught categories of "race" and culture

In postcolonial societies, development animates narratives of national progress and additionally often functions as a rationale for incorporating non-dominant groups into a "national" way of life. Inevitably these dynamics of postcolonial development operate through the imaginative and material power of gender and race (Fanon 1967). These racialized-gendered forms of postcolonial nation-building play out differently, depending on development goals and place-specific cultural hierarchies. In Ecuador, white urban women were offered as role models to racialized subaltern women at the same time as modernizing agricultural projects aimed to bring them into the market and reorganize their labor (cf. Hodgson 2001). In Turkey, the Anatolia Project targeted a region perceived to urgently require modernization particularly among its culturally marked (veiled, rural, religious, and ethnic minority) female population. Spivak's female figure "caught between tradition and modernity" appears here in Turkish form, as the project focused on women through "civil society participation, as well as particular efforts to reduce natality, [and] promote literacy" (Harris 2008).

Globalized representations of cultural difference circulating on the internet are changing the forms of representation of the South, the content of which remain persistently colonizing. Deliberate efforts to confront colonialist representations find it difficult to escape entrenched postcolonial associations and imagery. Since the 1990s, European non-governmental development organizations have rethought their publicity for Northern publics following new European Union guidelines. New representations comprise "positive images" of the Global South, highlighting agency and dignity, not vulnerability and risk. Yet racialized and exoticized female bodies remain central: women are represented hard at work, taking individual responsibility, in ways "consistent with the current neoliberal development consensus" (Wilson 2011: 328).

Postcolonial feminism has long critiqued the presumptions of Northern feminists to speak on behalf of Third World women (Mohanty 1988). Many utilize Gloria Anzaldúa's early insights into frontier identities as a tool to confront cultural essentialisms—whether in Northern policy views of "Third World women" as victims of "traditional" cultures, or in essentializing nationalisms. Mexico's indigenous women, alongside US Chicana and Arab women's movements, are struggling to insert a gender politics into anti-imperialist nationalism as much as against (neo-)colonialist power. Yet this resistance is also ambivalent as it "offers [women] spaces of resistance at the same time as it subordinates and colonizes their bodies in the name of identity and tradition" (Suárez and Hernández 2008: 20). Thus, although women struggle alongside men for national independence, this leaves an ambivalent legacy as cultural nationalisms and postcolonial identities construct an idealized femininity that exerts material and discursive force over women's bodies. In many cases too, "colonialism eroded matrilineal or women-friendly cultures and practices" leading women to think creatively about how to build a better present (Loomba 1998: 141).

Scholars explore how colonial difference is reproduced in an age of global development by critiquing race in development, and the privileges of whiteness. Following Said, Heron (2007) documents how Canadian women working in Africa construct an identity in relation to—and in opposition to—the aid recipients. As Canadian nationals, women reproduce their country's discourse that its aid (unlike that of the United States) is benign, while the construction of their identities in Africa perpetuate raced, classed, and gendered hierarchies of power in distinction to the local populations. By virtue of her race, class, and gender, the white feminist occupies a privileged place in a moral high ground over development and welfare, through establishing a public realm of power. Studies demonstrate white women's refusal to acknowledge their own power, and dissect their complicity in (neo-)colonialist geopolitics.

Taking these postcolonial insights further reveals how development languages and practices—in part due to GAD policy's focus on women—come to represent Third World *men* as more "feminized" than their Western counterparts. Colonialist discourses have historically considered men of color as less than fully rational and less than adult, leading to development's discourses of men as less capable of having "modern" attitudes (Loomba 1998). Global initiatives to provide opportunities to Andean indigenous groups positioned male indigenous leaders in feminized subordinate positions vis-a-vis the globalized male professionals in Washington (Radcliffe et al. 2004). In practice masculinities are constituted—as are femininities—within multiple spaces of differentiation and power; rather than a singular hegemonic global masculinity, diverse masculinities are constituted in local-global interplays (Davids et al. 2011). Postcolonial hierarchies and local-global gender constructions are routed via relations with women, themselves constituted through race-gender-nation, as illustrated forcefully in the case of Western female soldiers denigrating male prisoners in Abu Ghraib jail (Marchand 2009: 929).

In the Global South, development programs at the national level increasingly operate alongside the official recognition of multicultural populations. Multicultural populations thrown together within colonial boundaries created highly diverse internal development needs. Modernizing development treated these differences as a relic of the past, something to be overcome through incorporation into nationalist cultures and urban economies. The persistently poor development outcomes for ethnic and minority populations give the lie to this assumption. Recent development has reversed this model, granting "tradition" a renewed valence in development. This "ethno-development" tries to incorporate culturally distinctive qualities into growth-led projects; the gendered fall-out of these postcolonial adjustments is only now emerging. New legislation that reinvented customary landownership in Africa holds ambivalent effects for women (Whitehead and Tsikata 2003). In Ecuador, ethno-development positions indigenous women as cultural symbols of newly recognized communities, yet fails to clarify and enforce rights to resources, especially land (Radcliffe 2013). Reconfiguration of development around diversity hence does not erase hierarchies of gender-race-ethnicity, but reworks them.

Political economies of gender and postcolonial development

Postcolonial scholars ask critical questions about the interests served by policies for "growth," "empowerment," and wealth-generation for women in the Global South. Currently low-income women globally are offered an economic position through microcredit (see chapters by Aladuwaka and Huq-Hussain in this volume). In neoliberal development, the market also creates new channels for North–South connection such as fair trade (see chapter by Smith et al. in this volume). Wealthy consumers are encouraged to buy from fair trade networks, allowing women in the Global South to break relations of dependency and unequal trade. However, gendered scholars of these processes highlight the persistence of colonizing stereotypes. One chocolate bar advert recalled a long history of racist advertising, while sexualizing the black African woman producer (Wilson 2011: 328).

In the Global South, women are located at the juncture of multiple forms of power and oppression, not merely domestic patriarchs. With cutbacks in public services and privatization of public goods required for social reproduction, low-income women struggle to satisfy households' social reproductive needs as services and goods (water, gas, electricity) are privatized or subsidies removed. World Bank feminists base interventions on an intra-household negotiation model, "Dominant economic representations of social reproduction continue to rest upon a universalizing portrayal of the household economy and family life as mired in patriarchal tradition" (Bergeron 2011: 151). GAD hence reproduces negative representations of Third

World men, as noted above. Yet removing questions of social reproduction from wider political economic transformations means that women's rising workload remains invisible. Ignoring these labor costs and political economic factors, social development policy recently focused on the social reproduction of children, taking for granted women's unpaid labor contributions to community wellbeing. Such programs subordinate women's agency to caring for children, requiring regular child health checkups and school attendance in return for a minimal cash payment. Social reproduction is arguably one arena where the tension and disjuncture between gender and development policies, and women's grounded realities are most acute.

Whereas the current GAD intra-household model offers women the policy fix of the market, the evidence suggests that majority-world women create a *bricolage* of market, non-market, social reproductive, and self-exploiting means to survive. By excising questions about global income inequalities and domestic divisions of labor in the Global North, GAD models posit a neat separation of North from South, Third World women from First World women. Yet the international care work economy blurs this separation; Third World women are increasingly recruited into domestic care work for families in the wealthy world (see chapter by Yeoh et al. in this volume). Feminist GAD models ignore dynamics that push Third World women into transnational migration (Ehrenreich and Hochschild 2003). The care economy emerges out of and builds on colonialist power differentials and the low values accorded to (racialized) bodies of the majority world. Critical postcolonial accounts of care economies stress how women are pushed into migration by collapsing rural economies, underemployment, and inflationary living costs (Sylvester 2011).

Natural resource projects often "naturalize" the relation between women and resources (O'Reilly 2006: 961), a link that echoes Western narratives of colonized spaces as female. From a feminist political ecology perspective, struggles over the environment are not merely over resources' economic values but are embedded within power hierarchies and cultural-specific meanings. The colonialist discourse of "Third World women" being victims of change remains powerful, even as it fixes on emergent areas of policy concern such as environmental degradation and climate change. Globalized sustainable development discourses construct gender norms and performative repetitions of male–female, culture–nature dichotomies through which power is exerted in material ways. Foster (2011: 141) shows how sustainable development debates are underpinned by representations of "non-Western women as a homogenized, non-liberated, oppressed, and victimized mass [. . .] as particular victims of environmental degradation, armed conflict and violence." Linking with particular market-led policy models, sustainable development in turn justifies and legitimates neo-colonialist expansionism of Western economic interests, silencing the priorities of majority world women for food sovereignty and nutritious regional food production. Struggles over the environment and resources comprise contests around the social production of exchange and use values, gendered divisions of labor, racialized understandings of who uses resources "productively" or "efficiently," and multifaceted power relations.

Women's movements demanding change are highly place-specific, giving rise to diverse visions of development. These movements are not removed from the global landscape of power and development because they are often constituted at the interface between international gender frameworks (GAD, etc.), national political cultures, and local-regional women's struggles. Postcolonial perspectives bring crucial insights into these multiscale arenas of women's activism where development is defined, imagined, and fought for. Diverse meanings of empowerment arise in different cultural and political, geographical, and historical, contexts, often rejecting standard criteria for measuring empowerment, and demonstrating diverse different routes to empowerment (Sangtin Writers and Nagar 2006: 141). Postcolonialism questions a rigid boundary between Global North and South; "women of the 'North' and 'South' are not separate; gendered ideologies and practices are produced, sustained and challenged through inequalities across, as well as within, both 'places'" (Peake and Trotz 2002: 335). Movements in Chiapas, Mexico, illustrate women's activism operating at numerous scales simultaneously across South and North America (Speed et al. 2006). Global feminisms provide contradictory resources for Third World women's struggles; activists in the Ethiopian Women Lawyers' Association (EWLA) lobby against gendered violence through a strategic blend of global human rights discourses with "African features" (Burgess 2011).

Struggles for women's rights are hence shaped by postcolonial national identities, "global feminism's" ambivalent consequences, and the hybrid forms of agency and discourse innovatively put together by nationally embedded women.

Securitization and development from a postcolonial feminist perspective

Post-9/11 securitization of development impacts on GAD discourse and practice

Previous human development concerns—where gender had a presence, however constrained—have over recent years been displaced by a state-centered national security agenda. The new security agendas primarily use gender dichotomies as a means to mark boundaries of difference between Us and Them, Self and Other (Marchand 2009: 928–929). Whether in migration policies or resource extraction, security discourses manifest through differentiated impacts on women, men, and children in affected territories. Security discourses rework racial-gendered hierarchies, creating certain visible categories—such as young Muslim men in migration spaces—while making others less visible—such as female farmers displaced by biofuel production. In this sense, postcolonial feminist thought raises "questions about whose security and what kinds of security are important for GAD, thus challenging the current hierarchisation of security issues and the silencing or marginalizing of alternative views on security" (Marchand 2009: 931; Russo 2006).

The politics of aid to Afghanistan highlights how the security and development doctrine perpetuates colonial representations that represent Afghan gender dynamics as a dichotomy between deserving women, versus enemy men. "Afghan women's agency, power, and position within Afghan politics, society, community and kinship are marginalized from consideration in order to highlight their 'liberation' through modern sovereignty as an act of economic and political 'salvation'" (Fluri 2012: 37). Foreign aid workers self-identify as Western and modern, in opposition to the traditional and conservative Afghans. Hence, local women have contradictory spaces for pursuing their agendas. Educated women gain employment in aid programs but their expertise is ignored; middle-class Afghan women become ciphers for, rather than agents in development.

Postcolonial legacies under securitization mean that there is no singular response or experience among women of security and development issues. In Sri Lanka's civil war, uneven legacies of colonial language policy meant that some English-speaking women got jobs while others could not. Postcolonial feminist analyses highlight how women are not universally victims of war, and experience both positive and negative aspects according to class, location, religion, race-ethnicity, etc. The nature of women's political response is also not predetermined, ranging from motherist groups to feminist mobilization. In security and development contexts, a gender-specific policy is considered a "luxury" by military decision-makers, which further marginalizes women whose race-ethnicity distances them from marginalized gender experts

(Hyndman and de Alwis 2003). On the securitized Ecuador-Colombia border, diverse indigenous women are displaced from homelands and struggle to make their voices heard by the military and male-dominated ethnic organizations.

Speaking across cultures and power applies not only to development practitioners in new contexts of security. Equally challenging can be the creation and maintenance of peaceful dialogue between women during conflict and militarization, where women's development priorities may be shared, yet standoffs between masculinist ethno-nationalisms force divisions (see chapter by Ismail in this volume). Dialogue across Palestinian-Israeli difference offers one example, raising questions about what development can be done and by, and for, whom (Farhat-Naser and Svirsky 2001).

Conclusions

Feminism questions the positionality of knowledge, so too does postcolonialism—this chapter highlights the ongoing need for critical accounts of *whose* knowledge counts in the shifting development discourses and power relations, showing up the ways that North/South, male/female, racially unmarked/racially marked dichotomies and the power relations that (are) constituted in and through them operate. McEwan (2009) advocates the provincialization of north Atlantic and colonialist development. Northern subjects need to take responsibility for hierarchical relations and the knowledges upon which they rest. Similarly, Suárez and Hernández (2008: 18) advocate "spaces of collective identification with more inclusive practices and representations." These approaches offer useful accounts of racialized and subaltern women's struggles for rights, citizenship, and dignity in the context of securitized and neoliberal development, drawing on poststructuralist insights into power, difference, and resistance.

The challenges to postcolonial development research are similar to those of postcolonial development practice—how to challenge the appropriation of knowledge, and the need to take on a diversity of priorities. To acknowledge colonialist relations' enduring impacts on knowledge and practice is neither straightforward, nor innocent. Following Spivak's question "Can the subaltern speak?" Wood argues that attempts to assist "voiceless third world women authorizes, in new and equally problematic ways, the theory and practice of gender and development as a field" (Wood 2001: 430). Despite decades of women's struggles to insert difference into gender and development, the predominant representation remains a "third-world-woman-as-authentic-heroine, as a woman who is close to the earth, self-aware, self-critical, nurturing of cultures, community and family" (Wood 2001: 433). This chapter demonstrates the persistence of this representation across sustainable development, NGO publicity, and security and development. The challenge remains to "render a powerful field more consistently concerned to theorize development from the position of postcolonial people and their daily dilemmas" (Sylvester 2011: 200). A key element of this agenda involves deliberately turning the gaze away from the South to critically analyze the role of white femininities in development.

The chapter highlights how postcolonial/feminist accounts continue to have analytical purchase in relation to development even as its parameters, models, and priorities shift. Two major transformations will be of future interest. First, BRIC countries often claim to be free from colonialist baggage, yet their interventions will inevitably reconfigure racial, national, and gender politics with significant impacts for low-income women. Second, migrant trans-nationalism will entail important consequences for (urban) development as mobility grows, and programs attempt to capture remittances for development purposes (Raghuram 2009). While, on the one hand, transnational flows offer a "unique site of inbetweenness" through which to explore collective responses to non-aid funds, feminist postcolonial scholars caution against a celebration of these networks. Third, a number of countries in the South are now embarking on political economic reform distinct to Washington's neoliberalism, introducing new configurations of power constituted around gender-race-nation-location.

The "Third World" difference is increasingly a relational quality bound up in global media, racial difference, and multicultural societies, although development remains consistently defined via gendered and differential power relations. Postcolonial feminist insights suggest that development interventions will continue to be structured around national-racial-religious discourses and practices, positioning female beneficiaries in a way that reproduces power relations at the cost of women's agency.

Note

1   The term "female–male relations" is more neutral and has fewer associations with policy than "gender."

References

Anthias, F. and Yuval-Davis, N. (1993) *Radicalized Boundaries*, London: Routledge.

Bergeron, S. (2011) "Economics, performativity, and social reproduction in development," *Globalization* 8(2): 151–161.

Bhavnani, K., Foran, J., and Kurian, P. (eds.) (2003) *Feminist Futures: Re-imagining Women, Culture and Development*, London: Zed Books.

Burgess, G. (2011) "Ethiopian women activists at the global periphery" *Globalizations* 8(2): 163–177.

Chua, P., Bhavnani, K.-K., and Foran, J. (2000) "Women, culture, development: a new paradigm for development studies?" *Ethnic and Racial Studies* 23(5): 820–841.

Davids, T., van Driel, F., and van Eerdewijk, A. (2011) "Governmentalities and moral agents: male premarital sexuality in Dakar," *Globalizations* 8(2): 197–211.

Ehrenreich, B. and Hochschild, A. (eds.) (2003) *Global Women: Nannies, Maids and Sex Workers in the New Economy*, New York: Metropolitan.

Fanon, F. (1967) *Black Face, White Mask*, London: Pluto.

Farhat-Naser, S. and Svirsky, G. (2001) "Dialogue in the war zone: Israeli and Palestinian women for peace," in S. Perry and C. Schenck (eds.), *Eye to Eye: Women Practicing Development Across Cultures*, London: Zed Books, pp. 134–154.

Fluri, J. (2012) "Capitalizing on bare life: sovereignty, exception, and gender politics," *Antipode* 44(1): 31–50.

Foster, E. (2011) "Sustainable development: problematizing normative constructions of gender within environmental governmentality," *Globalization* 8(2): 135–149.

Harris, L. (2008) "Modernizing the nation: postcolonialism, postdevelopmentalism, and ambivalent spaces of difference in Turkey," *Geoforum* 39: 1698–1708.

Heron, B. (2007) *Desire for Development: Whiteness, Gender and the Helping Imperative*, Waterloo: Wilfred Laurier.

Hodgson, D. (2001) *Once Intrepid Warriors: Gender, Ethnicity and the Cultural Politics of Maasai Development*, London: Indiana University Press.

Hyndman, J. and de Alwis, M. (2003) "Beyond gender: towards a feminist analysis of humanitarianism and development in Sri Lanka," *Women's Studies International Quarterly* 31(3/4): 212–226.

Jacquette, J. and Summerfield, G. (eds.) (2006) *Women and Gender Equity in Development Theory and Practice: Institutions, Resources and Mobilization*, Durham: Duke University Press.

Kabeer, N. (2005) "Gender equality and women's empowerment: a critical analysis of the third Millennium Development Goal," *Gender and Development* 13(1): 13–24.

Kothari, U. (2001) "Feminist and postcolonial challenges to development," in U. Kothari and M. Minogue (eds.), *Development Theory and Practice: Critical Perspectives*, London: Palgrave Macmillan.

Loomba, A. (1998) *Colonialism/Postcolonialism*, London: Routledge.

Marchand, M. (2009) "The future of gender and development after 9/11: insights from postcolonial feminism and transnationalism," *Third World Quarterly* 30(5): 921–935.

McEwan, C. (2009) *Postcolonialism and Development*, London: Routledge.

Mohanty, C.T. (1988) "Under Western eyes: feminist scholarship and colonial discourses," *Feminist Review* 30: 61–88.

Momsen, J. (2010) *Gender and Development*, 2nd edition, London: Routledge.

Olowu, D. (2012) "Gender equality under the Millennium Development Goals: what options for sub-Saharan Africa?" *Agenda: Empowering Women for Gender Equality* 26(1): 104–111.

O'Reilly, K. (2006) "'Traditional' women, 'modern' water: linking gender and commodification in Rajasthan, India," *Geoforum* 37: 958–972.

Peake, L. and Trotz, A. (2002) "Feminism and feminist issues in the South," in V. Desai and R. Potter (eds.), *The Companion to Development Studies*, London: Arnold, pp. 334–338.

Perry, S. and Schenck, C.M. (eds.) (2001) *Eye to Eye: Women Practicing Development Across Cultures*, London: Zed Books.

Radcliffe, S. (2013) "Gendered frontiers of land control: indigenous territory, women and contests over land in Ecuador," *Gender, Place and Culture* www.tandfonline.com/doi/pdf/10.

Radcliffe, S., Laurie, N., and Andolina, R. (2004) "The transnationalization of gender and re-imagining Andean indigenous development," *Signs* 29(2): 387–416.

Raghuram, P. (2009) "Which migration? Which development? Unsettling the edifice of migration and development," *Population, Space and Place* 15: 103–117.

Russo, A. (2006) "The intersections of feminism and imperialism in the United States," *International Feminist Journal of Politics* 8(4): 557–580.

Sangtin Writers and Nagar, R. (2006) *Playing with Fire: Feminist Thought and Activism Through Seven Lives in India*, Minneapolis: University of Minnesota Press.

Speed, S., Hernández, A., and Stephen, L. (eds.) (2006) *Dissident Women: Gender and Cultural Politics in Chiapas*, Austin: University of Texas Press.

Spivak, G.C. (1993) 'Can the subaltern speak?', in P. Williams and L. Chrisman (eds.), *Colonial Discourse and Postcolonial Theory: A Reader*. London: Harvester, pp. 66–111.

Suárez, L. and Hernández, A. (2008) *Descolonizando el feminismo: Teorías y prácticas desde los márgenes*, Valencia: Cátedra.

Sylvester, C. (2011) "Development and postcolonial takes on biopolitics and economics," in J. Pollard, C. McEwan, and A. Hughes (eds.), *Postcolonial Economies*, London: Zed Books, pp.185–204.

Whitehead, A. and Tsikata, D. (2003) "Policy discourses on women's land rights in sub-Saharan Africa," *Journal of Agrarian Change* 3(1): 67–112.

Wilson, K. (2011) "'Race', gender and neoliberalism: visual representations in development," *Third World Quarterly* 32(2): 315–331.

Wood, C. (2001) "Authorizing gender and development: 'Third world women,' native informants and speaking nearby," *Nepantla* 2(3): 429–447.

6

GENDER AND RELIGION

'Gender-critical turns' and other turns in post-religious and post-secular feminisms

Maria Jaschok

Introduction

Carried by the momentum of Third World feminism and postcolonial studies, new schools of religious feminisms in an era characterized by Rosi Braidotti (2008) as a 'post-secular' feminist phase are paying ever more critical attention to the diversity of women's religious experiences as alternative forms of diversely embedded feminisms. Muslim women's experiences, values, choices and practices – constituting the main focus of attention in this chapter – are thus claimed by a number of prominent academics writing on Islamic feminism, foremost Saba Mahmood (2005) and Masooda Bano (2012), as purposefully negotiated rights to self-constitution. And yet a growing number of influential scholars writing on the gendered nature of Islam and Muslim culture have drawn our attention to the attendant ambiguities of such claims (see below), utilizing a wide range of what Ursula King (1995) characterizes as 'negative-critical' and 'positive-critical' approaches in the study of religion. As confirmed by the author's own research on the gendered nature of Islam and of Muslim organizations and leadership in China, while the case for close judicious examination of the intersection of gender and religion is regarded by an increasing number of academics as ever more pressing, there is also evidence in recent scholarship of the abiding instrumentalization of religion by political Islamists and other fundamentalist religious groups for legitimization of beliefs and policies that disempower women (Hélie and Hoodfar, 2012). King (1995) points to the need for continued attention to, and critique of, the global resurgence of patriarchy as religious mandate, while giving due recognition to the power of women who claim religion as a liberating force. In this 'post-secular' feminist phase, close scholarly attention is given to the diversity of women's religious experiences as alternative forms of diversely embedded feminisms. The main incentive behind Islamic feminist research today is the interweaving of women's perspectives and a faith position that subscribes to the Islamic doctrine and basic message, toward the activation of its 'just' and 'fair' principles and the production of gender-sensitive knowledge within an Islamic frame of reference.

The tension between secular/religious patriarchy and female dissenting agency forms the main thread of the second part of this chapter, featuring academic writing on female-led religious organizations in historical and contemporary Chinese society since the 1990s as an illustration of emerging post-secular feminism in scholarship in Asian contexts. Institutions of women's mosques and of female *ahong* (imams) in the Hui Muslim communities in central China evolved, developed and also disappeared in the course of a long and complex historical process as Islamic female leaders and their women's congregations negotiated for wider social space and a political voice.

Contextualizing 'gender-critical turn'

A feminist approach to the epistemological and ideological perplexities that have traditionally beset the relationship between gender studies and the study of religion is characterized for the purpose of this chapter as favouring a dual approach, entailing the search for faith-based emancipative values and rejecting discriminatory ideas and practices in the name of religion. Given an emphasis on the transformative potential of religion, the post-secular approach allows for production of alternative knowledge about minority-status women whose voices and stories, individually or collectively, have been undervalued by enlightenment feminists, assigning believing women to traditionalist or 'backward' stages of development. Rejection of the instrumentalization of religion to disempower women binds together feminisms in different religious traditions.

In her seminal work on the difficult relations between gender and religion, the scholar of cross-cultural studies of gender and religion, Ursula King, maps milestones of change as entailing a 'gender-critical turn'. Feminist studies (critical and transformative) and gender studies (more broadly conceived, more inclusive), she writes, have created paradigmatic shifts in religious study. Most crucial have been interrogation of the gendered nature of religious phenomena, the relationship between power and knowledge, the critical questioning of the authority of religious texts and institutions and involvement and responsibility of researchers undertaking such studies as gendered subjects. Yet there is no simple linear trajectory, and King warns of a persistent 'double blindness' affecting gender and feminist studies – remaining stubbornly 'religion-blind' – as well as religious studies that continue to be marked by 'gender-blind' approaches (King and Beatty 2005). The challenge to these entrenched mindsets, so King maintains, entails radical transformation of consciousness, knowledge, scholarship and social practices. Gender is so deeply embedded in religion, and religion so deeply embedded in gender regimes, that their mutually reinforcing constraint must be the subject of our most critical, analytical interrogation.

In noting the shifts in thinking that have occurred since the 1970s, gender and religious studies intersect with changes in feminist and postcolonial studies in engaging in critiques of Western and androcentric models of knowledge. Building on such critiques, women's voices as voices of emancipated faith emerge as research subjects and active agents in the production of knowledge. In turn, this has led to epistemological and ethical concerns about power, authority, representation, and to inclusiveness of representation in mainstream canons of scholarship and renewed calls for critical scholarly reflexivity in the process of research.

Cultural and academic discourses

Frances Raday (2003) refers to the fundamental shift in state culture that occurred during the time of the Enlightenment in eighteenth-century Europe when a dominant religious paradigm was replaced by secularism with all its philosophical, ideological, political and social implications. In consequence, 'All its [a secular-biased normative system's] premises, values, concepts and purposes relate to the homocentric world and to ways of thought freed from transcendentalist premises and from the jurisdiction of religious authority' (Yehoshua Arieli, quoted in Raday 2003: 1). The legacy of secularist dominance over ideas and movements of social and gender transformation was most markedly inscribed in the adversarial relationship of a 'gender-blind' tradition of the study of religion with the equally implacably hostile tradition of 'religion-blind' approaches by feminist academics to the study of women and gender.

Arguably, this first 'gender turn' in the study of religion came during the 1990s with a more wide-ranging application of critical gender theory to the traditions and study of religion, for example, with the publication of King's work that appeared to signal no less than a 'paradigm shift' in religious studies. The 'gender-critical turn' spawned a new field of inter/multidisciplinary study but also engendered critical re-readings of mainstream canons of scholarly literature, both in gender studies and in the study of religion. Illustrative of this development are the ever-growing number of academic conferences, special journal issues, general publications, student projects and academic reference works that bring to the fore a rich harvest of gender and religion studies.[1]

Carried by the momentum of Third World feminism and postcolonial studies, new schools of religious feminisms have been paying close scholarly attention to the diversity of Muslim women's experiences, values, choices and practices, which are claimed by Islamic feminism as diversely negotiated rights to self-constitution. Islamic feminism emerged from the 1990s, with Iranian feminists arguing in the wake of the Iranian Islamic Revolution – which enshrined strict Islamic gender norms at the core of Islamic legitimacy – that rights for women in the private and public spheres could be successfully derived and argued from within the Shi'ite juristic tradition. Through critical re-reading of Islamic texts and practices, scholars in the field of Islam and gender studies were noting the gap between the original message of Islam and its diverse patriarchal translations and applications, critiquing the resurgence of patriarchy as religious mandate and also evoking the power of women in claiming religion as a liberating force.

A further illustration of such a dual approach to the study of Islam comes from the Egypt-based Women and Memory Forum (WMF), a major initiative undertaken by Arab women academics. Its core mission, so the WMF website proclaims, is the production of alternative knowledge about women in the Arab world that would support justice, equal opportunities and reshape power relations within various social structures. Activities such as the recently convened international conference on Feminism and Islamic Perspectives: New Horizons of Knowledge and Reform, which took place in Cairo in 2012, have served the facilitation and dissemination of findings from the organization's Islamic feminist research projects. Women's perspectives are interwoven with a faith position that subscribes to the Islamic doctrine in order to serve the production of gender-sensitive knowledge within an Islamic frame of reference. As stated in the WMF website: 'Besides critiquing patriarchal discrimination, the ultimate aim is reform and reconstruction.'[2]

More radical still are developments that see critical interrogations by Islamic scholars of core assumptions of secular (Western) feminisms of the mandatory unfettering of women from the 'false consciousness' of oppressive forces (i.e., Islamic religion) predicated on notions of a 'pre-social self' that is constructed as existing independent of socially conditioning norms. The 'self' to be recovered and liberated, it is held, bears a strong resemblance to the paradigmatic Western model of emancipation. Saba Mahmood (2005), for instance, one of the most forceful critics of the privileging of 'universal' norms of enlightenment secularism over 'Islamic' norms, calls for the closer study of sociocultural matrixes by which trajectories of evolving individual and collective identities intersect and permeate each other. Western feminists are faced with thorny epistemological issues when Mahmood dissects what she holds to be questionable assumptions of the universality of standard norms of the liberation of women (rooted in Eurocentric, secular norms). Such universalized notions underlie, she says, a false and simplistic distinction by Western feminists of the necessary act of revolt against universally recognizable manifestations of patriarchy as the only expression of true *agency,* a precursor of the accomplishment of the liberation of women. Or they invoke, she says, *submission to* these same manifestations as the marker of a false consciousness that presents an insurmountable obstacle in the pathway to liberation. On the contrary, Mahmood maintains, there is no essential pre-social self to be shaped into emancipated womanhood in accordance with a recognizable feminist methodology of liberation through revolt and resistance. Granted this intellectual position, the question arises over the implications of context-contingency of concepts of 'subordination' or 'justice' for interpreters of women's histories. What might other voices, as, for example, belonging to women with religious convictions, tell about strategies for exerting agency and subverting patriarchal traditionalism?

Masooda Bano (2012) has studied the life and career choices made by educated, believing Pakistani women through the concept of 'the rational believer' as an agent of her fate. She maintains that highly privileged women from wealthy elite families in Pakistan, educated abroad as well as at home, make independent, rational choices for their own lives that reflect a preference for lives lived in piety and modesty. Far from suggesting disempowered passivity, the careful examination of the world of believing women, both Mahmood (2005) and Bano argue, reveals how women engage with norms and institutions that surround them, negotiating and shaping discourses to facilitate their interests and agendas. Women may choose to resist and subvert certain norms, and many do, but those who do not, so Bano (2012) writes in her study of Pakistani intellectual women, cannot be simply discounted as oppressed and disempowered victims.

Empowerment as directional

To think about 'empowerment' of women cross-culturally, particularly in societies in complex and uneven phases of development, scholars in the field of gender and development tell us, is to confront critical implications of criteria we employ when addressing empowerment as directional: from where? For whom? Towards what goal?

Advocates of feminist development approaches argue for alternative ways to 'globalization-from-above development' and an emphasis on the transformative nature of organizing out of local impulses. Modernization theorists argue that hegemonic and patriarchal orders are increasingly weakening and will disappear altogether. This is challenged by scholars who reject the belief in a unilinear process from a patriarchal to an emancipatory order, observing contradictory developments within nation-states and in transnational contexts. Religion is identified as a prominent force in tensions between *re*-traditionalization and *de*-traditionalization of gender regimes, bringing forth new regimes of justification and new semantics of legitimation. Within the reframing of theories of constitution of 'the social self', contingencies of agency and nature of preference, the 'post-secular turn in feminism' (Braidotti 2008) is opening up new areas of investigation in gender studies. For example, individual empowerment and social impact in relation to religiously informed activism is the subject of researchers investigating the gendered nature of religious philanthropy at local and transnational levels.[3]

In discussing the predicaments arising where religion or culture are exploited to accuse women of transgressing traditional markers of their assigned 'feminine' sphere of domesticity, feminist scholars draw our attention to the challenges for academics engaged in transformative research. They must not simply 'understand' what women in traditionalist contexts may have come to rationalize, Moller Okin says, 'but rather do so with a view to politicizing the deprived so that they can begin to ask new questions about their cultural norms, with a view to improving their situations'. The nature of such a serious engagement, she holds, must be one which calls upon *all* parties to imagine women's challenges and predicaments from various points of view in order not to 'standardize' or 'generalize' and 'level' but examine closely each unique situation before plans can be made and strategies developed (Okin 1995). It is only by respecting cultural specificities of local contexts and by acknowledging the vital significance of indigenous knowledge that research findings make for effective and meaningful development.

Religion as indigenous feminism – Hui Muslim women in China

Critical of Western modernization theory and its assumptions that Western modernization created the sole pathway to progress and modernity, the 'indigenous' is located in 'alternative geographies of modernities'. Women's 'habitus' contains the promise of 'alternative modernity' and of the transformative potential by which the indigenous becomes the 'feminist progressive.' This is the argument of Sarab Abu-Rabia-Queder, an Israeli Bedouin anthropologist of development. Her criticism of Western modernization theory on women is a critique of modern enlightenment models that, she says, are posing a false binary of 'public' space as modern, communicative, transparent and civic – and 'private' as closed, backward and inert in the face of change. Abu-Rabia-Queder argues for a postmodern approach and 'the embedding of local traditions in modern societies'. While the women's rights movements have historically targeted access to legal, economic and civil rights, the lack of cultural underpinnings of women's legitimacy to move within the public sphere would make a full realization of such rights difficult, even counterproductive. Abu-Rabia-Queder observes: 'Modernization gave them physical access to public space, yet they were still restricted culturally in this space,' concluding that 'for women whose modernist spaces limit their access to the public, the best solution is not to change the spaces already dominated by men, but to grant women their own separate but equal spaces' (Abu-Rabia-Queder 2006).

The rational choices made by believing women as explored in the studies of Mahmood and Bano are noted also in the author's own work among central China's Hui Muslim women. Here too not all women lack options, not all women are constrained in their agency by poverty and education, and not all women experience overt coercion in their life choices. On the contrary, the secularist nature of the Chinese communist revolution and its forceful developmentalist ideology would for a great many believers have offered temptation to disavow all religious piety and conduct. Religion, that is membership of all religions, it might be argued, is to this day tainted in public discourse by association with Confucian morality and thus by definition with women's cultural and political backwardness. It is thus not surprising that until recently narratives of women's alternative histories, weaving religious beliefs and customs into indigenous traditions and local ideals, were rare. Revisionist scholarship has redressed this gap as a new question has arisen: to what extent might indigenous, localized forms of women's practices, many religious in nature, be regarded as a source of transformative strength for women, that is, feminist in nature? Or are they compromised by their parochialism of ideal and bargaining powers?

The theoretical and empirical examination of the nature, role and impact of indigenous feminisms in relation to emancipatory development practices in diverse Muslim contexts was at the centre of a research framework that shaped the 'Women's Empowerment in Muslim Contexts' (WEMC), an international research consortium active between 2006 and 2010.[4] Indigenous strategies for women's empowerment may be understood, quoting from the WEMC interim report (Wee and Shaheed 2008: 7) as 'women's endeavours to assert their rights in their own socio-cultural context, with no attribution of indigenous identity to the women themselves'. Such an approach, it was reasoned, serves to highlight (ordinary) women's agency in the empowering processes; challenges the diktat of political Islamists worldwide wherever they deny women the right of entitlements to their own thought and action; and, moreover, serves to question certain modes and methodologies of development processes. These methodologies, criticized as infantilizing and disempowering, turn women into 'patients' rather than agents of development. The consortium's findings uncovered the wide spectrum of meanings attached by women in various Muslim contexts to notions of selfhood, progress, justice and approaches. This diversity of local ideas and practices was expressed, moreover, in the wide range of strategies by which Muslim women were found to address social issues and challenge perceived injustices at local level. For example, Hui Muslim women in China exploit indigenous traditions of female-led sites of worship and congregation to give legitimacy to public engagement with charitable and educational initiatives. Women *ahong* (imams), leading women's mosques in the relatively affluent Muslim communities in central China, have been found to organize development funds and mobilize direct aid in support of fellow believers in the more disadvantaged borderland areas of northwest China (see also Jaschok and Shui 2011).

At one time the idea of a 'modern religious Chinese woman' would have been considered an oxymoron, not only by the ideologues of the Chinese Communist Party but also by Chinese mainstream academics. Similar to the diffidence noted in discussions of international women/gender studies scholars, Chinese scholars in the field of women's and gender studies expressed a most profound ambivalence, if not outright hostility, to the role of religion in society and its impact on women's pathway to liberation. They wrote about religion as an 'opium of the female masses', criticizing Chinese women as all too gullible consumers of the false promises of otherworldly salvation. In the context of Communist Party-dominated master narratives of the liberation of the nation, and by definition of women, any alternative versions of salvation offered by women of religious faith constituted a most enduring affront to Maoist constructions of liberation. Maoist ideology of liberation was based on the dismantling of the very ties to which religious believers clung with such stubborn determination. After all, as students of Chinese women's history have argued, the Chinese Communist Party successfully mythologized 'women's liberation' into a cornerstone of its claim to legitimacy of government. Topics touching upon women's relationship with religious practices were left to scholars whose ethnic/religious 'minority' background, thus marginality, led to a convenient tokenism in academic institutions.

To engage in the 'double reading' of critical-negative and critical-positive gender perspectives (King 1995) is to examine words and practices of Islam in the intersections of state rhetoric, ethnic and religious policies and local contingencies. In their study of the historical presence of Hui Muslim women in organized Islam and of women as members of diverse local faith traditions in central China, Jaschok and Shui (2000, 2011) argue that these female-led religious organizations were able to use their historical social space to capitalize on, and thus widen, the liberating and emancipatory potential of religious practices for requirements of modern Chinese society.

This must beg the question as to whether such a postmodern approach might not make itself complicit in relations of power, whether religious/cultural/political or economic, which disadvantage women. Need remains for constant critical interrogation of the role played by 'traditions' and by patriarchal control mechanisms that reject emancipatory initiatives from within 'indigenous feminisms' as either 'endangering' to familiar norms and practices or as 'betrayal' of allegiances to the collective and destabilization of political status quo.

Chinese women's mosques: tradition as a site of contestation and change

China's Muslim population, comprising ten minorities, can be found in diverse geopolitical settings, widely dispersed, or in small clustered settlements, as is the case among Hui Muslims in central China.[5] Muslim populations also dominate larger townships and counties as is the case in the borderland areas of northwest China, where Hui and Dongxiang Muslims can be found. Women's mosques arose historically in widely dispersed Hui Muslim communities in central China where they continue to benefit from the greater openness to non-Muslim lifestyles in contemporary society. Living as citizens of a non-Muslim state, controlled since 1949 by the Chinese Communist Party, their daily interactions with non-Muslims contribute to conditions that make for their more ready access to the legal and social entitlements due to all Chinese women as well as for their more assertive stakes in rights associated with citizenship (however limited and circumscribed). Where Hui Muslim women and Dongxiang women live in closed communities, historically segregated from mainstream Han Chinese society, official treatment of these communities as 'autonomous' and 'self-governing' has given rise to 'zones of exemption' in respect to certain aspects of rights and entitlements applicable elsewhere.

An ever greater politicization of Islam and local inter-Islamic group rivalry add to the complexity of Muslim contexts in which women negotiate entitlements and constraints, rights and resources arising from their multiple embedded identities as Chinese, as members of their ethnic group, as Muslims and as women. The Chinese word for 'empowering', *fuquan,* suggests that power, understood as constitutional rights, is being *granted to* women entitled to these rights. Indeed, from this point of view, Muslim women in the most disadvantaged and powerless situations in borderland areas of northwest China would be considered 'empowered' by virtue of their identity as Chinese nationals. After all, China has some of the world's most progressive legislation concerning women and families. Yet their political, social and cultural participation in societal resources is constrained, qualified and even – where conservative Islamist traditions prevail in the so-called self-governing minority areas of China – tabooed as unseemly or *haram* (violating Islamic law).

Yet, the relationship of gender and religion is bound up with multifaceted meanings. For instance, as part of their study of the continued relevance of women's mosques in the twenty-first century, Jaschok and Shui (2011) explored how and why members of a given Muslim community perceive the institution of the women's mosque differently. Both dissenting and acquiescing perspectives came to the fore by which religious faith, political conformism, public role and uncertainties of the state's policies toward religion intersected with cultural connotations of gender, age and status. Increased contacts with Muslim countries through pilgrimages, and educational as well as financial transactions, have now made the unique institution of China's most independent women's mosques a matter of controversy even among those formerly sympathetic to women's mosques. Only a multi-voice narrative of perceptions, values and aspirations can challenge assumed religious homogeneity in positive or negative evaluations of the institution of the women's mosque and of popular support for its continuity. Interviewed Muslims contributed highly divergent understandings of the uses and significance of mosques in women's life and of the justification for their perpetuation (Jaschok and Shui 2011), revealing the importance of gender, age, education, professional and religious affiliations. Justifications for the existence of women's mosques differed even among those apparently united in support of the institution, ranging from statements that mosques are providing 'convenient facilities' for women's essential needs for scriptural education. Only women respondents invested 'their' space, i.e., the women's mosque to which they belong, with meanings of 'women's rights' and with a sense of 'entitlement' to gender equality (Jaschok and Shui 2011).

The advocacy for women's mosques may thus represent an enduring patriarchal 'walling in' of women, as well as standing for the justification of a 'modern' approach to women's rights to their own space, to their own female leadership and to equal standing in Muslim and in secular society. Believers create their sacred sites through their imagination, memories, actions and speech, with meanings derived from multiple sources and identities. Jaschok and Shui, in reference to the sociologist of religion, Kim Knott, suggested that to understand the significance of any religious site, we must comprehend its location in time and space and within its surrounding, everyday environment. Sacred space should be seen as 'inevitably entangled with the entrepreneurial, the social, the political, and other "profane" forces' (Jaschok and Shui 2000, 2011).

The case for religious literacy

Increasingly, the treatment accorded in mainstream development studies to religion has been subjected to critical appraisal making the case that little explanatory power may be derived from a secularist, developmentalist discourse that tends to essentialize religion either as a source of social progress or as an obstruction of social progress. Such weak religious literacy, it is maintained, first of all fails to understand the way that (Christian) religious values inform the Western researchers' own positionality and relationship with the subject matter. But lack of comprehension of religion as it plays itself out at the most local levels of belief and practice also leads to the failure to address central issues of agency and choice. And yet these concepts lie at the core of the study of social change. A multiplicity of standpoints and judicious approaches to religion that are both nuanced and context-specific is needed. The basic right of women to 'self-constitution' and to exercise of choice must be recognized. At the same time we must raise critical questions about the conditions under which choices are made and preferences expressed for given beliefs and practices.

The aforementioned international research consortium Women's Empowerment in Muslim Contexts (WEMC) incorporated researchers and research sites from three Muslim majority countries (Indonesia, Iran and Pakistan) and from one non-Muslim country (China). Selected communities in diverse Muslim contexts were studied for the conditions under which women's empowerment was negotiated by women themselves, subject to their own terms and aspirations, reconciling religious belief and local cultural traditions with the pressures and opportunities of modernity. Persuasive findings from studies across national and cultural boundaries illustrated many incidents of women's collective strength, of their capacity to formulate perspectives informed by individual and communal needs, and their resourceful strategies for changes to benefit all, while retaining deeply held religious convictions. Such findings, it was hoped, would serve to demystify widely held stereotypes of women's passivity and submission to dominant paradigms of Muslim femininity and of their unquestioning obedience to religious and secular patriarchies. On the other hand, the assertive agency by the believing women studied in the WEMC project, ready to organize for causes that women consider urgent, jars with the still ubiquitous lack of women's institutional presence in the religious sphere, or their minimal political representation in the public domain of society. More research is required. Remaining under-researched and under-explored, and thus in need of further examination in the field of Islam and gender, are the potent and manifold implications of women's inclusion in, and exclusion from the mosque, embodying the sacred and the profane, and the progressive and patriarchal dimensions of Muslim practices.

The current phase of what Braidotti characterizes as post-secular feminism provides a welcome and timely opportunity to consider some of the issues associated with 'gender-critical turns' that are bringing about major shifts in the relationship between gender and religion, allowing us to consider women's notions and practices of development from a faith position that need not preclude the exercise of agency, of considered choice and of preferred modes of modernity. The growing number of symposia, conferences and research projects in the area of religious gender studies make manifest the emphasis, shared with gender studies across disciplines, on voice, agency and respect for diversity of standpoint and conceptions of social justice and just development. Global economic and environmental crises as they are played out in a world order of distinctly demarcated religious affiliations and identities make the role played by faith-based development organizations and in particular by women's participation in these organizations as well as by women's religious philanthropy, individually or in institutional form, a subject matter of critical importance for scholarship (see chapter by Fiddiyan-Qasmiyeh in this volume). Although considerable scholarship is devoted to the subject of religious philanthropy, more critical attention is required to study the gendered implications of cross-cultural charitable transactions for the faith communities and their influence in shaping development discourse.[6]

Conclusions

'The clash between culture or religion and gender equality rights has become a major issue in the global arena. It is probably the most intractable aspect of the confrontation between cultural and religious claims and human rights doctrine' (Raday 2003: 3). Susan Moller Okin, cited above, argues that, while our ideas of justice are universalizable, we must yet be able to take note of difference. Not a 'substitutionalist' feminist practice is called for but an 'interactive' (or 'dialogic') feminism, both responsive and democratic. In the author's own study of women's lives and female-led Islamic organizations in the Chinese Muslim diaspora, observations relate to the difficulties of engaging in debates over women's agency, equality and empowerment in a context of Islam and diverse Muslim cultures. These are, as was pointed out above, in important respects due to, among other factors, feminism's historically hostile rejection of religion as detrimental to women's progress. And yet we are beginning to note a discernible, if uneven shift from post-religious to post-secular feminism.

The first international conference in Chinese studies to focus entirely on women/gender and religion took place in 2011 in Macau. It was a remarkable milestone in a long, personal but also more general academic trajectory that consistently had marginalized the study of religious women. Or, to use a Lori Handrahan term, had rendered women religious 'the invisible actor'. Invisibility, however, does not and has never equalled inaction. The editors of a book based on the Macau conference, *Gendering Chinese Religion: Subject, Identity, and Body* explain the aim of the publication to be a vigorous 'challenge [to] the "double blindness" in Chinese gender studies and Chinese religious studies by exploring from various "gender-critical" perspectives previously ignored gender patterns embedded in Chinese religious life.' Importantly, all the volume's contributions place women at the centre of analysis, 'exploring the formation of the female subject of religiosity and women's agency in negotiation with religious and social norms'.

When discussing the emerging work on women's constructions of faith within diverse faith traditions, on their aspiration and transcendence, and how women interpret intersectionalities of agency and structure, dissent and acquiescence, the author noted the consequences of a decentring or dislocating (borrowing from the terminology of subaltern study) of secularist feminist dominance. In the ongoing challenge to binary thinking in which enlightenment, modernity and progress are defined on the back of a religion-bound Other, two pertinent and interconnected facts may be noted. First, the recognition of the richness of the study of women, gender and religion (in China as elsewhere) as a subject matter of academic pursuit and, second, that the theoretical and empirical contribution made in whatever discipline and subject matter problematizes the relationship between women and religion. Although her reference is to Islam and gender in the Middle East context, Mounira Charrad represents a wider scholarly preoccupation in her explanation of the significance of interrogations of gender and religion, namely, 'first, to dismantle the stereotype of passive and powerless Muslim women and, second, to challenge the notion that Islam shapes women's condition in the same way in all places' (Charrad 2011: 417). In meeting the challenge, Charrad holds, the academic must be prepared to engage herself in a multiplicity of ways – whether this means facing up to the legacies of Orientalism, engage with a diversity of schools of feminist thought or problematize epistemological and methodological implications of the diverse nature of narratives for understanding women's agency and choice in the context of a volatile and fragmented world order. Due recognition to the power of women who claim religion as a liberating force, to return to Ursula King's (1995) exhortation cited at the beginning of this article, must ever entail a scholarly responsibility to probe and critique the implications for believing women of the global resurgence of patriarchy under the guise of a religious mandate.

Notes

1 Illustrative of major reference works published in recent years within the diversity of disciplines and topics that mark gender and religion: a six-volume set of *Women and Belief, 1852–1928* in the History of Feminism series (edited by Jessica Cox and Mark Llewellyn, London: Routledge); a six-volume *Encyclopedia of Women & Islamic Cultures* (edited by Suad Joseph, began publication in 2003, ongoing, Leiden: Brill); *The [Oxford] Encyclopedia of Islam and Women* (New York: Oxford University Press). Another important publication came with the freely accessible online academic journal *Religion and Gender*, which commenced publication in 2011, www.religionandgender.org.
2 See the Women and Memory Forum, Giza, Egypt, for the mission of Forum and its core intellectual and academic objectives and activities: http://fipcairo.wordpress.com.
3 For example, the ongoing research conducted by the anthropologist K.E. Kuah-Pearce whose early work includes 'The poetics of religious philanthropy: Buddhist welfarism in Singapore', in B.S. Turner (ed.), *Religious Diversity and Civil Society: A Comparative Analysis*, Oxford: Bardwell Press, published in 2008.
4 Entitled Women's Empowerment in Muslim Contexts: Gender, Poverty and Democratisation from the Inside Out (WEMC), the international Research Programme Consortium was funded by DFID, UK government, between 2006 and 2010. An interim report was published in 2008 (Wee and Shaheed 2008).
5 Rights enshrined in the Chinese Constitution and in the Law of the People's Republic of China on Regional National Autonomy, include independence of finance, independence of economic planning, independence of arts, science and culture, organization of local police and use of local language. Modelled on the Soviet Union, five autonomous regions, 30 prefectures, 117 counties and 3 banners were established after the communist takeover. Autonomous administrative areas, as well as the various rights granted to them, are affirmed by the government as a positive example of local self-rule in

ethnic areas, and an acknowledgement of minority self-determination unprecedented in Chinese history. They are seen as preserving the culture of minority peoples within a larger, stable Han Chinese society. Critics of this policy point to lack of autonomy, given government-appointed officials, and they draw attention to the real site of authority. While the head of local government comes from the minority population, the local Communist Party secretary does not. Since the 1950s, China's population has been classified in terms of ethnic, geographical, cultural and linguistic criteria; it comprises altogether 56 nationalities, with the largest constituted by the Han nationality (equated with Chinese culture per se).

6   Among them, J.A. Clark (2004) *Islam, Charity and Activism: Middle Class Networks and Social Welfare in Egypt, Jordan and Yemen*, Bloomington: Indiana University Press; P. Sundar (ed.) (2002) *For God's Sake: Religious Charity and Social Development in India*, New Delhi: Sampradaan Indian Centre for Philanthropy; also Kuah-Pearce, as cited in note 3.

References

Abu-Rabia-Queder, S. (2006) 'Between tradition and modernization: understanding the problem of female Bedouin dropouts', *British Journal of Sociology of Education* 27(1):3–17.

Bano, M. (2012) *The Rational Believer: Choices and Decisions in the Madrasas of Pakistan*, Ithaca, NY: Cornell University Press.

Braidotti, R. (2008) 'In spite of the times: the postsecular turn in feminism', *Theory, Culture & Society* 25(6): 1–24.

Charrad, M.M. (2011) 'Gender in the Middle East: Islam, state, agency', *Annual Review of Sociology* 37: 417–437.

Hélie, A. and Hoodfar, H. (eds.) (2012) *Sexuality in Muslim Contexts: Restrictions and Resistance*, London: Zed Books.

Jaschok, M. and Shui, J.J. (2000) *The History of Women's Mosques in Chinese Islam*, Richmond: Curzon.

—— (2011) *Women, Religion, and Space in China: Islamic Mosques and Daoist Temples, Catholic Convents and Chinese Virgins*, New York: Routledge.

Jia J., Kang X. and Ping Y. (eds.) (2014) *Gendering Chinese Religion: Subject, Identity, and Body*, New York: State University of New York Press.

King, U. (1995) *Religion and Gender*, Oxford: Blackwell.

King, U. and Beatty, T. (eds.) (2005) *Gender, Religion and Diversity: Cross-Cultural Perspectives*, London and New York: Continuum.

Mahmood, S. (2005) *Politics of Piety: The Islamic Revival and the Feminist Subject*, Princeton: Princeton University Press.

Okin, S.M. (1995) 'Inequalities between the sexes in different cultural contexts', in M.C. Nussbaum and J. Glover (eds.), *Women, Culture, and Development*, Oxford: Oxford University Press.

Raday, F. (2003) 'Culture, religion, and gender', *International Journal of Constitutional Law* 1(4): 663–715.

Wee, V. and Shaheed, F. (eds.) (2008) *Women Empowering Themselves: A Framework that Interrogates and Transforms*, Hong Kong: SEARC, City University of Hong Kong.

7

FEMINIST POLITICAL ECOLOGY

Rebecca Elmhirst

Introduction

Feminist political ecology has established itself as an influential subfield within gender and development studies and, from this, shares a broad commitment to understanding and addressing the dynamics of gender in relation to the natural environment and in the context of natural resource-based livelihoods. Feminist political ecology (FPE) offers an explicit emphasis on power and politics at different scales, directing attention towards the gender dimensions of key questions around the politics of environmental degradation and conservation, the neoliberalization of nature and ongoing rounds of accumulation, enclosure and dispossession associated with each of these. Its genealogy owes much to the wider field of political ecology, which seeks to understand 'the complex relations between nature and society through a careful analysis of . . . access and control over resources and their implications for environmental health and sustainable livelihoods' and explaining 'environmental conflict especially in terms of struggles over 'knowledge, power and practice' and 'politics, justice and governance' (Watts 2000: 257). For feminists working in political ecology, a key question has always been to ask in what sense is there a gender dimension to such struggles, and how might these intersect with feminist objectives, strategies and practices?

In one of the first contributions to this field, Dianne Rocheleau and colleagues set out a road map for FPE research and practice, by inviting political ecologists to extend their analysis of power to include gender relations, and to extend their consideration of politics to take in closer scales of analysis, thus complicating arenas of assumed common interest, such as 'community' and 'household' (Rocheleau et al. 1996). *Feminist Political Ecology: Global Issues and Local Experiences* offered a refreshingly open-ended and loosely configured framework that treated gender as 'a critical variable in shaping resource access and control, interacting with class, caste, race, culture, and ethnicity to shape processes of ecological change, the struggles of men and women to sustain ecologically viable livelihoods, and the prospects of any community for "sustainable development"' (Rocheleau et al. 1996: 4).

Since the publication of this landmark text in 1996, FPE has had to respond to new challenges in environment and development. There has been a pronounced intensification of economic reform programs that favour market-led approaches to natural resource management, while at the same time most rural populations are marked by heightened geographical mobility, as rural–urban and transnational linkages complicate and rework resource-based livelihood practices and institutions, often in gender-differentiated ways. Moreover, processes of environmental degradation (deforestation, desertification, climate change and urbanization) have intensified, bringing new and often enlarged shocks and stresses to livelihoods. Each of these has reconfigured patterns of natural resource use, heralding new forms of development intervention and environmental governance. Such interventions are themselves imbued with gendered power relations: they are inflected with gender discourses and assumptions that set in motion differentiated and unjust life opportunities and exclusions. At the same time, within gender and development studies more broadly, as Cornwall et al. (2007) have persuasively argued, there is disquiet that, amid efforts to mainstream gender into natural resource management interventions and into development policy more broadly, gender has lost its critical and politicized edge, having been institutionalized into a series of tools and techniques that are far removed from the wider emancipatory goals associated with the label 'feminist'.

It is in this context that there has been a renewed interest in the potential offered by FPE as a set of ideas, principles and practices that is well-placed to respond to the wider transformatory potential of feminism in gender and development studies, and that can address the kinds of challenges posed by global environmental change and neoliberal, marketized responses to these. Recent contributions to FPE suggest that this field is being remade to account for and represent the extraordinary conditions in which contemporary gendered lives and livelihoods are being reworked (Elmhirst and Resurreccion 2008). Debate is being invigorated at a time when social theory offers profound challenges to earlier gender and environment approaches, and as wider political economies of natural resource governance invite a vigorous analysis of gender and gendered socio-natures across multiple sites and scales. There is immense potential offered by FPE for gender and development studies as it is enriched and animated by its location at an exciting intersection of feminist theories of power and subjectivity, post-humanist debates concerning the relationships between human and non-human natures, and developments in the wider field of political ecology. This chapter provides an overview of recent work in this field, exploring the new conceptual terrain that underpins feminist political ecology and that is opening up new questions and concerns for consideration. The chapter begins by mapping out what counts as feminist political ecology before examining the specific ways this newly invigorated FPE contributes to gender and development studies in four related bodies of work: (i) analyses of gendered resource access and control, (ii) recent theorizations of gendered subjectivity and power, (iii) emerging debates around the relationship between human and non-human nature, and (iv) ideas around a feminist ethics of environmental care. Each of these areas is inspired by particular iterations of feminist theorizing, from feminist-inflected Marxist analyses of enclosure and resource access through to new materialist feminisms that are prompting a radical rethinking of the permeable boundaries between humans and nature, and in turn, raising new questions for gender and development studies more broadly.

Defining feminist political ecology

While 'gender and environment' forms an important theme within gender and development studies, relatively little work self-defines as 'feminist political ecology'. This is not to say that there is a paucity of research on substantive themes that are of interest in feminist political ecology. Research bearing a 'family resemblance' (Watts 2000: 271) to feminist political ecology may be found across a range of disciplines, focusing on substantive issues ranging from gendered resource access and property rights (water and land) to the dynamics of gender in policy discourses, collective action and social movements, much of which might be regarded as FPE but is not named as such.

This includes work on gender and resource access, where the gendered effects of land titling, alienation of common lands, green and corporate land grabbing, decentralized governance and resurgent 'traditional' or pluralistic tenure arrangements unquestionably deserve at least the attention they are getting. Similar debates around resource access are being heard with regard to water in both urban and rural settings in the context of neoliberalization processes. Analysis of gender dynamics in community-based institutions, gendered environmental knowledge and the dynamics of gender in policy discourses and within environmental departments of development agencies also connect with what Rocheleau et al. (1996) envisaged as FPE. Moreover, the field of environmental justice includes contributions that demonstrate the productive ways feminist thinking may be brought to bear on environmental research and activism.

Few of the contributions associated with these areas of concern carry the label 'feminist political ecology', reflecting in part some profound challenges that herald from within feminist research and praxis more generally, and that find parallels across development studies. In part, this reflects a lack of appetite for delineating the boundaries of feminist political ecology in a clear and unequivocal way, other than through a loose commitment to address the politics of gender and environment thought, practice and activism. However, it remains that in some instances, the label 'feminist' in gender and environment research carries unhelpful resonances and unwanted political meanings. This is particularly so in some contexts in the Global South, where both postcolonial critiques and new conservatisms (religious or otherwise) complicate everyday understandings of feminism's meaning and intent (Cornwall et al. 2007), in ways that may prove unhelpful in engagements with the varied participants and audiences of research on gender, development and the environment.

A second and related explanation for the relative invisibility of FPE as a label relates to the profound changes seen in social theory more generally. Since Rocheleau and her colleagues outlined the principles of feminist political ecology in the early 1990s, poststructuralist approaches to power, subjectivity and women's agency have grown in influence, placing the 'decentred subject' at the heart of many debates. In such work, 'gender' is destabilized as a central analytical category: instead, emphasis is given to an exploration of multidimensional subjectivities where gender is constituted through other kinds of social differences and axes of power such as race, sexuality, class and place, as well as through everyday, embodied practices of 'development'.

In gender and environment research, some of the negative unintended consequences of strategic essentialism – articulating a centred 'Third World woman' subject in order to build bridges between women globally and to seek women's inclusion in emerging sustainable development agendas – became all too clear in the 1990s. Such representational strategies were followed by a range of initiatives dubbed 'women, environment and development' that targeted women as a homogeneous and undifferentiated social category, charged with 'care' for degraded environments, resulting in the exacerbation of social and gender injustices in a number of documented instances (Leach 2007).

More nuanced approaches are now evident within the myriad activist networks that focus on environments and social justice, where gender is seen neither as analytically central nor as the end point of critique and analysis. Instead, people are conceptualized as inhabiting multiple and fragmented identities, constituted through social relations that include gender, but also include class, religion, sexuality, race, ethnicity and postcoloniality (see chapter by Radcliffe in this volume), as well as in multiple networks for coping with, transforming or resisting development. The importance of this kind of approach lies in its power to problematize naturalized and undifferentiated categories of people and social relationships (men, women, gender relations) and, critically in this context, relationships between people and the environment.

The question that remains is, can feminist political ecology articulated in this way be regarded as uniquely 'feminist'?

More recently, a new wave of 'material feminist theory' has grappled with such questions, suggesting that in line with a feminist relational ontology, there are always instances where 'bringing into view' women and gender transformations may be an ethical imperative. The epistemological questions being raised by this work are providing inspiration for renewing a 'feminist' perspective on environmental questions. As Karen Barad (2007: 394) puts it, 'intra-active practices of engagement not only make the world intelligible in specific ways but also foreclose other patterns of mattering ... therefore accountability and responsibility must be thought in terms of what matters and what is excluded from mattering'. The remainder of this chapter considers some of the avenues currently being pursued in FPE, and in which gender is shown to matter in important ways.

Gender politics of resource access and control

Drawing on political ecology's Marxist heritage, much FPE has centred on questions of resource access and control, with a particular emphasis on Global South contexts. Various studies share a concern to illuminate 'the crucial role of family authority relations and property relations in structuring the gender division of labour and access to rural resources' such as land and labour (Carney 2004: 316). Work has detailed the gender-specific impacts of ecological change and/or environmental interventions, insofar as these are shaped by existing household divisions of labor and differing resource rights of men and women. More recently, this kind of work has been taken forward in studies of the gendered impacts of nature's neoliberalization (as showcased in Resurreccion and Elmhirst 2008). A common theme is that men and women hold gender-differentiated interests in natural resources through their distinctive roles, responsibilities and knowledge within household/family divisions of labour. Gender is thus understood as a critical variable in shaping processes of ecological change and the pursuit of viable livelihoods (Elmhirst and Resurreccion 2008: 5).

In parallel with gender and development studies more generally, FPE has also considered household and community gender relations as a critical and often overlooked site for politics, particularly where environmental interventions have brought about gender conflict within households (and across conjugal partnerships) generating in turn ecological effects. Some of the most compelling accounts of such processes herald from African contexts. For example, Carney (2004) analysed the intra-household conflicts that arose following interventions to enhance the productivity of wetlands in Gambia. Irrigation schemes and horticultural projects had the effect of undermining women's customary access to rice land for income, and at the same time precipitated new demands by men on female labour, both of which were widely contested within households and communities (Carney 2004). Ethnographic work of this kind that emphasizes family (kinship) in negotiations over access to labour and resources is also revealing of the ways that gendered resource contestations are also struggles over meanings and identities, situated in specific cultural and historical contexts, and that gender categories are themselves negotiated and socially produced in the course of environmental struggles.

The importance of gender in family authority structures and conjugal relations is perhaps most clearly drawn in recent debates over gender differences in access to natural resources, particularly in settings where the ability to derive benefits from resources is contingent on social relationships that constrain or enable the realization of such benefits. In settings such as South Asia, hierarchical social norms and practices associated with the conjugal partnership mean women are very heavily dependent on male kin as a central conduit for access to resources (including land, labour and capital), and that this creates gender-specific vulnerabilities for those experiencing marital breakdown or widowhood. FPE has offered the conceptual tools necessary for revealing intra-household power dynamics of this kind and, in probing the assumed division between public and private spheres, has shown how gender discourses and practices cast through national and international policies bleed into the reproductive realm. For example, in a series of African case-studies reviewed by Razavi (2003), the impacts of recent changes associated with liberalization are a significant factor in the marginalization of women from natural resources. While such changes transfer kin-based resource access to the market (through individual title to newly commoditized land), they have reduced women's access to land as community members. At the same time, women's rights to land *as wives* has been established through household land titling, reworking the importance of marriage and, by extension, ways of 'doing gender' in different contexts (Razavi 2003). Within these kinds of FPE analysis, conceptual weight is given to the ways in which capitalism transforms and produces nature, and as such processes intersect with gender hierarchies at different scales, patterns of enclosure and marketization are seen as having important gender effects.

Politics and ecology: producing gendered subjectivities

In recent years, debates in FPE have been taken in new directions through the influence of poststructuralist theories of subjectivity. This kind of work explores how gender is constituted in different contexts as a component of multiple and complex subjectivities. The performance of masculinities and femininities construct and reconstruct the gendered subject through peoples' everyday practices. This approach allows masculinities and femininities to be regarded as in process: fragmented, provisional and wrought through the interplay of culture, class, nationality and other fields of power, and through regulatory frameworks such as normative heterosexuality. The emphasis on social construction challenges essentialist and binary views of relations between men and women that may overemphasize difference and opposition, and that may essentialize particular patterns of gendered disadvantage. Moreover, such an approach allows space to consider other kinds of gender relations that may be significant in peoples' lives beyond marital relationships; for example, seniority and status.

The significance of this view for gender, environment and natural resource management studies is that it shifts the direction and emphasis of analysis. Rather than seeing gender as structuring peoples' interactions with and responses to environmental change or shaping their roles in natural resource management, the emphasis is on the ways in which changing environmental conditions bring into existence categories of social difference including gender. In other words, gender itself is re-inscribed in and through practices, policies and responses associated with shifting environments and natural resource management, and whilst inherently unstable, through repeated acts, comes to appear as natural and fixed. Nightingale's (2011) work in Nepal shows how gendered subjectivities are defined and contested in relation to changing ecological conditions, becoming salient at particular moments through practices and discourses of gender associated with livelihoods, natural resources and wider development programmes and policies. This kind of work demonstrates the politics that underlie how gendered subjectivities are constituted. More recently, ideas have been taken forward through the concept of intersectionality in order to show how gender is but one of a series of modes of difference – including race, ethnicity, age, sexuality – that, in their interaction, constitute subjectivities.

Intersectionality is an approach for studying the relationships among various dimensions of social relationships and subject formation. Subjectivities (people's subject positions) are produced through the way axes of power (gender, race, ethnicity, class, sexual orientation, age, ability) intersect and emerge in relation to one another, rather than being based on stable or given understandings of social difference.

Feminist political ecology is contributing in important ways to intersectional theory by paying attention not only to how, for example, patriarchy and racism are imbued in shaping human-environment relationships (e.g., Mollett and Faria 2013: 117) but also to the role of nature in producing particular identities and bodies. In other words, gendered subjectivities emerge from the convergence of political economic structures and embodied everyday practices in *specific ecological contexts*. For example, Nightingale (2011) shows how the ecological materiality of space is crucial in producing (materializing) gender in Nepal. One aspect of her discussion is an empirical analysis of timber cutting, which is largely undertaken by Dalit (lower caste) men. Her discussion weaves around material and symbolic meanings attached to the forest and ideas about purity, but central to her discussion is how the materialities and ecologies of timber work (through men's everyday practices) are integral in redefining caste and in the reproduction of social inequalities. Her work shows how feminist political ecology can contribute and expand intersectional theories as nature plays a fundamental role in the constitution of gender, class, race and ethnicity through everyday practices of access to and control of resources.

A similar perspective has been adopted to examine how the dynamics of resource access and control link to normative and non-normative sexualities in Indonesia (Elmhirst 2011). In analysing the experiences of landless migrants and their efforts to secure land in forest margin areas, an intersectional feminist political ecology approach reveals how normative heterosexualities are in part produced through the materialities (labour requirements, positionings required to secure land) of access to particular kinds of nature at different times. In other words, the materiality of nature and its production through cultural meanings (around access) and work (everyday labour practices) mean that peoples' engagement with forest ecologies are also a dimension in the power plays that produce gendered subjectivity.

While much feminist political ecology work has centred on rural settings and the historical relationships that produce politicized agricultural environments (around women's access to land, for example), work in FPE also engages with the production of urban natures, and in particular, political ecologies of water access and the governance of sanitation (see chapter by Greed in this volume). This work draws on earlier Marxist-influenced analyses that trace the production of nature and city spaces through capital–labour relations and, more recently, through neoliberal policies that have further entrenched the commodification of water provision and brought about class-based water inequalities on a city-wide scale. New waterscapes are produced as people engage with the material things of the waterscape and in so doing, produce and reproduce power relations. Feminist political ecology is engaging with this work by extending it beyond the city, national or global scale to consider closer and more intimate scales where multiple social differences are reproduced through everyday and embodied water practices. Truelove (2011) describes how the materialities of water shortage in Delhi lead girls and young women to experience a constriction of their spatial mobility in ways that shape their life opportunities. She shows how a feminist political ecology view that centres on everyday experiences reveals the ways particular bodies bear the brunt of subsidizing and compensating for state water governance strategies, while at the same time, everyday practices around water reproduce multiple hierarchies of social difference. While there is an emphasis on gendered 'despair' in such work, others have suggested that a FPE analysis opens up space for focused attention to be paid to how the materialities and meanings of city waterscapes are produced through gendered practices, i.e., the unpaid reproductive labour of women as they seek to provision households with water. As Loftus (2007) suggests in his study of protests around water access in Durban, this in turn reveals the political possibilities that emerge through women's common concerns. What each of these studies of gender, environment and the politics of subjectivity demonstrate is how a FPE that is informed by theories of embodiment and intersectional subjectivity avoids an unhelpful retreat to simplistic understandings of gender divisions and singular gendered power relations in relation to the environment in a development context.

Materialist feminism and the agency of non-human nature

Recent FPE has provided a renewed focus on subjectivities in understanding the politics of human–nature relationships and, more specifically, gendered resource access and control. However, there is also a current of thinking that seeks to reconsider how 'nature' is understood, not as simply a backdrop against which social relations are played out, but through a 'post-humanist relational ontology'. Within new material feminisms, as explored in the collection edited by Alaimo and Hekman (2008), the modern nature–society dualism is replaced by ontologies that see phenomena as mixtures of human and non-human elements that interact in the form of assemblages. The importance of these kinds of ideas in FPE can be seen in three emerging threads within this subfield, all of which hold resonance in gender and development studies more broadly. First, a post-humanist relational ontology goes beyond the idea of bounded, interacting bodies to instead consider the flows between and through organisms, and between human and non-human natures. Those working in FPE have drawn on similar conceptualizations to analyse the metabolic flows associated with food, making important links between the ecologies underpinning neoliberal globalized food systems, production and consumption practices, and the more traditional feminist terrain of gendered bodies (Hayes-Conroy and Hayes-Conroy 2013). It is in this area of FPE that some key parallels with work in environmental justice emerge, where the seepages of pollutants and carcinogens across and between human and non-human natures are spatially uneven and associated with racialized and gendered processes of social and spatial marginalization.

A similar emphasis on the flows and interactions between human and non-human nature has been deployed to investigate the devastation of Hurricane Katrina, and in particular, the 'in-between' of 'natural', 'human-made', 'social' and 'biological' phenomena. Tuana (2008) invokes the agency of non-human nature to see through the eye of Katrina and consider the embodiment of levees, hurricanes and swamps as well as the embodiment of the women and men of New Orleans. The significance of this for FPE is how such ontologies challenge the privileging of the human subject and the centrality of human labour in understanding environmental change, and instead show the generative capacities of a pluralist non-human nature.

Finally, the interaction of human and non-human nature has been linked back to theories of intersectionality as those working in FPE begin to think about how the experiences of gender, race, ethnicity, class, age and so on often take shape through species-ist ideas of human-ness vis-à-vis animality (Hovorka 2012). Drawing inspiration from Donna Haraway's (2008) work on companion species, the focus in this line of work is on the doing and becoming of social identities across species boundaries. As Hovorka puts it: 'certain groups of humans become symbolically associated and materially related to certain other (non-human) species (and vice versa) – this process, together with hierarchical privileging and othering, reproduces the positionality and life chances of both humans and non-humans within society' (Hovorka 2012: 876). With particular resonance for gender and development studies, these kinds of ideas are put to work in a study of women chicken farmers in urban Botswana, where Hovorka examines the symbiotic relationship between women and chickens, understood through their mutually constituted positionality. After exploring the various ways in which women in Botswana are marginalized, this study considers a parallel process in which chickens are marginalized in relation to cattle, who occupy a privileged and valued position in terms of Botswana's development. What emerges is a picture of the shifting positionalities of women and chickens in symbolic and material terms in the context of agrarian restructuring, peri-urban development and official support for small-scale enterprises, all of which have brought a degree of empowerment to both women and chickens. In effect, feminist post-humanist thinking is used in an FPE context to consider the ways gender and species hierarchical arrangements work, in this case, in city spaces in urban Botswana.

A feminist ethics of care

In drawing on recent poststructuralist and post-humanist feminist theories, FPE is developing a suite of approaches for addressing critical environment and development challenges in the twenty-first century. The question remains, however, what animates a specifically 'feminist' politics in FPE? One pathway being pursued within FPE is to explore how a feminist ethics of care can be put to work to offer a post-capitalist alternative to neoliberal forms of natural resource-based development. Embracing a transnational perspective on environment and development and, in particular, the links between production and consumption worldwide, insights from FPE are being used to analyse practices of ethical consumption and cause-related marketing, and the ways in which such practices constitute the subjectivities of (women) consumers in the North, and human–environment relationships between and across the Global North and South.

While there is a sense in which some of these forms of ethical consumption draw on and reproduce forms of neoliberalism and do little to challenge material inequalities (including those associated with gender disadvantage), other possibilities for expressing an ethics of care are also being documented within FPE. Jarosz (2011), for example, examines the motivations of women farmers involved in community-supported agriculture in the USA and concludes that these are expressive of an 'ethics of care' that involves a sense of them nourishing themselves and others, nurturing people and the environment, as part of 'an ethical positioning that challenges the processes of privatization, unfettered capital accumulation, competition and discourses of personal responsibility for inequality and poverty, which construct individuals as neoliberal subjects' (Jarosz 2011: 308). While Jarosz is careful to avoid an essentialist connection between women and care for the environment and distant others, she suggests that, through their care work in community-supported agriculture, the women in her study reveal motivations that are not primarily economic, but rather are associated with social goals and desires to live a work-life that is satisfying and meaningful, and that express discourses of a post-capitalist politics and that might be interpreted as offering a radical responsibility to others. This renewed interest in ethics brings current approaches in FPE back to the agenda first set out by Rocheleau et al. (1996), in which the transformative potential of feminist activisms is seen as a key thread within this subfield.

Conclusion

Looking across all of these areas, what is clear is that FPE does not align closely with one, narrow analytical framework. While FPE broadly comprises an eclectic mix of theoretical positions drawn from feminist theory more broadly, there is a common commitment to presenting a re-politicized recognition of gender as an optic for analysing the power effects of the social constitution of difference: something Cornwall et al. (2007) suggest is a requirement in gender and development studies in general. From its beginnings in the early 1990s as a sub-field of gender and development studies, and through its engagement with recent poststructuralist, post-humanist and post-capitalist feminist theory, feminist political ecology demonstrates the myriad ways that feminist theorizations and new understandings of gendered subjectivity can be taken forward within and through the permeable boundaries of an open-ended *feminist political ecology*. As with political ecology more generally, this theoretical ecumenism is its strength when it comes to addressing some of the more pressing environment and development challenges of our time. Many important avenues are currently being explored as new feminist political ecologies are being articulated: the themes introduced here are just part of a continued flowering of this revitalized and important area of debate.

References

Alaimo, S. and Hekman, S. (eds.) (2008) *Material Feminisms*, Bloomington: Indiana University Press.

Barad, K. (2007) *Meeting the Universe Halfway: Quantum Physics and the Entanglement of Matter and Meaning*, Durham, NC: Duke University Press.

Carney, J. (2004) 'Gender conflict in Gambian wetlands', in R. Peet and M. Watts (eds.), *Liberation Ecologies: Environment, Development, Social Movements*, 2nd edition, London: Routledge, pp. 316–336.

Cornwall, A., Harrison, E. and Whitehead, A. (2007) 'Introduction: feminisms in development: contradictions, contestations and challenges', in A. Cornwall, E. Harrison and A. Whitehead (eds.), *Feminisms in Development: Contradictions, Contestations and Challenges*, London: Zed Books, pp. 1–20.

Elmhirst, R. (2011) 'Migrant pathways to resource access in Lampung's political forest: gender, citizenship and creative conjugality', *Geoforum* 42: 173–183

Elmhirst, R. and Resurreccion, B.P. (2008) 'Gender, environment and natural resource management: new dimensions, new debates', in B.P. Resurreccion and R. Elmhirst (eds.), *Gender and Natural Resource Management: Livelihoods, Mobility and Interventions*, London: Earthscan, Ottawa: IDRC, and Singapore: ISEAS Publications, pp. 3–22.

Haraway, D.J. (2008) *When Species Meet*, Minneapolis, MN: University of Minnesota Press.

Hayes-Conroy, J. and Hayes-Conroy, A. (2013) 'Veggies and visceralities: a political ecology of food and feeling', *Emotion, Space and Society* 6: 81–90.

Hovorka, A. (2012) 'Women/chickens vs men/cattle: insights on gender-species intersectionality', *Geoforum* 43: 875–884

Jarosz, L. (2011) 'Nourishing women: toward a feminist political ecology of community supported agriculture in the United States', *Gender, Place and Culture* 18(3): 307–326.

Leach, M. (2007) 'Earth mother myths and other ecofeminist fables: how a strategic notion rose and fell', *Development and Change* 38(1): 67–85.

Loftus, A. (2007) 'Working the socio-natural relations of the urban waterscape in South Africa', *International Journal of Urban and Regional Research* 31(1): 41–59.

Mollett, S. and Faria, C. (2013) 'Messing with gender in feminist political ecology', *Geoforum* 45: 116–125.

Nightingale, A. (2011) 'Bounding difference: intersectionality and the material production of gender, caste, class and environment in Nepal', *Geoforum* 42: 153–162

Razavi, S. (2003) 'Introduction: agrarian change, gender and land rights', *Journal of Agrarian Change* 3(1/2): 2–32.

Resurreccion, B.P. and Elmhirst, R. (eds.) (2008) *Gender and Natural Resource Management: Livelihoods, Mobility and Interventions*, London: Earthscan, Ottawa: IDRC, and Singapore: ISEAS Publications.

Rocheleau, D., Thomas-Slayter, B. and Wangari, E. (eds.) (1996) *Feminist Political Ecology: Global Issues and Local Experiences*, London: Routledge.

Truelove, Y. (2011) 'Conceptualizing water inequality in Delhi India through a feminist political ecology framework', *Geoforum* 42(2): 143–152.

Tuana, N. (2008) 'Viscous porosity: witnessing Katrina', in S. Alaimo and S. Hekman (eds.), *Material Feminisms*, Bloomington: Indiana University Press, pp. 188–213.

Watts, M. (2000) 'Political ecology', in T. Barnes and E. Sheppard (eds.), *A Companion to Economic Geography*, Oxford: Blackwell, pp. 257–275.

8

NAVIGATING GENDER AND DEVELOPMENT

Ragnhild Lund

Introduction

Any historical narrative is influenced by the situation and context of its author. Personal histories, experiences, and practices may inform people's meanings, values, convictions, and forms of expressions, and their context may help us understand how knowledge is made and how it is being reproduced and changed over time. At every juncture in a person's life course, insights into the strengths and weaknesses of their 'entanglements' (who they interact with and how) can be crucial to understand their capacity to act and how their interactions are situated. In my case, I had an opportunity to work for the Norwegian aid authorities as an undergraduate student in 1976, which formed my professional history and practices. I had written an exam paper on gender and development and was invited by a federation of women activists to assist the Norwegian Agency for Development Assistance (NORAD) in identifying how bilateral projects affected the situation of women. While working for NORAD in Oslo, on a six-month assignment, I also pursued my Master's thesis work at the Department of Geography, University of Bergen. It was this work that motivated me to study how resettlement programmes impacted female settlers in north-eastern Sri Lanka. NORAD staff facilitated my first visit to Sri Lanka in their newly opened Colombo embassy, introducing me to scholars and practitioners who worked on gender issues in the country. Since then, most of my professional work has been related to gender and development, primarily in Asia.

Through having long been part of the field of gender and development studies as both a researcher and practitioner, I have found various bodies of work important at various times and in varying contexts. However, it is difficult to view one's thoughts, ideas, and arguments in retrospect. The standpoint from where I perceive and interpret my previous work is different today from yesterday. Instead of trying to gain full understanding of what was important when a piece of work was written, or trying to trace previous misjudgements, I find it more interesting to assess only certain aspects of an earlier work, in an effort to identify major shifts of thought related to thematic choices, theory, and methodology. Such shifts may illuminate the development process of gender-related research.

In this chapter, autobiography is used as a method to understand how meanings of self are embedded in knowledge production related to gender and development. I present a brief history of the gender and development field by examining how I myself have navigated between practical and policy-relevant work and feminist research at several turning points in my professional career from 1976 to date. In reflecting critically on these turning points, my intention is to unravel how they have shaped my performances as a scholar but also how I have shaped them.

In the following, six turning points are identified, starting with the early stage of doing women's studies and development work, from the time I worked for NORAD and was a student (1976–1980); the time when the quest for a global structural view was raised in feminist circles and I myself worked as a junior scholar contributing to starting DAWN (Development Alternatives with Women for a New Era) (1980–1990); the point where researchers and practitioners in gender and development developed along different paths, and I wrote my doctoral thesis (1990–2000); the years of what geographers term 'the cultural turn',[1] when the gender and development field expanded into a wide variety of specific areas – in my case gendering post-disaster/post-conflict geography (2000–2010); and, more recently, to my present research on how mobility is gendered and my recent project on how researchers who have worked with gender issues for a long time think about gender today. In the concluding section, I identify what remain as key concerns in gender and development studies, as well as what seems to be changing.

Turning points

1976–1980: from WID to GAD

Prior to 1970, when Esther Boserup published her landmark book on women and development, it was thought that the development process affected men and women in the same way.

(Momsen 2004: 11)

Norwegian development assistance identified women as the poorest and most marginalized target group in 1977. My first consultancy was a sector-based examination of NORAD projects with regard to women (Lund 1977). It was a follow-up of the UN Women's Conference held in Mexico City in 1975, and the subsequent Chapter III of the UN Plan of Action, which concerned research, data collection, and consultancy work. The plan gave priority to all kinds of research and applied studies that could illuminate the situation of women at administrative and political levels. As a result, NORAD, in collaboration with women's groups and organizations, wanted to assess how much aid and what types of project activities benefitted women.

Methodologically, my consultancy included the study of letters, project documents and programmes, interviews with NORAD personnel, and a review of relevant literature. A total of 78 bilateral projects were examined in detail, looking at leadership and preparatory project work, participation and control, benefits for 'target groups', and implications for social change and project objectives. My findings were quite discouraging. I documented great ignorance about, even ridicule of, the issue of women in development (WID) among the respondents; NORAD projects did not have any policy formation related to sex and gender, and a sector-based organization of aid had rendered women's efforts invisible, although the agricultural sector had given some priority to income generation among women. The report suggested that the organization of NORAD routines and staff should be revised to integrate aspects of WID policies and strategies in its donor administrative structure.

The consultancy contributed to the initiation of several WID-related activities, plans, and campaigns. Norwegian aid directed various efforts towards strengthening the position of women (through bilateral and multilateral strategies, and mainstreaming gender in partner organizations). Official objectives were the alleviation of poverty for women, the achievement of equity between men and women, and the integration of women in all types of aid and productive activities, hence WID. I got involved in numerous consultancies in the African and Asian regions. Even though gender was later used as an analytical category, compared to WID, gender and development (GAD) policies never really took off in the organization and gender-specific interventions were basically targeted towards women only. These practical experiences shaped and sharpened my gender lens in the research that I conducted simultaneously for my Master's degree.

The earliest and most fundamental finding of both WID and GAD approaches was the marginalization of women in agriculture. My own master's thesis (Lund 1979), on the changing roles and functions of women in the 'modernization' of the Mahaweli resettlement area in Sri Lanka, documented how the traditional gender-based division of labour was altered, and women were placed in the situation of agricultural labourers for their household heads (husbands or fathers). First and foremost, this increased the women's dependent status, as well as their workloads. Second, women lost control over resources, such as land, which impacted on patterns of marriage, inheritance, and usufruct rights. Their access to new technology and inputs required for cash crop production was also limited. Third, the mobility and activity spaces of women had become more limited compared with those of men. Fourth and finally, the workload of women had increased more than that of men, expanding from housework, childcare, and subsistence-related activities, to include increasing participation in cash-crop production. The results of the study were presented in several publications during the period 1978–1983, and were heavily debated in Sri Lanka. Later, I used data from further visits to the Mahaweli area as a case study for my doctoral research.

Looking back at these formative years of my academic career, it is clear that close contact with practitioners and feminists influenced my positioning as a feminist geographer. From early on, I closely followed the discourses on practical gender needs (WID) and later strategic gender needs, power, and patriarchy (GAD). I realized that I was given an enormous chance to be part of what became a worldwide process, in which the Norwegian authorities contributed to shape gender policies.

1980-1990: towards a North–South divide

For Norway as a small country with a modest development aid involvement, all the elements of this tripartite consequence of research – the production of knowledge, of individual competence and of international research contacts – are essential.

(Skjønsberg cited in Stølen and Vaa 1991: viii)

In Norway, a research programme, Women and Development, was funded by Norwegian development aid authorities (1985–1990). I participated in the programme as a board member and later as a researcher. The overall aim of the programme was to promote and strengthen national expertise on women and development issues and to make this expertise available to development agencies. Three broad areas were identified: 1) production, work, and economy; 2) politics, power, and control; and 3) culture, knowledge and ways of life. In total, 20 Norwegian researchers from different fields participated and provided a wide range of approaches, locations (Asia, Latin America, Asia), and themes.

In a later anthology and largely inspired by actor-oriented approaches, most of the researchers analysed processes of change from 'below', uncovering the strategies of individuals or social groups in various sociocultural contexts characterized by rapid market integration. Some were concerned with continuity and change in women's and men's behaviour, others with discontinuities at the levels of ideas and perceptions of gender, while a few attempted to explore the intersection between the two levels (Stølen and Vaa 1991: 1). Emerging issues were the agency of women versus their victimization, a gender analytical perspective instead of a specific woman focus, in an effort to unmask gender biases in development interventions, the legal system, and technological innovations.

The study of young Muslim workers in the women and development programme that I conducted together with Merete Lie exemplifies the production theme. In the first stages of our work we critically examined the thesis of the new division of labour arising in the wake of restructuring of industrial production in Europe to Asia. We attempted to explore how Norwegian industrialization relocated to Asia impacted upon gender roles and whether there was any truth in the thesis of 'super-exploitation of women' (the exploitation of cheap female labour in appalling work conditions in global capitalist production), which dominated the feminist discourse. In an in-depth study of a local community in southern Malaysia, where some of the women were employed in a Norwegian factory, we found that women who had been previously protected and controlled had entered the labour market and to a large extent had replaced their fathers as the main income earners in their families. This provoked discontinuities in the meaning of gender, which had to be renegotiated and redefined without challenging the traditional authority structure. The women's identity and value still depended on how they conformed to traditional values, such as marriage, motherhood, and what, in their words, constitutes 'a good woman'.

Later visits to this study area over a period of more than 20 years led us to question the original findings. The middle-aged women that we met in the late 1980s were now mainly housewives and few had taken up work outside the village because, as they said, of the need to stay at home to look after children and the elderly. Instead, other young women, mainly foreign workers from countries such as Bangladesh and Nepal, constituted a large section of the industrial labour force in Malaysia. The factory we had studied had since moved to cheaper places for production (Mauritius, the Philippines, later China), thus exemplifying the footloose nature of industries in the era of globalized production.

In the same period, I was given the chance by the Christian Michelsen Institute in Bergen to assist Devaki Jain in formulating a Third World women's network of professional activists and scholars. The inception meeting of DAWN was held in Bangalore in 1984 and lasted for a full week. As a junior university lecturer, I clearly remember not being able to utter a word, as I was so taken aback by all the versatile, charismatic, and bright women present. In a later meeting in Bergen the group decided that Caren Grown together with Gita Sen would write a platform for action (Grown and Sen 1987). Since then, DAWN has made a huge difference by mobilizing Third World women worldwide, and it has provided a critical forum at all international women's conferences. It was a pioneer in raising Third World women's voices (albeit from an intellectual elite's point of view), in relating the processes of marginalization of women to inequality of power (emphasizing the role of access to resources, power, and the need to cut across caste, ethnicity, and culture), to structural adjustment programmes and development reforms, and to the relationship between the West and the rest (which raised criticism over dominant Western feminism).

1990–2000: researcher–practitioner distanciation

The 1990s represented a turning point in my career. I gradually turned away from practical and aid-related work on gender and development and instead took up a position as an associate professor at the Asian Institute of Technology in Bangkok. For three years (1992–1995) I taught a gender course in addition to more regular development studies courses. I met resourceful students from all over Asia, and some of them remain good friends. In the evenings I worked on my doctoral thesis on gender and place. This happened during a time when donor agencies started to gender mainstream[2] their organizations, gender and development specialists started to handle the gender and development agenda, and gender-related research (at least in my country, Norway) started to distance itself very much from donor debates and instead turned to explore deeper meanings of gender differences, roles, and relations.

Inspired by the works of scholars such as Vandana Shiva, Bina Agarwal, Janet Momsen, Pierre Bourdieu, Anthony Giddens, John Friedmann, Henri Lefebvre, James Duncan, Robert Chambers, and others, I grappled with feminization of poverty, (dis)empowerment, contextualization, science 'from below', and the role of women's agency. Drawing on my previous work in Sri Lanka and Malaysia, I constructed an analytical framework where I explored the links between gender and place in an effort to understand what constitutes social change from a gendered perspective (Lund 1993). Two spatial levels became important, the household and place. The household level was perceived as the smallest decision-making body, which divides familial and communal tasks according to custom by gender and age. It also provided the starting point from which to explain and solve conflicting human behaviour (as an arena of female subordination), gender-based division of labour, sexual relations, control over income and property, inheritance rights, children's education, and women's participation in civil and political affairs. While place hitherto had been understood as sites (land and territory), I took up the idea that place could also be analysed as a combination of spatial practices; forms of knowledge and hidden ideological content of codes, theories, and concepts; and abstract space as 'social imaginary'. This implied that relations between gender and place are context-specific (as social relations, their particular form cannot be generalized and is a product of particular socioeconomic, political, and ecological situations), neither gender or place are static relations (they change over time, and are intended or unintended products of socioeconomic and political decisions and actions), and gender and place represent a particular set of power relations (which find their expression in a range of institutions in society).

Some general conclusions were drawn from this work. While gender-related approaches saw the transformation of gender roles and identify gender as an integral part of development processes and social change, geography and development studies did not necessarily include gender. Against this background, it became clear that gender roles and gender relations are not formed by patriarchy alone. Other power structures, such as cultural (religion), economic (class), and political institutions, determine male and female ways of life. Hence, I saw a need for thicker contextual descriptions and less of a Eurocentric donor-driven and elitist development, which could provide more complex understandings (particularly with respect to my own discipline of geography).

In hindsight, the work described above represented a rupture with my own tradition of collaborating with practitioners. Instead, I turned back on my discipline of geography and, while critiquing it, I tried to formulate a focus that legitimized the study of gender and development. In a wider perspective, this work shows how thoughts at the time were already dealing with issues that were later termed 'intersectionality', culture, different means of power, feminization of poverty, and 'situatedness', to some of which I now turn.

2000–2010: multiple realities and situatedness

In drawing on culture as lived experience, a WDC [women, culture, development] lens brings women's agency in the foreground (side by side with, and within, the cultural, social, political and economic domains) as a means for understanding how inequalities are challenged and reproduced.

(Bhavnani et al. 2003: 8)

In the social sciences, the 'cultural turn' led to a multitude of research topics and challenged established approaches to development, which hitherto had focused largely on politico-economic aspects. Now emphasis was put on rights, social justice, the invisibility of reproduction, and women's political agency and situatedness. During this period, I carried out research on a broad range of topics: on children, reflecting on the participating child and its development through a gender lens; studying new faces of poverty and gendered livelihoods in Ghana; and studying empowerment, participation, and home-making in Sri Lanka. More importantly, I coordinated a small but effective forced migration research group, in which I myself continued with my engagement in development-induced displacement in Sri Lanka.

When the tsunami event of 2004 occurred in Asia, the focus of my work shifted to post-disaster recovery issues, which allowed me to return to how practitioners work. On 26 December 2004, the Indian Ocean tsunami devastated the coastlines of Sri Lanka. More than 35,000 people lost their lives, and it is believed that 500,000 people lost their homes. The disaster created new challenges for the people, the government institutions, and the humanitarian community in dealing with the shock and devastation and in finding ways to recover. A collaborative programme (2005–2008) between a Scandinavian NGO and our forced migration research group was developed to improve practices of recovery in the NGO. It encompassed a wide range of activities, such as action planning, developing a livelihoods strategy, monitoring routines in the field, and gender-sensitization.

The various project activities were based on the needs articulated jointly by the above-mentioned NGO and the university group. They should not be seen in isolation, but as parts of an integrated whole; for example, gender related to livelihoods, housing related to action planning, conflict related to how the organization worked, and also to how local people and leaders were approached and how they participated. Gender-sensitive approaches were developed in planning, in the formulation of a livelihoods strategy, in advisory services, and in participatory research. Community-based organizations and other local partners were targeted, primarily to identify bottlenecks in the operations and to stay loyal to the local mission of the NGO. The central administration of the organization was targeted to strengthen the awareness of how gender is a structural issue and embedded in the power structures in the organization. Valuable South–South learning was obtained through a visit to the Self-Employed Women's Association (SEWA) in Gujarat, India. The purpose of the training course and visit to SEWA was to create an opportunity for the local NGO staff to become exposed to and learn about the intervention strategies of the livelihoods component of SEWA.

The work of this collaborative project contributed to the accountability and transparency of practices and events within the NGO. For the university researchers – including myself, in my dual role as both coordinator and researcher – this experience contributed to deeper insights into the work and priorities of development agencies, and their interactions with different levels in their organization, and among stakeholders and other partners. However, some critical weaknesses influenced the success of the collaborative programme. Some related specifically to gender (my being a female coordinator and advising on gender issues within a male-dominated organization). Other weaknesses related to the fact that the project was initiated from the top, that the realities on the ground constantly shifted, that it was difficult to bridge differences in practitioner and academic understanding, and that a lot of knowledge in the local communities did not reach the main offices of the NGO in Colombo and in Scandinavia. Also the limited timeframe of the project activities prevented the achievement of significant social change.

Later, together with two of the academic project partners I wrote an article about our experience with gender-sensitization in post-crisis situations. The article analyses the relationships between partners in an NGO's attempt to mainstream gender in their post-tsunami projects (Attanapola et al. 2013). It shows the tensions between the requirements of quick recovery results and the aims of creating changes in society through gender mainstreaming. From interrogating the structural and individual challenges of working with gender in a post-crisis situation, it was found that different understandings of 'gender' prevailed within the NGO at different levels. For example, while the gender coordinator in the head office gave attention to both women's needs and structural constraints for changing existing gender relations, the field officers transformed their understanding of gender to encompass livelihoods, improvement and income generation in general. The fact that the various stakeholders understood 'gender' differently also resulted in different gender priorities. Arriving at a common understanding in this situation would have been useful, but we found that this did not happen. Albeit greater flexibility at the local level may have contributed to achieving more gender change, the organization did not succeed in its efforts to address how gendered structures are embedded in a larger political economy. Instead, it encountered well-known structural and organizational challenges and communication flows, which did not enable knowledge of gender to be co-produced between the various levels of the organization and achieve gender change (Attanapola et al. 2013: 83).

Looking back on all that we said and wrote on mainstreaming gender, the project clearly enhanced my understanding about the way ideas of gender mainstreaming have been circulated and how they must be better contextualized rather than uncritically adopted from the international aid organizations' blueprints on gender mainstreaming. This view fits with related discourses on gender mainstreaming. Sylvia Walby (2005: 453), for example argues, that:

> Tensions can arise as a result of actors seeking to mainstream quite different models of gender equality: based on equality through sameness; through equal valuation of difference; and through transformation. The intersection of gender with other complex forms of inequality has challenging implications for a primary focus on gender within gender mainstreaming.

In a recent article published in the *Emotion, Space and Society* (Lund 2012), I used my own fieldwork notes from this project as the source for understanding my emotions and the role of collaboration in my development as a feminist geographer. I think back on numerous meetings with those who were marginalized and suffering, how I dealt with feelings of guilt and hopelessness about not being able to be instrumental in changing injustice and malpractices and doing good. I realized that while the post-tsunami experience provided insights into biases among partners, for me personally the situation provided grounds for frustration that impacted on my research, as significant dilemmas arose regarding loyalty to the various project partners. Today, I find that the work on emotions expanded my understanding of collaborations among researchers and practitioners. Of particular concern became the divided loyalties – to the poor people on the ground or to the organization, which were not compatible, resulting in tension.

2010 – globalism and transnationalism

In the steadily growing literature on globalization, a gender perspective is missing, while in the literature on gender and globalization, the phenomenon of globalization is taken for granted and, thus, stays underexposed.

(Davids and van Driel 2007)

I have related to increasing globalism and transnationalism in two distinct ways: through a study of gendered mobility in China, India, and Laos, and through studies of rethinking gender and development in India and Malaysia.

Today, more and more people live on the move. This 'mobility turn' (of individuals, commodities, disease, organizations, and knowledge) characterizes the present development of the Global South (for we have moved from talking about the Third World to referring instead to the Global South). Jonathan Rigg (2012) argues that there is a need to develop a 'new mobilities paradigm' with a southern focus in the social sciences, where present forms of mobility and livelihoods are explored. Together with three Asian colleagues I have intended to fill in this gap through the research project Mobile Livelihoods and Gendered Citizenship: The Counter-geographies of Indigenous People in India, Laos and China (2010–2013). This study explores how the mobility of indigenous women and men in rural areas exposed to neoliberalist policies and structural reform are translated into new livelihoods. We argue that this process is largely gendered. It expands the current area of migration analysis by studying not just the relation between place of origin and place of destination, but the whole manner in which men and women manoeuvre differently through space.

As has been well documented by researchers such as Diane Elson, Saskia Sassen, Ruth Pearson, Naila Kabeer, Sylvia Chant, and others, the impacts on women's lives of major restructuring in the global economy have shown how the drive for profit has treated female labour as a commodity. Increasingly, women have to find paid work outside their home and community, leading to feminization of labour. Furthermore, global demands for agricultural and mineral products have pushed developing economies away from producing food and wage goods for their own people, resulting in migration and deteriorating traditional livelihoods, and increasing the burden of poor women and men seeking food, shelter, and other necessities. Economic restructuring is also intertwined with other structures of inequality, such as ethnicity and class.

The findings of the mobility study in China, India, and Laos concur with such findings. Subsequent to contextual and structural changes, such as patriarchy and new socioeconomic realities, we find that mobilities are gendered in new ways, which is amply demonstrated in the case of indigenous populations. Pushing women and men into different positions makes them challenge conventional gender norms. In Laos, there is a problem that relatives become dispersed due to resettlement; in India, tribal women organize themselves into work collectives and engage in wage work far away from their place of origin; in both Laos and China, agriculture is feminized because the men have moved out; and in both China and India, women become involved in new sectors of the economy as domestic workers, prostitutes, entertainers, agricultural labourers, and industrial workers, while men are pushed into mining and construction work. It is evident that rather than being an issue of power struggle within the confines of the household, women's access to new roles and spaces may become less dictated by local cultures and traditional social practices. Such a change has the potential to both empower and disempower women in mobility situations.

The final project I turn to is Revisiting Gender in Development: Complex Inequalities in a Changing Asia (2011–2014),[3] which aims to both revisit and rethink the concept of gender in a rapidly changing Asia. Starting from Cornwall's (2007) call to revisit the 'gender agenda', the overall objective of the project is to revisit gender as a concept that can engage simultaneously with change and continuities in today's developing regions and to ascertain whether gender as a social category has remained true to its earlier feminist promise of emancipatory and empowering outcomes for women and men, or whether it has instead blunted this possibility, as Cornwall suggests. Thirteen scholars from various European and Asian countries have come together to critically examine the concept of gender with respect to changing subjectivities; environment and resources; power, policy, and practices; justice and human rights

The sub-project Re-visiting Gender Activism – Tribal Women in Odisha, India with Smita Mishra Panda provides insights into ways in which activism through self-help groups (SHGs) and gendered practices intersect with other factors in indigenous people's lives and are intertwined with the body and body politics. Two major challenges for activist groups have an impact on indigenous women's lives in Odisha. First, most SHGs have contributed to make women visible in public and political spaces. However, most groups are unable to scale-up their activities due to patriarchal norms, social and political divides, and location. Second, the body spaces of the activists – both leaders and grassroots members – are caught in a complex web of politics of the Indian state, Hindu culture, and patriarchy, hence 'colonized'. Sadly, we find that the voices of the indigenous people are more marginalized and suppressed than ever despite their activism and the assistance given by local NGOs. The women themselves are of an opinion that they have come out in public space, are active, make their presence felt and bring about some changes for the better in their lives. We realize that there is a need to rethink gender by including the marginalization by other processes of oppression, such as ethnicity, class, location, and development policies.

Let me end this section with a brief note on the Malaysia study, which adds self-reflexivity to my work. The study Re-visiting Gender, Industrialization and Modernization in Malaysia with Zaireeni Azmi and Merete Lie is based on the long-lasting fieldwork (1988–2012) that we undertook in Johor Barhu with the aim of studying life course and livelihood strategies of women involved in foreign industrial production. During this period, our analyses have been affected by the general trends within feminist and gender studies, evidently such as the change from women's studies to gender studies and the warning against essentialism and dichotomization in gender studies. The field was revisited by a young Malaysian researcher during 2011–2012, with the intention to scrutinize the Western view of the previous findings, bring in a new gender lens, and provide fresh insights from the field. These data are used to deepen and illustrate our discussion as well as to throw insights into what changes and what remains. We find that our perspective is influenced by our changing theories, concepts, and research approaches, but at the same time our empirical findings make us question the very same theories as they are embedded in our own knowledge production and awareness.

We are currently about halfway through the Revisiting Gender in Development project. What strikes me when learning about the various sub-studies is that the agenda is no longer Western or Southern. Among the 13 subprojects, only four researchers come from the Global North, while the others are from the Global South and are critically grappling with research topics that are indigenous to them. It appears that our collaboration contributes to reduce a Eurocentric lens and instead build transnational knowledge.

Change and continuity in perspective

Unravelling what remains and what has been changing in my work on gender, I realize that juggling between doing development work and research over the years has given me a unique opportunity to learn how this field has been changing. Conducting the present scrutiny about my own work also makes me see myself change in this process towards becoming increasingly sensitive to the complexity of gender and being self-reflective.

Inquiries from 'below' have characterized my work throughout, through numerous field visits and ethnographic work in several Asian and African countries. Gradually, I became more sensitive to the changing power relations and contested notions of power of those 'below'. Equally important, I also questioned modernist gender and development 'top-down' definitions of development organizations and development programmes at an early stage. I continue to critically grapple with issues related to women's empowerment, as in the study of activism in India where such issues remain pertinent. What is new in my work is the examination of the interpretative power of gender in institutions, as in my post-tsunami work. This represents a break with my earlier works on gender mainstreaming, WID- and GAD-related consultancies. I now seek understanding of whose knowledge and voices count in research, as well as critically reflecting on how we work with gender issues in our collaboration with research partners. This reflects a development from focusing on the feminine subject and feminist methodologies to a critical examination of self in research and collaboration with research participants to conflate power differences.

In my projects I always perceived women as active agents of change, despite oppressive situations. Gradually, I co-opted a focus on gender, and increasingly tried to understand its complexity and situatedness. A challenge remains in that current gender and development approaches are inadequate to match the complexity of gendered lives. I have covered some ground with respect to *intersectionality* – the way gender is constructed for an individual depends on the gendered interactions the individual has with others as well as other identities or roles he or she may have, such as ethnicity, caste, class, and other oppressions, by putting emphasis on uniqueness and the significance of context. However, I have yet to come to grips with *performativity of gender* – how human beings behave in ways that consolidate an impression of being a man or being a woman – and *gendered entanglements* – how material phenomena (e.g., new technologies) in various ways are intertwined with gender, power, and knowledge.

The turning points that I have identified in this chapter thus coincide with related issues in other research circles. To some extent, I have contributed to existing discourses, and in other cases I have provided additional insights. What I find particularly intriguing is that the structural adjustments that we tried to understand in the 1980s still remain challenges for men and women in the present globalized and transnational society. This happens while research, activism, and political movements across borders have increasingly become transnational. Ideas are shared most effectively, fostering partnerships across the previous North–South divide. I will continue to explore how the complexities of gendered lives can inform the present unstable and mobile nature of gender.

Notes

1   The movement among scholars in the social sciences to make culture the focus of contemporary debates.
2   Mainstreaming a gender perspective is the process of assessing the implications for women and men of any planned action, including legislation, policies, or programmes, in all areas and at all levels.
3   Partly coordinated with Babette Ressurreccion and Philippe Doney, Asian Institute of Technology.

References

Attanapola, C., Brun, C., and Lund, R. (2013) 'Working gender after crisis: partnerships and disconnections in Sri Lanka after the Indian Ocean tsunami', *Gender, Place and Culture: A Journal of Feminist Geography* 20(1): 70–86.

Bhavnani, K.-K., Foran, J., and Kurian, P. (2003) *Feminist Futures: Re-imagining Women, Culture and Development*, London: Zed Books.

Cornwall, A. (2007) 'Re-visiting the gender agenda', *IDS Bulletin* 38: 69.

Davids, T. and van Driel, F. (eds.) (2007) *The Gender Question in Globalization: Changing Perspectives and Practices*, Aldershot: Ashgate.

Grown, C. and Sen, G. (1987) *Development, Crises, and Alternative Visions: Third World Women's Perspectives*, New York: Monthly Review Press.

Lund, R. (1977) 'NORAD-prosjektenes betydning for kvinnenes stilling, Oslo/NORAD (The significance of NORAD-projects on women's situation)', internal report.

—— (1979) *Prosperity through Mahaweli – Women's Conditions in a Settlement Area*, Master's thesis, University of Bergen.

—— (1993) *Gender and Place: Towards a Geography Sensitive to Gender, Place and Social Change*, doctoral thesis, Department of Geography, University of Trondheim.

—— (2012) 'Researching crisis – recognizing the unsettling experience of emotions', *Emotion, Space and Society* 5: 94–102.

Lund, R., Kusakabe, K., Panda, S.M., and Wang, Y. (eds.) (2013) *Gender, Mobilities, and Livelihood Transformations: Comparing Indigenous People in China, India and Laos*, London: Routledge.

Momsen, J.H. (2004) *Gender and Development*, London: Routledge.

Rigg, J.D. (2012) *Unplanned Development: Tracing Change in South-East Asia*, London: Zed Books.

Stølen, K.A. and Vaa, M. (eds.) (1991) *Gender and Change in Developing Countries*, Oslo: Norwegian University Press.

Walby, S. (2005) 'Introduction: comparative gender mainstreaming in a global era', *International Feminist Journal of Politics*, 7(4): 435–470.

PART II

Environmental resources: production and protection

9

INTRODUCTION
TO PART II

Both academic and practitioner views of women's engagement with the environment have shifted over the years, from a women and environment focus that portrayed women as having a natural, inherent, and altruistic relationship to nature to an emphasis on the gendered nature of access and control of resources, resource policies and politics, and management (see chapter by Elmhirst in Part I). This shift has reflected a more critical turn to approaches emerging from the field of political ecology that focus on "gender and the environment" rather than "women and the environment" (Elmhirst 2011; Jackson 1993).

If we look at the evolution of both the environmental movement and the field of environmental studies, the role of women has been substantial. Foremost is Rachel Carson, a marine biologist and author of *Silent Spring*. Carson is often credited with being the founder of the modern environmental movement. Many environmental activists were women who were keenly concerned about the health, livelihoods, and wellbeing of their families. In the United States, women pioneered movements to clean up toxic waste sites and protest nuclear power. Kenya's Green Belt headed by Wangari Maathai linked women's concerns about livelihoods to the health of their local environments and forest conservation.

With the dual rise of the feminist movement and the environmental movement, the 1970s marked the emergence of greater attention to women's relationship to the environment. One of the early and most significant writings that connected women to the environment in developing countries was Ester Boserup's 1970 book *Women's Role in Economic Development*. Boserup, a well-respected Danish agricultural economist who had written influential publications on agricultural growth, underscored the importance of women's work in agricultural enterprises. Her contributions highlighted the gendered nature of many crops and cropping systems, the role of women's customary rights to land in maintaining agricultural productivity, and women's overall economic importance in the household. This book provided a key intellectual underpinning for the women in development (WID) movement.

At the same time, ecofeminism was emerging in writings of women in both the North and South. Activists such as Vandana Shiva (1989) sought to portray women as both victims of environmental degradation and development, linking environmental exploitation to the exploitation of women. This notion of ecofeminism was rooted in the idea that women both depend more closely on environmental resources and are more inherently altruistic and caring about the environment. While Shiva's views of women's role in both environment and development now seem both simplistic and deterministic, ecofeminist views have influenced environmental policy. For example, ecofeminist thinking shaped the women, environment, and development (WED) approach that was adopted by small and large donors in the 1980s. This approach argued for programming that included women in environmental management, based on the idea of women's special relationship with nature (Leach et al. 1995). Other environmental movement leaders identified as ecofeminists. Wangari Maathai, the founder of Kenya's Green Belt movement and winner of the Nobel Peace Prize, believed that women should be involved in tree planting because they were the most aware of the loss of trees. This belief came from the fact that women were responsible for collecting firewood and experienced in nurturing seedlings, not because of their connection to nature per se. Maathai mobilized women to plant trees and became a powerful voice for sustainable development, peace, and democracy.

A commonly heard critique of ecofeminism is its essentialist nature that ignores historical and material relationships. Cecile Jackson (1993) put forth a more nuanced gender and environment perspective on conservation that moved away from women having an inherent relationship with the environment to one that is socially constructed. In this view, women's knowledges, rights, and responsibilities are influenced by issues such as class, culture, and ethnicity that, in turn, shape intra-household dynamics and relationships. Jackson's work foreshadowed what has become the current field of political ecology. The volume edited by Rocheleau et al. (2013), *Feminist Political Ecology*, further examined gendered knowledge and access to resources but also stressed the role of women in environmental organizations as grassroots activists and the gendered nature of environmental politics. Elmhirst elaborates on the field of feminist political ecology in her chapter in this *Handbook*.

Another important event in the formation of gender and environmental policy was the 1992 UN Conference on Environment and Development in Rio de Janeiro, commonly known as the Earth Summit. This summit, and Agenda 21, the conference's official document, highlighted the role of women in sustainable development. The conference mentioned the importance of women in areas ranging from biodiversity conservation (see chapter by Howard in this volume), climate change, and population policy. It also led to the founding of the Women's Environment and Development Organization (WEDO) as well as other international alliances. While many governments have not lived up to the commitments made in Rio, subsequent conferences have concentrated more on implementation. Women's groups at the World Summit on Sustainable Development (WSSD) conference held in Johannesburg in 2002 vocally asserted that sustainable development must take gender into account (Momsen 2009).

All the chapters in Part II of this *Handbook* largely focus on gendered access to resources. Several themes emerge from this section. Many chapters discuss how women's role in the productive and reproductive sphere are ignored and underplayed, with implications for resource access, particularly in the context of socio-economic and political change. Two chapters (by Kevane and Jacobs) describe the significant contribution of women to agricultural enterprises, both as farm managers in their own right but also as part of farm households where they are largely marginalized in terms of control and access to land. Jacobs' focus is on land redistribution and reform. Land distribution has been carried out all over the world in response to unequal access to land and to make land tenure more secure, particularly for the rural landless and poor. Many land redistribution programs operate at the household level. While single women and female-headed households have often benefitted, agrarian reforms frequently give formal rights to the male household head. This can sometimes disadvantage women by ignoring their land rights and important labor contributions. Kevane examines women's access to land in sub-Saharan Africa observing three trends: the rising value of land due to economic modernization, large-scale schemes commonly referred to as "land grabs", and government formalization of customary land rights. Several of these trends have worked to decrease women's access to land. Some regions of Africa are seeing a growing movement towards individualization of land rights and a counter-movement to enhance customary rights.

Hovorka's chapter about gender and livestock illustrates how women are regularly represented as the "silent guardian" who watches over animals, despite the fact that they play significant roles in livestock management. Women can benefit from livestock as assets that provide income-generating resources, saving and enhancing household food security. Yet they frequently face constraints in accessing productive inputs such as land, labor, and technical services.

Another theme that emerges from this section is the changing nature of gendered knowledge. Eriksen and Hankins consider the gendered dimensions of fire in eastern Australia and California where gender plays an important role in maintaining fire rituals and knowledge over time. Despite the fact that women and men's knowledge is often segregated, women have played a role as "keepers of knowledge" during times of upheaval. This forging of temporary generational knowledge crossovers allows knowledge to return to men during times of stability, maintaining their roles as cultural keepers of knowledge. Although wildfire management agencies have generally ignored indigenous knowledge, the authors argue that traditional understanding could be of value if incorporated into contemporary fire management, influencing ongoing debates about how to coexist with fire.

Howard explores the many roles that women play in the management of plant genetic resources, as housewives, gatherers, gardeners, herbalists, and plant breeders. With climate change, agrarian change, and privatization, this knowledge is in danger of being lost. Much of this depends on existing customary rights to plants and land; failure to recognize customary rights and decreasing access to land is the primary reason for the loss of plant genetic resources. Howard calls for more attention to the gendered rights of access to plant genetic resources, though research on this topic has been scarce. While treaties such as the Convention on Biological Diversity do mention women's role in conservation, as did the 1992 Earth Summit held in Rio de Janeiro, Howard suggests more concrete guidelines to incorporate gender into plant genetic resource management.

Two chapters nicely illustrate women's agency in managing natural resources. Kunze and Momsen's chapter shows how community programs that organize self-help groups and provide loans to women have created spaces for women to manage agrobiodiversity and create new forms of land cultivation. Women are leasing land and collectively farming, often growing valuable cash crop varieties and other crops that have social significance. Lahiri-Dutt's chapter illustrates how women have been important as political agents in mining struggles.

The role of technology emerges as another broad theme of these chapters. Verbrugge and Van Wolputte discuss how new small-scale mining technologies are shifting gender roles and relations in the mining sector in Tanzania. These new technologies are quite expensive and out of reach for most women, creating an unequal gendered playing field that is shifting mining gender roles and relations. Williams contends that, with technological changes and the rise of aquaculture, fisheries have entered a masculine phase. Recent fishing technologies have transformed the industry from "hunting to farming," yet half of fisheries workers are women in many different occupations. Many new technologies such as aquaculture projects ignore women and the social implications of new technology. Small-scale fishery declines have likewise negatively affected households; fisheries supply chains have squeezed out small-scale producers, which often include many women.

Two chapters demonstrate mining's "gender bias," referring to the dominant scholarship on mining that tends to focus almost exclusively on men, portraying women as having little influence except as poor and powerless wives and girlfriends or barmaids and prostitutes. Verbrugge and Van Wolputte's discussion of small-scale mines in Tanzania reveals how women have long been involved in all phases of gold extraction, from entering mines, processing materials, or providing services in mining communities. Lahiri-Dutt demonstrates how large-scale mining is changing women's labor. While women have historically provided labor in the mines themselves, the emergence of large-scale mining industries has relegated women to the surface, sex-segregating women's labor to offices and other above-ground services. Other factors make women's labor invisible in mining; employment and government records are fabricated to discount women's labor and erase the historic role of women in mining. Women furthermore play an important but underreported role in mining communities, providing services and supporting households that are essential for mining operations.

A final theme that many of the chapters address is the importance of scale in gendered resource access. Decisions made on the household and broader policy scale influence how gender intersects with resources. Several chapters argue that policies and programs ignore the role of women (Howard and Hovorka). Williams uses her personal experience as a researcher in fisheries management to consider the gendered nature of fisheries and aquaculture at different scales. She demonstrates how the broad fishery sector, from fishery operators to policymakers, is largely gender-blind. Gender groups rarely participate as stakeholders in fisheries networks, and, as the sector has become more industrialized, women's rights have been put aside. These changes marginalize women's role in production.

Several of our authors demonstrate the scalar mismatch of policies and local implementation of those policies. Kevane discusses how, at the national scale, many African governments are promoting land registration programs with a gender-neutral lens, yet the evidence on the ground is not promising for increased access. With land scarcity and the changing value of land at the local scale, many of the traditional institutions that provide women access to land are limiting access. Jacobs illustrates how land reforms are often enacted at the national scale to promote equity, but land redistribution can have the unintended consequence of increasing women's labor, particularly with agricultural intensification. This can increase men's control over women's labor and/or sexuality as increased labor needs increase control on women's fertility and reproductive capacity.

References

Boserup, E., Tan, S.F., and Toulmin, C. (2013) *Woman's Role in Economic Development*, London: Routledge.

Carson, R. (2002) *Silent Spring*, New York: Houghton Mifflin Harcourt.

Elmhirst, R. (2011) "Introducing new feminist political ecologies," *Geoforum* 42(2): 129–132.

Jackson, C. (1993) "Doing what comes naturally? Women and environment in development," *World Development* 21(12): 1947–1963.

Leach, M., Joeks, S., and Green, C. (1995) "Editorial: gender relations and environmental change," *IDS Bulletin* 26(1): 1–8.

Momsen, J.H. (2009) *Gender and Development*, London: Routledge.

Rocheleau, D., Thomas-Slayter, B., and Wangari, E. (eds.) (2013) *Feminist Political Ecology: Global Issues and Local Experience*, London: Routledge.

Shiva, V. (1989) *Staying Alive: Women, Ecology and Development*, London: Zed Books.

10

CHANGING ACCESS TO LAND FOR WOMEN IN SUB-SAHARAN AFRICA

Michael Kevane

Introduction

Although women work alongside men in farms across the breadth of Africa, they neither own nor control much land. Women work on fields owned by their fathers, their husbands, their brothers, and other males in their lineage. When they do control fields on their own, their ownership rights are limited. Even households headed by women are constrained by male lineage elders, and required to eventually pass on land to their sons rather than their daughters.

There has been relatively little improvement in this discriminatory and marginal position of women in land tenure. This is curious, because there has been rapid movement toward gender equality in many other domains (World Bank 2011). The halls of government have admitted more and more women in Africa. By early 2014, there were three female heads of state, and representation in legislative, executive, and judicial branches of government was increasing. All African countries have ratified the Convention for the Elimination of Discrimination Against Women (CEDAW) and many countries have ministries and other bureaucracies charged with promoting gender equality. Government policies and programs have been revised to be more gender neutral, and often to favor women, offering subsidies, education, training, and access to state jobs. Reforms of family law codes have eroded the marital and divorce privileges of men. In this regard, most countries in Africa have differed little from other countries around the world. Gender inequality remains a significant limit on human development and economic growth, but there has been much progress toward formal, legal equality. Progress in economic, social, and cultural equality has been slow, but discrimination is changing in character and intensity. Land tenure appears to be an exception to this positive trend.

The very slow movement towards gender equality in land tenure has perhaps been related to a context of comparatively broad change in land tenure generally. Change appears to have accelerated in the 2000s due to three related trends. First, continued high population growth and increases in prices of agricultural products on international commodity markets pushed the value of land higher. Second, agribusiness firms, both national and international, successfully lobbied for broad changes in national land legislation and investment codes to expedite establishment of large-scale farms and plantations. Third, growing urbanization, migration, and remittance income led to increasing differentiation in village economies, raising incentives for more secure private title. These forces have induced individual efforts to privatize and secure title, and governments to attempt to establish or improve land management systems. Governments have also tried to formalize customary tenure rules and recognize village and traditional social groups as landholding entities. These "induced" changes appear, by and large, to have reduced women's access to land. For example, the changes toward individual title and communal title, and the efforts of government to become more involved in land management, typically limit secondary use rights to tree and forest products. These secondary use rights are often the major form of access to land for women.

Partly as reaction to the induced institutional changes and the continued lack of progress toward gender equality in land tenure, policy initiatives of government and donors, such as the World Bank, have increasingly emphasized attention to gender-oriented reform of land tenure. One reform has been to legislate and enforce procedures of joint conjugal ownership over land upon marriage, with provisions for spousal consent over land use and transfer, and provisions for division of land upon the termination of marriage through death or divorce. Another reform has been to mandate that the local tribunals that resolve land disputes include women.

There has also been more attention to evaluating the effects of policy innovations, through the use of randomized control trials (RCTs). In a RCT, one group or region is "treated" with the policy change, and another "control" group remains under previous tenure institutions. If treatment and control groups are randomly selected from an appropriate population, comparisons between the outcomes for the two groups enable estimation of the effects of the programs. Despite the methodological advances of RCTs and increasing prevalence of large-scale tenure reforms, the bulk of knowledge about changing land access still comes from anthropologically-oriented small-scale studies of communities in rural Africa. These benefit from the researchers' close knowledge of community dynamics, and enable the voices and interpretations of persons involved to be communicated to the larger academic and policymaking community.

The plan of the chapter is as follows: a brief introduction to the male-biased land tenure systems prevalent across Africa, followed by a summary of anthropologically based understanding regarding how gender aspects of land tenure are evolving. Then a discussion of the results of recent RCTs designed to measure the impacts of policy changes on gender equality is followed by some concluding thoughts.

Gendered land tenure in Africa

A tenure system is a set of rules, processes, and institutions that determine access and control over land and how to resolve disputes over land. Disputes must be arbitrated before they culminate in violence, and arbitrators rely on precedent and consistency in their decisions. Land tenure systems are often in flux and contradiction, however, with changing and competing precedents, jurisdictions, and principles available to litigants. Moreover, litigants and their witnesses have interests in presenting different interpretations before an arbiter. There are, then, in most African societies, complex and varied discourses about land tenure. It is often hard to discern a "system" that fairly represents the ambiguity, intricacy, nuance, and dynamic nature of tenure.

Some persistent and common elements can be parsed from land tenure discourses in Africa (Doss et al. 2013; Kevane 2014). First, land tenure is a complex, plural affair with multiple jurisdictional venues. The national state, regional, and local government, traditional institutions such as chieftaincies, and ethnic and village traditions regulate land tenure. There are many arbiters of land tenure, and almost all are male. Second, most rural areas do not have government-maintained cadastres of agricultural land, nor do they have government officials managing written records of land tenure. That is, land tenure is largely an oral affair, consisting of memories of transactions that are not notarized or submitted to government officials. The paperwork of transactions and disputes is haphazard and contested, and often has the same standing as people's memories. Much of land tenure is local. Third, very little non-irrigated agricultural land is transacted through land markets, either as rental, sharecrop, or through sales. There are few arm's-length transactions. The bulk of non-irrigated land is allocated through a social process involving lineage elders and other local actors, and invokes discourses of community, ancestors, obligation, and responsibilities. Fourth, women are typically marginalized in the social processes of land allocation. A social process, by definition, involves a person's identity (in contrast to a market transaction, where identity is largely irrelevant to the transaction). In much of rural Africa, a person's identity as a woman rules out most of the "normal" processes for obtaining and keeping land. Women only have access through their husbands to small plots of land that they can farm on their own account, where they are socially recognized as the farm manager. To the extent that women have access to land directly, it is usually in the form of secondary rights to tree products, wild plants, stubble, and gleanings.

This static picture masks change, and several broad trends are clearly visible on the African tenure landscape. First, property rights to land are becoming more individualized as markets for land rental and sales at the individual level are emerging in many regions. Individuals with lineage authority are severing land from social networks. This is often outside areas that have seen official centralized schemes to transform tenure through cadastres and title registers. Private landholding and individual transfer are more often a spontaneous emergence and the informal privatization of land may involve stark expropriation of marginal or secondary holders of land rights. Women may benefit from this privatization, although there are no large-scale surveys demonstrating any significant shifts in the gendering of land ownership.

In most African countries, there has been perennial interest in reforming tenure to strengthen the institutions that secure individual and communal land. Three arguments are offered for tenure reform. First, informal tenure may result in an excess of land disputes, and the constant contestation over rights reduces the incentives of farmers to invest in improving their farmland. Why invest a lot if a politically connected neighbor can lay claim to your land? Tenure insecurity deters investments in land improvement, such as building terraces, planting trees, and applying manure. Second, banks may be reluctant to fund investments in land if the land cannot collateralize the loan. There may be too little lending for investment in agriculture because farmers do not have secure title. Third, tenure insecurity may lead farmers to continue farming extensively, instead of intensively, because they are reluctant to rent out land to others, for fear they might lose their rights as the person actually farming the land might gain the rights to ownership. Despite these apparently strong arguments, it is widely acknowledged that state efforts to effectuate programs of individualized land title have rarely been successful. Instead, large amounts of money are spent on registration campaigns, but then little attention is devoted to institutionalizing the individualized land tenure system, and so informal, local tenure rules quickly reassert themselves. The changes happening in the 1990s and 2000s have been interesting because they have not generally been the result of a top-down policy change. Rather the changes have emerged from decentralized decisions and processes. Nevertheless, evidence suggests that land titling programs have been and remain more favorable to men (Widman 2014).

Land tenure insecurity often affects women more severely than men because women are typically secondary rights holders. Insecurity and marginalization have direct effects on productivity and indirect effects by limiting opportunities to leverage assets. Goldstein and Udry (2008) showed that women were significantly less productive than men, in a large sample from rural Ghana, and that their lower productivity was due, in large part, to their insecure tenure. Because they had weak rights to land, women were reluctant to let land regenerate through fallow. Peterman (2011) estimated the empowerment effects of women having title to land, using data from a sample of households from rural Tanzania. She observed the same women over a 13-year period. Many of the women became formal co-title holders of land during the time period. The changes in women's titles and also in inheritance rights were significantly associated with women's employment outside the home, self-employment, and earnings.

This discussion of the link between tenure security and productivity is part of a broader debate on productivity in rural Africa. It is sometimes argued that women's access to land is not an important issue because women are less productive farmers than men. As just suggested, however, productivity may differ because rights and security of tenure are gendered. Moreover, many careful studies of productivity suggest that gender differentials are due to access to inputs, rather than efficacy of managing inputs. Women have less access to labor, fertilizer, pesticides, and timely credit. The differential access results in lower productivity.

A second broad trend is partly a reaction to the pressures toward individualization of tenure. In many areas, local jural groups are reasserting and strengthening their rights to communal ownership of land. Lineage and ethnic groups across the continent lay claims to land, and the failures of individual titling have led many policymakers to explore granting of communal tenure certificates. Traditional chiefs and earth priests recognize their powerful positions as arbiters of land conflicts within and between groups. These arbiters of land tenure see that if land becomes individualized, their role becomes an anachronism. With their allies, they form an influential body of citizens interested in maintaining notions of collective ownership. The ostensible justification for formalizing their control is that corporate groups, with legitimate leaders, well-understood processes for resolving disputes, and clear and inclusive membership would function like clubs. That is, they would share an interest in managing land effectively, and would be able to work out arrangements among members that would be mutually beneficial. They would reduce the likelihood and costs of tenure insecurity. The push to devolve authority to quasi-formal groups is attractive, not least because the simple devolution of powers has low costs for the national government.

Corporate groups in Africa, as elsewhere, exhibit considerable variation in their willingness and ability to provide public goods. Some are authoritarian, with leaders using violence, threats of sorcery, and discourses of exclusion to maintain their power. Many are dysfunctional, with competition among interest groups within the corporate group undermining sentiments of legitimacy and solidarity. Corporate ownership of land has been a concern to social movements that promote gender equality, because virtually all customary institutions in Africa are dominated by men and promote discourses of women as subject to the authority of their husbands and male lineage elders. Nearly all patrilineal groups exclude women from leadership or even voice in corporate affairs.

Changing gendered land tenure in African societies

Recent literature on changing access to land distinguishes between changes in rules and changes in the incidence of application of various rules of access. That is, sometimes some tenure rules change in ways that reduce access, but the incidence of women's access increases through use of other more favorable rules. There are two important processes of change that appear to be common to many African societies. The first process involves changing locally understood definitions or meanings about what property can be controlled by what persons. That is, social identities related to land are changing. A second process involves changes in the incidence of what kind of person women are, moving from traditionally married wives to cohabiting wives. That is, social identities relating to husbands (or, more broadly, to men) are changing.

Men and women negotiate definitions and meanings that are relevant to land tenure. These meanings are manipulated to favor one group over the other as the balance of power shifts from women to men. For example, sometimes particular crops are designated as male or female, or as individual or household crops, and control over land used for those crops shifts as control over the crop shifts. A number of studies of the effect of irrigated rice cultivation on women's access to land in Gambia drew attention to this process (Brautigam 1992; Carney 1988; Dey 1981). Farmers recognized both common and individual land rights. Women had controlled rice fields that they had cleared with their own labor. Their rights were well-defined: they controlled what they harvested but, more significantly, they controlled the right to transfer land, which they generally did, to their daughters. State-initiated irrigation projects changed the landscape. Rice land was cleared and developed by male-centered development projects. Local men were able to claim land, partly because newly irrigated plots were categorized as household property that came under the control of male household heads. Moreover, inputs and mechanized services were allocated overwhelmingly to men. Women's access to land was reshaped by redefining the meanings of the categories by which their access had traditionally been allocated. The new projects allocated some land to women, but that did not mean that women had the power to control the land. Irrigated crops turned out to be considered as household crops and hence under the control and management of men.

Another example of changing meanings concerns the language and discourses of land tenure disputes. Sometimes women have been able to deploy notions of fairness, rather than equality, in pursuing land claims. That is, in some localities women argue that denying them land is unfair, or is contrary to fulfillment of their social roles as mothers. In other regions, rhetoric of rights and responsibilities emanating from labor on the fields has enabled women to assert control rights over land. As men move towards cities and migrate for work, women left behind on rural homesteads begin to adopt rhetoric that their work on the land entails communal recognition in the form of rights to the land. Some of this change happens as hitherto marginalized discourses in local land tenure mesh with national discourses that are increasingly moving towards more formalized equality of rights, impelled in part by ratification and dissemination of the CEDAW treaty.

A second process that alters the terms of access to land rights is one in which the terms of marriage contracts change, and this affects the incidence of various kinds of marriage. In traditional marriages in many parts of rural Africa, the husband's lineage is obliged to give the wife a small plot of land for her to farm on a personal basis. The land is compensation, in a sense, for her loss of freedom to pursue her own economic projects. Men are under considerable social pressure to honor this commitment. As rural economies differentiate, however, and women are increasingly able to operate as small entrepreneurs, many women no longer insist on access to a personal field. Those women who do want to marry under the old rules, and obtain a plot of land, find that potential husbands are less willing to commit. There are many women in the marriage market who will no longer insist on controlling part of the land. As the terms of marriage shift from the traditional marriage to less formal marriages, closer to cohabitation, the incidence of access to land by women falls dramatically. The reasons for the changing character of marriage are complex, but across the continent seem to involve a common pattern of women seeking to emancipate themselves from the control of their fathers (who used to decide on their marriage partner). The loss of rights to land appears to have been the price paid for greater freedom in choosing marriage partners.

Randomized control trials in land tenure policy

Recent policy initiatives have emphasized large-scale low-cost implementation of tenure programs (such as certification programs, at both household and community level) and have included mechanisms to ensure more gender equity. These initiatives have responded to a perception that previous expensive and bureaucratic land titling efforts impeded, rather than facilitated, the emergence of individual property rights. The titling process often seemed to generate insecurity rather than tenure security, had perverse equity effects (especially for women), and had high implementation costs.

Assessing the impact of policy and programs to reduce gender inequality has been facilitated by recent adoption of a methodology of randomized control trials (RCTs) for land tenure interventions. In an RCT, some communities, households, or individuals, are randomly selected to participate in the intervention. Within this group, the nature or level of the intervention may vary (different programs, different costs). The non-selected group constitutes a proper control group for measuring the impact of the assessments. This methodology stands in contrast to the typical impact evaluations of the past: "pre and post" studies attributed all changes over time to the program, neglecting the often important effects of other factors that were changing over time; "participant vs. non-participant" studies neglected the importance of self-selection (by program designers or adopters). One of the virtues of RCTs is that they permit estimation of the secondary effects of changes in land tenure. For instance, increased access to land by women might increase their bargaining power within the household, and enable more household resources to be directed towards children's health and schooling. Observational studies have a very difficult time assessing these effects, because women with more bargaining power may have greater access to land, and so the causality runs from bargaining power to access and not from access to bargaining power. Moreover, other variables that are difficult to measure, such as inherent abilities, family background, and ethnic identity, may be influencing both bargaining power and access to land. RCTs can experimentally manipulate access to land, and so permit estimation of the effects.

One of the largest land tenure RCTs started in 2003 in Ethiopia. Communities were selected for a program to formalize land transactions. The program was not an official cadastre and titling program, but rather intended to improve recordkeeping and tracking of local land transactions. One of the key features of the program was that wives would be considered as equal legal actors in the land transactions. Under national law wives were co-owners with their husbands of marital property. The process was that a local committee for land use and administration was formed in each locality, supposedly by popular vote. The committee had to include at least one woman. The committee then issued certificates of land registration after public meetings. The certification program was envisaged as rapid and low-cost, in the sense that the goal was to deliver certificates to as many uncontested plots as possible. Indeed, more than 20 million certificates were delivered in two years from 2003–2005. A nationally representative large-scale survey revealed that the program had a mixed impact (Deininger et al. 2008). Many of the committees did not have women members and many apparently reneged on their responsibility to issue certificates jointly to husbands and wives. In Oromia region, for example, where all titles were required to be issued jointly, 58 percent of certificates were issued only in the husband's name. There was, however, some evidence of benefits to women. One analysis of the Ethiopia program suggested that female-headed households without access to family labor or cash to pay hired laborers benefitted by renting out their excess land holdings (Holden et al. 2011).

A similar land tenure regularization program was undertaken by the government of Rwanda in the late 2000s. Legal changes provided for compulsory registration of property and of transfers of land, and customary law was no longer given legal standing. Reforms protected property rights of women in land held by their husbands, provided they were in a legally registered marriage. Ali et al. (2011) evaluated a pilot program of land tenure regularization undertaken in 2007. The program intended to demarcate and adjudicate almost 15,000 parcels covering 3,500 hectares held by approximately 3,500 households. The pilot program chose to implement the program in areas where there were many former refugees, soldiers, and female-headed households. The project identified plots on aerial photos. If a plot of land was not subject to any disputes, as verified by neighbors, then the program entered the plot in a registry book with the names of owners and the heirs who would have claims to the land. The program issued a land certificate upon payment of a fee. In order to evaluate the program, the authors conducted a short survey with 3,500 households on both sides of the borders of four pilot areas in 2010, two and a half years after the government implemented the pilot. They assumed that households close to boundaries were similar, so that those not in the pilot program constituted a valid control group. The only difference between households on either side of the border was that on one side their plots were registered, and on the other side they were not.

The authors estimated that the intervention led to a large increase of investment in soil and water conservation, with an even larger increase for female-headed households. Moreover, the likelihood that married women claimed to have ownership over land increased dramatically because of the registration program. Respondents in the pilot program zone were also much more likely to indicate that daughters would inherit land. Women in traditional marriages (such as polygamous marriages) did not enjoy the same benefits as women in the monogamous marriages that were formally registered. The authors observed that the fees for obtaining land certificates were probably too high. Since there were externalities from widespread access to title (reduced conflict, better land management, more transfers to efficient users of land), it made sense for the program to offer certificates at very low prices, or even for free. Another problem identified in the paper was that the program did not have a clear plan for managing the land registry as households conducted transactions and land transfers, and passed on land to the next generation through inheritance.

Another form of land tenure project informs and empowers people about rights that they already have but do not exercise. Santos et al. (2012) evaluated a community awareness project undertaken in conjunction with Rwanda's land regularization effort. The project delivered in 2011 a series of public dialogue and awareness raising programs in Musanze District in Northern Province. The target populations included both local officials and the residents affected by the legal changes. The researchers selected a random sample of 355 landowning households, including monogamous households under various forms of marriage, polygamous house-holds, and households headed by widows. One third of the households were in villages outside of the program area, and constituted a control group. Overall, awareness of and participation in tenure regularization was extremely high, generally approaching 90 percent of the sample. The campaign was successful in changing men and women's perceptions of tenure security, with these rising substantially compared with the control group. There seemed to be no difference in wealth, gender, and ethnicity between participating and control households. The awareness campaign significantly affected responses to questions about whether sons and daughters were likely to inherit. However, wives in the awareness campaign area were less likely to be listed as having title, compared with the control group. The research further revealed a problematic ambiguity about the legal changes and their promulgation through awareness campaigns. Households were unclear about whether wives in polygamous households could share title with their husbands. Their marriages were not recognized by the state, which only recognized monogamous unions. Since husbands wanted to clarify inheritance for their children, in ways that would not lead to ambiguity about which children of which mothers would inherit, the husbands ended up in many cases having land registered in the name of their wife, rather than jointly. This outcome undermined, in local discourses, the land registration, since the local understanding was that the husband remained the owner, and the registration title was simply a document to satisfy the state. Down the road, one could imagine situations (divorce, death of the wife) where courts would likely have to make accommodation for local practice rather than strictly adhere to the letter of the land registration law.

There have been two RCT studies of the impact of programs to promote community ownership of land. Goldstein et al. (2013) presented preliminary results of the gendered effects of a randomized land tenure intervention in Benin that supported community ownership. The *Plan Foncier Rural* (PFR) program, funded by the United States' Millennium Challenge Corporation, organized information campaigns about land rights, surveyed land, developed village use plans, adjudicated disputes, and issued land use certificates. For an evaluation of the program, villages were randomly selected in 2008 from among those that applied. A survey of about 5,500 households in 2011, from 193 treatment villages and 98 control villages, measured a variety of agricultural and land use outcomes. Interestingly, the program appeared to have increased people's fear that one of their fields might be used by someone else. Perhaps because of that, there was a small increase in trees planted (planting trees is a way of reducing insecurity) and decrease in land loaned out to others (people perhaps feared they might not recover a field loaned out). There were few other changes, and no increase in average productivity. There appeared to be, however, an increase in trust in the village land committee. When the data were disaggregated by gender, the authors found evidence that the program appeared to be reducing conflict (perhaps because of awareness that there was greater scrutiny surrounding land issues). Women were somewhat more likely after the program to obtain land, and increased their inputs, in particular of fertilizer. Overall, then, the program seemed to be having more impact on women than men, though the results were very modest.

Knight et al. (2012) reported on an important initiative to experiment with methods to protect community rights to land. The study selected 20 communities in Mozambique, 18 communities in Uganda, and 20 communities in Liberia to implement community demarcation and title processes that were established in each country's land legislation. The communities were then randomly assigned to four different treatments. One group received monthly legal education sessions; a second group received the legal education and also regular visits by trained paralegals; a third group received regular assistance from lawyers; and a fourth group was the control group and simply received paper copies of relevant manuals and legislation. The communities were encouraged and supported in their efforts to establish a community land committee, demarcate community land, develop procedures for local adjudications, develop local land-use zoning plans, elect a legal local land authority, and proceed with steps to formalize community control vis-à-vis the national government. Although the sample sizes of communities were too small to draw robust conclusions, the findings, after two years, were suggestive. First, the process for establishing formal community rights was very slow and drawn out. None of the communities had established legal rights by the end of the second year of the project. Nevertheless, the authors documented that the process, in Liberia and Uganda, but not Mozambique, had high levels of participation and numerous positive effects in improving local civic accountability. The authors found that the education sessions with paralegal support appeared to be more cost-effective than the other interventions. Finally, the report found considerable evidence that a participatory process, if explicit about the need to incorporate women's land rights and supported by paralegals, could shift attitudes and community practices towards land tenure practices more favorable to women. The process in Mozambique, however, appeared to have had the opposite effect, undermining women's rights.

All in all, the results are mixed, in terms of gender outcomes, of these pioneering RCTs dealing with land tenure changes. The Rwanda trial found the largest effects for women; the other trials found small or non-significant effects, or even negative effects. It seems reasonable to conclude that there is still much to learn about what kinds of interventions will favor women's access to land and consequently lead to improvements in productivity, incomes, bargaining power, and well-being. A study on an RCT for urban land in Tanzania suggests that access to titling and subsidies to cover the cost of titling can have significant positive effects on co-titles that include wives (Ali et al. 2013).

Concluding thoughts

Women's ownership and control over land in African is limited and marginal. There are processes of change underway. Endogenous processes include changing patterns of marriage, and changing discourses concerning the gendering of agricultural crops. Exogenous processes include donor-funded land registration and awareness programs. National governments, pushed by narrow but sometimes effective women's movements, have revised land legislation to be more gender neutral. There is little evidence from early randomized control trials indicating that recent female-friendly land tenure programs have significant positive effects on women. This may be because programs are implemented in ways different from intended, or that local processes of land tenure favorable to men are deeply embedded. Limited efforts by government to intervene in local land tenure may be ineffective in these contexts. It may, of course, be too early to tell whether recent initiatives are having slow but steady impacts on women's access to land.

Overall, then, the trends of changing access to land for women in Africa are unclear. In many countries, observers are ambivalent about the eventual outcomes of the processes of change underway. Moreover, it remains the case that access to land is just one part of the broader set of determinants of gender equality in economic development. Women's access to labor (their own and through markets or social institutions) is often restricted. Government and private sector both continue to discriminate against women in provision of inputs such as fertilizer, credit, and insurance. Improved access to land will help, but not completely counteract, the unequal effects of other forms of unequal access.

References

Ali, D.A., Deininger, K., and Goldstein, M. (2011) *Environmental and Gender Impacts of Land Tenure Regularization in Africa: Pilot Evidence from Rwanda*, World Bank Policy Research Working Paper Series.

Ali, D.A., Collin, M., Deininger, K., Dercon, S., Sandefur, J., and Zeitlin, A. (2013) *The Price of Empowerment: Experimental Evidence on the Demand for Land Titles & Female Co-Titling in Urban Tanzania*, Working Paper 369, Center for Global Development.

Brautigam, D. (1992) "Land rights and agricultural development in West Africa: a case study of two Chinese projects," *The Journal of Developing Area* 27: 21–32.

Carney, J. (1988) "Struggles over land and crops in an irrigated rice scheme: the Gambia," in J. Davison (ed.), *Agriculture, Women and Land: The African Experience*, Boulder, CO: Westview Press, pp. 59–78.

Deininger, K., Ayalew, D., Holden, S., and Zevenbergen, J. (2008) "Rural land certification in Ethiopia: process, initial impact, and implications for other African countries," *World Development* 36(10): 1786–1812.

Dey, J. (1991) "Gambian women: unequal partners in rice development projects?" *The Journal of Development Studies* 17(3): 109–122.

Doss, C., Kovarik, C., Peterman, A., Quisumbing, A.R., and van den Bold, M. (2013) "Gender inequalities in ownership and control of land in Africa: myths versus reality," Discussion Paper, International Food Policy Research Institute.

Goldstein, M.P. and Udry, C. (2008) "The profits of power: land rights and agricultural investment in Ghana," *Journal of Political Economy* 116(6): 981–1022.

Goldstein, M., Houngbedji, L., Kondylis, F., O'Sullivan, M., and Selod, H. (2013) *Formalizing Rural Land Rights in West Africa: Evidence from a Randomized Impact Evaluation in Benin*, CSAE Conference 2013: Economic Development in Africa March 17–19, 2013, St Catherine's College, Oxford.

Holden, S.T., Deininger, K., and Ghebru, H. (2011) "Tenure insecurity, gender, low-cost land certification and land rental market participation in Ethiopia," *The Journal of Development Studies* 47(1): 31–47.

Kevane, M. (2014) *Women and Development in Africa: How Gender Works*, Boulder, CO: Lynne Rienner Publishers.

Knight, R, Adoko, J., Auma, T., Kaba, A., Salomao, A., Siakor, S., and Tankar, I. (2012) *Protecting community lands and resources: Evidence from Liberia, Mozambique and Uganda*, Washington, DC, and Rome, Italy: International Development Law Organization.

Peterman, A. (2011) "Women's property rights and gendered policies: implications for women's long-term welfare in rural Tanzania," *The Journal of Development Studies* 47(1): 1–30.

Santos, F., Fletschner, D., and Daconto, G. (2012) *Enhancing Inclusiveness of Rwanda's Land Tenure Regularization Program: Initial Impacts of an Awareness Raising Pilot*, paper prepared for the Annual World Bank Conference on Land and Poverty, http://landandpoverty.com/agenda/pdfs/paper/santos_full_paper.pdf.

Widman, M. (2014) "Land tenure insecurity and formalizing land rights in Madagascar: a gender perspective on the certification program," *Feminist Economics* 20(1): 130–154.

World Bank (2011) *World Development Report 2012: Gender Equality and Development*, Washington, DC: World Bank.

11

GENDER, AGRARIAN REFORMS AND LAND RIGHTS

Susie Jacobs

In many societies, landownership or landholding is emblematic of social belonging and is a marker of social status. This is a gendered phenomenon, since women are frequently excluded or marginalised from access to land on the same bases as men within their societies. Where they can access land, they often lack direct control over this and are constrained in making decisions over agricultural production. This acts as a powerful symbol of male domination and of the social construction of women's dependent status. Such lack of control affects women's livelihood security and their social status more generally.

This article explores gender relations and gendered inequalities in relation to land through a focus on agrarian and land reforms: i.e., the redistribution of land to landless or land-hungry households. This is particularly important given contemporary emphases on food security and sovereignty, and in face of growing food crises.

The first section of the chapter outlines women's substantial participation in agriculture; the second briefly defines agrarian and land reforms. The third section analyses the gendered implications of 'household' model reforms and the fourth examines collective models of reform, followed by a brief discussion of consequences of decollectivisation. The final section explores some contemporary mobilisations for women's land rights.

Background

Today, women's contributions to agriculture are increasingly acknowledged (FAO 2011), but this is a relatively recent development. For many years agriculture was treated as a male domain and references to farmers concerned men. This corresponds in part to the relatively hidden nature of women's roles, but also to their erasure from discourses concerning agriculture and peasantries. Due to contributions from 'gender and development' activists and academics since the late 1960s, however, it has become well-established that women across the globe labour within farms and agrarian households.

On both subsistence plots and land used for cash-cropping, women perform a wide range of tasks. The exact scope and types of work varies a good deal according to crops, soil type, size of holding and also according to sociocultural context and geographical region. Typically, they are responsible for sowing seeds, planting, weeding and other aspects of routine upkeep, preserving seeds, care of small livestock and processing crops, as well as fetching water and gathering firewood.

In China, for example, peasant women cared for children and for in-laws, cooked and prepared food – involving the difficult task of husking rice or millet – fetched water and fuel, cleaned the house, raised animals such as chickens and pigs, wove cotton cloth and sewed family clothing as well as making cloth shoes. Certainly in rice-producing areas women worked in transplanting seedlings and taking part in harvesting as well as everyday agricultural activity. In Lu village in the southwest, both in the 1930s and in the twenty-first century, women were primary agriculturalists, raising rice and beans as well as subsidiary crops.

There exist regional and social differences in the extent of women's agricultural participation. In sub-Saharan Africa, women have the main responsibility for agricultural production and they undertake the majority of agricultural work (see chapter by Kevane in this volume). In most of the rest of the world, it is men who are viewed as having responsibility for provisioning. However, women's work may be greater than or equivalent to men's in terms of effort and time spent (Dixon-Mueller 1985; FAO 2011). It has been noted for some time that official data often seriously underestimate women's agricultural labour (Dixon-Mueller 1985). Women's labour on farms often fails to translate into control over income, equal participation in decision-making or improvements in status. Sachs (1995: 129) notes: 'In fact, women frequently perform agricultural labour under men's direction or to increase male income.'

Women are often marginalised in landholding: this is particularly apparent in patrilineal and patrilocal societies. These exist, for instance, in much of Africa, including southern Africa; in north Africa and the Middle East; in west and much of south Asia and in east and parts of Southeast Asia, including China and Vietnam. In these, neither unmarried men nor women usually held land; nor did married women, although sometimes widows hold land on a temporary basis on behalf of sons. In the contemporary world, for instance in China, north Africa, south Asia and in parts of southern Africa, lineages are now rarely corporate or property-holding bodies; nevertheless, lineage principles are still important. Additionally, in customary law in sub-Saharan Africa, wives had (and often still have) the right to a 'garden' plot for cultivation of food, but access to this is up to the husband to allocate.

Although it is common to note that women own relatively little property across the world, it should be noted that estimates of women's landholding based on large-scale data sets are few and that differing criteria can be used to estimate male/female differentials in landholding and control over agriculture (Doss et al. 2013). Doss et al. find, however, that for Africa, women systematically own and control substantially less land than do men across all regions of the continent.

Women's secondary status may be reinforced ideologically, for instance through beliefs or taboos problematising their ability to cultivate land. An example is the traditional taboo on women ploughing in India. Bilateral kinship systems such as those in Europe and Latin America do hold the possibility of more egalitarian property relations. However, women nevertheless often experience a secondary and contingent relation to land and property. This may be emphasised or reinforced through other means – for example, domestic ideology, emphasis on men's roles as providers and devaluation of work habitually performed by women.

A number of factors – kinship relations, sexuality, aspects of culture, agricultural production on small farms and social policies – thus operate together and frame the social and economic positions of peasant and smallholder women.

Agrarian reforms – and land 'grabs'

Land redistribution has been carried out usually only after widespread and bloody struggles. Land and agrarian reforms can be divided into two types, according to unit of redistribution:

1) the 'household' model, redistributing land to individual households, and 2) the 'collective' model, in which a collective body – such as an agricultural production cooperative or collective – holds land. Collective model reforms were usually carried out by state socialist governments or movements (see discussion later in the chapter). The classic definitions of agrarian and land reform along household lines belong to the 'moment' of developmental states, especially after World War II. Redistributive land reform is usually considered to be a state-backed programme aiming to increase access to land and (sometimes) more secure tenure for the rural landless and poor. It also seeks greater democratisation in rural areas.

The reasons for household model land redistributions are economic as well as political and social. They include the reduction of rural class inequalities and quelling unrest of the landless or land-hungry. Most importantly, improved food security is a key aim, meant to be realised through greater effort and care and attention to production and through raising productivity on the land.

Increased democratisation for rural people, including lessening of landlord power and abuses, is an important rationale. More recently, environmental rationales for redistribution of land in small farm units have been recognised as a key concern in at least some cases.

With the advent of neoliberal economic policies from the 1980s, many states turned away from agrarian reform and to more market-oriented agricultural policies. However, several states in the Global South – such as Brazil, (contentiously) Zimbabwe and the Philippines – have responded to popular pressure by continuing or expanding redistributive land reforms

Recently, the landscape for land redistribution has been changing rapidly. A huge spike in commodity prices for rice, maize, soybeans and petroleum in early 2008 focused worries about food prices and security, particularly in arid and/or heavily populated countries in east Asia and the Middle East. What is now known as a 'global land grab' or large-scale land acquisitions (GRAIN, 2008; White et al. 2012) have affected sub-Saharan Africa in particular, but also extend to Southeast Asia, central Asia, Eastern Europe and elsewhere. The main investors in land include governments, transnational corporations and private equity groups. Detailed discussion of these dramatic developments is beyond the scope of this chapter, but it is worth noting that these threaten prospects for land redistribution and for *any* small-scale agricultural production since land grabs undermine rural people's existing rights, including any women may have.

'Reform' is not simply change; the term 'land reform' is today used particularly to refer to tenure reform and privatisation of holdings which often disadvantage the poor, including poorer women. The definition or usage of 'land reform' here to refer to state-backed redistributive reform, accords with the demands of a number of rural movements including La Vía Campesina, the umbrella organisation campaigning for food sovereignty (see below). And 'land grabs' affect land rights immediately and directly, while land titling and privatisation may undermine rights in the longer run. As noted, the need for land redistribution – already pressing – is likely to increase in future due to emerging global food crises.

Gender and agrarian reforms

Although there exists a large literature on land reforms, most of this does not study women's position or gender relations. This section presents an overview of 33 studies of gender and land reforms across Africa, Latin America, Asia and Eastern Europe that have paid attention to women's positioning and to gender relations within reforms. The countries in which studies were conducted include the following: Brazil, Burkina Faso (formerly Upper Volta), Chile, Ethiopia, Honduras, India, Iran, Kenya, Libya, Nigeria, Peru, the Philippines, Poland, South

Africa, Sri Lanka, Tanzania, Vietnam and Zimbabwe. These studies of gender and the household 'model' were carried out after land redistribution took place (see Jacobs 2010 for a summary). The case studies indicate that there have been both beneficial and detrimental aspects of land reform for married women.

Increase in food production and household incomes are key aims of land reforms and, where this has happened, single and married women often – although not inevitably – report that their lives have improved. Women in Andra Pradesh, for instance, saw stability and food security as an important marker of success of redistribution of land. Many land reform programmes have and continue to use a model of a nuclear family. Some wives experience this model as giving them more informal influence over the husband, partly because of increased distance from the extended family or lineage relatives.

Until very recently nearly all programmes allocated land titles or permits to men as 'household heads'. Men usually qualify for landholding as a result of gender alone, but women only on the basis that they support dependent children. Until recently, in many cases, the presence of an adult man in a household meant that he was counted as the 'household head'. This pattern is nearly universal and has served to entrench not only normative sexuality but also male privilege (Jacobs 2010). A number of negative repercussions have followed from this policy.

One is that women's workloads often increase with land redistribution, both because the household has more land to cultivate and because of pressure to work more intensively. There may exist as well pressure to bear more children to form a more self-sufficient farm unit. A lack of services (shops, schools, clinics, etc.), especially in early stages of programmes, is a common experience, and this affects women particularly as wives and mothers. Where wives hold customary land rights (e.g., wives' rights to a small 'garden' plot allocated by the husband), they may lose these because men's land rights are strengthened. It is also usual for wives to report a reduction in their own incomes, even while that of the 'household' rises. This is for a range of reasons: loss of marketing niches, loss of opportunities outside agriculture to earn incomes, and lack of equitable redistribution within households. Finally, wives often report a lessening in decision-making powers and autonomy due to their relegation to the role of 'housewife' as well as to the more continuous presence of husbands (this, because many schemes forbid off-farm work and migration). The fact that husbands are more continuously present means that they may have a stronger interest in the farm, but can also increase surveillance of wives' work and behaviour.

The situation of women-headed households has been ameliorated somewhat with regard to rights within land reforms, since many schemes now include female household heads as beneficiaries, although their numbers are typically small. But the lives of the majority of adult women who are married or live with male partners have not improved in any straightforward way, and may have deteriorated with respect to the ability to exercise rights or to make decisions with a degree of autonomy. Overall, within agrarian reforms husbands tend to gain power and influence as well as material benefits, but often at the expense of wives (Jacobs 2010). Although male-biased policy is an important factor, more structural factors also underlie.

In many smallholding or peasant households, family labour is crucial to household production as well as to its continuation as an economic unit. Even though today most rural people's livelihoods are diversified, where land reforms take place they increase the importance of household-centred agricultural production.

Several features encourage relations of domination over women – or patriarchal gender regimes – although much social variation also exists. One factor is the relative spatial isolation of smallholder communities and households. Another factor of relevance is the combination of production and domestic tasks in one household unit. In many societies, women's work in agriculture, craft production and processing of crops is a mainstay of production, and peasant men normally direct the labour of wives and daughters.

Peasant husbands and fathers also have an impetus to retain control over women's sexuality within peasant households since control of women's reproductive capacity ensures reproduction of the economic unit itself. Further, control over women's sexuality has symbolic importance for male identity within many peasant communities. A married peasant man may control little outside his household but he does usually have authority within the household and has a position of status as its head. Heidi Tinsman's (2002) study of the Chilean land reforms under Frei and then Allende (1965–1973), for instance, found that the process strengthened existing gender hierarchies, reinforcing married women's economic dependence on men and reinforcing peasant men's sense of authority over wives. Tinsman (2002: 184) writes that *peones* as new holders of land were particularly eager to display a reinvigorated masculinity by policing the parameters of feminine domesticity. Women's participation in political struggles was also experienced as demasculinising.

This outcome is likely to have been influenced by the historical period in which the land reform took place, but these assumptions sometimes continue in more contemporary settings.

Brazil: the land reform movement (MST) and gender

The example of Brazil indicates the difficulty of organising around gender issues within land reform movements, even in avowedly progressive settings. Brazil's 1988 Constitution is rare if not unique in granting women and men equality in land rights, including in agrarian reform. Nevertheless, land and property-holding is considered a male prerogative. Brazil also has perhaps the largest and best-known rural land movement in the world, the MST (O Movimento dos Trabalhadores Rurais sem Terra) or the Landless Workers'/Peoples' Movement – usually termed the 'Landless Movement' – with 1.5 million members or affiliates (MST 2013).

Over nearly four decades, the MST has organised numerous land occupations and has been successful in pushing land reform and agricultural sustainability onto the national agenda. The MST has also successfully promoted functioning land reform settlements or communities. It has been attentive to provision of services such as schools, clinics, marketing of produce and sustainable cultivation; these are initiatives that benefit all settlers, and particularly women.

Organisationally, the MST is meant to be democratic and participatory: all adult family members must join rather than simply the household head. Nevertheless, the MST has had an ambivalent relation to questions of the 'family' constitution of households and gender discrimination. This is despite acknowledging support from women in land occupations and emphasis on women leaders. It has a stated commitment to gender equity and to emancipatory ideals more widely, but this coexists with marginalisation of ordinary women and, sometimes, with silence about gender issues within encampments and settlements. One source of ambivalence may stem back to the MST's religious roots, which lay in 'ecclesiastical base communities' (CEBs), emphasising the need for economic redistribution, including of land, but also the links between landholdings and stable family life.

A number of feminists have critiqued the male bias of MST policy and informal practice (see Jacobs 2010 for sources; da Silva 2004). The summary below provides brief examples.

The MST tends to view farming units as unified households with a head, usually assumed to be the husband or father. Rules are often predicated on the existence of a nuclear family. For instance, residents of encampments must hold no other 'outside' job: this stipulation discriminates against the substantial numbers of single mothers who have occupied land. Married couples may decide that one spouse or partner should work outside the encampment but this strategy is not open to single parents with children. Male dominance within encampments is enacted in many ways: for example, behaviour and language, undercurrents of gendered hierarchy and the assumption that women mainly exist in a private, home-bound world.

Although in the late twentieth century and previously, only one person (usually the husband) could sign the official (INCRA) form, there now exists room on the form for the signatures of two spouses or partners – as is now common elsewhere in Latin America. The MST does support the policy of having two signatures; however, mobilisation for the change in policy was from feminists within rural trade unions rather than the MST itself.

Da Silva's study of a settlement in Santa Catarina state in the south (da Silva 2004) also stressed deep-rooted inequalities in the division of labour and assumptions about male land rights. The study found great discrepancies in the types of agricultural work that men and women undertook and in extent of cooperative membership. Moreover, the MST council in this case intervened in cases of sexual morality. In one example, a woman who had had an affair with a married man was asked to leave the settlement: the man, a movement leader, remained.

The MST has vacillated with regard to how much it decides to emphasise gender issues. Ambivalence has been evident in the public website, which has in various periods published and withdrawn material on gender issues. The present site acknowledges that inequalities exist within the organisation and lays out various aims concerning gender equity: 50 per cent participation by women in all MST education and training courses, and in leadership roles in national bodies, the guarantee of one male and one female coordinator in the community bases, and 'intensive discussion and study regarding the theme of gender in all MST courses and conventions' (MST 2013). The MST is a leading member of La Vía Campesina, and this accords with the latter's policy on equal gender representation at all organisational levels. The MST does attend to gender to some extent then, but tensions exist between priority given to gender and to class divisions.

The above examples suggest that, despite promises of democratisation and increased autonomy for small-scale producers, then, household model land reforms have usually privileged peasant men.

The next sections briefly discuss the collective model of agrarian reform, and gendered processes of decollectivisation.

Agrarian reform: the collective model

Following revolutionary upheavals in the twentieth century, societies such as the USSR, Cuba, China and Vietnam (among others) attempted collectivisation of agriculture either on a full or partial basis. Despite the usual understanding of land reform as being enacted through distribution of land to households, the greater part of land redistribution historically has taken place either through collectivist reforms or else the redistributions that have followed (see below).

Although collectives can take different forms, larger-scale collectives were favoured by state socialist governments for a variety of reasons not least that it was seen as imperative to follow the Soviet model. Collective farms dealt with the political issue of the peasantry, seen as individualistic and potentially petty capitalist – a suspect class. Collectivisation was also seen to have social advantages, especially for poorer peasants, by providing employment for most men and often for women as well. Collectives sometimes provided schools and clinics, benefiting all the collective, including women. Governments also saw advantages in the collective model, retaining state control over food production, which was hoped to ensure urban food security.

Collectivisation was also enacted because it was assumed that in agriculture as in industry, large units were the most efficient. The model also failed to realise that while some crops may be suited to large-scale farming, others are more effectively managed on a small and more flexible scale. An intricate interweaving of tasks characterises most farming. Within this process, the ability (of men and women) to combine different tasks is often crucial.

The assumption that agriculture was most efficient if organised in large industrial-like units was an error of great proportions, and one that was repeated even after the initial Soviet experience. For a variety of reasons related to the inefficiency of top-down approaches, and the detailed and complex care and attention needed for many crops, collective agriculture rarely proved capable of increasing production levels.

Collectivisation was not in the main undertaken for reasons to do with gender, but it nonetheless had profound gender implications. One reason for this was the payment of workers or collective members in work points, according to the types of work done and extent of contribution. Women were typically paid – as with wages – less than men. Various Sinologists, for instance, find that women were usually paid roughly 20–30 per cent less than men (Li cited in Jacobs 2010). However, the lack of private property and payment methods meant that women's work in fields was undertaken for the collective body rather than the family or husband. Collectives often provided an avenue for mobility for women, in that they were sometimes able to take on semi-skilled or skilled work – including, iconically, as tractor drivers. Although in practice women still suffered inequalities, this was meant to show that they were capable of skilled and heavy labour as were men.

Collectives also provided an avenue for independence for 'lone' women (widows, divorcées, single and never married women) who were often socially marginalised. There exists much testimony of support for collectives among such women; historical examples exist from the Soviet Union, as well as from Cuba, Bulgaria, Hungary, Uzbekistan and Vietnam (Jacobs 2010). Payments in work points lent visibility and public recognition to women's labour. This often met with resistance from husbands, who resented their relative loss of control. Collectives also usually had some type of representative body – including at times, women's committees – in which women could seek redress against household or other abuses.

Collectives were generally unpopular, including among married women. However, these tended to be more popular among women than men as long as they could ensure reasonable levels of production.

Wiergsma (1991) gives a powerful account, for instance, of peasant men's antagonism (especially, that of middle or self-sufficient peasants) and resistance to collectivisation in Vietnam. She highlights in detail means of resistance to and subversion of collectivisation policies, resulting in delayed implementation. Thus peasant households – and/or men within them – retained much control over agricultural plots and production.

Collectivisation was not disastrous in all cases, but it was in at least two. The USSR and China suffered large-scale famines as a result of the coercive manner of collectivisation or overly rapid reorganisation or (in China) violent punishment of cadres who reported poor harvests and mass starvation. At the same time, not all collectives or large cooperatives were failures. Collectives performed best when they were small- to medium-sized units that could exercise a degree of control over themselves, avoiding distrust between members and collective officials where mechanisms existed linking income to individual work effort and where crops are suited to large-scale farming.

The partial loss of control over women and their labour, then, differentiated collective model land reforms from the household model. There was much resistance, usually passive or semi-hidden to these policies; the vision was one of a small-scale 'peasant patriarchy' (Stacey 1983; Jacobs 2010) rather than a large collective in which wives earned work points.

After collectivisation

Land redistribution following the decollectivisations of the late 1980s and early 1990s involved large numbers of households, people and large amounts of land, although some remained within large units. The return of 'responsibility' (as it was termed in China) to the household or family usually resulted in increases in agricultural production. This also implied, however, a return of control over the farm as production unit to the husband/father as household head. The outcomes bore much similarity to those of land reforms enacted without intervening collectivisation. For instance, wives often experienced more pressure to bear children (especially sons) increased work pressures and loss of public and household recognition of their work. More positively, some women were able to exert more control over when to carry out farm tasks.

In Vietnam, wives were able to regain marketing roles – traditional for women in the region – when food markets and small businesses in general were reinstated. Vietnam differed from many other cases in that with decollectivisation, all adults were guaranteed a land allocation, and this included married women. Another legacy of the revolutionary period was that policy measures discouraged land concentration. Although women still suffered discrimination, their landholdings ensured some livelihood security. However, from 1993 the government rescinded individualised land allocations for newly born children, although their parents or households retained land rights. If the latter die intestate, daughters will receive equal inheritances to land. However, where fathers and mothers make wills – as most do – they are likely to favour sons in inheritance, particularly of agricultural land. This measure undermines the relative security that women had enjoyed and reinstates parental, particularly fathers' control over land.

What is to be done?

The example of gender relations within land reforms presented underlines the embeddedness of gendered and sexualised inequalities within peasant or small-scale agriculture. Deliberate and targeted policies are needed in order to avoid inequitable outcomes, even if those have been unintended.

Some issues can be addressed through legal or political change and within agrarian reform and titling programmes themselves. Where women face legal discrimination, the law should be amended to give them land rights equal with men's. For instance, a 2013 ruling in Botswana (de Lange 2013) upheld the rights of sisters residing in a family home to inherit their home, rather than a non-resident nephew. Justice Lesetedi and the court noted that social realities had altered over the past 30 years and that equality before the law should take priority over customary norms of male inheritance.

Where land is held communally it may be that mechanisms other than privatisation of land can be found: there are some signs that communal tenure can be reformed in a 'gender-friendly' manner. There also, however, exists much evidence of 'resistance to gender equity in land and property holdings' (Budlender and Alma 2011). It is crucial that women be able to claim the same kinds of property rights as men in their own groupings – i.e., communities, kin groups, households.

In contemporary situations in which many corporate and state 'land grabs' are taking place, it is easy to forget that women as well as men smallholders suffer from loss of lands and livelihoods. A recent NGO-led study of gender and commercial pressures on land (Daley 2011) noted the general lack of discussion of gender within the growing literature on land deals. Reviewing evidence from eight African and Asian countries, it concludes that although the rural poor are likely to be unfavourably impacted by land acquisitions, the impacts for women differ from those on men. Women are not only impacted differently: they are disproportionately likely to be affected in negative ways because of existing discrimination and disadvantage (Daley 2011: 57).

Historically, women have made most gains within agricultural programmes where there exists state backing for gender equitable measures – for example, rights for female-headed households, practical advice and mechanisms for adjudication of disputes between husband and wife. Requirements for signatures of both husband and wife on land permits or titles are an important step, but have proved difficult to enforce. Reforms to family law connected with property are also important. One of the most contentious areas concerns wives' rights to remain on land in case of divorce. It is common for wives to be ejected from land and to lose their livelihoods along with their marriages. This outcome points to a key factor: changes in legislation are insufficient in themselves. To be meaningful, these must be enforced (literally) 'on the ground' – through (for instance) educational campaigns to alter beliefs about women and landholding, through equitable legislation and through actual enforcement of laws. Women also need legal recourse when they are subjected to violence, which is frequently precipitated as a backlash to land claims. Most such social changes can be made only through political organisation. Initiatives such as La Vía Campesina's against violence against women (VAW) are therefore a key step.

Some campaigns for women's land rights take the form of direct action. For instance, the World March of Women participated in a world forum on food sovereignty in 2007. Marches for women's land rights have taken place in India and Brazil and smaller actions have taken place in South Africa. However, many initiatives are at the local or community level: for example, discussions with community elders, formation of women's cooperatives and many initiatives to list wives' names on permits or title deeds. However, the global arena is also key in leading movements. The importance of women's land rights is beginning to be recognised globally, not least by United Nations institutions, state aid agencies and many NGOs, as well as by local movements. Nonetheless, the struggle for gender-equitable land rights remains at an early stage; perhaps the disquiet that often arises from women's land struggles indicates their great importance.

There are some signs that non-married women in some areas are taking steps to acquire land, despite this being seen as 'non-normative' in most rural areas. For instance, in southern Africa a number of single women household heads have asserted demands for land, although these must usually be negotiated with traditional authorities (i.e., local chiefs). Recent laws both in Zimbabwe and South Africa have strengthened the role of traditional authorities in land allocations, and this can be problematic for gender equity. However, some traditional leaders have promoted rights for women within communities (see Budlender and Alma 2011).

A striking example of single (including divorced and widowed women; never-married women; wives fleeing domestic abuse) women's assertions of land rights is in the northwest Indian movement Ekal Nari Shakti Sangatham (ENSS) (Berry 2011). The movement explicitly challenges both women's dependent status and the necessity of (heterosexual) marriage in rural sectors, in that it demands resources and new forms of organisation enabling single women to subsist outside marriage. These demands include individual registration in local council registers, and ration cards, which are crucial markers of individual identity. They also call for access to a range of government programmes and resources, and the grant of two acres of state-held agricultural land to meet basic food needs.

What makes the ENSS particularly unusual is not only its organisation of non-married women, but also its demand for a new form of household or 'marital family' (*naya susural*) in which an older woman joins with a younger woman (usually, with dependent children) to form a viable farming unit. While this is in part simply a practical measure, such measures challenge the 'heteropatriarchal' (Berry 2011) basis of access to land in north India. Since women living outside the protected status of heterosexual marriage are automatically suspect, the new household relations are also intended to enhance single women's community status.

This case provides an attempt to construct women-centred households that also have an agrarian basis. These provide an interesting contrast to the consolidation of peasant households and control over wives indicated by mainstream household agrarian reform models.

Another example of rural women's movements comes from Bolivia. To date, little land redistribution has taken place (under the Morales government) but mobilisations are emerging. There, as in many other countries, women make up the majority of the agricultural workforce. Unusually, however, the most radical women's organisations are rural ones: the subcommittees of peasant unions such as Bartolina Sisas of the Confederación Sindical Unidade Trabajadores Campesinos de Bolivia (CSUTCB) and the newly formed National Confederation of Indigenous Women (Spronk 2012). Both have made agrarian reform a central plank of programmes since the mid-1990s, although they are in disagreement about the preferred tenure type in any future reform: that is, individual vs. collective titling (Spronk 2012: 262).

There thus exist signs of growth in rural women's movements. It is important as well, to acknowledge the obstacles faced: these remain substantial.

Women's household and farm labour are crucial to agricultural production systems, especially on small farms. Key to any equity within agrarian reforms or farming more broadly is that women's labour be recognised at household and community levels, as is beginning to happen at the international level. The example of agrarian reforms underlines the risks of marginalising women's contributions, and the need for public – including household and community – acknowledgement of their work. Rural women, like others, need secure livelihoods and increased autonomy. Equality of land rights are key.

References

Berry, K. (2011) 'Disowning dependence: single women's collective struggle for independence and land rights in northwestern India', *Feminist Review* 98: 136–152

Budlender, D. and Alma, E. (2011) *Women and Land: Securing Rights for Better Lives*, Ottawa: IDRC/ International Development Research Centre.

Daley, E. (2011) *Gendered Impacts of Commercial Pressures on Land*, Rome: International Land Coalition.

da Silva, C.B. (2004) *Relações de gênero e subjectividades no devir MST*, www.scielo.br/scielo.php.

de Lange, I. (2013) 'Victory for women's rights', *The Citizen*, 15 September, http://citizen.co.za/40221/victory-for-womens-rights.

Dixon-Mueller, R. (1985) *Women's Work in Third World Agriculture*, Geneva: International Labour Office.

Doss, C., Kovarik, C., Peterman, A., Quisumbing, A. and van den Bold, M. (2013) *Gender Inequalities in Ownership and Control of Land in Africa: Myths vs. Reality*, IFPRI Discussion Paper 01308, Washington, DC: IFPRI.

FAO (2011) *The State of Food and Agriculture 2010–11; Women in Agriculture: Closing the Gender Gap for Development*, Rome: FAO, www.fao.org/publications/sofa/en.

GRAIN (2008) 'Seized: the land grab for food and financial security', 24 October, www.grain.org/article/entries/93-seized-the-2008-landgrab-for-food-and-financial-security.

Jacobs, S. (2010) *Gender and Agrarian Reforms*, New York and London: Routledge.

MST (2013) 'About us: sectors – gender', www.mstbrazil.org/about-mst/sectors-gender.

Sachs, C. (1995) *Gendered Fields: Rural Women, Agriculture, and Environment*, Boulder, CO: Westview Press.

Spronk, S. (2012) 'Twenty-first century socialism in Bolivia: the gender agenda', in L. Panitch, G. Albo and V. Chibber (eds.) *The Socialist Register*, Pontypool: Merlin, pp. 255–265.

Stacey, J. (1983) *Patriarchy and Socialist Revolution in China*, Berkeley: University of California Press.

Tinsman, H. (2002) *Partners in Conflict: The Politics of Gender, Sexuality and Labor in the Chilean Agrarian Reform 1950–73*, Durham, NC: Duke University Press.

White, B., Hall, R. and Wolford, W. (eds.) (2012) *New Enclosures: Critical Perspectives on Corporate Land Deals*, special guest-edited issue, *Journal of Peasant Studies* 39(3–4).

Wiergsma, N. (1991) 'Peasant patriarchy and the subversion of the collective in Vietnam', *Review of Radical Political Economics* 23(3–4): 174–197.

12

EXPLORING GENDERED RURAL SPACES OF AGROBIODIVERSITY MANAGEMENT

A case study from Kerala, South India

Isabelle Kunze and Janet Momsen

Introduction

Rural women's mobilization in various movements has been growing for the last four decades in India and in other parts of Asia (Elmhirst and Darmastuti 2015). Adivasi or indigenous people's movements' protests initially occurred in India against caste and gender inequality. Basu (1995) described the movements as women's indigenous feminism, whereby women became conscious of their role and strength in improving the lives of their households. These grassroots movements sought to transform their everyday spaces. Adivasis have begun to make their presence felt as environmental change has eroded their livelihoods in many parts of India. They continue to be marginalized but have often been assisted in their protests by civil society, self-help groups, and local NGOs. As their land is increasingly alienated they look for multi-local livelihoods and become more mobile (Panda 2014). Only recently have such activities begun to take place in Kerala in southern India. This chapter looks at indigenous women's activism in Kerala as a response to land-use change and environmental degradation.

Wayanad is a hilly region in Kerala, South India, on the edge of the Western Ghats, 700–2,100 meters above sea level. It was designated a UNESCO World Heritage Site due to its remarkably high level of biological diversity and recognition as a biological 'hotspot'. *The Hindu* newspaper described the region's new status as a "reflection of India's concerted efforts to inscribe the world's hottest hotspot on the World Heritage List" (July 12, 2012). The ecological value of Wayanad's natural landscape characterized by paddy rice fields and forest is also essential for the livelihoods of the people living in the area. Unlike other parts of Kerala, Wayanad is largely "tribal" as 17.7 percent of the local population is made up of Adivasi communities, as indigenous population groups are called (Rath 2006). Paddy cultivation has a strong cultural meaning for Adivasis in particular because rice is their main staple food and, therefore, essential to maintain food security. However, the current agrarian crisis in India (Narasimha Reddy and Mishra

2009) as well as changes in land use and cultivation practices including the conversion from agricultural fields to land for housing and infrastructural development all contribute to rural diversification in farming households in India (Arun 2012). This trend influences current agrarian and gender relations amongst the Adivasis.

Determinants of changes in land use

Agricultural transition in Kerala is linked to changing cropping patterns such as the conversion from food to cash crops. In the past, Wayanad was mostly characterized by rice fields whereas today, agricultural fields are also used for plantain, ginger, arecanut, and vegetable cultivation. Overall, paddy cultivation has decreased as it is no longer profitable. Changes in demand for labor, including rising labor costs and labor shortages, further challenge the future of paddy cultivation. In addition, increasing variability in rainfall patterns negatively affects agriculture and particularly paddy cultivation, so that now high-yielding varieties of rice can be cultivated in only one season instead of two as in the past. Another dominant driver of land-use change is ongoing deforestation due to growing demand for land required for housing because of rapid population growth, infrastructural development, and the needs of the tourism industry in the region. The distribution of agricultural land is also influenced by changes in the social organization of indigenous communities in Wayanad. Increased education and migration has encouraged a shift from a joint to a nuclear family system leading to a reorganization of property rights from collective to individual ownership of land.

The Kuruma people

This case study focuses on the gendered dimension of land-use change among the Kuruma tribe, which is one of the landowning communities of Wayanad. The Kuruma consider themselves as traditional agriculturalists who have always been involved in paddy cultivation. Therefore, cultivation practices are also strongly embedded in the social organization of the Kuruma, whose sources of subsistence are mainly paddy rice and vegetable cultivation. In the past, subsistence agriculture served as a protection against poverty and starvation, whereas today both Kuruma men and women are involved in work outside the agricultural sector to generate additional income. Social changes associated with the trend to nuclear household structure as well as increased levels of education have altered the social status of the Kuruma women in particular. Elmhirst (2011) sees these changes as material feminisms linking nature and multi-local livelihoods. The introduction of community development programs such as the "Kudum-bashree" (women's self-help groups) enlarges opportunities for empowerment of poor women. Women meet in Kudumbashree groups not only for discussions about health and education, but also for organizing joint cultivation practices that offer a possibility for women to manage and use agrobiodiversity outside the masculine domain.

We explore how changes in land use influence women's ability to sustain food security while at the same time considering the gendered roles and responsibilities among indigenous communities on a household level. Crucial in this regard is to look at the ways in which community programs such as the Kudumbashrees offer new spaces for women to manage agrobiodiversity. The next part presents a short overview of contemporary work on gendered geographies in India, the phenomena of the gender paradox in Kerala, and the dynamics of the Kudumbashree movement in Kerala.

We used qualitative research methods including in-depth, semi-structured interviews, and focus group discussions with Kuruma women and men farmers from two different Kuruma communities living in two distinct locations. Field work was carried out from February–July 2011 and January–March 2012.

Gender, space, and development

Gendered geographies in India

Recent work within feminist geographies in South Asia calls for analytical frameworks that conceptualize the social and theoretical formations of space and place (Raju 2011; Raju and Lahiri-Dutt 2011). These frameworks aim at a better understanding of the spatial embeddedness of social relations that are shaped and created by notions of femininity and masculinity. Using a gendered geographical perspective means viewing gender as a social construct defined by behavioral norms for women and men. However, gender as a feminist concept is not fixed but fluid and changing over time. Crucial in this regard is sensitivity to difference among women from other subaltern locations. Thereby, the concept of intersectionality provides a useful approach to explore the ways in which gender is socially constructed in the South Indian context. It particularly considers differences in caste, class, gender, age, ethnicity, and identity in relation to the construction of social hierarchies.

A central and much contested debate is the binary separation into public and private domains. Unlike Eurocentric notions of the public/private binary, the "fluidity of binaries" (Raju 2011: 13) appears to be a unique feature in most Asian countries. Instead of reinforcing the private/public dualism, Raju and Lahiri-Dutt (2011) emphasize the socially constructed nature of space and place, while arguing that the boundaries between both are rather fuzzier than simply sexually segregated geographical locations. Moreover, the organization and meaning of space is a product of transformation and experience that differ among communities, caste, and gender. For example, the spatial segregation between genders among scheduled caste and tribal population groups appears to be less restrictive than for women from high castes due to different social norms and hierarchies between the genders. According to Raju and Lahiri-Dutt (2011), doing gender refers to the idea of situating gender and, therefore, is closely linked to a critical engagement with inequalities and imbalance of power. They offer an important perspective for the Wayanad context and claims that the intersection of space, power, and knowledge are experienced on an everyday basis. In addition, Raju (2011) explains that contemporary research on gendered geography in South India seeks to challenge patriarchy being manifested through formally and informally institutionalized structures in order to oppose women's essentialized enclosure within the paradigm of domesticity.

The "gender paradox" in Kerala

Unlike northern Indian states, Kerala has a tradition of a matrilineal system of inheritance in which familial property is passed on through females (Devika and Thampi 2007). It is one of the few states in India where the Gender Development Index (GDI) parallels the Human Development Index (HDI). As such, Kerala proves to have high indicators of social development referring to a high female literacy and women's greater enrolment in higher education compared to men, even with low per capita income (Sreekumar 2007). The North–South Indian controversies also form an interesting field of inquiry for gendered geographies. Women in Kerala have a longer life expectancy by as much as 18 years compared to women in Madhya Pradesh even though there is little difference in per capita income (Raju and Lahiri-Dutt 2011).

However, despite equal access to basic needs and material resources for women in the development planning process, research has identified two alarming trends that question the high development profile of women in Kerala: little participation of women in decision-making processes and increasing control over women's bodies through domestic violence (Erwér 2011; Mukhopadhyay 2007). This contradiction is interrelated with patriarchy established through formally and informally institutionalized structures (Raju 2011). Thereby, northern gender norms are being reproduced and fail to challenge the uneven distribution of power in gender relations in Kerala. Some scholars see this in relation to the Kudumbashree movement in which conservative gender norms and the gendered separation of space remain powerful (Devika and Thampi 2007).

The gender paradox debates in Kerala call for a critical engagement with the ways in which the "paradox" of Kerala has been socially constructed through the media, politics, and Keralite society itself. Sreekumar (2007) questions the mutual relationship between greater female literacy and women's emancipation as the comparison is based on social indicators that do not necessarily reflect progress made in gender and development. She examines Kerala as utopia and as dystopia. Utopia is explored through a discourse analysis of the Kerala model of development and tourism advertisements; dystopia is evaluated through discourses on AIDS and sexual violence. Whereas development discourses are often gender neutral, the tourism industry portrays the high status of women who enjoy a modern life in Kerala, which is portrayed as God's Own Country. The ideal woman of dystopia is vulnerable and does not fit into traditional gender norms. In Sreekumar's (2007: 49) view, utopia and dystopia are different but "equally hegemonic worlds." She critiques the gender paradox in Kerala as constituted by women who are marked by socioeconomic privileges and excludes the consideration of gendered differences amongst women from marginalized population groups (scheduled caste/tribes). This strongly relates to the hierarchical structure of Indian society as a whole in which those people belonging to scheduled tribes in particular are often left out.

The Kudumbashree: Kerala's poverty eradication program for women

The Kudumbashree Mission is a women-oriented poverty eradication program based on microfinance microenterprise launched by the government of Kerala in 1998. According to the government's program website (www.kudumbashree.org), the program aims at the empowerment of women in households living below the poverty line (BPL). Furthermore, it is considered to be one of the largest women-empowering projects in the country, having 4,054,000 members in 2013 and covering more than half the households in Kerala. As a response to current development concerns, the 11th Five-Year Plan of local state governments stresses agricultural production, local economic development, poverty eradication, and social equity as primary objectives. The Kudumbashree mission relies on participatory governance initiatives to support decentralization processes.

Overall, the program is based on the framings of development concerns and addresses issues of gender equity, participation, and poverty eradication (Williams et al. 2011). It encompasses three key organizational features: first, operationalization through women participants; second, self-organization in neighborhood groups (NHGs); and third, the group's relationship with the state. Men are generally not included in the Mission's overall structure and objectives. The Mission covers three different levels: the community development society (CDS) on the *panchayat* level, the area development society (ADS) on a ward level, and the neighborhood groups (NHGs) on a community level. All these are significant areas of interface of the CDS with the local government. Using a feminist analytical perspective, Devika and Thampi (2007) describe two historical reasons for successful implementation of the Kudumbashree in Kerala. First, it highlights the individualization of women in education. This includes employment on all levels with a relatively low domination of patriarchy. Second, the program builds upon domestic ideologies that favor unpaid labor. Instead of promoting gender equity, the Kudumbashree Mission rather fosters community development in which empowerment of the poor refers to the "creation of greater space and flexibility for the poor within the socio-political entrenched structures" (Devika and Thampi 2007: 44). This view on empowerment appears to be a useful viewpoint from which to further examine the potential of the Kudumbashree's offering of new space for women to manage agrobiodiversity and, therefore, be agents of development (Devika and Thampi 2007).

Gendered dimensions of land-use change in Wayanad

Kudumbashrees – a space in which women become agents of development?

Recent critical assessments of the Kerala government's poverty eradication program, the Kudumbashree (Arun 2012; Devika and Thampi 2007; Williams et al. 2011), offer fruitful insights to this case study on gender and land-use change in Wayanad. The Kudumbashree program is a useful way of exploring the interactive nature of public and private spaces. I use the example of the Kudumbashrees to reveal the ways in which women cope with agrarian change while using this space to manage agrobiodiversity. The Wayanad case study on the gendered dimension of land-use change stresses the importance of the Kudumbashree movement amongst the Kuruma. In three out of four villages, the microenterprise Kudumbashree groups are crucial for women's everyday life. In Wayanad, Kudumbashree neighborhood groups are called women's self-help groups (SHG). Several office bearers form the groups: a president who attends meetings with the local panchayat, a secretary who is responsible for minutes and pass books, a treasurer who manages the funds, a health worker who looks after houses and attends health classes in the panchayat in order to inform the group members accordingly, an income manager who attends classes on governmental money schemes, and a basic amenities worker who is in charge of housing and sanitation facility schemes provided by the government.

The ways in which the groups are socially organized is clearly gendered while reproducing traditional gender norms. Men have recently shown interest in becoming involved in Kudumbashree meetings for men. Unlike women's groups, men's Kudumbashrees receive little support from the government. Field research has shown that in all three Kuruma communities mainly women participate in the Kudumbashree groups. Usually, the groups consist of up to a maximum of 20 women members only. In one community, two men's Kudumbashree groups existed; however, one woman described these groups as unsuccessful because in her view, "men finish the money very easily and they don't save the money like women do. The secretary of that group took the money and ran away" (interview, June 13, 2011). This statement nicely illustrates a gender difference in dealing with savings that are originally meant to support the family.

For the Kuruma women, the groups serve four main purposes: first, they appear useful in providing opportunities to obtain loans required for health issues, children's education or marriages; second, they increase women's social and economic responsibilities; third, participating in the groups has increased women's mobility; and fourth, the Kudumbashrees foster women's empowerment. In order to become a Kudumbashree member, participants have to pay a weekly contribution of between 10–20 INR (£0.10–0.20 or US$0.15–0.30), which allows women to take loans which they must repay within two years. The interest for loans provided by a bank would be 9 percent whereas, being a Kudumbashree member, women need to pay only 4 percent interest, with the other 5 percent subsidized by the government. The microcredit program is particularly important for widows who have responsibility for their children's education. One senior woman states that the SHGs have improved her life because:

> No one from outside will give this much money as a loan. And their interest rates are also very high. I have two daughters and to marry them off, we have to have some money and similarly there are many things in the house that need money.
>
> *(interview, June 15, 2011)*

This directly leads to increased social and economic responsibility for women. A former group secretary highlights the link between the Kudumbashree movement and increased responsibility for women today.

> It was men who looked after all such things and now with the Kudumbashrees, women started doing most of the things at home. Especially taking loans are done by women. In the past, women were involved in household or kitchen work but now women are involved in many things including bank dealings.
>
> *(interview, June 13, 2011)*

The majority of the Kuruma husbands support women's participation in the group because through the units, women save money and help the family. One woman further clarifies that "if women have an opportunity to save this money then it helps in alleviation of poverty" (interview, June 13, 2011). This thought links to the main agenda of the Kudumbashree program stressed by the government of Kerala. Some women say that men also borrow money from women through the Kudumbashree groups. A tribal promoter described that through Kudumbashree, "responsibility of men is getting reduced and that of women is increased . . . Here women are helping men so they won't say not to go to Kudumbashree" (interview, March 8, 2012). This statement reveals increased responsibility for women to fulfil the needs of the family, which reinforces traditional gender roles and responsibilities. Nevertheless, most women participants support the view that the Kudumbashree mission brought lots of positive changes in women's everyday lives. For example, a young woman states: "I also learned many things through Kudumbashree. I learned how to deal with money matters, also banking process. I also learned how to behave with others. Now I am far better than I was" (Interview, February 9, 2012). For this young woman, the group helped her to overcome an inferiority complex because she has improved knowledge of managing monetary matters and enjoys more interaction with other women. A senior women from the same village argues that today, "women go for work outside and so get to know many things and do works and we earn money. Now women are almost independent" (interview, February 9, 2012). Overall, both statements highlight that improved access to training programs for poor women have reinforced self-confidence.

These insights are also related to another aspect that brought major changes to Kuruma women's everyday lives: increased mobility and participation in public life. A majority of the Kudumbashree participants appreciate that, instead of being restricted to the domestic domain only, the group offers space for social interaction outside the home. The shift in the social organization from the joint to the nuclear family system also results in greater mobility for women. In addition, greater freedom on how to use time is also linked to agrarian change. Due to the availability of rice mills, the processing of paddy demands less time than in the past, which allows for greater flexibility of activity. One woman said that today women enjoy "more freedom and get space for what they want to do and also to mingle with the rest of society" (interview June 23, 2011). This thought clearly emphasizes that the participation in the SHGs enables women to cross the domestic boundaries while entering a public space offered by the Kudumbashree movement. However, neighborhood groups usually take place in private places such as the home and, therefore, uncover the fuzzy boundaries between public and private spaces.

Mobility is closely related to the empowerment of women, which is—according to the official website—the overall objective of the Kudumbashree program. The study of the Kuruma in Wayanad stresses that from the women's perspective, the participation in the SHGs has helped women to "grow, to develop and to empower" (interview, June 14, 2011). The tribal promoter supports this view, arguing that "units like the Kudumbashree . . . gave opportunities for women to come forward and lots of good changes have come for women today" (interview, March 8, 2012). This thinking nicely links to Devika and Thampi's (2007) notion of empowerment understood as creation of greater space and flexibility for the Kuruma women within the existing social structures.

Despite the positive effects of the program on women's mobility and increasing social status within society, the structure of the groups reveals a generation gap. As stated by many women participants, the rules and regulations do not allow senior women over 60 years to participate in the program. In addition, the Kudumbashree excludes mothers with young children and reinforces the conflict between the group and domestic responsibilities including childcare. Furthermore, given the required weekly contribution of 10–20 INR, uneducated, extremely poor women are also excluded from the program because they do not have the economic means to become a group member. This contradicts the program's overall objective of poverty eradication for poor rural women. As men are completely excluded from the program, the Kudumbashree mission also fails to address its goal of promoting gender equity.

The Kudumbashrees and agrobiodiversity

Having looked at the social dimensions of the microenterprise community development initiative, the Kudumbashree also offer useful insights into how women cope with changes in land use. The meetings provide space for women to discuss any matters related to agriculture including the use of tractor, tiller, and other machinery or which crops and vegetables to cultivate. It has also been common to buy cattle through the Kudumbashree groups. However, due to shortage of available land, it has become increasingly difficult to feed them and, consequently, stock farming has decreased. In addition, the decline in cattle has also a religious dimension because, as explained by one member, "animals were being reared for meat . . . and my conscience did not agree with it" (interview, June 23, 2011). Even though dung can be used as fertilizer, the women in this particular group decided to move away from joint cattle farming.

One main joint task women do through the group is farming. Overall, the women find it easier to do group farming because "it is difficult to do cultivation separately and so we do it together in fields taken for lease" (interview, February 3, 2012). As such, Kudumbashree participants come together to grow paddy, ginger, and vegetables, which they either sell on the market or keep for themselves in order to maintain food security. If women decide to cultivate collectively, the group needs to apply for subsidies at the Area Development Society (ADS), which sends officials to the villages for approval of the subsidy. Once the application is accepted, the local government gives subsidies for cultivation. Vegetable seeds are provided by the "Krishi Bhavan" (a regional office of the Kerala Agricultural Department) but the choice of which rice variety to cultivate is decided by the women. Paddy rice cultivation is important for the Kuruma women and men farmers. They grow new and old rice varieties with the latter playing an important role in maintaining their tribal culture and traditions. Old rice varieties are necessary for catering for wedding celebrations and religious rituals. In one village, yam cultivation is an important additional source of income. One SHG member explains that, "we were doing yam cultivation and for that work is also less . . . and at the same time we can earn money also. We will get subsidy from the government to make masala powders or to have a mill or anything that we like" (interview, June 13, 2011). However, joint decisions over vegetable cultivation also create conflict between the young and older generations. Unlike the younger generation, older women prefer to cultivate yams.

Exploring the role of Kudumbashree among the three Kuruma communities has shown that the movement offers space for collective farming for women. This space is little influenced by patriarchal power relations due to the gendered separation of space. This fails to challenge the status quo in regard to traditional gender performances on the one hand but also provides opportunities for women to cope with agrarian change. Furthermore, the study demonstrates that the SHGs among the Kuruma are strongly related to agriculture, which stresses the interrelations between space and locality. Being agrarian Adivasi communities, women also successfully use the groups as a strategy to cope with changes in land use. Therefore, the case of the SHG offers a useful example through which to better understand the social construction of gendered spaces.

Multiple dimensions of land-use change

The reasons for changes in land use are constituted by agrarian and social changes occurring in Wayanad and among the Kuruma. One important driver for land-use change is the conversion from agricultural fields to land for housing. Population growth and infrastructural development, including the construction of roads and new houses, all contribute to increased demand for land. A crucial change for all the Kuruma communities is the replacement of thatched houses by brick houses. The construction of the latter is also subsidized by the government in order to strengthen tribal development. Further development projects include the construction of roads, of ponds in the fields and of wells for household water.

The shift in social organization among the Kuruma is also linked to changes in land use because more land for housing is required for nuclear families who wish to live in separate houses built for single households only. As a consequence, as outlined by one woman farmer, "there is not enough space to construct houses so even if we want to construct a house in the fields we have to get permission from government" (interview, June 14, 2011). Therefore, in order to prevent continued conversion of land, the government of Kerala has introduced a law that prohibits the construction of houses in paddy fields.

A second dominant trend leading to land-use change is changing cropping patterns associated with the shift from food to cash crops such as banana (plantain), yam, ginger, and arecanut. As a result, the *Ivayal* (paddy field) has become *kara* (elevated land). A 72-year-old village chief offers a historical perspective on changes in land use and argues that "now fields are less and if we cultivate other crops then paddy will not be sufficient" (interview, January 30, 2012). In addition, a woman farmer underlines this thought while pointing to the ongoing trend in shifting values in agriculture, which are linked to a changing landscape. She describes that "all those places were paddy fields earlier, now there are buildings everywhere. People are lazy and do not want to do agriculture now. They want to make money" (interview, June 23, 2011). In addition, a male farmer sees the shifting attitude towards agriculture based on economic rationality.

> The value of fields has changed and it is changing. Now nobody is interested in agriculture as earlier. Mainly because they have to spend lots of money for agriculture and they don't get the money back. Now people are looking for profit. That is why the shift came in agriculture.
>
> *(interview, February 15, 2012)*

This statement stresses that agrarian change in Wayanad is linked to the phenomena of de-agrarianization as farmers lose interest in farming activities as it is not profitable anymore. Changing cropping patterns are also linked to demographic changes in Wayanad. With the arrival of Christian settlers from the plains of Kerala, new crops such as arecanut have been introduced to Wayanad. However, according to a woman farmer, this change in agricultural practices negatively affected paddy cultivation. She explains that,

> Christian farmers are cultivating arecanut in the fields and along with it came water shortage. One of the reasons . . . is that they make canals for cultivating arecanut, so all the water is drained away. The level of the water table falls and . . . there is scarcity of water everywhere. So that is why paddy cultivation has decreased thereby leading to the cultivation of banana and ginger. There is less coffee and almost no pepper now and instead people are cultivating rubber trees.
>
> *(interview, June 15, 2011)*

This explanation highlights that the lack of water due to changing cropping practices is a major concern for Kuruma farmers. One woman clearly expresses her dislike of arecanut and explains that "there should not be any arecanut. It is very sad . . . all we could see was paddy fields. Now in between, everyone has planted arecanut. This has decreased the water capacity of the area and it is not a good change" (interview, June 14, 2011). This statement reveals that the lack of water can be interpreted as a result of changing cropping patterns. Central to water scarcity is also deforestation. Being described as another driver for land-use change, deforestation is closely related to population growth, increased demand of land for infrastructure development and water scarcity in the region. Rising temperatures in Wayanad and declining rainfall may have been caused by massive deforestation.

Overall, changes in land use in Wayanad have different causes and consequences for the Kuruma community groups. Agrarian change constitutes the conversion of agricultural fields and changing cropping patterns, which are both caused by population growth and changes in the social organization of the Kuruma.

Conclusion

Even though the Kudumbashree mission may reinforce traditional gender norms and binary divisions between public/private spaces, we suggest that this case study offers interesting insights into how rural, Kuruma women use that space to manage agrobiodiversity through collective farming approaches. Rather, the microfinance program presents one example of how boundaries between the private and the public spaces become fuzzy. Kuruma women either lease land or cultivate on their family's fields that could be seen as a private sphere but they also interact with the local government in the public. However, discussions and negotiations over the choice of crop or vegetables to be cultivated remain within a "safe" space among women only, in which the power of patriarchy is less dominant. Furthermore, as feminist approaches to the gender and development nexus have shown, gender identities and performances are shaped by shifts in the social and environmental context. The Kudumbashree program builds upon this thought and reveals that agrarian communities such as the Kuruma have transformed the Kudumbashree mission according to the local context and women's needs. Being traditional agriculturalists and considering the ongoing agrarian change in Wayanad, Kuruma women use the community program in order to maintain food security for their families. This reproduces the traditional gender norms associated with the caring notion of rural women on the one hand but also demonstrates the women's potential to become agents of development on the other (Devika and Thampi 2007). Therefore, the Kudumbashree's functioning among Kuruma communities in Wayanad can be seen as one aspect of empowerment of rural Adivasi women. As such, the Kudumbashree mission offers a new approach in rural development in which women in particular can perform as agents of development (Devika and Thampi 2007).

However, using a feminist perspective, the emphasis on women is highly problematic because it specifically excludes poor men from development initiatives. Furthermore, the impacts of the Kudumbashree on Kuruma men and women farmers are contradictory in nature. The Kudumbashree improves woman's mobility and fosters empowerment on the one hand but increases women's domestic responsibilities for sustaining the families' needs on the other. Consequently, Kudumbashree fails to address concerns associated with gender equity while reproducing traditional gender norms among the Kuruma.

References

Arun, S. (2012) "'We are farmers too': agrarian change and gendered livelihoods in Kerala, South India," *Journal of Gender Studies* 21(3): 271–284.

Basu, A. (1995) "Introduction," in A. Basu (ed.) *The Challenge of Local Feminisms*, New York: Routledge.

Devika, J. and Thampi, B.V. (2007) "Between 'empowerment' and 'liberation': the Kudumbashree Initiative in Kerala," *Indian Journal of Gender Studies* 14(1): 33–60.

Elmhirst, R. (2011) "Migrant pathways to resource access in Lampung's political forest: gender, citizenship and creative conjugality," *Geoforum* 42: 173–183.

Elmhirst, R. and Darmastuti, A. (2015) "Material feminism and multi-local political ecologies: rethinking gender and nature in Lampung, Indonesia," in R. Lund, P. Doneys, and B.P. Resurrección (eds.), *Gendered Entanglements: Revisiting Gender in Rapidly Changing Asia*, Copenhagen: NIAS Press.

Erwér, M. (2011) "Emerging feminist space politicizes violence against women," in S. Raju and K. Lahiri-Dutt (eds.), *Doing Gender, Doing Geography: Emerging Research in India*, New Delhi: Routledge, pp. 129–157.

Mukhopadhyay, S. (ed.) (2007) *The Enigma of the Kerala Woman: A Failed Promise of Literacy*, Delhi: Social Science Press.

Narasimha Reddy, D. and Mishra, S. (2009) *Agrarian Crisis in India*, New Delhi and New York: Oxford University Press.

Panda, S.M. (2014) "Exploring mobile livelihoods among tribal communities in Odisha. India: gendered insights and outcomes," in R. Lund, K. Kusakabe, S.M. Panda, and Y. Wang (eds.), *Gender, Mobilities and Livelihood Transformations*, London and New York: Routledge, pp.93–117.

Raju, S. (ed.) (2011) *Gendered Geographies: Space and Place in South Asia*, Delhi: Oxford University Press.

Raju, S. and Lahiri-Dutt, K. (eds.) (2011) *Doing Gender, Doing Geography: Emerging Research in India*, New Delhi: Routledge.

Rath, G.C. (ed.) (2006) *Tribal Development in India: The Contemporary Debate*, New Delhi: Sage.

Sreekumar, S. (2007) "The land of 'gender paradox'? Getting past the commonsense of contemporary Kerala," *Inter-Asia Cultural Studies* 8(1): 34–54.

Williams, G., Thampi, B.V., Narayana, D., Nandigama, S., and Bhattacharyya, D. (2011) "Performing participatory citizenship – politics and power in Kerala's Kudumbashree Programme," *Journal of Development Studies* 47(8): 1261–1280.

13

GENDER RELATIONS IN BIODIVERSITY CONSERVATION AND MANAGEMENT

Patricia L. Howard

General background

An emerging approach to biodiversity conservation that is ever more broadly expounded upon holds that the major custodians of the world's biodiversity are the people who depend directly upon it for subsistence. These "indigenous people" and "traditional farmers" have seen their livelihood systems disrupted and their cultural and biological diversity dramatically reduced by economic and cultural processes accompanying the expansion of markets and the adoption and imposition of exogenous cultural and economic value systems. The implications of such a perspective are that, in order to preserve biological diversity, the benefits should accrue principally to those who sustain it. The focus of conservation efforts, then, should be on maintaining the cultural and agro-ecological systems of small, "resource poor" and biologically rich farmers and rural forest dwellers, which is necessary not only to preserve existing biodiversity, but also to ensure its continual evolution *in situ*. In this perspective, local indigenous property rights, local biodiversity knowledge and management systems, and the direct participation of farmers and rural dwellers in the development and management of conservation efforts are a *sine qua non* of achieving biodiversity conservation.

These "custodians" include nearly three billion rural people in the least developed countries, of whom about 2.5 billion subsist from agriculture and feed a considerably larger number of people. About half rely on no- or low-input biodiversity-rich polycultures (e.g., agroforestry). Some 250 million live in and depend on forests, including 60 million indigenous people. About a billion regularly consume wild foods and 50 million depend on small-scale fisheries. People in the tropics use 18,000–25,000 plant species, *excluding* 25,000 herbal medicines. About 19 percent of the world's population live in "environmentally fragile" lands, where they must adapt to living with extremes. Biodiversity constitutes most of these households' wealth, the foundations of their cultural and material heritage, and the substance of the knowledge and practices that they pass on to future generations.

117

Much of this biological wealth is rapidly disappearing due to climate change, land-use change and habitat fragmentation, invasive species, overharvesting, and pollution. Agrobiodiversity (essentially, biodiversity that is directly useful to humans) is being lost particularly as agriculture becomes commercialized and high-yielding varieties displace local varieties of crops and animals. Loss of species and agrobiodiversity are major global concerns.

What was conspicuously absent until recently in debates around biodiversity that, by contrast, was given prominence in the 1996 World Food Summit and in the 1992 Earth Summit is the significance of women and gender relations for biodiversity conservation. The preamble to the Convention on Biological Diversity (CBD) recognizes "the vital role that women play in the conservation and sustainable use of biological diversity" and affirms "the need for the full participation of women at all levels of policy-making and implementation of biological diversity conservation." However, a Plan of Action to mainstream gender in the work of the CBD Secretariat was approved only in 2008, which led to the development of guidelines for mainstreaming gender in National Biodiversity Strategies and Action Plans. In 2012, the 10th Conference of the Parties to the CBD adopted Decision X/19, which promotes gender mainstreaming at all levels, but this is not legally binding on governments. The Global Plan of Action for the Conservation and Sustainable Utilization of Plant Genetic Resources for Food and Agriculture, signed in 1996, led to the legally binding International Treaty on Plant Genetic Resources for Food and Agriculture (ITPGRFA, signed in 2001), and refers several times to women's role in conserving agrobiodiversity and proposes actions to strengthen their capacity to manage these resources. However, according to the Food and Agriculture Organization of the United Nations (FAO), "Despite increased recognition of gender differences, and implications at the international level, little has been done to implement this knowledge in policies and programs for agrobiodiversity management and conservation" (FAO 2004).

The way that women are viewed in relation to biodiversity may significantly influence the ability to halt its erosion or loss, particularly of those species that are directly useful to humans. Against much thinking on the topic, women collectively hold the majority of knowledge about the world's plant genetic resources (PGR) because women's daily work requires more of this knowledge. Even when women's roles in rural development are widely recognized, their knowledge and management of PGR continue to be underestimated and undervalued, in part because this knowledge is documented and recognized only haphazardly. Women's motivations, needs, rights, and capacities as plant managers and conservationists are often not taken into account in conservation efforts. A large but fragmented and scattered body of scientific literature arising from fields such as ethnobotany, cultural anthropology, farming systems research, forestry, nutrition, and gender studies examines these issues (Howard 2003; Pfeiffer and Butz 2005).

Patterns in relief: women's knowledge and management of plant genetic resources

Woman the housewife

Women, in their roles as housewives performing domestic tasks, sustain an intimate relationship with plants that has fundamental implications for the use and conservation of PGR. What is associated with the kitchen (food processing, preparation, and preservation) and with the pantry (storage) is considered to be part of women's domain. While comparative statistics on plant use are not available, many case studies indicate that most plant varieties are cultivated or gathered for their domestic (medicinal, culinary, and nutritional) values. The kitchen and the pantry are quite possibly the most undervalued sites of PGR conservation.

Culinary traditions are a highly important aspect of cultural identity. Foods are valued not only for their nutritional content, but also for their emotional, ritualistic, spiritual, and medicinal values. Food is, in most cultures, also a fundamental constituent of exchange and hospitality, which are in turn basic organizing principles of many traditional societies. While men influence the idea of what constitutes an adequate meal or dish, women are the "gatekeepers" of food flows in and out of the domestic sphere. Culinary traditions are perpetuated by the careful transmission of knowledge and skills, particularly from mother to daughter. Most importantly, culinary traditions and preferences, and the post-harvest processes required to provide edible, culturally acceptable food have a marked influence on knowledge, selection, use, and conservation of PGR. This is well illustrated by the homegardens of women who migrated from a subsistence agricultural economy in the Yucatan Peninsula to a wage-labor, cash economy in Quintana Roo, Mexico. These homegardens are sites of *in situ* conservation not only of traditional crops, but also of elements of traditional Yucatan cuisine, which helps to preserve the cultural identity of immigrants in their new environment. Some 140 plant species were found in 33 gardens, mostly for culinary use (Greenberg 2003).

Food processing, preservation, and storage knowledge and skills are also essential to maintaining agrobiodiversity, and to meet nutritional and other cultural needs, but they go unrecognized. Like these, wild foods are also neglected since the number of species is large; many are very localized (grown or collected in small patches in homegardens, boundary lands, or between crops), and are mainly managed by women. They are usually cheaper and more accessible than exotic food, ecologically well-adapted, and high in nutritional content. Many are also toxic and require processing to be consumed. An example is the phi to (*Arisaema flavum*), a wild root that contains the same toxin (cyanogenic glycosides) that is found in bitter cassava. Practices passed down by women over generations are used to detoxify the phi to and convert inulin, a complex carbohydrate, to fructose and glucose, making it digestible. The detoxified phi to is made into bread, added to soup, or used to make raksi (distilled alcohol) (Daniggelis 2003). In the Western Division of the Gambia, for instance, Madge (1994) researched wild food plants and found that women's knowledge also often correlates with scientific knowledge. In fermentation, women stress the importance of heat, which reflects the degree of bacterial and enzymatic activity. Adding acid fruits reduce bacterial growth and shortens cooking time; this probably explains why velvet tamarind (bujaala), *Dialium guineense* is sometimes added to cooking water (Madge 1994).

The utility of agrobiodiversity for humans depends upon not only the demand for these resources related to culture and culinary traditions, but also on the knowledge of the properties of plants and the skills and technology available for processing, preserving, preparing, and storing plant materials. If these are lost, the use and knowledge of the plants concerned is also likely to be lost eventually, which in turn threatens the species that depend upon human intervention for their reproduction. The major reasons cited for the erosion of PGR are not usually related to the kitchen or the domestic sphere, but rather to factors related to production and the environment. But it is increasingly recognized that the nutritional transition and changing food habits that are related to urbanization, globalization (in general, "acculturation"), expanding food commodity markets, changing gender relations, and women's work result in the loss of local agrobiodiversity as well as worsening nutrition (Howard 2003; Johns and Sthapit 2004).

Woman the gatherer

Across the developing world, gathering provides a crucial source of food, medicine, and productive inputs, including fuelwood, construction materials, and inputs for handicraft production, as well as fodder for livestock and mulch for agriculture, without which soil fertility would be jeopardized. Most rural people engage in gathering, which can provide a substantial contribution to rural livelihoods, particularly in areas where there is abundant genetic diversity. Gathering in particular provides critical resources where populations are resource-poor and food supplies are short seasonally or during crises. The majority of those who gather wild plants across the world are women, who provide around four-fifths of total vegetal food collected, from a low of about half in East Eurasia to a high of 83 percent in North America—in all other regions women contribute more than half of all gathered food (Barry and Schlegel 1982). Gender relations affect the types of plants gathered or harvested, rights to access these resources, types and means of management, and the sustainable use of these resources (Howard 2003). The gender division of labor in gathering is often related to the type of plant and the space in which plants are gathered (Pfeiffer and Butz 2005). For example, among the Mauna in Papua New Guinea, women collect wild vegetables unless this involves tree-climbing (Hays 1974).

The explanations for the division of labor that generally proscribes hunting large game animals by women and thus confines them to gathering or hunting small game animals, has been sought in cultural phenomena such as proscriptions against women's use of men's weapons, spiritual concepts of incompatibility between procreation and killing, and taboos related to menstruation and animal blood. They also are related to the prestige of hunting, the status of meat as the most coveted food, and men's superior political authority in many foraging societies (Fedigan 1986; Brightman 1996).

The idea that plants growing in natural environments are "wild" may be mistaken: many "gathered wild plants" are not strictly either "gathered" or "wild," but are selectively managed and harvested, in an incipient process of domestication (Vodouhè and Dansi 2012) that begins with the preservation of wild plants, their protection *in situ*, cultivating and harvesting *in situ* or in fields or homegardens, and, finally, selection for specific traits. Many "wild" species are in process of domestication. Domestication is occurring in part because, across much of the globe, foraging resources are declining rapidly and, if wellbeing is not to decline, the amount of time and effort that must be invested to obtain these resources must increase, which puts a particular strain on women. Thus, women seek to conserve wild resources and also to reduce their labor inputs by domesticating those wild resources that are most important to them.

Woman the gardener

Homegardens are the oldest and most pervasive cultivation systems known and are found in all of the world's arable regions. They contain greater species diversity and ecological complexity than agricultural fields. Because of their spatial and cultural diversity, there is no universally accepted definition of homegardens, but they are considered to have certain commonalities. Over much of the developing world, they function as esthetic, social, and recreational spaces and provide secondary sources of food as well as medicines, condiments, fodder, building materials, and fuel (Eyzaguirre and Linares 2004; Kumar and Nair 2006). The land areas involved are small and situated near the home; they are managed with unpaid family labor and with no or low use of external inputs; many of the plants cultivated are traditional varieties known mainly to local people, and most of their produce is consumed at home. These same characteristics mean that they tend to be "invisible" to outsiders and their importance is underestimated. Yet a review of 82 tropical homegarden studies showed that women were the main gardeners in 94 percent of the cases in Latin America, 63 percent in Asia, and 65 percent in Africa.

Overall, 77 percent were primarily managed by women, 2 percent by men, and 21 percent by both sexes.[1]

PGR conservationists now focus more attention on homegardens (Galluzzi et al. 2010). Plant domestication often occurs in homegardens, just as it did at the dawn of agriculture; in part because, when wild species are threatened, they may be transplanted into homegardens (Vodouhè and Dansi 2012). Species and varieties that are not economical to grow in agricultural fields are planted in homegardens. As well, Niñez (1987) reported that the migration of the potato from South America to other parts of the globe occurred through homegardens. Homegardens are used to test new crop varieties; however, germplasm collectors and plant breeders generally do not collect specimens from homegardens, and thus underestimate total crop diversity and women's roles (Lope-Alzina 2007).

The knowledge required to manage tropical homegardens is specific and more complex in comparison with agricultural production. Their small-scale and intense management mean that different soil, water, and plant management techniques are used than in fields (Kumar and Nair 2006). Production is year-round, and greater species diversity requires more knowledge about growing requirements, plant associations, and the creation of plant communities in different microenvironments. Homegarden planning considers the outputs desired, which may include many hundreds of different items intended to fulfill a large number of household needs. Planning must combine an understanding of vegetative cycles, characteristics of perishability, processing, and storage, and timing and quantity of demand, including the need to meet nutritional and medicinal needs of households whose composition also changes over time (Finerman and Sackett 2003).

This complex knowledge is gained experientially as well as through lifelong interactions between homegardeners and their protégées that begin at a young age. It is transmitted largely among women and then principally among closely related kin (Hays 1974; Descola 1994; Greenberg 2003). Not only gardening knowledge is transmitted between mothers and offspring and women and their kin or neighbors, plant material is also transmitted as part of a "package" of cultural and physical capital that flows among women and between women and their offspring (Howard 2003; Delêtre et al. 2011).

Homegardens also reveal the ways in which women's social status, as well as social and cultural capital, is related to the agrobiodiversity they manage and the social relations entailed. Descola (1994) showed that, for the Achuar Amazonian Amerindians, it is a "point of honor" for women to cultivate large swidden gardens: so that "a woman who successfully grows a rich pallet of plants . . . demonstrates her competence as a gardener and fully assumes the main social role ascribed to women by proving her agronomic virtuosity" (Descola 1994: 166).

The status women gain through agrobiodiversity management is not confined to their skills or the beauty or diversity of their gardens. A recent study shows that, among Yucatecan Maya, homegardens are the principle source of goods for gift-giving and barter with kin and neighbors, and that women predominate as gift-givers and recipients. The agrobiodiversity that women manage provides the cement that binds people together in relations of solidarity and reciprocity.[2] In the village of Nenela, homegardens were the source of 62 percent of the gifts exchanged within the village (kitchens provided another quarter, and these also depend largely on produce from homegardens). Gifts were far more prominent as forms of exchange within the village (more than four-fifths of all transactions), whereas sales/purchases were the predominant form of exchange with outsiders (about two-thirds of the transactions). Women-to-women exchanges accounted for more than 90 percent of all within-village exchange, where gift-giving represented nearly 82 percent, and sales around 16 percent. The vast majority of homegarden planting materials that are not self-provisioned are acquired through gifts and barter, also mainly between women (Finerman and Sackett 2003; Delêtre et al. 2011). Such gift-giving and transmission of knowledge generate social and cultural capital for women that serve many different ends, such as access to labor, credit, and political power. Also, sales of homegardening products provide women with an important source of cash that can be generated on a regular basis throughout the year.

Women's status may decline as homegardens erode. Stavrakis (1979) noted that, in Belize, homegardening has lost considerable status in the villages she studied, with clear implications for women. Scholars have noted that women's traditional knowledge and management of indigenous vegetables and other local crop diversity may become valueless and their status reversed as such knowledge and production becomes increasingly associated with poverty and backwardness.

Commoditization subverts many of the dynamics that have maintained homegardens' value—diversity, independence and autonomy, local adaptability, home-centeredness, needs-centeredness, and multi-value-centered production. Several studies have discussed shifting responsibilities and benefits related to homegardens when these begin to generate substantial amounts of cash, even in contexts where homegardening is culturally strongly associated with women (Kumar and Nair 2006). A total of 77 tropical homegarden case studies reviewed reported on responsibilities for subsistence production, petty sales, or high-value sales. Of these, women alone were responsible for subsistence in 69 percent of the cases, whereas men were responsible for high value or major cash crops in 78 percent of the cases. As homegarden production becomes more lucrative, women's roles often shift. Markets are fickle things, being shaped at times by local needs and values, but increasingly by values and needs created thousands of miles away. Commoditization likewise has been a double-edge sword for many women: a route toward recognition, a source of power-as-money and of greater autonomy, and a means to deprive them of resources and force upon them responsibilities previously assumed by men, destroying complementarities that were a primary benefit of their unequal social standing.

Woman the herbalist

Despite the rapid spread of manufactured pharmaceuticals throughout the world, most people in developing countries still rely predominantly on plants for medicine. The World Health Organization estimated that 80 percent of the world's population use botanical medicines for primary health care. The aim of many ethnobotanical studies is to document the medicinal plant knowledge of indigenous peoples in order to prevent its disappearance as a result of biodiversity loss and cultural and economic change. Although the ethnomedicinal knowledge of local people is increasingly leading to commercial patents, most do not share in the benefits. Many medicinal plants are also disappearing rapidly due to the loss of tropical biotopes and overharvesting, as raw materials are exported, thus restricting local people's access.

Herbalists are specialists in treating illnesses through the use of botanicals and are frequently if not predominantly women. A very large number of studies demonstrate that women may have knowledge of dozens, if not hundreds, of different plant species used for medicine (Howard 2003; Pfeiffer and Butz 2005). However, ethnobotanists often fail to explore women's ethnomedicinal knowledge and therapeutic roles, as they make a beeline for the "shaman" or other "expert" healers, offices reserved for men in many cultures (Howard 2005). While women may hold most ethnomedicinal knowledge, they may be denied the power and status of the "expert," or they may find an avenue toward higher status in their communities.

Women's therapeutic roles and knowledge are everywhere essential to the health of households and, in most societies, lay women have a greater role in the knowledge and use of medicinal plants than men. Quichua women in the Ecuadoran highlands use more than 350 medicinal species and can detail the effectiveness of individual herbs for specific disorders, as well as the methods for preparing each plant for use in curing (Finerman 1989). Quichua mothers treated 86 percent of all family illness complaints and act as the first source of care for 75 percent of all ailments recorded; socialization processes in this community result in the exclusive transmission of curing skills and knowledge to daughters (Finerman 1989).

Gender should be explored as an integral aspect of local structures of power and prestige in which knowledge is held, generated, and expressed. Often, medicinal knowledge is held secret and is passed among women along kinship lines (Howard 2003; Pfeiffer and Butz 2005). Research on Mexican Chinantec men and women's knowledge of plants relating to women's health shows that sex is not in itself an indicator of plant knowledge (Browner and Perdue 1988). The Chinantec use a total of 111 herbal remedies to treat women's reproductive health problems and manage fertility. Men report nearly as many remedies as women, but their knowledge is quite differently structured. Women's knowledge tends to be shared along female lines, whereas men's knowledge tends to be idiosyncratic; while many women use the same plants for the same illnesses, men would report different plant used for the same illnesses, and women knew more remedies for menstrual hemorrhaging (Browner and Perdue 1988).

Woman the seed custodian and plant breeder

Women have very important traditional roles as seed custodians and as informal plant breeders. They are frequently directly responsible for crop production and as such consider all of those selection factors that formal plant breeders are coming to realize are critical to marginal farmers. Even more often, women are responsible for tasks related to seed management including selection, storage, preservation, and exchange. Informal seed exchange systems are often female domains, and include mechanisms such as the bride price, gift-giving, and kinship obligations, as well as market and barter transactions (Howard 2003; Delêtre et al. 2011).

The management and exchange of seed are also often considered to be women's responsibility, whether this occurs through kinship or other types of exchange obligations, barter or sale. In Gabon, for example, Delêtre et al. (2011) report that, upon marriage, women move to their husband's homes and often bring seed from their own villages with them. They also maintain close ties with their villages of birth and collect seed when visiting relatives. Not only in Africa, but also in Amazonia and Melanesia, seed exchange is often interrelated with marriage exchange.

Some explain the cross-cultural predominance of women's labor and decision-making in seed management in terms of the relation that this has with post-harvest and domestic work. Others suggest a cosmological explanation. For example, Zimmerer (1997) relates that, across much of the Peruvian Andes, women almost exclusively manage (select, store, and preserve) seed—and men are not allowed into women's seed storage areas. Tapia and De La Torre (1998) show that the explanation is found in Andean cosmology, where this relation between women and seeds is coherent with the tradition of Andean thinking in terms of a duality defined by the principles of masculine and feminine "seed." But cultural beliefs can also exclude women from certain aspects of seed management. Women's involvement in seed selection and storage may be seen as polluting, particularly due to taboos around menstruation and childbirth.

Women's varietal preferences are strongly related to dietary traditions, as well as food processing and storage technologies and conditions. Studies show that women have a broader set of selection criteria compared with men, and require a larger array of crop diversity (Zimmerer 1997). Farm-households use plant materials in diverse ways: for example, rice not only provides protein-containing seed, but also straw for thatching, mat-making and fodder, husks for fuel, and leaves for relishes. Breeding programs for cassava should make provisions for leaf as well as tuber yield, leaf palatability, and ease of harvest, otherwise formal breeders' varieties will continue to meet resistance from female farmers.

Women's influence on crop varietal selection is not restricted to women's crops or fields. Women influence selection and breeding where men and women manage farms together, and when men farm alone. Lope-Alzina (2007) found that among the Yucatec Maya, men and women "negotiate" not only which varieties to produce in men's fields, but how much of each to plant. Such decisions are also influenced by the varieties of the same crops that women produce in homegardens.

Women's rights to plant genetic resources

As is the case with property rights of all types, rights to plants are also differentiated by sex. It is usually assumed that whoever controls private land controls the biodiversity on that land, and that "common" land implies "common" rights to the plants found therein. This is not the case with homegardens where, while land may belong to men, women have usufruct rights and "own" the plants growing there (Howard and Nabanoga 2005). Gendered rights are not confined to domesticated plants, as historical research from North America attests: rights to wild plant harvesting grounds and territories were passed down from mother to daughter; violations of others' territories were punishable (Dick Bissonnette 2003; Howard and Nabanoga 2005). Price's research (in Howard 2003) in Northeast Thailand focused on rights to gather wild plant foods. Rights to harvest for domestic consumption carry different restrictions than rights to harvest for sale, and are related to the specific species and perceptions of their market value, taste, and rarity. Rights to collect these plants on other women's land are linked to matrilocal residence and inheritance and female kinship networks. Rights to plants establish "bundles of powers" for heterogeneous groups of users and knowledge holders within cultures, which also create differential entitlements to the benefits of their use (Howard and Nabanoga 2005).

There are very substantial dangers to the failure to recognize existing customary rights systems. Customary plant rights are a major institutional component for achieving the sustainable use and management of biological diversity and the equitable distribution of the benefits from such use, which are objectives of the CBD. Customary natural resource rights regimes have often helped to ensure sustainable management of both resources and their respective landscapes over the long term, as well as a distribution of resources that helps to ensure that all members of a society are able to meet subsistence needs.

The danger exists that outsiders will confer rights to the wrong groups, unwittingly usurping the rights of those who traditionally held them. The introduction of European land tenure systems in Africa and Asia often resulted in the usurpation of women's customary land rights and their transfer to men. Systems of rights can also change without any conscious attempt being made to do so. For example, in the Peruvian Andes, although women have traditionally had exclusive rights to potato seed, when modern varieties are introduced, men frequently receive and manage the new varieties, while women continue to manage the traditional varieties. What is not clear is whether this will eventually lead to the erosion of women's control over traditional varieties, but this is a possibility given the fact that women's control is rooted in cultural norms that are now *de facto* being challenged. If women's control over traditional varieties erodes and they are the main repositories of knowledge, then it is likely that this knowledge will erode absolutely.

Debates about rights to PGR and their outcomes cannot be considered as gender neutral since, while women constitute the majority of those gardeners, gatherers, herbalists, and plant breeders who have developed agrobiodiversity and identified useful plants, due to gender bias they are likely to be the last to have their rights recognized and therefore to benefit from related development or compensation schemes. Assuming that the rights or compensation given to "indigenous groups" or "farmers" will reach women is incorrect. The unequal distribution of resources between men and women across the globe that has been carefully statistically documented by the UNDP in its Human Development Reports attests to the fact that mechanisms of compensation that earnestly seek to benefit the provider of these resources and stimulate their continued conservation must carefully consider means by which the rights of women in particular can be respected.

Gender, biodiversity loss, and conservation

If women are the predominant managers of PGR, then researchers and development practitioners should consider the ways in which they may be specifically affected by or implicated in those processes leading to genetic erosion, such as the diffusion of modern varieties and increasing commoditization of plant resources, decreasing access to land, and changing consumption patterns. Gender relations are changing, as are women's incentives and management practices, which affect biodiversity management. Decreasing access to land is a primary reason for the erosion of the PGR that women manage. When land becomes privatized, women may lose access to forests and fields where they gather wild plants; when men turn to cash-cropping, women may lose access to land where they produce traditional crops.

Research methods, conceptual frameworks, and guidelines relating to PGR conservation are very important and must be further developed to specifically capture gender differences and permit examination of gender relations. More cooperation between PGR researchers and gender experts needs to happen to benefit both. In addition to the need to extend the type of research currently carried out by ethnobotanists, ethnoecologists, and anthropologists to deal with gender–plant relationships, some of the main gaps in current research and development interventions that need to be urgently addressed (without being all-inclusive) include:

- Gendered rights of access to PGR and related knowledge and causes and consequences of change in such rights, as well as analysis of politics around rights to PGR at national and international levels and their implications for different groups, including women specifically. The international and national political debates and claims, and the power bases that support these, should be examined critically to explain the exclusion of gender issues from the political agenda.
- Effects of changing labor demands on women's management of PGR, particularly those induced by male out-migration, decreasing resource access, and environmental degradation, as well as increasing demands for women's cash income earnings.
- Effects of changes in culinary traditions and post-harvest processes on PGR management and means to enhance the returns to women's labor and household wellbeing, and hence to promote PGR conservation in these activities.
- Effects of the introduction of modern varieties on men and women's PGR management. The analysis of factors affecting adoption, conservation, and use of local versus introduced

varieties should extend to women's perceptions of risks, benefits, and trade-offs as well as factors affecting this behavior.

- Evaluation of conservation programs and projects to determine their positive and negative impacts for women PGR managers and for the biodiversity that women conserve. This should include analysis of relations between induced change in PGR access and management and existing local biodiversity management and use systems, and the significance of gender in these relations.
- Analysis of the relevance of gender for local seed systems where the lack of knowledge about traditional seed supply and exchange systems is particularly acute, particularly how gender relations, kinship, and other exchange networks affect the functioning and resilience of these systems.
- Analysis of the economic and conservation value of homegardens and factors contributing to their enhancement, maintenance, or decline (including the trade-offs between these values), particularly considering the changing incentive structures that women face.
- Finally, research on women's status and PGR management and conservation is particularly scarce as is the importance of PGR for traditional social relations that contribute to social, economic, and cultural capital.

The global objective of future research, development, and conservation work should be to develop a sound body of comparative knowledge of socioeconomic and environmental change, gender, and PGR that links global and local processes. This work should be used as an input to the development of regional and global assessments under the aegis of the Intergovernmental Platform on Biodiversity and Ecosystem Services (IPBES), whose task is to "synthesize, review, assess and critically evaluate relevant information and knowledge generated worldwide by governments, academia, scientific organizations, non-governmental organizations and indigenous communities."[3] The First IPBES Expert Workshop on Indigenous and Local Knowledge Systems confirmed that

> Women and men commonly fulfill different, but complementary roles and responsibilities in relation to different components of biodiversity and sustainable use, resulting in different knowledge, needs, concerns, priorities and roles. For this reason, women may possess knowledge, not held by men, which can inform IPBES processes. To fulfill its operational priority to achieve gender equity in all relevant aspects of its work, IPBES should put in place mechanisms that ensure attention to gender specific-knowledge and gender balance in all components of its work.[4]

The fulfillment of this recommendation will depend on how much pressure different stakeholders can bring to bear and to what degree researchers and other scientists recognize that the aims of biodiversity conservation cannot be met without women's full and fair participation.

Notes

1 P. Howard, unpublished manuscript, Wageningen University, 2006.
2 Data come from a study by D. Lope-Alzina and P. Howard, unpublished manuscript, Wageningen University, 2013.
3 www.ipbes.net
4 Available at http://ipbes.net/events-feed/353-expert-workshop-on-indigenous-and-local-knowledge-systems-to-ipbes-2.html.

References

Barry, H. and Schlegel, A. (1982) "Cross-cultural codes on contributions by women to subsistence," *Ethnology* 21(2): 165–188.

Brightman, R. (1996) "The sexual division of foraging labor: biology, taboo, and gender politics," *Comparative Studies in Society and History* 38(4): 687–729.

Browner, C.H. and Perdue, S.T. (1988) "Women's secrets: bases for reproductive and social autonomy in a Mexican community," *American Ethnologist* 15(1): 84–97.

Daniggelis, E. (2003) "Women and 'wild' foods: nutrition and household security among Rai and Sherpa forager-farmers in eastern Nepal," in P.L. Howard (ed.), *Women and Plants: Gender Relations in Biodiversity Management and Conservation*, London: Zed Books, pp. 83–93.

Delêtre, M., McKey, D.B., and Hodkinson, T.R. (2011) "Marriage exchanges, seed exchanges, and the dynamics of manioc diversity," *Proceedings of the National Academy of Science* 108(45): 18249–18254.

Descola, P. (1994) *In the Society of Nature: A Native Ecology in Amazonia*, Cambridge: Cambridge University Press.

Dick Bissonnette, L. (2003) "The basket makers of the central California interior," in P.L. Howard (ed.), *Women and Plants: Gender Relations in Biodiversity Management and Conservation*, London and New York: Zed Books.

Eyzaguirre, P. and Linares, O. (eds.) (2004) *Home Gardens and Agrobiodiversity*, Washington, DC: The Smithsonian Institute.

FAO (2004) "Recognizing gender aspects in agrobiodiversity initiatives," www.planttreaty.org/content/recognizing-gender-aspects-agrobiodiversity-initiatives.

Fedigan, L.M. (1986) "The changing role of women in models of human evolution," *Annual Review of Anthropology* 15: 25–66.

Finerman, R. (1989) "The forgotten healers: women as family healers in an Andean Indian community," in C. McClain (ed.), *Women as Healers: Cross-Cultural Perspectives*, New Brunswick and London: Rutgers University Press, pp. 24–41.

Finerman, R. and Sackett, R. (2003) "Using home gardens to decipher health and healing in the Andes," *Medical Anthropology Quarterly* 17(4): 459–482.

Galluzzi, G., Eyzaguirre, P., and Negri, V. (2010) "Home gardens: neglected hotspots of agro-biodiversity and cultural diversity," *Biodiversity and Conservation* 19(13): 3635–3654.

Greenberg, L.S.Z. (2003) "Women in the garden and kitchen: the role of cuisine in the conservation of traditional house lot crops among Yucatec Maya immigrants," in P.L. Howard (ed.), *Women and Plants: Gender Relations in Biodiversity Management and Conservation*, London and New York: Zed Books.

Hays, T.E. (1974) *Mauna: Explorations in Ndumba Ethnobotany*, PhD dissertation, University of Washington.

Heywood, V. (1999) "Trends in agricultural biodiversity," in J. Janick (ed.), *Perspectives on New Crops and New Uses*, Alexandria, VA: ASHS Press.

Howard, P. (ed.) (2003) *Women and Plants. Gender Relations in Biodiversity Management and Conservation*, London: Zed Press and Palgrave Macmillan.

—— (2005) "Gender bias in ethnobotany: propositions and evidence of a distorted science and promises of a brighter future," www.researchgate.net/profile/Patricia_Howard2.

Howard, P. and Nabanoga, G. (2005) "Are there customary rights to plants? An inquiry among the Baganda (Uganda) with special attention to gender," *World Development* 35(9): 1542–1563.

Johns, T. and Sthapit, B. (2004) "Biocultural diversity in the sustainability of developing-country food systems," *Food and Nutrition Bulletin* 25(2): 143–155.

Kumar, B.M. and Nair, P.K.R. (eds.) (2006) *Tropical Homegardens: A Time-tested Example of Sustainable Agroforestry*, Berlin: Springer.

Lope-Alzina, D. (2007) "Gendered production spaces and crop varietal selection: case study in Yucatán, Mexico," *Singapore Journal of Tropical Geography* 28(1): 21–38.

Madge, C. (1994) "Collected food and domestic knowledge in the Gambia, West Africa," *Geographical Journal* 160(3): 280–294.

Niñez, V.K. (1987) "Household gardens: theoretical and policy considerations," *Agricultural Systems* 23: 167–186.

Pfeiffer, J.M. and Butz, R.J. (2005) "Assessing cultural and ecological variation in ethnobiological research: the importance of gender," *Journal of Ethnobiology* 25(2): 240–278.

Stavrakis, O. (1979) *The Effects of Agricultural Change upon Social Relations in a Village in Northern Belize, Central America*, PhD dissertation, University of Minnesota.

Tapia, M.E. and De la Torre, A. (1998) "Women farmers and Andean seeds," www.fao.org/sd/nrm/womenfarmers.pdf.

Vodouhè, R. and Dansi, A. (2012) "The 'bringing into cultivation' phase of the plant domestication process and its contributions to *in situ* conservation of genetic resources in Benin," *The Scientific World Journal* 2012(1): 176939.

Zimmerer, K.S. (1997) *Changing Fortunes: Biodiversity and Peasant Livelihoods in the Peruvian Andes*, Berkeley: University of California Press.

14

COLONISATION AND FIRE

Gendered dimensions of indigenous fire knowledge retention and revival

Christine Eriksen and Don L. Hankins

A fiery context

This chapter elucidates how gender is entwined in the spatial and temporal knowledge trajectories through which indigenous fire knowledge is retained and revived using a case study of eastern Australia and California, USA. Fire extends its roots far into the past of indigenous cultures worldwide, extending beyond basic domestic needs to responsible environmental stewardship. Fire has played a key role in the land stewardship practices of Aboriginal Australian and Native American women and men for millennia (Stewart et al. 2002; Gammage 2011). This includes cultural and gendered landscapes, such as indigenous sacred and ceremonial sites off-limits to women or men. However, a 'disconnect' between the past, present and future of both ecological and cultural aspects of fire underpins a tendency among many researchers, policymakers and practitioners to dismiss or ignore fire knowledge that is alive today among indigenous elders and cultural land stewards. This may be attributed to assumptions based on historic events, a lack of current burning and relatively low indigenous populations. Instead guidance is sought from archaeological, anthropological and ethnographic records from the past or from scientific models that project the future. An attitude also prevails that depicts historic use of fire by indigenous people as non-applicable in current-day environments due to environmental and demographic changes (White 2004). Yet, it is important to recognise that culture and knowledge are as dynamic as the environment is. From an applied standpoint, indigenous fire knowledge is fluid (for example, changing with past climatic events or gender-targeted genocide), and the ability to read the landscape to know how, when, why and what to burn comes with proper training. The concept of 'proper training', however, arguably plays out differently today from traditional[1] indigenous fire knowledge trajectories of the past due to the impact of history and politics. It is this marginalised political, technological and institutional position of indigenous peoples' knowledge in many 'developed' countries that makes this chapter relevant to a handbook of gender and development.

Colonisation introduced a new paradigm of law into indigenous cultures in many parts of the world, although it should be noted that colonial processes were uneven in time and space. Colonial interests in both Australia and the USA disrupted indigenous use of fire through the removal of people from their lands and policy prohibition (Gammage 2011; Vale 2002). Policymakers in other regions of the world are likewise using Eurocentric fire suppression policies, rather than developing and adopting fire management strategies adapted by indigenous people over many centuries to suit regional or local environments and agricultural practices. Indigenous use of fire, whether for resource harvesting, hunting, vegetation and soil regeneration or maintenance of communal areas, were instead seen as an 'evil', environmentally degrading practice, as such fires threatened both the property and the social hierarchies of rigidly ordered colonial societies (Eriksen 2007; Mistry 2000). Case studies on the environmental impact of indigenous burning practices in West Africa (e.g., Laris and Wardell 2006), research by Jay Mistry (2000) on policy and politics surrounding indigenous fire management in savanna regions globally, the persistent conflict over fire as a land managment tool in southern parts of Africa (e.g., Eriksen 2007) and insights into government-funded programme pressure on indigenous fire management in South America (e.g., McDaniel et al. 2005) have all concluded that conflicts between indigenous or local communities and official bodies over fire management often stem from global (Eurocentric) perspectives on environment and natural resource management having replaced local standpoints. This is despite the more recent acceptance of the role of non-equilibrium theory in ecosystem management, which highlights the value of fire in, for example, indigenous shifting cultivation systems and wildfire protection schemes (Eriksen 2007). The consequences of the continual dominance of Western environmental narratives over indigenous land management practices in many fire-prone regions have arguably been 'large-scale illegal burning, and the occurrence of catastrophic burns resulting in ecological and economic damage to land and property' (Mistry 2002: 308).

Indigenous burning practices are distinguished from the fire management of government agencies in the context of traditional law, objectives and the right to burn. At the core of indigenous eco-cultural fire processes is recognition of the interrelated and interdependent aspects of fire that follow the laws of the land (nature). Traditional law and lore are rooted in the landscape and stories that define a given culture (Black 2011). Gendered norms are interwoven into both law and lore. By 'lore' we refer to story, whereas indigenous law is coded in the lore. The landscape will convey its need for burning based on factors such as the accumulation of dead plant materials or the decline in resource conditions. Such knowledge may be encoded in the stories of a region. These stories may also convey the penalties for not following the laws of the land, as often depicted in Aboriginal fire paintings. The Aboriginal Australian governing principle is founded on the ontological concept of 'the Dreaming'. This concept reflects how indigenous people place their continuing practices of traditional action within a metaphysical context. Indigenous people do not generally separate the synergistic relationships between fire and other aspects of the physical and metaphysical. Rather those relationships are interrelated and interdependent. The scaling of the relationships extends from the individual to the universe, and is inclusive of the feedbacks within those levels. They may be separated at a community level to distinguish the responsibilities of, for example, a given gender, society or clan, but beyond that there is recognition of those relationships. Many examples of fire in the stories of indigenous people explain various aspects of fire knowledge from inter-specific gendered relationships to devastating fire. This knowledge forms how a culture interacts with fire and more specifically how, what, where, when and why burning occurs spatially and temporally for cultural and environmental reasons. This is evident, for example, in the dependent relationship between fire and water and its close connection to balanced gender relations:

When I go to gatherings in other places in the US, there's a really strict protocol around fire: the men handle the fire and the women handle the water. That balance is representative, but also practical, of a partnership that's always there whether it's recognised or not. Perhaps that more than anything else in contemporary US society is what's completely out of whack. Both the fire and the water now sometimes work in opposition because of human interference. So it makes me think to right things men and women have to be more in balance with each other or the fire and the water doesn't work.

(female Coast Miwok/Jenner Pomo cultural
practitioner, California 2011)

Changes in land tenure following colonisation have disrupted the continuous relationship between indigenous people and their homelands. Access to land is important to indigenous peoples' memory of land, self-identity and for their sense of belonging – i.e., it is a core element for both the physical and mental wellbeing of indigenous cultures. The land is not only the source of traditional law and lore, it is what defines many indigenous cultures; when the ties to the landscape are compromised, then so too is the culture (Brody 2002). Indigenous people struggle with mitigating many forced losses in the aftermath of colonisation from the loss of language to a loss or displacement of knowledge and culture linked directly to environmental stewardship. With colonisation, the indigenous obligations to burn as responsible environmental stewards were in many cases reduced from application at a landscape scale to memories, which have survived generations through oral tradition. To the dominant (white) society, indigenous fire knowledge, when not currently applied, has been relegated to a thing of the past. This negates the recognition of indigenous fire as a keystone process environmentally and culturally today. This is despite the tangible results of adaptive management frameworks that have empowered indigenous knowledge keepers to practice fire (e.g., Ross et al. 2011). In some regions of northern and central Australia, indigenous law and practice are still applied through fires ranging in scale from individual plants for food, fibre and other resources to fire at a landscape scale for hunting and environmental management purposes (Bliege Bird et al. 2008; Vigilante et al. 2009). In California this happens at a fine localised scale at present, although it was significant historically, occurring across scales from individual plants to the landscape (Stewart et al. 2002; Anderson 2005). These examples demonstrate a chain of knowledge from which to contrast indigenous and non-indigenous fire use and management practices.

More than two-dozen indigenous elders, cultural practitioners and land stewards have shared oral narratives with us over the past decade during participant observations at prescribed burns, fire knowledge workshops, fieldtrips with students, informal conversations and audio-recorded interviews in New South Wales, Queensland and California. In comparing indigenous fire knowledge and burning practices across geographical regions we run the gauntlet of scholarly criticism regarding the portrayal of all indigenous knowledge as being the same (Smith 2012). However, in addition to the regions' many ecological, climatic, colonial and pyro-geographical similarities, the approach is supported by the many similarities apparent in the narratives from both regions of indigenous eco-cultural fire processes.

Gendered dimensions of indigenous fire knowledge trajectories

Despite the impacts of colonisation, indigenous law has remained at the root of many Aboriginal Australian and California Native American communities through their continuing operation outside present-day laws established by the colonising governments of Australia and the USA. While applied skills in indigenous fire knowledge still exist in some, frequently remote, communities such as the Martu in Australia's Western Desert and the Karuk tribe in northern California, many indigenous people working with fire are today trained within the Eurocentric and patriarchal notion of fire*fighting*. Firefighting agencies and men are therefore likely to be their main source of fire knowledge – a potentially problematic knowledge-gender relationship also highlighted in the quote above. Fire among indigenous cultures is therefore a complex affair, which has been muddled by colonial laws, policies and practices that have largely placed Australian and American governance above indigenous. For instance, in the USA, only federally qualified individuals can burn on federal lands, which excludes most indigenous traditional practitioners from tending to their homelands as had been done for millennia. In place of indigenous-based fire knowledge, agency-derived policies established around the concept of fire suppression or fire*fighting* has become a societal norm, which today forms a baseline of sorts among both non-indigenous and indigenous people. From our experience, however, the knowledge of indigenous fire practices persists in varying formats among many indigenous women and men who are either cultural practitioners or land stewards within land and fire management agencies. Their employment or engagement with such agencies reflect the need for dynamism and change within a culture over time for its wellbeing and ultimate survival (Black 2011). Dynamism – spatially and temporally – is the crux of the story of how indigenous fire knowledge has been able to persist in New South Wales, Queensland and California. Although gender norms are interwoven into indigenous laws – in the most obvious form as women's laws and men's laws – the stories shared with us strongly indicate that the gender of specific indigenous knowledge-keepers is generationally dependent due to the impact of external social factors past and present. As exemplified below, a temporary generational crossover of gender roles and gendered knowledge has been forged to ensure the retention of indigenous fire knowledge and land stewardship, as well as ceremonial traditions more generally. The temporality of these crossovers can be indefinite.

Figure 14.1 illustrates a generalised view of this spatial, temporal and gendered trajectory of indigenous fire knowledge-holders, which has been derived from our work with California Native Americans and Aboriginal Australians in New South Wales and Queensland.

According to our research, the layering of Aboriginal east-Australian and California Native American eco-cultural fire knowledge has traditionally been the responsibility of fire-keepers – a role that according to tradition seems to have fallen mainly (but not exclusively) to men (White 2004). This finding correlates with the interpretation by Bill Gammage (2011: 160) of how Aboriginal Australians in 1788 managed what he terms 'the world's biggest estate': 'Fire was a totem. Whoever lit it answered to the ancestors for what it did. Understandably, fire was work for senior people, usually men.' More specifically, it seems that men in the pre-colonial period were the holders of the fire knowledge that was applied at a landscape scale. While women held some fire knowledge, the extent of their personal engagement with prescribed burning practices was in the context of finer scale burning for specific purposes, such as plant foods or basketry. Even though such gendered norms and gender roles were interwoven into indigenous everyday practices, this did not seem to preclude an understanding of the underlying knowledge by the other sex. M. Kat Anderson (2005) offers a rationale for the distribution of fire knowledge and responsibilities among California Native Americans based on traditional tasks men and women would engage in historically. The logic of multiple fire knowledge domains are also described by Bliege Bird et al. (2008) regarding different types of fire use for hunting by Martu women and men respectively in Australia's Western Desert. These fire-hunting practices correlate with the 'temporary activity-space relation' described by

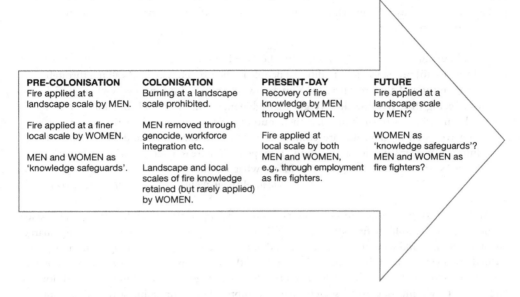

Figure 14.1 A generalised spatial, temporal and gendered trajectory of indigenous fire knowledge-holders in eastern Australia and California, USA (Eriksen 2014)

Francesca Gleeson (1993) as underpinning the complementarities and interdependence of apparent gender-specific activities and purposes of Australian hunter-gatherer uses of fire. Gleeson (1993: i) concludes: 'While most domestic and economic fire-related activities appear to be gender-specific, the only strict adherence to gender specificity is in ritual.'

That women in many places became the main carriers of fire knowledge is directly linked to the impact of external social factors, such as male genocide. However, recognition of whom – women or men – the knowledge and customs belong to traditionally remains with the intent of returning the knowledge to its rightful gender when time and space allow. Thus even when the practical connection to land has been hindered past and present, the cultural connection of indigenous laws to their source – the land – enables knowledge transfer across gender rather than knowledge prohibition caused by static gendered norms. An example of such dynamic transitioning of indigenous fire knowledge is the ways in which Aboriginal Australians and California Native Americans have been able to reconnect with land they are otherwise denied stewardship rights to through employment with wildfire management agencies. While agency burning practices may differ from traditional burning practices and outcomes, employment inadvertently opens up an avenue for the retention and fortification of elements of indigenous fire knowledge through interaction with (tribal) land – albeit in different contexts, such as agency wildfire fighting, and not always within one's own homeland.

The feeling of personal wellbeing recorded among male and female indigenous firefighters in our research is consistent with the correlation between indigenous burning practices and increased physical, mental and social health recorded by Burgess et al. (2005). This feeling of wellbeing through agency wildfire management, however, obscures the power struggles, contrasting cultural norms, rules and generational gendered fluidity that underpin the interaction between indigenous and agency fire knowledge. For example, the difference between agencies' emphasis on scientific notions of environmental 'thresholds' in comparison to Aboriginal perceptions of fire as a 'living thing' is problematic as differences in desired outcomes drive on-the-ground practices:

> When we drove out to Nullumbuy [in the Northern Territories] we actually saw an old car [with] probably eight traditional people in the car. They're just walking through the country and they're burning. You know, there was burning everywhere and I love seeing that. Because they've got management of that land and they know when to burn, what to burn, and what their outputs of that burning is. I think that we, even we as an agency, still are coming to terms with thresholds and all this sort of thing. 'Cause it's a scientific notion, I suppose. Whereas I see Aboriginal burning practices, it's a living thing, do you know what I mean?
>
> *(female Aboriginal ranger, New South Wales 2011)*

Another example is the current emphasis on equal opportunity within federal and state agencies, which result in fire knowledge and training opportunities in theory being shared equally with men and women of indigenous and non-indigenous heritage. Indigenous firefighters trained by agencies have been trained outside of the traditional rights and pathways of fire knowledge acquisition, which makes them privy to knowledge they might not have traditionally had access to within their tribal society. Agency approaches to firefighting thus contribute to the breaking of traditional rules surrounding what fire knowledge is shared with whom, defying cultural laws and practice, which could subvert the revival of traditional indigenous burning practices.

An example of cultural sensitivity (or lack thereof) that appears frequently in our research is the impact of wildfire fighting on indigenous sacred sites, women's and men's ceremonial sites and other areas of significance. Just as knowledge of fire has been retained and protected by indigenous people, so too has the knowledge of cultural sites. The laws governing access to such sites are often related to an individual's own role within their society. For some areas access may be linked entirely to gender or may be restricted to initiation into a given group. In modern society the implementation of prescribed fire and/or the suppression of wildfire may bring conflict with the traditional practices of a given group. Thus knowing where, when, what and how to burn is one set of attributes governing traditional fire, but knowing the deeper significance of the landscape is key to securing appropriate cultural context and sensitivity awareness. The need for knowledge of land is evident in the ways in which concerns for indigenous cultural heritage are seen as being thrown out the window in order to ease the logistics of firefighting operations. For example, when a helicopter used an Aboriginal rock art site as a landing pad, one Aboriginal firefighter felt the site was being 'desecrated'. Another Aboriginal firefighter spoke of her frustration with the lack of consideration for cultural heritage sites as part of the planning and incident management stages of firefighting operations. It is an interesting dilemma that agencies often overlook indigenous sacred ground, rights of access and cultural practices, in that if traditional burning practices were in place, then the right people would inherently be burning the places they were obligated to care for. However, since policy does not support such practice, the reality of having damaging fires scorch sacred ground is often only overcome by fire suppression by whoever is appointed by the agency to do so.

However, experiences to the contrary – of agency fire operations considerate of indigenous gendered landscapes – have been shared specifically by female non-Aboriginal wildland firefighters. At one fire, for example, the on-the-ground fire units were organised so only men would patrol the fire on a site sacred to Aboriginal men. Can this heightened awareness by some white female firefighters be explained by a greater sensitivity towards other minority groups given women's minority status within the male-dominated world of firefighting? The answer could be both yes and no. Bob Pease (2010) points out that, while awareness of experiences of oppression are much more common than consciousness of aspects of one's own privileges, members of dominant groups are at the same time conditioned by the normalisation of inequality. Privilege seems natural because processes of oppression are normalised in everyday life through habituated and unconscious practices. Many, therefore, do not recognise aspects of their own privilege as the cultural norms and bureaucratic institutions in which privilege is embedded legitimate it. Thus women within the male-dominated sphere of firefighting are continually reminded of how their gender is a source of discrimination through the habituated and unconscious practices of many male colleagues (Eriksen 2014). This may heighten their consciousness of other forms of oppression in their everyday lives. However, white female firefighters are simultaneously privileged by their race, which may alienate some indigenous women from this 'alliance' (Black 2011). Thus while indigenous employment with wildfire management agencies hold many opportunities and promises, its long-term effect on the retention and revival of indigenous fire knowledge is a critical unknown.

Conclusions

Gendered dimensions of Aboriginal east-Australian and California Native American fire knowledge retention and revival have been illustrated in this chapter through a temporal and spatial trajectory of fire knowledge-holders. This generalised trajectory reveals how gender is at the crux of the story of how fire knowledge has been able to persist over time in New South Wales, Queensland and California. By forging temporary generational crossovers of gender roles and gendered knowledge, the retention of indigenous fire knowledge and environmental stewardship has been ensured despite generations of externally imposed cultural hardship.

Men are presented as the traditional (but not exclusive) holders of fire knowledge that was applied at a landscape scale. While women held some fire knowledge, the extent of their personal engagement with prescribed burning practices was in the context of finer scale burning for specific purposes. As stated earlier, that women in many places became the main carriers of fire knowledge is directly linked to the impact of external social factors, such as male genocide during colonisation.

In discussing the pros and cons of employment with wildfire management agencies for indigenous fire knowledge retention and revival, the power struggles, contrasting cultural norms, rules and generational gendered fluidity that underpin interaction between indigenous and agency fire knowledge are highlighted. State and federal agencies approach to wildfire management contribute to the breaking of the traditional layering of indigenous eco-cultural fire knowledge and rules, for example, through the indiscriminate sharing of fire knowledge and training opportunities between female and male firefighters of indigenous and non-indigenous heritage. However, employment with wildfire management agencies is also an important element in the retention of indigenous fire knowledge through access to and caring for (tribal) land. Integration of cultural perspectives of fire provides indigenous peoples with the opportunity to engage with the restoration of healthy environments, despite the potential simultaneous subversion of traditional indigenous burning practices as cultural laws and practice are defied.

The retention and revival of indigenous fire knowledge through spatial, temporal and gendered trajectories of adaptation hold many lessons, which can aid ongoing debates on how to coexist with fire in wildfire-prone countries such as Australia and USA. Perhaps most importantly in light of indigenous knowledge systems is that in working together with Aboriginal Australian and California Native American communities, wildfire management agencies stand to gain through the protection and enhancement of a real asset at risk, namely the cultures that have shaped our landscapes since time immemorial. The continuing legacy of twentieth-century fire suppression policies acts against the laws of nature (including ecosystem processes) in many parts of the world. When indigenous people have not actively asserted customary law and applied fire to care for the environment, the laws of nature continue to play out through wildfires. Indigenous practice inherently has recognised the land 'speaking' its needs through wildfire. This recognition drives the implementation of indigenous prescription of fire. We believe a greater recognition of this traditional understanding of the environment could aid current struggles to manage the growing frequency of devastating wildfires if it is acknowledged by, and incorporated into, the practices of wildfire management agencies.

Note

1   By 'traditional' we refer to the time-tested knowledge and customary practice that still guide many indigenous societies.

References

Anderson, M.K. (2005) *Tending the Wild: Native American Knowledge and the Management of California's Natural Resources*, Berkeley: University of California Press.

Black, C.F. (2011) *The Land is the Source of the Law: A Dialogic Encounter with Indigenous Jurisprudence*, London: Routledge.

Bliege Bird,R., Bird, D.W., Codding, B.F., Parker, C.H. and Jones, J.H. (2008) 'The "fire stick farming" hypothesis: Australian Aboriginal foraging strategies, biodiversity and anthropogenic fire mosaics', *Proceedings of the National Academy of Sciences, USA* 105: 14796–14801.

Brody, H. (2002) *The Other Side of Eden: Hunter-Gatherers, Farmers and the Shaping of the World*, London: Faber and Faber.

Burgess, C.P., Johnston, F.H., Bowman, D.M. and Whitehead, P.J (2005) 'Healthy country: healthy people? Exploring the health benefits of Indigenous natural resource management', *Australian and New Zealand Journal of Public Health* 29: 117–122.

Eriksen, C. (2007) 'Why do they burn the "bush"? Fire, rural livelihoods, and conservation in Zambia', *The Geographical Journal* 173: 242–256.

—— (2014) *Gender and Wildfire: Landscapes of Uncertainty*, New York and London: Routledge.

Gammage, B. (2011) *The Biggest Estate on Earth: How Aborigines made Australia*, Sydney: Allen & Unwin.

Gleeson, C.T. (1993) *Gender and Australian Hunter-Gatherer Uses of Fire*, BA (Hons) thesis, University of Wollongong.

Laris, P. and Wardell, D.A. (2006) 'Good, bad or "necessary evil"? Reinterpreting the colonial burning experiments in the savanna landscapes of West Africa', *Geographical Journal* 172: 271–290.

McDaniel, J., Kennard, D. and Fuentes, A. (2005) 'Smokey the tapir: traditional fire knowledge and fire prevention campaigns in lowland Bolivia', *Society & Natural Resources* 18: 921–931.

Mistry, J. (2000) *World Savannas*, Harlow: Prentice Hall.

—— (2002) 'Savannas and development', in V. Desai and R.B. Potter (eds.), *The Companion to Development Studies*, London: Arnold.

Pease, B. (2010) *Undoing Privilege: Unearned Advantage in a Divided World*, London: Zed Books.

Ross, A., Pickering Sherman, K., Snodgrass, J.G., Delcore, H.D. and Sherman, R. (2011) *Indigenous Peoples and the Collaborative Stewardship of Nature: Knowledge Binds and Institutional Conflicts*, Walnut Creek, CA: Left Coast Press.

Smith, L.T. (2012) *Decolonizing Methodologies: Research and Indigenous Peoples*, Dunedin, NZ: University of Otago Press.

Stewart, O.C., Lewis, H.T. and Anderson, M.K. (2002) *Forgotten Fires: Native Americans and the Transient Wilderness*, Norman: University of Oklahoma Press.

Vale, T.R. (2002) *Fire, Native Peoples, and the Natural Landscape*, Washington: Island Press.

Vigilante, T., Murphy, B. and Bowman, D. (2009) 'Aboriginal fire use in Australian tropical savannas: Ecological effects and management lessons', in M.A. Cochrane (ed.), *Tropical Fire Ecology: Climate Change, Land Use, and Ecosystem Dynamics*, Berlin: Springer.

White, G. (2004) 'Restoring the cultural landscape', in National Wildland/Urban Interface Fire Program (ed.), *American Perspectives on the Wildland/Urban Interface*, Quincy, MA: FirewiseCommunities/USA.

15

GENDER AND LIVESTOCK IN DEVELOPING NATIONS

Alice J. Hovorka

This chapter explores gender and livestock issues within the context of international development. It surveys conceptual, theoretical and practical approaches, as well as empirical knowledge and available tools within existing literature and resources. It critically reviews and appraises the current state of gender, livestock and development and future trends in analytical and action-oriented realms. This chapter thus provides a comprehensive statement, commentary and reference point for students, researchers, scholars, practitioners and policymakers.

Domesticated animals such as cows, donkeys, goats, sheep and poultry play a vital role in national development agendas and people's everyday lives around the world. Livestock make important contributions in terms of agricultural production, foodstuffs, transportation and income generation for men, women and children especially in developing nations. Existing academic scholarship and development-oriented literature since the 1980s highlights the ways in which people's relationship with livestock is highly gendered. While often aggregated with issues related to gender and agriculture more broadly, literature specifically focused on gender and livestock clearly articulates the ways in which gender 'works' in terms of access to, use of and control over domesticated animals in developing countries.

Key gender and livestock issues

Existing gender and livestock literature primarily emphasizes three key issues. First, women play significant roles in livestock-related activities and their active participation yields numerous household benefits in terms of practical and strategic needs. Second, men's and women's roles associated with and access to livestock differ, with women affected disproportionately by constraints and obstacles impacting their productivity, self-sufficiency and empowerment. Third, livestock development research and projects must recognize the role of women, as well as gender difference and inequality in order to enhance women's participation and capacity in livestock management and to ensure successful operationalization of development interventions. Together these issues enrich understanding of gender-livestock dynamics in the developing world. A summary is offered below based on detailed overviews with comprehensive literature reviews from the FAO (Distefano et al. 2013; Köhler-Rollefson 2012; Okali 2011), Kristjanson et al. (2010) and Njuki and Sanginga (2013; including Kariuki et al. 2013 and Njuki and Miller 2013).

Women contribute to and benefit from livestock management

Women are key stakeholders in livestock management and in many contexts throughout the developing world they are the main keepers, labourers, caretakers and conservers of domesticated animals. It is estimated that women make up two-thirds of the 600 million poor rural livestock keepers found globally (Njuki and Sanginga 2013; Kristjanson et al. 2010: 1; Distefano et al. 2013: 6), while in urban areas women's livestock participation may range from one-third to more than half depending on the context. Women are, in many respects, 'silent guardians' of livestock management; although their role is at least equal to that of men, their contributions remain underestimated, undervalued and ignored (Köhler-Rollefson 2012: 7). Nevertheless, women's substantial involvement in livestock management includes rearing young animals, making breeding decisions, animal care and welfare, feeding regimens, waste disposal, milking or processing animal products, preparation for slaughter, food preparation, marketing and so on. For example, women in India provide 77 per cent of livestock care, while women in Pakistan take on major responsibilities with poultry, sheep and goats and 60–80 per cent of work involved with feeding and milking cattle. Women in Yemen are tasked with all sheep-related activities, as well as milk production and processing, while women in Vietnam are tasked with pig husbandry and are responsible for purchasing, rearing and marketing (Köhler-Rollefson 2012: 7–8, 21, 35).

In turn, women benefit from livestock in numerous ways. Specifically, livestock are an asset that women can access, through inheritance, marketing channels or collective action, and own more easily than other physical or financial resources (Njuki and Sanginga 2013). Livestock offer built-in capacity for capital growth through their reproduction, and for flexible mobility given that they may be relocated should a woman become widowed, divorced from their spouse or otherwise vulnerable to family members or others usurping their resources (Okali 2011: 2–3). Further, income-generation through the sale of livestock-related products, namely milk, eggs, meat, skin, hair/fur, manure, draught power and the animals themselves, offers women options for greater financial independence and budgetary control. By managing and owning livestock, and especially through the associated monetary benefits, women increase their household bargaining power and decision-making capacity, as well as their engagement within the broader community and market. Livestock also contribute to food security and women's roles in food production, processing and marketing. Domesticated animals provide draught power and fertilizer for agriculture and enable direct access to foodstuffs or cash for food purchasing, contributing to nutrition and dietary diversity for households and communities (Kariuki et al 2013). For example, women in southern Mexico earn 35 per cent of household income from sheep husbandry and the wool produced, while women's backyard poultry production in Afghanistan, often around ten hens producing 60 eggs per year, generates enough to cover household consumption and some surplus sale on the local market (Köhler-Rollefson 2012: 32, 35–36). Women in Mali diversify household diets through livestock contributions that may reach 78 per cent of cash income, while women in East Africa sell livestock and milk to purchase grain for household consumption (Kariuki et al. 2013).

Gender differences and inequalities within livestock management

Men and women play different roles – and thus have different circumstances and experiences – within the livestock sector on account of gender-based associations with, access to, use of, control over and interactions with domesticated animals. In China, women raise pigs and chickens around the homestead and men take cattle, sheep and goats to the mountain grasslands for grazing while in Tanzania women own poultry and rabbits and take on responsibility for milking, cleaning and feeding cattle, which are owned by men and increasingly incorporated into a male-dominated ranching system. The majority of pigs in India are kept in traditional scavenging and penning systems managed by women while larger stall-feeding units are operated and controlled by men (Köhler-Rollefson 2012: 16–17, 38). Women in sub-Saharan Africa control cow's milk when produced for household consumption yet they cannot sell it and keep the income while, similarly, women own and care for poultry yet cannot always take sole decisions over the use of birds or eggs (Kristjanson et al. 2010: 7).

Such differences are rooted in social norms shaping broadly conceived ideas of masculinity/femininity and male/female responsibilities, behaviours and expectations. In other words, gender–livestock relations are grounded in what it means to be a man or woman in a particular societal context. Gender roles, circumstances and experiences differ by livestock subsector, social class, as well as locale (e.g., household, community, country, region, etc.). At the same time, generalizable circumstances and challenges exist related to gender differences within livestock management, often reflecting inequality between men and women regardless of context (Distefano et al. 2013: 7). For example, women are more likely than men to be involved with livestock management featuring smaller animals, indigenous breeds, smaller scales of production, homestead-based, subsistence- rather than commercial-orientated, and informal rather than formal operations. Women tend to work with poultry, sheep, goats and other small animals (e.g., rabbits, guinea pigs, etc.), while less often exercise control over larger animals – although women do work with dairy cows (plus alpacas, camels, etc.). Women may prefer indigenous rather than improved or imported breeds because they are easier to care for, less expensive to purchase outright and adapt more easily to common property resources to which women are often relegated (Köhler-Rollefson 2012).

Women's smaller-scale livestock activities mean that their productivity levels in terms of income generation and product yields are often minimal compared to men; preference for small operations stems from limited access to land, capital and general constraints on time given women's productive, reproductive and community-based roles. Women tend to view livestock management as a source of household food security and a means of income to support education and health needs of family members, whereas men see livestock as entrepreneurial opportunities and additional income (Njuki and Sanginga 2013; Okali 2011). Women's livestock activities are often relegated to domestic spaces such as 'backyards' and 'homesteads' and feature less intensive agricultural systems than men (e.g., free-run indigenous chicken systems compared to vertically integrated broiler production) (Hovorka 2012). Within jointly owned and operated systems, women perform most of the routine, daily tasks associated with livestock management (e.g., feeding, cleaning, animal care) while men focus on specialized, occasional activities (e.g., infrastructure construction, marketing, vaccination) (Okali 2011). Given that women and men perform different yet complementary tasks in livestock management, they have different knowledge about them – men that herd animals may be more familiar with their group behaviour, while women who milk individual animals will know their temper, health status and reproductive behaviour (Köhler-Rollefson 2012: 17).

While women's participation in livestock activities contributes substantially to household livelihoods and community empowerment, women are often stifled in their efforts or marginalized within particular realms because of structural gender inequalities existing in society. Specifically, women face greater constraints than men in accessing land, labour, capital, natural resources, social networks, extension services, innovative technologies and marketing channels (Distefano et al. 2013; Njuki and Sanginga 2013). Lack of land is particularly detrimental to women's livestock-related activities – limiting grazing opportunities, investment collateral and increased production opportunities – while inadequate amounts of time, energy, feed and equipment undermine women's attempts to sustain or grow livestock operations. Women tend to have fewer financial resources in terms of savings, credit, remittances, inheritance and insurance on account of legal restrictions (e.g., dependents under the law), customary rules (e.g., male authority over budgets), lacking collateral or matching funds (e.g., on account of women's predominance in low-paying jobs or women's privileging of household expenditures on food, health and education particularly for children); financial constraints impeded infrastructure investment as well as operational funds (Distefano et al. 2013: 8). Women's relatively limited access to technologies, training and extension services is attributed to their lack of time or mobility to do so or social dynamics that render it unacceptable for women to participate in meetings or workshop sessions. Women often face marketing constraints whereby they may be unable to transport animals and products to market on account of limited means of transportation, time constraints or mobility allowances; women may also be constrained by low literacy and negotiation skills in such settings. In these instances, women's income generating is limited and (where involved) men conduct financial transactions and retain the income. Women's roles in livestock management arguably diminish as formal markets and commercial operations expand (Njuki and Sanginga 2013; Kristjanson et al. 2010: 22).

Women also face constraints related to decision-making, secure ownership of livestock, occupational health and safety risks, and gender roles. Owing to such resource constraints, women in turn hold limited participation and decision-making power relative to men in livestock management (Distefano et al. 2013: 9); women's lower status and inputs at the household level and beyond gives them restricted ability to state their opinions, shape local mandates and empower themselves. Further, the relative informality of livestock rights in many parts of the developing world mean that women's ownership may be challenged and women often hold more rights over animal-products than they do over the animals themselves (Njuki and Sanginga 2013). Also women's close contact with animals, often handling them or their products (e.g., raw meat and milk) on a daily basis, means women are at risk of animal-related illnesses such as salmonellosis, brucellosis, Q fever and leptospirosis. Women's reproductive household role means that they bear responsibility for nursing ill family members back to health should they become infected with animal-related disease (Distefano et al. 2013: 9). Finally, distribution of gender roles within the livestock sector means that both women and men are limited to those deemed socially acceptable – for example, should men wish to take on milking or women wish to take on slaughtering, these activities may be inaccessible and thus hinder self-actualization (Distefano et al. 2013: 7).

While animals are often among the few assets many developing country women can own, the relative insecurity of their rights to these animals, their marginalization into particular realms, the greater responsibility women may have for livestock-related tasks, and women's lack of access to vital physical, financial and social assets could make this livelihood activity less desirable than other options (Kristjanson et al. 2010: 21). Ultimately, the above noted constraints often prevent women, more so than men, from reaching their full potential within the livestock management sector and thus gender inequalities compromise the achievement of overall household security and community empowerment.

Gender essential to livestock development research, projects and policies

Addressing the gender gap in livestock management requires eliminating gender biases and inequalities in men's and women's associations with, access to, use of, control over and interactions with domesticated animals. Livestock development research, projects and policies can reduce gender gaps as long as they are designed to be gender responsive and promote women's empowerment (Njuki and Miller 2013). Past livestock interventions in which gender or women were absent led to considerable research on gender and livestock issues in the 1980s and 1990s; subsequent interventions became more appropriately focused on species (e.g., poultry, small ruminants, dairy cows) and approaches (e.g., participatory, group-based) aimed more directly at women's interests, needs and experiences (Kristjanson et al. 2010: 1). Today there exist numerous resources for development researchers and practitioners touting gender and livestock indicators, modules, checklists and approaches explicitly for the purpose of addressing gender inequalities and facilitating women's empowerment through livestock activities (e.g., Distefano et al. 2013; Njuki and Miller 2013; Njuki et al. 2011; Rota 2010; Kristjanson et al. 2010; Hill 2008). All emphasize the need for gender-disaggregated data and gender analysis within livestock research, as well as gender-based design, implementation and monitoring of development projects. Perhaps most innovative are three strategies for facilitating gender and livestock as a 'pathway out of poverty' (ILRI framework) suggested by Kristjanson et al. (2010: 21–22), namely securing livestock assets, increasing livestock productivity and enhancing participation in livestock markets. The associated literature review and suggestions for future research and action are insightful (and discussed further below).

Gaps and critiques in literature

The key issues summarized above predominate in existing literature on gender and livestock in the developing world, and enrich understanding of men's and women's relative circumstances and experiences with livestock-related activities. At the same time, there is not yet enough information about gender and livestock issues, with relatively little offered on women's roles and opportunities related to the livestock sector (Njuki and Sanginga 2013; Okali 2011; Kristjanson et al. 2010: 1). Further evidence is needed to substantiate gender–livestock dynamics (Okali 2011), which often vary from place to place, and in-depth exploration of causal relations, especially through quantitative methods, between gender and livestock production is necessary to fully appreciate and understand relevant issues (Kristjanson et al. 2010: 1). Specific empirical gaps and conceptual shortcomings are detailed below.

In terms of empirical or topical gaps, little evidence exists on gender and livestock ownership and acquisition, especially the extent of women's livestock ownership, which species are preferred or most important to them and the relative importance of livestock relative to other assets (Njuki and Sanginga 2013). Limited information exists on gender and livestock marketing, especially the factors influencing women's management of income and animal-products, the the differences across subsectors, and women's participation in market scenarios (Njuki and Sanginga 2013; Kristjanson et al. 2010: 22). Few insights exist on gender and livestock productivity, especially in terms of men's compared to women's production systems (Kristjanson et al. 2010: 22). Further, insufficient data exists on gender and livestock contributions to food security, especially in buffering households against food deficits and the relative impact livestock access has on men's and women's households (Kariuki et al. 2013). Finally, health- and environment-related elements of gender and livestock issues are lacking, especially the ways in which climate change, environmental degradation, zoonotic and emerging infectious diseases or food-borne illnesses impact men, women and livestock; negative impacts of livestock keeping (e.g., water contamination, waste disposal, feed production) must also be explored in terms of their gender impacts (Kristjanson et al. 2010: 1). It is important to recognize that human health is animal health and vice versa.

In terms of conceptual shortcomings, explorations of gender and livestock remain limited to particular social groups, production systems and geographic scales, and gender-livestock relations are viewed in somewhat static and simplistic terms. For example, literature focuses on the plight of poor rural women (Okali 2011: 1), rarely extending into different income categories of women (who are often marginalized relative to their male counterparts) or women residing in peri-urban and urban areas (indeed gender and urban livestock keeping remains woefully understudied). Literature also focuses on small-scale production systems and poultry, small-stock and dairying subsectors, rarely extending into larger scales of operation or subsectors in which women may not predominate but certainly participate (e.g., beef cattle). Gender and livestock issues are commonly considered within the household and only rarely in terms of the wider community, within markets and political economy, or in state and private institutional settings (Okali 2011: 1). Additionally, literature views gender–livestock relations as static – such that changing production systems, transitions between scales and subsectors, and highly networked operations are left unexplored – and simplistic or overgeneralized such that the diversity of circumstances and experiences, needs and interests of both men and women in livestock management are overlooked.

Gender and livestock dynamics in Botswana

The following case study of gender and livestock dynamics in Botswana reconfirms key issues raised in existing literature, in particular that women contribute to and benefit from livestock management and that substantial gender differences and inequalities exist. The case study also addresses gaps in existing literature. Broadly, it substantiates gender–livestock dynamics, offering in-depth exploration of causal relations to explain how and why gender differences and inequalities (re)produce themselves within human–livestock dynamics, and potential avenues for empowerment through livestock-related activities. Further, it explores circumstances and experiences of livestock keepers in often understudied realms: low as well as higher income producers, small- as well as medium-scale operations, commercial as well as subsistence production, urban as well as rural areas, and structural as well as household-level analysis. Ultimately, the case study illustrates gender–livestock relations as dynamic, complex and diverse.

Empirical findings are drawn from cumulative insights gained through my longstanding research program in Botswana, in particular several fieldwork sessions conducted between 1998 and 2010.

First, research conducted in October 2000 to September 2001 included semi-structured interviews with 48 male, 51 female and ten husband and wife joint owners of 109 agricultural enterprises in and around Gaborone, revealing the prevalence of women's participation in poultry production. Data collection and analysis highlighted socioeconomic, spatial and environmental variables shaping productivity levels. Second, research conducted in May/June 2004 and May 2007 focused specifically on the role of women in poultry production and chickens as pivotal species in livestock management. Data collection and analysis highlighted 72 participants' reasons for pursuing particular agricultural subsectors, namely poultry, as well as their perceptions of, attitudes towards, and actions regarding chicken compared to other livestock. Third, numerous field visits between 2008 and 2010 provided opportunities to investigate further structural issues, including national policy and programming efforts, politico-economic trends, and sociocultural norms, shaping livestock management issues in Botswana via through key informant interviews and secondary data.[1]

Livestock access, use and control reflect substantial gender differences and inequalities in Botswana, illustrated clearly through an example of men's association with cattle and women's association with chickens. Specifically, ownership and responsibility for cattle as symbols of social and economic status in Botswana fall to men as a birthright. Young boys are socialized into cattle rearing and management outside of the domestic sphere while girls are relegated to tending smaller animals, including chickens, pigs, sheep and goats, as well as subsistence crop production and preparation (Brown 1983: 378; DAPS 1997: 11; Kalabamu 2006; Van Allen 2007: 100). Sixty-two per cent of women do not own cattle as opposed to 32 per cent of men (Kidd et al. 1997: 54–56) and reports of women's participation in cattle production remain novel (e.g., Tsiane 2010). Cattle holdings by gender are men with 1,455,420 traditional cattle compared to 472,346 for women, and 93,042 commercial cattle (316 holdings) compared to 6,897 (29 holdings) for women. Another 43,309 cattle are within 152 commercial holdings that are predominantly men (CSO 2006: 160, 170). That women do not often acquire relevant skills or knowledge of cattle, affects their ability to access, interact with or control them without male assistance. Men have benefited greatly from associations with cattle: greater access to draft power for commercial food production, enhanced social status on account of quantity of cattle owned, and entry into exchange or income-generating from selling cattle on the open market. Without direct interactions with cattle, women are untimely with ploughing and planting plus incur costs of hiring tractors or labourers (Fortmann 1980: 6; Wikan 1984: 131; Mazonde 1990: 15; Mogwe 1992: 7; Kidd et al. 1997: 57). Women often borrow animals at the end of the ploughing season when (male) owners are finished using them, again risking low productivity levels. The lack of cattle as draught power also means that women do not have the opportunity to sell cattle in times of hardship (Mogwe 1992: 7). Women's associations with non-cattle species are explained in terms of socioeconomic status: for example, women can often only afford to purchase and keep sheep, goats and chickens (Kidd et al. 1997: 71; Hovorka 2006). Anecdotal connections are also made that, for example, poultry keeping is 'women's work'. As one male Batswana interviewed in 2004 notes: 'women have an advantage with poultry: they know how to nurse and look after babies. That chicken is like a baby, you have to raise it'; another male respondent commented that 'chickens are too delicate for men – they are more suited for women'.

Symbolic and material associations of men with cattle and women with chickens that occur in everyday realms become deeply entrenched on account of how these gender dynamics are reflected in national policy and programming efforts. First emerges the idea that successful Batswana men own and manage cattle, and thus women are relegated to other (smaller) animals. Second evolves the everyday practice whereby men's association with cattle provide them with benefits in terms of social status, economic contributions, labour, and food security. Women may benefit from small scale, subsistence-oriented association with chickens but are often excluded from the benefits that cattle provide. These gender differences and inequalities are further entrenched at broader scales given that cattle are central to Botswana politics, economy and culture. National agriculture policies and programs have long funnelled financial subsidies, land allocation and veterinary services towards cattle and thus towards men who hold primary control of them. The successful beef-oriented political economy and beef-focused dietary preferences are further evidence of symbolic and material privileging of masculinity in Botswana. Chickens have long remained relatively invisible and marginalized in this context, with little socioeconomic or political value given their association with women, subsistence, domestic and feminine realms (Hovorka 2012: 879).

Recently women's access to, use of and control over chickens has received a boost in Botswana. Specifically, agrarian restructuring and urbanization have given rise to commercial poultry production in urban and peri-urban areas of the country, in particular the capital of Gaborone.

First, the government of Botswana has notably redirected agricultural policy and planning since the mid-1990s, emphasizing diversification beyond dominant traditional realms of arable production and cattle ranching into new subsectors of poultry, dairy and horticulture. Paired with financial granting schemes, this government agenda has taken shape over the past two decades at the local level with burgeoning enterprises focused on intensive production, especially poultry farming. Second, urbanization trends have facilitated emergence of commercial agriculture in and around cities. Botswana's able-bodied agricultural labour force is found increasingly in urban areas, given the pull of jobs, services, and urban lifestyle and push from drought-prone, isolated, low-level investment rural areas. Yet urban Batswana continue to identify with an agrarian-based lifestyle and thus seek out farming-related opportunities. Further, financial grants available for local business development have been allocated to urban enterprises and this, together with government promotion of new agricultural subsectors, has facilitated emergence of agricultural enterprises in and around the city.

Women's land access, particularly to agriculturally conducive areas, has emerged through government financial grants promoting and supporting local enterprise development, which allow women to purchase, rent or use plots of land without dependence on male relatives. Grant provisions include easier eligibility terms, smaller matching contributions and larger monetary allocations to women compared to men. While some women gain access to commercially or agriculturally zoned land in and around Gaborone, others focus energies on their residential plots that in many cases they have sole ownership of or control over given their status as women heads-of-households. Residential plots hold numerous advantages for these women given direct water connections to homesteads, access to neighbours as consumers, as well as fellow producers with whom they share experiences of poultry rearing. Many women express a sense of pride and contentment associated with being landholders, reflecting their transformation into new spaces within the urban context.

Women are capitalizing on traditionally designated 'women's work' of chicken-keeping by taking on income-generating activities, often funded by government grants and loans, featuring poultry production. Chickens facilitate this because they are biophysically adaptable given their slight physique to various urban habitats, ranging from large-scale, industrial agriculture sites to smaller backyard areas. Chickens are also hearty and efficient in intensive enterprises given that 10–13 chickens can coexist in one meter squared of space, and grow quickly over a six-week period. Environmentally-conditioned houses ensure that chickens are consistently fed, watered and monitored (essential in Botswana's relatively harsh climate) yet require minimal care. Fixed-asset investments and operating costs are substantially lower in this agricultural subsector, particularly if enterprises are small-scale as is often the case for women in and around Gaborone. These factors offer women high and quick earnings from the poultry sector, something that other species cannot offer in the urban context. In particular, cattle are deterred from urban areas given their physical bulk and extended maturation cycles, as well as the lack of adequate grazing areas and prohibitive zoning regulations. Women's productive relationship with chickens in the urban context repositions them materially and symbolically. Income generated from chicken rearing is substantial and done so in circumstances where women can easily maintain their reproductive roles, in particular caring for their children and homestead. The resulting productivity means that women's ability to provide basic human needs and even lifestyle amenities to their households and their claims to land are reaffirmed through sustained production. Women's enterprises have caught the attention of larger-scale poultry producers within a vertically integrated agriculture sector. Contract farming has expanded with women and chickens providing much needed supply into the sector and given training, guaranteed market and continuous income in exchange. The visibility of women's chicken-based enterprises has enhanced their status within entrepreneurial realms; Ministry of Agriculture extension workers, who in the past focused primarily on men's activities, now feature women in short courses and routine visits to agriculture sites.

Poultry meat production is by far the largest and most substantial non-cattle-oriented commercial agricultural subsector in Botswana today. Sixty-six enterprises found in Greater Gaborone in 2000 generated more than 15 million kilograms of meat at a market value of 81,451,826 pula (US$16,290,365) from approximately 2,300,000 chickens. This represents a substantial increase in urban chickens from the previous year, when one million chickens inhabited the city, with trends continuing in the subsequent decade (e.g., approximately three million chickens in 2004). Given these financial and absolute numbers, chickens are now recognized for their productivity, adaptability and efficiency. However, they are noticeably absent from the government of Botswana's *Vision 2016* document published in 1997, which acknowledges and privileges beef cattle as fostering entrepreneurship, household economic prosperity and food security. Yet by the 2002 release of the ninth National Development Plan (NDP 9), chickens feature as the backbone of agricultural diversification and export strategies, and as a means of enhancing employment opportunities, technological innovation and household food security. National pride in chickens is prevalent in official realms praising the self-sufficiency and productivity of the sector, and sentimental attachment to chickens noted by ordinary citizens. Further, chickens are recognized as providing jobs and skills training for Batswana, and they are influencing food consumption trends by offering a healthy alternative to beef-oriented diets (Hovorka 2008).

Women's association with chickens via poultry production has provided numerous opportunities for empowerment in recent decades; however, empowerment has not been fully realized within this context of changing gender–livestock relations in Botswana. While individual women are empowered within Botswana's new commercial-urban context, only a certain kind of 'woman' is valued or offered the means to empowerment. Specifically, individual women are valued only when associated with chickens inside of a global capitalist system featuring intensive commercial production yet are not so recognized when within subsistence agriculture associated with the domestic, reproductive realm. This means that women moving into commercial livestock production is empowering because the realm itself is empowered not that women are empowered in their own right. Further, women as a social group remain subordinate relative to men in terms of freedom of choice (given relative access to rights and resources) and influence over others or within the livestock management context (given collective visibility, status and valuing) (Hovorka 2012: 881). Men tend to have greater access to capital and more extensive plots of agricultural land, which in turn allows them to generate greater quantities of produce across various subsectors. Women engage in small-scale production often constrained by access to small plots of land and dominated by chickens as the quintessential 'women's work'. Further, women's enterprises are couched with a vertically integrated poultry sector, dominated by the wealthiest producers, all of whom are men (Hovorka 2005). As contract producers, women are further threatened by small-scale male producers who wish to take over women's work in order to secure their own position as successful entrepreneurs and farmers in the city. Urban men's associations with cattle are stifled given the rural-base required, which means they travel to cattle posts during weekends and must leave stock in the hands of herd-boys during weekdays. Men are turning to chickens by establishing their own productive enterprises and subverting discourse to reflect 'poultry as a man's job', as one male Batswana stated, based on the now commercial (hence non-subsistence) orientation to human–chicken interactions in the urban context.

This case study of gender and livestock issues in Botswana illustrates the ways in which gender relations of power are (re)produced through men's and women's associations with particular animals, and through livestock management activities that are necessarily dynamic and complex. It is not enough then for researchers, development practitioners and policymakers to recognize women's contribution to and benefits from livestock-related activities. Revealing and explaining deeply entrenched gender differences and inequalities at household through national scales is vital to understanding the specific context, as well as informing development interventions to enhance gendered livestock management and its associated benefits and empowerment opportunities.

Conclusion

Livestock provide vital sources of food, labour, transport and income to individuals, households, communities and nations – that livestock-related activities and benefits are necessarily gendered requires both outright acknowledgement and in-depth exploration so as to ensure gender equity and advancement. Continued and enhanced development research and practice is warranted given the importance of domesticated animals and their management to men and women in developing contexts around the world. Investigations must continue to focus on the specific roles of women and men with respect to livestock, in particular the ways in which these roles are assigned, reinforced or transformed, and must continue to document the benefits (or lack thereof) accrued by those men and women best positioned to do so. Investigations must enhance focus on underexplored issues related to gender and livestock, in particular filling empirical gaps such as topics of ownership, marketing, food security, health and environment issues, or addressing conceptual gaps such as insights on various social groups, production systems, geographic scales or gender definitions. Thinking about and practicing such gender and development research involves both gender-disaggregated data collection and gender interpretation and analysis. On the one hand, researchers must collect information about the different experiences, needs, interests and access to opportunities and resources of both men and women with livestock to establish an accurate picture of a particular context, which is often complex and dynamic. On the other hand, researchers must ask why such gender and livestock dynamics occur by probing deeper to examine factors that create and influence differential opportunities and constraints for men and women at local, regional and global levels. In turn, gender and livestock development practice must capitalize on such in-depth understanding and translating it into specific action-items that address practical and strategic gender needs in that context.

Note

1   Excerpts of the case study below are reprinted from Hovorka (2012) with permission from Elsevier.

References

Brown, B. (1983) 'The impact of male labour migration on women in Botswana', *African Affairs* 82: 367–388.
CSO (2006) *Annual Agriculture Survey Report*, Gaborone, Botswana: Central Statistical Office.
DAPS (1997) *Gender Equity and Access to Economic Opportunities in Agriculture in Botswana*, Gaborone, Botswana: Ministry of Agriculture.
Distefano, F., Mattioli, R. and Laub, R. (2013) *Understanding and Integrating Gender Issues into Livestock Projects and Programmes: A Checklist for Practitioners*, Rome: Food and Agriculture Organization, www.fao.org/docrep/018/i3216e/i3216e.pdf.

Fortmann, L. (1980) *Women's Involvement in High Risk Arable Agriculture: The Botswana Case*, Washington, DC: USAID.

Hill, C. (2008) *Gender and Agriculture Sourcebook: Module 14 Gender and Livestock*, Rome: Food and Agriculture Organization, http://siteresources.worldbank.org/INTGENAGRLIVSOUBOOK/Resources/Module14.pdf.

Hovorka, A.J. (2005) 'The (re)production of gendered positionality in Botswana's commercial urban agriculture sector', *Annals of the Association of American Geographers* 95(2): 294–313.

—— (2006) 'The no. 1 ladies' poultry farm: a feminist political ecology of urban agriculture in Botswana', *Gender, Place and Culture* 13(3): 207–225.

—— (2008) 'Transspecies urban theory: chickens in an African city', *Cultural Geographies* 15(1): 95–117.

—— (2012) 'Women/chickens vs. men/cattle: insights on gender–species intersectionality', *Geoforum* 43(4): 875–884.

Kalabamu, F.T. (2006) 'Patriarchy and women's land rights in Botswana', *Land Use Policy* 23(3): 237–246.

Kariuki, J., Njuki, J., Mburu, S. and Waithanji, E. (2013) 'Women, livestock ownership and food security', in J. Njuki and P. Sanginga (eds.), *Women, Livestock Ownership and Markets: Bridging the Gender Gap in Eastern and Southern Africa*, London: Routledge, pp. 95–110.

Kidd, P.E., Makgekgenene, K., Molokomme, A., Molamu, L.L., Malila, I.S., Lesetedi, G.N., Dingake, K. and Mokongwa, K. (1997) *Botswana Families and Women's Rights in a Changing Environment*, Gaborone: The Women and Law in Southern Africa Research Trust.

Köhler-Rollefson, I. (2012) *Invisible Guardians – Women Manage Livestock Diversity*, Animal Production and Health Working Paper No. 174, Rome: Food and Agriculture Organization.

Kristjanson, P., Waters-Bayer, A., Johnson, N., Tipilda, A., Baltenweck, I., Grace, D. and MacMillan, S. (2010). *Livestock and Women's Livelihoods: A Review of the Recent Evidence*. Nairobi: International Livestock Research Institute.

Mazonde, I.N. (1990) *The Gender Issue in Botswana*, paper presented at the Annual Conference of African Anthropologists, Nairobi, Kenya.

Mogwe, A. (1992) *Country Gender Analysis: Botswana*, Stockholm: Swedish International Development Authority.

Njuki, J. and Miller, B. (2013) 'Making livestock research and development programs and policies more gender responsive', in J. Njuki and P. Sanginga (eds.), *Women, Livestock Ownership and Markets: Bridging the Gender Gap in Eastern and Southern Africa*, London: Routledge, pp. 111–128.

Njuki, J. and Sanginga, P. (eds.) (2013) *Women, Livestock Ownership and Markets: Bridging the Gender Gap in Eastern and Southern Africa*, London: Routledge.

Njuki, J., Poole, J., Johnson, N., Baltenweck, I., Pali, P., Lokman, Z. and Mburu, S. (2011) *Gender, Livestock and Livelihood Indicators*, Addis Ababa: International Livestock Research Centre, http://cgspace.cgiar.org/bitstream/handle/10568/3036/Gender%20Livestock%20and%20Livelihood%20Indicators.pdf.

Okali, C. (2011) *Notes on Livestock, Food Security and Gender Equity*, Animal Production and Health Working Paper No. 3, Rome: Food and Agriculture Organization, www.fao.org/docrep/014/i2426e/i2426e00.pdf.

Rota, A. (2010) *Gender and Livestock Tools for Design*, Rome: International Fund for Agricultural Development, www.ifad.org/lrkm/factsheet/genderlivestock.pdf.

Tsiane, L. (2010) 'Women venture into the cattle business', *Mmegi*, 10 June 10, www.mmegi.bw/index.php?sid=1&aid=2692&dir=2010/June/Wednesday2.

Van Allen, J. (2007) 'Feminism and social democracy in Botswana', *Socialism and Democracy* 21(3): 97–124.

Wikan, G. (1984) 'Development and women in Botswana', *Norsk Geografisk Tidsskrift* 38: 129–134.

16

FISHERIES AND AQUACULTURE NEED A GENDER COUNTERREVOLUTION

Meryl J. Williams

Introduction

The fishery sector, forever dynamic, is in a particularly active—even revolutionary—phase. Through history, fisheries have spread from harvesting fish in rivers, lakes and swamps, to coasts and eventually offshore and into deep-water ocean realms to the limits of polar ice. Farming fish in aquaculture is about to overtake wild fisheries in production. In the highly diverse fishery sector,[1] women and men fill a great diversity of roles that have evolved over millennia, along with the technology, use and economy of fish. Beyond evolution, however, contemporary fisheries and aquaculture have entered their most masculine stage, thanks to the dominance of commercial and industrial scale operations and international trade. Women's roles and those of many working men have been compressed, especially in public perceptions, into invisible—often exploited—support positions. A counterrevolution is needed to overcome the gender inequality and other social concerns that are costing the sector social and economic value.

The gender blindness of the present fishery revolution is the first hurdle to a gender counter-revolution. The blindness manifests as a blank wall of incomprehension and misperception from expert insiders, as illustrated by comments such as:

Why? Is there a problem with gender in aquaculture?
(retired male senior aquaculture professor, June, 2012)

Don't women have less trouble being recognized as equal in fisheries these days?
(male professor of natural resource economics, January, 2013)

Women have greater equality in aquaculture than fisheries, don't they?
(male senior government fisheries economist, January, 2013)

These comments are all from men with extensive international experience but similar views also come from women in mainstream positions.

Gender, gender equality and gender equity are neither understood nor considered important in current fishery sector norms, values and strategies. Groups with gender interests are weak and rarely participate as stakeholders in the sector's business, policy and advocacy networks. Over the past century, the sector has become firmly entrenched in the global economy and has attracted major technological and capital investment in production, processing and trade, helping to create the contemporary actors and networks that are highly masculine in personality. In most fishery pursuits, women are unrecognized and/or have negligible influence. The gendered impacts of sector change and the sector and social losses from gender inequality go unnoticed. The sector's bodies, such as fishery-specific management committees, commodity and professional societies, pay minimal attention to this gender bias, except to occasionally justify it on the grounds of the physical rigors of fishery work and women's lack of need to be involved. Although the overwhelming majority of the people engaged in the fishery workforce are in developing countries, and half of these are women, international development assistance agencies, national governments and civil society give little priority to the sector per se, and very little attention to women or gender issues.

In fact, all internal and external forces have had little impact on having the fishery sector mainstream, recognize and address gender inequality or behave in a gender equitable manner, although signs of hope exist. Inside the sector, researchers and advocates are slowly gathering facts and recommending directions, despite a weak and slow response from the sector mainstream. External factors, including global norms such as the Convention on the Elimination of all Forms of Discrimination Against Women (CEDAW), the Millennium Development Goals (MDGs) and national economic imperatives have created some stimulus for the fishery sector to examine gender. However, the agencies that implement global instruments, such as UN Women and the gender units of mainstream UN agencies, focus more on overarching themes and platforms, such as girls' education and ending domestic violence, than sector-specific programs. Even the main UN agency responsible for the fishery sector, the Food and Agriculture Organization (FAO), has been slow to address gender substantively, although it has made a start, as will be touched on below.

The present chapter starts with an overview of fishery sector change and its gendered nature, illustrating trends with examples from three specific change themes: fisheries management change, supply chain inequalities and legal status and assets. I then describe the struggle to get gender equity considered as a legitimate part of the sector agenda. Advocates, researchers and development assistance agencies have each made small and complementary contributions but against adverse trends. This part of the chapter draws on my personal experience as a researcher (originally a marine zoologist and statistician) and research manager in fisheries and aquaculture at sub-national, national, regional and international level, both in the mainstream sector and in helping organize and promote gender and fisheries activities in the Asian Fisheries Society and others. Finally, I reflect on what could be done, and the recent hopeful steps by some leading institutions, to begin a counterrevolution against the sector's near-blindness to its gender (and other) inequities.

The revolution creating gendered change in the fishery sector

Over the past half century, the hub of global fish production moved from developed to developing countries, which now dominate production and totally dominate in the numbers of fish workers. The direction of trade has reversed to flow predominately from developing country producers to developed country consumers. These changes did not occur overnight but the late 1980s and early 1990s were a major turning period.

Much fishery sector change is gendered in the sense that it arose from gendered processes in which most women and many low-ranking men had little decision-making power. They also experienced different outcomes compared to those of the powerful men and the few powerful women involved in fisheries and, as a result, found their needs inadequately addressed.

Change in the fishery sector

In 2010, according to FAO *State of World Fisheries and Aquaculture 2012*, global fish production was 154 million tons, eight times that of 1950, while the human population grew by 2.7 times. Forty percent of total fish production is traded internationally, making it the most traded food commodity by value, outranking coffee, tea, cocoa and sugar. Fish trade has increasingly directed fish processing into more centralized and mechanized modes, with quality controlled, value added and certified products. Prime market commodities—such as salmon, shrimp, catfish and tilapia—supply highly competitive markets in which cost/price pressures stress the profitability of farms and companies.

Aggregated global fisheries production (harvests from the wild) has been at a plateau for more than 20 years, although component fish catches vary annually. In 1950, world aquaculture production was negligible; today it nearly equals the production from fisheries. Aquaculture growth has been particularly rapid since the early 1980s, generating the partly appropriate narrative of "from hunting to farming fish." What is glossed over in this narrative is that "hunting" has also undergone revolution and that hunting also provides vital inputs to aquaculture, such as fish meal and fish oil in aquaculture feeds.

In the mid-1980s, developing countries' total production overtook that of developed countries and the gap has increased, building up large trade deficits in fish by the developed countries. Aquaculture expansion has been an important component, in extent, size and intensity. In Africa and parts of Asia, small-scale aquaculture was originally supported mainly by development projects, but after a patchy track record, full-scale commercial enterprises have tended to take over. Developing country fisheries also expanded, particularly in Asia but also in South America and Africa.

In terms of the numbers of fish workers, developing countries are even more dominant than they are in production. In fisheries, more than 97 percent of the 119 million fish workers are in developing countries (World Bank 2012). In aquaculture, workforce data are more limited but likely a similar or larger share of aquaculture workers are in developing countries, given Asian and other developing country dominance. FAO estimated 17.6 million fish farmers plus a similar number in the rest of the value chain, making for about 30 million aquaculture workers.

Even after decades of industrialization, the greater share of the global fishery workforce is engaged in small scale operations, many now mechanized, predominately in Asia and Africa. Formal statistics are weak and likely underestimate the total workforce, especially in small-scale fisheries, illegal fishing, inland fisheries and small scale aquaculture, and for women throughout (World Bank 2012). Gender disaggregated data are scarce.

Despite its masculine image, the best estimate is that nearly half of the workers in the fisheries branch of the sector—47 percent or 56 million people—are women (World Bank 2012). They fill many different occupations, ranging from powerful commercial and financial actors to the most menial workers, harvesters and cultivators. Women's participation in the fisheries workforce is highly variable, however, from more than 70 percent in Nigeria and India, to 5 percent or less in Bangladesh and Mozambique (World Bank 2012).

Developing country governments and private investors have given considerable fishery support in order to modernize the sector, enhance food fish production and gain foreign exchange from fish trade. Often, sustainability and social justice have been overridden in the quest to fulfill these other aims.

International development assistance agencies such as bilateral aid agencies and multilateral banks, and non-governmental organizations have given only minor support to the fishery sector, and that only from a few agencies. The low level of development interest is due partly to the small size of the sector relative to their other related interests, such as agriculture and rural development. These other larger sectors/themes are themselves subject to fluctuating development interest, largely depending on food prices. Although the overarching strategies of international development agencies, including gender strategies, help shape their assistance, they have rarely taken a strong position on gender inclusion in their fishery support. National fisheries development approaches have been left to the central government authorities such as ministries of finance and ministries of fisheries/agriculture.

Conversely, thematic gender programs in development assistance typically address thematic issues such as on violence against women, girls' education, reproductive health and HIV/AIDS, and rarely reach the fishery sector. Indeed, as a good example of the isolation of the fishery sector, the high HIV/AIDs risks in some parts of the fishery sector are only now being studied in depth, only after the alarm was raised in the 2001 Asian Fisheries Society Global Women in Fisheries symposium.

Fishery development assistance support has focused on natural resource management and technical matters, such as resource assessment, fishing, aquaculture and post-harvest technology, exploratory fishing, conservation, illegal fishing control, product certification and market access. As fish stocks declined from overexploitation, agencies withdrew or shifted their fisheries support, usually towards funding fisheries governance (e.g., co-management, community based and ecosystem approaches to management), conservation (e.g., marine protected areas) and aquaculture. Fisheries governance and conservation support, although aimed at community level and co-management, is overwhelmingly gender blind.

Many aquaculture development projects, focusing mainly on small-scale farmers, have also ignored gender. The projects that involve women as well as men have come up against systemic social and cultural issues in the new farming technology. These are location-specific, such as those illustrated for a large agricultural extension project run by the Danish International Development Agency and the Bangladesh Department of Fisheries (1997–2004). In a survey for the project, Mowla and Kibria (2006) found that women had less leisure time than men, but they still carried the burden of the household work and were much less likely to attend training programs (see Table 16.1). Indeed, aquaculture work often added to the women's burdens.

For post-harvest fish processing activities, several decades ago development agencies supported women's projects to improve small-scale processing technology, such as chokor ovens for smoking fish in West Africa. Today, development agencies focus on supporting certification and product safety standards to enable major commodities to enter global markets through capital-intensive value-addition.

Gendered change in the fishery revolution

The social impacts of fishery change have received little attention, by contrast with attention to the impacts of change on resource sustainability, the environment, production, profitability and trade. The sector change has benefited the owners of capital and markets, including fish

Table 16.1 Division of labor in integrated pond farming: activities and gender issues from a survey of 30 women and 30 men from two villages in the Patuakhali Barguna Aquaculture Extension Project in south and southeast Bangladesh coastal areas (from Mowla and Kibria 2006).

Activities	Gender	Issues
Training activities	More men than women (25 men, 15 women)	– Women felt too shy to attend mixed group training programs. – Women had multiple work burdens in the house. – Both men and women thought that most women were slow learners. – Religious prohibition.
Pond preparation	Men and women (27 men; 24 women)	– Women assisted male partners since the former were less confident in technical issues.
Preparing feed/ feeding/applying feed in feeding ring	Mostly by women and children (24 women; 10 men)	– Men and women considered this work the extension of household chores. – Both considered this was not men's work.
Making/repairing nets ("invisible" job)	Women more than men (24 women; 12 men)	– Traditionally, women do this in their leisure hours, while men take a nap or rest. – Men and women both considered women more dexterous with their fingers.
Stocking (fish)	Mostly men (28 men; 5 women)	– Men were more knowledgeable on species selection. – Women could not travel far from home to collect good fingerlings. – Women encountered cultural barriers on their mobility.
Marketing	Exclusively by men (29 men; 1 woman)	– Women had little marketing knowledge. – Women were less confident in the marketplace. – Society did not permit women to go to market and bargain with male buyers even at nearby market.
Credit or capital investment	Mostly men (28 men; 2 women)	– Men were better at investing money. – Women did not have easy access to credit while men had a better networking system.

farmers. Generally, small-scale fishers, women and fishing households have suffered losses. Of especial interest is that most of the fishery sector change has had different impacts on women and men, as I now illustrate with three examples.

Fisheries decline and new fisheries management regimes

Fisheries changes such as the collapse or decline of stocks and transitions to new management regimes produce gendered impacts that are rarely acknowledged or addressed in the ensuing adjustments. Indeed, this topic deserves major investigation in its own right, although isolated studies give glimpses of the impacts.

After a fishery declines, women in fishing households, and sometimes men, may migrate to seek employment elsewhere, or stay and work longer hours on additional crew duties to keep the reduced fishery income within the family. Women and men also cultivate practical, psychological acceptance and social support behavior in localized, culturally and gender-specific ways. The most vulnerable households are those in which several members are dependent on the fishery, say a husband who is a fisher and a wife who processes and sells the catch, as reported in studies in Samarang (Java) and Pantar Islands, Indonesia (see studies in Williams et al. 2012). After a crisis, even where the community remains intact, women tend to become less visible but more important in their community and household support roles. Also invisible are the impacts on workers such as male crew, many of whom are low-paid immigrants.

In countries such as Indonesia and the Philippines, local fisheries declines have dislocated communities, altered the local use of space and changed the gender divisions of labor in the supply chain. For example, in a Philippine fishing village, Turgo (2012) found that the localized fishing crisis changed who did what—such as men stepping into fish vending—and how the local fish market space became important for additional social and welfare purposes in addition to its trading house function. Where reef gleaning and inshore fishing is still carried out mainly by women, the lack of women's voice in community management initiatives—many supported by development agency projects—can cause them to be banned from their normal fishing grounds in the interest of conserving resources. For example, women lack a voice in the Pantar Islands, Indonesia, in the conservation-oriented Coral Triangle Initiative that has entered into regional and local governance.

Inequalities in changing fish supply chains

Inequalities, including gender, are pronounced in fish supply chains. A small but growing number of fishery value chain analyses show that women, with some exceptions, have smaller enterprises, little bargaining power, are more prone to be squeezed out and are less able to get investments to secure their businesses. The overall result is loss of value of the whole chain. For example, along the 150km fish value chain between Lake Selingué and Bamako, Mali, men manage the lorries that carry the fish and they control the traders association. Tindall and Holvoet (2008) found that small-scale traders had to take responsibility for all losses from the landing point to the wholesalers, leading many women, but relatively few men, to drop out.

As fishers and fish farmers trade their product into higher value fish markets controlled by capitalist systems, women may lose access to raw materials for processing and sale and play increasingly exploited and marginal roles. Nayak (2007) found this outcome in investigating the gender dimension of rapid fisheries industrialization in Gujarat and Tamil Nadu, India. Although growing female employment in fish processing factories is promoted as a compensating factor, this is not the case as the factories are located in other areas and their workers have a different demographic profile. For example, in India many processing workers are young women migrants from Kerala (Nayak 2007).

Factory line workers are a far cry from the earlier "herring lassies" of the United Kingdom who practiced under a high degree of autonomy or village women drying locally procured fish. Women's jobs in modern fish processing factories grant them little autonomy. Women rarely become factory supervisors and managers. Even on the processing floor, women's needs are little recognized. For example, the 2011 FAO *Technical Guidelines on Aquaculture Certification* do not address gender issues, working conditions and opportunities, focusing instead on the environment and food safety.

South Asian, and indeed Asian, women are the mainstay of seafood processing factories, just as they are in the garment, shoe and other factories. In seafood processing, workers receive higher wages and benefits than those in other sectors because they must be educated sufficiently to follow quality control procedures. Yet, women—mainly single women demanding few benefits—receive lower wages than men, have low bargaining power and are exposed to health and sexual harassment hazards. Where fish processing factories employ women migrants, their roles in the reproductive sphere are affected by the relocation and conditions that favor employment of unencumbered workers.

Legal status and access

Women and men's legal status often differ in fisheries and aquaculture. Women who fish may not be able to get licensed as fishers. For example, women divers in Japan and Korea are not permitted to register as fishers and join cooperatives in their own right (see Williams et al. 2012). Women's fishery support roles also are often insecure.

A common assumption is that aquaculture, being more modern, is more women-friendly than fisheries. Yet, opportunities in aquaculture depend on access to space, finance, physical assets and knowhow, leading typically to poorer outcomes for women. Except for well-educated women with significant assets, most women are active only in small-scale operations, as laborers or in processing and marketing. In some cases such as in Bangladesh and Nepal, access to aquaculture resources for women in development projects, although small, are set in motion by the temporary loss of local men who migrate to work in the cities or other countries.

In India, rights to coastal space for women's aquaculture suffer from unexpected gender bias (see study in Williams et al. 2012). In southern India, coastal mussel farming on the Malabar coast first prospered as a women's industry and "empowerment platform" supported by the self-help group movement of Indian government agencies and NGOs. However, open sea cage culture, needing larger investments, was developed by men. When mussel farming became profitable, men moved in and the women's groups found that the use rights over their grounds were not legally protected; by contrast, from the outset, the state protected the rights to the men's cage farming sites.

The struggle to get gender equity on the agenda

In this section, I present my first-hand observations from nearly four decades as a fishery science insider. Over that period, even the meager attention given to women and gender has waxed and waned.

In the 1980s, following the adoption of CEDAW, a very modest flurry of activity on women ensued, including by the FAO (1987). Harrison (1995) eloquently dissected how early and weakly designed women/gender efforts fizzled after failing to make practical contributions on policy and in field projects. I note that gender was such a marginal issue that the 1995 FAO Code of Conduct for Responsible Fisheries did not mention women or gender. These gaps have persisted even though the Code has been enhanced since by more than 40 topic-specific technical guidelines, international plans of action and conventions.

Despite lacking global attention and norms, a few national and regional gender efforts sprouted. In the 1990s, Dr. M.C. Nandeesha helped stimulate national women in fisheries work in India and Cambodia. The India effort was stimulated by the growth of women's employment in seafood processing and rapidly rising education levels of women that generated possibilities of their greater agency. In Cambodia, civil war had left a generation in which women outnumbered men, many households were headed by women, and national government policies stressed the economy-wide inclusion of women. Set up in 2000, the intergovernmental Mekong Network for Gender Promotion in Fisheries Development (NGF) grew out of the Cambodia efforts and subsequent formation of national networks in Cambodia, Lao PDR, Thailand and Vietnam. It is the world's longest running regional network but still struggles for support. In the main, national efforts have been difficult to generate and sustain.

The 2000 MDGs stimulated another burst of interest. The overall international development framework opened up important space in three ways: directly through sector projects that included or focused on gender, semi-directly via new national gender programs that reached out and included the fishery sector, and indirectly when sector actors, especially activists and researchers, recognized the opportunity.

In non-government circles, gender advocacy is mainly interpreted as women's advocacy. One exception is the International Collective in Support of Fishworkers (ICSF), the global body leading small scale fisheries advocacy, which has a comprehensive policy position on gender in the "quest for sustainable, equitable and gender-just fisheries." ICSF's position derives from a "feminist perspective . . . to reshape gender relations by questioning the dominant discourse and those who set its terms. Gender issues thus focus not only on women, but on the relationship between men and women, their roles, rights and responsibilities, while acknowledging that these vary within and between cultures as well as by class, race, ethnicity, age and marital status" (http://wif.icsf.net/). In developing countries, some small scale women's fisheries organizations have been created, such as Masifunde Development Trust in South Africa. Despite these and other examples, women/gender organizations are scarce and struggle to sustain themselves.

Development agencies have made some attempts to incorporate women or gender activities in fishery assistance. However, fishery project planning more frequently ignores gender or, if included, assumes that gender-sensitive development means finding work, preferably paying work "that women can do," or small scale technologies, or women-only group enterprises. Although such approaches can be useful, unless women are in control of the programs they simply added to women's work burdens, as shown in the above example from Mowla and Kibria (2006). The failures of agencies to mainstream gender often start with the failure to carry out prior gendered contextual analysis. And the need to do this is a time and cost impediment to projects.

In the mid-1990s, the focus on small scale fisheries and poverty alleviation placed any gender emphasis onto poor women and, in small scale fisheries, conflated "women" with "poverty," despite the different drivers and subjects of problems in each case. Fisheries experts subsequently considered poor women the primary gender issue in development, and this development focus remains to the present, thus framing gender as having a welfare or charity focus for poor women.

Professional fisheries and aquaculture societies, populated by members from the biophysical sciences and economists, have paid minimal attention to gender/women. One major exception is the Asian Fisheries Society (AFS) that started to pay attention in the mid-1990s and has held regular women/gender symposia since 1998. In 2012, the International Institute for Fisheries Economics and Trade (IIFET) held its first dedicated gender session, addressing gender and fish value chains (http://genderaquafish.org/events/iifet-gender-sessions-2012-dar-es-salaam-tanzania/).

Research on women/gender in the fishery sector has helped move the agenda forward but is still in an immature and early stage of conceptual development. Further, although well-intentioned, many studies and projects suffer from weak designs and methods. Those studies that are based on strong social science methodologies stand out among lay efforts of scholars trained in other science disciplines.

"Gender" research has been almost exclusively on women. Many studies are narrowly conceived and not framed within gender-based frameworks. For example, they tend to use empowerment only in its economic sense, ignore the role of patriarchies or fail to differentiate among feminisms. They also overlook gendered power relations, intersectionality and gendered use of space, among others. Women are assumed to lack power and agency, regardless of social or economic class, and all households are treated similarly, regardless of religion or culture.

Since studies on women dominate, the corollary applies: few studies and actions concern men and masculinities. Yet, masculinities are also highly diverse in the fish sector and male experience of fish sector change is no less gendered than female experience. The rare studies on men cover a spectrum. On one end, studies focus on the patriarchal dividend by which men in fisheries own most of the licenses, power, capital and rights and on how men's roles and power affect women's outlook. For example, Overa (2003) found that male support is critical to the success of powerful women entrepreneurs in three different ethnic communities in Ghana. At the other end are the few studies on the actual experiences of men in fishing, although almost never men in aquaculture.

The 2011 FAO Workshop on Future Directions for Gender in Aquaculture and Fisheries (FAO 2012) exposed differences among experts from academia, applied research and development agencies as to whether gender should be addressed more broadly or specifically focused on women and their oppression.

Women and gender studies can be roughly grouped into those that address: sociocultural issues, development themes and applied development. Sociocultural studies are not intended to inform development action but rather to reveal processes, roles and relationships. Studies by anthropologists and historians delve into the cultural and social practices in fisheries or, more rarely, aquaculture, examining fishing cultures, use of traditional fishing gear, methods and pre-existing resource management practices. The high walls between academic disciplines, however, meant that the valuable knowledge gained from these studies was ignored by fisheries scientists who are typically biologists, oceanographers and mathematicians.

Studies on development and its processes include gendered critiques and analyses of development practices and challenges such as corporate economic globalization driven by laws, policies and programs that are gender blind but not gender neutral. Biswas (2011) critically reviewed how development paradigms that favored market-based solutions shifted the focus of studies on women in fisheries. She found that the studies moved away from those on women's labor in the economy to studies on women's survival and livelihood. Studies using the unclear concept of "community," paradoxically, rarely included women in decisive roles. The shifts redirected activists into weaker mediator roles.

The third category of studies, using an applied development perspective, is the dominant form used by fishery sector researchers. Many of the practitioners are fishery experts from fields such as biology and economics who are interested but not expert in the gender dimension. They apply social science methods, drawing from many disciplines including sociology, geography, economics, extension and marketing. Their underlying premises are typically that women and gender equity can be instrumental in improving fisheries/aquaculture performance or household food security and nutrition. Many of the studies in the Asian Fisheries Society gender/women in fisheries symposia fall into this category. The studies describe gender divisions of labor, women's control of assets and gendered value/supply chain analysis.

A new stream of fishery relevant applied development research is on climate change, its trajectory and impacts. Many fisheries researchers, including gender experts, have "followed the money" and shifted their fields of research to climate change. In climate research, gender/women's vulnerability and agency are now subjects of inquiry.

A more recent movement, exemplified by the emerging gender work of the Consultative Group on Agricultural Research, is the "gender transformative approach," which its supporters advise should be added to the ongoing efforts to achieve gender equity. The transformative approach intends to address systemic barriers and inequalities in society and its institutions.

Building the road to gender equity

In the subject matter of the main fishery networks, gender has not achieved legitimacy and people concerned with gender have not been able to translate this concern into legitimate inclusion. Turning this bleak reality into a new one in which gender and gender equity is a normal and substantive consideration will need new networks—more active gender networks—that take the responsibility to go beyond current more acquiescent efforts in advocacy, research and projects.

Those of us who are involved have been naively expecting a spontaneous awakening of understanding and interest in gender for social justice and practical, instrumental ends, such as the economy and food production. Reflecting on what we have achieved on gender in the Asian Fisheries Society (AFS) gives only a little satisfaction. One achievement is that women/gender symposia and their outputs are now well accepted by the mainstream of the AFS after an interesting start. In 1995 we started with the non-threatening women in fisheries photo competition. In 1998 we held our first symposium. Just before the symposium, the women and men who organized it were subjected to the repeated jocular and mocking question: "can men attend too?" Happily, many men did attend and contribute, and when the event and the proceedings were seen as serious, this question vanished and never returned. A second achievement is that, by the continuity of our events and products, and in creating a small, informal infrastructure for information exchange using social media (see http://genderaquafish.org/), AFS has created a gender focal point for fishery sector researchers, advocates, and the public that is, *de facto*, global.

These achievements are minor, however, and handicapped by the small number of funded projects and minimal research underway, resulting in thin material on which to draw. Most active experts are doing gender on the side of their primary jobs.

Weak human capacity on gender could be one of the greatest constraints to fishery sector progress. The field has only a small number of gender experts, and in fact has lost some of its stars to other, better supported gender domains. In addition, major fishery sector institutions lack gender programs, staff policies and know-how, and the enabling environment lacks policies, funding, training and education facilities.

The bleak reality does, however, contain several important signs of hope.

FAO has started to address gender. In 2011, FAO in its *State of Food and Agriculture: Women in Agriculture – Closing the Gender Gap for Development* report predicted how closing the agriculture productivity gap between women and men would increase world food production. Sector approaches, of necessity, will take an instrumental view of gender. "Gender equality thinking should not focus just on the numbers of women and men in fish supply chains. The overall aim should be how to empower women and men in supply chains to boost overall productivity," said Gifty Anane-Taabeah (Ghana), a young women and the final panelist on "Overcoming Gender Equalities in Fish Supply Chains," at the IIFET 2012 conference.

Important leading fishery institutions, such as FAO, ICSF, the Network of Aquaculture Centres in Asia (NACA), the fishery institutes of the Indian Council for Agricultural Research, and the AquaFish Cooperative Research Support Program (USA supported) have embarked on active gender strategies and programs. Paradoxically, public opinion is starting to exert pressure through major fish marketers over social justice on fishing vessels and in processing factories. Such social issues have been overlooked while the corporations grapple with the complexities of environmental and resource sustainability under the pressure of environmental NGOs.

As they respond to demands for greater gender and social responsibility, fishery sector institutions, public and private, will need to learn "on the job" how to change from being strongly masculine, driven only by the economy, and become human-centered. Textbook solutions are not available but shared learning and practice can accelerate progress.

More gender champions are needed in sector networks. The existing isolated gender in aquaculture and fisheries networks should collaborate to build critical mass, for example, to persuade key sector forums of their need to address gender, and to track and share developments. Although development assistance agencies interested in fisheries are few, their formal or informal networks and relevant projects should also be encouraged to participate. External gender and development networks should be encouraged to support the fishery sector, as happens in Cambodia, but the sector should not rely on this happening.

Conclusion

The fishery sector has undergone rapid economic and structural change in a gender-blind manner, forcing many women and men into marginal roles and undermining the social and economic returns—a veritable revolution. Calls for greater gender equality by gender activists and researchers have received only minor mainstream responses in terms of norms, policies and action. Yet, the drivers of change and inequality persist, creating tough challenges. A gender counterrevolution now is needed, not to return to some notional ideal past, but to counter the gender-blindness of the revolution the fishery sector is pursuing. We have some glimmers of hope for the counterrevolution, but these need to be really ignited and incited to spread.

The time for wishfully thinking that, somehow, the fishery road towards gender equality would be built has gone. If the current core of committed gender actors does not take up the challenge, positive change will not happen by itself. For mobilization and campaigns, the committed actors will need to sign up key sector leaders who understand the issues and carry wider legitimacy.

Because of the nature of the sector, the people in it and their networks are idiosyncratic and heterogeneous. This means that only the people and institutions of the sector can instigate the revolution and find solutions to the gender problems. Although coalitions with actors from other networks—for example, agriculture, environment—may be useful, to achieve legitimacy the fishery sector needs its own gender solutions. Ideas need to be translated into solid action. The global Code of Conduct for Responsible Fisheries needs to show leadership by addressing gender equality.

Institutions cannot deliver gender programs without gender expertise and workplaces that are fair and gender-sensitive. Gender and fish sector experts are scarce and competition for their services is strong. Many recruits will be new and junior staff, entering institutions where their fellow researchers are senior, experienced biotechnical experts. Senior managers will need to deliberately support the new gender experts in such challenging professional settings.

Existing fisheries and aquaculture staff need educating and retraining on gender. For their part, educational institutions—already taking in more and more women in traditionally male fisheries and aquaculture courses—should be preparing all students, regardless of their specializations, with "101" gender courses.

Acknowledgements

First, thanks are due to my late and deeply missed colleague and mentor, Dr. M.C. Nandeesha (1957–2012), who first ignited my abiding interest in women/gender in fisheries. Second, please note that, due to the procedures mandated by the publisher for quoting sources, many individual papers could not be referenced in the text. To meet the limits on numbers of references, individual papers by Ria Fitriana, Holly Hapke, Sun-ae II, Cristina Lim, Ramachandran C. Nair, Marilyn Porter, Sunila Rai, Zuzy Anna and their co-authors were not mentioned explicitly but can be found in Williams et al. (2012) and in searchable information on http://genderaquafish.org/; for example, all the proceedings and reports from the five AFS women/gender symposia, and the 2012 IIFET gender session report by Hillary Egna and myself. Unless otherwise sourced, most statistics are taken from the FAO *State of World Fisheries and Aquaculture 2012* publication. I am also deeply grateful to Janet Momsen for the opportunity to contribute and to Leslie Gray and Anne Coles for their substantial editorial assistance.

Note

1   In this chapter, the term "fishery sector" is used to refer to the whole system for fish production and its movement through the supply chain to the market. The two main branches of the sector in terms of production are: *fisheries* involving the production of fish and other aquatic organisms by capture from natural stocks, and *aquaculture* that covers raising fish through culture in ponds, cages, on rafts and hanging lines in the water.

References

Biswas, N. (2011) "Turning the tide: women's lives in fisheries and the assault of capital," *Economic & Political Weekly* 46(51): 53–60.

FAO (1987) *Women in Aquaculture: Proceedings of the ADCP/NORAD Workshop on Women in Aquaculture, 13–16 April 1987*, Rome: Food and Agriculture Organization.

—— (2012) *Report of FAO Workshop on Future Directions for Gender in Aquaculture and Fisheries Action, Research and Development*, FAO Fisheries and Aquaculture Report. No. 998, Rome: Food and Agriculture Organization.

Harrison, E. (1995) "Fish and feminists," *IDS Bulletin* 26(3): 1–18.

Mowla, R. and Kibria, M.G. (2006) "An integrated approach on gender issues in coastal fisheries projects in Bangladesh: problems and challenges," in P.S. Choo, S.J. Hall and M.J. Williams (eds.) *Global Symposium on Gender and Fisheries*, Penang, Malaysia: WorldFish Center and Asian Fisheries Society, pp. 21–27.

Nayak, N. (2007) *Understanding the Impact of Fisheries Development on Gender Relations in Fisheries: The Importance of Reorienting the Focus of Fisheries Management Strategies Towards a More Life Centered and Gender Just Perspective*, Kerala, India.

Overa, R. (2003) "Gender ideology and manoeuvring space for female fisheries entrepreneurs," *Research Review NS* 19(2): 49–66.

Tindall, C. and Holvoet, K. (2008) "From the lake to the plate: assessing gender vulnerabilities throughout the fisheries chain," *Development* 51: 205–211.

Turgo, N. (2012) "'Bugabug ang dagat' (rough seas): experiencing Foucault's heterotopias in fish trading houses," *Social Science Diliman* 8: 31–62.

Williams, M.J., Porter, M., Choo, P.S., Kusakabe, K., Vuki, V., Gopal, N. and Bondad-Reantaso, M. (eds.) (2012) "Gender in aquaculture and fisheries: moving the agenda forward," *Asian Fisheries Science* 25S: 1–13.

World Bank (2012) *Hidden Harvest: The Global Contribution of Capture Fisheries*, World Bank, Report No. 66469-GLB, Washington, DC: World Bank.

17

GENDER *IN* AND GENDER *AND* MINING

Feminist approaches

Kuntala Lahiri-Dutt

Introduction: explaining 'in' and 'and'

The branch of economic geography that focused on extractive industries has analysed why certain industries develop in certain locations, and theorized about these locations, but gender did not enter into these locational conversations. Gender thus turned into an 'overburden', to use a mining term, which is disposable to the main purpose of mining. This chapter steps away from such avowed gender neutrality and illuminates the extractive industries as gendered places of work at various scales. In addition, it also illuminates the industrial production and processing of the minerals as a gendered labour process having gendered effects on communities. It draws on the growing body of literature that has been built by feminist researchers over the years, literature that had been lying hidden, unexplored and untapped, under the heavy weight of hegemonic masculinist knowledge that was propagated by conventional geography. In this context, it is imperative for feminists to ask two central questions to expand the discourse. The first question is why there exists a need to consider mineral extractive industries such as mining, and the second is through what approaches we should study mining.

My objective is to present the main strands of feminist arguments arising from approaches that are based on both political ecology and political economy. Thus the chapter presents an overview of the 'industrial' studies of labour by feminists coupled with feminist writings on the 'gender selective impacts of mining' emerging primarily from studies in less affluent nations where mining has been expanding rapidly since the 1980s. Indeed, the masculinity of mining as an industry cannot be divorced from the gendered impacts of the industry; a masculine workplace with masculine labour processes are intimately linked to situations where women bear disproportionately heavy burdens of environmental degradation and social disruptions caused by mining activities.[1]

Images of mining that have conventionally been invoked when dealing with extractive industries continue to be masculine, illustrating not just the dirty, dangerous and manual nature of the work as it used to be but also a new kind of masculinity that is associated with power that emanates from the capital that such industries represent. The space constraints of this chapter do not permit an in-depth exploration of these representations. Suffice it to say that it is possible that the equation of modern mining with the 'death of nature' by eco-feminists such as Merchant (1990) came from an association of women and nature, and the obvious analogy of mining violating a sacred mother's womb. Such gendered symbolism about mining can also be detected in the work of anthropologist Mark Taussig (1980) in interpreting the changing belief systems surrounding mining in Latin America. Taussig documented that as Bolivian peasants were transformed into industrial proletariats, their incorporeal mother earth and the benevolent spirit of *pachamama* changed into the idol of an unpredictable male devil that inhabits the underground and who, instead of being worshipped, is feared by all.

Research on mining has explored less of such symbolic domains and more of the *real* or the absolute space that minerals and mining occupy. The thematic areas that have been explored by conventional research included the 'special' nature of mining settlements that defy any spatial modelling, the widespread in-migration into mining areas leading to rapid and widespread urbanization, and the 'place identity' that extensive mining of minerals such as coal creates – whether in mining camps or over extensive regions. As a feminist researcher, one cannot help but observe that each of these areas of research can be gendered. Indeed in recent years, feminist contributions to labour geographies and geographies of economic work have brought to light women's key roles in building the mining industry and constructing coherent mining communities.

This chapter summarizes these feminist contributions under four broad (and somewhat overlapping) heads: mining as a gendered industry or a workplace (or women and men *in* mining); the working-class struggles by mining workers as gendered, which studies the gender interstices of home and work in mining communities; and mining as a global/national agent of capital accumulation and dispossession of the poor in the contemporary world. At the end of the chapter, I offer some possible directions for future research on mining.

Women as economic agents: mining as a gendered industry

Influential authors such as Lewis Mumford (1934, 1967), who were familiar with the extractive industries that came into existence in Europe and America after the industrial revolution, saw these industries as an ultimate expression of modernity – a metaphor of the modern world – representing capitalist exploitation of labour and the gross commodification of nature by the hegemonic corporate enterprise. The extraction of mineral commodities was instrumental to capitalist exploitation as coal and iron ore formed the backbone of modern industries that built themselves on the manual labour performed in the mines (and factories). Humphries (1990) offers a Marxist explanation, pointing out that the rural poor in late eighteenth- and early nineteenth-century England enjoyed non-legal but insuperable customary and common rights to local agricultural resources. The loss of these resources through enclosures changed women's economic positions adversely within the family as their dependence on wages (and wage earners) increased.

There is no doubt that women have been in the mines with men, as part of a family labour unit, from early times. In this regard, there was hardly any difference between peasant production strategies in farming and mining. Early treatises such as *De Re Metallica* (1556) portrayed women in ancient mines as taking on the tasks of breaking and sorting of ores, hauling and transporting them, participating in the smelting and processing activities, and sometimes even undertaking the physically demanding job of working the windlasses. Describing early mining in Europe, Vanja (1993) outlines women's contributions in preindustrial mining of an artisanal nature in Europe, where women worked in small pits and smelted ore.

The gold rushes that mark the history of mining were also gendered; women not only accompanied male prospectors into the new gold fields, providing a range of services, but also prospected on their own. In the American West, Zanjani (2006) shows that many women prospected for gold during major phases of gold rushes. When similar rushes took place in the colonies, women were not far behind. Similar stories of women gold prospectors can be found in Australia or in the Brazilian Amazon, where some women accumulated significant wealth as prospectors as well as service providers.

Interestingly, early modern or capitalist industrial mines also saw the participation of women in large numbers. In the mines in Belgium until the early twentieth century,

> numbers of women working underground in Belgian coal mines actually grew . . . [and] Belgium's women coal-miners earned some significant portion of the public respect and reverence elsewhere given so readily to male coal-workers.
> *(Hilden 1993: 110–111)*

Women coal mine workers came to be known as *hiercheuse*, a proud title connoting the feminine version of *mineurs*, the male miners. Hilden further goes on to narrate that while the male miners wore hard leather helmets, trousered suits and wooden clogs, the *hiercheuses* were dressed in white linen suits of knee-length trousers and a buttoned, long-sleeved jacket. Groups of women worked in shifts, going down the mine at 5am, and returning at about 9pm, loading between 60–70 carts during these hours.

One might expect that European colonial powers would also hire women when they were establishing mines in the colonies, but that was not the case. In contrast, early Indian colliers initially employed women as part of a 'family labour' system that kept the workers, often drawn from local tribal communities, tied to the modern industrial mines (Lahiri-Dutt 2012b). With respect to South African collieries, Alexander (2007) has shown that the colonizers hired male black men and kept them in barracks strictly separate from those of the white workers. The job division in the pits meant that *kamins* (female workers) performed quite different tasks from those performed by the *coolies* (male workers) in India (Lahiri-Dutt 2007). The *kamins* worked not only as the 'gin girls' (from the term 'engine') to lift coal from the shafts, but also as loaders of coal (cut usually by their male partners – father, brother or husband) in both surface and shallow underground or open cast mines that were locally known as *pukuriya khads*. A similar pattern of sex-based labour division existed elsewhere in modernizing Asia as well; Nakamura (1994) describes the *naya* ('stable') system of work in Japanese coal mines:

> A working pattern in which a married couple worked as a unit, with the husband (*sakiyama*) digging out the ore and the wife (*atoyama*) assisting him by carrying away the coal, became widespread [during the early twentieth century]. Married women comprised most of the female workforce in the coal mining industry.
> *Nakamura (1994: 15–16)*

The *naya* system was an indirect labour management system with the *naya* chief as an agent, who recruited the miners, and then made them live in the bunk-houses he provided, and supervised their daily lives. Monies borrowed from him kept the miners in bondage.

However, in spite of this long and impressive history, the issue of women's labour in the mines remains an area fraught with controversies owing to four main reasons: sex-segregation of jobs that pushed women into relatively lower-status jobs compared to those performed by men; the marking of tasks as male and female, which created spatial segregation *within* the mine or even the pit; the prevalence of a family wage that meant women hardly ever received the full recognition for their component of labour; and the various protective legislations, largely initiated by the International Labour Office during the 1920s and 1930s, around women's labour in mining. Women's labour in mines within a strict sex-based division of tasks was (and still remains) subject to gender ideologies – situated in the home as well as those propagated by the state.

The relationship between gender ideologies and sex-based division of labour has meant that women's labour as part of the family remains largely invisible. John (1980) perceives that since the male members utilized the labour of their female relatives, those women who worked as 'pit-brow lasses', women who worked at the brow of coal pits in early modern collieries in England, were not usually recorded in colliery accounts. The maintenance of masculine dominance in the household also meant that the division of able-bodied and the disabled was not sharply drawn and the male was presented as the head even when physically ill because the central concern to mining families was to ensure the physical wellbeing and continued wage-earning capacity of male breadwinners (Forestell 2006).

The relationship between state gender ideologies and economic imperatives meant that women were encouraged to join the mining industry at the convenience of the state. For example, Ilič (1996) has documented that despite the fact that the Soviet Labour Code included a ban on the employment of female labour in underground work, women continued to work extensively and in increasing numbers throughout the interwar period in response to the need for additional labour.

These factors made women's labour invisible. Clearly, official employment records do not show the full extent of women's participation in mining. When government officials visited an asbestos mine in South Africa in 1950, they found the records showed that the mine employed 100 males but made no mention of women. A month later, a health inspector, also from the government, visited the same mine and noted that there were 102 male employees and 40 women (McCulloch 2003). Such discrepancies occurred because of women's invisibility as workers, but not necessarily as individuals with health-related problems.

In retaliation for the invisibility of women mineworkers in government records, feminist labour historians have resorted to alternative sources of documentation such as oral histories. For example, Gier and Mercier (2006) and Tallichet (2006) have used oral histories of women mineworkers in the Midwestern regions of the USA, and in central Appalachia respectively in order to unearth their hidden histories.

Women as political agents: active in mining struggles

It is important to note that women are not only workers in the mining industry. They are also political agents who are active in mining struggles. Throughout the history of modern industrial mining development, they have fought (and continue to do so) against exploitation, either standing side-by-side with men or in the wings by providing critical support and logistical assistance. Their efforts have helped in the maintenance of solidarity as part of the working class. Mining and gender literature also uncovers the histories of women who helped the male (and female) workers to protect both their families and their collective interests. The archetypal images of women protesting against capitalist exploitation in mines are contained in Emil Zola's 1885 masterpiece *Germinal*, set in the context of the coalminers' strike in northern France in 1860 and in particular, that of the inspirational title character in Maxim Gorky's 1906 novel *Mother* about Russian factory workers. It is important to remember, however, that the struggles of industrial mineworkers that such images represent are somewhat different from the contemporary struggles portrayed in the social movements and resource conflicts literature. I will address this later in the chapter.

While women supported workers' struggles against capitalist exploitation and mine closure, and even confronted management directly over issues such as food and housing, they also adopted an impressive number of strategies to ensure their own position. Their strategies often pitted gender against gender and even occasionally transcended class lines. Women activists, even though working for auxiliaries, sometimes acquired considerable power by refashioning the rhetoric surrounding motherhood and through the deployment of strategies to organize themselves (Waldron 2006). Another strategy involved closer self-observation of women and their relationship with the trade union activists.

The crucial roles women played in miners' strikes were invaluable, as they not only supported men's struggles, but also took up activist roles themselves to be the 'voice' for their own concerns. Describing the more recent strikes in the Bowen Basin in Australia in the 1980s or the dispute with the Australian mining company BHP in 2001, Murray and Peetz (2011: 288) comment that 'without this support [that women gave to striking miners and to each other] it is questionable whether any of the actions would have succeeded'.

Similar stories have been documented of miners' strikes in other countries. The key areas that emerged include not just the unpaid work by wives and the economic support they provided to families through the crisis by adopting diverse economic activities, but also their active political roles as protesters representing class interests.

The supportive roles played by women in mining struggles are legendary. However, some feminists have critiqued this view based on the conflicts in class and gender identities that never make women an integral part of mining unions anywhere in the world. Indeed, the politics of socialism, in which miners have historically played a central part, tends to generalize and present a fictitious character of 'the miner's wife' who supports men's struggles in solidarity. Trade unions were most often insensitive to the specific needs of women as workers and held working-class interests as representing the interests of all workers, irrespective of their gender. Moreover, the ideal of dual unionism (separate unions for female and male workers) meant that women's unions were always lower in status, poorly staffed and had a largely ineffective voice. Even in India's coal mining industry, which spearheaded struggles against the exploitation of working classes in India, male camaraderie contributed to the marginalization of women workers in coalmines (Lahiri-Dutt 2012a).

Mining corporations are aware of this social need of mining communities to conform to gender ideologies and, being gendered organizations, they exploit this social need and themselves push women into domestic spaces. In the past, in mining towns of the northern Rhodesian copper belt, mining corporations encouraged more mineworkers to live with their wives at home to enhance the 'stability' of the labour force, while women wrestled with their 'male protectors' for access to family incomes (Parpart 1986). Although a woman's potential prosperity largely depended on her husband's wages, higher wages alone did not guarantee a woman's financial position. In contemporary times, they take up other strategies such as setting up women's clubs to create and reinforce a strong occupational identity that overall emphasizes women purely as reproductive agents by placing them as part of the home and the community.

Women as reproductive agents: mining wives building up the community and family

Rhodes (2006) lived in mining camps as the wife of a mining engineer. Based on her personal experiences, she shows how unpaid labour by wives – at home and in the community – helps sustain a flourishing social life around mines. Her observations are supported by the earlier observations of Robinson (1986) who documented that the managers' wives in a mining town in Indonesia were expected to take on leadership and welfare roles in the community through involvement in the Association of Inco Families, an organization in which their position paralleled that of their husbands. In other words, they were constructed as an auxiliary branch of the capitalist enterprise of the mining company. Broadly speaking, this illustrates the meaning of *mining culture*, a culture that attributes domesticity to miners' wives by socially constructing and locating women within the home in mining communities. These women can then be described as 'the hewers of cakes and drawers of tea', and relegated to their place at home for men to gather in union halls (or local pubs) in order to form their class-solidarity (Gibson-Graham 1994).

Mining settlements are not organic and can be seen as 'remote' and a kind of exception to ordinary settlements (Pattenden 2005). More importantly, the communities who inhabit these settlements comprise of new migrants, building up a 'new' community, whose very existence is ruled by the state, the company and labour organizations – all male-dominated institutions. Klubock's (1998) study of a copper mine community in Chile shows how transnational mining capital and the state ignore the complexities of gender within the community in monitoring men's and women's roles and behaviours, thus engineering a shift to the male head-of-household model. Anthropologist June Nash (1979:12–13) put women in Bolivian mining communities within the context of the home as the wife of a male miner, subjected to the limitations of the house, to the dominance of the man whose needs she must dedicate herself to and to almost continuous childbearing: 'Male and female roles are dichotomized in the mining community, and there is still a mystique about women not entering the mine.' Women in mining communities are therefore seen as belonging to the working class *because* of their men, the 'male contoured social landscape' burgeoning with tacit as well as overt support from the company.

This statement continues to be true broadly of all contemporary mining settlements. Not just the larger *company town* but also the camps meant for mineworkers living closer to the pits are highly structured along class hierarchies. In these settlements, company hierarchies are reproduced within and between social spheres to provide an informal instrument of subjugation for those occupying a lower status at work. Managers' wives have relatively higher social and economic status than the wives of staff members and much more so than the local or indigenous women, who, in turn, then develop a heightened awareness of class and wealth. Also, in multinational operations, race too can play an important role in segregating places according to the complex intersection of gender, race and class (Lahiri-Dutt 2013).

Undoubtedly, the mining communities are where everyday lives are performed and gender is constructed and enacted through the processes of daily production and consumption. Feminist geographical interest emerges from the fact that gendered people from these communities occupy the interstices of home and work. The contrasting female identities – of the domestic woman versus the political woman (or the economic woman) – become a key force in reinforcing normative gender roles by ascribing certain tasks to the individuals from each sex. The majority of domestic women shoulder onerous maternal and domestic burdens of unpaid domestic and emotional work, and accept male subordination to fit the picture. This is fast becoming the norm. To what extent women have a degree of autonomy in producing their identities is still unclear (Hall 2001).

Women as victims: dispossessed by mining

In contrast to the historical literature, contemporary feminist studies have primarily used the lens of political ecology (see chapter by Elmhirst in this volume) to analyse the literature on mining. The reason for this may possibly be attributed to the fact that, since the 1980s, most new large mining projects have been established in less affluent countries. It is also possible that this selective focus is reminiscent of the bias expressed by Mumford and Merchant that I referred to earlier in this chapter.

Collectively, this genre of research has emphasized that the introduction of large-scale mining affects women disproportionately. The impacts on women are more severe than that on men. These impacts range from the erosion of the physical subsistence base to social and cultural changes such as notions of authority and interpersonal power equations both at home and within the communities. Women are affected both by lack of access to assets and resources, as well as increased cash flowing into local economies, which predominantly finds itself into the hands of men. The gendered impacts often cut across class and race, but women (and men) who are already at a disadvantaged position within the socioeconomic hierarchy are more adversely affected by large-scale mining than their more privileged counterparts. Again, the gender-selective impacts have been noted both in more economically developed countries such as Canada as well as in less developed countries with smaller economies. For example, Hipwell et al. (2002) noted three broad categories of gendered impacts of mining on the indigenous populations in Canada – those affecting women's health and wellbeing, those affecting women's work and traditional roles, and those affecting the (unequal) distribution of the economic benefits from mining. Equally important to their discussion were the loss of traditional autonomy and the changes in the productive roles of women.

In less affluent countries, the family unit's food security primarily burdened already fast depleting subsistence bases combined with environmental degradation (Ahmad and Lahiri-Dutt 2007). Degradation in the physical environment in mining areas occurs due to the contamination of air, dust and water from activities related to mineral extraction and processing. Positioning the problem in the environmental justice paradigm, Bose (2004) observes that the alienation or loss of productive agricultural lands, reduced access to water sources, the loss of livelihood opportunities and the loss of forest cover negatively affect women more than men. Bhanumathi (2002) also observed the decreased ability of women to work on remaining land as men tend to migrate out of villages to cities in search of cash incomes.

In her early ethnographic work on the political economy of development in a mining town in Indonesia, Robinson (1986:12) observes that 'the fundamental change in Soroako has been the loss of the village's most productive agricultural land to make way for the mining project'. Therefore, the loss of productive land rapidly transforms a large proportion of male villagers dependent on wage labour for the company reducing them to a semi-proletariat. Generally, women tend to become less active economically due to a sudden change in the production systems, relations and spatial orientations according to Rothermund (1994). Consequently, women sometimes resort to a variety of activities in the informal sector, or become more home-bound than before. As cultural expressions of gender change, both women's autonomy and personal freedom that they may have enjoyed earlier either become restricted or decline altogether. As the community changes from peasant agriculture to wage labour for their subsistence, women generally experience a decline in their economic independence.

Mining projects are highly capitalized; their heavy equipment, offices, roads and other infrastructure, salaried workers and their residences introduce a new cash-based economy into a rural area. Often these cash incomes actually play significant roles in reducing women's autonomy. A variety of social and cultural impacts occur when large numbers of migrants disrupt the social fabric of a mining area. The loss or changes to traditional culture removes some of those older cultural norms that might have attributed some authority and power to women. Thus, the loss of older values often more adversely affects women than men; the effect is particularly evident in the devaluation of women's productive work at home and by undermining their status as decision-makers and landowners. In the Pacific islands, a sudden influx of mining revenues within local communities generally marginalizes women. In general, the monetized economy introduces a different 'mining culture'. This external culture is reinforced by the mining company that attributes new notions of authority to men, putting women either in lower status jobs or rendering them invisible by involving men in decision-making processes relating to issues of compensation and agreement-making. When a mining corporation negotiates with the community, the latter is most often represented by traditional leaders who are usually male elders. This happens because, being gendered organizations themselves, the mining corporation personnel are usually also male who carry with them false assumptions about who might be the 'head of the family', or how household resources are allocated among the members. Usually, the land belongs to men and, more often than not, men receive monetary compensation for the loss of land, making it difficult for women who have little formal political authority to be able to influence how the mine would shape their autonomy in decision-making and, in a broader perspective, their lives.

One interesting dimension in determining the winners and losers of the mining projects among women is their geographical distance from the mining operation. In general, physical proximity to the mines leads to the direct experiencing of noise and vibrations, and the visibility of gigantic machines arouses fear and a sense of insecurity creating a heightened sense of negative impacts for women according to Lahiri-Dutt and Mahy (2007).

Mining's introduction of a cash-based economy also affects women indirectly. The extra cash earned by men is spent on sexual promiscuity, on pubs, karaoke bars and brothels that pop up overnight in the most remote places. Lahiri-Dutt and Mahy (2007) observe that, generally speaking, the negative gender impacts of mining are related to the shifting power equations within communities and families. These shifts can be specific to a particular person or groups, and they also include the increased cost of living, the lack of direct employment opportunities in the mine for women, environmental impacts and women's lack of decision-making power at the community level. All these translate into a dependency on male relatives.

Reflecting on the literature that relates to the impacts of mining on women, one observes that it has reaffirmed the role of women as victims of multinational capital that destroys the environment and adversely affects the lives of the poor. Of the several victim figures, the most convincing and lasting one has been that of 'the prostitute' who has been turned into one by mining projects and who makes a living selling sex around the mine site. Again, as a metaphor, this has been interpreted as equal to the vandalization of nature by mining, thereby causing the degradation of women by degrading the land. Indeed, the attitudes towards both sexuality and towards women change with the rapid influx of money and new social-cultural values (Emberson-Bain, 1994). For example, while many societies in Papua New Guinea incorporated long periods of male sexual abstinence, there is evidence that in mining towns this is being eroded. MacIntyre (2002) notes that mining communities report a growing incidence of alcoholism, rape and other forms of violence against women and an increasing incidence of teenage pregnancy. However, the theoretical positions of much of this evidence have been questioned in recent years (Lahiri-Dutt 2012b).

This discussion is the derivative of the highly contested literature on women, environment and development (WED), which emphasizes the affinity of women with their environments. This scenario leads to the danger of depoliticizing the environmental and community politics in mining and creates a dualism between women and men with separate spheres and spaces of production.

Conclusion

As can be seen, I have depended very heavily on historical material in writing this chapter. Tentatively, I can only speculate and offer some broad and general views for this paucity in the creation of new knowledge on gendered perspectives on mining and vice versa. In my view, one of the reasons lies in how most people see mining: quintessentially as the 'other', as a human endeavour that is 'not natural' and one that competes with farming for land; that is physically remote and scattered; that is associated with chance or luck that does not yield itself easily to modelling; and finally, one that represents a 'special' kind of human project in its disregard for preserving nature/the environment.

From my position at the margin of margins, I can only indicate the possible directions of future research on gender *in* and gender *and* mining. The body of knowledge that developed in the context of modern industrial mining remains invaluable and will remain so, but tomorrow's feminists would do well to engage more with four main sets of theories: postcolonial feminist perspectives that critically reflects on power relations (see Chapter 5), intersectionality, feminist political ecology (see chapter by Elmhirst in this volume), and related but not quite the same, gender and development (GAD) theories. They would explore how mining impacts not just on all women as a homogenous group but how gender creates advantages as well as disadvantages selectively. They would explore deeply how gender can be integrated within the mining 'project cycle', that is, from the stage of exploration to that of closure and – by engaging with the mining communities – companies and policymakers would offer new solutions to the continued invisibility of women and gender. They would explore what race and age means for women and men in mining camps and towns, thereby focusing on how gendered bodies of individual men and women perform certain kinds of work, how masculinity (and femininity) of the enterprise is transmitted into the communities, and the gendered social lives within these communities, as well as the implications of the intricate sexually-based division of labour within mining organizations.

Acknowledgement

I prepared this chapter during my stay as a Senior Visiting Fellow at the Asia Research Institute (ARI) of the National University in Singapore (NUS) I thank both ARI and NUS for providing me with an engaging academic environment that immensely facilitated this writing project.

Note

1   A caveat is better mentioned up front: this paper does not explore informal mineral resource extraction that has been a central plank of my recent feminist critique of mining (Lahiri-Dutt 2012a).

References

Ahmad, N. and Lahiri-Dutt, K. (2007) 'Engendering mining communities: examining the missing gender concerns in coal mining displacement and rehabilitation in India', *Gender, Technology and Development* 10(3): 313–339.

Alexander, P. (2007) 'Women and coal mining in India and South·Africa', *African Studies* 66: 201–222.

Bhanumathi, K. (2002) 'Mines, minerals and PEOPLE, India', in I. MacDonald and C. Rowland (eds.), *Tunnel Vision: Women, Mining and Communities*, Fitzroy: Oxfam Community Aid Abroad.

Bose, S. (2004) 'Positioning women within the environmental justice framework: a case from the mining sector', *Gender, Technology and Development* 8(3): 407–412.

Emberson-Bain, A. (1994) 'De-romancing the stones: gender, environment and mining in the Pacific', in A. Emberson-Bain (ed.), *Sustainable Development of Malignant Growth? Perspectives of Pacific Island Women*, Suva: Marama Publications.

Forestell, N. (2006) '"And I feel like I'm dying from mining for gold": disability, gender, and the mining community, 1920–50', *Labor Studies in Working-Class History of the Americas* 3(3): 77–93.

Gibson-Graham, J.K. (1994) '"Stuffed if I know!" Reflections on post-modern feminist social research', *Gender, Place and Culture* 1(2): 205–224.

Gier, J. and Mercier, L. (eds.) (2006) *Mining Women: Gender in the Development of a Global Industry, 1670–2005*, New York: Palgrave Macmillan.

Hall, V.G. (2001) 'Contrasting female identities: women in coal mining communities in Northumberland, England, 1900–1939', *Journal of Women's History* 13(2): 107–131.

Hilden, P.P. (1993) *Women, Work and Politics, Belgium, 1830–1914*, Oxford: Clarendon Press.

Hipwell, W., Mamen, K., Weitzner, V. and Whiteman, G. (2002) *Aboriginal People and Mining in Canada: Consultation, Participation and Prospects for Change*, Working Discussion Paper, North-South Institute.

Humphries, J. (1990) 'Enclosures, common rights, and women: the proletarianization of families in the late eighteenth and early nineteenth centuries', *The Journal of Economic History* 1(3): 17–42.

Ilič, M. (1996) 'Women workers in the Soviet mining industry: a case study of labour protection', *Europe-Asia Studies*, 48(8): 1387–1401.

John, A. (1980) *By the Sweat of their Brow: Women Workers at Victorian Coal Mines*, London: Croom Helm.

Klubock, T.M. (1998) *Contested Communities: Class, Gender and Politics in Chile's El Teniente Copper Mine, 1904–1951*, Durham: Duke University Press.

Lahiri-Dutt, K. (2007) 'Roles and status of women in extractive industries in India: making a place for gender sensitive mining development', *Social Change* 37(4): 73–64.

—— (2012a) 'The shifting gender of coal: feminist musings on women's work in Indian collieries', *South Asia: Journal of the South Asian Studies Association* 35(2): 456–476.

—— (2012b) 'Digging women: towards a new agenda for feminist critiques of mining', *Gender, Place and Culture: A Journal of Feminist Geography* 19(2): 193–212.

—— (2013) 'Gender (plays) in Tanjung Bara Mining Camp in Eastern Kalimantan, Indonesia', *Gender, Place and Culture*, https://crawford.anu.edu.au/pdf/staff/rmap/lahiridutt/2012/gender-plays.pdf.

Lahiri-Dutt, K. and Mahy, P. (2007) *Impacts of Mining on Women and Youth in Two Locations in East Kalimantan, Indonesia*, report available from http://empoweringcommunities.anu.edu.au.

MacIntyre, M. (2002) 'Women and mining projects in Papua New Guinea: problems of construction, representation and women's rights as citizens', in I. Macdonald and C. Rowland (eds.), *Tunnel Vision: Women, Mining and Commuters: An Anthology*, Melbourne: Oxfam Australia, pp. 26–29.

McCulloch, J. (2003) 'Women mining asbestos in South Africa', *Journal of Southern African Studies* 29(2): 413–432.

Merchant, C. (1990) *The Death of Nature: Women, Ecology and Scientific Revolution*, San Francisco: Harper.

Mumford, L. (1934) *Techniques and Civilisation*, New York: Harcourt Brace and Company

—— (1967) *The Myth of the Machine, Vol 2: The Pentagon of Power*, New York: Harcourt Brace Jovanovich.

Murray, G. and Peetz, D. (2011) *Women of the Coal Rushes*, Sydney: University of New South Wales Press.

Nakamura, M. (1994) *Technology Change and Female Labour: Manufacturing Industries of Japan*, Tokyo: United Nations University Press.

Nash, J. (1979) *We Eat the Mines and the Mines Eat Us: Dependency and Exploitation in Bolivian Tin Mines*, New York: Columbia University Press.

Parpart, J.L. (1986) 'Class and gender on the copperbelt: women in northern Rhodesian copper mining communities, 1926–64', in C. Robertson and I. Berger (eds.), *Women and Class in Africa*, New York and London: Africana Publishing Company, pp. 141–160.

Pattenden, C. (2005) *Shifting Sands: Transience, Mobility and the Politics of Community in a Remote Mining Town*, unpublished PhD thesis, School of Social and Cultural Studies, University of Western Australia.

Rhodes, L. (2006) *Two for the Price of One: The Lives of Mining Wives*, Perth: Curtin University Press.

Robinson, K. (1986) *Stepchildren of Progress: The Political Economy of Development in an Indonesian Mining Town*, Albany: State University of New York Press.

Rothermund, I. (1994) 'Women in a coal mining area', *Indian Journal of Social Science* 7(3–4): 251–264.

Tallichet, S.E. (2006) *Daughters of the Mountain: Women Coal Miners in Central Appalachia*, Pennsylvania: Pennsylvania State University Press; Urbana and Chicago: University of Illinois Press.

Taussig, M.T. (1980) *The Devil and Commodity Fetishism in South America*, Chapel Hill, NC: University of North Carolina Press,.

Vanja, C. (1993) 'Mining women in early modern European society', in T.M. Safley and L.N. Rosenband (eds.), *The Workplace Before the Factory: Artisans and Proletarians, 1500–1800*, Ithaca and London: Cornell University Press, pp. 100–117.

Waldron, C.M. 2006. '"We were not ladies": Gender, class and a women's auxiliary's battle for mining unionism', *Journal of Women's History*, 18(2): 63–94.

Zanjani, S. (2006) *A Mine of her Own: Women Prospectors in the American West, 1850–1950*, Lincoln and London: University of Nebraska Press.

18

JUST PICKING UP STONES

Gender and technology in
a small-scale gold mining site

Hannelore Verbrugge and Steven Van Wolputte

Introduction[1]

In May 2013, the Australian-South African International Mining for Development Centre (IM4DC) organized an African Women in Mining and Development study tour.[2] This initiative was not the first nor the only one of its kind, and the past decade witnessed a growing interest in the role of women in large-scale mining (LSM) and, especially, artisanal and small-scale mining (ASM) in the world's mining hotspots, such as South Africa, Australia, Ghana or Papua New Guinea. This suggests that international development initiatives only recently noticed the important role women play in mining.

The reason for this may be the heroic image of muscled miners armed with headlights and pneumatic drills, covered in sweat and dust, which dominates the Western imagination when it comes to mining (see chapter by Lahiri-Dutt in this volume). This image is, no doubt, a recent product. Reforms in nineteenth-century industrializing Europe were geared towards domesticating, in every literal sense, working-class women active in mining and in heavy industry, such as textile mills. Yet, it partly explains why there was no place for female miners in popular representations and in development initiatives until recently, except in their derived capacities as worrying mothers, wives or girlfriends, or as "poor, powerless, and invariably pregnant, burdened with lots of children, or carrying one load or another on her back or head" (Win 2007: 79).

This skewed image of mining as a predominantly male activity does not hold true today nor did it earlier. Throughout the world, women are, and always have been, involved in digging up precious minerals and other materials. Yet, as the Women in Mining Conference, held in Papua New Guinea in July 2013 observed, "benefits tend to be captured by men".[3] This also means that women's role in and around the mine, and their agency, is often misunderstood and ill-represented (see Bryceson et al. 2013; Lahiri-Dutt, this volume).

Another misconception is that mining in the Global South is something of the recent past. If we restrict ourselves to Africa, we can safely assume that, for instance, iron ore was mined as early as 1,000 BC. But gold was mined as well. African gold, for instance, fuelled the economies of Ancient Egypt, Greece and Rome. Likewise, the invention of banking in thirteenth-century Italy was to an important extent sparked by the gold imported from present-day Ghana (the former Gold Coast) via trade routes across the Sahara, thus predating the Portuguese "discovery" of West Africa by more than a century.

A third preliminary remark is that, in general, "mining" is but a generic term for an activity that might vary considerably depending on the kind of mineral or deposit, the form it is found in, the geomorphological context, the available infrastructure, the availability of (cheap) labour, the technology one has access to, and so on. Mining bulk minerals such as iron ore, copper or bauxite requires a different approach than excavating diamonds, silver or gold, whose value is measured not in tons, but in carats or milligrams. Bulk minerals are often processed on the spot, as transport is expensive, whereas precious stones and nuggets are cheaper and easier to transport, trade or smuggle. Excavation can take place through shafts, or in open mines; sometimes, underground mining requires the use of explosives, sometimes the mineral (such as uranium) demands that special safety precautions are observed, and sometimes the extraction technologies (such as, in the case of gold, the use of cyanide and mercury) imply a health and environmental risk. These material aspects should be considered in their relation to – local and (inter)national – structures of power, histories (for instance, of colonization), existing expectations with regard to land, gender and authority, worldviews, the economic situation, consumption patterns or labour and tax legislation – and then especially how people deal with it: knowing how to circumvent or bend the law also constitutes a "technology". All these elements have their impact on the minute particulars of mining sites across the globe, on the kinds of risks involved, the division of labour, the need and use made out of specialized technology, the need for capital, kinds of labour, and the like (see Ali 2006; Kumah 2006).

Given this complexity it is all but impossible to provide an outline of women's roles in mining around the world, or even in, say, Africa. A first reason is that the particularities of minerals and ores, of technologies, of labour and gender laws (and how they are dealt with), of cultural and political factors, and of geomorphology get in the way of a bird's eye overview. So, for instance, in northern Ghana, women make up 50 per cent of the mining workforce; their job mainly consists of loading trucks with the rocks and with ore crushed by the men. They receive a daily wage (US$1.66–3.33), with work being predominantly a family, not an individual affair. In neighbouring Burkina Faso, women are more involved in the actual mining process, even though it remains unknown if they also reap the benefits from this greater participation (Yakovleva 2007). In South Africa, however, women are historically far less involved in mining because of labour laws forbidding women to work underground, even if this legislation has been altered by the post-apartheid government.

A second reason is that also scholarship in the social sciences is skewed by "mining's gender bias".[4] Following Ferguson's (1999) seminal work on the expectations of modernity on the Zambian copper belt, authors have focused on the push and pull factors of labour migration under colonial rule, and then especially on the exploitation of (male) labour under colonialism. Scholars concentrating on the present direct their attention towards mining in the unruly borderlands of the postcolonial state, towards the ambivalent relationship between mining and "occult economies", or towards mining and the forging of contemporary masculinities. Obviously, these publications are not without merit. However, most have in common that they focus almost exclusively on male miners, middlemen and traders, and on men as main agents of consumption and modernity. In reaction, recent scholarship has started to also include women in the picture (see Bryceson et al. 2014; Hinton et al. 2003).

Therefore, this chapter presents an anthropological case study that focuses on an artisanal and small-scale gold mine in Makongolosi, a small mining town in Mbeya Region, in southwest Tanzania. Its main focus is on how women miners conduct their daily lives in and around the mine, and on how technologies introduced over the past decade or so not only impact upon their day-to-day activities, but also affect gender relations in and around the mine. These changes, however, must be set against the background of relative recent reforms inspired by the liberalization of the mining sector throughout Africa.

The remainder of this chapter is divided into three parts. First we sketch the context of small-scale gold mining in Tanzania and Makongolosi. In this, we consider gold – despite its particularities – as representative for the entire mining sector in Tanzania (and Africa): often the gold mining sector has proven to be a frontrunner in the neoliberal race for minerals and resources (even if in July 2013, its price on the world market dropped considerably). Second, we turn to an ethnography of female miners there, focusing especially on their daily lives in and around town. In the third and final section, we turn to a brief discussion of our findings, and their implications for women's development.

Case study: context

> With the discovery of dry-blowing, alluvial miners moved from the creeks to gold-bearing ridges and new fields opened up, the most famous being Makongolosi and Golding Hill. In 1933, 10,000 Africans were employed at the Lupa goldfields. The miners were very young and poor. There were problems with water, housing and a high spread of venereal diseases occurred.
>
> *(Dar es Salaam National Archives, 2012: letter from*
> *J.F. Millard, 27 July 1938)*

Gold can be produced either through large-scale mining (LSM) operations, or through artisanal and small-scale mining activities (ASM). The former refer to "major industrialized practices involving substantial capital investments" (Jønsson 2009: 9). These mining methods are based on modern capital-intensive technology and thus require a limited workforce. Currently, LSM activities in Africa are responsible for about one-third of the world's annual gold production.

ASM activities, in contrast, range from "poverty-driven, informal and seasonal activities carried out by individuals or groups of people using pickaxes and hammers and chisels to reach a few meters below the surface to formal operations with several people involved using mechanical drilling and hoisting to reach more than a hundred meters below the surface" (Jønsson 2009: 9). Large-scale and small-scale mining operations, however, are not strictly separated. They, for instance, have particular markets in common. Also, miners quite easily move from formal employment to self-employment and back (see Bryceson et al. 2014). The fact that ASM activities are usually situated in the informal or shadow economy and that over the past decade they have been increasingly branded as "illegal", however, makes it difficult to estimate their importance in numerical terms.

Makongolosi is located 100 kilometres south of the regional capital Mbeya and has been a gold mining settlement ever since 1932, when British companies started to excavate the Lupa goldfields in southwest Tanzania and used Makongolosi as a processing site. An old storage tank and broken-down water basin some five kilometres from the current centre testify to this. During this period (1934–1938) job opportunities within the Rungwe and North Nyasa districts (present-day Mbeya and Songwe) were very limited. Therefore, many Nyakyusa men and women came to the Lupa goldfields to find employment.

The Second World War, however, brought about a decline in gold production. This led to the abandonment of many artisanal mining operations, and throughout the 1950s production kept falling sharply (Lissu 2001). In the course of the 1980s, however, following the liberalization of the mining sector in Tanzania – itself an outcome of the Economic Recovery

Programs and Structural Adjustment Programs (SAPs) imposed by the International Monetary Fund and World Bank – Makongolosi experienced a second gold rush. As liberalization implied shutting down the state-owned mines in Tanzania (in 1986, some 50,000 government employees in the sector were retrenched), many semi-skilled workers tried their luck in the artisanal or informal mining industry, causing the exponential growth of the latter (Drechsler 2001; Lissu 2001), catapulting Tanzania to the fourth place on the list of gold-producing countries in Africa, behind South Africa, Ghana and Mali. It ranks sixteenth in the world, with gold accounting for 90 per cent of Tanzania's mineral exports and constitutes the country's main export. Tanzania thus exemplifies the boom Africa's mining industry has experienced over the past 20 years.

As causes of this mineral boom we can, first, refer to the rising prices of old and new mineral resources and their growing strategic importance, making mining profitable even in more challenging circumstances. Gold reserves are indeed depleting: as Ali (2006: 456–457) observes, 80,000 tons are circulating above the surface, with only an estimated 50,000 tons remaining to be mined.[5] Second, throughout the continent various initiatives have been taken to create a more "investor-friendly climate", including privatizations and the implementation of new mining laws as a consequence of, among others, the SAPs imposed by the World Bank. A third reason is that Africa's mining industry profited from the introduction of new technologies, making mining more cost-effective, but also more dependent on capital investments.

A first difference with the first gold rush in the 1930s, however, is that at present the percentage of women involved in mining activities is much higher, constituting nowadays about half of the estimated 800,000 inhabitants in Tanzania's ASM towns. According to the 2012 Population and Housing Census, Makongolosi has 33,406 inhabitants. Women constitute 49.6 per cent of the mining population, men 50.4 per cent. As gender proportions have levelled out, Makongolosi should no longer be considered as a rush site (like it was in the 1930s and 1980s), but as a mature mining town. According to the most recent census, migration to Makongolosi has increased from 3.7 per cent in 2002 to 9.6 per cent in 2013. Although significant, this increase is still significantly lower than the influx of miners during the gold rush in the 1930s and 1980s. Also, it keeps pace with the steady increase of the gold price on the international markets after the historical low in the 1990s.

A second difference is that women nowadays may be involved in all phases of the mining process, from bringing it up from underneath the soil to marketing their produce. This means, among others, that the threefold distinction Burke (2006: 26) made with regard to women miners in Asia – women miners ("involved in the actual extractive process either below ground in large mines or in small-scale artisanal mines"), surface workers ("involved with sorting, crushing or preparation of the ore") and women "who are members of the mining community" – does not hold true in an African or Tanzanian context. At best, a distinction can be made between women miners and female mining service providers (such as gold sellers, food vendors, barmaids and so on), in the understanding that the boundary between these two roles is not very strict and women move back and forth in the course of their career. To complicate matters, most women are also involved in (subsistence) farming, growing sunflower, sesame, ground nuts, maize, millet and tobacco. This diversification follows a seasonal cycle: during the dry season they prepare the land and work the surface pits; during the rainy season they engage in alluvial mining and in farming. The distinction between farming and mining is, in other words, not that rigid: an estimated 70 per cent of the population is involved in farming, although not always on a full-time basis. At the same time, however, nearly everyone in town is involved in mining in one way or another, with new mining sites opening up every day.

These quantitative and qualitative differences between women's position in the mines around Makongolosi in the 1930s and the present can partly be explained by referring to labour laws and recruitment practices. Under British rule, as elsewhere in the colonized world, labour recruitment focused almost exclusively on men, especially when it came to hard labour. Mining, however, when available, has become – as Yakovleva (2007: 31) notes for Ghana – an "indispensable non-farm income-earning opportunity". In a similar vein, Drechsler (2001: 67) cites the deterioration of subsistence farming, low demand of public and private employment, lack of trading commodities, high inflation, a high birth rate and extended families to explain why women have engaged in gold mining themselves, rather than contenting themselves with the role of service providers. Moreover, it is easier for women to enter ASM in comparison to other sectors as this occupation, in general, is easily accessible and requires only a small initial capital investment. This, however, makes women in particular vulnerable to fluctuations in the price of gold, and to competition from labour-saving technologies or capital investments.

Figure 18.1 Picking up stones

Apart from this, one also has to take into account women's aspirations and speculations about the future. These considerations play a pivotal role in their decision to join the mining profession. In general, they perceive mining as a good way to accumulate wealth and thus as an opportunity to, for example, start up their own businesses or invest in their other activities.

Mining in Makongolosi: a day in the life of two women miners

Makongolosi is, both literally and figuratively, a crossroads. Built around a T-junction that makes up the town's centre and serves as its main road along which trucks loaded with tailings and buses filled with people make their way to the north, it also is a place where miners' pathways converge with those of gold brokers, business people, barmaids and other opportunists and fortune-seekers, to diverge again somewhat later.

> During the day, Makongolosi is quiet. Most mining sites are far, and away from the T-junction. Most men in town are hard rock miners, digging for stones in deep underground shafts far from the city centre. They usually work in teams, and remain absent for longer periods of time. Women, in contrast, work closer to town, in abandoned mining pits at about four to six kilometres from the T-junction, where they look for gold-bearing rocks. Some four to six sites are still active. They resemble moonscapes, with deep pits and high heaps of rocks and sand, between which one sees a dozen or more ladies shovelling sand and crushing rocks. Children lie or play in the shadow of a bush or a rare tree. Work is hard, and conversations are few. Wielding a pickaxe or shovel demands skill, strength and concentration, and the only sounds breaking the silence are those of the pickaxes hitting a stone, a toddler crying, or the murmurs of disappointment when their hard work remains fruitless. Yet, these miners do enjoy each other's company, especially during lunchtime or on the long walk from and to town. A few hours before sunset, they return to town, to shop and prepare dinner. Then, the town comes to life. Motorbikes flash by, honking their way through traffic. Street vendors sell their wares and bars are suddenly crowded with returning miners and passing-by truck drivers. Things go quiet again only after midnight.
>
> *(field notes, 21 November 2013)*

Mining gold does not correspond to the image of "striking a vein" or "finding" a nugget in your pan. Mining gold is hard, and often unrewarding, work. This is particularly evident in the surface mines around Makongolosi, especially during the dry season.[6] This section follows two women, Mama Vumi and Mama Salome, from the mining site back to town, where they try to sell their gold. These women miners have to negotiate a highly competitive world, where all the key positions (from claim holders to gold buyers and the owners of machinery and technology) are male.

In and around Makongolosi people distinguish between two kinds of gold. On the one hand, there is the "real gold" (*kikole*, pure gold, larger fragments or nuggets). On the other hand, one finds amalgamated gold (*dhahabu yakukamatisha*), obtained by pulverizing gold-bearing rocks and extracting the gold with mercury. At the time of writing, the former is valued at TZS 52,500 (US$32.58) per gram, while the latter changes hands for about TZS 45,000 (US$27.92). This distinction corresponds to another one people make. Men, and then especially the younger ones, leave to "go and do mining" (*ninaenda kuchimba*). Women miners, in contrast, say they "go to pick up stones (from the ground)" (*ninaenda kuokota mawe*). This means that both women and men are involved in mining, albeit in different ways. "Real gold" is usually found by men, often with the aid of metal detectors: they descend into the shafts, or are the first ones to mine the open surface sites around town. Women, in contrast, are not allowed to work in the shafts or underground. They are considered "unclean" (*wanawake wachafu*, an allusion to menstruation) and their presence may prevent the men from striking gold. Therefore, in Makongolosi, women can only work the surface mines that were abandoned by the men. Also, women do not team up with male miners, with family, or with people they are close to. Unlike, for instance, small-scale mining in Ghana, women in Makongolosi tend to work alone.[7]

> I'm mining myself. I'm carrying stones. I send them to the crushing machine. They crush it and they "are putting together" with mercury. Then, I get money. Don't wait for a fellow woman to help you. Women they are saying: "If I help you, what can I get?"
>
> *(interview with Bibi, miner, on the way to the mining site near Kitete, 14 October 2013)*

Women, moreover, are confined to surface mines at walking distance from town, where their older children spend most of the day in daycare or with friends or neighbours. Here, they enter into competition with the older men, who deem the work in the shafts too demanding. However, because of Makongolosi's rapid expansion and the influx of miners from the rest of Tanzania and its neighbouring countries, and because of the recent introduction of metal detectors (see below), women miners have to walk longer distances "to find good stones".

> Mama Vumi and Mama Salome each choose a pit that seems unmanned. They kick off their shoes and immediately start looking for gold-bearing "red stones". Suddenly Joram, the claim holder, appears. His attitude and tone are blunt. He starts shouting at the women, demanding that they pay TZS 1,000 [US$0.62] each for working his claim. Mama Salome sweet-talks him and hands him TZS 1,000. "For the both of us", she says. He takes the money but assures her that he will be back for the remaining 1,000.
>
> *(field notes, mining site Mwanzo mgumu ("Every beginning is hard"), 26 October 2013)*

Not all claim owners demand payment: it depends on how much gold they speculate the miners to find.[8] To find out, they petition the Mining Officers of the Resident Mines Office (RMO, a local extension of the Tanzanian Ministry of Energy and Mines) to come and inspect their land. Or they see how many people show up every day. Perhaps paradoxically, the more people are working their claim, the more gold it must hold and hence the higher the price they can charge.

After being collected, the rocks brought home from the site need to be crushed to free the gold fragments encased in them. Mercury is then used to amalgamate these tiny fragments into larger ones. These are taken to gold brokers who, in turn, sell it to other middlemen. On a normal day, women working at the surface sites around Makongolosi collect between half a bag and one full bag of rocks, which they bring to the crushing site every four to five days. Makongolosi has quite a number of these enterprises. They charge their clients about

TZS 3,500 (US$2.17) per bag to make use of their ball mill, a diesel-powered machine – only operated by men – that grinds the rocks into a fine powder (*unga*, "flower"). Crushing sites demand an enormous capital investment; start-up costs range between TZS 9,000,000 and TZS 12,000,000 (US$5,584.91–7,446.55) not taking into account maintenance and fuel of the ball mill and engine.[9]

Crushing sites also offer facilities to "wash" (*kuosha*) the powder from the ball mill. Washing reduces the powder from the ball mill to a handful of gold-holding sand that, next, is panned with the mercury provided by the owner to amalgamate (*kukamatisha*) the gold. Crushing sites usually offer to buy the women's produce and, especially on calmer days, there is a fierce competition between their owners. Besides the obvious (what they charge their clients, and the price they pay for their gold) they try to attract clients by offering them extra facilities (such as free motorbike transport or providing "good" or "kind" service).

> Six in the morning. Mama Vumi and Mama Salome arrive at the crushing place, with two bags each. Yet they are told to wait, despite the ball mill's first come, first serve policy. Hungry from the walk, they decide to take breakfast at a nearby café. Eleven o'clock. Mama Vumi and Mama Salome are called to the ball mill. Finally. The young operator boy opens up the lid and a huge pile of dust and powder pours out. The two women fill up the machine with Mama Salome's stones. Forty minutes later, the whole ritual is repeated. When they are in the process of washing their precious powder, a young woman starts a discussion with the ball mill operator boy. "Why are you offering us only TZS 5,000 [US$3.11] per 0.1 gram of gold?" she asks. He routinely replies that "The price of gold has dropped". She then starts telling all the other women in the mill of this other place she had heard about, where they are offering TZS 5,800 [US$3.60] per point.
>
> *(field notes, 16 October 2013)*

Miners can sell their gold to different buyers in town. Usually, however, they sell to the owner of the mercury they used. Obviously, selling gold is a tricky business. First of all, miners need to know who is offering the best price. But they also need to dodge the repertoire of tricks at the disposition of their buyers, such as tampering with the weights or scales, throwing gold on the ground, lying about the current gold price and the like. So even if women miners do shop around for buyers, they prefer those with whom they have a good relationship, one they can trust or one that may offer his help in times of need, someone with whom they have some manoeuvring space.[10]

> *Nimeporola!*, Mama Vumi shouts. "I have missed the money".[11] Both women extracted 0.4 grams from their two bags of stones, but Mama Salome sold her gold on the same day she came back from crushing her stones, after she heard that one particular crushing place offered a higher price, and managed to sell her gold at a rate of TZS 5,800 [US$3.60] per 0.1 gram. Mama Vumi, in contrast, had waited to sell until the very next day, only to find out that overnight the price of gold had dropped to TZS 5,000 [US$3.11] "per point". Twelve days before they only got TZS 5000 from their two bags. This time, they had more luck: after subtracting all costs, Mama Salome ended up with TZS 9,200 (USD 5.72), while Mama Vumi earned TZS 6,000 [US$3.73]. The following day, they decided to take two days off.
>
> *(field notes, 17 October 2013)*

Technology: gender matters

Ball mills were introduced in Makongolosi about a decade ago, and according to the miners, it greatly facilitated their work. In the words of Baba Eva: "Before, there were no ball mills. People were using iron clubs to crush the ores. Many people died because of this, because it was very heavy work" (interview, 2 October 2013). However, this newly available technology places women in a vulnerable position. For one, women miners have become dependent upon the entrepreneurs who were able to finance the purchase of a ball mill. Also, miners feel exploited as they deem the prices they are charged with as way too high. What adds to this is that, as of 2012, ball mill owners started to sell their tailings (the left-over debris of the crushing machine) to Arab, Chinese or Zimbabwean-owned tailing companies that mushroomed around Makongolosi. The sale of tailings is much more profitable than what they earn from crushing the miners' rocks, leading the latter to ask the question why they have to pay for crushing at all.[12]

The past few years saw a second technological innovation, in the form of metal detectors. These devices are in high demand; also, they are quite expensive and require skill and expertise to handle. Access to them is thus effectively limited to a limited number of (male) experts, or to those who can afford to hire them: the price to hire a metal detector is about TZS 10,000 (US$6.20) per gram of gold extracted at the site, which boils down to one-fifth of the current price people get for their gold. Access to this technology is thus limited to those who already are somewhat better-off than most women miners are. In practice, this means that over the past year women have been pushed out of the more easily accessible surface mines close to town, and have to walk longer distances in order to execute their profession.

Figure 18.2 Washing the sand

Obviously, women miners can also hire one of these specialists to survey a particular area. This, however, requires great negotiation and speculative skills (see below) on the miner's behalf. As the operator wears headphones, one is never sure whether or not he is telling the truth. The detector may or may not work, and the operator may or may not lie about his findings by keeping the information to himself, and come back later to dig the soil. Hiring such a specialist is thus shrouded in secrecy and suspicion, also because they are reported to resort to witchcraft to "blind" others looking for gold. These operators have access to a secret knowledge of, first of all, their expertise and, second, about their findings. The suspicion, however, is mutual: when a miner goes to sell her gold, the operator will join her to make sure she will not rip him off. What this means is, first of all, that the introduction of metal detectors is unfavourable to women miners, who find themselves in an unbalanced power position vis-à-vis these specialists. It also reveals, secondly, the importance of speculation in the lives of these miners (and, by extension, in Makongolosi).

Analysis/discussion

As Mohammed Banchirigah (2006: 167) notes, development practitioners and policy makers often overlook the fact that the poverty that in sub-Saharan Africa drives people towards ASM is largely the result of reforms, such as privatization (as in Tanzania) or mechanization (such as the ball mill). Moreover, similar reforms have, in many cases, devalued smallholder farming leading to diminishing returns and driving people to do so-called illegal mining. Local structures of authority and power, and existing gender roles (such as, in Makongolosi, the idea that women miners are "unclean" and hence may jeopardize the finds of their male counterparts), may exacerbate these inequalities, but they do not cause them.

Our case study also illustrates that the introduction of new technologies affects women and men in different ways (for an overview, see Bray 2007). The huge capital investments required to purchase a ball mill or, to a lesser extent, a metal detector effectively put these devices out of reach for women miners. What is more, women miners have become dependent on these technologies that, for various reasons, are in the hands of specialized experts, usually men, who possess the knowledge (and means) to operate them, and who benefit most from Makongolosi's gold fever. Because of this growing dependence on technology they do not master (such as the metal detectors) nor control (as miners, for instance, depend on the mercury provided by the owner of the ball mill), Hilson and Pardie (2006: 109) have referred to these technologies as "agents of poverty".

Obviously, the bigger the discrepancy between the dream of being "lucky" (*kupata bahati nzuri*) and everyday reality, the more attractive this dream becomes, and most women miners are committed to the art of speculation. For most in the mining industry, "modernity" (the access to "modern" commodities and amenities, health care, education) is a gamble, a bout of luck that bears little resemblance to the project of "development" as envisaged by development professionals and state officials. For most of them, modern life is as occult as is witchcraft.

Finally, despite these constraints and despite the observation that, across the globe, mining is notoriously male, any analysis must also take into account the agency and various strategies deployed by women miners around the world (see Hinton et al. 2003). These strategies depend on the minute particulars of context and life-world, including the materiality of minerals (such as "real gold" and "rocks"). Important here is to also take into account a more holistic approach to women at mining sites. Many, for instance, are drawn into the mining profession as it offers them an escape from patriarchal control (see Bryceson et al. 2013: 53). Therefore, it is important to look into how and where women's lives intersect with that of men, how their lives take shape across divides of seniority and authority, but also how their lives are changed by forces and power structures at the national and international level, by state (de)regulation and international speculation.

These aspects, however, remain underinvestigated. Much more research is needed to understand how gender relations affect women's involvement in mining; how it affects women miner's personal autonomy, alters family structures and practices, and how it bears on relatedness and sexual relationships. Women on mining sites do not constitute a homogenous group. They are barmaids, business women, migrants and mining professionals, often in combination with other jobs; they are young or old, have different upbringings and backgrounds, and different expectations of what the future may bring. The interplay of these elements, and their bearing on gender relationships, is another venue for further research.

Acknowledgments

The research for this chapter was made possible by a VLADOC scholarship from the Flemish Interuniversity Council (VLIR-UOS).

Notes

1 Data for this case study were gathered as part of Verbrugge's ongoing doctoral research, through semi-structured interviews and informal conversations with 36 women miners and 76 female service providers, and through participant observation. Besides long walks, hanging out in bars and chatting, this also included working side-by-side with women miners in and around the shafts of Makongolosi. All conversations, meetings and field visits were conducted in Kiswahili and in the presence of a research assistant, Salima Benson Wanderage, with whom a daily ritual of discussing passed events and experiences was established.
2 http://im4dc.org/activity/african-women-in-mining-and-development-study-tour.
3 http://go.worldbank.org/8KI2XV6VD0.
4 http://go.worldbank.org/8KI2XV6VD0.
5 Other sources estimate the total volume of gold ever mined at 172,000 tons, or even 2.5 *million* tons. The estimates on how much gold still can be mined, however, remain remarkably constant at about 50,000 tons.
6 During the rainy season, most people turn to alluvial mining.
7 In 2009, for instance, in Ghana a mine shaft collapsed, leaving 15 people dead, including the owner. Thirteen of the victims in what, for eight months, was Ghana's biggest mining disaster were women: http://news.bbc.co.uk/2/hi/africa/8356343.stm.
8 In Tanzania, all land is owned by the state. Formally, one needs a Primary Mining License in order to mine a plot, but people only rarely possess such certificates. In reality, both women and men are allowed to keep land, even though only very few women do own pits, as they lack the authority to "protect" it.
9 The crushing site consists of a ball mill containing 200 to 500 iron balls in different sizes, a gearbox and an engine. Other items needed to start up a crushing site are timber, barrels, chains and bolts to fasten the ball mill, a ball-mill operator and access to water (personal communication with owner of a crushing site, 30 November 2013).
10 There are no female gold buyers in Makongolosi. Some women do buy and sell gold, but always on their husband's or son's behalf.
11 This Kisafwa exclamation has become a common expression (also in Kiswahili) to express one's disappointment for not finding (lots of) gold. The opposite expression in Kisafwa is *Nimebongoa*.
12 Recently, a Russian leaching firm has also found its way to Makongolosi. Leaching is a process whereby gold is extracted using cyanide.

References

Ali, S.H. (2006) "Gold mining and the golden rule: a challenge for producers and consumers in developing countries", *Journal of Cleaner Production* 14(3–4): 455–462.

Bray, F. (2007) "Gender and technology", *Annual Review of Anthropology* 36: 37–53.

Bryceson, D.F., Jønsson, J.B. and Verbrugge, H. (2013) "Prostitution or partnership: wifestyles in Tanzanian artisanal gold-mining settlements", *Journal of Modern African Studies* 51(1): 33–56.

Bryceson, D.F., Fisher, E., Jønsson, J.B. and Mwaipopo, R. (2014) *Mining and Social Transformation in Africa: Mineralizing and Democratizing Trends in Artisanal Production*, New York: Routledge.

Burke, G. (2006) "Opportunities for environmental management in the mining sector in Asia", *The Journal of Environment & Development* 15(2): 224–235.

Drechsler, B. (2001) *Small-Scale Mining and Sustainable Development within the SADC Region*, Research Report 84, London: International Institute for Environment and Development, http://commdev.org/files/1798_file_asm_southern_africa.pdf.

Ferguson, J. (1999) *Expectations of Modernity: Myths and Meanings of Urban Life on the Zambian Copperbelt*, Berkeley: University of California Press.

Hilson, G. and Pardie, S. (2006) "Mercury: an agent of poverty in Ghana's small-scale gold-mining sector?", *Resources Policy* 31: 106–116.

Hinton, J.J., Veiga, M.M. and Beinhoff, C. (2003) "Women and artisanal mining: gender roles and the road ahead", in G. Hilson (ed.), *The Socio-Economic Impacts of Artisanal and Small-Scale Mining in Developing Countries*, Oxon: Taylor & Francis, pp. 161–203.

Jønsson, J.B. (2009) *Golden Livelihoods: Organizational Practices, Strategies, and Trajectories of Small-Scale Gold Miners in Tanzania*, PhD thesis, University of Copenhagen.

Kumah, A. (2006) "Sustainability and gold mining in the developing world", *Journal of Cleaner Production* 14(3–4): 315–323.

Lissu, T. (2001) "In gold we trust: the political economy of law, human rights and environment in Tanzania's mining industry (work in progress)", *Law, Social Justice & Global Development Journal* 2: 1–60, www2.warwick.ac.uk/fac/soc/law/elj/lgd/2001_2/lissu1.

Mohammed Banchirigah, S. (2006) "How have reforms fuelled the expansion of artisanal mining? Evidence from sub-Saharan Africa", *Resources Policy* 31(3): 165–171.

Win, E.J. (2007) "Not very poor, powerless or pregnant: the African woman forgotten by development", in A. Cornwall, E. Harrison and A. Whitehead (eds.), *Feminisms in Development*, London: Zed Books, pp. 79–85.

Yakovleva, N. (2007) "Perspectives on female participation in artisanal and small-scale mining: a case study of Birim North District of Ghana", *Resources Policy* 32(1–2): 29–41.

PART III

Perspectives on population and poverty

19

INTRODUCTION
TO PART III

On March 26, 2014, the United States Census Bureau calculated the world's population at 7.159 billion people. The annual rate of population growth peaked at 2.2 percent in 1963 and has since slowed down, falling below 1.1 percent in 2012. This decline is in part a response to urbanization, as children are more expensive to maintain in terms of housing, food, and education in cities (see chapter by Chant in this volume). At the Third International Conference on Family Planning in November 2013, held in Addis Ababa, family planning was seen as a universal human right and as part of women's emancipation. The conference report, released with more than 70 donors and countries, pledged to ensure that 120 million more women and girls would gain access to family planning by 2020. However, this emphasis on contraceptive provision with its focus on access to services and methods that meet women's needs, in support of the slogan "full access, full choice," prioritizes technical availability rather than rights and personal decision-making. Greater attention needs to be given to women's lack of agency in specific cultural contexts.

A corollary of the fall in birth rates is an increase in "the old age dependency ratio"—the ratio of old people to those of working age. This is particularly true in the many countries where, along with falling birth rates, health services and rising living standards have enabled more adults to survive into old age. Between 2010 and 2035 the old age dependency ratio in China will more than double from 15 to 36 and Latin America will see a shift from 14 to 27. This will have a gendered component as well, as women generally outlive men and tend to experience poverty more acutely. The major exceptions to this general greying of the population in the developing world will be south Asia and Africa where fertility is still high (*Economist*, April 26, 2014: 22).

Reproductive rights and sexual health are another important population concern. Twenty years after the first conference in Cairo, in April 2014 the 47th United Nations Commission on Population and Development ended with a recommitment to continue placing women's and girls' reproductive rights and sexual health at the core of their program. The Commission also expressed deep concern over the pervasiveness of gender-based violence and the need to prevent and eliminate such harmful practices as early and forced marriage and female genital mutilation (FGM). Finally in 2014 the first arrest has been made of a doctor carrying out FGM in Britain. Yet efforts to make legal abortion available everywhere did not succeed at the Commission meeting (RESURJ 2014).

Clearly, policies to improve the status of women and children must include education, particularly for girls, at their core. The World Bank Report 2012 states "investments in health and education-human capital endowments-shape the ability of men and women to reach their full potential in society" (World Bank 2011: 108–109). Actions taken to achieve Millennium Development Goal 3's target of eliminating gender disparities in primary and secondary education by 2005 and in all levels of education no later than 2015 have been very successful. The World Bank Report indicates that less than one-fifth of inequality in educational outcomes stems from gender but that, especially at secondary level, family wealth is the main differentiating factor. Even where boys and girls are both equally enrolled in school, girls tend to drop out earlier in order to take over domestic duties for their mothers. While in school, girls often have to take time out to collect water, consequently easier access to water usually improves educational returns for girls. Teenage girls also may have to take time off school during menstruation if the school toilet facilities are inadequate (see chapter by Sultana in this volume). Children may also face sexual harassment from male teachers and others and "the experience or even the threat of such violence often results in irregular attendance, dropout, truancy, poor school performance and low self-esteem, which may follow into their adult lives" (UNESCO 2014). Where the journey to school is long and inconvenient, girls may be less likely to go than their brothers. We found this in a village in Quintana Roo in Mexico, where children had to travel by bicycle, not thought suitable for girls, or stay in town to go to secondary school. Girls without relatives to stay with tended to drop out of school and the main wish for mothers, but not fathers, was to have a secondary school in the village (Momsen 1999). However, a recent and ominous response to the greater presence of girls in school in mainly Moslem areas is the Boko Haram movement in northern Nigeria, which has destroyed schools and kidnapped girls, and attacks on schoolgirls in Afghanistan and Pakistan by the Taliban as in the case of Malala (Yousafzai and Lamb 2013).

On the other hand, female participation in tertiary education has grown faster than that of males so that in many countries, especially in Latin America and the Caribbean, women outnumber men in universities today. But there are still distinct differences in subjects studied, with women enrolling predominantly in education, arts, and social sciences while men tend to study sciences and engineering. Boyden et al. (in this volume) show that parental aspirations for the education of their children were biased towards boys in Ethiopia and India by the age of 12 and towards girls in Peru and Vietnam. By the age of 15 these aspirations had been transmitted to the children with higher educational aims shown among boys in Ethiopia and India and among girls in Vietnam.

The five chapters in this section explore gendered aspects of population in the Global South, addressing economic aspects, family structure, gendered aspects of children's lives, and the commercialization of reproduction. Poverty is one overarching theme of most of the chapters in this section. Varley's chapter addresses gender and urban poverty. Today more than half the world's population lives in cities and urbanization has been most rapid recently in the Global South affecting gender roles. Cities in the developing world have grown as hubs of modernization and services rather than the industrialization that typified the cities of the Global North during their periods of rapid growth a century and more ago. Rural migrants fled to cities such as São Paulo in Brazil making it the biggest city in the southern hemisphere by the 1960s when it overtook Rio de Janeiro and became the industrial hub of Brazil (Henshall and Momsen 1974). Cities allowed rural migrants new opportunities, although few succeeded as well as Luiz Inácio Lula da Silva who, through leadership in a Paulista labor union, eventually became the 35th president of Brazil. As president, he set up the Bolsa Familial, a cash transfer program, which has done a great deal to reduce inequalities due to poverty in Brazil. Urbanization produces both material and normative changes: it limits the feasibility of adherence to conventional gender patterns, offers new opportunities, and leads to broader value transformations. At the same time, city life brings with it problems attributable in part to gender norms or changes that are experienced differently by women and men.

Sylvia Chant takes apart the complexity of the diverse meanings, causes, and processes of gender and poverty while Boyden et al. suggest that children are mainly seen in relation to the poverty of their mothers. Chant shows that poverty for women is a problem of multiple deprivations, not just of income but also of lack of assets and agency leading to vulnerability, violence, and loss of dignity and self-respect. Policy interventions aim microfinance projects and cash transfer programs (CCTs) at poor women as a simplistic and often inadequate way of empowering women (see chapter by Huq-Hussain in Part VII). Both Barrow and Chant talk about pathologies, such as instability and problems of childrearing, of non-standard households especially the female-headed household widely found among Afro-Caribbean peoples. However, they both see such families as resilient and not necessarily poorer than male-headed households. Yes despite normative patterns of women as household heads, and generally being better educated than men, patriarchal power is still strong in the Caribbean (Momsen 2002). Barrow provides a very clear example of the influence of colonialism on attitudes to family structure. Early ethnographic studies by white, male, foreign researchers, as part of an imperial project for social reconstruction and the moral uplift of black populations out of poverty and backwardness, saw female-headed households and multiple sexual partners as alien and undesirable. Barrow further notes that "Commentators were especially shocked at illegitimacy levels as high as 70 per cent along with low marriage rates, unstable conjugality and absentee husbands and fathers." She identifies an ongoing contradiction between so-called black family breakdown and new social policies facilitating divorce and abolishing the distinction between children born in and out of wedlock.

Another theme that emerges from these chapters is the new technologies that are changing fertility patterns and decisions. Varley describes how women in urban areas can more easily access family planning services or escape an unhappy marriage. The downside of this, though, is that urban women have access to ultrasound technology enabling female feticide in countries where son preference is culturally important, as in China, India, and South Korea. The proportion of missing girls at birth is still high in China and India and in Armenia, Azerbaijan, and Georgia, but in South Korea urbanization and education have reduced such gender discrimination (World Bank 2011).

The spread of assisted reproductive technologies (ARTs) has allowed a new transnational surrogacy movement as described by Schurr and Fredrich. The outsourcing of industrial production to the Global South that led Varley to suggest that factory women are sometimes thought of as tainted by modernity is now being followed by outsourcing of reproductive processes. Schurr and Fredrich describe how women of lower economic and social status in the Global South provide ova and surrogate services for wealthier women in search of inexpensive fertility treatments. Drawing on a case study from India, they examine how transnational surrogacy can be both a development strategy improving the livelihoods of poor women and at the same time a new arena for exploitation and commodification of women's bodies. They argue for geographically situated research that produces policy relevant knowledge to help regulate and monitor this new market.

Finally several chapters examine the gendered effects of urbanization, migration, and modernity. Boyden et al. consider the gendered effects of urbanization and modernity on children drawing on a 15-year study of childhood poverty in Ethiopia, India, Peru, and Vietnam. Gender differences in access to education vary from country to country. Yet information and communication technologies present new learning opportunities for the young and expand their horizons. Varley points out that older women living in cities will soon outnumber those in rural areas and widowhood will be an increasingly common experience in the world's poorer cities. In the Caribbean the commonality of migration and transnational families reinforces the role of grandmothers as carers and contributes to increasing problems of masculinity, especially among boys, as shown by Parpart (in Part I).

Urbanization produces both material and normative changes: it limits the feasibility of adherence to conventional gender patterns, offers new opportunities, and leads to broader value transformations. Ann Varley considers the changes to gender roles brought by life in the city. Movement to the city tends to weaken kin control and encourage the formation of nuclear families and Varley suggests that this may make it easier for non-heterosexual relationships to exist. However, acceptance of such relationships varies regionally and Barrow points out that the Caribbean region continues to outlaw same-sex sexuality, as in parts of Africa, and reinforces heteronormativity as the bedrock of morally acceptable family life.

Overall, this section takes a global view of the impact of life stage on adaptation to change by women and men and looks at some of the new directions being taken in search of improving living standards in the Global South.

References

Henshall, J.D. and Momsen, R.P. (eds.) (1974) *A Geography of Brazilian Development*, London: Bell and Hyman.

Momsen, J. (1999) "A sustainable tourism project in Mexico," in M. Hemmati (ed.), *Gender and Tourism: Women's Employment and Participation in Tourism*, London: UNED-UK, pp. 142–148.

—— (2002) "The double paradox," in P. Mohammed (ed.), *Gendered Realities: Essays in Caribbean Feminist Thought*, Jamaica: University of the West Indies Press, pp. 44–55.

RESURJ (2014) *Public Statement on UN CPD 47 "20 years after Cairo. Women: Continue to Reproduce But Never Mind about Sexuality"*, www.awid.org/layout/set/News–Analysis/ Announcements2/RESURJ-Public-Statement.

UNESCO (2014) *School-Related Gender-Based Violence in the Asia-Pacific Region*, Bangkok: UNESCO.

World Bank (2011) *World Development Report 2012: Gender Equality and Development*, Washington, DC: World Bank.

Yousafzai, M. and Lamb, C. (2013) *I am Malala: The Girl who Stood Up for Education and was Shot by the Taliban*, London: Weidenfeld & Nicolson.

20

GENDER AND POVERTY IN THE GLOBAL SOUTH

Sylvia Chant

Introduction: the complex terrain of gendered poverty, and the need for a multidimensional perspective

That poverty in general is increasingly acknowledged as multidimensional, with UNDP's recently-launched Multidimensional Poverty Index (MPI) comprising ten variables relating to education, health and living standards, has driven to a substantial degree feminist scholarship. The latter has consistently highlighted in recent decades that gendered poverty is not just about income, and gender is not just about women (Chant 1997, 2007; Jackson 1997).[1]

Despite a multiplicity of shorthand terms used to describe the complex nexus between gender and poverty, 'gendered poverty' is perhaps one of the most apposite insofar as it gestures towards the need for an holistic gaze on the diverse meanings, causes, processes and outcomes embroiled in different forms of privation among women and men, and to their intersections with a host of social, economic, demographic and other variables (Chant 2010: 3).

Although strong evidence for persistent and ubiquitous gender differences in access to material resources such as earnings means that income must remain a key element in gendered poverty analysis, whether it should be pre-eminent is another matter. As Gerd Johnsson-Latham (2004: 26–27) has posited, income may actually be a less 'robust' indicator of women's disadvantage than factors such as access to land, agency in decision-making, legal rights within families, vulnerability to violence, and (self-)respect and dignity. For instance, on the matter of the most tangible issue here – land – gender gaps are more marked than in earnings. While women workers worldwide may still only receive an average 75 per cent of male wages, often on account of restricted employment options (see Figure 20.1), this pales into relative insignificance given that a mere 15 per cent of landowners globally are estimated to be female.

Bearing in mind caveats in data both on earnings and land and property ownership, the particularly parlous position of women in respect of the latter is due to discriminatory inheritance systems, and to the fact that their statutory rights are frequently compromised by lack of knowledge, plural legal codes, and sociocultural emphasis on the prerogatives of males and their consanguineal kin over land and property (see Chant 2013 for discussion and references). Even in Latin America, where women fare rather better in terms of asset acquisition than elsewhere in the world, in only two out of 11 countries that formed the basis of a comparative

Figure 20.1 Informal woman street trader, Greater Banjul, the Gambia © Sylvia Chant

regional study by Carmen Diana Deere, Gina Alvarado and Jennifer Twyman (2012) – namely Nicaragua and Panama – has gender parity been achieved in homeownership (Deere et al. 2012: 525).

Many women in developing countries can only find work in the informal economy, such as ambulant vending, which is marked by long hours and low remuneration.

The primacy routinely accorded to income in poverty assessments plays down other practical and potentially measurable – as well as subjectively interpreted and experienced – dimensions of poverty such as 'overwork', 'time deficiency', 'dependency' and 'powerlessness'. As a growing body of research in different geographical contexts has revealed, the latter may be as relevant, if not more so, to women's perceptions of disadvantage, and to the 'trade-offs' they are able to make between different aspects of privation (see Chant 2007 for discussion and references). There is also a spatial element to gendered poverty, concisely summarized by Sarah Bradshaw (2002: 12) who stresses that women's poverty is 'not only multidimensional' but is also 'multisectoral [and] . . . is experienced in different ways, at different times and in different "spaces"'. Such spaces include not only the labour market and the realm of asset and property ownership, but the home, the neighbourhood, the city, the legal environment, the social policy arena, the political economy and territories of war, conflict and natural disaster (Bradshaw 2002; Chant 2010). Moreover, mobility between spaces, both physically and virtually, is often heavily circumscribed among women on account of restrictive norms governing female propriety, male-biased transport systems, and a gendered digital divide, with mutually reinforcing effects on income and other forms of poverty (see Chant, 2013 for discussion and references).

It is, of course, important to recognize that gendered poverty is also experienced by men and boys. As for women, this is differentially impacted by the diverse influences of age, 'race', nationality, sexuality, class, household headship and composition, marital status, fertility and family status, urban versus rural provenance and residence, migration within and across national borders, availability of public as well as private assets, labour market possibilities in the formal and informal economy, and state social transfers. Many of these variables not only intersect with gender and poverty, but with one another, and are associated with myriad processes and outcomes contingent on whether analysis is filtered through the conceptual lens of monetary poverty, capability poverty, social exclusion and so on; whether methodology is based on 'objective'/quantitative or more 'subjective'/qualitative/participatory measures; and whether the subject and scale in question are personal, domestic, local, national or global (Chant 2010: 2).

Having provided a brief introduction to some of the complexities of thinking through the nexus between gender and poverty, in the sections that follow, I concentrate on two interrelated issues that have dominated the field. The first is the 'feminization of poverty', which has become a common shorthand in academic and policy circles to describe women's economic disadvantage, but which remains problematic on numerous counts. In critiquing the 'feminization of poverty' I pay particular attention to the increasingly widely contested assumption that female-headed households are the 'poorest of the poor', as well as to alternative formulations of the 'feminization of poverty' that play down a focus on income in favour of greater emphasis on gender-biased labour and time inputs in household livelihoods. The second issue is the mounting tendency for poverty alleviation initiatives to target women, which is discussed with particular reference to conditional cash transfer (CCT) and microfinance programmes.

The 'feminization of poverty'

The term 'feminization of poverty' is generally attributed to the sociologist Diana Pearce (cited in Chant 2007), who used this to refer to the increased concentration of income poverty among women in the USA, especially among Afro-American female-headed households, between the 1950s and 1970s. Following gradual uptake of the term in a variety of geographically disparate studies, in the mid-1990s the 'feminization of poverty' began a new lease of life as one of the most widely generalized and circulated truisms of international development. At the Fourth United Nations World Conference on Women in 1995 the (albeit erroneous) pronouncement that women accounted for 70 per cent of the world's poor, and that this proportion was rising, ignited support for addressing 'the persistent and rising burden of poverty upon women' under the auspices of the Beijing Platform for Action (BPFA). Since this time the 'feminization of poverty' has effectively been regarded as a global phenomenon, and associated with three apparently intuitive (if inadequately substantiated) notions. Reflecting the legacies of Pearce's original time- and place-specific concept, these are, first, that women are poorer than men; second, that the incidence of poverty among women is increasing relative to men over time; and third, that growing poverty among women is linked with the 'feminization' of household headship (Chant 2007).

Each of these elements is open to question on conceptual and/or empirical grounds. The first assertion – that women are poorer than men – is static, and therefore anomalous within a construct whose very nomenclature implies dynamism. As summarized by Marcelo Medeiros and Joana Costa:

> In spite of its multiple meanings, the feminisation of poverty should not be confused with the existence of higher levels of poverty among women or female-headed households ... The term 'feminisation' relates to the way poverty changes over time, whereas 'higher levels of poverty' ... focuses on poverty at a given moment. Feminisation is a process, 'higher poverty' is a state.
>
> *(Medeiros and Costa 2006: 3)*

While the latter is inscribed in the second tenet, namely that there is a tendency for poverty incidence among women to be increasing relative to men, it is important to recognize that this too is intrinsically vague because no time period is specified. Besides this, in practical terms it is immensely difficult to establish whether gender gaps in poverty are widening over any interval except perhaps the past one or two decades. This is not only due to a dearth of sex-disaggregated panel data especially in developing countries, but the fact that even where data are available on income poverty, no consistent trend emerges in the direction of mounting economic privation among women. Through a rare and painstakingly detailed quantitative study on eight countries in Latin America between the early 1990s and early 2000s, for example, Medeiros and Costa (2006: 13) concluded that there was 'no solid evidence of a process of feminisation of poverty' (see Table 20.1). This interpretation was drawn not only on the basis

Table 20.1 Trends in the feminization of poverty, selected Latin American countries

Country (period)	Overall trend in poverty	Feminization of poverty	
		Women–Men	Female–Male headed HH
Argentina (1992–2001)	increased	no	yes
Bolivia (1999–2002)	stable	no	no (except for females without children)*
Brazil (1983–2003)	decreased	no	no
Chile (1990–2000)	decreased	no	no
Colombia (1995–99)	increased	no	no
Costa Rica (1990–2001)	decreased	no	no (except for females with children)
Mexico (1992–2002)	decreased	no	yes
Venezuela (1995–2000)	increased	no	no

Source: Adapted from Medeiros and Costa (2006, 12, Table 20.1)

* The exception occurs when comparing female-headed households without children with couple-headed households without children.

NB: 'no' stands for a rejection of the feminization of poverty hypothesis and 'yes' for the opposite.

of per capita income figures for women and men in general and according to male- and female-household headship, but also took into account the incidence, severity and intensity of poverty (see Table 20.1).

Interestingly Medeiros and Costa's research indicated that in Latin America the presence of young children was more likely to place households at greater risk of poverty than female household headship. This is massively important in light of the fact that the third and perhaps the single most widely cited tenet of the 'feminization of poverty' is that rising female poverty is connected with increases in households headed by women – a notion confirmed by repeated, not to mention categorical, statements that female-headed households are the 'poorest of the poor' (see Box 20.1).

Most generalizations about the links between female household headship and poverty are suspect because they are based only on assumptions, or on statistical scrutiny that falls far short

---

Box 20.1 Assertions about female-headed households as the 'poorest of the poor'[3]

the global economic downturn has pressed most heavily on women-headed households, which are everywhere in the world, the poorest of the poor.

(Tinker 1990: 5)

Women-headed households are overrepresented among the poor in rural and urban, developing and industrial societies.

(Bullock 1994: 17–18)

One continuing concern of both the developing and advanced capitalist economies is the increasing amount of women's poverty worldwide, associated with the rise of female-headed households.

(Acosta-Belén and Bose 1995: 25)

What is clear is that in many countries women tend to be over-represented in the ranks of the 'old' or structural poor, and female-headed households tend to be among the most vulnerable of social groups.

(Graham 1996: 3)

the number of female-headed households among the poor and the poorer sections of society is increasing and . . . they, as a group – whether heterogeneous or not – are more vulnerable and face more discrimination because they are poor and also because they are man-less women on their own.

(Bibars 2001: 67)

Households headed by females with dependent children experience the worst afflictions of poverty . . . Female-headed households are the poorest.

(Finne 2001: 8)

Households headed by women are particularly vulnerable. Disproportionate numbers of women among the poor pose serious constraints to human development because children raised in poor households are more likely to repeat cycles of poverty and disadvantage.

(Asian Development Bank 2003: 11)

of the kind used in Medeiros and Costa's rigorous analysis. For example, in some instances female-headed households are posited as poorer than their male-headed counterparts on the basis of comparisons of unweighted aggregate household incomes, which, given the generally smaller size of women-headed households, makes them 'a visible and readily identifiable group in income poverty statistics' (Kabeer 1996: 14). A further problem is that female-headed households are frequently taken as a proxy for women. Bearing in mind Lampietti and Stalker's (2000: 2) caution that 'Headship analysis cannot and should not be substituted for gender and poverty analysis' (see also Deere et al. 2012), the situation of women within male-headed households is missed, even if their economic predicament may be worse on account of 'secondary poverty' deriving from the inequitable allocation of household resources. As Medeiros and Costa (2006: 9) point out, even where per capita income is the basis of household poverty measurements, one cannot assume that income is distributed equally among members.

While not denying that some female-headed households, especially in societies marked by extremes of gender discrimination, may find themselves worse-off than their male-headed counterparts, the automatic assumption that they are the 'poorest of the poor' bears witness to a widespread pathologization of 'non-standard' (non-patriarchal) family units that often flies in the face of empirical evidence (see Chant 1997; Safa 1998).

In another vein, an overwhelming (if implicit) emphasis on income poverty in the construction of female-headed households as the 'poorest of the poor' goes against the grain of women's subjective experiences of privation. As has been attested to in a number of studies, female heads may well be 'income poor', but also 'power-rich' (see Chant 1997; Kabeer 1996).

Calls to acknowledge not only the multidimensionality of women's poverty, but the critical part played by women's subjective experiences of, and reactions to, poverty, have led to various attempts to deepen and diversify interrogation into the 'feminization of poverty' such that it more accurately depicts trends in female privation and offers a more meaningful basis for analysis and action. For example, although it remains difficult to establish a generalized tendency towards a feminization of income poverty, one recent development emerging from in-depth comparative field research has been the notion of a 'feminisation of responsibility and/or obligation' (Chant 2007).[4] Among the factors emphasized in the latter concept are growing gender disparities in the range and amount of labour invested in household livelihoods, a persistent and/or growing unevenness in women's and men's capacities to negotiate gendered obligations and entitlements in households, and an increasing disarticulation between women's and men's investments/ responsibilities and their personal rewards/rights (Chant 2007, 2008). Viewed from the perspective of these gender disparities, the 'feminization of poverty' may be more accurately attributed to the actual (and idealized) majority position of male-headed households than a rise in the numbers of households headed by women (Chant 2007). In male-headed households, for instance, an increasingly observed tendency for women to cope with a mounting onus of household survival arises not only because they cannot necessarily rely on men and/or do not expect to rely on men, but because a growing number seem to be supporting men too, including financially. This underlines the argument that poverty is not just about the privation of minimum basic needs, but of opportunities and choices. Although female household heads could conceivably be seen as an extreme case of 'choicelessness' and 'responsibility' – insofar as they have little option other than to fend for themselves and their dependents, and on potentially weaker grounds given gender discrimination in society at large – this needs to be qualified: a) because female-headed households do not necessarily lack male members; b) free of a senior male 'patriarch', their households can become 'enabling spaces' in which there is scope to distribute household tasks and resources more equitably, and c) women in male-headed households may be providing for spouses as well as children as an increasing proportion of men seem to be stepping out of the shoes of 'chief breadwinner' into those of 'chief spender' (Chant 2008: 188).

In short, it appears that conventional portrayals of the 'feminization of poverty' have simplified concepts of both poverty and gender to the extent that some individuals in need are missed, those individuals most associated with the phenomenon (particularly female heads of household) are scapegoated as key to the 'problem', and issues of major importance in feminized poverty, such as gendered power relations, within, as well as beyond, the household are ignored.

Leading on from this, while it might be desirable to retain the term 'feminization of poverty' given its historic expediency in drawing attention to women's privations and mobilizing resources for gender-sensitive policy interventions, it is necessary to ensure that the term is fleshed out more explicitly, and is understood to encompass a broader range of feminized privations than income alone (Chant 2007: 338–339).

Gender, poverty and policy interventions

With these provisos in mind, it is perhaps no surprise that routinized 'feminization of poverty' thinking tends 'to have translated into single-issue and/or single group interventions which have little power to destabilize deeply-embedded structures of gender inequality in the home, the labor market, and other institutions' (Chant 2008: 185). While nation-states, non-governmental organizations (NGOs) and bilateral and multilateral agencies are often genuinely well-intentioned in their bids to promote gender equality and 'female empowerment' as part of poverty reduction, a number of pitfalls and unintended consequences are in considerable evidence. Indeed, despite widespread consensus on the fact that empowerment is a complex and highly contested term, many 'empowerment' interventions boil down to increasing women's access to material resources as a route to widening their choices. Yet this itself is probably not enough, as revealed by debates around conditional cash transfer (CCT) programmes and microfinance schemes.

Conditional cash transfer (CCT) programmes and women

Most CCT programmes allocate stipends to women in exchange for 'co-responsibility' in respect of a series of health- and education-oriented duties designed to capacitate upcoming generations to exit poverty. In the Mexican programme *Progresa/Oportunidades*, for example, 'co-responsibility' requirements for cash transfers and food hand-outs are contingent upon women ensuring their children's school attendance and health (including taking them for medical check-ups and participating in health workshops), as well as undertaking voluntary labour such as the cleaning of community facilities (see Molyneux 2006).[5]

Directing cash transfers to women signals recognition that women are prudent with money, that they are good managers, and in a sense provides official endorsement for women to assume greater control of household income. In turn, and arguably much more positively, cash transfers can yield appreciable intergenerational gendered dividends such as affording young women unprecedented opportunities for schooling and labour force entry, as discussed in the context of young indigenous Mexican women by Mercedes González de la Rocha (in Chant 2010). As also noted in evaluations of World Bank-sponsored CCT trials in Malawi and Tanzania, girls tended to stay in school and rates of HIV-prevalence declined due to a fall-off in risky sexual behaviour – mainly because cash transfers reduced their needs to engage in unsafe transactional sex (Harman 2010).

By the same token, one major concern about CCT programmes is their instrumentalist reliance on women as a 'conduit of policy' (Molyneux 2006), whereby resources channelled through them are expected to translate into improvements in the wellbeing of their children and other family members.

Such instrumentalism is not only grounded in essentialist stereotypes such as 'female altruism', but also plays a part in perpetuating them. Leaning on women's financial prudence, and/or 'voluntary' unpaid labour in the context of assumptions about their 'traditional' proclivities as wives and mothers usually adds to, rather than alleviates, already heavily feminized burdens of time, labour and financial management (see Bradshaw 2008; Chant 2008; Molyneux 2006; Tabbush 2010).

In turn, in the process of depending heavily on mothers, and making little effort to involve fathers in any unpaid volunteer work, such programmes have often 'built upon, endorsed and entrenched a highly non-egalitarian model of the family' (see Molyneux 2006; also Bradshaw 2008; Tabbush 2010).

Microfinance schemes and women

With regard to microfinance, women also represent a disproportionate share of 'beneficiaries' in the belief that giving women a 'helping hand' with their businesses will help them (and their households) exit poverty. At a world scale an estimated 84 per cent of microcredit clients are female (Daley-Harris 2006, cited in Chant 2010), and in the Gambia and the Philippines specifically, more than 90 per cent of microfinance loans made by government and international agencies in recent years have been directed to women (Chant 2007: Chapters 4–5).

The granting of loans to women who have historically been excluded from bank credit is deemed to afford them welcome opportunities to embark upon and/or strengthen their entrepreneurial ventures, to enhance personal wellbeing and economic status, and thereby to challenge gender unequal norms.

Yet despite these positive outcomes, a number of downsides are also apparent. For example, as highlighted by Supriya Garikpati (in Chant 2010) in relation to the Self-Help Group (SHG) Bank Linkage Program in India, many women's loans end up invested in assets that are primarily controlled by husbands, or are used for household production or consumption, neither of which help women with loan repayments. Indeed, in order to fulfil their loan obligations women are often forced to undergo the 'disempowering' process of having to work harder as wage labourers, while also experiencing a growing gendered resource divide within their households. Related tendencies, including the reinforcement of women's responsibilities for household expenditure with no expansion of their rights, have also been noted in other contexts (see Cons and Paprocki 2010 and Goetz and Sen Gupta 1996 on Bangladesh; Mayoux 2001 on Cameroon), compounding what Linda Mayoux has famously referred to as a 'feminization of debt'.[6]

Another concern raised about microfinance schemes is their common emphasis on group loans. In the context of Bolivia, Kate Maclean (in Chant 2010) contends that these often help microfinance institutions (MFIs) to guard against risk of default and ensure their own survival. Yet, to borrow Maclean's expression, 'capitalising on women's social capital' (which refers to utilising relationships of female reciprocity), may not be in the best interests of women insofar as this can exert immense pressure upon them to sacrifice individual wellbeing to keep up collective repayments. Indeed, the question of whether poor women are necessarily inclined to join forces in their bids to exit poverty is one in which general assumptions are ill-advised.

Questions of collective versus individual interests aside, although microfinance can make some inroads to reducing poverty – mainly because it allows women to 'smooth-out' fluctuations in household income – it is arguably ineffective at providing a pathway out of gendered poverty. Beyond paltry attention on the part of implementing institutions to the gendered structures of constraint that act to limit women's personal autonomy, common obstacles include the small size of loans, lack of specialist expertise and training in enterprise development, and the fact that there is often inadequate dynamism in the local or even macroeconomic contexts in which women operate to allow for truly successful business ventures (see chapters by Bibars, Casier, Mohamed and Sweetman in Chant 2010). Thus, while, on balance, many MFIs may enjoy high rates of on-time repayments and rich returns on their investments, including capital growth, whether an escape from poverty, and advances in women's wellbeing, form part of this scenario are less assured (Chant 2010). The issue of whether microfinance provides a route towards 'female empowerment' also remains a moot point, with Fauzia Mohamed (in Chant 2010), inter alia, stressing the need for a range of ancillary interventions such as reforms in education and legal systems.

Leading on from this the underlying premises of both microcredit and CCT programmes alike have been called into question. As persuasively argued by Stefan de Vylder (2004: 85), the pursuit of gender equality has normally been regarded by the GAD community as an end in itself from a human rights perspective; yet from the vantage point of stakeholders whose primary interest is alleviating poverty in general, the incorporation of women into poverty reduction programmes may be driven less by imperatives of social justice than by goals of pragmatism and efficiency.[7]

Missing men

The fact that men and gender relations remain largely absent from contemporary policy responses to gendered (women's) poverty goes against the grain of men's frequently significant presence in women's lives, and mounting recognition that their exclusion can seriously prejudice the success of gender projects and policies (see Chant and Gutmann, 2000; Cornwall and White 2000).

In Costa Rica, for example, lone and partnered mothers on low incomes have been the exclusive client group under at least two flagship 'gender and poverty' programs since the mid-1990s, notably the 'Comprehensive Training Program for Female Household Heads in Conditions of Poverty' (*Programa de Formación Integral para Mujeres Jefas de Hogar en Condiciones de Pobreza*) and 'Growing Together'(*Creciendo Juntas*) (see Chant 2007: Chapter 6). Despite the fact that both these schemes have nominally been 'empowerment'-orientated and have comprised modules in *'formación humana'* (human development/training) aimed at sensitizing women to gender and human rights, the exclusion of men may not just have perpetuated, but exacerbated, prevailing male tendencies to curtail women's power and liberties (Chant 2007).

Additional dangers of excluding men include the fuelling of gender rivalry or hostility, with evidence indicating that growing pockets of male social, educational and economic vulnerability can manifest itself in violence in the home and in the community, in drug or alcohol abuse and other forms of disaffected behaviour (Chant and Gutmann 2000; Molyneux 2006).

Missing the point?

Despite the best-intentioned efforts of even the most rounded programmes designed to alleviate poverty and empower women, the narrow focus on income, and on women alone, leaves much to be desired. Instead of women finding themselves 'empowered' to strike new deals within their households as their education, skills, and access to economic opportunities expand, more often than not they simply end up burdened with more obligations (Chant 2007). In the Gambia, for instance, where, as a result of dedicated policy initiatives to address gender inequality and poverty, young women are beginning to enjoy increased access to education and employment, familial claims on their newly acquired human capital act to deplete the possibilities of personal mobility. Many daughters not only continue helping out in the home, but are also expected to study hard and, when they do find work, to use the bulk of their earnings to subsidize their parents' or brothers' expenses. Later on in life, they often find themselves in a similar situation with husbands (Chant 2007: 169; see also ECLAC 2004: 29 on Latin American countries). Leading on from this, men's individualism, not to mention their perceived entitlements to the fruits of female labour, continue to be tolerated. As mentioned earlier, women's duty towards others is rarely questioned, which is partly to do with an aforementioned resilience of culturally condoned expectations of female altruism and servility – a phenomenon that in itself is probably owed to some degree to the tendency for GAD policies to focus only on women (Molyneux 2007: 4). Yet if we are to acknowledge that poverty and human rights are integrally linked, women's rights to stand up for themselves and to negotiate social expectations of their roles in the family are fundamental. As affirmed by the Asian Development Bank:

> poverty is increasingly seen as [a] deprivation, not only of essential assets and opportunities, but of rights, and therefore any effective strategy to reduce poverty must empower disadvantaged groups, especially women, to exercise their rights and participate more actively in decisions that affect them.
>
> *(ADB 2002: xvi–xvii)*

Conclusions

It is vital that the alleviation of poverty and advances in gender equality feature as policy priorities in the twenty-first century, and there is encouraging evidence that this is occurring at an unprecedented scale. However, despite widespread rhetoric around 'empowering' women through involving them in poverty reduction initiatives, in practice there are worrying signs that they may be further disadvantaged in the process. In the name of making anti-poverty programmes more gender aware there has been a unilateral focus on women to the exclusion of men, and a bias towards women's condition (incomes) as opposed to position (power) (Johnson 2005). Compounded by accompanying tendencies to capitalize on essentialist notions of women's altruism (ECLAC 2004: 55; Molyneux 2006), a palpable reluctance on the part of policymakers to engage in what Cecile Jackson (1997: 152) terms 'intra-household "interference"', and a continued subordination of gender equality to the priorities of poverty reduction and economic growth, women's burdens appear to have intensified rather than lightened.

Notwithstanding the importance of income in helping women to challenge other aspects of inequality, boosting women's economic status is unlikely to go very far on its own given that the '"feminization of poverty" is . . . an issue of inequality that extends to the very basis of women's position in economic relations, in access to power and decision-making, and in the domestic sphere. It is emphatically not addressed in a sustainable manner, solely by measures to improve the material conditions of women' (Johnson 2005: 57; see also Jackson 1997).

The reiteration of 'smart economics' thinking in the World Bank's flagship World Development Report 2012 on Gender Equality and Development is unlikely to do much to counteract the seemingly inexorable tendency for gender equality and 'female empowerment' to be sacrificed to the larger goals of poverty reduction and economic growth. Aside from the fact that WDR 2012 concentrates its social policy attention to CCTs alone, as well as neglecting 'serious engagement with the gender biases of macroeconomic policy agendas' (Razavi 2011: 11), it also (re)affirms that 'greater gender equality is also smart economics, enhancing development productivity and improving other development outcomes, including prospects for the next generation and for the quality of societal policies and institutions' (World Bank 2011: xii). And even where 'smart economics' is not articulated as such, it is thinly veiled by the type of pronouncements that appeared on the World Bank's website in the run-up to WDR 2012, including 'The Bank recognizes the importance of gender equality for poverty reduction and development effectiveness' and 'One rationale for policies aimed at improving gender equality is that such policies ... will yield a large dividend in terms of economic growth'.[8]

If *gendered poverty* is to be tackled more effectively in future, it is crucial that the 'feminization of poverty' is recast to adapt to the multiple deprivations encountered by women subjectively, as well as objectively, and that more empirically informed, imaginative, holistic and less instrumentalist ways are devised to ensure the attainment of comprehensive, and sustainable, gender justice 'on the ground'.

Acknowledgements

Thanks are due to Marcelo Medeiros and Joana Costa for their kind permission to adapt Table 20.1 from their 2006 paper.

Notes

1 This chapter draws heavily on previous publications, especially Chant (2007, 2008).
2 The incidence of poverty measures the proportion of the poor in a given population and is the most commonly used indicator when assessing poverty differentials between women and men, or between female- and male-headed households. The intensity of (income) poverty is measured by the aggregated difference between the observed income of poor populations and the poverty line, while the severity of poverty refers to 'some combination of the incidence and intensity of poverty and inequality among the poor' (Medeiros and Costa, 2006: 20n).
3 References to the authors cited in Box 20.1 can be found in Chant (2007).
4 Another variant on this terminology is the 'feminization of family responsibility' coined by Lynne Brydon (in Chant 2010), and an important theoretical antecedent to Chant's formulation was Saskia Sassen's (2002) concept of a 'feminization of survival'.
5 An extended version of Molyneux's (2006) paper, which I also draw on in this chapter, is Molyneux (2007) *Change and Continuity in Social Protection in Latin America: Mothers at the Service of the State*.
6 The 'feminization of debt' was most explicitly detailed in Mayoux (2002).
7 For an extended discussion of the different paradigms underpinning microcredit programmes see Mayoux (2006).
8 These and other Bank-stated rationales for 'smart economics' can be found in Table 2 in Chant (2012).

References

ADB (2002) *Sociolegal Status of Women in Indonesia, Malaysia, Philippines, and Thailand*, Manila: Asian Development Bank, www.asianlii.org/asia/other/ADBLPRes/2002/1.pdf.

Bradshaw, S. (2002) *Gendered Poverties and Power Relations: Looking Inside Communities and Households*, Managua: ICD, Embajada de Holanda, Puntos de Encuentro, https://eprints.mdx.ac.uk/4031/1/genderedpoverties.pdf.

—— (2008) 'From structural adjustment to social adjustment: a gendered analysis of conditional cash transfer programmes in Mexico and Nicaragua', *Global Social Policy* 8(2): 188–207.

Chant, S. (1997) 'Women-headed households: poorest of the poor? Perspectives from Mexico, Costa Rica and the Philippines', *IDS Bulletin* 28:3, 26–48.

—— (2007) *Gender, Generation and Poverty: Exploring the 'Feminisation of Poverty' in Africa, Asia and Latin América*, Cheltenham: Edward Elgar.

—— (2008) 'The "feminisation of poverty" and the "feminisation" of anti-poverty programmes: room for revision?', *Journal of Development Studies* 44(2): 165–197.

—— (ed.) (2010) *International Handbook on Gender and Poverty: Concepts, Research, Policy*, Cheltenham: Edward Elgar.

—— (2012) 'The disappearing of "smart economics"? The World Development Report 2012 on gender equality: some concerns about the preparatory process and the prospects for paradigm change', *Global Social Policy* 12(2): 198–218.

—— (2013) 'Cities through a "gender lens": a golden "urban age" for women in the Global South?', *Environment and Urbanisation* 25(1): 9–29.

Chant, S. and Gutmann, M. (2000) *Mainstreaming Men into Gender and Development: Debates, Reflections and Experiences*, Oxford: Oxfam.

Cons, J. and Paprocki, K. (2010) 'Contested credit landscapes: microcredit, self-help and self-determination in rural Bangladesh', *Third World Quarterly* 31(4): 637–654.

Cornwall, A. and White, S. (2000) 'Men, masculinities and development: politics, policies and practice', *IDS Bulletin* 31(2): 1–6.

Deere, C.D., Alvarado, G.E., and Twyman, J. (2012) 'Gender inequality in asset ownership in Latin America: female owners vs household heads', *Development and Change* 43(2): 505–530.

de Vylder, S. (2004) 'Gender in poverty reduction strategies', in G. Johnsson-Latham (ed.), *Power and Privileges: Gender Discrimination and Poverty*, Stockholm: Regerinskanliet, pp. 82–107.

ECLAC (2004) *Roads Towards Gender Equity in Latin America and the Caribbean*, Santiago de Chile: Economic Commission for Latin America and the Caribbean, www.eclac.org/publicaciones/xml/7/14957/lcl2114i.pdf.

Goetz, A.M. and Sen Gupta, R. (1996) 'Who takes the credit? Gender, power and control over loan use in rural credit programmes in Bangladesh', *World Development* 24(1): 45–63.

Harman, S. (2010) *Why Conditional Cash Transfers Matter for HIV*, London: City University, www.city.ac.uk/intpol/policy-briefs/harman-conditional-cash-transfers-HIV.html.

Jackson, C. (1997) 'Post poverty, gender and development', *IDS Bulletin* 28(3): 145–155.

Johnson, R. (2005) 'Not a sufficient condition: the limited relevance of the gender MDG to women's progress', in C. Sweetman (ed.), *Gender and the Millennium Development Goals*, Oxford: Oxfam, pp. 56–66.

Johnsson-Latham, G. (2004) 'Understanding female and male poverty and deprivation', in G. Johnsson-Latham (ed.), *Power and Privileges: Gender Discrimination and Poverty*, Stockholm: Regerinskanliet, pp. 16–45.

Kabeer, N. (1996) 'Agency, well-being and inequality: reflections on the gender dimensions of poverty', *IDS Bulletin* 27(1): 11–21.

Lampietti, J. and Stalker, L. (2000) *Consumption Expenditure and Female Poverty: A Review of the Evidence*, Policy Research Report on Gender and Development, Working Paper Series No.11, Washington, DC: World Bank, Development Research Group/Poverty Reduction and Economic Management Network.

Mayoux, L. (2001) 'Tackling the downside: social capital, women's empowerment and micro-finance in Cameroon', *Development and Change* 32(3): 435–463.

—— (2002) *Women's Empowerment or the Feminisation of Debt? Towards a New Agenda in African Microfinance*, paper presented at One World Action Conference, London, 21–22 March, www.oneworldaction.org/Background.htm.

—— (2006) *Women's Empowerment through Sustainable Micro-finance: Rethinking 'Best Practice'*, Discussion Paper, Gender and Micro-finance website, www.genfinance.net.

Medeiros, M. and Costa, J. (2006) *Poverty Among Women in Latin America: Feminisation or Over-representation?* Working Paper No.20, International Poverty Centre, Brasilia, www.ipc-undp.org/pub/IPCWorking Paper20.pdf.

Molyneux, M. (2006) 'Mothers at the service of the new poverty agenda: Progresa/Oportunidades, Mexico's Conditional Transfer Programme', *Journal of Social Policy and Administration* 40(4): 425–449.

—— (2007) *Change and Continuity in Social Protection in Latin America: Mothers at the Service of the State*, Gender and Development Paper 1, Geneva: United Nations Research Institute for Social Development.

Razavi, S. (2011) *World Development Report 2012: Gender Equality and Development: An Opportunity Both Welcome and Missed (An Extended Commentary)*, Geneva: UNRISD, www.unrisd.org/80256B42004C CC77/(httpInfoFiles)/E90770090127BDFDC12579250058F520/$file/Extended%20Commentary%20 WDR%202012.pdf.

Safa, H. (1998) 'Female-headed households in the Caribbean: sign of pathology or alternative form of family organisation?', *The Brown Journal of World Affairs* 5(2): 203–214.

Sassen, S. (2002) 'Counter-geographies of globalisation: feminisation of survival', in K. Saunders (ed.), *Feminist Post-Development Thought*, London: Zed Books, pp. 89–104.

Tabbush, C. (2010) 'Latin American women's protection after adjustment: a feminist critique of conditional cash transfers in Chile and Argentina', *Oxford Development Studies* 38(4): 437–451.

World Bank (2011) *World Development Report 2012: Gender Equality and Development*, Washington DC: World Bank.

21

AT HOME IN THE CITY?

Gender and urban poverty

Ann Varley

Introduction

Cities have long been the poor relation of gender and development. They have suffered from an inverse 'urban bias', whereby rural problems are seen as more pressing (Masika et al. 1997). In a world where more than half the population is now urban, however, neglecting cities is more difficult to justify than ever. In this chapter I ask what difference city life makes to gendered poverty, and what difference gender makes to urban poverty.

When we ask such questions about urbanization, gender, and poverty in developing regions, we find echoes of ideas and experiences from elsewhere. For instance, the belief that women who move to the city to find work are immoral (a threat to the patriarchal order) has frequently accompanied industrialization. When a young migrant to Beijing reports that people in her village believe 'girls who go out will become bad' (Hairong 2008: 45), her comment resonates with the notion that 'mill girls' in Victorian Britain were not 'respectable' (Gordon 2002: 37). When men advertising for marriage partners in Sri Lanka stipulate that workers from Colombo's garment factories need not apply (Tambiah 1997: 29), they express much the same concern as lonely hearts advertisers in Mexico who sought wives free from 'the vices of modernity and the big city' (Rubenstein 1998: 68).

Women working in the Sri Lankan factories are known by the derogatory term 'Juki girls',[1] but some garment workers escape this branding: 'They are assumed to be "good" because they are still living in their villages' (Lynch 2007: 155). Such distinctions constitute a gendered variation on the binary representations of country and city identified, for Britain, by Raymond Williams (1973: 1):

> On the country has gathered the idea of a natural way of life: of peace, innocence, and simple virtue. On the city has gathered the idea of an achieved centre: of learning, communication, light. Powerful hostile associations have also developed: on the city as a place of noise, worldliness and ambition; on the country as a place of backwardness, ignorance, limitation.

Such similarities point to the need to reject assumptions of incommensurable difference between Global North and South. They are not, however, evidence of uniformity or convergence (Varley 2013). Similarities conceal cultural and historical specificities. The context of globalization and foreign investment in Sri Lankan garment factories adds a national(ist) significance to the distinction between 'good girls' and 'Juki girls' that would have no equivalent for Victorian 'mill girls' in Britain. The move to the city is also gendered in different ways. The motivation for young women in post-Mao China to 'go out', for instance, arises in part from ideological and material changes to rural life that have undermined the self-worth women could once gain through agricultural labour and 'speaking bitterness' (Hairong 2008: 46; Jacka 2006: 57). Such changes, restricting women's activities to 'moving around the stove', leads young migrants to speak consistently of life in the Chinese countryside as 'inert' and 'meaningless' (Hairong 2008: 46). As one put it, 'If I had to live the life that my mother has lived, I would choose suicide' (Hairong 2008: 25).

The difference a city makes

Few people would think of the provision of 'urban' services such as water, sanitation, and waste disposal as a key *gender* policy, but an analysis conducted for the *World Development Report 2012* suggests that such a description might not be inappropriate.[2] The analysis focuses on 'missing women': excess female mortality, calculated by comparing 'the mortality risks of women relative to men in every country and every age' with the corresponding ratios for high-income countries (World Bank 2011: 120). Early childhood emerges as one of the most dangerous times, accounting for about 10 per cent of the 'missing women'. The analysts then undertake a similar exercise comparing the statistics for high-income countries today with the historical record. While European countries and the United States faced similar problems to today's poorer countries at the start of the twentieth century, the excess female mortality in early childhood had virtually disappeared by the 1930s (World Bank 2011: 125). The dramatic change is explained by investments in clean water and sanitation. It was a reduction in infant mortality occasioned by the resulting decline in infectious diseases that led to the disappearance of this excess female mortality (World Bank 2011: 126).[3] Similar outcomes are observed today. Where countries such as Argentina, Bangladesh, China, and Vietnam have extended water and sanitation networks, they have experienced declines in infant mortality and, as a result, in excess female mortality in early childhood. By contrast, in West Africa, where the share of urban households with piped water has *fallen* in recent decades, the decline has been much slower (World Bank 2011: 127, 138).

Living in urban areas, with better provision of water and sanitation, is, then, one reason for more children surviving early childhood, and benefits girls even more than boys. But it is not only after birth that girls benefit from urbanization. The girls who are never born – a result of sex-selective abortion – account for nearly one-quarter of the world's missing women (World Bank 2011: 120). The experience of South Korea shows that the problem can be reduced by urbanization and the spread of urban values. The male–female sex ratio at birth first increased dramatically despite an impressive record of social and economic development, including increases in education and employment for women; but it has declined since the mid-1990s (Chung and Das Gupta 2007). This turnaround is a result of material changes associated with urbanization that have rendered the roles sons play in their parents' lives less crucial. Urbanization makes it more likely that daughters (who would previously have moved to their husband's village) might still be able to provide help, but also that sons might no longer live nearby; reduces community pressure on sons to provide support; and creates alternative sources of support, such as faith-based social networks, in which women play key roles. Such factors have contributed to a decline in the drive to have sons that has now become generalized (rural groups are currently displaying the fastest decline in son preference) (Chung and Das Gupta 2007: 778).

Urbanization has not had a similar impact on son preference elsewhere in Asia, but the Korean example is striking because it illustrates how material differences can promote normative change. Rural and urban are not separate worlds. Cable TV in rural India led to increases in women's autonomy together with a decline in the share of women reporting son preference and the belief that it is acceptable for a man to beat his wife. It also reduced the differences between rural and urban women by between 40 and 70 per cent (Jensen and Oster 2007: 3, 43).

The Korean experience also introduces an important theme linked to urbanization: a decline in patrilocality, the custom of newly-weds living with the husband's parents. This practice has been widely associated with mothers-in-law behaving in a domineering and sometimes violent fashion towards young wives, leading for example to ante- and post-natal depression in their victims (ICRW 2000; Kishor and Johnson 2004; Roomruangwong and Epperson 2011). The practice declines with a shift to urban residence, because of the alternative housing and income-earning options in the city. That city life makes a big difference was made clear to me during a group interview with women in the Mexican city of Puebla in 1995. In the countryside, one woman declared to the murmured agreement of others, you have no option: you *have* to go and live with your in-laws. Participants in a later study, in Guadalajara, mostly reported a poor relationship with their mother-in-law. Living with the in-laws, said one woman, was like 'scratching yourself with your own nails' (Varley 2013: 136).

The ability to escape early marriage, unhappy relationships, or overbearing in-laws is, then, an important reason for some women to migrate to urban areas, despite the particularities of urban poverty and the double marginalization poor women face, through gender and through their material and social circumstances. Migration is not always a woman's own decision, but even in regions where more men than women have historically migrated to cities, the proportion of women moving on their own has grown (Masika et al. 1997). Some women have little option but to leave, for their own safety, or because they need to support their children, whether or not they are married. In parts of Africa and Asia, in-laws deny widows inheritance of the means of support: something experienced by almost two-fifths of the currently unmarried women who had migrated to one of Africa's largest informal settlements, in Nairobi (COHRE 2008: 31). If these women's husband had died of AIDS, disinheritance and loss of his income could be accompanied by hostility from relatives wary of infection. Parallel research in Ghana found that some wives in polygamous marriages had been forced to migrate to Accra to seek a living when their husband failed to support them (COHRE 2008: 36, 98). It is also probable that women preferring relationships with other women will find this somewhat less difficult in cities (Jolly 2010).

City life is commonly perceived as changing not only women's lives but also gender relations. The greater quantity and variety of employment opportunities is one reason, despite women's association with the informal sector and low earnings. Although changes in women's education and lower fertility rates may be more significant drivers of women's employment than urbanization, the fact that female labour force participation rates in Latin American have risen faster for married than for single urban women points to the importance of normative change (Chioda et al. 2011: 51, 60–82). Kin-based social networks may be weakened in cities, for better or for worse, but there are alternatives, such as rotating savings and credit associations, in which women often play a key role. The *World Development Report 2012* argues that it is the combination of income generation and social capital that is crucial, giving urban women more say about how many children they bear and providing resources with which to combat domestic violence (World Bank 2011: 95).

It is not only migration that challenges the association of women with the home and 'inside' (Davin 1996: 27). Attitudes to women's mobility outside the home are likely to change, particularly when they are contributing to household income. Research with married couples in Ethiopia found that the proportion of women needing to ask their husband for permission to work was twice as high in rural locations as in the capital, and the percentage of urban men believing that wives should hand over their income only half that for rural men (Tarazona and Munro 2011).[4] Although commuting may involve long, tiring, and sometimes dangerous journeys on crowded public transport, the 'legitimacy' of such movements helps erode male control over women's mobility. In countries where many women need male permission to visit a clinic or hospital, fewer do so in urban areas (World Bank 2011: 176). Women interviewed in the capital of Odisha State, India, cited being able to 'go out of house to do marketing, shopping, and other household work such as paying . . . bills' or enjoying their 'husband's support and permission to go outside of the house' in response to questions about what they understood by being 'powerful and free' (World Bank 2011: 94). Younger women compared themselves with their mothers, who 'never came out of their houses, so they did not even know what was happening outside' (Muñoz Boudet et al. 2012: 58).

The changes in women's lives often find expression in changes to their style of dress and appearance. Such bodily refashioning reflects enthusiasm for new experiences. When asked about the main changes to their personal lives since arriving in Beijing, one-quarter of domestic workers surveyed mentioned 'learning to dress well', and one-half 'knowing a lot of new things that [they] didn't know before' (Hairong 2008: 161). Demeanour, gesture, voice, and willingness to meet someone's eyes can all give physical expression to such a sense of change (Haque 2000).

The emphasis on new ways of presenting the self is not just skin-deep consumerism. Yan Hairong (2008: 161) identifies what is at stake in her discussion of disparaging remarks made by migrant women about the appearance and behaviour of their male counterparts in Beijing, such as:

> You still see them squat there eating their meals at construction sites, baring their chests, and watching TV together in large crowds. It's as if they still live a village life in the city. That's why some of us migrant women are not attracted to this kind of male migrant.

Women have more to lose in marriage than men, as they may find their liberty curtailed by both husband and in-laws (Hairong 2008: 162). Urban men are reluctant to marry rural women, so these women want to marry migrants with more liberal attitudes. They 'demand more modernity of their potential marriage partners', such that 'their quests for personal development and new self-expression do not mesh well with courtship' (Hairong 2008: 162).

Across a wide range of countries, demands for equality are more likely to be voiced by urban than rural women, and less negatively received by urban than rural men (Muñoz Boudet et al. 2012: 34–37). City life fosters such changes, presenting practical limits to the feasibility of adherence to established gender norms; but pragmatic adaptation to circumstances also leads to broader and more sustained normative transformations.

Cities, in short, represent more opportunities for women to earn a living; a weakening of social control by kin-based networks; and greater likelihood of normative change. This observation in no way denies the conditions of urban poverty and insecurity in which many people find themselves living, which may even lead to an urban health 'penalty' (Hawkins et al. 2013; Ruel et al. 1999). Such conditions often represent a greater immediate burden for women, given the likelihood that they undertake more domestic labour and spend more time around the home, on both reproductive and productive activities (such as homeworking). They also add to the stresses of parenting (Meth 2013). What life in the city nonetheless means to women may, perhaps, be illustrated by the responses of residents of two informal settlements in Nairobi to questions about how satisfied they were with their location (Mudege and Zulu 2011). Their views were by no means unanimous. Many expressed a desire to return to rural areas, but having family in Nairobi, not being the decision-maker or not owning land or a house outside the city meant some women felt they had no option but to stay. Women were, nonetheless, less likely than men to say they wanted to leave Nairobi when older, mostly because they could make some sort of living and enjoyed some independence in the city. Two quotations make the point:

> Things are better than they were before. I have an income which is helping me to sustain myself, my mother and the children . . . Now I can do my own things at my own pace without any interference. I can take care of myself without having to depend on anyone.
>
> *(Mudege and Zulu 2011: 225)*

> I would say I am satisfied because I managed to educate my children . . . nobody has interfered with me, no one has stolen from me or beaten me, and no misfortune has befallen me.
>
> *(Mudege and Zulu 2011: 227)*

Living in here

Comparing rural and urban life almost inevitably predisposes us to talk about migration and migrants, but most cities are growing more through natural increase. It is important, then, to ask what 'living in here' means.[5] Life in the city also brings problems attributable in part to gender norms or that are experienced differently by women and men. This section therefore considers a number of gender and development issues associated with urban living.[6]

Finding a home in the city

Urban residence has generally increased women's housing options, reducing reliance on one's relatives or in-laws for accommodation. In East Africa, for instance, urban women have better access to land and housing than their rural sisters (Benschop 2002: 15). Women are nonetheless at a considerable disadvantage, compared to men, in securing access to a home in the city.

Discriminatory inheritance practices are often discussed as a rural practice, but in Tanzania the problem 'reaches deep into urban areas and firmly entrenches male dominance' (Hughes and Wickeri 2011: 836). Widows are often disinherited by their in-laws. Childless widows are the most vulnerable, subject to 'near-certain expulsion from the marital home' (Hughes and Wickeri 2011: 837). Widows' ability to resist such practices is limited (for example, by the need for in-law support if they are to remarry) (Hansen 1996). Inheritance has been described as the major urban housing issue for poor women in African and Islamic countries, and the problem has been aggravated by AIDS deaths (Peters 1996; Sweetman 2008); but women elsewhere can also be disinherited by close relatives (Varley 2010).

Women face particular problems in gaining access to credit, because informal sector work means many are unable to prove a regular income, and they are therefore often excluded from low-income housing schemes. Even if they are able to access land to build on, they face major barriers to building a home as they lack appropriate skills as well as lower incomes. This is one reason why women-headed households tend to be overrepresented among the tenant population, although rental accommodation in central areas may also be attractive because of access to employment opportunities and services (Gilbert and Varley 1991: 114). In our survey of 405 low-income households in Guadalajara, Mexico, for example, female-headed households comprised 27 per cent of tenants compared with 17 per cent of all households surveyed, and 39 per cent of tenants in a central area with a lot of rental accommodation. A study in Botswana found that women living by themselves dominated the rental market in Gaborone (their children were with relatives in rural areas) (Datta 1996). But women face disadvantages even as tenants: suspicion of single women's morality means landlords are reluctant to rent to single women (Gilbert and Varley 1991: 169; Hughes and Wickeri 2011: 859). As participants in our Guadalajara study put it, 'a widow or a woman separated from her husband needs to look out for herself much more . . . [she is] fair game for all and sundry'. Women without male support are also less able to resist rent hikes, arbitrary eviction, or landlord opposition to their running a business from home. *Letting* accommodation has, however, served as a good way for women of relatively modest means to support themselves (Datta 1995; Hughes and Wickeri 2011; Varley 1995).

Some women are able to build their own home, with the help of paid labour, although ensuring that the workers do not cheat them may be more difficult for them than for a man, as women in Guadalajara noted. They may also join in. One woman told us how, every afternoon, she would go to help the bricklayer, until her fingertips were 'running with blood, which is why I value my *rinconcito* [little corner]'.

Critics of the impact of property titling in informal settlements in recent years have suggested that it is bad for women, because it collapses overlapping rights in the home into ownership in the husband's name. Creating a public record of ownership can, however, help defend women entitled to half the marital property against a husband seeking to sell their home (Varley 2007). Women in different parts of the world regard joint title, in particular, as protecting their home and helping to guarantee their personal safety (ICRW 2006; Varley 2010). In urban India, titling may even prove to be 'women's best chance of becoming property owners' (Datta 2006: 292).

Urban services

Planners are often accused of gender bias. They fixate on a model of breadwinning that matches the classic pattern of factory or business districts rather than the environments where women often work (Sweetman 1996). Alternatively, they simply prioritize male activities – providing, say, sports-oriented playgrounds without considering girls' leisure activities (Peters 1996). When such bias interacts with gendered limitations on women's activities, the combination can be fatal.

Urban transport and social amenities provide one example. Men generally have first call on private transport, making women dependent on public transport or walking. A study of street vendors in Hanoi found that whereas more than two-thirds of women vendors carried their goods from door to door, almost three-fifths of the men used a bicycle (UN-Habitat 2004: 5). Women's childcare and domestic responsibilities mean they make trips to a variety of locations, rendering them relatively more seriously affected by accessibility restrictions in the peripheral areas that tend to be the worst served by public transport. Making multiple short journeys means they may have to use more expensive informal transport such as shared taxis. Researchers in Durban, South Africa, found that where women work outside the locality, they face higher overall travel times (and costs) than men, making it particularly important for them to enjoy good access to social amenities, such as health clinics (Venter et al. 2007: 674).

Poor services can threaten women's physical integrity. Inadequate sanitation in informal settlements is a case in point. It obliges people to use makeshift alternatives, including defecating in the open air. Women in urban India and Africa prefer to do so in the dark; but this exposes them to attack (Amnesty International 2010; COHRE 2008). One woman in Delhi, where squatter settlement residents use a nearby urban forest described as the 'jungle', recounted one experience to Ayona Datta (2012: 122–123):

> I went to the jungle for toilet, there were three men there . . . And they were trying to touch my chest, and I turned myself as I passed them . . . So I kept waiting there for them. When they came . . . I hit them with a stick. I was so angry . . . Then some other men came to my help and they beat them up too.

Middle-class efforts to protect the forest led to construction of a wall separating it from the settlement. People had to use ladders and piles of bricks on the forest side as steps, leading to nightly injuries as sentries along the wall removed the bricks or beat up residents trying to cross it (Datta 2012: 130–131).

An urban future

Older women are often marginalized in feminist scholarship, including work on gender and development (Varley 2013). In just a few years' time, however, older people in cities will outnumber those in rural areas in all regions (UNFPA 2007: 29).[7] As with the urban population in general, then, it becomes ever more difficult to justify neglecting the older population in cities; and there is a particular gender dynamic to ageing. Women's greater life expectancy and tendency to marry older men, plus men's greater likelihood of remarrying, mean that older widows far outnumber older widowers, and widowhood is set to become an increasingly common experience in the world's poorer cities.[8]

The effects of urbanization are often depicted in gloomy terms, suggesting that older people will be left to fend for themselves. There is, however, no consistent or substantial difference between their living arrangements in urban and rural areas of developing countries. In some countries they are more likely to live alone, or to live with adult children, in urban than in rural areas; in others, vice versa (United Nations 2005: 69–72). In addition, levels of wellbeing and support cannot be read directly from living arrangements (Palloni 2001). In Buenos Aires, children may be a burden rather than a resource (Lloyd-Sherlock and Locke 2008); and living with relatives does not guarantee wellbeing. For instance: the balance of power between mothers- and daughters-in-law can shift in later life, and even lead to abuse of the older woman (Varley and Blasco 2003). The effects of this shift are exacerbated by historical changes, including urbanization, that undermine the mother-in-law role. Rising suicide rates among older women in Taiwan have been interpreted as reflecting their resultant loss of status (Hu 1995: 208). Rates are higher in rural areas, but younger women in low-income areas of Chennai, India, associated the shift in power relations with city life: 'here it is the daughter-in-law who lies on the bed ordering her mother-in-law around' (Vera-Sanso 1999: 577).

Partly for such reasons, some older women say they prefer to live alone ('to avoid burdening relatives'). Men are unlikely to express similar views, perhaps because it is a fairly unlikely prospect, demographically, so not one they think about much. However, Mexican census data show that while older women are more likely to live alone, the reverse applies for *unmarried* older people: men are then more likely to live alone (Varley and Blasco 2003: 529).[9] The same applies elsewhere in Latin America (Palloni 2001). This can be explained partly in terms of women's domestic skills meaning they continue to be seen as useful to relatives, whereas men may be dismissed as 'useless' once they cease to provide; but also of family resentment of men's prior neglect or mistreatment of their wives or children (Varley and Blasco 2000). In this context, gender comes to work to men's disadvantage, and it also does so beyond the family/household. Women's neighbourhood-based social networks are more likely to endure than men's work-based ones, particularly when men worked elsewhere in the city. Consequently, male community leaders in an informal settlement in Nairobi observed that:

> [I]t is the women who come together to form groups that can assist them in times of need . . . Elderly men do not live for so long. Because they worry a lot they die before their age.
>
> *(Mudege and Ezeh 2009: 252)*

Afterword

Elizabeth Pritchard concludes her analysis of how postmodern feminist theorists reproduce the Western rhetoric of mobility as progress that justifies development interventions by asserting that:

> Feminists cannot and should not avoid telling spatial stories that narrate women's escape from their confinement to certain places – but we can do so in a more self-critical manner. We can tell of such maneuvers only in the guilty context of colonialism.
>
> *(Pritchard 2000: 62)*

Pritchard's caution lends a postcolonial slant to the critique of dualistic oppositions of country and city. Closure, stasis, and tradition, on the one hand, and mobility, openness, and development, on the other, can easily be mapped on to the representations of rural and urban discussed by Raymond Williams (1973). Migration is literal displacement or dislocation; and women's migration to the city chimes with the postmodern fondness for mobile *subjectivities*. To acclaim urban life as emancipatory thus reproduces a (gendered) understanding of development as heroic liberation from the confines of tradition (Gibson-Graham 1996; Pritchard 2000; Scott 1995).

We should not, however, ignore the real constraints on women's lives or simply invert the valorization of rural and urban, dwelling on images of 'drudgery and dire poverty' (UN-Habitat 2004: 3) that perpetuate hierarchies of difference between cities of the Global South and North (Robinson 2006). (Post)development rejection of 'closure' and celebration of openness run the risk that 'fixation on a "way out" forbids the sketch of a "way in" to somewhere else' (Pritchard 2000: 60). In other words, fear of reinscribing a Western valorization of mobility and development in our representations of the urban should not prevent us from attending to gendered injustice. Nor should we forget that citizens use Western discourse for their own gendered purposes: for better or for worse, but not *always* for the worse.

Notes

1   'Juki' is the name of a Japanese sewing-machine brand.
2   Remarks by Ana María Muñoz at the Report's UK launch, Chatham House, London, 23 November 2011.
3   Girls are 'more robust than boys', but particularly so 'for perinatal conditions relative to infectious diseases', such that a decline in the share of deaths caused by infectious diseases increased girls' advantage over boys and led to the disappearance of excess female mortality (by comparison with today) (World Bank 2011: 126).
4   Women and men in Addis Ababa were consistently less conservative in relation to eight normative beliefs (Tarazona and Munro 2011: Table 4).
5   The title of Mexico's contribution to the 2008 Venice Architecture Biennale. Curated by Javier Sánchez, the exhibition sought 'in here' alternatives to the isolated extra-urban housing estates that have appeared in Mexico (and elsewhere) in recent years.
6   The issues addressed reflect my own research interests; they do not pretend to be comprehensive.
7   Not currently the case for Africa or Asia.
8   Women outnumber men by 3:1 among unmarried people aged 60 or over in less developed regions (calculated from United Nations 2012).
9   One-quarter of unmarried men aged 60 or over lived alone in 2000 compared with one-sixth of unmarried older women in larger localities, although there were still over two such women living alone for every such man doing so. The gap is slightly narrower in larger localities.

References

Amnesty International (2010) *Insecurity and Indignity: Women's Experiences in the Slums of Nairobi, Kenya*, London: Amnesty International.
Benschop, M. (2002) *Rights and Reality: Are Women's Rights to Land, Housing and Property Implemented in East Africa?* Nairobi: UN-Habitat.
Chioda, L., with Garcia-Verdú, R. and Muñoz Boudet, A.M. (2011) *Work and Family: Latin American Women in Search of a New Balance*, Washington, DC: World Bank.
Chung, W. and Das Gupta, M. (2007) 'The decline of son preference in South Korea: the roles of development and public policy', *Population and Development Review* 33(4): 757–783.
COHRE (2008) *Women, Slums and Urbanisation: Examining the Causes and Consequences*, Geneva: Centre on Housing Rights and Evictions.
Datta, A. (2012) *The Illegal City: Space, Law and Gender in a Delhi Squatter Settlement*, Farnham: Ashgate.
Datta, K. (1995) 'Strategies for urban survival? Women landlords in Gaborone, Botswana', *Habitat International* 19(1): 1–12.
—— (1996) 'Women owners, tenants and sharers in Botswana', in A. Schlyter (ed.), *A Place to Live: Gender Research on Housing in Africa*, Uppsala: Nordiska Afrikainstitutet.
Datta, N. (2006) 'Joint titling – a win-win policy? Gender and property rights in urban informal settlements in India', *Feminist Economics* 12(1–2): 271–298.
Davin, D. (1996) 'Gender and rural-urban migration in China', *Gender and Development* 4(1): 24–30.
Gibson-Graham, J.K. (1996) *The End of Capitalism (As We Knew It): A Feminist Critique of Political Economy*, Minneapolis: University of Minnesota Press.
Gilbert, A.G. and Varley, A. (1991) *Landlord and Tenant: Housing the Poor in Urban Mexico*, London: Routledge.
Gordon, L. (2002) *Mill Girls and Strangers: Single Women's Independent Migration in England, Scotland, and the United States, 1850–1881*, Albany: State University of New York Press.
Hairong, Y. (2008) *New Masters, New Servants: Migration, Development, and Women Workers in China*, Durham, NC: Duke University Press.
Hansen, K.T. (1996) 'Gender, generation, and access to housing in Zambia', in A. Schlyter (ed.), *A Place to Live: Gender Research on Housing in Africa*, Uppsala: Nordiska Afrikainstitutet.
Haque, G.T.M. (2000) *Lived Experiences of Empowerment: A Case Study of a Vocational Training Programme for Women in Bangladesh*, PhD thesis, University College London.
Hawkins, K., MacGregor, H., and Oronje, R. (2013) *The Health of Women and Girls in Urban Areas with a Focus on Kenya and South Africa: A Review*, Brighton: Institute of Development Studies.

Hu, Y.-H. (1995) 'Elderly suicide risk in family contexts: a critique of the Asian family care model', *Journal of Cross-Cultural Gerontology* 10(3): 199–217.

Hughes, K. and Wickeri, E. (2011) 'A home in the city: women's struggle to secure adequate housing in urban Tanzania', *Fordham International Law Journal* 34(4): 788–929.

ICRW (2000) *Domestic Violence in India: A Summary Report of Four Records Studies*, Washington, DC: International Center for Research on Women.

—— (2006) *Property Ownership and Inheritance Rights of Women for Social Protection: The South Asia Experience: Synthesis Report of Three Studies*, Washington, DC: International Center for Research on Women.

Jacka, T. (2006) *Rural Women in Urban China: Gender, Migration, and Social Change*, Armonk, NY: M.E. Sharpe.

Jensen, R. and Oster, E. (2007) *The Power of TV: Cable Television and Women's Status in India*, Working Paper 13305, National Bureau of Economic Research, Cambridge, MA.

Jolly, S. (2010) *Poverty and Sexuality: What Are the Connections? Overview and Literature Review*, Stockholm: Swedish International Development Cooperation Agency.

Kishor, S. and Johnson, K. (2004) *Profiling Domestic Violence: A Multi-Country Study*, Calverton, MA: ORC Macro.

Lloyd-Sherlock, P. and Locke, C. (2008) 'Vulnerable relations: lifecourse, wellbeing and social exclusion in Buenos Aires, Argentina', *Ageing and Society* 28(8): 1177–1201.

Lynch, C. (2007) *Juki Girls, Good Girls: Gender and Cultural Politics in Sri Lanka's Global Garment Industry*, Ithaca, NY: Cornell University Press.

Masika, R., with de Haan, A. and Baden, S. (1997) *Urbanisation and Urban Poverty: A Gender Analysis*, Report No. 54, BRIDGE, Institute of Development Studies, University of Sussex.

Meth, P. (2013) '"I don't like my children to grow up in this bad area": parental anxieties about living in informal settlements', *International Journal of Urban and Regional Research* 37(2): 537–555.

Mudege, N.N. and Ezeh, A.C. (2009) 'Gender, aging, poverty and health: survival strategies of older men and women in Nairobi slums', *Journal of Aging Studies* 23(4): 245–257.

Mudege, N.N. and Zulu, E.M. (2011) 'In their own words: assessment of satisfaction with residential location among migrants in Nairobi slums', *Journal of Urban Health* 88(2): Supplement 219–234.

Muñoz Boudet, A.M., Petesch, P., and Turk, C., with Thumala, A. (2012) *On Norms and Agency: Conversations about Gender Equality with Women and Men in 20 Countries*, Washington, DC: World Bank.

Palloni, A. (2001) 'Living arrangements of older persons', *Population Bulletin of the United Nations. Special Issue 'Living Arrangements of Older Persons: Critical Issues and Policy Responses'*, 42/43: 54–110.

Peters, C. (1996) 'Chris Peters talks to Catalina Trujillo about her work for the UN Agency Habitat', *Gender and Development* 4(1): 53–56.

Pritchard, E.A. (2000) 'The way out West: development and the rhetoric of mobility in postmodern feminist theory', *Hypatia* 15(3): 45–72.

Robinson, J. (2006), *Ordinary Cities: Between Modernity and Development*, London: Routledge.

Roomruangwong, C. and Epperson, C.N. (2011) 'Perinatal depression in Asian women: prevalence, associated factors, and cultural aspects', *Asian Biomedicine* 5(2): 179–193.

Rubenstein, A. (1998) *Bad Language, Naked Ladies, and Other Threats to the Nation: A Political History of Comic Books in Mexico*, Durham, NC: Duke University Press.

Ruel, M.T., Haddad, L., and Garrett, J.L. (1999) 'Some urban facts of life: implications for research and policy', *World Development* 27(11): 1917–1938.

Scott, C.V. (1995) *Gender and Development: Rethinking Modernization and Dependency Theory*, Boulder, CO: Lynne Rienner.

Sweetman, C. (1996) 'Editorial', *Gender and Development* 4(1): 2–8.

—— (2008) *How Property Titles Make Sex Safer: Women's Property Rights in an Era of HIV*, Oxford: Oxfam.

Tambiah, Y. (1997) 'Women's sexual autonomy: some issues in South Asia', *Options* 9(1): 28–32.

Tarazona, M. and Munro, A. (2011) *Experiments with Households in Four Countries*, London: Economic and Social Research Council and Department for International Development.

UNFPA (2007) *State of World Population 2007*, New York: United Nations Population Fund.

UN-Habitat (2004) *Gender, Culture and Urbanization*, paper prepared for World Urban Forum, Barcelona, 13–17 September, HSP/WUF/2/11.

United Nations (2005) *Living Arrangements of Older Persons Around the World*, New York: Department of Economic and Social Affairs, United Nations ST/ESA/SER.A/240.

—— (2012) *Population Ageing and Development*, New York: Department of Economic and Social Affairs, United Nations.

Varley, A. (1995) 'Neither victims nor heroines: women, land and housing in Mexican cities', *Third World Planning Review* 17(2): 169–182.

—— (2007) 'Gender and property formalization: conventional and alternative approaches', *World Development* 35(10): 1739–1753.

—— (2010) 'Modest expectations: gender and property in urban Mexico', *Law and Society Review* 44(1): 67–100.

—— (2013) 'Feminist perspectives on urban poverty: de-essentializing difference', in L. Peake and M. Rieker (eds.), *Rethinking Feminist Interventions into the Urban*, London: Routledge.

Varley, A. and Blasco, M. (2000) 'Exiled to the home: masculinity and ageing in urban Mexico', *European Journal of Development Research* 12(2): 115–138.

—— (2003) 'Older women's living arrangements and family relationships in urban Mexico', *Women's Studies International Forum* 26(6): 525–539.

Venter, C., Vokolkova, V., and Michalek, J. (2007) 'Gender, residential location, and household travel: empirical findings from low-income urban settlements in Durban, South Africa', *Transport Reviews* 27(6): 653–677.

Vera-Sanso, P. (1999) 'Dominant daughters-in-law and submissive mothers-in-law? Co-operation and conflict in South India', *Journal of the Royal Anthropological Institute* 5(4): 577–593.

Williams, R. (1973) *The Country and the City*, London: Chatto & Windus.

World Bank (2011) *World Development Report 2012: Gender Equality and Development*, Washington, DC: World Bank.

22

CARIBBEAN KINSHIP RESEARCH

From pathology to structure to negotiated family processes

Christine Barrow

The evolution of Caribbean kinship research is complex and contradictory. Centred almost exclusively on Afro-Caribbean[1] families and households in poor, rural communities since anthropology and social work took root in the region, the trajectory takes in matrifocality, male marginality, female-headed households and transnational kinship networks. In this discursive, thematic review, we interrogate these and other conceptualisations of Caribbean families proposed by key researchers across the region. En route, we explore the intersection of ideology, theory and ethnography, as deficit models of family "dysfunction" and "breakdown" were replaced by praise-songs of strength and resilience; as household structure and function gave way to kinship ideology and culture; and as imperialist policies to reconstruct families in line with nuclear normativity were replaced by the idea of family as private space beyond state intervention. Today, kinship has fallen off the radar of Caribbean scholarship. We propose a revitalisation using the interpretive frame of *negotiated family processes*.

The foundation: family pathology and reconstruction

The study of Caribbean families dates back to the 1940s, to the post-emancipation imperial project for the social reconstruction and moral upliftment of black populations out of poverty and backwardness. To this end, and precipitated by the riots that spread across the Caribbean during the late 1930s, the West India ("Moyne") Commission was established. It drew attention to the social ills of family life, attributing "illegitimacy" and "promiscuity" to the absence of male household headship and financial support. Support for this family pathology came from a band of social welfare experts subsequently dispatched by the Colonial Development and Welfare Office in London, the most influential among them being Thomas Simey. Such stigmatisation persisted for decades and was not confined to the Caribbean. In the US, the influential Moynihan Report of 1965 emphatically enforced the stereotype of African-American families as "highly unstable" and as a "tangle of pathology".

In the Caribbean, colonial ideology backed by its own brand of welfarism set the stage for deficit modelling. Perhaps more than any others the world over, Afro-Caribbean families became the target of censure, riddled as they were seen to be by an array of social and moral faults (Barrow 2001). Commentators were especially shocked at illegitimacy levels as high as 70 per cent. along with low marriage rates, unstable conjugality, single mothers and absentee husbands and fathers. Judged against the Eurocentric ideals of Christian marriage, monogamy and nuclear co-residence, black families were "disintegrate" and "loose" and conjugal unions, although occasionally faithful and enduring, were more often "promiscuous" and "transitory" (Simey 1946: 15–16, 84). Intensifying the moral panic was the "culture of poverty" thesis claiming that, since the family was the cornerstone of society, any deviation from the nuclear model, and its male–female division of family labour, would perpetuate a plethora of social problems. Illegitimacy and promiscuity were breeding grounds for juvenile delinquency, lawlessness and poverty. Family reconstruction then, was the prerequisite for social order, development and civilisation. However, despite harnessing the state, the church, the school and other moral authorities, the experiments in social engineering that followed were doomed to failure. Examples range from the notorious Mass Marriage Movement in Jamaica, a campaign designed to eliminate the evils of promiscuity, to gender-segregated education – for girls in basic reading and writing and the feminine refinements of housewifery and French conversation, while their future husbands were prepared for the world of work.

Early anthropology: household structure and function

The next phase, arguably the heyday of Caribbean family studies, dated from the mid-1950s through the 1960s and was charted by anthropologists. The two seminal studies by Edith Clarke (1970 [1957]) in Jamaica and Raymond Smith (1971 [1956]) in British Guiana (now Guyana) were complemented by a host of others across the region. Their strength lay in meticulous ethnography; their weakness in the theoretical model of structural functionalism that biased and prematurely foreclosed interpretation and analysis.

Intensive community-based fieldwork, over long periods spent conducting in-depth, participant observation complemented by household surveys, generated a deeper, contextualised understanding of Caribbean family life. These early researchers were concerned to show how family structures and roles were shaped by community conditions – social, economic and demographic. Clarke and Smith, for example, each selected three communities for cross-comparison. Clarke attributed marriage or faithful concubinage, parental responsibility and kinship solidarity in Orange Grove and Mocca to relative economic prosperity, community organisation and neighbourliness, household economic cooperation and residential stability; while in her third community, Sugartown, the seasonality of the sugar industry, population mobility and poverty generated "casual concubinage", "promiscuity" and absentee fathers. Raymond, (or R.T.) Smith[2] traced a causal pathway between economic stagnation, male unemployment, male marginality and matrifocal family structure. For Michael (or M.G.) Smith, working in Carriacou, demography was the decisive factor. The highly skewed sex ratio of more than two women to every man as a result of high male migration had generated a system of "extra-residential mating", with leftover women as "keepers", lesbians, lovers and prostitutes (Smith 1962).

True to their model, this generation of anthropologists sacrificed history, not that there was much Caribbean historiography at the time to shed light on family patterns in the past. It was the contemporary context that was all important in shaping kinship; families were a structural and functional response to socioeconomic conditions, here and now. In this, they connected with the debate that had raged between protagonists Melville Herskovits, who attributed Caribbean family structure to African origins that had survived, though not intact, and

E. Franklin Frazier for whom slavery had destroyed African culture, leaving only insignificant, soon to be forgotten "scraps of memories".[3] They weighed in on the side of Frazier despite his blatant pathologising. Either that, or they dismissed history altogether. R.T. Smith (1971 [1956]: 228) cautioned that "the prior task of sociological analysis may be side-stepped when historical factors are prematurely introduced as 'explanatory' devices". For him, historical analysis was out of the question because informants' memories were faulty, and also unnecessary since there had been so little change in family patterns over the years (Smith (1971 [1956]: 112). Clarke (1970 [1957]: 21) went so far as to dismiss the search for the roots of Caribbean family, either in slavery or African culture, as "dangerous" and "sterile". And so, "synchronic analysis" was privileged and history relegated to the domain of practical impossibility, unscientific speculation and academic eccentricity. Precise tracing of familial and cultural roots to African origins may well lead down a blind alley, but to deny the history and politics of a Caribbean kinship legacy is quite another matter, one to which we return to below.

Detailed ethnography reduced, but did not entirely eliminate, the ethnocentrism and deficit modelling of earlier studies. Centring the "domestic group" as their unit of analysis, household and family were lumped together and much effort devoted to the construction of elaborate classification systems. Edith Clarke's typology of "residential groupings", for example, identified six types and a further five subtypes (Clarke 1970 [1957]: 117). In the process, however, the idea and ideal of the nuclear norm persisted. Family labels juxtaposed conjugal and household forms in line with the nuclear standard and labelled "Christian", "faithful" and "stable", against others perceived to be "incomplete", "sub-nuclear" and "denuded".[4] Visiting unions were "not real", merely "semi-conjugal" (Smith 1971 [1956]: 109–110, 185). The preoccupation with household typologies – generating seemingly unlimited lists of types, but still forcing families into boxes where they did not belong – was described by R.T. Smith (1988: 8–10), on reflection, as "the refinement of error". Adding to the futility of the exercise was the reality that Caribbean families, like those elsewhere, are not places of residence, but sets of inter-household, intercommunity and transnational relationships and practices. But, for the time being, contextual interpretation was trumped by theoretical misconceptions and the ethnographic convenience of households as units of analysis.

Community-level cross-comparison of "the lower-class Negro family" to the exclusion of class-race analysis also generated assumptions of a nuclear norm in the higher ranks of society. In an acrimonious, Smith-versus-Smith, debate models of pluralism and creolisation were pitted against each other: M.G. claiming that each sociocultural segment of Caribbean society had its own familial and cultural characteristics; R.T. that these were creolised and shared across class and colour lines. In research that followed, R.T. Smith (1988) proved his point by showing that matrifocality, visiting relationships, illegitimacy and "outside" women and children featured at all levels of Afro-Caribbean society, from poor to elite.

Nuclear normativity also presumed roles for men as household heads and breadwinners and for women as wives and mothers, home-based and engaged in housework and childcare. Mothers cared and nurtured: fathers supported and disciplined. But in the Caribbean context of matrifocality and male marginality, coalitions of women as grandmothers, mothers and daughters constituted the family, while men, as often as not, lived elsewhere and were "on the fringe of affective ties that bind the group together" (Smith 1971 [1956]: 223). As fathers, they were seen to play little or no role beyond that of biological procreation: "Children derive practically nothing that is of importance from their fathers, they do not suffer if they never even see their father" (Smith 1971 [1956]: 147). Hence the thesis, reflected in the title of Clarke's book, that it was "my mother who fathered me". Nuclear modelling also assumed that, on reaching adulthood, individuals would shift residence and function from the family of origin to the family of procreation. But Caribbean men appeared not to be making the transition as they should; they remained attached to their mothers in relationships marked as "exclusive and often obsessive" (Clarke 1970 [1957]: 164).

This generation of anthropologists may not have managed a total break with hegemonic modelling, but they were the first academic scholars of the region to provide a firm foundation for Caribbean kinship scholarship. In the main they denounced and distanced themselves from colonial social welfarism. Clarke, however, in her position as Jamaican social administrator crossed the line into policy and proposed legal and social reforms to ensure that irresponsible men acknowledged paternity and provided child support.

Extended family networks and empowered women

Arguably, the studies that followed during the 1970s represented less of a new conceptual phase, than a reconfiguration using the same functionalist model. They did, however, provide new insights. Firmly denouncing perceptions of black family breakdown and dysfunction, they replaced the nuclear norm with extended family networks, reimaged women as powerful, resourceful survivors, and celebrated family strength and resilience.

No longer were "domestic groups" the unit of analysis. Replacing them were extended kinship networks reaching far beyond the four walls of the household. Family boundaries were perceived as elastic and open, generating an expansive family culture of love and support; friends and neighbours became fictive kin embraced and addressed as "Ma", "Sis" or "Papa"; and migration was culturally constructed not as departure and family fragmentation, rather as the creation of transnational networks and strong and supportive ties of familyhood across the globe.[5]

Caribbean feminists joined the debate by contesting the hegemonic stereotype of submissive wives and mothers, dependent on male partners. Their favourite target was Peter Wilson's concept of "respectability" drawn from his work in Providencia to describe a feminine value system as against that of "reputation" for men (Wilson 1969). Respectability built on the notion of a "double standard of sexual morality" and presented Caribbean femininity as centred on modesty, obedience and fidelity, at home and in church. It was denounced and replaced with images of women's sexual assertion, economic autonomy and empowerment in the public domains of work and community politics. Black women were re-scripted as superwomen, multitasking as household heads, breadwinners, caring labourers and homemakers, empowered by motherhood and coping resiliently by devising ingenious survival strategies. And so, Caribbean matrifocality was rewritten as matriarchy. In this, there were parallels with black feminists in the US who challenged depictions of women as "welfare mothers", and countered their white sisters' ideas of families as a source of oppression by elevating them as sites of women's power and endurance, and political resistance to racism and patriarchy. The challenge to Eurocentric femininity was timely and critical, but Caribbean family and gender realities were all but buried in this ideological celebration. And the potential of Wilson's work to lift the discourse from a functionalist male–female division of labour and place to contextualised gender analysis was lost.

Although this feminist discourse was grounded in functionalism and continued, therefore, to disregard the past, a parallel Caribbean historiography emerged. Enslaved women in Barbados, for example, were depicted as the backbone of the plantation labour system, and older generation women "as matriarchal figures . . . empowered with immense moral and social authority in slave communities" (Beckles 1989: 123). They resisted efforts to transform themselves and their families. Marriage and respectability would have curtailed their economic activities and sexual strategies for survival and betterment (Beckles 1989: 147). A gender ideology that centred women's personal and economic autonomy emerged to thwart the grand colonial project to civilise the ex-slave population by reconstructing families and domesticating women from field labourers to wives and mothers. The married woman, at home with the children putting dinner on the table for her working husband, remained an illusion of empire.

In contrast to the revisioning of Caribbean woman during this phase of Caribbean family studies, if anything, notions of "male marginality" and "reputation" became further entrenched. The symbolic space of masculinity was outside the home and family and centred on sexual freedom and prowess with many women and "children all about" as proof of virility.

During this phase also, Caribbean family structure was reinterpreted as an "adaptive response" to social, racial and economic conditions. In his study of village life in Trinidad, Hyman Rodman overturned the perception of Caribbean families battered out of shape by adverse circumstances. For him, adversity generated strength and resilience and it made better sense to see family patterns, not as "problems", but as "solutions" to socioeconomic conditions (Rodman 1971: 197). Other researchers joined in by describing family patterns as "flexible", "elastic", "fluid", "malleable" and "adjustable" and by rewriting extended family networks, fictive kinship, flexible conjugality and child shifting as "adaptive survival strategies".[6]

There is much to appreciate in this work, in particular the elimination of deficit modelling and normative portrayals of women. But it seemed, at times, that the objective was more to challenge stereotyping than to interrogate Caribbean kinship. Although the concept of "adaptation" introduced personal choice and flexibility in contrast to functionalist role pre-scription for social stability, this sometimes went too far, implying that individuals operated in a social and cultural vacuum. Rodman (1971: 195), for example, claimed that individuals, including his so-called "circumstance-oriented man", either "stretch" social values or ignore them altogether and "react to circumstances pragmatically rather than normatively; they are neither guided nor hampered by allegiance to any set of values". In addition, the realities of gender and family life were overshadowed by praise-songs of resilience and empowerment, although a few scholars continued to remind us that female household headship in the context of patriarchy and unequal gender relations, more often than not, placed women among "poorest of the poor", struggling to manage the double burden of care and support and to make ends meet (Massiah 1983). Caribbean researchers also warned of the plight of children "shifted" from one home to another or "left behind" when their parents migrated; childcare shared within kinship networks did not always operate in their best interests (Barrow 2010).

Caribbean family and kinship studies had reached a crossroads with contradictory appraisals of family dysfunction versus resilience, and polarised profiles of women. Controversy also centred around fundamental conceptual questions. For example, should the family system be characterised as matrifocal or matriarchal? On this issue, R.T. Smith entered the debate by reiterating matrifocality, and affirming matrifocal families as mother-centred, not female-headed or matriarchal (Smith 1988: 7–8). Another pivotal question was whether the apparent empowerment of women in the family was transferred to the public domain. On this note, Patricia Anderson (1986: 319) pointed to the contradictions between a "gender ideology of male dominance" and "women's independence from male control"; and Janet Momsen (1993) articulated the paradox of patriarchy and a domestic ideology of female subordination and dependence coexisting with female-headed households and women's educational achievement and economic autonomy. A third central issue was whether male marginality reflected the reality of Caribbean men and their families. Patricia Mohammed and Althea Perkins (1999) added to the growing evidence contesting the concept by pointing out that models of empowered motherhood further discredited men. Already marginalised, they were positioned even further towards the edge of family life by the celebration of motherhood and female-centred kin networks. On a more promising note, the extended family framework has enabled researchers to move beyond nuclear categorisations of men as co-resident fathers and husbands to capture the significance of visiting fathers in the lives of their children (Roberts and Sinclair 1978: 54–55) and of other men who as grandfathers, uncles, older siblings and stepfathers were also socially fathering children (Barrow 1998: 344–349). It may be that, in the Caribbean culture of shared mothering and the informal adoption of children within extended families, social fathering is delinked from biological fathering to a greater degree than in nuclear family contexts. However, also centred in today's discourse is "hyper-heterosexuality", a concept that fosters notions of "disruptive", "harmful", "predatory" masculinities by demonstrating intersections with child sexual abuse, sexual coercion, rising rates of HIV and STI, gender-based violence and homophobia.

Adding further complication was evidence from other ethnic groups that shattered conceptual foundations: if families were structured in response to prevailing economic conditions, why were Indo-Caribbean families living in similar conditions patriarchal, not matrifocal? The answer lay in culture and history.

Introducing culture and history: family meanings and negotiations

The most recent phase of Caribbean family studies has replaced regimented standpoints around structures and roles with an interpretive approach to family relationships and family processes. Rather than viewing families as the outcome of adversity, scholars rewrote Caribbean kinship as a system of meaning and action embedded in culture and shaped by history.[7]

Rather than impose theoretical misconceptions and mismatched concepts, they adopted a more open, dynamic approach. Replacing the study of family groups as formal structures of kinship obligations with a clear division of labour and place, their *emic* approach centred the meanings of lived experiences of Caribbean family life; focusing on people's contextualised understandings and assumptions about their own families. Although there is little evidence of transatlantic communication between researchers, synergies are evident with, for example, Kerry Daly's understanding of family in the UK as "a province of meaning" and Anthony Giddens' concept of families as ways of living and relationships as the search for intimacy (Daly 2007: 72–73; Giddens 1992).

To their already well-established repertoire of qualitative methods, Caribbean scholars charted genealogies and added case studies and life histories that captured informants' own definitions of family and reflections on family experiences as expressed in their own words, thereby removing structural preconceptions. Interpretive ethnography opened the way to the familial worlds of men, women and children through meanings attached to masculinity and fatherhood, femininity and motherhood, childhood and family.

Researchers challenged presentations of fatherhood, motherhood and childhood. They contested male marginality with narratives testifying to the deep, primordial meanings around fatherhood – to expanded responsibilities of fathering from economic support and discipline to embrace care and nurturing – and to men's daily involvement in "fatherwork" whether or not they live with their children (Barrow 1998; Brown et al. 1993). Although men are still seen to be occupying "outside" space, they are also expected, with maturity, to "settle down" and assume familial responsibilities (Barrow 2010: 31; Chevannes 2001). Unpacking the meaning of fatherhood, Mindie Lazarus-Black (1995: 51–60) demonstrated unequivocally that mothers do not and have never fathered children. Motherhood is non-negotiable and never-ending, while the degree and duration of fathering is much more a matter of choice. Jack Alexander

(1978) interrogated culturally shaped meanings around "friendship", "love" and "marriage" in Jamaica, and another study, conducted in Barbados, addressed retrospective narratives of childhood that reflected an evolution of consciousness as individuals rethought and renegotiated their experiences as "outside" or "adopted" children or "left behind" by migrant parents (Barrow 2010).

With this approach, developed for research in the UK by Janet Finch and Janet Mason (1993) and centred on *negotiated family processes*, ideas around individual choice can be revisited. Individuals make decisions about their family behaviour in accordance with motives, interests, resources and power, but these are negotiated against a set of family moral codes, not activated in a cultural vacuum of anything goes. A family ideology of shared principles and meanings shapes family life by setting out how each family member ought to perform. There are clear ideas about the actions and practices that constitute "good" mothering and fathering, and what makes a "good" childhood, in different cultural and historical contexts. But these gendered cultural and moral scripts leave room for individual agency and negotiation. Family obligations are not fixed, unconditional or inescapable, but open to modification according to what is considered appropriate in the circumstances. Even the iconic principles of "good" motherhood may be reconsidered. In specific circumstances, of violence for example, unconditional mother-love may run out and take second place to self-preservation.

The concept of negotiated family processes also challenges the representation of families as extensive kinship networks that embrace and support all. Caribbean peoples, like others elsewhere, distinguish between *kin* as structured and fixed by blood and affinity, and *family* as chosen and flexible. It follows that individuals can be and are evicted from the family circle. Those who fail to deliver according to the moral codes of family life or who threaten family reputations by deviant behaviours may be excluded. Individuals may deny fathers who neglected them as "outside" children or mothers who migrated and "abandoned" them (Barrow 2010). Social exclusion denies family membership and rights to gay men, persons living with HIV and other stigmatised groups. Emerging from this approach is a very different conversation around family morality, one that is understood as a process of negotiation – and renegotiation – between mothers and sons, fathers and daughters, with grandparents and cousins, rather than imposed by prescribed roles framed by nuclear normativity and targeted by deficit modelling (Barrow 2010: 48–54).

Caribbean kinship studies have also broken community boundaries. No longer seen to be structured solely by local conditions, families are situated at the intersection of colonialism and capitalism, racism and patriarchy, and are impacted by global declarations, national laws and social policy, and mass media messaging. At the same time as we acknowledge remarkable cultural continuity in Caribbean kinship over the generations, despite extensive economic and social transformation (Roberts and Sinclair 1978; Smith 1988), there is evidence of inter-generational change as childhood, mothering, fathering and conjugality are being renegotiated and reshaped. While there is resistance to the United Nations Convention on the Rights of the Child (UNCRC) as alien and as subverting traditional norms of child socialisation, its principles are filtering into local thinking, albeit slowly, to improve parenting practices, and also to promote children's rights to contact with both parents and their participation in decisions affecting their lives. While strong social pressure for "good" mothering persists, today's mothers and grandmothers are negotiating a different balance between selfless childcare and their own pursuit of career goals and leisure (Mohammed and Perkins 1999). And, as mentioned, the rituals of "good" fathering have changed over the last generation or so. Caribbean men contrast their own commitment to fathering with the "neglect" of their own fathers (Barrow 1998: 350–351). Conjugal relations appear to be shifting from "segregated" to "joint" (Smith 1988:

134–148) as partners seek love and intimacy, although conjugal tension persists. Young women view their partners' infidelity as betrayal, while for their own mothers and grandmothers it was expected, even condoned. A central question amid all this flux is how today's men are negotiating the tensions between hyper-sexualised masculinity, conjugal intimacy and caring fatherhood. It may be that men, previously criticised as stuck in their families of origin as sons to their mothers, are rebalancing affection and support towards their partners and children – in other words, paradoxically, towards nuclear family relationships. Each of these issues presents a fascinating locus of gendered tension and renegotiation for future research on Caribbean families.

A way forward

Caribbean scholars have, in the main, abandoned the preoccupation with structure, function and adaptation, with binary models of family breakdown versus resilience, and with crude deficit modelling and structural engineering, in favour of a focus on kinship relations and gendered meanings rooted in history and culture. The focus for future studies proposed in this chapter is negotiated family processes – on how people create families through agency, action and interaction. Sexualities, partnerships and marriage, childbearing and childhood socialisation, and support for the elderly and vulnerable are all matters for family debate and negotiation. The approach exposes the reification and privileging of the nuclear norm and appreciates families as sites of diversity, complexity and ambivalence; of tension, conflict and chaos, as well as consensus and order. So *The Family* as a thing, with a structure and given set of roles, is replaced by family as a set of practices and negotiated processes. The research focus shifts from what the family is to what the family does.

All this is not to deny the critical importance of more policy-related research on the "dark side" of family life. Research must continue to debunk the myth of care, support and safety for all at home by revealing families as sites of patriarchal power and oppression for children, women and some men. Neither is it to sidestep the ongoing contradictions between a persistent public pathology of "black family breakdown" – currently bemoaning extended family fragmentation – and the expectation that families will act as universal shock absorbers in times of economic crisis. In family politics too, the rhetoric driven by fundamentalist faiths and other moral authorities around structures of marriage, heterosexuality, monogamy, co-resident fathers and extended family unity coexists with laws, and to some extent social policies, that have been reformed to recognise common law unions, to facilitate divorce and to abolish the distinction between children born in or out of wedlock. On the other hand, the Caribbean region continues to outlaw same-sex sexuality, if anything, hardening against sexual autonomy and diversity by reinforcing heteronormativity as the bedrock of morally acceptable family life. Furthermore, the reconfiguration of family as private space beyond state intervention and community sanction has also given exploitation and violence space to persist with impunity behind the closed doors of the household. And anecdotal evidence suggests that legal and social service procedures privilege the preservation of family unity above the delivery of social justice.

The study of Caribbean kinship has become unfashionable and fragmented into subgenres – sexualities, masculinities, motherhood, "unattached" youth, gendered child socialisation patterns and children's rights being among today's favourites. But Caribbean men and women have always constructed their families against hegemonic norms, and reflections on this diversity and on the rich heritage of scholarship across the region may stimulate a revival through the interrogation of family ideologies and processes, as well as the formulation of more effective, less judgemental family policy and programming. It may also rescue Caribbean research from relative isolation and encourage engagement with diasporic conversations around intimacy, gender politics, morality, modernity and transnationalism in family life.

Notes

1   This is not to say that those of other ethnic groups have been omitted. For example, a rich scholarship on Indo-Caribbean families has emerged from Trinidad and Guyana, the main destinations for indentured labourers from India. Constraints of space prohibit inclusion here, but see Barrow (1996: Chapter 6).
2   We follow the established distinction between the two founding fathers of Caribbean kinship by referring to Michael Smith as "M.G." and Raymond Smith as "R.T.".
3   For more detailed discussion and references, see Barrow (1996: 3–8).
4   For more detailed discussion and references, see Barrow (1996: Chapter 2).
5   For more detailed discussion and references, see Barrow (2010: 21–26).
6   For more detailed discussion and references, see Barrow (1996: Chapter 3).
7   For more detailed discussion and references, see Barrow (1996: Chapter 4).

References

Alexander, J. (1978) "The cultural domain of marriage", *American Ethnologist* 3(1): 17–38.
Anderson, P. (1986) "Conclusion: women in the Caribbean", *Social and Economic Studies* 35(2): 291–324.
Barrow, C. (1996) *Family in the Caribbean: Themes and Perspectives*, Jamaica: Ian Randle Publishers.
—— (1998) "Masculinity and family: revisiting 'marginality' and 'Reputation'", in C. Barrow (ed.), *Portraits of a Nearer Caribbean: Gender Ideologies and Identities*, Jamaica: Ian Randle Publishers, pp. 339–359.
—— (2001) "Contesting the rhetoric of 'black family breakdown' from Barbados", *Journal of Comparative Family Studies* 32(3): 419–441.
—— (2010) *Caribbean Childhoods "Outside", "Adopted" or "Left Behind": "Good Enough" Parenting and Moral Families*, Jamaica: Ian Randle Publishers.
Beckles, H. (1989) *Natural Rebels: A Social History of Enslaved Black Women in Barbados*, London: Zed Books.
Brown, J., Chevannes, B. and Anderson, P. (1993) *Report on the Contribution of Caribbean Men to the Family: A Jamaican Pilot Study*, Jamaica: Caribbean Child Development Centre, University of the West Indies.
Chevannes, B. (2001) *Learning to be a Man: Culture, Socialisation and Gender Identity in Five Caribbean Communities*, Jamaica: University of the West Indies Press.
Clarke, E. (1970 [1957]) *My Mother who Fathered Me*, London: George Allen and Unwin Ltd.
Daly, K. (2007) *Qualitative Methods for Family Studies and Human Development*, California: Sage.
Finch, J. and Mason, J. (1993) *Negotiating Family Responsibilities*, London and New York: Tavistock/Routledge.
Giddens, A. (1992) *The Transformation of Intimacy: Sexuality, Love and Eroticism in Modern Societies*, Cambridge: Polity Press.
Jones, A., Sharpe, J. and Sogren, M. (2004) "Children's experience of separation from parents as a result of migration", *The Caribbean Journal of Social Work* 3: 89–109.
Lazarus-Black, M. (1995) "My mother never fathered me: rethinking kinship and the governing of families", *Social and Economic Studies* 44(1): 49–71.
Massiah, J. (1983) *Women as Heads of Households in the Caribbean: Family Structure and Feminine Status*, Paris: UNESCO.
Mohammed, P. and Perkins, A. (1999) *Caribbean Women at the Cross-Roads: The Paradox of Motherhood among Women of Barbados, St. Lucia and Dominica*, Jamaica: Canoe Press.
Momsen, J. (1993) "Introduction", in J. Momsen (ed.), *Women and Change in the Caribbean: A Pan-Caribbean Perspective*, London: James Currey; Jamaica: Ian Randle Publishers; Bloomington: Indiana University Press, pp. 1–11.
Roberts, G. and Sinclair, S. (1978) *Women in Jamaica: Patterns of Reproduction and Family*, Millwood, NY: KTO Press.

Rodman, H. (1971) *Lower Class Families: The Culture of Poverty in Negro Trinidad*, London: Oxford University Press.

Simey, T. (1946) *Welfare and Planning in the West Indies*, London: Oxford University Press.

Smith, M.G. (1962) *Kinship and Community in Carriacou*, New Haven and London: Yale University Press.

Smith, R.T. (1971[1956]) *The Negro Family in British Guiana: Family Structure and Social Status in the Villages*, London: Routledge and Kegan Paul.

—— (1988) *Kinship and Class in the West Indies: A Genealogical Study of Jamaica and Guyana*, Cambridge: Cambridge University Press.

Wilson, P. (1969) "Reputation and respectability: a suggestion for Caribbean ethnography", *Man* 4(1): 70–84.

23

GENDER, DEVELOPMENT, CHILDREN AND YOUNG PEOPLE

Jo Boyden, Gina Crivello and Virginia Morrow

Introduction

Most of the world's children live in developing countries, and children and young people are a major focus of intervention in international development. Yet despite the success of recent feminist critiques in shifting focus within academic and policy circles from 'women' to 'gender' and foregrounding questions of power relations and equality, children and young people are marginal in gender and development debates, and the gender-development nexus continues to be explored mainly through the lens of adult experience. When children do appear, the focus is often on their relationship to women/mothers and to processes of female poverty and the attention given to girls rather than gender per se, emphasis being given to their victimization through diverse 'harmful practices' and to the women they will become (see, for example, UNICEF's *State of the World's Children* reports 2007, 2008 and 2009, concentrating, respectively, on 'Maternal and newborn health', 'Women and children – child survival' and 'Women and children – the double dividend of gender equality'). Although gender in childhood and youth receives some consideration within social anthropology, social geography and childhood studies, international development engagement with this literature has been negligible (see chapter by Barrow in this volume).

We suggest that international development needs to pay greater attention to gender in childhood and youth. We make four key points: 1) that the international development discourse is neglectful of children's gendered experiences; 2) that gender in childhood is influenced by intra-household forces, sociocultural context, institutional structures and economic pressures; 3) that gendered disparities do not always conform to conventional assumptions; 4) and that the gendering of childhood is affected by various 'modernizing' influences that tend towards universal understandings of child wellbeing, such as rights-based approaches that in turn often underplay gender differences among the young.

The chapter draws on research from developing countries, and particularly on recent findings from Young Lives, a 15-year study of childhood poverty being carried out with 12,000 children in Ethiopia, India (in the state of Andhra Pradesh), Peru and Vietnam (see www.young lives.org.uk). Young Lives provides some of the most up-to-date data on children in developing countries and, as a longitudinal study, can trace changes over time. Further, the countries were chosen to reflect a range of cultural, economic, geographic, political and social contexts and therefore echo several trends that occur more generally in the developing world.[1] They differ on international indicators of gender inequality, and according to the 2011 Human Development Report, Vietnam was ranked 48th, Peru 72nd and India 129th out of 176 countries on the Gender Inequality Index (UNDP 2011: 139–142).[2]

How is gender in childhood framed in the literature?

This section traces how gender within childhood has conventionally been framed in international development. Overall, it is accepted in policy circles that children and youth should be prioritized in development processes. Research from developmental and behavioural sciences conducted largely in industrialized countries has provided evidence concerning the factors influencing positive development in the young. This research conceptualizes childhood as a life-phase shaped mostly by universal biological criteria. Recently having been taken up by development economics and applied to children in the developing world, this research emphasizes the factors that affect human capital formation during childhood, and links children's competencies with adult outcomes. Highlighting the education and health 'investments' required to increase future economic performance, it proposes early childhood as the most critical window of opportunity for intervention (Heckman et al. 2006). However, human capital approaches focus on individual characteristics, and tend to disregard power relations and sociocultural factors in explaining differential outcomes for boys and girls.

Concern about the loss of developmental potential in childhood because of poverty and inequality and a desire to maximize human development potential has led to an emphasis in international development on measuring children's progress, with a focus on quantitative research in which gender is operationalized as a discrete variable. For the first time, a global coalescence to achieve the Millennium Development Goals (MDGs) – which set targets related to poverty, gender equality and enrolment of children in primary school – has required systematic efforts to track changes in children's wellbeing over a 15-year period. But the limited availability of data that simultaneously disaggregates gender and age masks the gendered and generational dynamics of poverty (Jones et al. 2008: 1), concealing power relations, as well as the responsibilities and perceptions of children, especially when data are collected at the household level. Scholars and activists are calling for the post-MDG development agenda to prioritize the identification and tackling of global inequalities, and the lens on gender in childhood is also increasingly focused on disadvantage, especially the problems faced by girls. This provides an opportunity to deepen understanding of how gender power operates in particular contexts, but will require better integration of quantitative and qualitative research approaches.

Attention to the 'improvement' of the human condition and the elimination of poverty often leads to a 'problematizing' approach to gender in childhood, where being a girl or boy is narrowly framed in relation to deficiencies or threats requiring intervention. The spotlight is commonly on children facing extreme risk and deprivation. The ordinary worlds of boys and girls are frequently overshadowed by categories of children requiring special protection – 'street children', 'orphans', 'child labourers' and 'trafficked children', for example. These categories are based on universal approaches to child protection with scant regard to how vulnerability is construed and experienced in local contexts (Myers and Bourdillon 2012). If gender is considered at all, the received wisdom is that girls are almost inevitably more disadvantaged and at greater risk than boys. Hence, a distinct narrative has emerged in the past five years, centring on the 'empowerment' of girls. Influential organizations including the UK Department for International Development (DFID), the World Bank and several UN agencies have rallied behind 'girl-led development'. In 2008, the Nike Foundation, in collaboration with the NoVo Foundation, United Nations Foundation and the Coalition for Adolescent Girls, launched 'The Girl Effect', which is marketed as a 'revolution' to end global poverty, led by adolescent girls (www.girleffect.org).

Much of this discourse is based on normative understandings of gender. International organizations advocate attitudinal, behavioural and policy change to bring about gender equality to safeguard girls' rights and to protect girls from harmful practices (Boyden et al. 2012). Girls' problems and sexualized risks permeate the priorities of child protection within developing countries, such that major international agencies have concentrated on protecting female minors from marriage, circumcision, sex-work, trafficking and domestic work. Although 'girls' are the focus, there is no guarantee that the interventions aimed to improve their lives are informed by contextualized gender analysis. This raises questions about the evidence base of interventions, and the extent to which universal prescriptions to address child and youth poverty reflect young people's social worlds, including gendered experiences of poverty and livelihoods. Even if girls are prioritized, it is not possible to fully understand their situations without also paying attention to experiences, roles and perspectives of boys and young men, and constructions of masculinity (see chapter by Parpart in this volume).

At the same time, there are parallels between the way girls and women are cast within development discourses, as 'instrumental' to realizing the goals of economic growth, investments targeted at women and girls being described as 'smart economics' (World Bank) and 'intelligent design' (The Girl Effect). The rationale is economic, and the 'cost of excluding girls' is calculable: for example, in Uganda, 85 per cent of girls leave school early, seemingly resulting in $10 billion in lost potential earnings. By delaying child marriage and early birth for one million girls, Bangladesh could potentially add $69 billion to the national income over these girls' lifetimes (The Girl Effect). However compelling this narrative, development is not just about economics, and gender is not just about females (Jones and Chant 2009).

The dominance of normative assumptions about gender in childhood and the concentration on measuring the effects of gender inequality raises concerns about the neglect of the phenomenological dimensions of gender, and the way gender mediates boys' and girls' evolving identities and experiences. Since the 1980s, a growing number of studies in developing countries have focused on the diverse social and cultural constructions of childhood, and on children's perceptions, roles and responsibilities, the constraints on their agency and everyday experience. Bringing together perspectives from anthropology, sociology and human geography, the views and interpretations of young people are sources of data, and the methods are qualitative. This research demonstrates that constructions of gender in childhood vary across time and space. Young people are active in creating their gender identities, and even where gender 'gaps' may not be significant, gender mediates young people's experiences, including their experiences of poverty.

Gender differences among children and young people

Here we trace some trends in gender in developing countries, from birth to early adulthood, describing patterns of survival, school attendance, children's work activities, lifecourse changes, and risks and reputation.

Birth and survival

The number of males per 100 females, or the human sex ratio, differs greatly between countries and regions. In many countries, female disadvantage in survival during childhood has declined dramatically in recent decades, and life expectancy at birth has increased for both boys and girls. However, India and China have an abnormally high human sex ratio of 106, and in India the ratio has increased over the past century (Bhattacharya 2013). A combination of factors explain this – female infanticide was (and to an extent continues to be) practised, girls received less medical care than boys, and female children tended to be neglected. In India, ultrasound foetal scanning has led to a rapid increase in female foeticide, which is more prevalent in urban areas and among well-educated women, although this does vary across the country, and is more common in the north-western states. The preference for sons in India and China tends to be explained by patrilineal kinship systems where married daughters become part of their husbands' families (Bhattacharya 2013). In India, this is exacerbated by the continuing and spreading practice of dowry payments; legislation has been ineffective in eradicating the practice and sex determination tests and female foeticide have been marketed with slogans such as 'Spend Rs. 500 now, save Rs. 50,000 later' (Bhattacharya 2013: 127).

Proverbs are indicative of cultural values favouring boys, such as in Andhra Pradesh where there is a saying: 'It is better to be born as a tree in a jungle than to be born a girl,' and another: 'Bringing up a daughter is like watering a plant in another's courtyard.' In Vietnam, where the number of male births compared to female births has risen, sons are essential to their parents because they carry on family lineage and names and perform ancestor worship; they are expected to take care of parents in old age; and having a son improves women's status in the family and confirms men's status in the community.

School attendance

The huge effort expended on meeting the MDGs means that enrolment in primary school is now the norm globally and the emphasis on gender equity has resulted in girls' enrolment increasing at a faster rate than that of boys, narrowing the gender gap at primary level. The United Nations (2012) reports that, overall, gender parity in primary schooling has been achieved worldwide although gender inequalities persist in some regions and countries. Sub-Saharan Africa has the highest rate of girls out of primary school, at 26 per cent; girls also account for 70 per cent of the total share of primary-age out-of-school children in Northern Africa. In Southern and Western Asia, the percentages are 55 and 65 respectively. In Northern Africa, between the 1990s and 2000s, the attendance gap between poor girls and boys was reduced by more than half. Despite progress in closing gender gaps at primary level, large differences in attendance persist between rich and poor in all regions. Moreover, there has been little progress since the 1990s in closing secondary attendance gender gaps (United Nations 2012).

Within Young Lives, analysis of preschool access for children aged between three and five years found small differences between boys and girls (compared with socioeconomic differences), which were often not statistically significant (the largest being a 5 percentage point difference favouring boys in rural Peru, much smaller than other socioeconomic related gaps) (Woodhead et al. 2013). Nevertheless, gender differences are more evident during middle childhood, as children become progressively integrated into the household economy, undertaking domestic tasks that are differentiated along gender lines that reflect adult patterns. Typically, children balance schooling with a range of economic activities. Boys usually spend more time doing paid or unpaid work on family farms, while girls spend more time caring for others and on domestic tasks. That said, sibling composition and birth order have been shown to be as important, or more important, as gender in shaping roles and time use in some countries (c.f. Punch 2001).

While gender parity in primary school enrolment was achieved in all four Young Lives countries by 2009, children's trajectories beyond primary school reveal different patterns. At age 15, in Vietnam, boys were less likely (72 per cent compared to 80 per cent girls) to be in school in 2009 and the disparity is intensified by poverty, as only 40 per cent of poor boys were in school at age 15, compared to 52 per cent of poor girls. In Andhra Pradesh, more girls than boys had left school at this age, yet poor girls were more likely to be in school than poor boys. This may be the result of policies that aim to keep girls in school. However, the quality of schooling that children receive varies and, overall, using the results of vocabulary and maths tests as a proxy for quality, Young Lives has found that boys perform better in Andhra Pradesh and Ethiopia, while girls perform better in Vietnam. In Ethiopia and Andhra Pradesh, girls may miss school for several days each month during menstruation, due to the lack of sanitation at school, and in Andhra Pradesh, girls report distance to school and inadequate transport as reasons for discontinuing. A rapid growth in low-fee private schools in Andhra Pradesh has further favoured boys as parents generally spend more on their sons' education and by 2009, 39 per cent of boys and 23 per cent of girls aged eight were in a private school. Although unregulated, private schools tend to perform better than government schools.

Overall, gender differences are more pronounced in Andhra Pradesh and to a lesser extent Ethiopia than they are in Vietnam and Peru (Woodhead et al. 2013). Gender disparities in childhood are often interpreted as evidence of entrenched discrimination. However, decisions affecting girls and boys are regularly made in light of restricted options and resources, and frequently reflect calculations of the short- and longer-term risks and advantages of various alternatives, bearing in mind boys' and girls' employment, marriage and social mobility prospects. For example, in Andhra Pradesh the opportunity costs of schooling are higher for girls than for boys, because girls are less likely to obtain well-paid jobs or to contribute to the natal home when they marry, which may be at a relatively young age. As a result, parents tend to discourage their daughters from remaining at school beyond the acquisition of basic skills. In contrast, investments in boys' education are made with the expectation of future returns, so boys are more likely to be encouraged to remain at school and/or enter private schools. As one mother explained: 'We wanted to stop [our daughter] from further studies . . . Will she give us money once she starts working? Who will she give it to? We won't make anything from her. She is better off working here . . .'. Such gendered practices appear to be transmitted through the generations in some communities, producing a systematic bias against girls. Lower educational aspirations of parents for their daughters at age 12 may be linked to girls having lower educational aspirations for themselves, as well as lower scores in both math and verbal tests by age 15 (Dercon and Singh 2013).

Although gender disparities can be discerned, other factors associated with household wealth and levels of education of caregivers, rural or urban location, ethnicity, language or caste, explain greater gaps between children's outcomes (Dercon and Singh 2013). This is apparent even in Andhra Pradesh, where gender difference is most pronounced. The difference between access to private schooling for boys and girls is smaller than that between children in rural and urban areas, or between children with more and less educated mothers, or between children from traditionally deprived caste and tribal groups and the other castes (Dercon and Singh 2013). Often, the effect of gender disparities is most evident when they co-occur with and exacerbate other processes of socioeconomic, cultural and/or spatial differentiation.

Children's work and time-use

Gender norms influence children's roles and responsibilities, how they are treated, their experiences and their aspirations, especially as they enter middle and late childhood. But within the context of children's lived experience, it is impossible to separate gender out from other dimensions of their evolving identities and from the multiple hierarchies that they navigate. In her ethnographic study in rural Bolivia, Punch (2001) finds that children's roles and time-use were not determined solely by gender, and that household composition and sibling birth order also explain the assignment of work tasks. Gender is one among several considerations in this respect. Thus, gender cannot be reduced to something that individual boys and girls 'have'; rather, gender mediates young people's experiences, including their experiences of poverty.

Heissler and Porter (2012) find that in Ethiopia the nature and amount of work done by children is affected less by levels of household poverty than by economic shocks, such as illness and death in the family. But they also establish that girls and boys are affected differently by household circumstances, with girls working more overall and on domestic tasks when their mothers are ill or absent. Boys appear to work less in larger households, while girls work more if they have sisters, and if they have younger brothers. Both boys and girls seem to be affected by adverse events generally, but boys work more when the household has more livestock. A shortage of children may lead to the blurring of gender roles. Although this ensures that all the tasks necessary for household maintenance are accomplished, it can be a source of considerable discomfort for the boys and girls concerned (Heissler and Porter 2012).

Youth – risk and reputation

Gender differences become more pronounced during life-course transitions such as puberty, leaving school, marriage or commencing work, and at times of crisis. Research demonstrates the uneven impact of modernization on young people's transitions, and there is a growing discourse on the failures of development as evidenced by the problem of youth unemployment and unfulfilled aspirations. Studies examining the challenges in the transition to social adulthood tend to focus on the experiences of young men. Jeffrey (2010) shows that many young men in Uttar Pradesh, India, acquire copious qualifications, but, in a competitive labour market where opportunities are structured by caste and other political considerations, are unable to find jobs that require their skills, and remain in a limbo of student-hood. Young men in a diversity of contexts globally articulate their frustrations in terms of 'waiting', 'boredom', 'timepass' and being 'stuck'.

Jones and Chant (2009) explore the 'crisis of youth' in the Gambia and Ghana, and question whether the confidence placed in education and employment is justified. They emphasize that despite efforts to increase girls' school enrolment and employment, social norms continue to dictate that the bulk of reproductive work is done by young women, thereby constraining the amount of time they can dedicate to schooling, and to making contacts that are relevant to securing paid work. Similarly, a study of male Palestinian youth living in a refugee camp in Jordan finds that those young men who successfully displayed the dominant form of masculinity in the camp (*maukhayyamji*) risked marginalization from wider Jordanian society, including through access to jobs and resources (Hart 2008). The transition from childhood to adulthood is therefore not a straightforward process, and young people in the developing world struggle to meet the gendered social expectations placed on them within the context of limited opportunity.

Young Lives research also finds that young people actively pursue education in the hope that it will enable them to secure respectable jobs (Camfield 2011; Crivello 2011; Morrow 2013). However, many are disappointed by the outcome and resist taking on menial labour for risk of social shame; this phenomenon seems to affect both girls and boys. Gendered social expectations also constrain boys and girls alike. Young men and women share many concerns – for example, contributing to household welfare, securing a good marriage, earning a 'good name' – and these concerns are shaped by gender norms. Social risks also influence perceived options, especially when neighbourhood environments are deemed to be dangerous. For example, families living in a shanty town on the outskirts of Lima worry about the presence of gangs and drug dealers and the prevalence of violence. Parents are especially resistant to letting their daughters work outside at this age, although some girls earn money by working from home, sewing or making jewellery. Susan is sympathetic to her mother's insistence that she not leave the house alone: 'Because I think there's a lot of danger on the streets . . . the thing nowadays is robbing . . . they can steal something from you.' Girls see themselves as and are construed as potential victims of violence. A parallel desire is to protect boys from being drawn into 'bad company'. One mother said, 'it's about what he might get himself involved in, because here . . . it's where the gang members smoke [drugs] . . . so there are times you can't even let the kids leave [the house]. And if they do, you have to be watching them.'

Gender-differentiated social risks are also a concern for young people in Addis Ababa. Young women articulate everyday fears of rape and harassment, and general physical insecurity. Such fears are instrumental in constraining their mobility, and girls seek work close to home, for example, assisting mothers in laundering, or food preparation for sale, whereas boys find jobs in the local market. In short, not only are young people's livelihood opportunities differentiated by gender, but differing moral judgements are made about boys' and girls' choices, respectively, even when these choices are constrained by poverty. These examples underscore the moral values that shape how boys and girls negotiate their everyday sociospatial environments.

In Andhra Pradesh and Ethiopia, making a 'good' marriage for daughters is a priority, with chastity a criterion of eligibility. In some communities, the social standing of households depends on the conduct of unmarried female members, and any breach is a social risk for the individuals involved and for their families. Particularly in Muslim communities in Andhra Pradesh, the need to limit the interaction between unmarried girls and unrelated males inhibits girls' movement outside the home, especially after puberty. This often results in the weakening of their social networks, as they spend less time with school friends – male and female – and more time with siblings and relatives within the home (see Boyden and Crivello 2012). Earning a good name for oneself is also important for boys, and by age 16, boys are often concerned about how to maximize their chances of a good marriage. For example, Salman described how formal qualifications would improve his marriage prospects. When initially interviewed, he was not interested in marrying an educated girl, but his views changed over time. He worried that eligible girls will prefer to marry an educated boy, not 'this boy who drives an auto, or rickshaw'. This reflects how he sees himself. Boys' greater access to mobility may be a resource for material and social gain, but it is not a guarantee, and girls may be excluded from such opportunities due to rigid gender norms.

With the growing focus on schooling for children, children's migration for education is a key strategy for accessing better facilities, and gender norms influence decisions regarding child migration. In rural Peru, young people and their families look to the city as a place of opportunity, and migration is increasingly considered an inevitable element in the pursuit of wellbeing. A mother in Peru explains, 'more and more, we're more alone, it's not like it was before where you're with your children. This is the change with children, they begin to leave, and one feels . . . more alone.' Where girls' moral reputations are strongly guarded, family networks are often crucial to their migration and to overcome restrictions on their mobility.

For older children and youth, mobility is also a resource for accessing post-secondary education, and for pursuing livelihoods. Their options are structured by gender processes, such as the gendered nature of personal networks, household responsibilities and labour markets. For example, constraints on youth mobility in Lusaka, Zambia, are described as 'getting stuck in the compound' and accessing other parts of the city depend on young people's households, skills and resources and their use of the city (Hansen 2005). Young women lack job contacts and economic means relative to young men, and cannot afford public transportation. Mobility and material inequalities are thus deeply gendered processes.

The gendered impacts of modernity: continuity and change?

Despite the persistence of longstanding gender norms and hierarchies, the experience of childhood in developing countries is changing rapidly due to the restructuring of economies, advances in technology, infrastructure and communications and the spread of basic services. Internal and international migration are on the rise, and the experience of childhood is increasingly urban; it is estimated that by 2050, seven in ten people will live in cities and towns. New information and communication technologies (ICT), including low-cost internet, present new opportunities for the young and are tools that can help improve their living conditions and expand their horizons. But access remains uneven across the Global North and South, between rural and urban locations, and between male and female users. Studies suggest that technological innovation does not automatically alter fundamental gender and age-based asymmetries in society. Modernizing influences may represent welcomed social change, but increasingly accessible communication tools such as mobile phones and internet can also represent risks to established gender relations; modernity does not therefore guarantee greater freedoms for girls.

At the same time, schooling has fast become one of the defining features of modern childhood, and increasingly shapes children's lives, changing their use of physical space and time, their relationships with and responsibilities towards others, and their aspirations. Earlier, the transition to adulthood was achieved through the assumption of increased gendered responsibilities and by means of marriage and parenthood. Today, schooling plays an increasingly decisive role in shaping social transitions and children's futures. Importantly, public education systems are invariably structured along lines of age and grade and most attempt to minimize gender distinctions, for example through coeducational provision at primary, and sometimes secondary, level, as well as shared school curricula. Although biases still exist in education, shown above, mass schooling may gradually give rise to a less gendered conception and experience of childhood that applies across economic and sociocultural groups as well as geographic localities. However, ensuring gender neutrality in schooling is far from straightforward and may in practice inhibit girls' attendance, for example when shared toilets infringe notions of gender propriety.

Among the Young Lives countries, overall, gendered experiences and expectations of childhood seem to have shifted most noticeably in Ethiopia (Boyden et al. 2012). As part of a wider commitment to socioeconomic development, the government has been keen to change attitudes and practices towards children, with the intention of achieving greater gender parity and eliminating customs that are deemed to be harmful to girls in particular. Government intervention has ranged from comprehensive services, such as the development of education and health systems, to measures intended to modify gendered practices, such as female circumcision and (female) marriage below the age of 18. These measures have had a dramatic impact on gendered expectations even within one generation. Thus, while the majority of mothers and grandmothers in Ethiopia have been circumcized, had limited or no schooling and were already in arranged marriages by the time they reached puberty, many of their daughters and granddaughters are not circumcized, are remaining at school to secondary level, staying unmarried until their late teens or early twenties, and aspiring to professional careers. Nowadays, most girls express a desire to delay marriage and to bear less children than women of their mothers' and grandmothers' generation, or at least to marry later than their mothers. Yet gendered social transitions in childhood and youth remain important and are marked by rites of passage, for example in Andhra Pradesh at menarche, when girls and their families celebrate becoming a 'big girl'. In Peru, many girls celebrate their fifteenth birthdays with a *quinceañera* ceremony. In Ethiopia, boys celebrate getting their first goat or small plot of land.

Moreover, in all four Young Lives countries there are weaknesses in school systems, ranging from limited availability and accessibility, especially in rural areas, to poor school quality and high costs of attending. Uncertainties about the value and relevance of schooling lead to contingency planning by boys and girls; depending on perceptions of the relative benefits of school and work at a particular moment, they may move in and out of school. Families may also strategize to maximize benefits to the household, so that an elder or younger child, a girl or a boy, or maybe a child that is brighter academically than her/his siblings, is singled out to remain at school while others work.

Behavioural and attitudinal change in respect of gender is not a linear process, but reflects a complex mix of shifting and sometimes competing, or even contradictory, understandings. Generational differences in views about gender may lead to conflict within families. New ideas give rise to new concerns about children's wellbeing, some of which are gender-specific. For example, attending school can compromise traditional conceptions of appropriate behaviour in girls, especially following puberty when their social identity becomes more sexualized (Boyden and Crivello 2012). In Ethiopia, where marriage is critical to girls' successful transition to adulthood and is customarily arranged by the couple's parents, delayed marriage may be seen as a social threat to young females, heightening the possibility of promiscuity, sexually transmitted infections, abduction, single parenthood and abandonment by partners. Similarly, while the majority of parents appear to view completing education as the best way for girls to secure their futures, others regard extended schooling as a potential risk, because girls are less likely to marry while at school. In other words, although education is considered an investment, it is but one of many social expectations that young people must negotiate, and different generations may have different views about what matters most for individual children and/or their families.

Conclusions

Gender differences in childhood are often overshadowed by the research and policy focus on human capital. Gender is a vital element in how boys and girls experience the world and their identities. Gender mediates and structures everyday life, influencing how children and young people are incorporated within their households and communities. This point is often eclipsed by a tendency to study 'outcomes' in poor children, disconnected from everyday experience. Thus, gender differences grow during middle and later childhood, shaped by changing expectations of girls and boys, which are in turn framed by the socioeconomic circumstances of the household as well as by perceived social risks and opportunities. Gender also underpins young people's evolving moral identities. However, gender is but one form of social power and intertwines with other social factors and, as we have seen, poverty intersects with gender and remains a powerful constraint on boys' and girls' trajectories.

In conclusion, gender is dynamic and gender disparities in childhood sometimes confound common assumptions. Gender shapes children's lives in crucial, although complex, ways. A balanced picture is needed, which recognizes the ways some inequalities develop progressively through childhood, while others can open up through specific life events, and yet others are amplified as children face key transitions. Gender inequalities offer a clear example of these processes. Global and local processes of economic, social, cultural, and environmental change are creating new opportunities and constraints. Finally, policies aimed at reducing gender-based difference need to engage with the context that influences parents' and children's choices, as well as discrimination (and the way different forms of gender discrimination interconnect with other forms of discrimination based, for example, on age, race/ethnicity and class). Policies need to take account of how programs may impact differently on boys and girls, and the gendered nature of poverty across the life course.

Notes

1   Peru, Vietnam and India have experienced consistent economic growth since the study began, but high levels of social and economic inequality persist. Ethiopia remains a low-income country, despite averaging 11 per cent growth for six years up to 2009. While gender norms appear relatively stable in some areas in these countries, gendered experience is affected by ever-changing forces that affect individuals and groups at various societal levels. Household factors such as wealth and security of livelihood are often decisive, as are macro-level mechanisms, such as media representations, institutional structures and political-economic processes (Woodhead et al. 2013).
2   Data were not available for Ethiopia. The Gender Inequality Index reflects women's disadvantage in reproductive health, empowerment and the labour market.

References

Alanen, L. (2009) 'The generational order', in J. Qvortrup, W. Corsaro and S. Honig (eds.), *The Palgrave Handbook of Childhood Studies*, Basingstoke: Palgrave Macmillan.

Bhattacharya, P. (2013) 'Gender inequality and the sex ratio in three emerging economies', *Progress in Development Studies* 13(2): 117–133.

Boyden, J. and Crivello, G. (2012) 'Political economy, perception and social change as mediators of childhood risk in Andhra Pradesh', in J. Boyden and M. Bourdillon (eds.), *Childhood Poverty Multidisciplinary Approaches*, Basingstoke: Palgrave Macmillan.

Boyden, J., Pankhurst, A. and Tafere, Y. (2012) 'Child protection and harmful traditional practices: female early marriage and genital modification in Ethiopia', *Development in Practice* 22(4): 510–522.

Camfield, L. (2011) '"From school to adulthood"? Young people's pathways through schooling in urban Ethiopia', *European Journal of Development Research* 23(5): 679–694

Crivello, G. (2011) 'Becoming somebody: youth transitions through education and migration in Peru', *Journal of Youth Studies* 14(4): 396–411.

Dercon, S. and Singh, A. (2013) 'From nutrition to aspirations and self-efficacy: gender bias over time among children in four countries', *World Development* 45: 31–50.

Hansen, K. (2005) 'Getting stuck in the compound: some odds against social adulthood in Lusaka, Zambia', *Africa Today* 51(4): 3–16.

Heckman, J., Stixrud, J. and Urzua, S. (2006) 'The effects of cognitive and noncognitive abilities on labor market outcomes and social behavior', *Journal of Labor Economics* 24(3): 411–482.

Hart, J. (2008) 'Dislocated masculinity: adolescence and the Palestinian nation-in-exile', *Journal of Refugee Studies* 21(1): 64–81.

Heissler, K. and Porter, C. (2012) *Know Your Place: Ethiopian Children's Contributions to the Household Economy*, Young Lives Working Paper 61.

Jeffrey, C. (2010) *Timepass: Waiting, Micropolitics and the Indian Middle Class*, Stanford, CA: Stanford University Press.

Jones, G. and Chant, S. (2009) 'Globalising initiatives for gender equality and poverty reduction: exploring "failure" with reference to education and work among urban youth in the Gambia and Ghana', *Geoforum* 40: 184–196.

Jones, N., Holmes, R. and Epsey, J. (2008) *Gender and the MDGs*, ODI Briefing Paper, www.odi. org.uk/sites/odi.org.uk/files/odi-assets/publications-opinion-files/3270.pdf.

Morrow, V. (2013) 'Troubling transitions? Young people's experiences of growing up in poverty in rural Andhra Pradesh, India', *Journal of Youth Studies* 16(1): 86–100.

Myers, W. and Bourdillon, M. (2012) 'Introduction: development, children and protection', *Development in Practice* 22(4): 437–447.

Punch, S. (2001) 'Household division of labour: generation, gender, age, birth order and sibling composition', *Work, Employment & Society* 15(4): 803–823.

United Nations (2012) *The Millennium Development Goals Report: Gender Chart*, www.unwomen.org/wp-content/uploads/2012/12/MDG-Gender-web.pdf.

UNDP (2011) *Human Development Report 2011: Sustainability and Equity: A Better Future for All*, New York: United Nations Development Programme.

Woodhead, M., Murray, H. and Dornan, P. (2013) *What Inequality Means for Children: Evidence from Young Lives*, Oxford: Young Lives.

24

SERVING THE TRANSNATIONAL SURROGATE MARKET AS A DEVELOPMENT STRATEGY?

Carolin Schurr and Bettina Fredrich

Summary

This chapter engages with the transnational market of gestational surrogacy resulting from the innovation and spread of assisted reproductive technologies (ARTs) and reproductive services. In this new transnational market, women of lower economic and social status—often situated in the Global South—provide reproductive materials such as ova and surrogate services for elite women who come to the Global South in their search for inexpensive fertility treatment. Mapping the geographies of transnational surrogacy, this chapter first discusses the spatial expansion of surrogacy markets within the Global South. Second, drawing on India as a case study, it is questioned to what extent transnational surrogacy can be considered a development strategy both for national developing economies and for individual women in the Global South. The chapter concludes discussing the implications of these new markets for feminist research.

Transnational surrogacy—have dystopias turned into everyday reality for women in the Global South?

The distinction between the genetically worthy and unworthy is likely to increase [. . .]. As I envision it, most women in a reproductive brothel would be defined as "nonvaluable" [. . .]. Certainly women of color would be labeled "nonvaluable" and used as breeders for the embryos of "valuable" women.

(Corea 1985: 276)

Once embryo transfer technology is developed, the surrogate industry could look for breeders – not only in poverty-stricken parts of the US, but in the Third World as well. There, perhaps one tenth the current fee could be paid women [. . .]. Using Third World women as surrogates [. . .] would benefit them because the women would earn money with which to raise their other children.

(Corea 1985: 215)

Box 24.1 Glossary of key terms

Assisted reproductive technology (ART): A treatment for infertility that involves the laboratory handling of eggs, sperm, or embryos.

Egg donation: The process of fertilizing eggs from a donor with the infertile women's partner sperm (or from a donor) in a laboratory dish and transferring the resulting embryos either to the infertile intended mother or to their surrogate.

Gestational surrogate: A woman who carries and delivers a baby for an infertile couple and is not genetically related to the child. Eggs are removed from the infertile woman (or from a donor) and fertilized with the partner's sperm (or from a donor), and the resulting embryos are placed into the uterus of a gestational surrogate, who will carry the baby to delivery.

In-vitro fertilization (IVF): A process in which eggs are surgically removed from a woman's ovaries and combined with sperm in a laboratory dish to bring about fertilization. The resulting embryos are transferred to the uterus of the woman being treated or can be frozen, donated, or discarded. Also known as *test-tube baby*.

Source: Thompson 2005: 309–316

In 1985, journalist Gena Corea published the cited statements in *The Mother Machine*, which can be considered one of the earliest feminist engagements with new reproductive technologies and the effects they may have on women in the Global South. At the same time, Margaret Atwood (1985) wrote her dystopian novel *The Handmaid's Tale*, picturing a science-fictive society in which young, fertile handmaids' only social function is to bear children for the wives of the ruling elite men. With their work, both Corea and Atwood have aimed to call public attention at an early stage to the potential risks the development of new assisted reproductive technologies (ART) such as in-vitro fertilization and gestational surrogacy (see Glossary) will bring about—depending on the context—for non-white, non-elite women living in scarce conditions.[1]

Studying the fast growing market of transnational surrogacy in 2012, it is striking how precisely Corea and Atwood have foreseen the effects ART might have on the bodies and lives of women in the Global South. As recent ethnographic studies have shown (Hochschild 2012; Pande 2008, 2009a, 2010a, 2010b), the accommodation and surveillance of gestational surrogates in Indian surrogacy hostels resembles the "reproductive brothel" (Corea 1985: 276) and the "thrilling" of the handmaids (Atwood 1985) to a great extent. Pande's (2009a: 382) ethnographic narratives about an infertility clinic in Anand come close to the dystopias pictured by Corea and Atwood:

> The clinic runs several hostels where the surrogates can be kept under constant surveillance during their pregnancy—their food, medicines, daily activities can be monitored by the doctor or her employees. All the surrogates live together, in a room lined with iron beds and nothing else. [. . .] The women have nothing to do the whole day [. . .] while they wait for the next injection.

These kinds of reproductive "assembly-line techniques" (Corea 1985: 276) applied in private fertility clinics in India have turned the country into "the mother destination" (Akanksha

Infertility Clinic 2012). The clinic in Anand advertises that the center has more than 50 surrogates pregnant at the same time for Indian and foreign couples. As the example from India shows, in this "biomedical mode of reproduction" (Thompson 2005: 250), the reproductive labor of non-Western women is exploited to produce babies for the global market.

Mapping the transnational geographies of surrogacy

As the introduction highlights, recent techno-scientific innovations in biomedicine that have succeeded in reproducing human reproductive cells (sperms, oocytes) in-vitro have dramatically changed reproductive practices (Franklin 1993; Franklin and Ragoné 1998; Raymond 1993; Thompson 2005). New technologies such as artificial insemination, in-vitro fertilization, and the cryopreservation of sperm, ovarian tissue, and embryos have made it possible to transfer and circulate reproductive materials through space and time (Parry 2008, 2012) for use in an increasingly transnational market of assisted reproduction (Ikemoto 2009; Martin 2009; Nahman 2010). The trade of reproductive materials and services has 'sped up' (Harvey 1989) and extended its geographical reach in the last decade (Panitch 2012; Spar 2005; Whittaker and Speier 2010). Simultaneously, national legislations regulating the use of reproductive technologies either do not exist or do not keep up with the fast transnationalization of the ART business. As Sengupta (2012: 30) highlights,

> just as Western multinational corporations have found it profitable to shift their production through "outsourcing" to developing countries, both to capitalize on the cheap labor power and minimal legislation and social protection measures, so has the reproductive industry found both its resource base and new markets in those countries, leading to outsourcing of reproduction.

The enormous price differences for surrogacy between countries of the Global North and the Global South make 'outsourcing' surrogacy to the bodies of women of the Global South an attractive option.

The money a surrogate mother receives varies according to where she offers her services (see Figure 24.1). While a surrogate mother in California earns on average $20,000 (Spar 2006: 87), surrogates in Ukraine receive an average of $8,000 (Martinović 2010), in India $7,000 (Sengupta 2010: 49), and in the emerging market of Guatemala an estimate of just $3,000 (Rotabi 2010) for their services.[2] The identified hotspots in the global bioeconomy of surrogacy such as India, Guatemala, or Ukraine attract international 'clients' for various reasons: the ART-friendly legislations, the fact that they offer surrogacy for same-sex couples and single women and men, their endless supply of egg donors and surrogates, little waiting time for treatment, and their good-quality infertility treatment with cheaper costs for procedures, and the ease of travel and visa requirements (Sarojini et al. 2011; Spar 2005; Thompson 2011). India, for example, has turned into the 'Cradle of the World' (Ghosh 2006) due to the country's legislation,[3] which can be considered "one of the friendliest laws on surrogacy in the world" (Pande 2010a: 295) and its high-class medical facilities that provide low treatment costs. As a consequence, a "new geography and economy of the body is mapped on the globe" (Whittaker and Speier 2010: 365) with the identified locations as global hubs for transnational surrogate services. These new geographies link women of the Global South and the Global North in new ways when encounters take place between Western elite women and their non-Western surrogates in clinics in the Global South that are supplied with "First World" technologies and staffed with health personnel often trained in the Global North.

Figure 24.1 The most important markets for transnational surrogacy

Commercial surrogacy as a development strategy?

When eggs and sperm are offered in a transnational market and surrogate contracts are made across national borders, revenues are generated in this transnational reproductive market. While eggs and sperms are now widely available in Western markets, wombs for rent are most easily found in non-Western countries. Spar (2006: 94) highlights that "surrogacy is a tiny piece of the baby business" (which further includes adoption and IVF) in Western markets. For example, in the American market, the US$27 million revenue of the surrogacy business represents less than 1 percent of the US$2.8 billion infertility market. In India in contrast, experts estimate that the surrogacy industry is currently worth around US$445 million with huge potential to grow. It does not come as a surprise that the Indian state considers the transnational surrogacy industry a "pot of gold" (Maranto 2010). In the meantime, India's health market has turned into an elite market for women from across the world who take advantage of low-cost fertility treatments, while at the same time a large number of Indian women are denied basic health care. As Sengupta (2011: 314) shows, only 17.3 percent of Indian women have had any contact with a health worker. Neoliberal reforms in the 1990s that cut down the public expenditure for health care to 1 percent of GDP have not only dismantled public health facilities in many parts of the country but have also resulted in the privatization of health care. This prepared a fertile ground on which reproductive tourism has been able to thrive. As a consequence, skilled health personnel are absorbed by the booming private clinics, which have an estimated share of 68 percent of all hospitals, leaving especially rural parts without any qualified health workers (Hazarika 2010). Satpathy and Venkatesh (2006: 33) calculate that already half of the local community health centers lack gynecologists. The internal brain drain further compromises women's access to health care as today only 17.9 percent of these local health centers have a female doctor, a precondition for many women to access health care due to cultural norms (Sengupta 2011: 314). In sum, while the growing international surrogacy industry might be considered a development strategy in the eyes of neoliberal consultants and private investors, this growing market in India menaces the living and health conditions of many marginalized, rural, and poor Indian citizens.

Paradoxically, in the light of widespread poverty and poor public provision of basic social services, surrogate services are pictured as an individualized development strategy for rural and poor women and their families—just as foreseen in Corea's introductory quote. Pande (2009b: 150) shows in her study of 42 Akanksha surrogates that 34 of them live at or below the Indian poverty line. The fee earned for their surrogate services equals roughly five times their normal household income. On the basis of interviews with surrogates, Pande (2010a) outlines that surrogates justify their decision for surrogacy as a necessity as the quote from a Indian woman who is pregnant for a couple from Washington evidences:

> Who would choose to do this? I have had a lifetime worth of injections pumped into me. [. . .] But I know I have to do it for my children's future. This is not a choice; this is *majhoori* [a necessity]. When we heard of surrogacy, we didn't have any clothes to wear after the rains – just one pair that used to get wet and our roof had fallen down. What were we to do? If your family is starving . . .
>
> *(cited in Pande 2010a: 301)*

In fact, most surrogates in India offer their reproductive services to improve their family income in order to provide health treatment to other family members, to replace or supplement an un- or underemployed husband's income, to pay for their children's education, to buy a home, or to start a small business. As Sengupta (2012: 36) argues, "in the absence of employment opportunities with which they can earn a living wage, this [surrogacy] seems to them a good option" to improve the living condition of their families. Regarding lacking employment options, many women in the Global South consider the money earned through surrogacy as some kind of "jackpot" (Bhatia 2009). Gestational surrogacy, however, puts these women's bodies at a high risk: gamete transfer, embryo implantation, pregnancy, and birth are risky medical maneuvers that might be complicated through ectopic pregnancy, fetal reduction, and cesarean sections that accompany gestational surrogacy (Smerdon 2008; Zelizer 2010). Further, the question remains whether any amount of money can recompense the emotional challenges transnational surrogates face during the pregnancy and especially when handing the baby over to the commissioning parents (Hochschild 2012).

Conclusion: surrogacy between feminist discourses of empowerment and exploitation

Given the development of new reproductive technologies and their resulting markets of assisted reproduction and gestational surrogacy, how can feminist scholars respond to the issues raised in this chapter? Banerjee (2010) identifies two dominant arguments in feminist debates around transnational surrogacy. On the one hand, based on choice and self-determination as key concepts of second wave feminism, feminists have advocated a kind of "reproductive liberalism" (Raymond 1993). It is hoped that women can use these new technologies to their own advantage when making self-determined decisions about their own bodies and reproductive capacities. For example, women who provide their reproductive services on this market are enabled to improve their economic and hence their social position in the family and lesbian or single women using sperm donors may realize their desire of having a child outside of heterosexual patriarchal family structures. While these hopes have certainly become true to some extent, fertility clinics and surrogacy agencies have since instrumentalized feminist discourses of choice and self-determination to justify women's right to sell their eggs or rent their wombs.

On the other hand, especially Western feminists argue that surrogacy is an exploitive practice that turns women's bodies into a commodity. More recent work in this strand highlights that the choices of women to become surrogates need to be contextualized within current global inequalities (Bharadwaj 2012; Vora 2009; Whittaker and Speier 2010). In short, the dilemma of surrogacy consists of the fact that "surrogacy contracts, although expanding their options numerically, seem to coerce women into making a choice they do not prefer, yet cannot refuse because the price of refusal is too high" (Damelio and Sorensen 2008: 274). In a similar vein, Pande (2009b: 160) argues that (Eurocentric) portrayals of surrogacy as either exploitation or choice do not incorporate the reality of transnational surrogates in the Global South who define their surrogate service not as work but as *majboori* (compulsion). As Sengupta (2006) argues, feminist research needs to go beyond a binarist response that judges surrogacy as either right or wrong. While reproductive liberalism does not acknowledge that surrogacy is not a choice but a *majboori*, the call for legally prohibiting transnational surrogacy is problematic as well. It is not only unrealistic in neoliberal times of expanding transnational markets but might even obstruct the creation of safer spaces for transnational surrogates. Feminist research is challenged to produce knowledge about how to make the transnational spaces of surrogacy safer. Linking geographically anchored and ethnographically situated case studies to broader economic, cultural, and political processes might be a first step to produce policy relevant knowledge that helps to regulate and monitor this new market in the long run.

Notes

1   We would like to highlight the difficulty of terms such as "non-white," "non-elite," women of the Global South as these social categories homogenize and essentialize diverse subjectivities and living conditions. While we do not aim to fix the subjectivities of women in the Global South, we use these terms in Spivak's sense of strategic essentialism to refer to structural power relations between individuals and regions.
2   It is important to take into account that in the US up to 50 percent of the cost the commissioning couple pays goes to the surrogate, in India most of the money is appropriated by the sperm banks, clinics, and lawyers (Qadeer 2009).
3   See Ministry of Health and Family Welfare, Government of India, Indian Council of Medical Research, 2010. The Assisted Reproductive Technology (Regulation) Bill & Rules – draft, New Dehli, www.icmr.nic.in/guide/ART%20REGULATION%20Draft%20Bill1.pdf.

References

Akanksha Infertility Clinic (2012) *We Care for Your Emotions*, www.ivfcharotar.com/about.php.

Atwood, M. (1985) *The Handmaid's Tale*, Toronto: O.W. Toad Limited.

Banerjee, A. (2010) "Reorienting the ethics of transnational surrogacy as a feminist pragmatist," *The Pluralist* 5(3): 107–127.

Bharadwaj, A. (2012) "The other mother: supplementary wombs and the surrogate state in India," in M. Knecht, M. Klotz, and S. Beck (eds.) *Reproductive Technologies as Global Form: Ethnographies of Knowledge, Practice, and Transnational Encounters*, Frankfurt: Campus Verlag, pp. 139–160.

Bhatia, S. (2009) "Investigation: surrogate baby delivered every 48 hours," *London Evening Standard*, May 20, 2009.

Corea, G. (1985) *The Mother Machine: Reproductive Technologies from Artificial Insemination to Artificial Wombs*, New York: Harper & Row.

Damelio, J. and Sorensen, K. (2008) "Enhancing autonomy in paid surrogacy," *Bioethics* 22(5): 269–277.

Franklin, S. (1993) "Postmodern procreation: representing reproductive practice," *Science as Culture* 3(4): 522–561.

Franklin, S. and Ragoné, H. (1998) *Reproducing Reproduction: Kinship, Power, and Technological Innovation*, Philadelphia: University of Pennsylvania Press.

Ghosh, A. (2006) "Cradle of the world," *Hindustan Times*, December 23, 2006.

Gupta, J.A. (2006) "Towards transnational feminisms," *European Journal of Women's Studies* 13(1): 23–38.

— (2012) "Reproductive biocrossings: Indian egg donors and surrogates in the globalized fertility market", *International Journal of Feminist Approaches to Bioethics* 5(1): 25–51.

Harvey, D. (1990) *The Conditions of Modernity: An Enquiry into the Origins of Cultural Change*, Oxford: Blackwell.

Hazarika, I. (2010) "Medical tourism: its potential impact on the health workforce and health systems in India," *Health Policy and Planning* 25(3): 248–251.

Hochschild, A.R. (2012) *The Outsourced Self: Intimate Life in Market Times*, New York: Metropolitan Books.

Ikemoto, L.C. (2009) "Reproductive tourism: equality concerns in the global market for fertility services," *Law and Inequality: A Journal of Theory and Practice* 27: 277–309.

Maranto, G. (2010) "They are just the wombs," *Biopolitical Times*, 6 December, www.biopolitical times.org/article.php?id=5497.

Martin, L.J. (2009) "Reproductive tourism in the age of globalization," *Globalizations* 6(2): 249–263.

Martinović, I. (2010) "Ukraine: Die Baby-Fabrik," *Der Biber*, October 18.

Nahman, M. (2010) "'Embryos are our baby': abridging hope, body and nation in transnational ova donation," in J. Edwards, P. Harvey, and P. Wade (eds.), *Technologized Images, Technologized Bodies*, New York: Berghahn Books, pp. 185–201.

Pande, A. (2008) "Commercial surrogate mothering in India: nine months of labor?" in K. Kosaka and M. Ogino (eds.), *Quest for Alternative Sociology*, Melbourne: Trans Pacific Press, pp. 71–88.

—— (2009a) "It may be her eggs but it's my blood: surrogates and everyday forms of kinship in India," *Qualitative Sociology* 32(4): 379–404.

—— (2009b) "Not an 'angel', not a 'whore': surrogates as 'dirty' workers in India," *Indian Journal of Gender Studies* 16(2): 141–173.

—— (2010a) "'At least I am not sleeping with anyone': resisting the stigma of commercial surrogacy in India," *Feminist Studies* 36(2): 292–312.

—— (2010b) "Commercial surrogacy in India: manufacturing a perfect mother_worker," *Signs: Journal of Women in Culture and Society* 35(4): 969–992.

Panitch, V. (2012) "Surrogate tourism and reproductive rights," *Hypatia* 28(2): 274–289.

Parry, B. (2008) "Entangled exchange: reconceptualising the characterisation and practice of bodily commodification," *Geoforum* 39(3): 1133–1144.

—— (2012) "Economies of bodily commodification," in T. Barnes, J. Peck, and E.S. Sheppard (eds.), *The Wiley-Blackwell Companion to Economic Geography*, Malden: Blackwell, pp. 213–225.

Qadeer, I. (2009) "Social and ethical basis of legislation on surrogacy: need for debate," *Indian Journal of Medical Ethics* 6(1): 28–31.

Raymond, J.G. (1993) *Women as Wombs: Reproductive Technologies and the Battle Over Women's Freedom*, San Francisco: Harper.

Rotabi, K.S. (2010) "Human rights and the business of reproduction: surrogacy replacing international adoption from Guatemala," *RH Reality Check*, May 20, http://rhrealitycheck.org/article/2010/05/20/human-rights-business-reproduction-surrogacy-begins-replace-international-adoption-guatemala.

Sarojini, N., Marwah, V., and Shenoi, A. (2011) "Globalisation of birth markets: a case study of assisted reproductive technologies in India," *Globalization and Health* 7(1): 27.

Satpathy, S.K. and Venkatesh, S. (2006) "Human resources for health in India's National Rural Health Mission: dimension and challenges," *Regional Health Forum* 10: 29–37.

Sengupta, A. (2010) "The commerce in assisted reproductive technologies," in: S. Srinivasan (ed.) *Making Babies: Birth Markets and Assisted Reproductive Technologies in India*, New Dehli: Zubaan, pp. 44–55.

—— (2011) "Medical tourism: reverse subsidy for the elite," *Signs: Journal of Women in Culture and Society* 36(2): 312–318.

Smerdon, U.R. (2008) "Crossing bodies, crossing borders: international surrogacy between the United States and India," *Cumberland Law Review* 39(1): 15–86.

Spar, D. (2005) "Reproductive tourism and the regulatory map," *New England Journal of Medicine* 352(6): 531–533.

—— (2006) *The Baby Business: How Money, Science, and Politics Derive the Commerce of Conception*, Boston: Harvard Business School Press.

Thompson, C. (2005) *Making Parents: The Ontological Choreography of Reproductive Technologies*, Cambridge, MA: MIT Press.

— (2011) "Medical migrations afterword: science as a vacation?" *Body & Society* 17(2–3): 205–213.

Vora, K. (2009) "Indian transnational surrogacy and the commodification of vital energy," *Subjectivity* 28(1): 266.

Whittaker, A. and Speier, A. (2010) "'Cycling overseas': care, commodification, and stratification in cross-border reproductive travel," *Medical Anthropology* 29(4): 363–383.

Zelizer, V. (2010) "Risky exchanges," in M. Bratcher Goodwin (ed.), *Baby Markets: Money and the New Politics of Creating Families*, Cambridge: Cambridge University Press, pp. 267–277.

PART IV

Health, survival and services

25

INTRODUCTION
TO PART IV

As the MDGs acknowledge, improved services are vital components to achieving holistic development. Women are deeply involved in provision and use of services relating to health, water, and cleanliness. Their crucial roles, concerns, and needs are often distinct from those of men, yet men tend to dominate decision-making in both the traditional and modern sectors. Women have particular health needs related to their reproductive role and specific requirements for personal hygiene (see chapters by Parfitt and Greed). They are traditionally responsible, not only for the day to day health and care of the young and the elderly, but for providing their households with water for a range of domestic purposes (see chapter by Sultana). From a development perspective this means accessing safe and sufficient water to maintain health and a suitable standard of living. Worldwide, women are normally responsible for the cleanliness of the home and often for that of the community, including communal washing and sanitation facilities (chapters by Greed and Thomas-Hope). While medical services have a specific curative function, women's hygiene-related, routine tasks are vital in protecting people against disease. Looking beyond health benefits, Joshi (2013) illuminates the contribution cleanliness makes to welfare. For the poor to "feel clean" is to achieve self-respect (for dirtiness is associated with poverty). Realistically accessible services that provide privacy and security enable poor women to have dignity. For women, a clean home and clean neighborhood environment are particularly important, since it is here is that they may spend most of their time (Joshi 2013; chapter by Greed in this volume).

The subjects addressed in these four chapters, on health services, water governance, sanitation, and waste management, are thus interrelated in complex ways. The authors focus on different facets, each contributing to gender and development. Parfitt, a trainer of nurses and midwives, identifies the patriarchal barriers that prevent women attaining better health, emphasizing the need to improve their access to services. Sultana uses a case study to show how mutual miscomprehension prevents water professionals from realizing well-intentioned plans to raise women's status by involving them in decision-making over new water points. Greed, a rarity as a female sanitary engineer, exhorts women to press for sanitary facilities that are women friendly in design and location. Thomas-Hope, also looking ahead, presses for continuing employment opportunities for women as waste management becomes more mechanized.

While problems of malnutrition and infection still exist and remain important in the developing world, particularly among the poor and young children, the so called "chronic diseases" such cancer and diabetes associated with obesity, common in developed countries, are also becoming more widespread. These partly reflect changes in ways of life that involve a reduction in exercise and the increased availability of attractive, cheap processed foods, rich in sugar and fat, associated with urban life. The result is a "double burden" of disease that complicates the "traditional" medical map. Parfitt provides international comparisons, and adopts a lifecycle approach for considering the different health issues of children, adolescents, adults, and the elderly. Using examples from her own research, she shows how, while poverty, deprivation, and marginality affect the health of both men and women, widespread social discrimination and even sexual violence adversely influence not only women's health but also their future welfare. Moreover, patriarchy and associated cultural norms often prevent women from accessing available medical services. Yet when able to do so, women use these services more than men and, although suffering more morbidity, have lower mortality.

Sultana uses her experience of water provision in Bangladesh to make two important points. She describes how women, the very people responsible for family water provision, are typically excluded from the participatory processes intended to ensure that user perspectives are incorporated in the design, development, and management of new community water supplies. Development professionals often fail to understand the strength of local patriarchy. Villagers, both men and women, connive to subvert their efforts. Men tend to focus on the technical aspects, leaving women to sort out problems of access in a status ridden society. In this respect, however, her Bangladeshi example is perhaps rather more pessimistic than that of some of the authors in *Gender, Water and Development*, who have shown that gender equality has improved, at least somewhat, in the course of participatory projects to improve water resources (Coles and Wallace 2005).

Then, unlike most social scientists writing on water provision, she valuably shows the importance of location, of the spatial distribution of sub-surface potable water, and how this in turn affects the ability of women to access it. The arsenic that poisons many tube-wells is scattered and unevenly distributed. Moreover, more women suffer arsenic poisoning than men probably due to poorer nutrition, and, if women show symptoms, their chances of marriage are adversely affected (Momsen 2010; chapter by Greed in this volume). Gender and geology, and the physical environment more generally (the rocks, soil mantle, geomorphology, and the seasonality and variability of the rainfall), are thus intimately and perhaps surprisingly related (Coles 2005).

In development perceptions, sanitation is very much the Cinderella, the disregarded add-on in "water and sanitation." Yet, as Greed points out, inadequate sanitation is not only common but impedes attainment of other development goals, including health and education. Again it is women and girls who have most need of better, more appropriate services and who, in both the developed and developing world, are least likely to have their needs considered. Their requirements are not merely physiological but also social. Sanitation is now big business; managers and especially engineers are overwhelmingly men; the default design has been automatically male. There is discrimination against the poorer sections of the population—not only against women. It is impossible to provide "flush loos" for all but a privileged minority in most countries. The environment is again important: technology is now emphasizing designs that make no use of water, an increasingly scarce commodity in many areas and also designs where recycling can reduce future pollution. It is for women as campaigners and cleaners to ensure that the resulting products are appropriate.

In many developing countries, particularly in large urban areas, waste management, like sanitation, has lagged behind—this time behind growing consumption with its associated debris. And women, as with water, tend to have responsibility for dealing with domestic waste, including feces. The cleaning of neighborhoods may be an informal women's task or a local authority responsibility. Thomas-Hope shows that disadvantaged women as well as men play a part in the often hazardous and unhygienic recycling of waste from garbage collection points and dumps. Waste-picking provides an economic opportunity for the urban poor. In this low-tech informal sector, men tend to control access to higher value waste, such as metals. While women are left to sort the less financially rewarding materials, this may still provide an important livelihood strategy for women who have few other opportunities. Thomas-Hope's recommendations are a plea that, as waste management becomes more industrialized, women should continue to have the chance to be involved, rather than being further marginalized as so often happens with mechanized innovation. This can be possible as Fredericks (2009) has shown in her study of women garbage workers in Senegal. Here the employment of women has varied with political vagaries, but at one stage they were even driving trucks.

References

Coles, A. (2005) "Geology and gender: water supplies, ethnicity and livelihoods in central Sudan," in A. Coles and T. Wallace (eds.), *Gender, Water and Development*, Oxford and New York: Berg.

Coles, A. and Wallace, T. (eds.) (2005) *Gender, Water and Development*, Oxford and New York: Berg.

Fredericks, R. (2009) "Wearing the pants: the gendered politics of trashwork in Senegal's capital city," *Hagar Studies in Culture, Polity and Identities*, 9: 119–146.

Joshi, D. (2013) "Apolitical stories of sanitation and suffering women," in T. Wallace and F. Porter, with M. Ralph-Bowman (eds.), *Aid, NGOs and the Realities of Women's Lives*, Rugby: Practical Action Publishing.

Momsen, J. (2010) *Gender and Development*, London and New York: Routledge.

26

GENDER AND HEALTH

Barbara Parfitt

Introduction

Gender differences in health among men and women are thought to be the result of a combination of biological, sociocultural, and psychological influences. Women live longer than men despite suffering higher levels of morbidity. In both developed and developing countries women use health services, when able to access them, more frequently than men and suffer more clinical signs of illness. Men are reluctant to seek medical advice although often engaging in lifestyles that expose them to greater threats (Hankivsky 2012).

The key issue surrounding gender and health is the inequalities that are found between men and women (Philips 2008). Inequalities have been related to outcomes that include infant deaths, mortality rates, morbidity, disability, and life expectancy. The link between inequity and health is irrefutable and directly linked with the sociocultural norms of a given society. These norms often leave women in a vulnerable position exposing them to a number of unnecessary health threats. Powerless women suffer from neglect, violence, and abuse. Lack of education and ignorance of much basic life-enhancing knowledge, and a failure to utilize resources available to them, affect their quality of life, but do not necessarily reduce their overall lifespan. Studies exploring women's role in society show that sociocultural determinants are more important than the economic position of the individual in determining health status. For women their role within the family and society and their relationship particularly with the dominant males in their family is the most important factor.

In this chapter I will examine issues of gender and health in the light of the limited evidence coming from the developing countries. I discuss gender and health using a lifecycle approach considering differing health issues evident from birth, infancy and childhood, adult, and old age that affect men and women. I will address the anomaly of women living longer than men but having greater morbidity and issues surrounding access to health services.

Longevity and the lifecycle

The physical experience of health or ill-health throughout the lifecycle is compounded by a person's gender. When health care is needed but is delayed or not obtained then people's health worsens. For women who frequently experience a delay in the provision of appropriate health care coupled with deprivation and social prejudice, ill-health is made worse through the unnecessary suffering imposed by their circumstances. In many developing countries, the combination of being a woman and being poor puts women not only at greater risk overall than men but also directly impacts upon their quality of life.

Different factors influence the experience of each individual throughout their lives and at each stage of their growth and development from infancy to old age. Individuals, both males and females, will respond not only to the biological effects of continued growth and aging but also to the social demands and cultural determinants that influence behavior. Differences in morbidity across the lifespan of men and women are reflected in demographic trends with the continued decrease in infant mortality and fertility rates expected to change the balance of males and females across the world. In addition to disease patterns other factors are also at play in this situation. Variables such as ethnic group, social and cultural beliefs and behaviors, social disruption and disasters also need to be taken into account. Deprivation is also a major influencing factor to consider. Studies that have confirmed the positive relationship between women and life expectancy have also shown that deprivation coupled with social inequality is important (Worrall-Carter et al. 2012; Husain et al. 2012).

The length of time that an individual lives is influenced by a number of different factors across their lifespan. In almost every country in the world, women, despite experiencing more general illness and higher rates of morbidity throughout their lives, are reported to live longer than men with three out of four women surviving into older age. The increased morbidity in women is often accounted for by gynecological and obstetric causes compounded in many countries by cultural attitudes and traditions and cultural control within patriarchal systems. But why, with a higher morbidity rate, women should live longer than men has yet to be understood. The longer lifespan of women is thought by some to be due to the hormonal benefits that women have, while others suggest that men's statistics are skewed by the high mortality experienced through risk-taking and war. It is difficult to explain the phenomenon simply in biological terms as biological and or sex-related explanations are rarely the only ones.

Much of the mortality in the past was the result of pestilence and famine experienced equally by men and women while there is now an increase in degenerative and manmade diseases. Major benefits have resulted from scientific advances as shown by the immunization of young children against diphtheria, polio, and tetanus, along with the abolition of smallpox, protection against hepatitis B, and a reduced incidence of typhoid, rickettsia infections, and rubella but there are still situations in some countries where boy children are immunized while girl children are not. This is not a new phenomenon. In the early 1970s during one of the last epidemics of smallpox in the remote Hindu Kush mountains it was noticeable that the majority of people who died in the community were women and girls. This was because the male members of their families had not permitted them to receive the vaccination against the disease. Programs to ensure that children are immunized have had a major effect on reducing the incidence of many preventable infectious diseases among both girls and boys but there are still pockets of resistance among some rural populations. Recently in Bangladesh many families refused the rubella immunization for girls because it was rumored that this was a plot to make them sterile, and similar reactions have occurred in northern Nigeria and Afghanistan and Pakistan in relation to the polio vaccine where those delivering the vaccine, usually women, have been attacked.

But the developing world is experiencing what the World Health Organization (WHO) terms as a double burden of disease and ill-health (World Health Statistics 2012). HIV/AIDS, tuberculosis, malaria, and other infectious diseases continue to cause high levels of mortality and morbidity and the incidence of diseases such as heart disease, diabetes, and congestive pulmonary disease is increasing. Pneumonia, diarrhea in children, coupled with problems associated with poor nutrition and poor management of maternal and child health compounds this situation and places a greater burden of illness and ill-health on women.

Gender differences during the lifecycle

Pregnancy, birth, infancy, and childhood

Maternal mortality is not simply an issue of public health but the consequence of multiple unfulfilled rights. A woman suffering from chronic malnutrition, who lives in a slum without access to safe water and sanitation and who does not have an education, is at a much higher risk of dying during pregnancy or childbirth. The same woman is at an even higher risk of dying if she is aged between 15 to 19, has suffered female genital mutilation, an early or forced marriage, gender-based violence or sexual exploitation. She would be more exposed if she has HIV/AIDS or if she is discriminated in her private and public life because she belongs to an indigenous group or because of her race, or for being an irregular migrant worker. In order to ensure that vulnerable women and girls in remote rural parts of a country have access to family planning, skilled attendants at birth and access to emergency obstetric care without delays, public policies must address broader human rights issues, rather than simply deliver a set of technical interventions. A failure to do so, might continue to condemn millions to be neglected in the fulfilment of the MDGs.

(Navanethem Pillay, United Nations High Commissioner for Human Rights, Speech at International Round Table exploring human rights and maternal mortality, Geneva, September 2, 2010)

In the developing countries things are improving for women but maternal and infant mortality rates remain comparatively high and many developing countries particularly in South Asia and parts of Africa will struggle to achieve their Millennium Development Goals by 2015 (OECD 2013). This state of affairs is not directly related to the intrinsic nature of women but rather because society has failed them. The knowledge and skills to reach out to these women and provide the necessary attention that they need is there but due in many cases to sociocultural and economic pressures, the women and their families do not receive the help that they deserve.

The birthing experience, both for women and their babies, remains a major health hazard. WHO reports that in 2010, 287,000 women died while pregnant or giving birth and 3.1 million newborns died in the neonatal period. The highest incidence of infant and maternal mortality (IMR) occurs at birth and the majority of infant deaths occur in the first 24 hours. Poor management at delivery due to the lack of skilled birth attendants leads to many unnecessary neonatal deaths; these deaths are then compounded by dangerous cultural practices that can endanger the infants' lives. This is especially the case for women from impoverished societies where there is little health infrastructure or support and where adequate antenatal and postnatal care is not provided.

Newly born infant girls are often at a greater risk because of selective lack of attention and care. The reasons for this are not fully evident but it is partly due to the burden that having a girl in some societies lays on the family, especially because of the dowry system, which can impoverish a family and leave them in debt for many years to come. The overriding perception in many communities is that girls are not only unable to contribute economically to the family but also are a serious liability. There are other reasons for preference for a boy that are culturally defined and difficult to explain. For centuries boys have been the preferred child and valued above girls. To have a boy child gives the mother status and position in the family and the community. To fail to have a boy child can lead to divorce, separation, and abandonment of the woman as well as exposing her to physical violence. Even before birth, discrimination against a girl fetus may occur. Concerns exist about female infanticide and forced abortions for, with access to ultrasound, women in South Asia as well as in parts of China and Africa are often believed to abort unwanted female children. In India, for example, selective abortions rose to nearly 4.5 million in the 2000s with an estimated total of 12.1 million from 1980 to 2010. As a consequence there is an imbalance of the child sex ratio leading to social and cultural instability (Jha et al. 2011).

The medicalization of the normal aspects of a woman's lifecycle is also said to be harmful. When the natural processes of the reproductive system without complications are reduced to health problems unnecessary medical intervention will result. Deliveries by caesarean section globally have increased almost five times over the past ten years from 2.6 percent to 12.2 percent with half of facility deliveries being caesarean sections. Specifically caesarean rates have risen exponentially within certain countries such as China and the USA where there is a 40 percent rate of sections, and in many developing countries it is over 25 percent. Although evidence of the direct relationship between IMR and caesarean sections is not strong, there is an indication that suggests that a more conservative and holistic approach to maternal welfare during the antenatal, natal, and postnatal period improves survival rates and reduces the need for caesarean sections. There is also evidence to suggest that much of the rise in caesarean rates is not the result of clinical need but rather a combination of pressure from the medical profession and lack of information given to women to ensure realistic choices. However many women in developing countries do not have the luxury of a choice for, even if a caesarean section is needed, they may be unable to reach a health center offering caesarean sections due to lack of transport or insufficient funds to pay for transport and the cost of hospitalization. Also in some cases they or their male family members do not trust the system or the care given by an obstetrician so needlessly go without care, putting themselves at greater risk. In other circumstances the legislation for supporting women is outdated and fails to provide for appropriate interventions.

A unique situation arose in Tajikistan where Family Health Nurses (FHN) had been trained to manage normal deliveries in the home. The law on delivery was still unchanged from the Soviet period which forbade women to deliver at home insisting that they attend the local maternity clinic and where a severe penalty was applied to anyone assisting them in a home delivery. Following the withdrawal of the Soviet government, transport to the clinic was no longer available and a woman who was about to deliver but lived more than 15 kilometers away from the Centre was unable to travel. The FHN who was called was unable to assist the women who delivered at home because she was very afraid that in doing so she could be arrested and sent away from her family. The baby died while the FHN waited outside of her house in a very distressed state.

(author's fieldwork, Tajikistan, 2007)

The maternal mortality rate (MMR) in Bangladesh has recently decreased from 322/100,000 live births to 194/100,000 live births from 2001 to 2010. The Department of Health and Family Welfare in Bangladesh identify the reason for this decline as largely due to improved education for girls, a reduced number of early marriages, and improved access to direct obstetric care. The focus in the Millennium Development Goals on improving education and opportunities for women has led to policy changes that focus on women's rights and these are beginning to impact upon the status and wellbeing of women in the country. However 87 percent of women in the rural areas still deliver without the assistance of a skilled birth attendant or midwife. For comparison, in Africa MMRs range from that of Botswana at 160/100,000 (2010) to that of Chad recorded at 1,100/100,000 (OECD 2013).

In a review of selected global risk factors, being underweight in childhood continues to be a major risk factor in developing countries for girls from deprived backgrounds. Girl children are frequently neglected and high child morbidity rates for female children are often the result of the lack of value placed on girls with poor nourishment, nurturing, and lack of adequate health care given at an early age. Boys from affluent families in the developing world are often identified as obese and in the Asia Pacific region there are an increasing number of both girls and boys who are obese. This is the result of overindulgence and the introduction of fast foods available to middle-class families. Worldwide, 40 million children under five were overweight in 2010 (WHO 2012).

Adult health

Adults, both men and women, suffer from increasing rates of chronic disease. Comparisons between the developed and the developing countries show significant changes taking place. Developing countries, which previously had lower rates of heart disease, stroke, and diabetes, now show a higher incidence of these conditions, while those in the developed world are improving. In situations where there is deprivation, the main causes for poor health include childhood and maternal underweight, unsafe sex, unsafe water, inadequate/lack of sanitation, poor hygiene, poor nutrition, and respiratory disease as a result of indoor smoke from solid fuels and pollution in the environment.

Occupation is also an important factor as in many developing countries health and safety regulations rarely exist and people are exposed to many risks in the workplace. Many of the newer industries emerging in developing countries, funded through world trade, have led to serious threats to safety. As women are the ones primarily being employed in these new industries they are most at risk. A number of major accidents have occurred in Bangladesh in the clothing industry. Twenty-seven people died in Ashulia in 2010 from a factory fire and in 2012, 2,000 people, mainly women, died in the collapse of the factory building in the Rana Plaza. On the other hand, the development of opportunities for work for women, especially in South Asia and China, has given women an opportunity to earn money and contribute to their families' income.

Danger threatens the security, health, and wellbeing of women worldwide. Often it is a result of violence instigated by cultural and traditional beliefs. The most dangerous countries for women have been identified as Afghanistan, with the physical dangers of rape and murder but also child marriage and poor education opportunities; the Democratic Republic of Congo, where 1,152 women were raped every day in 2011; Pakistan, where women face violence and discrimination and where one fifth of murders of women are the result of so-called honor killings; India, where the incidence of gang rape has recently been publicized, where sex discrimination is present before birth, and where 100 million people—mostly women and girls—were reported as being involved in sex trafficking in 2009; and Somalia, where 95 percent of women face female genital mutilation between the ages of four and eleven.

In many countries, road accidents remain a major cause of injury and death. In developing countries, careless and reckless driving, poor infrastructure, and unlicensed drivers with poorly maintained vehicles contribute to the high toll of accidents. In the developed world, excessive alcohol consumption is often responsible. Alcohol is not only responsible for road accidents but is also associated with violence and suicide as well as many diseases, in particular cardiovascular disease, cirrhosis of the liver, and cancer. Traditionally it has been men and youths who have been the heavy drinkers. However, those women who drink excessive amounts of alcohol during pregnancy can affect the normal growth and development of their child that may lead to mental and physical handicap.

WHO reports that rates of obesity among adults has doubled since 1980. Obesity reduces life expectancy for men by almost six years and for women by almost seven. In 2008 more than 1.4 billion adults over 20 years old were overweight. A study reviewing global nutrition rates for men and women identified that in countries where there was overweight and obesity or underweight and malnutrition these were experienced equally by both men and women. Lifestyle behaviors such as unsafe sex, tobacco, and alcohol have been identified as the leading causes of the global burden of disease, contributing to poor health in both developed and developing nations. Smoking is a major health risk and, although the rates among male smokers are reducing in the developed world, in the developing world smoking continues to be a major practice for men, increasing their risk of heart disease and emphysema.

Men and women experience different diseases partly due to their biological makeup but also as the result of different sociocultural behaviors. For example in parts of sub-Saharan Africa there is a higher incidence of HIV and AIDS than in other parts of the world, with many exposed to the AIDS virus. The reason why African men and women are more susceptible to the HIV virus is not fully understood. Women and men are exposed to HIV infection through close proximity to each other and sexual activity with an infected partner. HIV is transmitted via bodily fluids, especially semen and vaginal secretions. It has been shown that one of the main underlying causes of its proliferation is the behavior of men and women and the cultural norms that influence their understanding of the disease. This behavior can put whole communities at risk and the rate of transmission in parts of Africa is directly influenced by norms that prevent women from having control over their own sexual practices. Women in many societies, especially young women, cannot refuse to have sexual relationships with their partners, as they are at risk of being beaten and abandoned. A UNAIDS Report indicated that approximately 850,000 young women between the ages of 15 and 24 were newly infected with HIV in 2011 alone. Women can pass on the disease to their children during childbirth. However, treatment is becoming more widely available in the developing world and the disease is no longer spreading as fast.

The most common cancer suffered by women the world over is breast cancer and the incidence is increasing in the developing countries. It was estimated in 2004 that 519,000 women died as a result of breast cancer. Some 69 percent of all breast cancers occur in the developing countries. The overall improved survival rates for all cancers that women have over men may be the result of the biological response of men and women to cancer. Survival is also influenced by lifestyle choices such as smoking and excessive alcohol consumption. It could also be due to the different health-seeking behavior patterns of men and women, as it is acknowledged that women seek health care more frequently and sooner than men, and also women take health advice more seriously. This may account for their higher recorded morbidity. They engage with more health-promoting behaviors than men and generally have a healthier lifestyle pattern. Being a female is the strongest predictor of preventative and health-promoting behavior. Men, apparently regardless of ethnicity, are reluctant to consult a doctor. Because of the socialization of men, they often do not show pain or want to appear to be weak. Consequently they do not seek help as long as they are able to work. Undiagnosed hypertension is a major health hazard and men also regard overweight and obesity as women's problems with the result that they fail to ask for treatment or advice (Peate 2011).

The gender gap in health status between men and women appears to become narrower during periods of economic development when women are earning. High income, working full-time and caring for a family, and having social support are more important predictors of good health for women than men. In contrast other studies have shown that women's longevity decreases to match that of men when they are working (Philips 2008). When there are periods of economic constraint, when women are not earning as much as men, women retain their advantage over men, living longer with greater longevity. The reason for this may be that work that gives women greater financial independence may at the same time expose them to greater stress levels perhaps influencing an increase in alcohol intake and smoking, which lead to adverse health outcomes and also risks of abuse, violence, and discrimination.

Older age health

With declining fertility and improvements in life expectancy there is a rapid increase in numbers of older aged people in both the developing and the developed countries. This is creating a major burden on society as fewer people are earning to support the elderly dependent population. The number of people aged 65 or older is projected to grow from an estimated 524 million in 2010 to nearly 1.5 billion in 2050, with most of the increase in developing countries. It is predicted that by 2050 more than 30 percent of China's population will be over 60. Children and young adults who have grown up in poverty and ill-health in developing countries will be entering old age with a major burden of illness that will challenge the already inadequate health service available in many countries. These services barely provide the necessary support for the current generations and a health service that provides for the increase in older aged people will be a major challenge for most developing countries By 2030, WHO states that among the 60-and-over population, non-communicable diseases will account for more than 87 percent of the burden of disease. Over the next 10–15 years, older people in every world region will suffer more death and disability from non-communicable diseases such as heart disease, cancer, and diabetes than they do from communicable diseases. Because women live longer than men, women constitute the major group affected by these circumstances.

Health issues for the elderly, however, are rarely confined to a single health problem. As people get older, they suffer from several chronic health conditions at the same time. Compounding this situation, many elderly people, particularly women, experience discrimination, abuse, neglect, and poverty. Traditional family support systems often cannot cope with the extended life of older aged relatives with the result that they are increasingly marginalized. Mental health is also a key issue for the elderly. Loneliness leading to depression is often the consequence of neglect and separation from family and friends and community. Dementia and Alzheimer's disease are more prevalent in the developed world but this is largely due to undiagnosed cases in the developing world and fewer numbers of adults surviving into old age; they are expected to be an increasing problem in the future. Compound this situation with the level of discrimination experienced by women at present, then the prospects for older aged women in the future are poor.

Gender inequity and the sociocultural impact on health

Domestic violence is evident in both developed and developing countries. The health consequences of violence for men and women are different and more profound for women than for men. Violence against men is generally committed by other men, while men are the ones who are violent against women, normally those to whom they are closest such as wives, girlfriends, daughters, and other family members. Abuse by men as a form of neglect coupled with ignorance also leads to health consequences for women. The actual act of violence, especially sexual violence, is not the only consequence experienced by women. Women are invariably blamed for sexual violence against them and the associated shame both for themselves and their family will have a long-lasting effect and in some cases is impossible to live with.

> A 13-year-old girl was kidnapped on her way home from school. She was gang raped by four young men and later dumped by the side of the road. She was brought to the clinic where rape was confirmed and serious damage was identified to her genital area. The incident was reported to the police. Her family took her home as her father did not give permission for the necessary referral or treatment to be given. The family were ashamed of what had happened. The young girl developed a vaginal urinary fistula which became infected. She did not go back to school, she could not be married off, and she committed suicide this year. The young men who were local were never brought to justice although everyone knew who they were.
>
> *(author's fieldwork, Bangladesh 2012)*

The health effects of violence against women and on women are manifold. In addition to bruises and breaks, disfigurement from acid-throwing, and dishonor from rape and trafficking, such events will lead to the end of marriage prospects and young women essentially becoming social outcasts. Drug abuse and murder are also not uncommon with the mental trauma caused by sustained abuse affecting women's general health, reducing their personal self-esteem and confidence, and leading to a poor quality of life and an inability to carry out any form of paid or unpaid work. They become victims and as such they are weak, subordinate, and passive. Ignorance of their rights makes it impossible for women to question their circumstances and the behavior that gives rise to the abuse is often strengthened by other women who maintain the status quo by reinforcing male authority on younger women in order to gain some power and control of their own. The ignorance of young girls about their own body and sexuality is also often an additional cause of exploitation leading to unwanted pregnancies, sexual abuse, and neglected genito-urinary health problems.

Harmful practices that lead to health problems and are inflicted as a result of social or cultural influences raise ethical and medical questions. Cultural determinants have a major influence on practices that can bring harm to both men and women. For young men in all countries there are rites of passage that have to be undertaken for acceptance in the community. Subcultures such as inner-city gangs, political and extreme religious groups can require from young men behaviors and practices that put their lives at risk. Young women are subjected to other dangerous cultural practices, one of which is female circumcision, which is both a medical and a sociocultural problem. There are dangers inherent in this practice as poor procedures can lead to long-term genital damage. There are also ethical questions as to whether the practice should be allowed at all. Two distinct points of view are put forward. One is the position of harm reduction: by teaching practitioners safe procedures and providing facilities for surgery, the harmful effects can be reduced. The other view favors abolition and rejects any intervention in the belief that the practice threatens the equality of women and is primarily a human rights issue and as such must be stopped. The argument against harm reduction is that this is a health problem that can be prevented and merely treating it will not solve the inherent problem. The United Nations has banned any kind of procedures for female circumcision. In the meantime, however, many women are suffering. The dilemma of whether it is better to insist on a complete banning of the procedure or whether to adopt a 'harm reduction' strategy is not easy to solve. Similar discussions are held regarding access to abortion, which in some countries is a criminal offence, leading to illegal abortions that threaten the lives of many women.

Education can overcome gendered health disadvantages that lead to poorer health among women. But in some cases there is a backlash against the education of girls from many male-dominated societies. In these cases, girls continue to be abused with resentment from men spilling over into violence. They are also more vulnerable as they have little protection once in the co-educational school system; the abuse often takes place in the schools with many young girls subjected to sexual violence from the boys and from the male teachers in order to achieve the grades they require. Many families discontinue their daughter's education and arrange their marriages at a young age as they see this as a way to protect them from exploitation and sexual harassment. The daily papers in Bangladesh regularly report on girls who have been stalked and abused while attending school (Chisamya et al. 2012). A South Asian newspaper reported the incident of a young married man cutting off the hand of his new wife because she was studying.

The Grameen Caledonian College of Nursing in Bangladesh was set up to provide an opportunity for the daughters of the Grameen Bank Borrowers, young women from disadvantaged rural backgrounds, to undertake a combined nursing and midwifery program. Training these young women as professional nurse-midwives, future leaders, and change agents has already begun to make a difference not only in their own lives but also in the lives of their families and their communities as they spread their knowledge and demonstrate how young women previously destined for early marriage, early pregnancy, and possibly early death can now make such a difference. Among the first graduates from the college in 2012, some have been appointed to work with the joint government, BRAC (formerly Bangladesh Rural Advancement Committee) and USAID program training skilled birth attendants (SBAs), while others are working with Grameen Communication to provide comprehensive health assessments in the rural villages and offer advice and referral when necessary. Even at this early stage, it is expected that there will be improvements in both maternal mortality and infant mortality in the near future. The presence of Skilled Birth attendants has been shown to make a difference in a study carried out in Matlab, a district of Bangladesh. The study showed that over the past 30 years, maternal mortality fell by 68 percent in this service area and by 54 percent in the government service area. In the service area the speed of the decline was faster after 1990, when the skilled-attendance strategy was introduced ($p = 0.09$). Although using skilled birth attendants was successful and was shown to make a significant difference there was also an acknowledgement that the uptake was lower than expected. The need to persuade families to access maternal services for women is a slow process and requires major shifts in cultural attitudes of families particularly in the rural areas (Chowdhury et al. 2009).

A young married woman of 17 years was expecting her first baby. She was examined by the community health worker (CHW/SBA) who had some anxieties about the position of the fetus. She was advised to go to the clinic for further examination. She did not attend. She was seen again when she was almost at term and the CHW/SBA once again advised her to attend the clinic for her delivery feeling that she might need assistance. The young mother was very distressed and anxious but her father-in-law, mother-in-law and husband had forbidden her to attend the clinic and would not allow her to seek help. She and the baby both died during labor.

(author's fieldwork, Bangladesh 2012)

Access to health services and the medicalization of health provision

Poor people in the developing countries have less access to quality health care than the wealthy. For women geographical accessibility and availability of health care services are major issues. They are less so for men who have greater mobility. When services are centralized in large cities, those living and working in that area can normally gain access, but for rural and suburban dwellers the cost, inconvenience, and dangers of travelling to the hospital or clinic may prevent them from attending except in cases of emergency. Available accessible clinics and health care facilities have been shown to encourage increased attendance especially for women. The type of facility providing the health services also impacts upon usage (Peters et al. 2009). For example, tertiary services providing specialist care that are normally located in cities favor both men and the better-off, while primary health and preventative services favor poorer communities and women. Primary health care services generally encourage usage from both women and the poor and the difference between different economic groups is less evident than for those provided in tertiary institutions. Women are often materially reliant on men and as a consequence are dependent on them for decisions as to what health care they can receive. Poverty and access go hand-in-hand and women are disadvantaged most in a situation of extreme poverty.

Acceptability of services is also a major determinant of access to health care. The failure to recognize and adapt to cultural practices often leads to women refusing to visit local clinics. Many health care services are not acceptable to women, often due to the lack of female physicians, resulting in an inequity in the provision of health services especially for poor women. In many Muslim countries, the failure of women to seek medical care early because the services are unacceptable has led to a high incidence of undiagnosed breast cancer and poor uptake for family planning. Little investment is made to provide services for women's health, antenatal, delivery, and postnatal care. The poor continue to deliver without a skilled birth attendant leading to higher rates of maternal and infant morbidity and mortality. In some countries obstetricians see female attendants or midwives as non-essential to the modern birth process. The medical model in a patriarchal society ensures that women conform to their social role but in doing so in many cases they expose themselves to higher risks.

Conclusion

In this chapter I have tried to highlight some of the key discussions that focus on gender and health. Cultural, psychological, and social factors that influence health and disease and lead to gender discrimination are often ignored because the emphasis is placed on sex and biological factors. Gender is one of the most important factors to influence health-related behavior both for men and women. For women, their gender identity is highly significant as it has a greater influence on their morbidity and mortality rates. Women are affected by the reproductive process and the associated hormonal influences; childbearing is a major factor influencing their overall health status. Lifestyle is an important factor for men, giving rise to a higher incidence of some diseases such as heart disease and chronic pulmonary disease. There is some evidence that as women adopt the lifestyles of men, they take on the same patterns of health and are subject to the same disease patterns.

The pattern of disease is changing with an increase in chronic diseases such as diabetes, cancer, heart disease, and chronic pulmonary disease. Infectious diseases are no longer the main cause of death, although in the developing countries they still remain significant and give rise to the double burden of disease. The anomaly of women suffering higher incidences of morbidity but generally living longer than men is still not resolved but the indications are that the higher morbidity rates are influenced to a degree by biological and genetic differences but also by a combination of sociocultural factors.

There continues to be a general picture of discrimination against women both inside and outside the family reinforced by the level of ignorance and lack of health education that women possess. The vulnerability of women and the extent of abuse that they encounter both psychologically and physically within society is largely due to the sociocultural factors that influence behavior and this has a significant effect on their health. The lack of access to health services due to poor, inadequate, or inappropriate provision as well as to neglect, direct violence, and abuse that leads to physical and mental health breakdown all contribute to gender inequality in health experienced by women.

References

Chisamya, G., DeJaeghere, J., Kendall, N., and Aziz Khan, M. (2012) 'Gender and education for all: progress and problems in achieving gender equity,' *International Journal of Educational Development* 32: 743–755.

Chowdhury, M.E., Ahmed, A., Kalim, N., and Koblinsky, M. (2009) 'Causes of maternal mortality decline in Matlab, Bangladesh,' *Journal of Health, Population and Nutrition* 27(2): 108–123.

Hankivsky, O. (2012) 'Men's health and gender and health: implications for intersectionality,' *Social Science and Medicine* 74: 1712–1720.

Husain, N., Cruickshank, K., Husain, M., and Khan, S. (2012) 'Social stress and depression during pregnancy and in the postnatal period in British Pakistani mothers: a cohort study', *Journal of Affective Disorders* 140: 268–276.

Jha, P., Kesler, M.A., Kumar, R., Ram, F., Ram, U., Aleksandrowicz, L., Bassani, D.G., Chandra, S., and Ban, J.K. (2011) 'Trends in selective abortions of girls in India: analysis of nationally representative birth histories from 1990 to 2005 and census data from 1991 to 2011,' *Lancet* 377: 1921–1928.

OECD (2013) *Health Data: Statistics and Indicators*, www.oecd.org/dac/theoecdandthemillennium developmentgoals.htm.

Peate, I. (2011) 'Men and cancer: the gender dimension,' *British Journal of Nursing* 206: 340–343.

Peters, D.H., Garg, A., Bloom, G., Walker, D.G., Brieger, W.R., and Hafizur Rahman, M. (2009) *Poverty and Access to Health Care in Developing Countries*, Baltimore: Johns Hopkins Bloomberg School of Public Health and Institute of Development Studies.

Philips, S.P. (2008) 'Measuring the health effects of gender,' *Journal of Epidemiology and Community Health* 62: 368–371.

WHO (2012) *Global Burden of Disease Study 2010*, Geneva: World Health Organization.

World Health Statistics (2012) *World Health Statistics 2012*, www.who.int/gho/publications/world_ health_statistics/2012/en.

Worrall-Carter, L., Edward, K., and Page, K. (2012) 'Women and cardiovascular disease: at a social disadvantage?' *Collegian* 19: 33–37.

27

RETHINKING COMMUNITY AND PARTICIPATION IN WATER GOVERNANCE

Farhana Sultana

Introduction

The international community has recognized water as a human right, but water continues to be unequally available and accessible across the globe. The Millennium Development Goals (MDGs) have pointed to the importance of increasing safe water provision, which is also a gender issue, since it is women and girls throughout the developing world who labour to provide water for their families. Development goals can be hindered when insufficient water constrains people's abilities to live healthy and productive lives. As a result, water governance has become important in development policies and projects in the Global South.

Community and participation have permeated development discourses and practices in recent years. This is particularly prominent in the water sector, where there has been a shift from state-led, technocratic water management programs to an increase of 'participatory' and 'community-based' water management. Emanating from the participatory development models of the 1970s and 1980s, where civic participation and ownership of development endeavours were seen to result in better outcomes, participatory community-based water management projects have become popular among states, international donors, and NGOs. Thus, there have been changes in the governance of water resources away from state-controlled management to a focus on community-based institutions and reduced direct state responsibility.

Despite their popularity, there are several problems with discourses and practices of participation and community in water governance. Gendering such approaches further exposes both conceptual and practical limitations of the concepts. Water governance approaches that generally view communities as homogeneous entities can overlook complex realities where access to and control over water resources vary by multiple, interlocking, and hierarchical systems of differentiation. Similarly, participation involves processes of inclusion, exclusion, negotiation, and resistance, which are insufficiently understood or addressed. This chapter demonstrates that not only are gender and class relations important in assessing how notions of community and participation must be extended, but that geographical location and nature/water play important intersecting roles in local water management projects and institutions. The analysis is based on extensive research conducted in areas of rural Bangladesh that are facing acute drinking water crises from arsenic contamination of groundwater sources (Sultana 2007). There are an estimated 35 million people in Bangladesh consuming poisoned water, as naturally occurring arsenic (from aquifer sediments) is present in groundwater pumped up by tube-wells that are widely used for both drinking water and irrigation purposes. While the introduction of tube-wells was deemed a development success, as 'safe' groundwater reduced mortality and morbidity from pathogen/ microbial surface water sources historically used for drinking water, the discovery of arsenic and subsequent poisoning have resulted in drinking water crises in recent years (Smith et al. 2000). Arsenic's unpredictable spatial heterogeneity in the aquifer (due to minor differences in Holocene deposits) is reflected in the spatial heterogeneity of contaminated tube-wells (i.e., there can be contaminated and uncontaminated tube-wells within a few hundred feet of each other, and entire villages may have only contaminated tube-wells, or have few uncontaminated ones, which may be spatially clustered at a variety of scales and often be quite dispersed in pattern on the landscape). Arsenic primarily occurs in the shallow aquifer, where the vast majority of the tube-wells can affordably draw water; a few rich households can afford to drill into the deep aquifer, where there is little or no arsenic. As a result, there is great hardship in accessing safe drinking water and great conflict over safe water in many areas. In order to address the water crisis, arsenic mitigation projects have been implemented by organizations throughout Bangladesh to provide safe/alternative water supplies. These projects generally mobilize discourses of community and participation in local water management institutions. This chapter discusses some of the failings and challenges in the conceptualizations and practices of community and participation, whereby water management strategies produce inequalities and differences in safe water usage, access, and exposure to arsenic.

Community and participation in water governance

Scholars have debated the controversial issues surrounding the ways that community and participation have been conceptualized, mobilized, and deconstructed in natural resources management and development literatures (Agrawal and Gibson 2001). Despite critiques of exclusions, captures, and marginalization, the considerable staying power of notions of community and participation in development policies has resulted in a proliferation of community-based and participatory projects throughout the Global South. In the water sector, creating water user committees as part of community-based water management plans are common, whereby the committee is responsible for representing communities/villages in managing water infrastructure and decision-making at the local scale. Committee members often are assumed to have common interests and goals, overlooking social difference and heterogeneity of communities as well as environments. While development planners may acknowledge the problems that exist, project implementation often operates by treating communities as territorially defined intact wholes within the remit of the projects. However, notions of community being inherently egalitarian are problematic (Cornwall 2003). The social and power relations that play out in water management can challenge the notions of democracy and equity that are increasingly embodied in national water development policies uncritically espousing community and participation (Mosse 2003). Thus, while notions of community in water management may be externally defined by implementing organizations (e.g., local or extra-local NGOs, donors, states), the projects are implemented through local power relations, where different people with various strengths and weaknesses based on their structural position in village society will negotiate their positions within such projects vis-à-vis the costs and benefits in the context of their overall lives and livelihoods. As a result, it is important to look at the ways that community institutions operate in creating boundaries, exclusions, inclusions, and regulation.

The second popular discourse, related to that of community, is participation. Community members are expected to participate in projects in order to enhance equity and efficiency, as well as to feel greater ownership towards projects, which is also expected to lead to better water management and improved overall development. Participation invokes notions of inclusion, of people's abilities to make decisions, and to voice opinions/concerns that are heard (Agarwal 2001; Cooke and Kothari 2001). As such, participation hints at deliberative democracy. While notions of participation have become hegemonic in development discourses, they conceal processes of unjust and illegitimate exercises of power. Agarwal (2001) further argues that participatory institutions are often socioeconomically inequitable and perpetuate unequal relations of power. While locally accountable representatives can be sufficient if everyone cannot participate, their accountability is often a problem as there can be elite capture and corruption. Downward accountability may be lacking, although there is meant to be greater sharing of powers and resources with all the members who are meant to benefit from a project. Networks of relationships of reciprocity and livelihoods may also mean that people make decisions to support dominant institutions and not to challenge them (Cleaver 2000). Traditional notions of participation in village life are often worked out through patronage systems and kinship structures. It is within such unequal set-ups that participatory water management projects often embed themselves and thereby perpetuate cycles of inequality. As a result, participation is a process that involves conflict and consensus, within broader historical factors and constraints, and is not just a mechanism to facilitate project success or a set of techniques, although this is primarily how it has been treated in a majority of development projects.

Gendering and spatializing community and participation

Furthering these debates are contributions made by scholars who have looked at either the gendered dynamics of community/participation, and/or the spatialities of the processes of community/participation. I argue that it is important to look at these issues simultaneously. By undertaking a gender analysis, Cleaver and Elson (1995) expound the view that community water management schemes may not be equitable and may lead to further marginalization of poorer women in accessing water. Gendered analysis allows for understanding structural inequalities in community and household resource use and allocation. Women's and men's involvement in community projects have to be assessed in terms of their decision-making powers and the benefits accrued to them in various forms. If the beneficiaries/participants are conceptualized along certain criteria, then groups of people may be targeted, for example 'women'. In such instances, it is likely that women of any background can be assumed to be representative of the different groups of women, and differences between women in a locality get overlooked or obscured in the project. Blindly assuming that having rich or elite women participate in the project leads to 'gender mainstreaming' can be problematic, as exclusions and privilege may become institutionalized.

While adding women to a project may seem to address gender concerns stipulated in project documents, it does not necessarily address power issues between men and women, and among different women. Gendered subjects experience simultaneous processes of inclusion and exclusion based on other intersecting social processes, and thus it is not possible to generalize across all women or even men (Agarwal 2001). Social relations of class, kinship, marriage, and household relations can all complicate the ways that people experience exclusion and inclusion. There may be a range of different lines of connection and differences that situate women differently from each other, and the myth of female solidarity thus does not hold up to the ways that women may choose to pursue different desires, connections, and needs (for example, not all women in a neighbourhood may be similarly exposed to contaminated water or have similar water needs). In water management, however, some more clear patterns of exclusion do emerge, vis-à-vis men excluding women in decision-making roles, and men and women of wealthier households excluding people of poorer households from accessing their safe water sources. What is evident is that it is not just women but many poor and marginalized men are also excluded, which is often not captured by only focusing on women.

Participation is often portrayed as increasing overall 'empowerment' of women, but recent evidence suggests that many women are disempowered and marginalized in the process (Cornwall 2003; Agarwal 2001). The critical assessment of how participation is conceptualized and a gender perspective on who participates, in what capacity, to what effect, and with what means, is important in understanding the outcomes of participatory management institutions being set up as the solution to water governance problems as well as achieving problematic notions of 'empowerment'. For instance, women's participation in the process of planning or decision-making regarding water resources, generally seen as a male domain, is constrained by gendered responsibilities (both productive and reproductive), time, costs, as well as local norms of what is deemed appropriate gender behaviour. Agarwal (2001) posits that seemingly participatory institutions can exclude people through 'participatory exclusions', which can individually and interactively constrain a woman's participation in water resource management. These are: rules of entry, social norms of women's behaviour and actions (e.g., speaking in public for a gender division of labour), social perceptions of women's abilities, entrenched territorial claims by men, personal endowments and attributes of women (e.g., education), and household endowments and attributes (e.g., class).

Furthermore, participation is a spatialized process, taking place in specific spaces and places, which are symbolically gendered. As a result, spatialized subjectivities can discourage people from speaking in public, and people may perform differently in different spaces. For instance, when meetings take place in bazaars or market places, it is more difficult for women to attend meetings (as these are gendered spaces for men). Public space and decision-making in participatory development projects in many places also exclude women largely due to notions of appropriate feminine behaviour as well as practices of *purdah* (varied practices of veiling and seclusion that curtail women's mobility as well as public behaviour). Given that participation activities are largely conducted in public spaces, or what are perceived to be public activities of decision-making and sharing opinions, notions of femininity and masculinity can be challenged when women and marginalized men are involved. This results in both women and men being uncomfortable with projects that attempt to have participatory planning sessions or public committee meetings. These gendered subjectivities and identities are shifting, contested, and rethought in development projects so that they make sense to each individual in what it means to be a 'good' man or woman, husband or wife, son or daughter, within the contexts of other factors, experiences, and goals in their lives (Sultana 2009). Thus, women's mobility and autonomy, as well as decision-making powers, are spatially challenged in addition to the sociocultural ideologies of their capacities and rights to participate in decision-making fora. Such participatory exclusions can be powerful in highly unequal and patriarchal settings.

Greater attention to both gendered identities and agency is thus important in understanding how and why women and men participate in water management projects or not. Thus, women can manoeuvre through patriarchal structural forces in resisting, challenging, and reproducing power relations that operate in the ways that participation plays out in water management. Heeding subjectivities of femininity and masculinity that are associated with activities of participation help explain why different people relate to community participation in the ways they do. Partaking in water projects is bound up with sensitivities beyond the 'rational' water user that is assumed in participatory development projects, where water users are expected to automatically want to participate and do so with unified and collective identity. This is generally not the case. People display varying opinions and agency in the ways that water projects function in their locality, and what it means in their own access to safe water. Such realities are not just socioculturally defined, but also inflected by various understandings of water contamination and relations to water. This is where a closer attention to nature/water comes to make a difference.

Arsenic mitigation and water management projects in Bangladesh

Nearly all of the arsenic mitigation projects studied in this research were promoting community-based participatory water management options in order to address the acute drinking water problem. It was found that local people were less involved in the actual selection of technology (such as deep tube-wells, dug-wells, rainwater harvesting technologies), conceptualization, fund mobilization, and decision-making about management, than in site selection, collection of fees, and construction of infrastructure. While such involvement varies across different projects, what was noticeable is that there is greater investment and attention given to technology and physical infrastructure and less to the social organization and management institutions by the implementing organization. The main criteria by which most arsenic mitigation projects proceed are: explicit interest from the local community in having water technology options, agreement to invest in costs and construction (typically some 5–15 per cent of capital costs and 100 per cent of maintenance costs), and commitment to self-regulate user access and control. However, there is usually little follow-up beyond the physical construction and initial fund collection by the implementing organizations. At the village level, arsenic committees or water user committees are often set up as the local management institutions for the projects. In some instances, these committees are set up ad-hoc by implementing agencies in order to get arsenic mitigation projects started up; in other instances, existing village groups are used.

In the majority of cases, people are asked to form groups on their own, or recommend others to join, and this type of group formation is generally understood to form the 'community' for the project, as well as those who will 'participate' in it. In many projects, since users are usually seen as household units, it is the household head whose name is on the list of users. Since this is most likely to be a male head, the committees end up involving men. Thus the rule of entry ends up discriminating against the actual users (i.e., the women and girls who fetch the water or manage it at the domestic level). Community projects also require certain skills such as literacy, numeracy, organizational, networking, and people skills, which can also be lacking in many areas, or taxing on the few who have the skills. Thus there are limits to local capacities, as the numerous community projects ongoing in any given area can often involve the same people, thereby overburdening them with work and contributions of time and resources for the various projects.

The notions of participation and community among implementing officials largely consist of following guidelines that project documents have articulated, often under assumptions that spatially clustered arsenic-afflicted households have equal water needs. In order to have the quickest and optimal outcome in terms of project delivery schedules, officials generally contact the village leaders or elders and work through them. Very few officials felt the need to operate otherwise. However, all projects articulate goals of participation of the poor and equitable water use. When community groups are formulated along traditional lines of kinship or power hierarchies, there is a tendency for marginalization and elite capture, which have been identified to be serious problems in implementing community water projects. Similarly, since project requirements often required involving some women in water committees, this was often accomplished in instrumentalist ways (where any woman was placed on the committee irrespective of their water problems) or subverted (where fake names were included or the actual women noted were not really involved at all).

Members of committees are supposed to be engaged in the implementation of arsenic mitigation projects in their area, or at least to be responsible for the running (operation, maintenance, fee collection) of the water project with which they are affiliated. A caretaker is sometimes chosen from among those on the committee and is responsible for looking after the maintenance of the water technology and ensuring only official members obtain water from the source. However, several projects involved the implementing organization freely donating water technology without formation of committees or groups. These organizations generally did not require financial contributions and expected that local people will sort out management and maintenance issues on their own. Many of these donations were seen on the land of influential and wealthy households, although some were also specifically targeted and given to households with arsenicosis (arsenic poisoning) patients, or where there was a large cluster of contaminated tube-wells. Given the difficulties in mapping the exact location of arsenic in the aquifer, the criteria that organizations often follow involve looking at contamination levels in individual tube-wells, and the spatial clustering of unsafe tube-wells (which is a product of both natural distribution of arsenic in the aquifer and local geology, as well as human settlement patterns and historical placement of tube-wells). Government attempts to inform people about arsenic levels in their tube-well water have been to paint red those tube-wells with contamination levels above 50 micrograms/litre (the Bangladesh government standard of allowable arsenic levels) and paint green those tube-wells with arsenic levels deemed safe for human consumption. These act as visual markers of safe/unsafe water sources, in a binary colour system that does not inform people of the seasonal variations in arsenic concentration nor the actual concentration of arsenic in their water. When more than 80 per cent of the tube-wells in a village are identified to be contaminated and painted red, external organizations are more eager to implement projects and have greater access to donor funding.

The siting/location of a community water option is one of the critical issues that also requires negotiation between people and underscores various power relations that exist. Usually if wealthy households want to donate part of their land to install a community water option, they are instructed that they must allow access by other users. However, when the option is located inside the *bari* (homestead consisting of a cluster of huts of families in the same kinship structure around a common courtyard), especially close to the dwelling huts, there are greater access restrictions imposed by the landowner. Often people will debate and negotiate which spot is the best one for installing a community water source, but it is frequently overtaken by powerful families who dictate where the location should be (often donating their land or more money in order to control the project).

In places where community water projects were operating relatively well, the general opinion of those involved in the project was the need to increase the number of water options available, reduce the number of households dependent on each option, reduce the costs involved, and configure better ways to share the water. However, among most water user group members, there was general satisfaction that they had somewhat better access to a safe water supply, even if they had to pay for it. But a majority of the women involved in these projects did raise complaints that the water sources are often not maintained, that the people on whose land it is on tend to monopolize the source and often treat it as their personal source, and that there are crowding and time factors involved, as well as conflicts and arguments at/over water sources (Sultana 2009). While outright denial of safe water may be less common, at what cost (both literally and figuratively) water is fetched is an important factor for many households. Water–society relations are also inflected by gender sensitivities in that, even if a household's water source is contaminated with arsenic and deemed unsafe, concerns of women/girls venturing far or into public spaces to get safe water from community sources often result in families continuing to consume their own unsafe water. This can happen even if the household is officially within the reach of a participatory community water project, thereby undermining the goals of a project to provide safe water to all project users. Men do not participate in fetching domestic water, as it is seen to be an unmanly job, and thus gender norms and practices can challenge the success of water projects (see Sultana 2009 for further elaboration).

Reworking and renegotiating community and participation in water

Not all households feel the burden of safe water scarcity or arsenic poisoning in the same way in any given locality. In the territorial/spatial delineation or 'catchment area' of any community water project, there may be owners of safe/green tube-wells, which complicates notions of the continuous presence of needy households in the project area. Such safe water sources can also offer alternative water options to other households in the vicinity who can also opt out to get water from such water sources rather than from the community projects. The heterogeneous presence of arsenic in the aquifer results in a discontinuous presence of contaminated tube-wells, thereby reducing the desires by many who have safe water access/ownership to invest any interest, money, and time in the success of community projects or participate in them. Dynamic social power relations of inclusion/exclusion as well as uncertainties and fragmentations in nature come to undermine community projects. As a result, the very discourses of participation and community that are supposed to bring people together to enhance equity and efficiency of projects may not come to fruition as expected, and are implicated in the very relations people have with water (i.e., their relative location to contaminated parts of the aquifer, having a tube-well that is deep enough to draw out safe water from the deep aquifer, or living in areas where there are no safe tube-wells at all). Yet on the other hand, there are many areas that have acute problems with no safe water sources, but a community project may be quite far away or beyond the financial or social resources of the people in the locality to participate in existing water projects.

In areas where severe arsenic contamination has resulted in community water projects being introduced formally, a variety of social power relations complicate the ways that community and participation are understood and practised. A common theme regarding the management and operation of the community water projects was that many people did not know about community water projects properly, especially about their management mechanisms. Almost everyone had heard about the community arsenic mitigation projects and group formation in their area, especially if they were being implemented by an active NGO, and especially during early phases when information was disseminated. But the majority of people did not know exactly how the community-based projects functioned or how this was defined. Often, the prevalent notion was that the person on whose land the option is was fully responsible for it and that others did not need to be involved (i.e., they thought it was private property). In general, those not affiliated with projects were less aware of user committees or, if they were aware, most were not members. Generally, the rural elite and elders were key decision-makers in user committees. In some instances, committee meetings were called and people informed of the water issues and concerns, but this was more a rarity than a norm. As a result, few people knew about or attended community meetings regarding arsenic mitigation projects and water management decision-making. Very few community projects actually had functioning user committees where people actively participated and felt communal ownership of the water infrastructure. Furthermore, different notions of community and participation operating in water projects of different organizations in the same locality can further complicate the issue (e.g., for some it means making a financial contribution, for others it may mean becoming a member of the project group and sharing in decision-making). Often the same people are representing the community in the different projects, and the diverse approaches and modalities of operation under the same rhetoric can create confusion. In the end, what was apparent is that overall the men were more interested in getting technologies and the financial aspects of arsenic projects, with the assumption that issues of access, use, and conflict would be borne by women (see also Sultana 2007).

Instant validation of participation is seen when some people show up at any meeting – even at various stages of the meeting, or if they leave at various times, whether they have actually listened to the discussion, or said anything – as physical presence is generally understood to imply participation. Usually, on paper, a large number of people's names are included as committee members, but there are usually few who attend meetings or feel they are actually members of the project; also, there is little, if any, recordkeeping of meetings and how and what decisions were made. While the flexibility of rules can allow for faster action, it is also open to different interpretations and control by the powerful. In most of the community projects, few formal meetings were held as people didn't feel that there was a need if the water option was functioning properly, fees were collected, and there were no major conflicts to resolve. Costs of meetings and participating were factors that influenced this: these can be in terms of time spent, loss of income from loss of time, overcoming social barriers, and perceived risks to upsetting existing social hierarchies. As community participation involves time, those who are marginalized or poor usually cannot afford that kind of time (compared to the rich who, for instance, are generally less involved in agricultural wage labour and have opportunities to free up time for project work). Also, a sense of abandonment and powerlessness can further complicate the participation of those who feel marginalized at meetings. Thus, not everyone can or wants to participate, or at least not in the ways that are articulated in development and water governance plans. In some instances, people said they didn't want to challenge existing authority as they wanted to benefit from projects in whatever way they could. Meaningful participation may not result even when people are able to attend meetings, due to existing power structures and social norms on who can speak, when, and how. Many poor households thought that some financial or labour contribution is sufficient participation, and that decision-making should be left up to others. Nonetheless, the majority of the people were interested in having some voice and sharing their opinions in the ways that water management affected their lives, but were less certain how to enact this. For households with contaminated water sources, it was a critical concern, whereas for those with easier access to alternative safe water sources there was less concern.

What is notable was that just setting up a committee and having meetings does not address issues of subordination, marginalization, or vulnerability (Cleaver 2000). The rights of excluded and marginalized peoples cannot be redressed by sitting in at meetings or being formal members of water groups, although it can ensure water security to some extent (which is important for household reproduction and livelihood needs). Thus, people make trade-offs between maintaining power structures for overall livelihood needs and having access to safe water, whether they participate or not. Thus, meaningful participation does not come to fruition because of the problematic way it is conceptualized, implemented, and circumvented. The meanings of participation in different contexts and spaces and in relation to other people present greatly influence how people understand and operationalize participation. The public nature of the spaces for participation and decision-making often reinforce social norms concerning who can and cannot speak up. Most women and many poor and marginalized men often do not feel comfortable speaking in public spaces. Being seen to voice an opinion that may challenge existing power structures or ideas about water management is often deemed to be risky by those who need to maintain various kinship and social networks for their livelihoods. Further-more, class mentalities often position less powerful or poorer men at a disadvantage as they are often expected to go along with more powerful or wealthy patrons. While many may be resigned to this arrangement and accept the outcomes, some did wish they had more voice. Thus, it is seen that even within predominantly male groups, there are differences by class, age, and education. Invocations of differences through gender, class, location, literacy, and religion are common in community projects, and influence how people experience 'participatory exclusions' (cf. Agarwal 2001).

Gendered participation and decision-making activities for women were generally curtailed by age, marital status, education, and socioeconomic class positions across households with similar exposure and experience with arsenic contamination. Which women are allowed to participate or appreciated when they do participate also varies in patriarchal settings. Men often will listen to more senior and wealthy women if they have some history of influence or power in the village. Younger women find it more challenging. But if they are very educated and able to communicate with men, they are given space to speak. Nonetheless this is often looked down upon as well, as such women are out of the ordinary and seen to destabilize social norms if they speak up too much or against any older man. In general, older educated women who have played some leadership roles (e.g., schoolteachers) are more respected and participate more in decision-making fora. Similarly, hired female labour may fetch water for the wealthier households, but it is the households' more powerful women who are in a position to participate at decision-making fora; similarly, younger women (especially daughters-in-law) who actually procure drinking water are largely left out of water management institutions and decision-making processes as older women from the household may be involved, if at all. Thus, different women in different social locations can have very different opinions and experiences about water and water management.

In many community projects, women were not aware of their rights and roles, or even membership in such institutions. Often their names existed only on paper, they attended no meetings, or were not informed of meetings nor asked for their opinions. In most cases, the water user committee consisted of only men, or mostly men with a few token women. Even if women were asked to attend meetings, they mostly just listened in, rarely giving their opinions in public, and they were not given sufficient assistance or encouragement to attend. There is a general sense that women's role is limited to deciding where to fetch the water from, and they are less involved in how to improve the access, control, and managerial aspects of water governance. While most women felt that they should have more decision-making powers, and expressed interest in voicing their opinions and having more decision-making capacity, a majority were not willing to challenge the norms and authority of their husbands, fathers, brothers, or elders in order to do so. Such constraints need to be viewed within the broader context of women's lives, as well as local geological contexts, as women in households without arsenic problems are less willing to engage in gender equality in water projects compared to those facing greater challenges in accessing safe water. Women can resist, accept, and create different meanings out of notions of participation; this can be done directly and indirectly, actively and passively. Women are more likely to share their concerns with those in similar subject-positions than with women in general or with men (although this varies depending on the conjugal relationship and household structure). Women are also likely to make strategic alliances with men in their households in order to push forth their agendas for safe water in public fora; thus, there is control, domination, negotiation, and cooperation that can be brought to bear on how households and members in the household participate in water projects. As a result, irrespective of whether women participated or not, they remained overwhelmingly circumscribed by their gendered positions of subordination and were not able to challenge or change power relations through water management projects.

Conclusion

Nuanced and critical explorations of participation and community in development discourses and practices explain why certain development endeavours fail or succeed. Gendering such analyses exposes inequities and marginalizations that are covert and overt. Simultaneous attention to social heterogeneity (gender, class) as well as natural heterogeneity (arsenic deposits, safe aquifers) helps to clear space for a better understanding of complex nature–society relations in the context of development. Water governance strategies, and development projects more broadly, thus have to critically address the gendered outcomes and dynamics of any interventions, as the very attempts to solve a problem may end up exacerbating suffering in unexpected ways.

Despite the problems of discourses of community and participation discussed above, such notions continue to have enormous staying power in development projects, as policymakers, project officials, and local elites buy into the various understandings of what community and participation mean, and how different benefits can be reaped from mobilization of such polyvalent terms. Community-based water management can simultaneously further stratify communities, as well as create and congeal communities for specific projects, where the spatiality and heterogeneity of nature play a critical role in the ways that institutions are crafted and operationalized. People are more aware of what community means, or is supposed to mean, and can work through the processes of such development interventions vis-à-vis their own relationship with (un)safe water. Thus, a community reflects its internal stratification at such formalizing moments of interventions, where differences in needs, abilities, power, and influence become evident, as do the locational differences and relations people have to a spatially heterogeneous nature (one which is both benign in providing safe water, as well as harmful in providing poisonous water, depending on where and at what depth one draws water from). Differences are thus reinforced through arsenic mitigation and water management institutions, in that relations of domination and control tend to further marginalize those who do not have access to safe water, meaningful participation in water management institutions, and information about arsenic and mitigation (including arsenic's distribution, safe water options, impacts on health, and health management). As such, the creations of differences that are gendered, classed, and geographical (in relation to access to safe water sources as well as in relation to where contaminated aquifers are) are reinforced by the very notions of participation and community that are expected to reduce such differences and promote egalitarian and democratic water institutions.

Acknowledgements

Substantially revised and abstracted from: Sultana, F. (2009) 'Community and participation in water resources management: gendering and naturing development debates from Bangladesh', *Transactions of the Institute of British Geographers* 34(3): 346–363. Reprinted with permission from John Wiley & Sons.

References

Agarwal, B. (2001) 'Participatory exclusions, community forestry, and gender: an analysis for South Asia and a conceptual framework', *World Development* 29: 1623–1648.

Agrawal, A. and Gibson, C. (2001) *Communities and the Environment: Ethnicity, Gender, and the State in Community-based Conservation*, New Brunswick: Rutgers University Press.

Cleaver, F. (2000) 'Analysing gender roles in community natural resource management: negotiation, lifecourses and social inclusion' *IDS Bulletin* 31: 60–67.

Cleaver, F. and Elson, D. (1995) 'Women and water resources: continued marginalisation and new policies', *The Gatekeeper Series of International Institute for Environment and Development's Sustainable Agriculture Programme* 49: 3–16.

Cooke, B. and Kothari, U. (eds.) (2001) *Participation: The New Tyranny?* London: Zed Books.

Cornwall, A. (2003) 'Whose voices? Whose choices? Reflections on gender and participatory development', *World Development* 31: 1325–1342.

Mosse, D. (2003) *The Rule of Water: Statecraft, Ecology, and Collective Action in South India*, Delhi: Oxford University Press.

Smith, A., Lingas, E., and Rahman, M. (2000) 'Contamination of drinking-water by arsenic in Bangladesh: a public health emergency', *Bulletin of the World Health Organization* 78: 1093–1103.

Sultana, F.(2007) *Suffering for Water, Suffering from Water: Political Ecologies of Arsenic, Water and Development in Bangladesh*, unpublished PhD thesis, University of Minnesota.

—— (2009) 'Fluid lives: subjectivities, water and gender in rural Bangladesh', *Gender, Place, and Culture* 16(4): 427–444.

28

GENDER EQUALITY AND DEVELOPING WORLD TOILET PROVISION

Clara Greed

Introduction and contents

This chapter discusses the importance of achieving adequate and appropriate levels of toilet provision for women in the developing world. First, the chapter outlines the global sanitation situation, with particular reference to the nature of the problems encountered in the developing world. The main global players and toilet providers are identified, namely international governmental bodies, commercial toilet companies, development agencies, but also community and user groups. The extent to which women's "different" needs have been taken into account is discussed with reference to the enduring influence of colonial inheritance of patriarchal toilet standards, and the impact of modern, scientific, and mainly Western toilet standards on the developing world. The implications of not taking gender considerations into account are discussed with reference to health and education. It is argued change cannot be achieved by setting unenforceable and unrealistic international standards. It is proposed that changes in attitude, resource allocation, and gender awareness may be achieved through toilet-education programs, greater involvement of toilet user groups, more women in the sanitary engineering professions, and respect for and engagement with women within local communities. There is also a need for toilet issues to be integrated into the mainstream development agenda and not treated as a low-status subcategory of water and sanitation.

The chances of toilet availability

Of the approaching seven billion people in the world, around two billion lack toilet provision, but women generally have less provision than men. Water supply, toilets, and hygienic disposal and recycling of human waste are the absolutely necessary basics to achieving world health and development (Black and Fawcett 2008). This is particularly important when more than 50 percent of the world's population is now urbanized, but a third of that number live in slums, shanty towns, and unofficial settlements lacking the basics in terms of water and sanitation. More people in the world have mobile phones than toilets. Western sit toilets are popular amongst the emerging middle classes and have an aura of "progress, science, and modernization" about them. But the majority of the world's population still squats rather than sits, often on the bare ground, and most of the world's toilets do not have a flush and are not connected to an infrastructural sewerage system. For that matter, most people in the world do not use toilet paper, and most women do not have access to sanitary protection products.

But one must avoid geographical or, for that matter, gender determinism to explain the world's toilet problems. It is true that the main problems are found in the Global South, but not everyone is affected equally, as one's chances of having access to a toilet are not just shaped by gender or "nature," but also mitigated by "class" and income. For example, in high density mega-cities of South America, it is not necessarily lack of water supply or infrastructural services that prevents everyone from having toilet provision and running water in their homes. Rather it is a matter of being able to afford to be connected to the system, not how close one is to the pipes, as in many countries you have to pay for privatized "public" services (Mara 2006). Likewise lack of provision in India is not universal, but mitigated by class and income. One can readily find examples of women laboriously carrying water pots on their heads, when just across the road or behind a wall one finds a hotel with a swimming pool and fountain, or a new middle-class housing area with all mod cons.

The issue of racial discrimination adds another dimension on top of gender. Race has long been a major factor in shaping toilet provision in other countries, particularly where there has been a history of apartheid and segregation. For example, Barbara Penner has highlighted the racial aspect of toilet provision in the USA (Gershenson and Penner 2009). In 1961 in Jackson, Mississippi, a black woman, Gwendolyn Jenkins, was arrested for her attempts to desegregate public toilets, by trying to use the white women's toilets. We never hear of her, but she was the toilet equivalent of Rosa Parks. This all may seem distant history but in post-apartheid South Africa, there remains a marked toilet divide, between the situation in the deprived, predominantly black, shanty towns and the prosperous white cities.

The levels and types of global toilet facilities are shaped by both international and local standards, and by both historical and current attitudes. During the colonial era many developing countries were subject to European toilet standards and design principles, particularly in emergent, middle-class urban areas. For example the British Standard 6465 on Sanitary Installations (first developed in 1875 and last revised in 2009) (BSI 2009) was widely used as the basis for provision right across the British Commonwealth (which comprises two billion people). This standard was, until relatively recently, drafted by all male committees of engineers, who effectively stipulated twice the level of places to pee for men as against women, because men were provided with copious urinals in addition to cubicles (Greed 2004). As Michelle Barkley, an architect (one of the few women, along with two others and myself, on the current committee) commented, the British Standards thus exported gender inequality and toilet queues to the rest of the world. The principle of lesser provision for women is still pervasive and taken for granted as "normal" within the male-dominated world of sanitary engineering. The lack of adequate female toilet provision is a shared global experience, endured by women within affluent countries, and amplified and rendered starker within the developing world. Overall there has been little guidance or agreement as to global standards or what should be done in the developing world. The International Code Council (ICC) based in the United States produced in 2010 "Global Guidelines for Practical Toilet Design" (www.prweb.com/releases/2010/10/prweb4676224.htm). But this was much criticized for being too Western, and American, in perspective, and stronger on plumbing details than social considerations.

Following independence some erstwhile colonies have written their own toilet standards. For example, Malaysia (previously Malaya) has updated its standards and has recognized the need for greater gender equality. Indeed, a restroom revolution has taken place across the

Far East and in some such countries one now finds a 2:1 ratio of toilet provision in favor of women. However, in some cases, it was found that only the cubicles were included in the calculations, and not the urinals, giving the men a higher level of provision of places to pee: such are the unquestioned forces of patriarchy and the male urinary prerogative. The World Toilet Organization (the "other" WTO), led mainly by Chinese and Japanese toilet experts, commercial interests, and campaigners, initially sought to establish global toilet standards and, to their credit, the WTO has prioritized gender parity. But having discussed the issue fully, the WTO decided that cultural and development differences were too great to provide one definitive global toilet standard. Rather than producing specific standards, the WTO sees its role as assisting and training people in developing countries who are getting involved in toilets. For example it holds annual conferences when papers are presented on all aspects of toilet provision, and it also runs the World Toilet College in Singapore, mainly aimed at mid-career managers, engineers, and community workers who have gradually got into toilets and want to improve their knowledge and understanding.

The main players in the toilet arena

Rather than looking for definitive standards, one needs to identify and appreciate the influence of the main players in the international toilet arena to understand "who gets what, where, and why" in the global toilet stakes. The main players in shaping toilet policy, provision, and standards comprise international governmental bodies, commercial toilet companies, development agencies, but also community and user groups. All the provider groups are overwhelmingly male-dominated with a predominance of male technical experts and professionals, reflecting the lack of women in the fields of engineering, plumbing, and architecture. Women are more likely to be found in the community, campaign, and user groups.

First, international governmental bodies and organizations shape global toilet provision: but significantly generally in an indirect manner. All the main international organizations state they have a major commitment to water and sanitation issues, as is the case with the UN, UNESCO, WHO, UNICEF, the World Bank, World Urban Forum, UN Habitat, and so forth. Many produce detailed statistics on the lack of water and the need for better sanitation, such as www.unwater.org/statistics_san.htlm and www.unicef.org/wash/index_statistics.htlm. But, on closer investigation, one finds that while there are many high-level policies and pronouncements of good intentions, there is little detailed material or substantive policy on the earthy topic of toilets. "Toilets" become "sanitized" and, for that matter de-gendered, within the generic, and rather impersonal topic of "sanitation." Furthermore, there does not seem to be much joined-up thinking in relating the importance of dealing with toilet issues, in achieving clean water supplies, and improving world health. As will be explained, improving toilet provision will also contribute towards the alleviation of poverty, enabling children to attend school. Likewise, the provision of reliable water supply locally will free women from having to walk miles every day to collect water, enabling them to undertake other economic activities.

But, the provision of adequate toilets, especially for women, is fundamental to the achievement of the Millennium Development Goals (MDGs), especially Goal 3: promote gender equality and empower women, and Goal 7: ensure environmental sustainability (see www.un.org/millenniumgoals). In particular Goal 7c says, "halve, by 2015, the proportion of people without sustainable access to safe drinking water and basic sanitation." Many development agencies do not believe these have been adequately achieved and that many years of additional effort are needed. Since the MDGs have fallen behind schedule, a new set of Sustainable Development

Goals (SDGs) is being introduced, which are more about the processes and methods of achieving the MDGs and so do not mention sanitation per se.

A second set of major players in the global toilet arena are the commercial toilet companies along with their retinues of designers, manufacturers, scientists, and technical experts. Restroom provision is big business, comparable to the global motor car or tobacco industries, in terms of firms seeking and securing new markets in newly developing countries. Many of the companies involved are North American or European such as Armitage Shanks, Franke Sissons, Interpublic, WallGate, Twyford, and JC Decaux. But there is an increasing Japanese presence in the form of Toto and Fujitsu, the latter being a major construction conglomerate in the Far East, better known as a computer manufacturer in the West. Such companies are extremely male-dominated across the board, at professional, technical, managerial, and trades levels. However, Toto (of electronic talking toilet fame), which has a major presence throughout the world under localized brand names, undertakes ergonomic and sociological research to understand the needs of all types of toilet users including women, the elderly, and people with disabilities. In addition to toilet hardware manufacturers, chemical corporations such as Unilever are very active in the developing world promoting toilet products such as Domestos bleach, which is marketed as killing all known toilet germs (thus apparently neutralizing leakage into groundwater systems).

A third category of toilet players are the plethora of aid agencies and NGOs such as Oxfam, Tear Fund, and Water Aid, all of whom have got increasingly involved in toilet issues. Many seem to have come into dealing with toilets from an initial concern with drinking water and thus with polluted wells (Water Aid 2012). While women are increasingly found in many areas of "development planning," and there are some notable women in Water Aid, I would still argue that women are more likely to be found in the caring professions, working in nursing, nutrition, missionary and aid work, social care, and community development. So many women—as health professionals—are dealing with the results of poor sanitation rather than addressing the reasons for disease such as polluted water courses and groundwater, which in turn are caused by poor sanitation and lack of toilet provision. Most of the engineering, science and construction specialisms in development agencies are still male-dominated. Even well-intentioned toilet campaign groups are fairly male-dominated. But feminism is beginning to penetrate the world of sanitary engineering! It has taken years for women to get into senior positions in the construction professions whereby they might exert an influence on "what is built, and for whom." But it is very difficult to be taken seriously in the world of sanitary engineering as a woman, particularly if one lacks the technical background.

Meanwhile women have been active in community and grassroots organizations, both in the developed and developing world, challenging the policies, views, and official standards that affect their lives. This is the fourth set of players, mainly representing toilet users rather than providers. In particular a new generation of young girls is becoming a force to be reckoned with, increasingly involved in toilet campaigning within developing countries (see http://plan-international.org/girls).

The effects of lack of toilet provision on health

Why bother to do anything about the toilet situation? Everyone, not just the poor, is affected by toilet inequality. In highly urbanized situations, as in the South American mega-cities, rich and poor often live in close proximity, luxury apartments across the road from shanty town development (Burdett and Sudjic 2012). Flies and other vectors are no respecters of class or income and so rich people could be "eating other people's shit" as the flies fly over from the cess pits of the shanty towns and land on the food plates of the rich. Secondly fecal pollution seeps into the ground water, and is found in rivers and drinking water sources. This situation is caused by both outdoor defecation and by leaks and ground penetration from poorly constructed pit toilets, composting toilets, cracked pipes, and poorly maintained sewerage systems.

Jack Sim, founder of the WTO, notes that in India 90 percent of surface water is contaminated by shit. In Africa 80 percent defecate in the open, while worldwide more than one billion do so. Sixty percent of Africans do not have access to a toilet, and some will find other solutions including using "flying toilets" (that is wrapping excreta in plastic bags and throwing it away), a custom particularly prevalent in Nairobi, Kenya (CLTS 2011). Poor drainage and surface sewerage, along with lack of clean water for hand-washing and basic hygiene, has other health implications too. Sixty percent of the world population, including 80 percent of children, have worms and intestinal bugs, with 1.5 billion people worldwide having round worms alone (Greed 2006; Roma and Pugh 2012). Therefore, "I care a shit" was the motto of WTO conference in 2012, which was held in Durban, South Africa.

Much of this may appear to be generic rather than gendered in nature, but women account for at least half of the population of areas affected, but they face additional problems in view of unequal and inadequate female toilet provision, which, for example, are likely to result in them being more likely to use the bush (often under cover of darkness when evaporation rates are lower), thus increasing the rates of surface defecation. Also women are often more vulnerable to water-borne diseases as are their babies and children and therefore they have more at stake than men in terms of dealing with the consequences of lack of basic sanitation and toilet provision. The list of water-borne, feces-related diseases is limitless, including at least 30 major developing world diseases, with cholera, dysentery, and diarrhea being major players, and poliomyelitis, botulism, typhoid, giardiasis, and more recently SARS being subjects of concern.

The effects of lack of toilet provision on education

Lack of school toilets for girls undermines the chances of achieving the MDGs, especially in relation to gender equality and education. To illustrate this I refer to the 2012 WTO Organization Conference in Durban. There were far more women in the audience, more women speakers, and indeed entire classes of schoolgirls attending, when normally women are in the minority at WTO conferences. A group of schoolgirls came on the stage and gave their heart-rending toilet testimonies, explaining how the lack of school toilets for girls affected their educational prospects. The only option was to go in the bush or walk home at midday, often several miles. Eighty percent of schools only have pit toilets and many lack adequate facilities for female pupils and women teachers. Fears of snakes, wild animals, rape, and loss of privacy were explained in relation to having to go in the bush. (Schoolgirls were more vulnerable than boys because of additional personal safety fears, plus snakes and animals may be more conscious of female bodies because of the hormones they emit.)

In particular, girls may have to stay away from school every month when they are menstruating because of lack of school toilets (Roma and Pugh 2012). Around a quarter of all women of childbearing age will be menstruating at any one time, and every month up to five days will be lost in terms of school attendance. Fifty percent of girls do not continue with school following the onset of menstruation because of lack of toilets. Significantly, in spite of this specifically gendered problem arising at secondary school level, the MDGs only refer to increasing gender equality at primary school level (Goal 2).

When school toilets are provided they should be designed to provide girls with privacy. They should be clean and dry to enable them to change sanitary protection hygienically. But do the schoolgirls need Western disposable sanitary pads and tampons? Saskia Casteltain, who has been working with the UN on menstrual hygiene, argues that if every woman in the developing world used Western sanitary pads and tampons it would create such a pile of waste for disposal that it would be far higher than the disposable nappy mountain. Washable pads that can be made by local women and can be recycled are essential. The sewerage system (if there is one) cannot cope with the plastics, chemicals, bleaches, and non-biodegradable components found in many Western sanitary protection items. In fact, most women in most of the world do not have access to or cannot afford tampons and sanitary towels, so they are likely to use rags, leaves, wool, inter alia. In many cultures women will not use tampons or anything else that goes inside them, such as the Moon Cup, because of virginity and adultery laws and fear of infection, or they may be unable to use them because of female circumcision.

The open discussion of menstruation at the Durban toilet conference was a real breakthrough in the global toilet agenda, breaking down another toilet taboo. Amanda Marlin, a health expert working on sanitation in Africa, made the very interesting point that women are disadvantaged if their needs and existence are never recognized, that is if there is no empirical evidence collected on women. Indeed lack of data is itself a sign of discrimination. Women are just plain invisible. There are parallels with the British and American public toilet situation in that women's needs are so often ignored as there is no gender differentiated data on their needs, or the male is taken as the "norm" and women are just an irritating addition, best ignored, or seen as an extra expense. From this male-mindset flows all sorts of problems such as women being charged for toilets and men not, queues for women's toilets, and—in the developing world—a disregard for women's modesty, privacy, fear of attack, and lack of consideration of menstruation issues.

Recommendations

There are two main aspects to providing better toilets for women: first, the design of the toilet itself, and, second, dealing with the excreta to prevent the spread of disease and water pollution, both of which aspects actually benefit everyone else too. Some of the issues and trends discussed apply to the world as a whole and some specifically to the developing world. Some of the recommendations draw on Western practices and some on those preferred in the majority world.

First of all women need at least equal levels of provision as men, a situation that does not even exist in most developed world countries, let alone in the developing world. Queuing remains one of the most visible forms of gender inequality, still pervasive in the West in public toilets, theatres, cinemas, and stations. In Britain and many other Western countries, public toilet provision has been reduced being a soft option in these times of government cutbacks. But in North America around 20 states of the USA now have "potty parity" and attempts are being made to make this a federal-level requirement (Anthony and Dufresne 2007). In France, the new socialist government is taking toilet equality more seriously, influenced by the philosophical ideas of Lefebvre, regarding "la droit à la ville" that is "the right to the city." So it was declared in 2012 that all public toilets in France, and already in Paris, are to be free for everyone, residents, tourists, public transport users; albeit many towns do not have much provision to start with. The developing world usually copies toilet trends established in the developed world, but it is yet to be seen whether they will be inspired by Western gender equality movements and by the Far East restroom revolution, or will succumb to the Western trends towards toilet privatization and closure, which disadvantage women even more.

Not only is it important to get more women's toilets relative to men's provision overall, there is also a need to think more carefully about their design, management, and maintenance. As stated, most people squat to use the toilet. In the nineteenth century the export of "sanitary ware" from Britain was a major earner, as we convinced the world it was old-fashioned, and even un-Christian, to squat. Many women and men prefer squat toilets and some have great difficulty using Western toilets, attempting to squat on the seat, especially if they find the design is too high for them to use. The sit/squat debate is one of the major issues discussed by WTO members. Japanese toilet experts have designed modern high-tech squat toilets and these are preferred, for example, by elderly women who cannot "go" using sit toilets after decades of their muscles being used to the squatting position. From a development "sites and services" perspective, squat toilets are simpler to install and more cost-effective.

Personal hygiene is vital to reduce the spread of disease. In much of the world, toilet paper is unknown and people wash themselves after defecating. This is not only the case with Muslims (around a quarter of the world's population) but many other religions and ethnic groups too. So water needs to be supplied within the modesty zone of the cubicle or the sheltered part of a rudimental toilet building. In contrast to our secular, scientific society, in the rest of the world religion still matters a great deal. Religion is not just for special holy days and there are rules that cover every aspect of daily life. There are all sorts of toilet rules and regulations, about which hand you should use to do your toilet ablutions, which foot you should enter the toilet door with, and, depending on the religion, quite complex rules about where, when, and how you should urinate and defecate, not necessarily in the same building or receptacle. There are many important Hadiths (wise religious sayings) on toilet behavior and practice, which are important to Muslims, who are not only found concentrated in the Middle East but in all sorts of other countries across the globe. Indeed Muslim toileteers, including women architects and engineers, have had a major input in revising toilet design and levels of provision in Malaysia and Indonesia and increasing gender equality.

Women especially need to wash their hands after they use the toilet and also to have facilities for cleaning themselves during menstruation. At least 1.5 billion people lack access to clean water for drinking, but where it is available it is more important to use some of it for hand-washing than to waste it on flushing systems. Many women prefer to use jugs and bowls of their own to carry out their ablutions so they need taps that have adequate space under them to stand a jug or a vessel, and therefore Western washbasins (sinks) and taps (faucets) are not much use. Soap is also unheard of in many locations but often local people have their own recipes from soap wort, oils, and other substances, and such cultural habits need to be respected or else they may not wash their hands at all. Simple design features such as a shelf for women to use when changing their menstrual protection should be provided away from contamination.

Many toilet rules are there to maintain the purity of the toilet user and the surroundings. Unfortunately, women are often seen as unclean simply because they menstruate. Many women do not "go" in public and may wait until is dark before venturing into the bush or jungle to relieve themselves. Many are fearful of being seen as immodest, sexual, or in Western parlance "asking for it," and may therefore "hold on for hours." All this has a negative effect on their bladders, bowels, and general wellbeing, and may lead to incontinence and other medical conditions. In many developing countries, childbirth is more frequent than in the West, so pregnancy needs to be taken into account in designing the internal dimensions of toilet blocks and the frequency of provision. Many developing countries have populations where children and young people predominate, and so good toilet habits, hand-washing, and, for that matter, teeth cleaning should all be instilled and provided for. In contrast there are far less elderly people than in the West, although this is gradually changing with women taking the lead.

The toilet block needs to be designed to take into account women's modesty and cultural preferences. Unisex toilets are a disaster in many developing countries because they offend religious conventions and make women feel very unsafe (as they do in the West). A classic blunder is to create "gaps" between the floor and the walls in order to facilitate ventilation within toilet blocks, but this may result in women being seen from the outside when they squat down to use the toilet. It is preferable, especially in schools, to locate the girls' toilets some distance from the male toilets and to ensure there is good visibility and that women do not have to pass the entrance to the male toilet to reach their own. In many village and domestic situations, it is better to have single family toilets outside the house, rather than impersonal rows, so people can develop a sense of pride in their toilet.

Women are the main cleaners of toilets throughout the world. Granted low-status men take on this role too, including untouchables, *dalits* who are night soil collectors in India, and "house boys" in affluent African households. But at the domestic private house toilet level for those who have their own toilets and in the case of community toilet blocks in villages and barrios in the developing world, it is mainly women who do the "shit work." Women are also the main bearers of water (see chapter by Sultana in this volume) and deciders as to its allocation for cooking, washing, and toilet cleaning. To carry water is a low status role for women, but once men get involved transmitting water by pipes and sewers it becomes a well-paid honorable profession. But to be fair I have come across just a few well-intentioned and gender-aware sanitation engineers and they can make a huge difference as other men take them seriously when they talk about gender and other social toilet factors. For example Trevor Mulaudzi (Trevor the Toilet), runs the Clean Shop (www.thecleanshop.co.za), which is a toilet cleaning organization that puts a lot of effort into improving the hygiene in school toilets in South African townships and rural areas. He has given up a lucrative career in mining engineering to do so. It is important to get men more involved in toilet issues, water collection, and sanitation.

In the conference exhibition hall at Durban, I discovered a contraption that looks like a garden roller, which is in fact a pull-along barrel for collecting water, is proving popular with men (called the Hippo). Collecting water was always women's work in the villages, but since the introduction of this "machinery" apparently men had volunteered to collect water too. But in successful village toilet projects it is always older women, as community and family matriarchs, who are the best people to employ to manage the toilets as they can draw on their social networks, relatives, and local respect to get the job done. Building a toilet is only half the battle, it has to be cleaned, maintained, and serviced, and it has to be attractive and acceptable enough for people to want to use it and maintain it as their own (George 2008).

The second big issue is how the excreta is disposed of or recycled. In much of the world there is no mains drainage, no sewerage system, and water is too expensive just to flush away. Many highly urbanized areas still do not have infrastructural services especially not in the shanty town areas. Waterless "Enviro" toilets are being introduced in South African schools, for example, at Eqinisweni Primary School, Durban, there is a philanthropic community project interestingly supported by Domestos. In the waterless heart of Australia, solar power is being used to break down toilet waste into fertilizer. In India, the Sulabh International Service Organization (www.sulabhinternational.org), pioneered by Dr. Bindeshwar Pathak, has developed a three-pit system for the recycling of human excrement, based on one active pit beneath the moveable toilet cubicle, one pit where over several months the excreta turns into compost, and one where the resultant fertilizer can be used. There are many other versions developed in different countries but the most important factor is to make sure that the urine is separated from the feces and that the latter is treated with respect because of the potential transmission of pathogens (Roma and Pugh 2012). The pit used must be water-tight. Waste must not seep into the surrounding soil, groundwater, and should not run downhill into the local pond or river.

Since sanitation is such a major issue, especially toilet provision, it is projected that one billion toilets are needed worldwide. But, it is not just a matter of building more Western toilets, particularly in countries where there is no sewerage system or water supply to service the toilets. Dry toilets, ecological pit toilets, composting toilets, solar decomposition, compound, and low-tech toilets are the way forward. But in all of this it is essential that women feel safe and comfortable using these new types of toilets and that they are consulted about the changes. "Teaching toilets" in a manner that does not set threatening fixed standards but rather encourages participants to understand the issues is evidenced in a web-based program supported by the WTO (www.sustainablesanitation.info). This was established at Linkoping University by Dr. Jan Olof Drangert, a toilet expert who gives particular attention to sustainability issues, and to the siting and design of toilets so that pollution does not soak into the groundwater, while using simple sustainable technologies so that local people can manage and maintain the toilets without the benefit of Western experts.

Water-based sewerage systems are not necessarily the best solution. The West is very backward in using millions of gallons of water to get rid of its human waste, with old-fashioned sewage works heavily dependent on chemicals. A range of useful outputs can be achieved if dry excreta are treated carefully, such as fuel, fertilizer including nitrates, recycled purified water, ammonia, and electricity. For example fertilizer is being created out of urine by Professor Christopher Buckley at Natal-KwaZulu University in South Africa. This requires a special design of toilets that separates the urine from the feces, and this has gender implications as women do not use urinals, but apparently women's urine mixed with menstrual blood is also useable, whereas feces are not. More adventurous is the creation of electricity from urine, achieved by a process developed by scientists at the University of the West of England, Bristol, which also produces the by-products of drinking water and fertilizer. This has already received Gates Foundation Sanitation Program Funding (Ieropoulos et al. 2011). I have been invited to contribute an ergonomic design aspect of the project to make the human-machine interface, more user-friendly for ordinary village women and men. Bill Gates says he wants to do for the toilet what he has done for the computer: put one in every home on the earth. Approaching two billion people lack access to affordable electricity, and most use biomass such as wood, peat, and dried dung for cooking. For example, 60 percent of the population of sub-Saharan Africa has no electricity. Therefore solving two problems at once—by generating electricity from human excreta—seems ideal.

Such programs increase the chances of putting toilet issues at the center, not the edge, of the sustainability agenda, while linking toilets positively to economic development, clean water creation, and the reduction of environmental pollution. But it is still difficult to get toilet issues on the development agenda, and particularly to get women's needs taken seriously, including inadequate provision and menstruation issues. But one can argue the need for toilets from a variety of perspectives. These include the business case, as poor toilets are such a blockage to progress on development, education, and health. Toilets are potential sources of wealth, in the form of fuel, fertilizer, and electricity. Toilets are the missing link in creating sustainable, prosperous, accessible, and equitable countries and cities in both the developing and developed world. While development agencies are now more aware of these issues, one still has to work hard to change the attitudes of the heavily male-dominated, technically minded "global plumbing fraternity" of toilet providers, designers, and policymakers. This is why, in this chapter,

I have highlighted male experts who are supportive of improving women's toilet provision. But we need to train up our own troops. Young feminist women would do the world a great service if they chose a career in sanitation, services engineering, or plumbing, so that in the future they are in a position to shape toilet provision in favor of women.

References

Anthony, K. and Dufresne, M. (2007) "Potty parity in perspective: gender and family issues in planning and designing public restrooms," *Journal of Planning Literature* 21(3): 267–294.

Black, M. and Fawcett, B. (2008) *The Last Taboo: Opening the Door on the Global Sanitation Crisis*, London: Earthscan.

BSI (2009) *British Standard 6465 Part 1: Sanitary Installations – Code of Practice for the Design of Sanitary Facilities*, London: British Standards Institute.

Burdett, R. and Sudjic, D. (2012) *Living in the Endless City*, London: Phaidon.

CLTS (2011) *Taking Community-Led Total Sanitation to Scale with Quality*, Nairobi: Community Led Total Sanitation.

George, S. (2008) *The Big Necessity: Adventures in the World of Human Waste*, London: Portobello Press.

Gershenson, O. and Penner, B. (eds.) (2009) *Ladies and Gents: Public Toilets and Gender*, Philadelphia: Temple Press.

Greed, C. (2004) *Inclusive Urban Design: Public Toilets*, Oxford: Architectural Press.

—— (2006) "The role of the public toilet: pathogen transmitter or health facilitator," *Building Services Engineering Research and Technology* 27(2): 1–13.

Ieropoulos, I., Greenman, J., and Melhuish, C. (2011) "Urine utilisation by microbial fuel cells; energy fuel for the future", *Physical Chemistry: Chemistry Physics Journal* 14: 94–98

Mara, D. (2006) "Modern engineering interventions to reduce the transmission of diseases caused by inadequate domestic water supplies and sanitation in developing countries," *Building Services Engineering and Technology* 27(2): 75–85.

Roma, E. and Pugh, I. (2012) *Toilets for Health: A Report of the London School of Hygiene and Tropical Medicine in Collaboration with Domestos*, London: Unilever.

Water Aid (2012) *Sanitation and Water for Poor Urban Communities: A Manifesto*, London: Water Aid.

29

GENDER, POLLUTION, WASTE, AND WASTE MANAGEMENT

Elizabeth Thomas-Hope

Introduction

Waste management and sanitation are two of the basic services that have been a part of the global debate on sustainable human development. The reason for this is that, despite many decades of development, large proportions of the rural and urban populations of low- and middle-income countries lack adequate provision for those basic services.

Waste

First, let us consider what is meant by waste and waste management. Arguably, every product ends up as waste. Waste generally refers to discarded, unwanted material with zero value in its original market; but from a different perspective, waste is the refuse or residual matter that is unwanted, or the undesired material left over after the completion of a process. This definition must be qualified by the consideration that uselessness and lack of value are relative in many respects—dependent, in general, on differences of culture, socioeconomic status, and gender. Some items that are waste to the rich can be usable and valuable to the poor, or some items regarded as junk by women may be found useful by men and vice versa.

Waste occurs as a solid, liquid, or gas. When released as a liquid or gas, waste is referred to as effluent and emission, respectively. In terms of *solid waste*, most manufactured products are destined to become waste at some point in time, with the volume of waste production roughly similar to the volume of resource consumption. Solid waste refers to all construction debris, commercial, and institutional wastes (including medical wastes), street sweepings, and domestic refuse. In some poor parts of cities in developing countries, the solid waste management system also handles ash from incinerators and stoves, human wastes such as night-soil, septic tank sludge, and sludge from sewage treatment plants, all of which fall under the definition of solid wastes as provided in Chapter 4 of Agenda 21 (United Nations 1992). Post-consumer waste is the waste produced by the end-user (for example, the garbage disposed of by households). This is the waste that most involves the public both in terms of its generation, collection, disposal, and recycling. Therefore, this is the most visible form of waste, but it is a very small component of the overall quantity. For example, in the process of mining and industrial production, the volumes of waste produced are very much greater than post-consumer waste. In mining for some metals, such as gold, the volume of waste can be 500,000 times the volume of metal extracted. For each gram of gold produced, 500kg of mining waste is generated, containing other heavy metals that may pollute the atmosphere in their powdered form. These manufacturing wastes are by far the greatest output of many industrial production systems. *Liquid waste* is contaminated water or other fluids resulting from industrial or domestic activities. Such waste can cause serious pollution of aquifers, waterways, marine environments, and soil. The contaminants vary from bacteria to heavy metals (for example, lead, aluminum, and mercury).

Waste management

Waste management includes all intervention in the generation, collection, disposal, treatment, and reuse or recycling of discarded material. Responsibility for material regarded as waste, as well as for household and community cleanliness, and removal and management of the waste, are all socioeconomic-specific, status-specific, as well as gender-specific in many cultures. Gender and socioeconomic status are relevant to the overall management of waste in a number of ways. The poor and the most disadvantaged among the poor, including women, are the groups to whom household and community waste management are relegated, including human and animal waste management, street sweeping, and the maintenance of public spaces, separation of waste at source, reuse of waste materials, collection, transport, and disposal of solid waste from households and businesses. Women are usually associated with responsibility for the cleanliness of the home and for the health of the family and, in most cultural contexts, this is extended to responsibility for or special interest in cleanliness of the community. However, there are points at which control of the waste management process switches over to men, a shift that again reflects the gendered positions, roles, and balance of power between women and men in the specific cultural context (Scheinberg et al. 1999).

Improperly managed waste leads to contamination of the environment—air, water bodies, and soil—with the consequent negative impact on the health and wellbeing of ecosystems, commercial and domestic plants and animals, and—either indirectly through the food chain or directly through contact, inhalation, or ingestion—on the health of humans. The impact on humans is, therefore, contingent on the societal context and the levels and nature of inequity in relation to particular types of waste to which people are exposed in various ways—at the workplace or home and places of residence.

Pollution

The extent to which waste pollutes the environment partly depends on the volume and the location of the material that is discarded or emitted. Waste produced in the wild is reintegrated through natural recycling processes, such as dry leaves in a forest decomposing into soil. In an urban setting these same wastes, such as vegetable and animal matter, may become problematic. Accumulated solid waste and the pollution caused by inappropriate disposal is one of the most conspicuous aspects of inadequate environmental management in the cities of poor countries, especially in the poorest parts of those cities characterized by high levels of unplanned urbanization. As part of the goal to address the issues of sanitation and waste management, Agenda 21 Chapter 7 stresses that one of the parameters of sustainability in urban areas is effective management of its wastes (United Nations 1992).

Pollution by waste also depends on the characteristics of the waste. For example, some wastes are biodegradable (such as plant or animal remains) but if left exposed, they attract vermin and insects and the proliferation of bacteria. Other wastes are non-biodegradable (such as most plastics and metals). Additionally, some wastes have hazardous properties, whether toxic, inflammable, infective, corrosive, or radioactive. Industrial wastes are potentially the most hazardous. Much industrial waste consists of chemicals that are highly toxic, and must be treated properly to ensure that they do not contaminate the groundwater. The waste from medical laboratories, clinics, and hospitals also contains many hazardous elements. These include microbiological components containing living cells, such as viruses and bacteria, which can be extremely harmful (and even life-threatening) to humans. Medical waste also refers to anatomical waste as well as contaminated materials and equipment used in surgical and laboratory procedures, sharps in the form of needles, scalpels, and other instruments. Other forms of medical waste are discarded pharmaceutical and chemotherapy substances. In countries where the waste management system is not strictly controlled, workers come into direct contact with any or all of these hazardous wastes.

Waste and development

In developing countries, the generation of large quantities and varied types of waste is a major consequence of development and modernization. More and more varied types of waste are produced with increasing and new forms of consumerism, chiefly that of the rich. Therefore, the composition of the waste stream changes with increasing modernization and Westernization as significant changes in consumer products and packaging occur. In general, disposable items, chiefly plastics, as well as electronic, synthetic, and chemical materials increase. For example, in Côte d'Ivoire, previously used biodegradable wrapping (banana leaves) for certain goods is largely substituted by non-biodegradable plastic bags. "In some places [Côte d'Ivoire and Mauritania], all you can see for miles and miles, are fields of discarded plastic bags" (Burland, cited in Scheinberg et al. 1999: 10). At the same time, some of the greatest challenges to the management of waste are felt most keenly in the poorest countries. This is part of the paradox of development—factors that create the most intransigent problems currently facing the developing countries are invariably those that derive from development itself (Thomas-Hope 1998). This irony is based on the growing gap between the patterns of growth and modernization in the developing world, on the one hand, and the capacity to pay for, plan for, and effectively manage solid waste as part of an integrated national or municipal system, on the other.

The connection between waste management, poverty, and gender is most intense in urban areas and especially in large cities. This means that the relationship or connection between gender and waste management are contingent upon the circumstances and levels of poverty that occur and, linked to this, the nature of urbanization that exists in any specific case. For while urbanization, poverty, and waste management are connected to each other, differentiated gender roles relate to all three. It is important to examine the current gendered distribution of roles and the productive uses of waste-derived resources so that the access of women, in particular poor women, to resources in waste management is recognized. Once the roles of these women are acknowledged, their contribution to the overall environmental management needs to be strengthened, and not reduced or removed, by modernization or privatization in development planning.

Gender and waste management in the context of poverty and urbanization

Most developing countries are characterized by a large lifestyle gap between the socioeconomic elite and the poor, and the waste produced by the rich is largely disposed of by the poor. The pollution that the waste creates would be considerably higher, and alternative solutions of disposal would have to be found by the authorities if it were not for the poor who engage in the collection and disposal or reuse of the waste generated. Thus waste becomes directly and indirectly an economic resource for the poor—both men and women. Further, the collecting, disposing and/or reusing of waste materials form part of a strategy of economic survival that brings with it positive implications for wider environmental management. In large cities of developing countries, it has been observed that these activities—in both the formal and informal systems—absorbed 1–2 percent of the workforce. The urban poor thus render important services to the society in terms of waste management and reduction in levels of pollution, which to a large extent go unrecognized by that society.

Service-based waste management activities

Waste management activities that are related to payment for a service are usually based on the removal of waste, litter, latrine sludge, or excreta. Such services are not only very badly paid, but also pose health and safety risks to those engaged in them, which are not compensated.

The collection and disposal of waste in urban areas is expected to be undertaken by municipal authorities or private companies on contract to the authorities. In the formal system, men are chiefly employed by the waste management authority to drive the trucks and collect garbage at the curb-side, on account of the need to lift heavy loads. In developing countries, the tipping of garbage containers from street level into the trucks is generally not mechanized and requires heavy lifting for which men may be more appropriate. However, the driving of the trucks could potentially be carried out by either women or men and, although this job still goes primarily to men, in some countries a few women have also been employed in this capacity. In general, women are employed for the lower-paid work of sweeping the streets and edges of the major urban roads; and women appear to be more skilled and efficient in carrying out these tasks. In Hyderabad, India, it is reported that men earned nearly twice as much as women in the recycling sector, based on the argument that men did hard labor, namely carrying the waste, while women were employed in sorting (Snel, cited in Scheinberg et al. 1999).

Despite this gendered bias in the formal waste management sector, research findings in Haiti's capital city, Port au Prince, showed that in jobs such as supervising and weighing solid waste at the depots, the performance of females compared favorably with that of the male employees (Noel 2013). Second, unlike male workers, the women were more punctual in arriving at work and more reliable in working the agreed number of hours. This was not an isolated case. An example can also be cited from Ouagadougou, the capital of Burkina Faso, where the community leaders insisted that men be hired for the waste collection work, as women had to "stay home and look after the children." Yet, the men proved to be unreliable and careless workers who quit the job as soon as they had other work. Women were then hired "as an experiment." The women performed the waste removal work to everybody's satisfaction, motivated by their desire to make full use of this rare opportunity to earn regular money, albeit under very difficult conditions (Arsens, cited in Scheinberg et al. 1999).

Commodity-based waste management activities

Commodity-based economic activity involves the trading of items or materials for a price. The payment is based entirely on the value of the item or material, that is, its commodity value. In the case of waste materials, the commodity value is derived from the original value-added on the item being disposed, and the activity of recycling or recovery is based on capturing this residual or retained value.

The unofficial collection of waste from the formal urban sectors is a common part of the resource of the informal economy. The waste that is picked or collected is reused or recycled within both the formal and informal economies. There is intense competition among the urban poor for collection of waste of higher value and, in general, women find it very difficult to get hold of the more economically valuable material. Not only do men tend to take greater risks in retrieving material in seemingly chaotic situations where trucks are offloading and tractors compacting surface material in the same location, but access to the higher priced items is aggressively controlled by men. There is evidence from many countries of distinct and generally accepted gender divisions in "rights" to certain materials. To this end, there are gender differences in the concentration of women or men at different points in the waste stream. Where there are waste transfer stations (sidewalk depots), men tend to congregate there as they can get first access to the loads; and women are concentrated at the landfill, which is the end of the chain of waste-picking opportunity. Many authors have noted that most of the thousands of waste-pickers on any urban disposal site were adult women and children as young as nine years of age.

Where there are few or no transfer stations, then women on the dumpsite are usually outnumbered by men. For example, in the case of Kingston, Jamaica, a study conducted by Meiklejohn (2010) found that the ratio of men to women on the disposal site was five-to-one. This numerical dominance of males was matched by the level of male control that occurred. The disposal site was regarded as a man's domain and women were merely tolerated there. Fights and scuffles frequently broke out at the disposal site and sometimes these involved women. The lead man, who assumed control over the collection of specific material, would send in gangs of two or three other men— his "soldiers"—to collect the metal and glass, thus preventing others from accessing such items. Anyone could collect from the "bottom waste," which means the items left after the metals and, to some extent, glass, had been removed. So it was largely from this residual pool of resources that the women collected.

It is a general phenomenon in the informal urban waste management systems of poor countries, that the clear differentiation in access to the higher-value items is mainly along lines of gender, with men collecting metal and women collecting food items as well as clothing and other lower-value items, such as plastic containers. The items are recycled by the pickers or sold, in the case of metals, to other agents in the chain of metal recycling. In Port au Prince, Haiti, 94.7 percent of the male scavengers in a study conducted by Noel (2013) collected metal items, compared to a mere 5.6 percent of the females. Some 92 percent of males collected bottles and 11 percent of the women did. By contrast, 66.7 percent of the women collected foodstuff and 13.2 percent of the men; 33.3 percent of the women collected clothes as compared with 7.9 percent of the men; 13.2 percent of the men collected wood and 22.2 percent of the women; 5.3 per cent of the men collected old tires and none of the women. In cultures where men do the trading, one may expect to find traders of recyclable materials to be men. In East and West Africa, where women do the trading, one may also expect to find junk-shop owners who are women. Prices for commodities are not usually set locally, but are set by the world marketplace, particularly the prices for paper and metals. The local prices then reflect the global cost discounted to allow for the collection, handling, storage, and transport of the materials to the nearest or most advantageous buyer. Because of this connection to the global commodities trade, even small, localized commodity-based enterprises operate in a high-risk environment. There is the potential to realize high revenues at the level of junk-shop owners and recycling workshops, but there is also high risk of losing. This often results in a situation in which waste pickers or itinerant buyers are contingent labor, without fixed employment, and in addition, with low and uncertain rates of return on their labor (Muller and Scheinberg 2003).

Despite the gender inequalities, those persons picking waste from the waste stream, both men and women, were able to survive economically on account of that work. Studies have shown that some of the women recycled material collected from the disposal sites and were able to sell these. For example, textiles and clothing were washed and made up or, in the case of reusable clothes, were sold as they existed or refashioned and then sold. Many of the persons sorting and picking from the dump were also involved as hucksters or informal traders. Of the sample of women in Meiklejohn's study (2010) at the Riverton disposal site in Kingston, Jamaica, 63 percent relied for their sole income on sorting and recycling of materials sourced from the dump; 10 percent had independent sources of income (most likely support from partners or the fathers of their children), and 23 percent had other means of income-generating work—including small shops or stalls, itinerant vending, or livestock rearing (also on the disposal site). Funds obtained as a result of the waste-picking and recycling activities served as the start-up finances for many of the entrepreneurial activities in which they were additionally engaged. Women, who used the waste to build capital for other entrepreneurial activities, were eventually in a position to reduce and even stop going to the dumpsite.

Waste-picking therefore presents a real economic opportunity for the urban poor, especially those living near the disposal sites. One of the reasons for this is the vicious cycle of poverty in very poor families. It is known that in most cities at least 30 percent of all households are headed by women who are the sole income earners, and that the majority of very poor families are headed by women, few with skills of competitive value in the labor market. In such households, the income that children can earn is also absolutely necessary. Therefore, if mothers are working at the waste dump, they will train their daughters in waste-picking, many of whom will continue their working life in this activity. It has been found in many situations that the average amount earned by a female waste-picker is significantly more than the legal national minimum wage.

Value-added aspects of the informal waste management sector

City landfills and disposal sites or dumps provide more than just direct sources of economic survival but, also, are the source of items which contribute to the social wellbeing of those engaged in the sorting and picking activity. This, allows many women to create some possibilities of economic autonomy.

In many cities, the waste-pickers are resident in the adjacent informal settlements and their dwellings are constructed from materials recovered from the disposal site, including plywood, corrugated metal, and sheets of plastic. Some 87 percent of the dwellings in Kingston's Riverton Meadows, adjacent to the disposal site, were found to be constructed partially or entirely from material recovered from the dump (Meiklejohn 2010). Additionally, household equipment and kitchen utensils collected from the dump provided many households with all the items they possessed. Some of the female waste-pickers also relied entirely on the waste for items of clothing for personal use. In the same settlement, around 42 percent of the total of 20 small-to medium-sized commercial operations (shops and bars) were owned and operated by women, based on start-up funds obtained from picking at the disposal site. Livestock rearing also provided an economic base for many households. There were approximately 12,000 pigs in the community, and the women owned around 48 percent of the pig operations. This was a lucrative source of income on account of the demand for pork and the fact that there were negligible overhead costs in running the pigpens and maintaining the stock. All food for the pigs was derived from the disposal site.

The study of the Riverton waste disposal site and adjacent settlement showed that of the women interviewed, only 17 percent had ever worked in the formal labor force and more than half the sample were single heads of their households. The economic vulnerability of these, and many other women in similar circumstances, led to their dependence on a male partner and increased their exposure to the risk of domestic violence. Most had been victims of physical, sexual, or psychological abuse. It is undoubtedly the case that informal work in waste management provides a means whereby many poor men and women can reduce their levels of poverty. This is especially important for women because their work options are generally fewer than those of the men on account of their childbearing and rearing responsibilities. Additionally, their reduced poverty through earning some income, and the ability to acquire clothing and other needed items from the dumpsite, enhanced their general quality of life and in so doing reduced their vulnerability in terms of the dominance and violence of their male partners and other men. Therefore, the work on the disposal sites played an important part in improving the roles and relationships of women in the household and their local community. It also contributes to the ability of women to afford the clothing and meet the miscellaneous costs incurred in sending their children to school.

Conditions of work in waste management

Despite the material advantages derived by the urban poor in working in waste management, because of the nature of wastes (as previously described), this work invariably exposes individuals to potentially hazardous and unhygienic conditions. In the cities of poor and middle-income countries, the work is usually intense and conducted under harsh environmental conditions. The unhealthy conditions of disposal sites are from time to time exacerbated by fires spontaneously sparked by combustible waste material or deliberately lit by workers to control vermin and odors. These fires invariably burn for several days, or even weeks, at a time. Exposure to excessive heat or cold, with the only protection being the shelters that the workers construct for themselves, together with the noise of heavy machinery constantly on the move—trucks dumping and tractors flattening the waste—create a chaotic and hazardous environment. Additionally, there are usually a number of animals on the waste disposal sites of cities in the developing countries. These are chiefly pigs, dogs, and cows along with egrets (large white birds), that forage for food. Under these conditions, vermin proliferate and animals and people pick through the accumulated and decomposing garbage. Men and women sustain injuries of various kinds, most commonly cuts, bruises, and bites. Gastroenteritis and skin disorders are also common. Few women in the informal waste sector use or wear any protective devices or clothing and are often unaware of the extent of risk of illness to which they are exposed.

Where the management of waste presents an ongoing and increasing challenge to the health of the people and environmental conditions of countries, in particular poor developing countries, then policy and planning require the involvement of all stakeholders. This includes private commercial interests, both small and large, government agencies, non-governmental organization (NGO) sectors, and consumers, both rich and poor. But it also includes the equitable involvement of women and men at all critical stages in the overall waste management process. Women are among the most vulnerable population groups worldwide and are in many cases subjected to inappropriate service system designs, including solid waste collection systems, due to their male bias. The poverty and substandard working and living conditions of women solid waste workers are often due to their exploitation by middle men, which draws attention to the importance of ownership and/or management of the waste operations.

Policy and planning from the perspective of gender, waste management, and development

Planning is crucial to effective and acceptable solid waste management. But protecting the poor and promoting gender equity are invariably not major objectives of waste management policy. Instead, solid waste management policy and planning are driven by notions of modernization that focus on the technological dimensions of systems to the exclusion of the social. For example, in South Africa, planning for the sector largely focused on the technical issues of waste disposal with little or no attention paid to the social and economic aspects of households (Poswa 2004). Yet, the irony is that planning is also usually intended to promote the improvement of management frameworks and systems that exist in the urban centers of developing countries. These are the very centers that house large numbers of their country's poor for whom environmental health and environmental sustainability are critical issues.

Modernizing the waste and recycling sector

Modernization of the waste management system is now widely conceptualized in terms of what is called "integrated waste management." One of the main characteristics of this shift towards modernizing waste management systems is the inclusion of recycling and composting. This is added to the generally accepted formal waste management activities of collecting and disposing of human and solid wastes in a safe manner, and keeping the streets and public spaces clean (Muller and Scheinberg 2003). But integrated waste management is more likely to be seen as a move to the modernization and, thus, the mechanization of all stages in the process, namely, residential or commercial waste collection, industrial or urban cleaning, and separation of recyclables from the waste stream. While these efforts are to be welcomed, there is the danger in developing countries that new schemes for managing waste materials may be insensitive both to the poor and to gender differences, thus with the potential to disturb and even destroy fragile livelihoods. Therefore, it is important in the effort to modernize waste management systems that planners should devise strategies that can strengthen the position of the poor. In terms of gender, planning should strengthen women's access to resources in the waste stream and their roles in recycling waste-derived resources. Additionally, planning should focus on improving the environmental conditions under which women and men work, and ensuring equitable rewards for their labor.

If the goal is to create socially productive, sustainable livelihoods for women and men in an equitable way, as Muller and Scheinberg (2003) suggest, the two basic questions that emerge are: whether the modernization of the urban waste sector, especially the development of waste removal and recycling systems, can offer opportunities for improved livelihoods; and what the economic characteristics and gender dimensions of those livelihoods are. It would be interesting to see if women are permitted to work with compost. As material for compost is often traditionally not recovered, this activity may currently be considered gender-neutral, and provide substantial opportunities for women. Further pragmatic questions include how gendered roles and responsibilities and skills can contribute to improving waste management so that its practices are sustainable and environmentally sound. At the same time, when introducing new technology for waste collection, disposal, or recycling, gender-related questions of capacity are critical. Do women, too, have the managerial expertise required for working in the formal system? Do women, as well as men, have equal access to the necessary training? Can women, as well as men, continue with related income earning activities, such as sorting the waste? How does the new technology affect the health of women and that of men? Does it create equal risks or offer equal protection against health risks? Leaving such issues to the existing forces of competition and inequality in a society will tend to reinforce, or even increase, women's socioeconomic disadvantage. The experience has been that when women seek to move from waste picking or scavenging to the status of micro-entrepreneurs, their access to credit and family support tends to be less than that of the men, so they are more likely to be handicapped from the start. Yet, one could look at this positively, and hypothesize that intensive on-the-job training on the waste disposal sites equips both women and men for careers as buyers or dealers in waste, having become part of a social network that makes a business out of waste and controls the waste flow in the city.

Solid waste management services form the core of municipal services and cannot be sustained without community cooperation and participation in all operations. The new demands for access to community services on an equitable social (including gender) basis requires a paradigm shift in the approach of those organizations that control the waste management system. Efforts are needed in the planning for change in the social and institutional structures within which roles and relations are embedded, challenges are experienced, and aspirations created.

The importance of partnerships in ownership

The involvement of all stakeholders in waste management is important for its success. But the responsibilities, power, and interests of the people involved are not all the same. The gender question is whether women-led companies are in the same position as those that are men-led to deal with the three types of clients: the private sector, households, and government officers. As was observed in Dar-es-Salaam, Tanzania, in 1998 the first contracts for street cleaning were issued to several companies led by women, and one led by a man (Muller and Scheinberg 2003). In contrast is the situation in many Southeast Asian countries. Although women are traditionally responsible for the household waste and sweeping the streets and take pride in keeping the environment clean and tidy, as soon as any of these tasks become paying jobs, men secure the jobs and ultimately dominate the structures and decision-making systems. The volunteer unpaid work of women at the household and community level remains on a lower status, taken for granted and not quantified or appropriately rewarded (Hayes, cited in Scheinberg et al. 1999).

Women traders in most situations undoubtedly face obstacles, as was illustrated in an example from Ghana where women had a long tradition of small- and large-scale trading, including the sale of bottles for reuse. Nevertheless, the expansion of their operations was limited by such factors as the lack of financial support and of appropriate markets. Further, the women also lacked the required managerial skills due to low levels of education, if any at all. They had limited access to technology because they were only experienced in the recovery and sale of the material that could still be used, not in recycling the glass to create new products. Despite the challenges, when waste pickers have organized themselves into cooperatives they have been known to improve their incomes, their working and living conditions. At the same time, they have contributed to solving the problems of insufficient collection and inappropriate disposal of solid wastes in developing countries. Further, the experience and skills acquired by women in waste management, even in informal work, potentially can open up opportunities. For example, in Belo Horizonte City in Brazil, waste pickers formed an association, supported by a religious NGO (Dias, cited in Muller and Scheinberg 2003), and when the City Cleansing Department started a selective handling and treatment system of solid waste several years later, they found the waste-pickers' association to be an obvious and useful partner.

Conclusion

The inclusion of gender-sensitive concerns in waste management must be seen not merely as a means of improving the conditions and empowerment of women in developing countries, but as a means of achieving gender equity as an integral aspect of development overall. The issue of the relationship between gender and development, as suggested by van Wijk-Sijbesma (1998), may be taken as a starting point for theoretical, also ideological and political activities in the area of gender equality, as well as the very prerequisite for development.

This perspective is based partly on the widespread evidence that gender permeates the interaction between people and the environment and all utilization of the environment. The relationship between a human being and the environment is shaped by gender roles, responsibilities, expectations, norms, and division of labor, from which follow the differences in attitudes to the nature and natural resources. For example, as van Wijk-Sijbesma (1998) states, women are more likely than men to recycle, buy organic food, and be concerned about energy efficiency. It is also based on the objective of incorporating a gender perspective in development programs so that in promoting equity between women and men in society, people—both women and men—will be empowered to become major participants in their own development. Bisht (2005) concludes that this would necessarily include the improvement in efficiency, poverty alleviation, social development, and environmental sustainability.

Policy approaches to waste management require good governance, involving all stakeholders as an essential aspect of development. In the already mentioned chapters of Agenda 21 relating to women, the provisions relating to waste management include:

- the impact of waste (especially hazardous waste) and pollution on the health of women;
- the inclusion of women's organizations (including grassroots organizations) in the promotion, information, and education on recycling, mobilization of communities in reuse of waste as a resource;
- the promotion of gender equity in decision-making and policy formulation relating to natural resource management, waste management, and environmental protection, as well as policies at all levels;
- the consideration of gender roles in planning and implementation of all modernized or integrated waste management systems.

All the above mentioned issues reflect the importance of designing appropriate solid waste systems that are equitable and affordable by all sectors of the population, especially those that were historically disadvantaged (Poswa 2004). The goals are defined based on the global goals in the area of gender equity, such as equal distribution of power, impact, and resources between women and men. Finally, one may posit that a gender-sensitive approach in waste management, a clear commitment to gender equity and the empowerment of women in the modernization of the waste management system, are critical in all new initiatives in urban services and environmental protection and sustainability. As pointed out by Muller and Scheinberg (2003), the emphasis should be on promoting gender equity as part of the solution, not the problem; and as an essential prerequisite for development.

References

Bisht, M. (2005) *Sanitation and Waste Management: A Perspective of Gender and Diplomacy*, New Delhi: Institute of Social Studies Trust.

Meiklejohn, A. (2010) *Women Sorters at the Riverton Landfill: An Exploratory Analysis of Informal Waste Recovery as a Sustainable Development Approach to Waste Management in the Kingston Metropolitan Area*, unpublished MSc thesis, University of the West Indies.

Muller, M. and Scheinberg, A. (2003) "Gender-linked livelihoods from modernising the waste management and recycling sector: a framework for analysis and decision making," in V. Maclaren and N. Thi Anh Thu (eds.), *Gender and the Waste Economy Vietnamese and International Experiences*, Hanoi: National Political Publisher, pp. 15–39.

Noel, C. (2013). "Solid waste workers and environmental management in Port-au-Prince, Haiti," in E. Thomas-Hope (ed.), *Environmental Management in the Caribbean: Policy and Practice*, Cave Hill, Barbados: University of the West Indies Press, pp. 103–121.

Poswa, T.T. (2004) *The Importance of Gender in Waste Management Planning: A Challenge for Solid Waste Managers*, Durban, South Africa: Durban Institute of Technology Department of Environmental Health.

Scheinberg, A., Muller, M., and Tasheva, E.L. (1999) *Gender and Waste: Integrating Gender into Community Waste Management: Project Management Insights and Tips*, Gouda, the Netherlands: Urban Waste Expertise Program UWEP Working Document 12.

Thomas-Hope, E. (1998) "Introduction," in E. Thomas-Hope (ed.), *Solid Waste Management: Challenges for Developing Countries*, Kingston, Jamaica: University of the West Indies Press, pp. 1–10.

United Nations (1992) *Agenda 21: United Nations Conference on Environment and Development*, Rio de Janeiro, Brazil, June 3–14, www.un-documents.net/agenda21.htm.

van Wijk-Sijbesma, C. (1998) *Gender in Water Resources Management, Water Supply and Sanitation: Roles and Realities Revisited*, Delft, the Netherlands: IRC International Water and Sanitation Centre.

PART V

Mobilities: services and spaces

30

INTRODUCTION
TO PART V

In an age of globalization, movement of people both within countries and between countries has been increasing. The proportion of women in these migrations has also been growing, although it is hard to obtain precise and recent figures. Lund et al. (2014) see this as a "mobility turn" for society with mobility as a capability translated into different livelihood strategies. Mobility and immobility, particularly the changing, flexible relationships between those who move and those left behind, and those who return represent relative degrees of resilience and stability, insecurity and fragility. These relationships inevitably change and can be remarkably flexible depending among other things on the purpose of the migration, its timing with regard to the lifecycles of the families concerned and new means of keeping in touch, whether by the internet or cheap transport links (Coles and Fechter 2008). Mobility is clearly a gendered phenomenon that depends on context. Today almost as many women as men move internationally but the gender balance of mobilities varies across continents as Kofman and Raghuram note. The chapters in this section follow several themes.

The provision of care in its multiple forms underlies much of the movement of women both within the Global South and from South to North (Momsen 1999). This has a long history worldwide as shown by Yeoh et al. It has often been conceptualized through the framework of global chains of care (Hochschild 2000) and seen as a counterfactual of globalization. Women find employment as domestic workers, nurses, nannies, and caregivers in households and in hospitals and residential homes for the elderly (Yeoh et al.). Women also move for other reasons: as agricultural workers, family migrants, petty traders, entertainers, and sex workers. Not all these moves are by choice as discussed by Townsend et al. and Samarasinghe. Some may be forced migrants trapped into working as sex slaves as described by Samarasinghe or as domestic workers unable to leave their places of employment. Kofman and Raghuram argue that many of the models of migration for care work have been based on South to North mobilities. This frequently impacts on families left behind in the South, as migrant workers take over care work for families in the North where more middle-class women are working outside the home and public provision of services such as preschools and elder care homes is coming under financial pressure. The growth of transnational care relationships and mobilities between and within emerging economies may alter the economic status of both migrants and their families.

Some migration is linked to the push factors of environmental degradation and disasters (Foresight 2011). In the cases of islands such as the Maldives, rising sea levels may eventually force the transfer of the whole nation's population to another country. Sometimes people may become trapped in places increasingly vulnerable to climate change such as rural areas of declining soil fertility as in Uganda but in other places such as Kenya similar environmental problems lead to migration (Foresight 2011: 96). In the face of extreme climatic events, predicted to become more frequent, such as Hurricane Katrina, migrants may become refugees. In these cases, the speed, nature, and scale of migration challenge the delivery of services, the sustainability of resource use and increase risks to human security and rights. Migration in response to environmental issues is likely to be increasingly movement from rural to urban areas and to impact food production as loss of rural labor leads to less cultivation. However, when disasters hit, migrants from these areas may be a vital source of economic relief for those left behind. Abbasi shows how the use of ICTs enables the transnational family to keep in touch and to send remittances more easily.

The migrant worker is seen as an agent of development bringing benefits such as remittances, transnational flows of investments, and the sharing of knowledge through circular migration as described by Torres. In her report on women left behind in Mexico, Torres sees them as agents of development playing a central role in the reproduction of migrant labor, managing remittances and overseeing local projects funded by money transfers. The so-called feminization of labor in Asia, especially from the Philippines, is part of a state development strategy as described by Yeoh et al. However, when migrants return home they may face problems of reintegration into the family as noted by Torres. In Nepal women who migrated young were often unable to obtain citizenship on return as to do so required support from male family members, which was often not forthcoming as some had been prostitutes and were seen to have brought shame to the family. Without citizenship it was hard for return migrants to find employment as Townsend et al. document.

Poverty is a theme running through this section. Mobility is seen as a way of overcoming unstable livelihoods at home for the unskilled. Swain shows that tourism reverses the flow with tourists coming from the rich North to the impoverished South on a short-term basis and possibly bringing economic benefits to the hosts, although much tourist spending remains with Northern tour operators and airlines. Pro-poor tourism (PPT) emerged in the 1990s as an approach to tourism aimed at providing social and cultural as well as economic benefits to the poor (Scheyvens 2011). It works through strategies such as employment, and encouraging the poor to supply goods such as food or souvenirs to tourists (Torres and Momsen 2004), helping locals to operate enterprises aimed at tourists, encouraging voluntary giving by tourists, and the establishment of improved infrastructure that benefits both tourists and local people. Furthermore, by bringing them into contact with tourists it exposes local women to a more cosmopolitan worldview and may encourage visitors to help them to migrate to the North. Tourism may also result in negatives, however, through the commoditization of culture and of bodies through sex tourism and trafficking. Yet production of souvenirs and cultural performances for tourists may also be a way of preserving this culture. Yeoh et al. illustrate how transnational domestic work has been both a driver of national development and a household livelihood strategy. Yet, domestic work can also result in a downward spiral of poverty through debts, emerging from unfair labor practices and unseen costs associated with obtaining visas and jobs overseas.

Only recently have researchers looked at the problems of those left behind. It is often said that women are more reliable in sending back remittances, as noted by both Torres and Yeoh et al., which can become a major source of foreign exchange for countries such as Mexico and the Philippines, as well as support for families left behind. But it is usually women and children who are left behind hoping that the migration of the male head of the household will overcome the poverty at home. In many cases, as Torres indicates, the remittances slowly reduce or end as migrant husbands start second families in the foreign location. She sees the "immobile" as the most affected by migration with the burdens falling on the women and the absence of fathers affecting children's upbringing. It has been argued that women left behind become more empowered and independent but they may also become vulnerable, dependent on remittances, and afraid to take any decisions. Remittances may be used to fund further migration, as in the Caribbean, rather than being invested in local resources and local capacities. Where it is women who migrate, children may be brought up by adult relatives too old or too busy to cope or fathers who may not see it as their role to provide day-to-day care.

Space is a constant theme, whether as a migrant domestic worker in a foreign home or for those left behind. For women left behind their use of local space becomes crucial and reflects age and mobility, and ethnic divisions as illustrated by Ismail and Sun for Sri Lanka and China. Studies of ethnic divisions in social spaces allow us to highlight the gendered nature of spaces. Newbury and Wallace (2014: 8) argue that "Looking at who accesses which spaces and how particular spaces come to be populated by particular actors allows the interaction of power, voice and agency, within clearly defined arenas to be explored." Cornwall (2002) sees space as creating as well as circumscribing possibilities for agency. She identifies three different types of spaces: closed spaces that are hard to enter, especially for women; invited spaces created by external agencies often designed for women to participate; and claimed spaces created by people themselves from or against power holders. Sun adds to this taxonomy of space a temporal aspect of spaces used for occasional activities or regular actions and Ismail shows how language influences use of village spaces by different groups. Without men, however, there appears to be more flexibility of use of space. In both China and Sri Lanka, women use certain spaces and activities to provide places for gossip and mutual support in addition to the more formal customary organization of space.

References

Coles, A. and Fechter, A.-M. (eds.) (2008) *Gender and Family among Transnational Professionals*, London and New York: Routledge.

Cornwall, A. (2002) "Making spaces, changing places: situating participation in development," IDS Working Paper No.170.

Foresight (2011) *Final Project Report*, London: Government Office for Science.

Hochschild, A. (2000) "Global care chains and emotional surplus value," in W. Hutton and A. Giddens (eds.), *On the Edge: Living with Global Capitalism*, London: Jonathan Cape.

Lund, R., Kusakabe, K., Mishra Panda, S., and Wang, Y. (eds.) (2014) *Gender, Mobilities and Livelihood Transformations: Comparing Indigenous People in China, India and Laos*, London: Routledge.

Momsen, J.H. (1999) *Gender, Migration and Domestic Service*, London: Routledge.

Newbury, E. and Wallace, T. (2014) *The Space Between: An Analytical Framework of Women's Participation*, Maynooth: Trocaire.

Scheyvens, Regina (2011) *Tourism and Poverty*, London: Routledge.

Torres, R.M. and J. H. Momsen (2004) 'Challenges and potential for linking tourism and agriculture to achieve pro-poor tourism objectives' *Progress in Development Studies* 4 (4): pp 294–319.

31

TRANSNATIONAL DOMESTIC WORK AND THE POLITICS OF DEVELOPMENT

Brenda S.A. Yeoh, Shirlena Huang, and Yi'En Cheng

Introduction

As a key aspect, if not the foundation, of social reproduction,[1] domestic work—whether paid or unpaid—is highly gendered as women's work. When paid for, domestic labor is not only classed, but often racialized as well. While long ignored in academic research, largely because it takes place within the home, domestic service has garnered growing attention especially since the 1990s as a result of at least two developments: first, the recognition of feminist arguments that negotiations within the reproductive sphere are as critical as, and cannot be separated from, those within the productive sphere; and, second, as domestic service work has taken on a transnational dimension with large numbers of women from the world's less developed nations migrating to work as domestics in developed countries as a result of global economic restructuring. Feminist research into waged domestic labor has identified it as a site of multiple exploitations by gender, class, race/ethnicity, and nationality almost universally, whether in the West (such as Canada, the US, or the UK), Latin America, the Middle East, Southeast and East Asia, and elsewhere, with the state (of both receiving and sending nations) often implicated in this exploitation. While receiving nations are often criticized for the lack of legislation protecting migrant domestic workers, sending nations are censured for their focus on the overseas remittances from migrants to bolster national budgets and development efforts at the expense of the women and their families. Despite their often marginalized positions and bleak working situations, female migrant domestic workers are vital contributors to the maintenance and economic wellbeing of their families and even local communities.[2]

This chapter aims to contribute to an understanding of the relationship between "gender" and "development" as refracted through the critical lens of domestic service work in the context of Asia. After a brief discussion of the history of paid domestic service, we focus on contemporary understandings of domestic service as a form of (lowly) paid work predicated upon post-1980s migration. We then review the scholarly literature to gain insight into the relationship between transnational forms of domestic work and "development" at three interrelated scales: countries and regions; the "family"; and individual migrant domestic workers.

The growth of paid domestic service as (migrant) women's work

Although domestic service has had a long history, its gendering as women's work came about only after the eighteenth century, when large-scale industrialization resulted in the separation of work into public and private spheres. Prior to that, both men and women, usually drawn from villages and rural areas, worked in domestic service (such recruitment of domestic servants through internal migration was often kin-based). Indeed, men were common as valets, butlers, houseboys, and other indentured servants in Europe, Africa, Asia, and America, albeit with some regional variations. For many of the rural–urban migrants (both male and female), paid domestic work was a "bridging occupation," or the first step towards seeking other forms of paid employment in urban centers as opportunities for agricultural work became scarce with mechanization. In Europe, structural shifts in the economy brought about by the Industrial Revolution created employment for men in waged labor in the productive sectors, but not for women. The "cult of domesticity" that accompanied industrialization and economic development in the West crystallized gender roles by idealizing the home as women's domain, and led to men's withdrawal from domestic work; at the same time, employment opportunities for women in domestic service opened up as middle-class women, aspiring towards becoming the ideal wife and mother, displaced the undesirable aspects of domestic drudgery to domestic servants. As other countries industrialized, the practice of transferring the more demanding and demeaning aspects of household work to female domestic workers, whether local or foreign, was replicated in the West and Asia. Today, as in the early industrial period, this form of gendered labor substitution that accompanies economic development simply excuses men from the domestic sphere and does nothing to dispel the idea that reproductive labor is women's work, as female surrogates take over domestic chores that society has deemed are the wife's responsibility.

Transnational flows of men and women for domestic service can be traced back to the eighteenth century when indentured servants and black slaves were transported to modern Europe and colonial America. The transnational movement of indentured servants was also found in Asia with *mui tsai* (meaning "little sister" in the Cantonese dialect) brought from China to British colonies in Asia in the early twentieth century to meet a shortage of domestic workers. Not all historical movements for transnational domestic service, however, were tied to indentured servitude. For example, propelled by shortages of work and prospective husbands, single women from Europe moved within the region or to the United States, Canada, and Australia to work as "wet-nurses and maids" from the mid-nineteenth century onwards (see Momsen 1999 for examples). In Asia, migrant groups of domestic workers included Hylam cooks, houseboys, and *amahs* from China, as well as *ayahs* from India, who moved to British colonies to work (see Huang et al. 2006 for further details).

Today, the figure of the female transnational domestic worker constitutes an important aspect of global migration flows, as a result of the shift of reproductive work from the household to the market due to economic restructuring. As many have observed, not only has global economic restructuring intensified the demand for paid reproductive labor in the world's industrialized countries as more women have entered the labor force, but the increased labor force participation of skilled women in these developed economies is being sustained on the backs of migrant women from low-growth countries seeking work as domestics. Unlike their better-off female employers employed in the productive sphere, paid domestic workers are positioned on the blurred boundary between home and work. Often perceived as labor that women require little training to do, and coupled with its location within the home, paid domestic service is not only devalued and poorly paid but subject to potential exploitation and abuse. When performed by migrants, paid domestic labor is further devalued and commoditized as reproductive labor that is not only unskilled and lowly, but not work that locals wish to do. As with earlier times,[3] cultural differences between domestic workers and their employers result in the former becoming the object of gendered, racialized, and nationalized subjectivities, exacerbating the already uneven employer–employee power relations as host societies subject the migrant "Other" to stereotype profiling. Qayum and Ray (2010: 112–113) in fact argue that while employer–employee relations may vary among contemporary societies, the power relations, hierarchies, and domination/dependency—which are "channeled through the discourse of class and culture distinction" and are the "hallmarks of domestic service"—are sufficiently universal that we can identify a "global culture of servitude."[4]

The cross-border transfer of female reproductive labor from poorer labor-surplus regions in the Global South to provide household and care services to the middle and upper classes of the more well-off Global North is now widely understood as being part of a global "care chain" linking a series of women and their families who depend on other women for paid or unpaid care work to fill the "care deficit" they have each left behind upon entering the labor market. Currently, such care chains link Eastern Europe to Western Europe, North Africa to Eastern Europe, Central and Latin America to North America, South and Southeast Asia to the Middle East and East Asia, and the Philippines to the rest of the world. Parreñas (2012: 271) reminds us that the care chain is also gendered in the sense that men are more often the recipients of care than the care-givers; in addition, the care chain is not color-free but often predicated on a "'racial division of reproductive labor', one in which women of color usually perform menial tasks." While these global care chains transfer social capital from the poorer to the richer countries and receive economic capital in return, it is also important to note that the economic value of the labor declines as it moves down the care chain, thus resulting in diminishing returns for each subsequent woman.

On the one hand, post-1980s structural shifts in the global economy may be lauded as having opened up overseas employment opportunities for women whose home economies are "struggling, stagnant or even shrinking," thereby providing an important source of livelihood for the women as well as their families and communities; on the other hand, the women's participation in these global labor circuits also acts as much, if not more, as a fount of "profit-making" for institutions such as labor agents and remittance agencies involved in perpetuating these large scale, brokered cross-border circuits of international labor migration, and provides an important means for sending governments to accrue foreign currency (Sassen 2000: 503, 523). By transferring the care deficit from wealthier to poorer countries and unevenly redistributing women's reproductive labor, the global care chain can be regarded as reflecting and contributing to the uneven geographies of development in the face of globalization, with the burden fallen onto women's shoulders. As such, rather than allowing women to rise above their present situation, the care chain can also be conceived as a metaphorical chain that ties migrant women to the occupational ghetto of transnational domestic work, and a lifetime of traversing the cross-border domestic labor circuit to earn remittances because, as many scholars of transnational domestic work have noted, female migrant domestic workers' earnings are not only a matter of household survival but also often of national security. The next section of the paper goes on to consider more specifically the relationship between gender, transnational domestic work, and the politics of development at the level of the nation, the household, and the individual worker.

Transnational domestic workers and the national development discourse

In response to the global demand for waged domestic labor, governments in developing countries in parts of Asia have increasingly promoted overseas labor migration as a development strategy to address issues of poverty, domestic unemployment and underemployment, as well as to grow foreign exchange income through remittances. The so-called feminization of labor migration in Asia in recent decades is in part a response to growing but gender-segmented demand for domestic and care labor in the more developed countries, as well as initiatives on the part of the state in less developed countries such as the Philippines, Indonesia, and Sri Lanka to promote labor migration—largely low-skilled—as a development strategy.

As a key Southeast Asian example, the Philippine state has been integrally involved since the 1970s with the formulation of an overseas employment program crafted to take advantage of the employment opportunities created by the oil boom in the Middle East. Since then, the Philippines has positioned itself to become a major source country, supplying workers with a range of skill levels to more than 100 countries (Yeoh et al. 2005). Making up 11 percent of the country's 94.8 million-strong population, the 10.5 million Filipinos abroad in 2011 not only represent a far-flung social network facilitating flows of information and material goods but also, through remittances, a significant and reliable source of revenue accounting for 10 percent of the country's gross domestic product, outweighing in magnitude the annual flow of foreign direct investment into the economy. State promotion of the deployment of Filipinos as overseas contract workers (OCWs) is wedded to a discourse of the OCW as "national hero," an appellation that stems from the firm belief that the economic remittances generated by OCWs serve as a key resource for national development. It is accompanied by the development and elaboration of an institutional and legal framework governing all phases of migration, from pre-deployment to onsite services, and the return and reintegration of migrant workers. More recently, in the light of the Philippine state's globalizing aspirations, there has been a shift to a more intensive marketing of the Filipino as a "global worker" underscored by the target to send out a million overseas contract workers, a move that might abate the state's resolution to focus on the protection of migrants' rights as it races to meet the target.

Other Asian countries have also taken steps to facilitate the outflow of labor migrants. For example, the Indonesian state has employed a specific strategy of providing unskilled workers (including domestic workers) at relatively low cost to give it comparative advantage over the Philippines, an undertaking which has been aided by a "migration industry" that has been given liberty to proliferate, and that has both increased the supply of women in Indonesia and expanded the market of their labor overseas. Similarly, the Sri Lankan government set up structures such as the Sri Lanka Bureau of Foreign Employment (SLFBE) to facilitate the outflow of labor migrants through legal channels, with the aim of minimizing exploitation and corruption in the recruitment process.

In fact, state measures often operate from a paternalistic stance and focus on protecting the rights of *women*, rather than that of workers or migrants, in order to manage female transnational migration. In India, the practice of having the husband (if married) or parent of the domestic worker accompany her to the office of the Protector of Emigrants to obtain clearance for emigration as a domestic worker essentially removes the responsibility of the domestic worker from the state to the accompanying family member. This attitude of the government is ironic given that it is often the failure of patriarchal structures at home that drives Indian women to migrate in the first place; as such, passing the welfare of these women to their families is problematic. It is in Bangladesh that we find the strongest tensions between economic motivations to send women abroad as domestic workers and moral (generally paternalistic) imperatives to protect female migrants predicated on gendered discourses of vulnerability and/or sexualized discourses on the protection of virginity, presumably under the religious justification that women's mobility presents a threat to both decency and chastity (Yeoh et al. 2005).[5]

The repositioning of the "migrant worker" as an "agent of development" in nationalist discourses in many parts of the developing world is also congruent with the recent swing towards a "new optimism" about migration and development championed by international organizations such as the International Monetary Fund, the United Nations, the World Bank, and the Global Forum on Migration and Development. Much has been written since the 1990s about the benefits of migration including transnational financial flows towards and investments in countries of origin, the multiplier effects of remittances, and the sharing of knowhow through circular migration, although there is also a growing literature arguing that migration, remittances, and transnational engagement are not panaceas and "cannot substitute for proper development strategies" (Gamlen 2010: 20). Within this larger migration and development discourse, specific information about the remitting behavior of migrants working in the domestic work sector is hard to come by, and at best piecemeal. For example, the average earnings of Filipina women migrant workers are lower than those of men, and men remit more money in terms of magnitude back home compared to women. Yet there are others who argue that women migrants remit more frequently and are more "reliable" remitters, hence providing a steadier stream of funding for household use. For example, Jampaklay and Kittisuksathit (2009) found that 25 percent of the women sent more than 50,000 Thai baht (approximately US$1,560) during the period of 2006 and 2007, as compared to 17 percent of the men in their research on migrant worker remittances in the context of Cambodia, Lao PDR, and Myanmar. Research on Indonesian domestic workers in Singapore, Malaysia, and Hong Kong show that despite domestic workers' low earnings, it is found that they are better savers and remit a greater proportion of their earnings back home than male migrant workers. In general, the evidence across the developing world suggests that migrant men—who usually earn more than women—tend to send larger sums, while women are more consistent and reliable remitters who send home a larger proportion of their earnings. Scholars have argued that this observation may be attributed to the greater moral obligation to remit placed on women, and that migrant mothers who leave family members behind are expected to willingly practice self-sacrifice for the sake of their families and especially their children (and in contrast, migrant fathers do not face the same moral censure for neglecting their children).

Finally, it is instructive that feminist scholars have also questioned the "gendered" rhetoric of the migration and development discourse by troubling the unnatural divide between "productive" (e.g., capital investments) and "unproductive" (e.g., subsistence needs, children's education) uses of remittances. Mahler and Pessar (2006: 45), for example, point out that:

> many policy makers have stressed the "productive" uses of remittances and how to promote them. In their view the vast majority of remittances are spent by recipients on "unproductive" purchases such as food, shelter, clothing, and education. The development project is to increase the percentage of these moneys that are saved, not spent, so that the capital can be invested. Gender seeps subtly into a seemingly neutral notion of "productive" versus "unproductive" uses of remittances.

In the current more optimistic thinking about migration and development, using remittances for consumption is thought to create demand for local products thereby fuelling growth through multiplier effects, as well as contributing to safeguarding the welfare of the family (Gamlen

2010). In the next section, we turn our attention to the household level in examining the role that transnational domestic work plays as a livelihood strategy in sustaining family wellbeing.

Domestic work migration as household livelihood strategy

The increased volume of transnational domestic work migration originating from developing Asia in the past two decades is an increasingly significant driver of contemporary social transformation of the "family/household" in sending communities. Households across Southeast Asia that have resorted to this non-traditional, gender-differentiated form of migration as a livelihood strategy have to contend with the pressures of change stemming from at least three areas: the potential rearrangement of gender roles and redefinition of gender identities for the migrant and left-behind family members; the everyday politics over the use and management of remittances; and the consequences of indebtedness and failed migration on the family.

As a "non-traditional" form of labor migration, domestic work migration involving "migrant mothers" immediately triggers off the need to adjust arrangements and relationships of care for millions of left-behind children growing up for part or all of their young lives in the absence of their mothers, and under the care of "left-behind fathers" and/or surrogate caregivers. While the available research on how care deficits are dealt with in families with migrant members in sending countries at the Southernmost end of the global care chain is still limited, recent migration studies using Southeast Asian cases examine the durability of the woman–carer model. First, current research on migrant mothers emphasizes the resilience of gender ideals surrounding motherhood even under migration in the transnational context. While mothering at a distance reconstitutes "good mothering" to incorporate breadwinning, it also continues maternal responsibility of nurturing by employing (tele)communications regularly to demonstrate transnational "circuits of affection." While migrant mothers actively worked to ensure a sense of connection across transnational spaces with their children through modern communication technologies, it has also been argued that "long-distant mothering" is an intensive emotional labor that involves activities of "multiple burden and sacrifice," spending "quality time" during brief home visits, and reaffirming the "other influence and presence" through surrogate figures and regular communication with children (Sobritchea 2007). Second, the research thus far suggests that the care vacuum resulting from the absence of migrant mothers is often filled by female relatives such as grandmothers and aunts or eldest daughters and, in some cases, female domestic helpers. The continued pressure to conform to gender norms with respect to caring and nurturing practices explains men's resistance to, and sometimes complete abdication of, parenting responsibilities involving physical care in their wives' absence. These studies conclude that the "delegation of the mother's nurturing and caring tasks to other women family members, and not the father, upholds normative gender behaviors in the domestic sphere and thereby keep the conventional gendered division of labor intact" (Hoang and Yeoh 2011: 722). Third, more in-depth studies combining quantitative and qualitative analyses, however, have begun to reveal a more complex picture of more flexible gender practices of care in sending countries. Even in the context of Vietnam with its strong patriarchal traditions, Hoang and Yeoh (2011) argue that Vietnamese men struggle to live up to highly moralistic masculine ideals of being both "good fathers" and "independent breadwinners" when their wives are working abroad, by taking on at least some care functions that signified parental love and authority while holding on to paid work (even if monetary returns are low) for a semblance of economic autonomy.

Another strand of scholarship focusing on domestic worker migration as a household livelihood strategy trains the spotlight on the "family" as a site of cooperation and conflict over the management of resources, thereby raising important questions concerning the way gender and intergenerational relations in the family mediate the use and distribution of remittances. Research on the cultural expectations in relation to remittance-sending suggests that these familial obligations may serve to entrap women migrants more so than men. Basa et al. (2011: 12) document the experiences of Filipino migrant domestic workers in Italy and "how the pressure to provide remittances is locking women even further into the global care chain, with not only economic, but social and cultural consequences." Other studies of household dynamics in labor-sending countries in Asia show that remittances are often under the control of women—the wife (when the migrant is a married man) or the grandmother and eldest daughter (when the migrant is a married woman). Apart from the fact that men are often stigmatized as bad money managers, the general preference for women to manage remittances also derives from the widespread perception of them as more altruistic spenders whose control of family resources is more likely to be associated with the enhancement of collective wellbeing. In some contexts, however, the deployment of remittances tends to run along gender lines (i.e., sons to fathers and daughters to mothers) rather than being strictly a female activity. Yet in other cases, remittances sent back by female migrants are often channeled to the male members of the family, such as sponsoring their brothers' education or marriage, or facilitating their migration projects. The question of whether remittances have the effect of transforming or further entrenching gender ideologies and relations across transnational space remains an open one. The control over remittances can also present a source of intra-family conflict, driving a wedge between husbands and wives, parents and children, nuclear families and more distant relatives. The conflict may extend to wider familial networks when migrants and their spouses are reluctant to share the economic benefits of migration outside their nuclear unit, thereby rejecting relatives' claims for mutual access to assets and undermining deeply held patterns of family, kinship, and caste associations.[6]

A third area of work that has hitherto not attracted significant attention but which is crucial in considering the developmental impacts of domestic work migration relates to issues of indebtedness and failed migration. Much of transnational labor migration in Asia is arranged by commercial brokers who are known to charge exorbitant fees that may drive migrant families into debt. Studies suggest that large amounts of debts incurred in order to finance migration may pressurize migrant workers to go to great lengths to hang on to work in host countries, such as overstaying their visas, running away from legal employers, or resorting to crime. For example, a 2008 World Bank study estimates that 30,000 Indonesian domestic workers in Malaysia run away each year to escape the debt that would be deducted from their salaries over a period of six months. In Southeast Asia, debt repayment features as a significant part of remittance use, while migration-related indebtedness is an important factor explaining why international remittances fail to convert into upward socioeconomic mobility for many migrant families. In the worst case scenario of "failed" migration, the burden of debts may lead to a downward spiral and dire consequences for family wellbeing.

Scholars have variously argued that migrant women working in domestic service are poorly paid, experience job insecurity given the transient nature of contractual work, and undergo "de-skilling" due to the non-recognition of qualifications obtained from sending countries. Despite these vulnerabilities that transnational domestic workers as individuals face as part of the commodified global care chain, the number of households in developing Asia turning to domestic work migration as a viable strategy for poverty alleviation and upward social mobility has been increasing. In the longer term, the sustainability of such a livelihood strategy at the household level would depend on, inter alia, the degree of flexibility in the household division of reproductive labor as care arrangements are forced to change, the equitable management and use of remittances to promote the welfare of household members, and the ability to avoid the debilitating effects of debt entrapment in the course of migration. We turn next to examining the impact of transnational domestic work on the individual migrant.

Self-development and migrant rights

The prevailing discourse around the migration–development nexus tends to position the individual migrant as an "agent of development," where development is usually defined in narrow economic terms. In this vein, the current consensus is that migration is intrinsically "good" for migrants, as migratory moves are often associated with material gains. For example, the UNDP shows that "migrants from the poorest countries, on average, experienced a 15-fold increase in income, a doubling of school enrolment rates and a 16-fold reduction in child mortality after moving to a developed country" (quoted in Gamlen 2010: 18).

Critics of the dominant discourse have in turn argued that there is a need to broaden our understanding of "migration and development" to include how migration impacts on social developmental issues including education, health, social welfare and security, political participation, and how these are linked to the democratization of human relations (i.e., non-discriminatory policies and attitudes in terms of class, race, ethnicity, nationality, and gender). Adopting such an understanding involves changing the attention that is focused on the scales of national development and household dynamics to foregrounding the personal development of the individual migrant herself. In fact, even as transnational communities and diasporas contribute economically towards families, communities, and other developmental processes in sending countries, there is no automatic guarantee that, at the individual level, the migrant's personal development is enhanced by migration. As Gamlen (2010: 18) observes, "ensuring that individual migrants 'win' from migration requires streamlining migration controls in destination countries, whilst bolstering national and international mechanisms for protecting migrants' rights." The provision of proper legal status, channels for family reunification, and regulations that protect the labor rights of migrant workers are crucial to enhancing the benefits of migration for individual migrants. For transnational domestic workers in particular, acknowledging domestic work as equally worthy of protection under labor laws along with other non-domestic sectors, and the establishment of standardized and transparent contractual agreements between employers and domestic workers are fundamental to the promotion of their welfare and rights. International organizations and advocacy groups are also increasingly championing the redefinition of "social rights" to better reflect the changing realities faced by migrant domestic workers. For instance, the right to access social protection such as health care, whether short-term or long-term or within host country or upon return, has been a concern for Migrant Forum Asia, given that social security mechanisms for domestic workers are almost non-existent in the ASEAN region.

Additionally, a broader conceptualization of "rights" also leads to a consideration for migrant workers' "right to self-development." Indeed, many domestic workers are able to secure a better life through the acquisition of both work-related skills such as job training and language learning, as well as personal skills such as financial planning and budgeting. These skills not only have the potential of helping migrant women benefit financially from their overseas work but also build up their social capital and bargaining power.

However, the notion of migrant rights is highly contested in the context of Asian nation-states. In the case of Singapore, migrant workers' rights are negotiated over an ambivalent ground, where anxieties exhibited by Singaporeans over the ostensible invasion of foreigners into territory of citizenry are met by an emerging human rights consciousness within an embryonic civil society landscape. In recent years, the latter is marked by several small but progressive steps undertaken by local civil society organizations to improve the working condition of domestic workers, such as the standardization of the employment contract and a successful change to make a weekly day-off mandatory. However, progress in advocating for human rights is handicapped by the fact that migrant domestic workers in Singapore are not organized in any formal collective form to negotiate relations between employer and employee, while public assemblies are illegal without a permit. This is in contrast to the situation in Hong Kong, where transnational domestic workers are covered under the Employment Ordinance and their basic rights such as a minimum wage, rest days, paid annual leave (as well as paid leave for sickness and maternity) specified in a standard employment contract. Hong Kong's more vibrant civil society also features a number of migrant worker organizations and coalitions that regularly address the injustices suffered by foreign domestic workers, and the workers themselves are actively engaged in protest campaigns and victory marches.

Given the multiple marginalizations that transnational domestic workers experience in the course of migration and work, feminist scholars have been generally wary about overstating the emancipatory value of transnational migration for this group of women. Indeed, for migrant domestic workers who often have limited material resources and few civil rights, the prospect of self-development and the building up of new subjectivities and political spaces to act often runs up against state and society's rules of marginality and exclusion. The lack of political will on the part of most governments to deal with the rights and civil liberties of major groups of migrant women such as domestic workers implies that the role of non-state actors such as NGOs and faith-based groups is crucial in encouraging migrant women to participate in welfare-oriented or advocacy work, forge cross-national alliances and networks, and pursue transnational activisms to negotiate for human rights and better working conditions.

Conclusion

Over the past three decades in Asia, domestic work migration has grown as a highly gendered phenomenon in response to globalizing pressures and the new international division of reproductive labor. While there are multiple intersecting dimensions at work in shaping the phenomenon, we highlight the persistence of "gender" in framing discourses around migrant women's contributions to nation-building; in transforming household politics over the negotiation of care arrangements and the use of remittances; as well as in the creation of both constraints and opportunities for women who become global domestic workers.

By paying attention to the gender dimensions of domestic work, we are better placed to avoid the danger of uncritically positioning the transnational domestic worker as an agent of development within migratory life-worlds, which continue to be shaped by patriarchy, paternalism, and paradoxes. Indeed, it can be argued that transnational domestic work is, in fact, characterized by the exploitation of a paradox of intimacy and distance—the worker is located in proximity to the family but at the same time distanced from its members on the basis of class, gender, and other hierarchies. Other feminist scholars have observed that the phenomenon of the "maid trade" as a retrogressive, non-modern form of domestic work has further entrenched the gendered nature of domestic work and, as a consequence, the often unquestioned patriarchal underpinnings of society. Aguilar (1996: 5), for example, argues that "the plentiful supply of migrant workers from the Philippines [and other labor-exporting countries] has made possible the retention of archaic, slave-like forms of domestic labor in Singapore [and other labor-importing countries], which has propped up customary patriarchal relations in family and society."

At the same time, it is also important not to represent transnational domestic workers as mere victims of globalization. Instead, as this chapter illustrates, rethinking the links between domestic work migration and development needs a consideration of the politics of "development" at three interrelated scales below the "global." Importantly, the developmental potential of domestic work migration should be examined not just in relation to state-level discourses about migration and development, remittance generation, transfer and usage but also in terms of how this form of transnational migration features as a livelihood strategy in changing household dynamics and sustaining the wellbeing of left-behind family members of the migrants. A more holistic approach to "development" should also go beyond the household level and focus on empowerment and access to rights for the individual migrant domestic worker. This entails a shift in perspective in considering not just the economic developmental impact of migration on sending communities and nation-states (as is the case in the current preoccupation with migration as a generator of remittances), but to also take into account the conditions that shape recruitment processes, work conditions, and residency regimes that influence the quality of the migration experience for the domestic workers themselves.

As a form of "development," domestic work migration inextricably links the fates of individuals, households, and nation-states. More needs to be done in connecting women's labor migration and the roles they play in development across sites and scales, the multiple forms of contribution (economic, social, cultural, and political) that women make to nation-states and households, as well as the spaces for self-development of the women themselves.

Notes

1   The range of activities that fall under social reproductive labor is now generally understood to include both the physical aspects of household chores (such as cooking, cleaning, and laundering), as well as the emotional aspects (such as the care of adults and children, socialization of children, and maintenance of community ties).

2   While there is widespread acknowledgement of the gendered dimensions of remittances, including the greater contribution to household income of remittances sent home by men due largely to gender inequality and segmentation in global labor markets, this does not negate women's smaller but sustained and regular contributions to support their families.

3   For example, Indonesian servants in the Netherlands Indies were deemed dirty, lazy, and unreliable by their employers. More generally, Dutch and British families in colonial Southeast Asia had deep xenophobic anxieties about their Asian domestic servants (Huang et al. 2006).

4   In comparing domestic service in Kolkata and New York, Qayum and Ray (2010: 113) note, for example, that all households in Kolkata that can afford a domestic will employ one, with the relationship between employer and servant "infused with the rhetoric of love," while in New York, domestic workers are not considered essential and are employed "on the assumption of workers' rights and empowerment." Nonetheless, in both cities, the relationship is one of "domination, dependency and inequality."

5   The material presented in the first four paragraphs of this section is adapted from Yeoh et al. (2005).

6   See Yeoh et al. (2005) for a review of some of these studies.

References

Aguilar, F.V. (1996) "Filipinos as transnational migrants: guest editor's preface," *Philippine Sociological Review* 44(1–4): 4–11.

Basa, C., Harcourt, W., and Zarro, A. (2011) "Remittance and transnational families in Italy and the Philippines: breaking the global care chain," *Gender & Development* 19(1): 11–22.

Gamlen, A. (2010) *People on the Move: Managing Migration in Today's Commonwealth*, London: The Ramphal Centre.

Hoang, L.A. and Yeoh, B.S.A. (2011) "Breadwinning wives and 'left-behind' husbands: men and masculinities in the Vietnamese transnational family," *Gender & Society* 25(6): 717–739.

Huang, S., Yeoh, B.S.A., and Abdul Rahman, N. (2006) "Economics: paid domestic labor, Southeast Asia," in S. Joseph (ed.), *Encyclopedia of Women in Islamic Cultures, Vol. 4, Economics, Education, Mobility and Space*, Leiden: Brill, pp. 226–229.

Jampaklay, A. and Kittisuksathit, S. (2009) *Migrant Workers' Remittances: Cambodia, Lao PDR and Myanmar, Bangkok*, ILO Regional Office for Asia and the Pacific, http://ilo.org/wcmsp5/groups/public/---asia/---ro-bangkok/documents/publication/wcms_111543.pdf.

Mahler, S.J. and Pessar, P.R. (2006) "Gender matters: ethnographers bring gender from the periphery toward the core of migration studies," *International Migration Review* 40(1): 27–63.

Momsen, J.H. (ed.) (1999) *Gender, Migration and Domestic Service*, London and New York: Routledge.

Parreñas, R.S. (2012) "The reproductive labour of migrant workers," *Global Networks* 12(2): 269–275.

Qayum, S. and Ray, R. (2010) "Traveling cultures of servitude: loyalty and betrayal in New York and Calcutta," in E. Boris and R.S. Parreñas (eds.), *Intimate Labors: Cultures, Technologies and the Politics of Care*, Stanford, CA: Stanford Social Sciences, pp.101–116.

Sassen, S. (2000) "Women's burden: counter-geographies of globalization and the feminization of survival," *Journal of International Affairs* 53(2): 503–524.

Sobritchea, C. (2007) "Constructions of mothering: female Filipino overseas workers," in T.W. Devasahayam and B.S.A. Yeoh (eds.),*Working and Mothering in Asia: Images, Ideologies and Identities*, Singapore: NUS Press, pp.173–194.

Yeoh, B.S.A., Huang, S., and Rahman, N. (2005) "Introduction," in S. Huang, B.S.A. Yeoh, and N. Rahman (eds.), *Asian Women as Transnational Domestic Workers*, Singapore: Marshall Cavendish Academic, pp. 1–17.

32

CARE, WOMEN
AND MIGRATION IN THE
GLOBAL SOUTH

Eleonore Kofman and Parvati Raghuram

Introduction

Women, who are almost universally predominant in caregiving, have been significantly affected by recent reductions in funding and provision of care – they have become incorporated into both the formal and informal labour markets as caregivers in new ways. At the same time, the rising labour market participation of women has also resulted in substantial labour shortages in unpaid care provision that women had often provided, intensifying demand for paid caregivers. This demand is increasingly being met by migrant female labour. As a result, the nexus between migration, gender and care provision has become a key issue of concern in public and social policy (Razavi 2007). It has also led to a rich vein of theorizing of the linkages between North and South that draws on detailed and valuable empirical studies of care relationships in some Northern and Southern countries and the role of migration therein.

International migration is deservedly a significant driver for the analysis of care regimes. The number of female migrants globally was estimated at 94.5 million (or 49.6 per cent of total) in 2005. The share of women among migrants in Southern countries was about 38.9 million (or 51 per cent) in 2005, compared to 46.2 million (or 51 per cent) in the high-income countries belonging to the Organisation for Economic Co-operation and Development (OECD) and 8.7 million (or 40 per cent) in the high-income, non-OECD countries (Ratha and Shaw 2007). The provision of care in its myriad forms underlies much female migration. Thus, large numbers of female migrants move to provide care in a range of contexts and sites. They find employment as domestic workers and as care professionals, such as senior carers, nurses and social workers, and facilitate the care of children, adults, disabled and elderly within households, in residential homes and hospitals. Women also move for other reasons – as family migrants, petty traders, agricultural or manufacturing workers, sex workers and entertainers, and in a range of other professionalized occupations. However, the mobility of these women also leaves care gaps to be filled in the areas they leave behind.

Hence, care demands are both being created and met through women's employment, highlighting the complex causal relations that tie together migration, gendered labour and care regimes. Initially the relationship between gender and care was the focus of feminist economics, the sociology of work and social policy (Razavi 2007). More recently the transfer of labour from the South to the North has captured the attention of migration theorists who explore the nexus between the three, especially through the concept of global chains of care (Hochschild 2000). Large-scale migration for care purposes characterizes South to North (OECD and non-OECD countries such as Hong Kong Special Administrative Region, Saudi Arabia, Singapore and the United Arab Emirates). Conceptualization and models have been developed with a primacy of this form of South–North migration in mind and have, therefore, incorporated Southern countries' experiences selectively. As a result, much of the analysis has focused on the impact of transfers of care services on wealthy countries in East Asia, Europe and North America, i.e., countries that are incorporated into global circuits of care and ignoring the implications of regional dimensions and particularities of different places. In particular, the changing landscape of development and the role of gender therein has been underexplored.

Using the examples of selected countries in the South, especially middle-income countries, this chapter aims to unsettle some of the assumptions that underlie existing analysis and to lay out some questions that might need to be addressed to make questions of care in the South reflect the diversity and dynamic of migratory systems, gender regimes and welfare arrangements. The middle-income countries pose interesting questions as they are tied into global circuits of care in distinctive ways, have different kinds of care (relating to their income status). Several of these countries serve as poles of migration and have begun implementing active social policies and/or intervened in the provision of care; others are countries largely exporting care labour or countries where care systems have been overwhelmed by the HIV/AIDS pandemic.

The main aim of this chapter is to analyse the implications of the diversity and dynamic of Southern countries for analysing migration, gender relations and care provisioning in the countries of the Global South. In developing these implications, it should first be noted that there is inadequate empirical data to achieve any kind of comprehensive understanding and that we can only hope, in the context of this chapter, to suggest some partial insights. Second, although this chapter refers to the Global North and the South, both are highly heterogeneous in terms of their welfare regimes, wealth and migratory patterns and this diversity extends beyond that considered in this chapter. Rather our analysis suggests future lines of study.

The chapter begins with an overview of Southern migration patterns, especially those of women. The second section looks at the theoretical frameworks through which migration and care have been analysed, in particular global chains of care, and explores some of the conceptual gaps that exist. The chapter then explores what acknowledging the diversity and dynamism of Southern countries and the resultant variations in care mean for theorizing care.

Women and migration in a diverse 'South'

Although most research focuses on South–North migratory flows, migrants from the South are as likely to migrate to other countries of the Global South as to the richer countries of the North (Ratha and Shaw 2007). Only about two in five migrants from the South reside in the high-income OECD countries. Some 20 million (or 13 per cent) are estimated to reside in high-income countries outside the OECD – among them Hong Kong (China), Saudi Arabia, Singapore and the United Arab Emirates.

South–South migration, on the other hand, accounts for about half of all international migration from the South. In Africa, the proportion of emigrants (relative to population) moving to low-income countries is higher than that to middle-income countries. While emigrants from low-income countries are more likely to migrate to neighbouring countries, those from middle-income countries are more likely to move to high-incomes ones. However, even migration between areas of similar income levels can help families diversify income sources and thus reduce risk (Ratha and Shaw 2007: 19).

South–South migration is also overwhelmingly intraregional (except in South Asia). Regional wage differentials have led some countries such as Mexico and Turkey to become both origin and destination countries, while others have become 'migration poles'. The major middle-income migration poles are Argentina and Venezuela in South America, Jordan in the Middle East, Malaysia and Thailand in Asia, the Russian Federation and parts of Eastern Europe. These migration poles constitute a diverse group of countries with very varied histories and types of migration. In former colonial lands, such as Argentina and South Africa, large-scale migrations from Europe, regional systems of migration with neighbouring countries and internal rural–urban migrations coexisted in the twentieth century. Political repression and transition, such as apartheid and post-apartheid in South Africa and the dictatorship and severe economic crisis in Argentina, also led to emigration, especially of skilled migrants.

Furthermore, in many countries of the South, there are high levels of both internal and international migrations. Globally, internal migration is far more significant than international migrants, much of it accounted by urbanization. In many countries of Africa, Asia and Latin America rural–urban migration accounts for 40 per cent of urban growth. In some cases, migrants move internally before emigrating; in other cases they move to fill the vacuum left by international migrants.

As stated in the introduction, the proportion of women in migration flows globally has increased in the last few years. By 2005, women formed 53.4 per cent of migrants in Europe, 50.4 per cent in North America and 45.5 per cent in the Global South. However, the international migration of women is not new. Latin America was the first region where the number of women migrants equalled men (Staab 2004). On the other hand one of the most feminized flows of migrant labour is that from the Philippines. The country has sent approximately 1.5 million overseas foreign workers throughout the Asian region – many of whom find employment as domestic workers.

Spatial patterns of female migration are influenced by a range of factors, particularly labour demands. For example, the entry of women into the workforce in some middle-income countries has created a growing need for domestic workers. Environmental disasters (Ratha and Shaw 2007) and difficult economic conditions following economic structuring have also led to outflows of migrants. Seasonal migrations to meet agricultural labour demands and cross-border (or in the case of the Caribbean islands, inter-island) trading occur across many parts of the South. Political conflicts, as in Latin America, too have generated outflows. Finally, regulations and regional agreements that operate to facilitate interregional mobility have also shaped migration, although different agreements vary in the extent to which freedom of movement, residence and settlement are given to citizens of participating countries. For instance, the Economic Community of West African States (ECOWAS) has provided freedom of movement but so far the right to set up and establish businesses has not been extended. There is increasing movement between the countries that are part of the Mercado Común del Sur, or Mercosur (Brazil, Argentina, Paraguay, Uruguay). In 2002 an agreement on residency for nationals of the member countries was signed. It allows temporary and permanent migrants to receive the same treatment as the nationals of the country in which they are resident. Argentina was the first country to implement the agreement and in 2006 regularized half a million migrants.

However, most agreements are operating through a mode of labour circulation without offering settlement. Even where regional flows have been opened up, they can be selective. For example, within the North American Free Trade Association (NAFTA), free movement between Canada and the United States is limited to those with college degrees but there are quotas for migration from Mexico. Similarly, the free movement of people within the Caribbean Community and Common Market (CARICOM) was initially restricted to those with university degrees, and then extended to artists, media persons, sports persons and musicians.

Concerns over the vulnerability of female migrant women centre on their possible exploitation as workers, migrants and as women. Unorganized sectors such as domestic work are, in some cases, being brought into the remit of bilateral agreements and memorandum of understanding in order to improve the conditions of migrant workers. However, some sectors, such as sex work, remain unrecognized, although they offer job opportunities for migrants. Limited efforts to address these sectors have come from the specificity of women's experiences. Thus, in an attempt to control the spread of HIV and other infections, migrant women, who are at greater risk to such diseases, may, ironically, gain recognition.

The available data and literature indicate that migrants who travel to other Southern countries enjoy much lower increases in income, are more likely to be irregular, are subject to greater risks of exploitation, and are more likely to be expelled than are those who migrate from Southern countries to the North (Ratha and Shaw 2007). Nevertheless, if the benefits from South–South migration are limited, it is also likely that many South–South migrants are poor, or are forced to migrate because of war or ecological disaster. Even small increases in income can have very substantial welfare implications for people in such circumstances. Differences in country incomes are likely to be much greater, on average, for migrants travelling outside their home region than for intraregional migration, partly because larger income differentials are required to overcome higher costs associated with travelling over greater distances (geographic and cultural).

In sum, Southern migration has been marked by the overwhelming importance of regional and internal migration, and diversity of destinations, outcomes and experiences of migration. Yet, despite this empirical evidence, as we shall see the analytical frame for analysing migration has largely privileged particular forms of South–North migration and has left a conceptual legacy that needs questioning.

Theorizing care, gender and migration

The empirical basis for most theorization of care, gender and migration has rested on the experiences of selected Northern and Southern countries. Drawing on the groundbreaking work of Rhacel Parreñas (2001) on the experiences of migrant Filipina women moving to the US and Italy and the internal or even regional international migration that it sparked to fill the care deficit in the migrant's own homes, Arlie Hochschild (2000: 131) suggested that the migrants were tied together in a global chain. Defined as 'a series of personal links between people across the globe based on the paid or unpaid work of caring' (Hochschild 2000: 131), the chain acts to abstract labour (physical and emotional) upwards in a process that leaves the South as a global commons (Isaksen et al. 2008). Women's migration reconstitutes the division of labour among women such that extended female kin absorb some of the caring activities. Thus, 'other mothering', defined by acts of nurturing and caregiving rather than biological relationships between mother and children, becomes significant in assisting in the raising of children and looking after other family members who require care.

The concept of global care chains has rapidly become influential in theorizing care globally. Its global reach was dependent on the theoretical base offered by Sassen's analysis of global flows of labour (Sassen 2000). Drawing on dependency theorists Sassen suggested that such domestic workers were part of the (counter)flows of globalization.

However, while the care chain analysis has usefully provided a global framework for understanding different migratory movements, current analyses tend to be premised on a narrow range of relationships, institutional arrangements and care regimes. In the North the analysis has centred around the restructuring of public services and on rising female employment, while the Southern literature has focused on the reconfiguration of households (and particularly their transnationalization), especially where such changes have been driven by demands for migrant labour and the ensuing female migration from South to North. The differences among the Northern countries and those among Southern countries are not adequately explored. Moreover, the concept of global chains of care has primarily focused on transnational mothering relationships and on the international aspect of migratory movements. Other familial relations and the cascade of effects generated by migratory movements or the implications of gender and care regimes in different regions of the South (Kofman and Raghuram 2009) are side-lined. Finally, although empirical studies on care differences are mounting, their differential incorporation into global chains has not been adequately investigated. This section suggests a number of ways of broadening and deepening the analysis of global chains of care in the South.

Southern countries: the different roles of internal migration and international migration patterns in shaping care

Care chains have become important conceptual tools because, unlike in the late eighteenth century, today's migrant workers perform domestic work not as a stage in the life course prior to marriage, but as a full-time and lifelong occupation that may involve their movement from one country to another. Mature-age migrants who have their own caring responsibilities are increasingly engaged in this work full-time. Moreover, domestic work has become more clearly defined and associated with the private space of the household so that more of the workers, including migrants, entering such fields are women.

Existing literature identifies differences in the empirical patterns in these chains. First, female-led migration, although growing rapidly, is still numerically smaller than male-led migration. Thus, while in countries such as Thailand, it is primarily fathers who leave children behind, in the Philippines and in Sri Lanka it is mostly women who do so. Second, there appear to be class differences in how families respond to migration: in working-class families other women usually took over the work; in middle-class families fathers usually relied on paid domestic labour, other kin and family members such as older daughters. Third, there are variations in the extent to which women completely withdraw from caring after migration or indeed who picks up this labour afterwards. Migrant women frequently maintain their emotional concerns and advice from afar, sustaining an active, although distant, transnational mothering. As a result, mothers may contest the myth of the male breadwinner but retain the myth of the female homemaker. Men also do pick up caring responsibilities in many instances. In a detailed study, Asis (2006) suggests that in about half the cases where the mothers have migrated, children identified men as the primary caregivers. However, in South Asia, there appears to be much less redistribution of childcare – men rarely take up the responsibility for caregiving but rather, appear to need care when women migrate.

The implications of these (and other) empirical variations on the analysis of care have yet to be fully understood. For instance, much of the analysis of global care chains has focused on the experiences of international migrants. Yet, internal migration is very important in shaping care arrangements. Moreover, in many Southern countries, internal migration far outweighs international migration in influencing care regimes and may be independent of international migration. This pattern of internal migration is particularly notable in larger countries such as India and China.

Even internal migration may not be the primary source of care workers in highly differentiated societies. For countries such as India, migration is given little importance and international migration simply does not appear to be nationally significant. Rather internal variations within a place based on caste and class are far more important than differences between places. The global care chain analysis is, on the other hand, premised on a hierarchy of places wherein rural–urban difference are a lower order difference than internal–international ones. Enlarging the empirical scope beyond the countries on which the traditional care chain analysis was based leads us to suggest the need to interrogate the wider significance of differences within a place as opposed to those between places in understanding care.

Class, the lynchpin for understanding emigration in care chain analysis, needs to be contextualized and understood intersectionally with other axes of difference such as race, caste and ethnicity. The effect of climate change, so well-analysed in some of the migration and development literature, may also provide unexpected insights into the analysis of global care.

Finally, the 'South' is a geopolitical category with inheritances from colonialism, postcolonial economics and political affiliations. All these shape care regimes and migration patterns. For instance, the meaning and delivery of care in India is influenced by its histories in social work, in missionary activity and the role of the church in shaping philanthropy, care provision and the care of carers. On the other hand, migration into India is overwhelmingly influenced by India's historic colonial boundaries, its role in the war of independence in Bangladesh and the nature of the boundaries that therefore exist between the two countries. Together this has meant that 3.74 million of the 6.16 million migrants into India are from Bangladesh and significant proportions of women from this group are employed as domestic workers.

Theoretically the lines of connection between the North and the South seem to be trumped by those across the South. However, a historical analysis shows that these South–South relations were themselves conditioned by colonialism, i.e., previous rounds of North–South relations. Thus, it is not that North–South relations are not important but that the influence of the North in migration may be different from that which is usually analysed. In terms of scale the contemporary influence of those relations (impact of colonial borders, migration systems as in Africa and of decolonization) may indeed be even more important than the circuits and new connections theorized in global care chain analysis.

Thus the North–South relations analysed in care chain analysis may not be the most significant form of migration in the South. Indeed migration may itself not be that significant.

Dynamic migrations in the South: middle-income migration poles

Most theorizations of care chain draw on structuralist accounts of global inequalities and implicitly on the core-periphery theory offered by dependency theorists. However, in a critique of dependency theory Wallerstein (1979) offered an analysis of world systems which gave a more dynamic and contradictory role to the semi-periphery or the middle income countries. The shift to a discourse of Global North and Global South has tended to occlude the more complex stratifying effects of contemporary globalization. For instance, the differences among the Northern countries and the many countries in the middle position, e.g., the European Union accession countries and certain countries in Asia and Latin America, get short-shrift in care chain analysis. Instead the hierarchy of places implied in the global care chain analysis has had a polarizing effect on thinking about care. In this section, we therefore briefly examine the position of middle-income countries as migration poles in which care in the domestic sector relies to varying degrees on female migrant labour.

Migrant care worker flows are significant in some middle-income countries. Examples of movement from low- to middle-income Southern countries within a region include those to Argentina, Chile, Costa Rica, Jordan, Malaysia, the Russian Federation, Venezuela and parts of Eastern Europe. The analysis of variations in experiences of migration, gender and care in these middle-income countries has been limited but has much to tell us. Compared to the low-income countries, their labour markets are more formalized, there is an increasing – although uneven – level of social protection, and care-related measures have been implemented in an attempt to reduce class and regional inequalities. And where democratic systems have emerged, as in Argentina and South Africa, new immigration policies have been introduced. In both these instances, the effect has been to expand and entrench regional systems of migration. And to varying degrees care work is undertaken by a mixture of migrants and non-migrants. So too does the emigration, often of skilled workers, raise issues of care for those left behind, and in particular of older parents.

As previously mentioned, the sources of migration, including female migration, have in the twentieth century been multiple. Internal migration was a significant factor in large countries and was the primary source of migrants in earlier times. Moreover, large numbers were involved in care work in some form. However, these proportions may well be changing with economic growth as there appears to be a negative relationship between the proportion of own country nationals working as domestic workers and the country's economic performance. This also resulted in an intellectual legacy whereby appropriation of value and of effect of caring relations were primarily analysed as an effect of rural–urban difference. Over time the regional dimension has also become important.

Middle-income countries have become placed as migration poles in these circuits raising interesting questions about regional – in contrast to global – care chains. The shortening of the chain to a region may influence care arrangements, making it possible for women to bring their dependants with them and reconstitute families in destination countries. Latin American research suggests that many women have taken their children with them so that only 7.6 per cent of Bolivians and 21.7 per cent of Paraguayans have left their children behind. Hence, both the spatial circuits of care chains and the role of care in the lifecycle can be very different than in the classic description of chains of care based on the Filipino experience. Proximity may only be one of the factors in the reconstitution of families in the countries of immigration. Immigration policies may also be significant in enabling family members to join migrants.

What is also interesting is that, in Latin America through the 1980s, young women were coming to the cities to do domestic work but once they got married they would move into small businesses. Hence, the sectors now encompassed within care were sites of sectoral mobility and often also provided a springboard for class mobility. To explore this mobility we need to trace the movements of individual workers into and out of other occupational sectors or, indeed, into and out of countries.

Finally, the emerging economies marked by rapid growth rates, such as China and India, also need a different form of analysis. Some of these countries are marked by a rapid rise in income inequalities and where this is accompanied by extant dependence on informal care the personal services sector, particularly occupations such as domestic work, can show sharp rises. This may spur migration into these countries where little existed before. The economic changes to individual countries and regions within a country can be a barometer for care demands.

The significant point here is that the countries that are low-or middle-income may actually alter in their economic status and this dynamic can fundamentally influence how they interact with global and regional chains of care.

Not all middle-income countries have the same issues relating to gender, migration and care. For instance, the South African case has a different set of gender, migration and care configuration much of it underwritten by apartheid and post-apartheid changes. During the apartheid period, race influenced, if not determined, who cared for whom. Moreover, many Zambian and Zulu men were employed in South Africa as male domestic workers but they were accommodated into the gendered care divisions by infantilizing them. Irrespective of age they were called 'boys', fossilized in a state of youth and thus accommodated within the dominant ideology. The incongruence between the ideology and the practice of sex-stereotyping of tasks was thus overcome.

The dismantling of apartheid has not overcome the legacy of extremely polarized class and racial divisions. It has also had lasting effects on household structures with a high proportion of fathers living apart from the family and women providing care. At the same time, with changes in its restrictive migration policies, migration has become more Africanized and feminized. However, the absence of entry routes for low skilled work in the 2002 New Immigration Act (amended in 2004), and based on skills and qualifications, privileges male occupations. The cross border pass for citizens of countries with borders with South Africa for tourism, education, business, medical treatment and visiting relatives does not permit work. This means that migrant domestic workers from neighbouring countries are largely undocumented. International migrant domestic workers are also in competition for domestic work with South African women (migrant and non-migrant) who still undertake it. It is not clear how these forms of migration relate (if at all) to the longer-distance migration of South Africans to Europe and the US.

Although the above section only provides a schematic account of some of the challenges that middle-income countries pose for thinking about care chains, what is clear is that the global is more varied than is currently accounted for in global care chain analysis. Middle-income countries may replicate the global chain, may short-circuit it or may interrupt longer-distance mobility. These are all challenges to dominant debates in gender and development today.

Who cares and for whom?

There are four different ways in which migration of people affects gendered care, although in practice a mix of all four may occur in any context. Care issues arise when:

1   people migrate as care providers;
2   people migrate and leave some care responsibilities behind;
3   people migrate and bring some care responsibilities with them; and
4   people migrate and have either daily or emergency care requirements, particularly as they get older.

Most current theorizations are concerned with the first two aspects, whereby migrants move as care providers and leave some care responsibilities behind. However, it is worth remembering that these issues of care have a longer history than the recent theorizations of care chains. In Africa, for instance, there is a long history of circular migratory strategies that involved mothers and grandmothers sharing childcare responsibilities in rural areas while taking turns to do domestic work in towns. Sharing arrangements have also existed between co-wives, with one wife staying in the town with the migrant husband while the other wife goes back to the village to look after the older generation and to oversee the harvest, for instance. Transnational family strategies have also been common in the Caribbean, which means that Caribbean families are resilient to shifts in care practices due to adult female migration. Nevertheless, what might be considered new is the geographical spread of these phenomena to different parts of the world – the extent to which these care relationships now stretch across international borders and the larger numbers of people involved in such global care regimes.

The third form of care redistribution and migration occurs when many migrants also bring some care responsibilities with them, especially if they are internal migrants, where they are international migrants moving as families (e.g., refugees or ecological distress migration) or where there are regional agreements or weak boundary controls over who migrates. Normative notions of family often mean that while migrants may be allowed to bring children with them they are rarely allowed to bring elderly relatives such as parents, except as visitors. Yet, older people too are left behind and need to be taken into account in the examination of responsibilities and arrangements for care resulting from migration. Even child migration faces restrictions: in some countries, such as the UAE, migrants have to earn a minimum income before they are allowed to sponsor children. Moreover, women may also face harsher rules around family reunification than men. Thus, in the UAE, women are not allowed to sponsor children unless they are employed as doctors, nurses or teachers.

Importantly, people who migrate also have care requirements themselves – either daily or emergency care requirements, particularly as they get older. Yet this aspect is far less considered than the others discussed above. In much South–South migration, such care is complicated by the fact that most migration is temporary or circulatory, albeit over long periods. We might ask whether migrants go back when they themselves require care, and what this means for the countries to which they return. Similarly there is the issue of children who migrate on their own as paid carers as in many parts of Africa and the provision of schools for such migrants that take account of their working lives. This is an issue that has long had purchase in the South, as migrant child workers have generally played a dominant role in delivering domestic work as well as care of the elderly. Hence, several children's NGOs, for instance, have spent time trying to meet the care needs of the children, who themselves act as paid carers in employing households.

In addition, countries that have become dependent on the emigration of domestic workers will in the future need to take account of the care needs presented by ageing care workers who are distanced from their own families. Thus, while trans-migrant practices may bind domestic workers to those whom they have to help, and to those who care for their children, the dissolution of transnational relations provides an even more frightening scenario. The welfare of the domestic workers themselves, especially in their old age, may not be assured through these relations, so that they are left without social protection in both destination and origin countries.

Conclusion

In this article we have examined the significance of gender, migration and care in the Global South and highlighted the different dimensions of care rearrangements that ensue as a result of gendered migrations. Most of the burgeoning literature on global chains of care has focussed on the transfer of physical and emotional labour from the South, normally low income countries, to households in the North. The theoretical framework and empirical studies have also primarily been based on selected countries of emigration and immigration.

We argue that the heterogeneity of the Global South warrants far more recognition and inclusion in discussions of global chains of care. The incorporation of migrant labour in such countries reflects the legacy of past migrations and recent socioeconomic change and political transition. Thus, these countries act as catalysts for a potential chain rather than its supplier.

These insights also lead us to ask how exactly the analytical model of chains of global care, which rests upon the idea of cascading chains, plays out in a diverse and dynamic South. At the moment we have little evidence of the consequences of care redistribution for care rearrangements of migrants moving within the South or to the North. There are several possibilities. Most of the focus is on the way in which migrants from a poorer area, often assumed to be rural, migrate to replace the care needs generated by the migrant. This would be an example of a connecting and cascading chain. However, as we have noted in this chapter, internal migration may operate quite separately to international migration. However, care requirements may also be filled by non-family labour that is remunerated but is sourced from the neighbourhood or elsewhere within an urban area. This would be an example of the marketization of care labour within the household. Besides, caring labour may be performed by family members, especially 'other mothering', and remunerated to varying degrees, whether through the equivalent of a salary or the payment of expenses; for example, the school fees of their own children. The first and second variant might be combined or change over time. Neither, however, represents a direct and linking connection in a migratory chain. Furthermore much of this literature focuses on the children who are left behind; very little pays attention to older people left behind and the care arrangements resulting from the departure of children. From some countries, such as India and China, the children who migrate are not only skilled but also work in skilled sectors. This too may contribute to the marketization of care in the home country, especially in instances where all the children have left. This set of related issues would benefit from a great deal more empirical research in a diversity of places with different institutional configurations of care.

References

Asis, M. (2006) 'Living with migration: experiences of left-behind children in the Philippines', *Asian Population Studies* 2(1): 45–67.

Hochschild, A. (2000) 'Global care chains and emotional surplus value', in W. Hutton and A. Giddens (eds.), *On the Edge: Living with Global Capitalism*, London: Jonathan Cape.

Isaksen, K, Hochschild, A. and Umadevi, S. (2008) 'Global care crisis: a problem of capital, care chain, or commons?', *American Behavioral Scientist* 52: 405–425.

Kofman, E. and Raghuram, P. (2009) *The Implications of Migration for Gender and Care Regimes in the South*, UNRISD, www.unrisd.org/unrisd/website/document.nsf/%28httpPublications%29/9C17B4815B7656B0C125761C002E9283?OpenDocument.

Parreñas, R. (2001) *Servants of Globalization: Women, Migration and Domestic Work*, Palo Alto: Stanford University Press.

Ratha, D. and Shaw, W. (2007) *South–South Migration and Remittances*, Working Paper No. 102, World Bank, Washington DC.

Razavi, S. (2007) 'The return to social policy and the persistent neglect of unpaid care', *Development and Change* 38(3): 377–400.

Sassen, S. (2000) 'Women's burden: counter-geographies of globalization and the feminization of survival', *Journal of International Affairs* 53(2): 503–524.

Staab, S. (2004) *In Search of Work: International Migration of Women in Latin America and the Caribbean: Selected Bibliography*, Santiago de Chile: Women and Development Unit, ECLAC.

Wallerstein, I. (1979) *The Capitalist World-Economy*, Cambridge: Cambridge University Press.

33

GENDER, POST-TRAFFICKING AND CITIZENSHIP IN NEPAL

Janet G. Townsend, Nina Laurie, Meena Poudel and Diane Richardson

Introduction

The United Nations High Commission for Refugees (UNHCR 2013: 1) estimates that 'Statelessness is a massive problem that affects an estimated 12 million people worldwide'. In many parts of the world many people, women in particular, are still non-citizens, so that basic legal, political and/or welfare rights have not yet been attained. Exclusionary policies lie at the root of many statelessness situations. We argue here that citizenship has widely been constructed in a male image, because for women access to citizenship has historically been linked to motherhood in the context of heterosexual marriage. A new literature linking the discourses of citizenship, development and sexualities has appeared, but such debates must be situated within particular socioeconomic and geopolitical contexts: case studies are needed.

Our case study is with a number of women in Nepal: 'returnees' or 'survivors', women who have been trafficked, usually for sexual purposes, to India or beyond and have managed to return to a difficult situation. The Nepal census of 2011 reported a population of 26,849,041 (Central Bureau of Statistics 2012). Nepal still has problems of statelessness, particularly regarding citizenship. Even though it achieved in 2007 the largest reduction of statelessness the world had seen, issuing nationality to 2.6 million people after the end of the civil war (1996–2006), Nepal may still host some 800,000 people whose nationality is not confirmed and who cannot, without a citizenship certificate, access important government services or secure their human rights (UNHCR 2013). From May to July 2013, all political parties advocated the granting of citizenship along with voter registration for elections in November for a new Constituent Assembly (to write a new Constitution for Nepal and also to act as government). This included a massive campaign to encourage younger people.

In this chapter, we shall introduce our project and describe our case study: the problems of stigma for returnee trafficked women as a hindrance to livelihoods and citizenship, and the complexities surrounding livelihoods and professionalisation. We shall not deal with trafficking itself, rather our research examines livelihood opportunities for trafficked women on return to their home country. This was a collaborative project between Shakti Samuha – a Nepali non-government organisation (NGO) working with trafficked women, Newcastle University, UK,[1] and the International Organization for Migration in Nepal.

We shall see that many women who return to Nepal from trafficking, to India, the Middle East or Southeast Asia are rejected by their families and their communities. If they do not already have citizenship, this may mean that they are stateless since recommendation by a male relative has traditionally been necessary. Again, we shall see that those who lack citizenship may be excluded not only from education, health, legal assistance and training but from many jobs. In practice (Richardson et al. unpublished), this is still likely to be the case even though 'mother *and* father' (but not '*or*', so that mother alone cannot recommend) are technically now named in law for the recommendation. Returnee women therefore have a desperate need for livelihoods, but very few are available without the support of their families. Unfortunately, the emphasis of Western donors in anti-trafficking is on the roles of the police and the courts, so the vital long-term goal of lobbying for systematic and structural changes in society gets overlooked. We shall not develop this further here. We shall conclude that comprehensive change *from within* Nepali society is needed not just to reduce trafficking but to support the rights of returnees (compare Bennett 2008: 218). We shall argue above all that citizenship and livelihoods are fundamental to these rights.

There has been little research on post-trafficking livelihoods, or on their place in development. Our goals were to secure knowledge well-grounded in the lives of women returnees, and to emphasise the voices of these women. This agenda follows our partnership with Shakti Samuha, which is now one of the leading anti-trafficking organisations in Nepal and the only NGO in South Asia, and possibly the world, founded and run by returnee trafficked women (GAATW 2007).[2] Shakti Samuha are deeply proud of their foundation story as we see here. In 1996,

> 148 Nepalese girls and women were rescued from slavery in Indian brothels during widespread police raids . . . These women were then locked away in remand homes in India, where conditions were as bad as – if not worse – than prison. The Nepalese government was reluctant to bring the women back to Nepal, claiming they would bring HIV into the country with them.[3]

Shakti Samuha also reports that NGOs in Nepal took the lead in returning and rehabilitating the girls. Sadly, even in these rehabilitation centres, the women's treatment did not help to restore their self-esteem and basic human rights. Only after months had passed and they had training in their rights did they realise they were not to blame for being trafficked. The women felt it was time to claim their rights, so they set up Shakti Samuha.

Shakti Samuha has built up counselling, support and training, grounded in its founders' personal experiences. They feel that their approaches are more appropriate in design and delivery than those of other NGOs. Having been trafficked themselves, they claim to know better what is needed. They work to challenge discourses of victimisation, seeking rights of citizenship and to chosen livelihoods for trafficked women. To a degree, by organising themselves and other trafficked women, they have changed many Nepali trafficked women's understanding of themselves and their society (Richardson et al. 2009). We need to remember, 'a woman's life is richer than her trafficking experience' (Rapier-Moore 2012: 231).

We explore through returnee women's livelihood strategies the intersectionalities of sexuality, gender, citizenship and development. Nepal elected a Constituent Assembly in 2007 to rewrite the national Constitution, which includes citizenship as a fundamental right. Our project's collaborative partnership and the emerging findings have been used to lobby the Constituent Assembly process, organising workshops and meetings to discuss the challenges facing returnee women without citizenship. The Assembly was disbanded by the Supreme Court in May 2012, leaving the revised Constitution incomplete and the future of citizenship uncertain pending fresh elections (eventually 19 November 2013).

Case study

This case study is based on analysis of interviews given generously by returnee trafficked women and by stakeholders in anti-trafficking organisations. To reach the experiences of the trafficked women themselves and respect the sensitivity of the issues, our research methods were qualitative. We conducted 46 interviews with returnee trafficked women from urban and semi-rural locations identified in the past by government for high occurrences of trafficking. Overall, we selected women with differing degrees of engagement with NGOs and social movements and who self-identified as returnee trafficked women. They range from 17 to 44 years of age, from eight caste/ethnic social groups and four religious backgrounds. They represent different local home regions, returnee routes and timeframes of return; ages of being trafficked/returned; ethnicities and castes; length/number of trafficked journeys; and access to citizenship.

Nepal's census of 2011 enumerated 125 ethnic/caste groups and 123 languages. Of the Hindu population, Brahmins and Chhetris were in 2001 officially clustered as upper castes and were widely distributed across the country. Their language is Nepali, and in 2001 they were 39 per cent of Nepal's population. Dalits, also Hindu (once 'untouchables' and still suffering discrimination) are officially clustered as lower castes, and live in the hills/mountains and the terai/lowlands. In 2001 they were around 40 per cent of the population. Ethnic/indigenous groups in our sample are mainly Buddhist Tibeto-Burman. They include the indigenous groups of: 1) the Magar, along the southern edge of the Himalaya, following Buddhism/Hinduism; 2) the Tharu, indigenous to the Terai, the southern foothills of the Himalayas across into India, almost all Hindus; 3) the Tamang, in the entire mountainous region, mainly Tibetan Buddhists; 4) the Sherpas, also Tibetan Buddhists, in the most mountainous area; 5), the Rai, in the eastern foothills, Hindu/Buddhist. The actual pattern is extremely complex, and in many areas religions may shade into each other; households may even call on different priests for different rituals.

We started recruiting and selecting subjects from NGOs and then used snowballing techniques to reach women without current, direct NGO contact. In our first analysis of an initial set of 37 interviews, professionalisation (see below) emerged as a theme, so a further nine were conducted specifically with returnee trafficked women involved as activists/professionals in anti-trafficking. We held a further 14 stakeholder interviews with other activists, key personnel in NGOs, donors and government. Through these interviews we explored discourses and emerging policies on trafficking and citizenship in Nepal and internationally, and tracked the evolution of debates in the Constituent Assembly. The interviews examined returnee women's own experiences and assessments of the strategies that enable women to exit (or not) from trafficking and prostitution, including passing as 'migrant workers'. They gave us their evaluation of the significance of a range of factors including local contacts, social capital and skills-training. Interviews were taped and transcribed in Nepali (or in a few cases ethnic dialects), and then translated into English. Interview transcripts use the idiom of the original translation as we wish to recognise that Nepali English is one of many forms of global English that is spoken. All names are pseudonyms.

Kathmandu, the capital, is where government policies are formed and decisions taken; it is the hub for domestic and international transport; it is here that the majority of returnee women settle. The three locations selected outside Kathmandu are all extremely poor. Meena Poudel, the Nepali member of the team, conducted the 37 initial interviews with returnee trafficked women, given the sensitive nature of the discussion and her fluency in most Nepali languages and dialects. Other members of the team then took part in interviews with the returnee trafficked activists/professionals, and the stakeholders (donors, civil servants and anti-trafficking NGOs other than Shakti Samuha). A leading theme of the interviews was stigma (see Poudel 2011) where the attachment of even a suspicion to a woman of having been trafficked is a serious constraint, which may lead to social isolation, loss of confidence, depression and deprivation. Materially, the stigma may painfully limit possible livelihoods, our second theme, through her being rejected for work as impure. Our third theme is the professionalisation of work, which might provide opportunities, but where social exclusion and limited opportunities can be problems. Our fourth is citizenship, where government officials may reject a woman suspected of having been trafficked.

Stigma

In Nepal, *izzat*, the honour of a man or a family, is in the hands of its women; 'women's sexuality . . . can compromise the pedigree of an entire household or lineage' (Rankin 2004: 148). Nepal was the world's last constitutionally declared Hindu state but after the movement for democracy the Nepali Parliament amended the constitution to make Nepal a secular state. Questions of honour still surround trafficked women on their return, across the very diverse Nepali cultures. The 2011 census reports the Nepalese population as 81.3 per cent Hindu, 9.0 per cent Buddhist, 4.4 per cent Muslim, 3.0 per cent Kirant/Yumaist (an indigenous religion) and 1.4 per cent Christian.[4] In the (Hindu) Newar culture[5] a household's honour accrues not only by maintaining social investments but by managing the sexuality and ritual purity of its women (Rankin 2004). In a Hindu upper-caste settlement studied by Lynn Bennett (1983) for ten years, the patrilineal system stressed the sacredness of daughters and other blood-related women, and ritual purity dominated women's lives, from confinement to an outdoor shed during menstruation to being responsible for most of the ritual activities of the household. Most Tamang, on the other hand, are Buddhist. Nepali views on purity vary greatly between ethnic groups and religions, but we found that Hindu norms applied widely to *trafficked* women, so that many Hindu, Tamang or Christian post-trafficked women suffer deeply from stigma. In most places, a woman who is trafficked, however unwillingly, is labelled as a prostitute (and in some cases as an 'AIDS carrier'; in our interviews, nine women said they came back from trafficking HIV positive). It is as if many post-trafficked women are rejected by their families, their communities and their government. We discuss their occupations in Nepal under 'Livelihoods?' below. We sought to avoid their revisiting the trauma of their varied experiences abroad, partly because ESRC would have required us to have had specialised help on hand if questions caused distress.

One woman said that people think trafficked women are bad, *chhada* (out of control), and may 'spoil' other women. Another woman, Nisha, was asked whether she could work and feed herself since returning from trafficking, but said no, there were problems: 'For example, the boss at work place trying to misconduct with us; not providing us salary . . . trying to do jabarjasti [force to have sex] and balatkar [rape].'

Stigma surrounds trafficked women with sources of pain. They are clearly victimised in many different ways. For some, even though society may remain difficult, a shift in their attitude helps.

Dimi: I see myself as a trafficked woman but sometimes I forget the fact . . . Five years ago I saw no way in front of me. I was in a condition of dying, felt like already a dead person. For me, life was over; life was finished . . . Now I think trafficking is not a big issue, just your attitude, just your thinking. I am a human being more than a trafficked woman. Now I feel I am something and I can do something in life.

Dimi put this achievement down to Shakti Samuha's kindness, counselling and training. Most of our interviewees valued the Shakti Samuha representatives very highly for their warmth and effectiveness, first in refuges, later in counselling, training and finding employment.

One escape from stigma may be marriage, which may also lead to citizenship and possibly livelihood (Richardson et al. 2009). Many women interviewed, however, live with situations of extreme abuse from their husbands, which raises questions about how sustainable marriage is in the long term as a way of coping with stigma.

Ramila: My husband is aware of it [her trafficking history]. He because of this knowledge also gives me much torture [emotional in tone]. He knows it. I told him prior to marriage.

Livelihoods?

Returnee trafficked women are one of the most stigmatised, marginalised, excluded and vulnerable groups in Nepal, yet rights to sustainable livelihoods have become a rallying point for lobbying activities. Individual women look to the labour market as some escape from stigma and poverty. NGOs play an important role in skills training. Many of these skills are traditional jobs for women, such as cooking, carpet making, sewing or tailoring. From our interviews, however, these jobs seldom generate enough income to provide sustainable livelihoods. One problem is that many of the women have had little education.

Many were fully occupied in childhood. At the age of eight, Usha provided casual labour for other farmers, carrying water, cutting grass for feed, cleaning up the cow dung, taking the goats out for grazing. Other poor women have similar backgrounds, but exclusion after trafficking adds greater difficulties. As Usha said: 'Back then, it was not common to send the daughters to school. It was only for boys then . . . it was out of question with no place to stay and no food to eat . . . how can we think of education?"

Poverty is a key factor in the lives of post-trafficked women. For commentators such as Lynn Bennett (2008: 218), poverty reduction will not be achieved without culture change, which is very slow and ebbs and flows. In 1983, Bennett reported on her anthropological work in the 1970s, the goal of which was to interpret the Hindu perception of women in one village. This is now a classic study and gives her a very valuable perspective. For Bennett now, bedrock Hindu values continue to structure the worldview of most Nepalis, of all religions. She calls instead for more emphasis on Hindu egalitarian strands, for human rights to be reflected in law and practice and for corruption to become unacceptable. She emphasises that specific, desired cultural changes cannot be induced by anyone and certainly not by outsiders (Bennett 2008), thereby highlighting further the work of national activists and NGOs such as Shakti Samuha who both support individual women and provide an important lobbying role in the country. For Shakti Samuha, livelihoods and citizenship are central to solutions.

Today, some NGOs provide start-up loans for women launching their own businesses, including 'non-traditional' occupations such as driving 'tempos' (moto-rickshaws) or working as plumbers or electricians. They also provide basic business training. It appears that women can earn more in jobs such as driving or working as security guards, but some non-traditional occupations are better options than others. These are typically male occupations and may present workplace prejudice for women. A plumber or electrician needs to go into people's homes to work, and to build up a client base. For a returnee woman, this threatens challenges for confidence and personal safety, and for outsiders, Nepali or foreign, it is not easy to know where danger lies. For example, while a traditionally female workplace, carpet factories have historically been a source of recruiting young girls into trafficking (Samarasinghe 2008).

Similarly, computing, a 'modern' job in which skills training is being provided by NGOs, seems alarming to some, as Jyoti explains:

> *Jyoti*: We also learnt computer for four month in Pokhara Vocational Training Centre where I had received training [on the advice of Shakti Samuha] but computer I could not practice.
>
> *Question*: Why?
>
> *Jyoti*: Because computer I think was something, easy skill for men because if you go to computer places, you see many men practising and few women, but if you see beautician you see many women and very few men. Also what I feel is even if I do computer work with a group of men/boys I fear that I will be unsafe.

Personal safety is indeed an issue. Returnee women feel seen as more available for sex than others:

> *Nirmala*: I had to struggle a lot to get work; I used go to the offices [including Coca-Cola] directly to look for a job. They used to make direct proposals that if I stayed there for the night, he would get me a job. I thought instead of making money that way, I would rather return back to the village. I did not pay attention to their advances.

Other poor women also suffer from sexual harassment in employment, but returned trafficked women seem to feel rendered more vulnerable by their experiences.

Returned trafficked women refer to themselves as survivors, not victims. Preeta, a senior manager at a major international foundation with wide experience and understanding, has outspoken views on the strength of many returned trafficked women:

> We have to realize: we don't have to train people on mechanics, electronics or main-stream job. We have to depend on outside institutions to do that. We have to have links with mainstream technical institutions, colleges and vocational training centre. We should be able to push them to accept those girls. We need to move these young people or girls to start getting training in institutions which are absorb in the mainstream rather than giving skills inside the home or inside the shelter [refuge] ... *A woman who has been into a sex industry and who has worked as prostitute or has been trafficked has much skills stronger than any of us do because she survived the extreme circumstances and probably knows better* [emphasis added].

Professionalisation?

One answer for livelihoods is professionalisation. Some members of Shakti Samuha become staff in that NGO, with work from cooking or cleaning to counselling, awareness-raising or running hostels to serving on the executive board of Shakti. Kathmandu has a remarkable concentration of NGOs, including many anti-trafficking, which are mainly charitable, not developmental. Donor support (from bilateral, multilateral and charitable sources) has led to a growing demand for staff.

For members of Shakti Samuha, working for their NGO is a popular option, as Nisha shows.

> *Nisha*: I think to continue the work, what I am doing now. I think on working on awareness against trafficking, organizing, as much as possible involving teens, other communities wherever we are working, publicizing about trafficking, trafficking and violence, making this success and working in trafficking continuously . . . It is difficult for filling stomach. It is expensive in place like Kathmandu, after having baby there is so much of expenses involved with baby, have to pay for room rent, that is creating so much of difficulty . . . Whatever opportunities Shakti Samuha is providing for affected [trafficked] women, to members of the organizations and they are providing much facilities, support for income generation to them, job support for them, if they don't have any provision keep them in their hostel and manage food and clothes, if they want to go to home make a conducive environment for reintegration, it does all these, here they do counselling, there is so much, there are friends, and feel relaxed. Staying in hostel, providing trainings, jobs before they leave. I feel satisfied in this.

Nearly all the staff of Shakti Samuha are returnee trafficked women, including the executive, which is elected every two years. Some have gone through exacting training schemes. Some professional jobs requiring specialist training, such as finance, have been filled from outside the organisation, creating tensions. When we asked Shakti members if they would be willing to work for other NGOs, they replied that while other anti-trafficking NGOs also employ returnee women, they employ them as staff, while Shakti Samuha employs them as members with voices in the organisation.

Professionalisation is a vexed question, involving questions of power. Shakti Samuha has a substantial staff and a commitment to training and educating them, as far as possible, to their full potential. For Shakti Samuha, education to become a professional is a strategy for overturning discriminatory social hierarchies. They reject the neoliberalism of many of Kathmandu's NGOs, where hierarchies of knowledge are produced and discrimination and social exclusion reproduced. The members of Shakti Samuha argue that they value merit. As a test of the outcome, we can look at the caste/ethnicity of our returnees. Fourteen of 46 were upper caste, ten Dalit, eleven Tamang, four Magar, three Rai, three Tharu, one Sherpa. Of the executive board members in 2012, five were hill ethnic/indigenous, two Dalit, two Terai ethnic and one upper caste. There may be no reproduction of social discrimination. If only Shakti Samuha's tenets were more widely held, non-neoliberal professionalisation would be an option for many more returnee women.

As a part of our collaboration, we asked Shakti Samuha what other dimensions of support they would like to emerge. They asked for training in research methods so that they could do their own research; this was subsequently provided through a modular course delivered by Nepali specialist academics to Shakti's executive. Kamala is an executive member:

> *Kamala*: Research has been very important to understand the *jeevan ra jagaat* (life and the world), social world we live in. We have been involved in research work through consultancy and all but it is very important to take the research training once in life.

I have realized this importance of research training after I took it. I don't know whether I would be able to sustain my livelihoods being a researcher or not but taking this training is very important. Shakti has been doing research funded by other donors and recruiting researchers for us. But this is us doing research for ourselves and it is very important to analyze our social world from our perspectives.

The final stages of the training programme involved the design of a free-standing research project by Shakti Samuha, in progress at the time of writing, focused on the social impact of the legal process. Very few cases are actually filed in court against traffickers. What is the women's experience? Trafficking survivors are the researchers in this project and will share the feelings of survivors who have gone through legal cases. Ethics are a major concern and have been considered very thoroughly. It is thought that there are so few cases because of the fear and reality of harm to the survivors. Safety is the first consideration. This enquiry will follow the World Health Organization's Ethical and Safety Recommendations for Interviewing Trafficked Women, which highlight 'do no harm' strategies to avoid re-traumatizing trafficked women.

Shakti Samuha's leading goals are citizenship and livelihoods for its members and other women. Citizenship emerged as a core focus in our collaborative work.

Citizenship

Young men and women may apply for citizenship from the age of 16 (Richardson et al. unpublished). Both can acquire citizenship through recommendation, but in practice young women need fathers or other male relatives to recommend them. Many were trafficked too young to be 'citizens' then. On return, many male relatives refuse to recommend them for citizenship, feeling that these women have dishonoured the family. In our sample, just over two-thirds of the women did not have citizenship when they were trafficked. Of those who did have citizenship, almost all obtained it with the support of their fathers, husbands or male relatives, at least nine before trafficking and 14 after. Six of our sample did not have citizenship at the time of the interviews. There is little tradition of registering of births and the border with India is open, facts that present problems of evidence of birthplace. At present a Nepali woman's citizenship (if any) allows her to live, work, vote and spend money in Nepal, but does not allow her to pass those rights to her children. DFID and the World Bank (2006: 26) call for equal rights to citizenship for women and men in Nepal. Lynn Bennett (2008: 218, 222) discusses the proposed policies of these agencies from the same project.

Citizenship is highly valued by returnee women, and they are vocal on the subject. It is difficult to find a room to rent without it. This represents a 'class' divide in access to livelihoods. 'There is no work without citizenship' (Sabita and others). In practice, many women work without citizenship, but much of that work is unskilled, say in agriculture, cleaning or brickworks. Citizenship is likely to be needed to access government services, including health, education and training, obtaining a marriage certificate, registering a birth, the right to vote and the right to a passport.

> *Rupa*: He can say that my children are not his. In that situation marriage certificate is useful [citizenship is needed for marriage certificate or certificate of birth].

Citizenship is also fundamental in supporting women who are attempting to claim property rights. Uma, a returnee who is seeking such rights, said: 'If you don't have Nepali citizenship you don't have identity.' In Nepal, most inheritance systems emphasise patrilineal descent and patrilocal residence (less so among the Tamang). This effectively places restrictions (that do not apply to men) on a woman's rights to inheritance of her and her husband's property, or, if unmarried, from her natal family (Rankin 2004; Samarasinghe 2008). These laws on property rights are important for women seeking citizenship from male relatives who have a vested interest in them remaining non-citizens, not least where denying a woman proof of citizenship may be 'justified' as being for the collective good through maintaining the family honour (Richardson et al. unpublished). Citizenship has been a major field for advocacy for Shakti Samuha, involving its paid staff and members who volunteer.

Question: How does stigma attached to being trafficked shape access to citizenship?

Sabita: It affects all our lives . . . in my case, my family supported me so it [citizenship] happened, but in other friends' case, it is really difficult.

With Shakti Samuha, we organised two workshops with activists in Kathmandu. The first, in February 2011, brought together 80 participants including leading anti-trafficking NGOs, Nepali government representatives, donors and key members of the Constituent Assembly, including the head of the Fundamental Rights Committee, who used case study material from the workshop in press releases. Thirty trafficked women survivors also attended. Extensive media coverage followed this event, including interviews on Nepali TV stations (which are distributed to the diaspora internationally through the web) and radio station interviews, as well as print articles in the Nepali press. The second workshop, a highly successful research seminar, was held on 4 November 2011 to disseminate preliminary findings from the analysis of the interviews conducted with returnee trafficked women. This was opened by the Minister for Women, Children and Social Welfare and attracted more than 100 participants, including several members of the Assembly and senior policymakers. More media coverage followed. On 11 March 2012, we took part in a DFID UK Anti-Trafficking Stakeholder Meeting in London, along with representatives from six DFID pilot projects and civil servants from DFID and the UK Treasury.[6]

Nepali feminist, human rights and anti-trafficking activists challenged citizenship provision in the then-proposed new Constitution, arguing that citizenship for women and men should be granted based on birth in the country of Nepal, with the proviso that, until such changes occur, mothers or fathers should be able to confer citizenship by descent. Children born abroad to sexually exploited mothers but now living in Nepal should be granted citizenship based on state endorsement. Our findings indicated that this is a model of citizenship that many post-trafficked women thought should be adopted.

Even for women who do have formal citizenship, the patriarchal nature of Nepali society (Rankin 2004; Samarasinghe 2008; Richardson et al. unpublished) renders them unequal citizens in many respects. They also demand equal rights to inheritance, property and land, education and health care, non-discriminatory laws on travel and migration and much more legislation and action to eliminate violence against women. At least one woman in our sample had been raped and impregnated by police when she went to report her trafficking: very large changes are needed.

In 2011, Nepal's Supreme Court ruled in support of provision that mothers as well as fathers could confer citizenship on their children but in practice there would appear to be resistance to implementing this. With regard to naturalisation and the requirement for foreign men to live 15 years in Nepal to achieve it, it is suggested that this reflects geopolitical securitisation fears about the open border with India and the potential that changing citizenship laws could make Nepal vulnerable to Indian interests.

Elections

Elections for the government and Constituent Assembly that took place on 19 November 2013 attracted more than 70 per cent of the 12 million eligible voters to cast their votes. In 1990/1991, nine women were elected out of 205 seats; in 2006, 30 women out of 240; but 2013, ten out of 240. In 2008, the Maoists (now United Maoists), won the largest number of votes, but this time they failed to secure an outright majority. Pushpa Kamal Dahal, better known as Prachanda, their leader, claimed a 'conspiracy' and called for a recount, but it was refused. (He lost his Kathmandu seat, but won a seat in Siraha, southern Nepal.)[7] No party had a clear majority, so a government was not expected to be formed for several weeks.[8] Without the United Maoists, who had threatened to boycott the CA, the CA process and change agendas were endangered. Analysts said the moderate politicians who triumphed would struggle to form a stable government and write a new constitution.[9]

Conclusion

At the time of writing, any improvement in the access to citizenship of returnee trafficked women and of millions of other Nepali women without passports is delayed or ended. But the draft constitution was saved in the CA secretariat and will, we hope, be reopened after the recent CA election when a new Constituent Assembly is resumed. This research has convinced us of these women's need for citizenship. Most young women in Nepal in general, especially rural women, need far more assistance than they currently get from the government to be able to improve their opportunities in life and to avoid trafficking. Action is needed especially in access to citizenship, education, work, safe migration and a radical reduction of violence against women. Systematic, structural changes in caste, class and gender are necessary to defeat trafficking and all forms of violence against women: without them, the police and the courts will not succeed. The above applies to millions of Nepali women. The needs of returnee women, as we have seen, are greater, and more individual support is required as many lack the help of their families and communities. At the same time lobbying and advocacy on all these fronts are vital.

One confirmation of Shakti Samuha's success in caring, advocacy and campaigning is an international award. Shakti Samuha was selected for the internationally recognised Ramon Magsaysay Award 2013 by the Ramon Magsaysay Award Foundation, Manila, Philippines. Established in 1957, this award is described as Asia's highest honour and is widely regarded as the region's equivalent of the Nobel Prize. The formal conferment of this year's Ramon Magsaysay Awards was at the Cultural Center of the Philippines on 31 August 2013. The slogans of this foundation, set up by the Rockefeller Foundation, are 'Greatness of Spirit. Leadership. Asian Solutions.'

Acknowledgements

To Shakti Samuha and all our interviewees, from Shakti Samuha and stakeholders, as well as to our advisory group for its critical support and to the International Organization for Migration (Nepal).

Notes

1   Post Trafficking Livelihoods in Nepal: Women, Sexuality and Citizenship; funded by the UK Economic and Social Research Council (ESRC) from September 2009 to April 2012, grant reference: RES-062–23–1490, www.posttraffickingnepal.co.uk.
2   Global Alliance Against Traffic in Women, www.gaatw.org.
3   www.shaktisamuha.org.np
4   www.indexmundi.com/nepal/demographics_profile.html.
5   Newars occupy a complex position in the caste system (Rankin 2004: 79–80).
6   www.posttraffickingnepal.co.uk.
7   BBC News, 21 November 2013, www.bbc.co.uk/news/world-asia-25030107; *Mail Online*, India, 24 November 2013, http://newsr.in/n/India/74w4fcbsy/Prachanda-wins-by-thin-margin-in-Nepal. htm.
8   11 BBC News Asia, 28 November 2013, www.bbc.co.uk/news/world-asia-25135595.
9   12 *The Financial Times*, 5 December 2013, p. 6.

References

Bennett, L. (1983) *Dangerous Wives and Sacred Sisters: Social and Symbolic Roles of High-Caste Women in Nepal*, New York: Columbia University Press.
—— (2008) 'Policy reform and culture change: contesting gender, class and culture change in Nepal', in A. Ahmad Dani and A. De Haan (eds.), *Inclusive States: Social Policy and Structural Inequalities*, Washington, DC: World Bank, pp. 197–224.
Central Bureau of Statistics (2012) *Final Report of National Population and Housing Census 2011*, http://cbs.gov.np/wp-content/uploads/2012/11/National%20Report.pdf.
DFID and the World Bank (2006) *Unequal Citizens: Gender, Caste and Ethnic Exclusion in Nepal*, Nepal: DFID and the World Bank.
GAATW (2007) *Respect and Relevance: Supporting Self-Organising as a Strategy for Empowerment and Social Change*, Bangkok: Global Alliance Against Traffic in Women, www.gaatw.org/index.php?option= com_content&id = 666&Itemid = 73.
Poudel, M. (2011) *Dealing with Hidden Issues: Social Rejection Experienced by Trafficked Women in Nepal*, Saarbrucken, Germany: Lambert Academic Publishing.
Rankin, K.N. (2004) *The Cultural Politics of Markets: Economic Liberalization and Social Change in Nepal*, London: Pluto Press.
Rapier-Moore, R. (2012) 'Revisiting feminist participatory action research', in K. Kempadoo with J. Sanghera and B. Pattanaik (eds.), *Trafficking and Prostitution Reconsidered: New Perspectives on Migration, Sex Work and Human Rights*, second edition, London and Boulder, CO: Paradigm Publishers, pp. 231–248.
Richardson, D., Laurie, N., Poudel, M. and Townsend, J. (unpublished) 'Becoming citizens in the "New" Nepal'.
Richardson, D., Poudel, M. and Laurie, N. (2009) 'Sexual trafficking in Nepal: constructing citizenship and livelihoods', *Gender, Place and Culture* 16(3): 257–276.
Samarasinghe, V. (2008) *Female Sex Trafficking in Asia: The Resilience of Patriarchy in a Changing World*, London and New York: Routledge.
UNHCR (2013) *Stateless People*, www.unhcr.org/pages/49c3646c155.html.

34

FEMALE SEX TRAFFICKING

Gendered vulnerability

Vidyamali Samarasinghe

Introduction

Human trafficking generally involves a movement of people from one location to another, within a country or across borders.[1] While, the movement itself may be voluntary or involuntary, a fundamental definition of human trafficking is based on the premise that the trafficked person is often held captive by the perpetrators of trafficking and forced to work under exploitative conditions. At the core of the definition of trafficking is the recognition that "trafficking" involves several sets of actors and stakeholders, divided along the lines of supply and demand. While human trafficking is not a new phenomenon, it acquired a new urgency during the latter part of the twentieth century, partly due to the unprecedented cross-border movements triggered by forces of globalization. The stronger commitment by the global community to combat human trafficking, slavery-like practices, and transnational criminal activity has resulted in the enactment of the *UN Protocol to Prevent, Suppress and Punish Trafficking in Persons, Especially Women and Children*, which is supplemented by the *UN Convention and Interpretive Notes on the Trafficking Protocol* (taken together henceforth referred to as the UN Protocol).

Female sex trafficking flows demonstrate an intricate web of networks that connect most countries across the globe as source, transit points, and destinations. While statistics are notoriously unreliable of this clandestine activity, the International Labour Organization estimates that 21 million people are in forced labor, of which 4.5 million—accounting for 22 percent of the total—are victims of forced sexual exploitation primarily as a result of trafficking. Furthermore, 98 percent of forced sexually exploited victims are women and girls.[2] An estimated one-half of all cross-border female sex trafficking involves females from South and Southeast Asia, about one-quarter involves females from the newly independent states (NIS) of Eastern Europe. The rest comes from Latin America and Africa. It is reported that female sex trafficking rakes in an estimated income of nearly US$34 billion.[3] As Shelley (2010) demonstrates, about half of this comes from industrialized countries followed by Asia, NIS, Latin America, the Middle East, and North Africa (half of that from sub-Saharan). Facilitated by the internet, the location, marketing, and purchasing of women and girls for the purpose of sex trafficking is fast, convenient, global, and—for the customer—anonymous as well. While it prospers as a lucrative industry, female sex trafficking is regarded as a deviant, illegal enterprise, which should be eliminated altogether.

Focus

It is the contention of the study that the understanding of female sex trafficking requires a gendered analysis of both the global issues and country-specific sociocultural and economic factors that produce the enabling factors leading to female sex trafficking. The chapter is divided into four mutually complementary sections. In the first section, I briefly outline the main features of global female sex trafficking protocols with an overview of the significant contentious issues arising from the UN protocols. In the second section, I examine the gendered impact of globalization and how the processes of "feminization" of migration and work have created enabling conditions for trafficking of women and girls to the commercial sex industry. In the third section, using three case studies from Asia I analyze how place-specific socioeconomic and political differences produce different forms and structures of female sex trafficking. In the final section, I briefly review the impact of opposing economic, ideological, and political interests of the numerous stakeholders and actors in initiating globally implementable anti-sex trafficking policy measures.

Female sex trafficking: protocols, debates, definitions, and flows

The commercial sex sector, into which women and children are trafficked, includes not only standalone brothels and female streetwalkers, but also sex-specific work in massage parlors, strip clubs, bars (including karaoke bars), escort services, as well as mail-order-bride purchase systems. The female sex trafficking victim, by definition, has not entered the trade on her own free will. She is held captive, forced to perform sex acts for clients, and is exploited by the pimps, traffickers, and brothel owners who claim the money from her clients while the victim remains mostly unremunerated. The fundamental premise of female sex trafficking incorporates all sexually specific activities, which are non-consensual. A female trafficked into the sex trade is at the bottom of the hierarchy in the sex industry. The clients, brothel owners, recruitment agencies, pimps, entertainment industry entrepreneurs, hotels, travel groups, bar owners, marriage bureaus, military establishments, and criminal groups who form the demand sector is overwhelmingly male.

For the purpose of this study I use the following definition of human trafficking as defined in the UN Protocol, Article 3, as

> Trafficking in person shall mean the recruitment, transportation, purchase, sale, transfer, harboring or receipt of a person by threat or use of violence, abduction, force, fraud, deception or other forms of coercion, abduction, of fraud, of deception, of the abuse of power or of a position of vulnerability or of the giving or receiving of payments or benefits to achieve consent of a person having control over another person for the purpose of exploitation.

The adoption of the UN Protocol introduced an internationally recognized definition, which was expected to be the basis for other international legal instruments dealing with the subject of human trafficking in general and sex trafficking in particular. Its Article 3(a–d) defines the concepts of "trafficking," the issues of "exploitation," "consent," and "child."[4] While it is important to have a clear definition of the concept of trafficking, including specifically associated processes that form the structure of trafficking, the UN protocol has been critiqued on many grounds (Scarpa 2008). For example, the human rights advocates are concerned that its emphasis on criminal prosecution is likely to detract initiatives aimed at protecting the human rights of the victim. In particular, in the case of cross-border sex trafficking flows, law enforcement seems to give more precedence to the illegal immigration issue over attempts to ascertain whether the female was actually trafficked into the sex industry. Consequently, a sex trafficked victim may be deported before she is granted access to the human rights protection embedded in the UN Protocol guidelines on anti-trafficking initiatives. Furthermore, the issue of how to define "exploitation" and "consent," in relation to prostitution, the basis of sharp disagreements between the abolitionists and pro-free choice sex worker groups[5] during the drafting of the UN Protocol, continues to be a contentious issue between these two highly visible and vocal groups.

Differentiation in power dynamics along gender lines becomes somewhat complex in analyzing prostitution,[6] within which female sex trafficking is located, since not everybody would agree that female prostitution in itself is tantamount to a gendered sexual exploitation of women and girls. It is argued that some of the women who engage in prostitution, who have entered the trade willingly, have exercised a "choice" and individual "agency," and they sell sex, not their bodies. In particular, while abolitionist groups insist that female prostitution in all its forms are exploitative and, therefore by extension, that prostitution is synonymous with female sex trafficking, pro-free choice sex work groups define trafficking of women into the sex industry as the exploitation of labor using coercion and force. While both groups agree that violence, coercion/force, and sexuality are basic premises of female sex trafficking they sharply diverge on the fundamental issue of the gendered analysis of prostitution.

Gender, class, and place in female sex trafficking

In female sex trafficking, the gendered dichotomy between dominance and vulnerability is clearly manifested in the male dominant power of the demand side as against the vulnerability of the trafficked victims on the supply side. The vulnerability of the trafficked victim arises from her gendered position, which is impacted by the historical and contemporary socioeconomic and political situations. Society determines the role of women and men in society and the less than equal position women hold in relation to men is best illustrated in the way how, in general, females who either voluntarily or involuntarily join the ranks of the supply side of the commercial sex industry are depicted as the "face" of the industry, and openly stigmatized as perpetrators of deviant, immoral behavior. For the most part, males who are customers and clients escape such social censorship.

Female sex trafficking is also dichotomized along class and spatial lines. The supply side is financially poorer compared to the better-off demand side. This line of separation is reflected in the relatively poorer girls and women who are lured into sex trafficking flows on the one side, and the better-off customers who buy their services on the other. This pattern is also reflected in the flow of trafficked victims from poorer societies, communities, and countries to service male customers in richer societies, communities, and countries. There is also a reverse flow of customers and traffickers from relatively wealthier countries moving to poorer source regions of supply. Analysis also illustrates that violent conflicts have provided fertile breeding ground for female sex trafficking. Furthermore, a new dimension has been added to the existing military deployment repertoire on female sex trafficking with the first UN-sponsored peacekeeping mission deployed in Cambodia in 1991. The soldiers in the multination peacekeeping missions are widely accused of increasing the demand for female prostitution and sex trafficking victims (Samarasinghe 2008; Whitworth 2004; Enloe 2000).

Female sex trafficking and feminization of migration and work

The current forces of globalization have triggered an increase in cross-border flows in female sex trafficking due to several mutually complementary factors. First is the "compression of time and space" conceptualized in the globalization discourse, where it is argued that geographical space is increasingly fragmented from place and connected to other social spaces across the globe, while simultaneously time horizons get shortened (Harvey 1990; Mittleman 2000). It has created more enabling conditions, including the expansion of the global reach of information technology, relatively unfettered flow of capital across borders, and ease of travel for people internationally. Second, for debt ridden poorer countries of the Global South, globalization has opened a new space for export of its labor to the Global North in an effort to earn foreign currency. Thus, international migration has become an iconic activity of globalization. Third, the globalization processes, both in terms of movement and work opportunities, have created new transnational constructions of gender that have greatly increased the attractiveness of women workers leading to a combined process of "feminization" of "migration" and "work." Women's work in the reproductive sphere, which had been hitherto largely uncounted and therefore invisible to the public eye, has been socially reconstructed to support the emergence of a gendered globalized workforce. In the current globalized economy, poorer women from the South have provided a key source of labor, especially for export-led manufacturing industries such as garments and electronics, and for domestic labor in the "international maid trade."

Together the processes of feminization of labor and migration have created, according to Saskia Sassen (2000), circuits of "counter-geographies" of globalization. As Sassen explains, global societies have witnessed a growing presence of women in a variety of cross-border circuits that have become a livelihood for people, profit-making ventures, and a source of foreign currency earnings for a number of countries in the Global South. While these circuits are diverse, they have one common feature: they are profit and revenue-making ventures developed "on the backs of the truly disadvantaged" (Sassen 2000: 503). While female sex trafficking clearly fulfills all the characteristics of "counter-geographies of globalization," it also displays the unfortunate dimensions of illegality of a set of activities that nurture an underground economy based on exploitation, stigma, and humiliation for the trafficked women and girls. Thus, a darker side of globalization is created when the very processes of globalization that enable legitimate work for poorer women from the Global South is turned into an exploitative, stigmatized exercise of trafficking women into the sex trade as well (Samarasinghe 2008). Female sex trafficking is perhaps the most significant unintended consequence of the forces unleashed by globalization.

A key component of globalization is the phenomenal upsurge in the efficiency of information technology. The internet has been highly effective in transcending borders to connect clients to traffickers and criminal groups. It has thousands of sites that advertise the availability of women, especially those from South, Southeast Asia and the newly independent states of Eastern Europe. The potential client can access the services of commercial sex while retaining his or her anonymity. The transfer of money takes only few minutes and easily transcends borders (Samarasinghe 2003).

Place-specific differences

The interaction of "place" and "space," a core theme in the study of human geography, is gender-specific and has produced different experiences for women. While the current forces of globalization have had a decisive impact on the general form and structure of both the supply side and demand side of female sex trafficking, country-specific sociocultural histories and economic conditions create diversity in the form and the nature of female sex trafficking experiences. The three case study examples, Nepal, Cambodia, and the Philippines (Samarasinghe 2008), illustrate that, while in all three countries gendered norms of patriarchy dominate women's lives, "place"—based on specific sociocultural, economic, historical, and political norms—creates different spaces in relation to female sex trafficking.

Nepal is designated as a source country of sex trafficking victims (USDOS 2012). It is one of the poorest countries in the world sharing a 1,500km porous land border with a larger, relatively better-off India, which also has designated red light districts in the urban slums of Mumbai (Kamathipura) and Kolkata (Sonagachchi). According to the current estimates of the Nepali Census Bureau, nearly 40 percent of its population is under the age of 15 years. Trafficking of girls from Nepal to India cannot be explained away by simply projecting the image of a young girl, clandestinely abducted and forcefully taken across the border to nurture the openly legal sex industry in the Indian cities of Mumbai and Kolkata. While abducting/kidnapping is part of the female sex trafficking pattern, many trek across the border willingly in search of waged work: some may even be aware that they will be recruited as prostitutes.

As clearly shown in Ruchira Gupta's award-winning documentary *The Sale of the Innocents*, while a Nepali parent was willing to sell his daughter to the filmmaker who posed as a recruiter of young girls as prostitutes to the brothel, it was also the young teenage daughter's shy, smiling willingness to accompany the trafficker to the unknown,[7] which together give a glimpse of the complex socioeconomic and cultural dimension of the processes, causes, and outcomes of place-specific dimensions of female sex trafficking. One of the fundamental issues that emerges with regard to female sex trafficking, especially in Nepal, is a disconnect between the international law and local society norms. There is an international agreement that child prostitution and child trafficking of any kind is a violation of the rights of a child, and such activities are designated as criminal.[8] Certain cultural values and traditional practices in Nepal, however, specifically in terms of arranged marriages of minor girls bring out significant issues that run counter to the legal definition and its associated conceptual meaning of child and childhood. The socially sanctioned sexual activity in marriage for young teenage girls, especially among poor rural communities in Nepal, also becomes an unintended license for traffickers to act as prospective bridegrooms to hoodwink unsuspecting parents and recruit their young daughters to work in brothels in India.

Globalization is generally associated with forces that are not "rural-friendly." Neoliberal market forces generally concentrate on locating economic and service activity in urban areas or focusing on large-scale market oriented cash crop cultivation. Neglect of rural areas impacts negatively on those who survive on small-scale rural subsistence agricultural pursuits. In fact in Nepal there is a notable presence of women in subsistence agriculture, exacerbating feminization of poverty in rural areas. This has triggered a migration flow of young girls and women to the Kathmandu valley in search of work, especially in homes as domestic workers, in rug factories, and to the commercial sex industry to India and the Gulf states.

Relatively free movement of people from Nepal to India was a pre-globalization phenomenon. However, the issue of female sex trafficking from Nepal to India burst into the public arena in 1996 when the government of Nepal refused to accept 200 young Nepali female victims of sex trafficking that the Indian government attempted to repatriate to Nepal. Most of them were also infected with the HIV/AIDs virus.[9] Widespread reports of abuse of trafficked girls and women in brothels in India prompted the government of Nepal to impose harsh controls on the migration of unaccompanied young women.[10] The continuing negative impact of the structural socioeconomic and cultural norms on women and girls in Nepal, the contiguous porous borders, and ease of travel combine to create conditions for a steady flow of trafficked females from Nepal, especially to the brothels in India. And, as noted in a report from Nepal, "traffickers always fish in the stream of migration."[11]

Is the structure seen in Nepal duplicated in Cambodia? While, as in the case of Nepal, Cambodia is deeply patriarchal, nearly three decades of violent conflict within its borders and the impact of globalization have created different spaces for female sex trafficking in Cambodia. It is a source, destination, and transit point for sex trafficking flows. Cambodia shares porous borders with Thailand, Laos, and Vietnam. While Cambodian women and girls are trafficked across the border to feed the brothels of Thailand, it is also a destination point for young girls and women moved along the Mekong River from Vietnam to the brothels in Cambodia along the Thai border. Cambodia is also known for internal trafficking, mainly from rural source regions to serve the brothels of Phnom Penh, Battambang, and Poipet by the Thai/Cambodian border, and Sihanoukeville, a seaside tourist resort.

Women's sexual vulnerability in times of war has a long recorded history. A new twist to militarization and sex work was added when, for the first time, the United Nations deployed a multinational peacekeeping force of about 17,000 soldiers to Cambodia under the auspices of the newly formed United Nations Transitional Authority to Cambodia (UNCTAC). As Enloe (2000: 99) notes, "along with landmine removal and voter education UNTAC brought an upward spiraling of prostitution. UN did not supply the pimps, but it did supply the customers." Many factors contributed to the enticement/coercion of young Cambodian women and girls into the sex industry to service the UNTAC soldiers. Cambodia was a fragile, war-weary country, left with a larger postwar female population due to the long brutal war that had taken the lives of a large number of males. Many women were forced to leave their villages (often landmined and dangerous) for the city to find ways to survive. The arrival of the UNTAC soldiers socialized to expect sexual services from local women created a space to coerce vulnerable Cambodian women and girls into the sex industry. Clearly, many adult Cambodian women willingly participated in the sex industry before and during the UN peacekeeping mission. However, taking note of the persistent cultural stigma against female prostitution in Cambodia, the increasing demand created by the UN peacekeepers, without a doubt, had to be met by trafficking as well. A display in the wax museum in the cultural village of Siem Reap, which depicts important events in the history and culture of Cambodia, prominently features the wax figures of a UN peacekeeper with a young Cambodian female prostitute, bearing testimony to the association of the UN peacekeeping mission with the sex industry in Cambodia. While the number of female prostitutes in Cambodia dropped from an estimated high of 25,000 at the height of the UN peacekeeping mission to around 17,000 when the soldiers left in 1994, it has been noted that economic liberalization that soon followed brought in its wake new incentives for a rejuvenation of the commercial sex industry, especially in Phnom Penh and Battambang.

The newly opened export-led manufacturing, especially the garment industry, attracted a large number of younger women migrating from the rural areas to Phnom Penh in search of employment. Not all of them could find employment in the new industries and the growth in female migrants to Phnom Penh has seen a concomitant rise in sex workers often trafficked into the industry. Alongside the growth of export-led manufacturing industries, foreign tourism—clearly identified as an important avenue for generating much needed foreign currency earnings—also triggered a steady flow of men seeking commercially available sexual services. Cambodia also gained notoriety as a supply base of young Cambodian and Vietnamese virgins for sex tourists. The demand for girls younger than 15 years of age and young virgin women represented approximately 50 percent of the total request of Western and Japanese customers

(Samarasinghe 2008). Internet sites, with their wide global reach in advertising in the sex trade in Cambodia, give a distinct boost to sex tourism and the associated risk of trafficking of girls and women into the trade to satisfy the growing demand.

The story of female sex trafficking in the Philippines clearly shares some of the fundamental characteristics of female vulnerability in patriarchal societies common to Nepali and Cambodian cultures. But it also illustrates differences resulting from its location and the diverse socioeconomic, political, and historic paths it has taken. Filipinas are better educated than the Nepali and Cambodian women. Unlike the cases of Nepal and Cambodia, the Philippines, as a group of islands, do not share porous land borders. Consequently, the movement of Filipinas into sex trafficking flows is focused on migration overseas.

Trafficked Filipinas cater to a multitude of demand situations in the commercial sex trade. Filipinas provided sexual services voluntarily and involuntarily to US male military personnel when they were stationed in Clark and Subic Bay American military bases. Female sex trafficking in the Philippines is also directly linked to voluntary overseas migration, with or without legal travel or employment documents, in particular to serve in the entertainment industry in Japan. This dimension of sex trafficking of Filipinas brings into sharp focus the issues of globalization. It showcases the issues of feminization of work and migration, which has provided spaces for overseas trafficking of females for sexual exploitation, mostly under the guise of legitimate overseas contract work. The Philippines has also become a major exporter of mail-order brides (MOBs), who are also known to run the risk of becoming victims of sexual exploitation. The Philippines continue to nurture a vibrant local female sex industry catering to the demand of local Filipino men as well as foreign male tourists.

Philippine society demonstrates sharp contradictions in terms of gender inequality. On the one hand, it has some women in the highest positions in government, business, and academia, and, on the other, women are plagued by lower salaries, stereotyped into appropriate women's jobs, subjected to unfair labor practices, and exposed to recruitment into certain activities that may be illegal and often stigmatized. Daughters in the Philippines are socialized to accept responsibility of catering to the needs of the family both financially and caregiving. Often, the impact of limited local economic opportunities, gender role socialization, and family dynamics, reinforced by pressure from parents to provide for the family may increase the risk for women becoming vulnerable to trafficking, especially into the commercial sex industry overseas. In response to the outcry that the Filipina entertainers and hostesses in Japan are forced into sexual activities in Japan, both the Japanese and Philippines governments have severely tightened the visa regulations for Filipinas seeking to enter Japan as entertainers. Rhacel Parreñhas (2011) insists that migrant entertainers are not trafficked persons, or individuals coerced to do hostess work, but instead they are labor migrants. While she does show that coercion, force, and indentured servitude are not unknown in the industry, she cites them as severe structural restraints on labor, not sex trafficking.

Strategizing for anti-sex trafficking policy: local and global contexts

The *UN Protocol on Trafficking* has set forth guidelines for anti-trafficking strategies based on "prevention," "protection" of victims, and "prosecution" of perpetrators. Individual member states are expected to formulate anti-trafficking policies using the guidelines, popularly known as the "three Ps." First, in reviewing the different scenarios of female sex trafficking in Nepal, Cambodia, and the Philippines, it becomes very clear that any initiatives to combat female sex trafficking has to take into serious consideration the place-specific socioeconomic and political issues that, in the first place, have created different spaces of female sex trafficking among individual countries. Second, in formulating cohesive, global anti-sex trafficking initiatives, it has to be recognized that female sex trafficking issues at the policy level are also simultaneously mired in the complexities of international migration issues in terms of the movement associated with trafficking. Third, the ideological debate on defining all prostitution as exploitation of female sexuality or identifying it as a form of work plays a decisive role in formulating country-specific policies to combat female sex trafficking.

Strategizing to stop female sex trafficking has to address the complexities that cross-border migration brings to the discourse. The overwhelming majority of persons caught in cross-border sex trafficking are young women and girls. They face a double jeopardy, i.e., being victims of sex trafficking as well as being (often) illegal immigrants. Women and girls who become victims of cross-border sex trafficking are often reduced to depersonalized bodies, with no voice at any point in the international migration process, i.e., source, transit, or destination countries. Migration policies and practices adopted by poorer source countries and richer destination countries affected by female sex trafficking often illustrate opposing goals that are driven by socioeconomic and political needs of respective countries.

Overseas employment fulfills important needs of poorer countries of the Global South. They create spaces to ease local unemployment problems and such labor exports have additionally become "cash cows" in the form of remittances. Remittances have become the human face of globalization, accounting for an estimated $610 billion by 2014, of which $467 billion will flow to developing countries of the Global South, and remittances outstrip development aid by a wide margin (Ratha et al. 2012). Consequently, poorer countries in the Global South will thus have little interest in controlling outward migration of women or men, be it legal or illegal. Furthermore, many of the female sex trafficked victims from poorer countries of the Global South are also likely to be undocumented. They often do not have the resources to track the flow of undocumented migrants who leave the country, when such countries have porous borders with relatively richer destination countries.

While the poorer source countries are, at best, fumbling with migration issues in relation to anti-female sex trafficking policies, in richer countries of the Global North, migration offences take center stage in trafficking related and law enforcement practices. The UN Protocol emphasizes the need to treat migrant trafficked persons as "victims" of a crime rather than criminals who have violated the immigration laws of the country. To institute "protection" measures in anti-trafficking policies, first the "victimhood" has to be established. This poses an immediate dilemma for law enforcement of the destination country. If she has entered the country as an illegal immigrant, the state has the legal right to take her into custody for violating immigration laws. Furthermore, in destination countries where prostitution is illegal, for a woman who is also undocumented, the immediate response of law enforcement is to charge her on both counts and often deport her as well, as reported in the case of the US (Bernstein 2010).

In an effort to facilitate the prosecution of traffickers and to ensure at least temporary protection for victims of sex trafficking, many destination countries—such as the US, the Netherlands, Belgium, and Italy—offer temporary visas to victims of trafficking, mainly in order to support prosecution cases against the traffickers. However the provision of such a visa, known as a "T" visa in the US, is inconsistent and fraught with challenges (Samarasinghe 2008). In general, despite the ratification of the UN Protocol and the ILO Convention on Migrant Workers by source, transit, and destination countries of sex trafficked women and girls, there is a serious lack of convergence of migration-related policies on sex trafficking between supply-based source countries and demand-based destination countries.

As noted earlier, the discussion on anti-sex trafficking initiatives has evoked a sharply divided debate between the abolitionist advocates of prostitution and free-choice sex work advocacy groups. The contrasting policy initiatives adopted among neighboring countries of Europe offer some insights on the impact of this debate and the limitations of country-specific laws in combating an activity that has a cross-border reach. Adopting a free-choice ideology on prostitution, Netherlands legalized prostitution in 2000. The Trafficking in Persons Bill adopted by the Netherlands made a distinction between forced and voluntary prostitution, with provision for legal penalties for those charged with trafficking. It also included provisions to protect the human rights of the free choice prostitutes plying their trade. Germany has a similar policy. At about the same time, Sweden, in contrast, adopted an abolitionist ideology on prostitution and introduced new legislation to criminalize the purchase of commercial sex, while decriminalizing prostitution itself. The model employs a law enforcement strategy where the goal is not to punish the female prostitute, who is explicitly identified as the more vulnerable partner exploited by the dominant male client. It is the client who is apprehended and charged. Finland has a similar policy. However, as Shelley's (2010) comprehensive review of human trafficking illustrates, neither the Netherlands nor Sweden has been able to make any significant inroads into combating sex trafficking. While clients from Sweden and Finland can go to a country where prostitution is legal such as the Netherlands or Germany, which is likely to lead to an increase in demand for sex workers in those countries, such a process in turn has the potential to boost sex trafficking flows. As Shelley notes, Europe remains a main magnet for female sex trafficking especially from the Balkans. In the case of Finland, Finnish customers reportedly travel overseas, especially to the "East," identified as Russia, the Baltic States, and the Far East, in search of commercial sex (Martilla 2003). While resources for law enforcement are also very limited, there is also little effort to reduce demand in Europe for trafficking of women and girls from poorer countries of Europe and also from Africa into the commercial sex trade (Shelley 2010).

Conclusions

The fundamental premise of female sex trafficking rests on gender subordination resulting in the vulnerability of females and the appropriation of female sexuality by the dominant males. Within the parameters set out by that premise, place and time create different manifestations of female sex trafficking. Each of the countries we briefly examined illustrates differential impacts of globalization and also produces a different mix of factors, resulting in significant differences in the structure and form of female sex trafficking. Female sex trafficking flows transcend state borders, which creates deeply embedded tensions between source and destination countries. Furthermore, the definition and analysis of female sex trafficking have created a passionate and contentious feminist debate on prostitution/sex work, which has injected a different set of complications for developing a harmonized, globally cohesive set of anti-female sex trafficking initiatives.

While it is well recognized that sex trafficking includes the sexual exploitation of boys and men, it is also widely acknowledged that female sex trafficking accounts for the largest single proportion of all segments of human trafficking. It invokes most outrage for different reasons. It tarnishes the image of source countries for not being able to "protect" their girls from male predators. It involves the sexual exploitation of females leading to a subversion of female dignity. The continuing social stigma associated with prostitution in general sharply highlights yet another form of gender-based discrimination inflicted on girls and women.

Notes

1 The US Department of State (USDOS) observes that "'trafficking in persons' and 'human trafficking' have been used as umbrella terms for the act of recruiting, harboring, transporting, providing or obtaining a person for compelled labor or commercial sex acts through the use of force, fraud or coercion." It adds that "human trafficking can include but does not require movement" (USDOS 2012: 33).
2 International Labour Organization, *Forced Labor Statistics Factsheet June 01, 2012*, www.ilo.org/global/-the-ilo-/newsroom/new/WCM.
3 2007 ILO estimates quoted in Shelley (2010: 7).
4 "Exploitation" as defined in the UN Protocol on Trafficking 2000, Article 3(a) "Shall include, at a minimum, the exploitation of prostitution of others or other forms of sexual exploitation, forced labor or services, slavery or practices similar to slavery, servitude or the removal of organs." According to Article 3(b) of the UN Protocol, "consent" of a victim of trafficking in persons to the intended exploitation set forth in paragraph 3(a) above. Article 3(d) of the UN Protocol notes that the "child" shall mean any person under the age of 18 years (see note 8 below). See also Samarasinghe (2008: 21–23).
5 Abolitionist advocacy group defines all forms of prostitution as exploitative. Pro-free choice sex worker advocacy groups use the term "sex work" in place of "prostitution" and argue that some sex workers are in the sex trade by choice and is a form of employment.
6 I use the term "prostitution" and not "sex work" because, in my fieldwork in Nepal, Cambodia, and the Philippines, it is the term used to identify those who are trafficked for the purpose of sexual exploitation.
7 According to footage of the video, which portrays an actual incident, the teenage girl who is accompanied by her father is told that she will be working in the commercial sex sector. She smilingly agrees.
8 ILO's work on combatting child trafficking has been reinforced by the adoption of the ILO Convention #182 on *Worst Forms of Child Labor*. Under the provisions of this convention any person under the age of 18 years of age is a child and a minor and cannot legally give "consent." The *UN Protocol on Trafficking* and the Trafficking Victims Protection ACT (TVPA) of October 2000, legislated by the US Congress, which is the basis for the *Annual Trafficking in Persons Report* published by the US Department of State uses the same definition for "child" and the legal validity of "consent."
9 The young trafficked victims were repatriated to Nepal due to the intervention of Nepali women's organizations.
10 The 1950 Open Border Agreement between India and Nepal allows citizens of India and Nepal to travel freely across the border without passports or visa. The government of Nepal amended Section 12 of the Foreign Employment Act in 1998 to prohibit foreign employment for girls and women without the express written permission from their guardians and the government. Confronted with harsh criticism from human rights groups and women activists, the age restriction for female travel was finally lifted in 2006.
11 Spotlight, "Girl Trafficking: Immoral Trade Feeds on Violence and Trafficking" Nepalnews.com, 22(24), December–January 2003.

References

Bernstein, E. (2010) "Militarized humanitarianism meets carceral feminism: the politics of sex rights, and freedom in anti-trafficking campaigns," *Signs* 36(1): 45–71

Enloe, C. (2000) *Maneuvers: The International Politics of Militarizing Women's Lives*, Berkeley CA: University of California Press

Harvey, D. (1990) *The Conditions of Modernity: An Inquiry into the Origins of Cultural Change*, Oxford: Blackwell.

Martilla, A. (2003) "Consuming sex: Finnish male clients and Russian Baltic prostitution," paper presented at Gender and Power in New Europe, European Feminist Conference, Lund University, Sweden, August 20–24.

Mittleman, J.H. (2000) *The Globalization Syndrome*, Princeton, NJ: Princeton University Press.

Parreñhas, R.S. (2011) *Illicit Flirtations: Labor, Migration and Sex Trafficking in Tokyo*, Stanford, CA: Stanford University Press.

Ratha, D., Mohopatra, S., and Silwal, A. (2012). *Migration and Remittances: Fact Book, 2011*, Washington, DC: World Bank, Migration and Remittances Unit.

Scarpa, S. (2008) *Trafficking in Human Beings: Modern Slavery*, Oxford: Oxford University Press.

Samarasinghe, V. (2008) *Female Sex Trafficking in Asia: The Resilience of Patriarchy in a Changing World*, New York: Routledge.

—— (2003) "Confronting globalization in anti-trafficking strategies in Asia," *Brown Journal of World Affairs* 10(1): 91–104

Shelley, L. (2010) *Human Trafficking: A Global Perspective*, Cambridge: Cambridge University Press.

Sassen, S. (2000) "Women's burden: counter-geographies of globalization and the feminization of survival," *Journal of International Affairs* 53(2): 503–524.

USDOS (2012) *Trafficking in Persons Report*, Washington, DC: US Department of State.

Whitworth, S. (2004) *Men, Militarism and the UN Peacekeeping: Gendered Analysis*, Boulder, CO: Lynne Reinner.

35

TOURISM AND CULTURAL LANDSCAPES OF GENDER IN DEVELOPING COUNTRIES

Margaret B. Swain

Introduction

Tourism, or leisure travel and the industry that serves it, is built on attractions to differences or similarities, and is profoundly gendered for both service providers and tourist consumers (Swain 2002). I focus here primarily on providers, acknowledging that hierarchies of power and synergies of meaning, or interactions of ideas, among tourists and other tourism stakeholders shape these gender landscapes. Tourism is frequently used as an engine for economic development policy in developing countries. As such tourism work has been evoked and critiqued as a harbinger of gender equity, when women's income earning is equated with empowerment (Ferguson 2011; Tucker and Boonabaana 2012). This chapter interrogates such equations, drawing from global assessments by development agencies, feminist theory in tourism research, and case studies. I will address the proposition that "gender issues and women's participation must be mainstreamed for sustainable tourism to be a reality."[1] Specifically, it is argued here that cultural landscapes of gender, the systems of ideas about women's and men's relative worth, identities, and roles on global and local scales must be understood and negotiated to find a way forward, to enable equitable and sustainable tourism.[2]

Referring to International Labour Organization (ILO) findings, Baum (2013: 12) notes that "notions of gender and gender equity lie at the heart of an understanding of sustainability within tourism, particularly in a developing context." Sustainable tourism here does not simply mean ecotourism or green practices, rather it refers to the social, economic, and environmental sustainability of tourism as a means for development. In the late twentieth century, tourism—often claimed as the world's largest industry—was labeled "smokeless" to imply that work conditions and environmental impacts were considerably better than factory work. Research has shown that this is far from the case, raising numerous questions about the sustainability of tourism as a development strategy and its role in promoting more gender equitable societies.

Cultural landscapes of gender—including divisions of labor and consumption, livelihood potentials, exploitation and advantage, sexuality, aesthetics, and environmental knowledge—shape the potential of tourism for development. Some of these issues facing women, men, and children were addressed in the 1999 United Nations Commission on Sustainable Development report on *Gender and Tourism: Women's Employment and Participation*, using case studies, although its analysis was hindered by a lack of global statistical data. The 2010 *Global Report on Women in Tourism*, produced by the United Nations World Tourism Organization (UNWTO) and UN Women, combines quantitative analysis of data from the ILO with qualitative case studies. It notes issues of low status jobs and gender stereotyping, calling for "decent work," free from exploitation. This study also highlights a critical lack of information on the informal sector where much of tourism work is located, from craft production to home hospitality.

Sex tourism and trafficking is mentioned, but not addressed in the 2010 UN report. Ideas about gender and sexuality articulate exploitative aspects of tourism. Many developing nation destinations cannot successfully combat tourism practices based in gender inequalities and violence against women without broader societal transformations as well. Shifts are needed towards women's political, social, and economic empowerment to create gender equity locally and globally, as reflected in the UN's Millennium Development Goal 3 (MDG3) to promote gender equality and empower women. Gendered policies for tourism development should promote poverty reduction, gender equality, and women's empowerment. There are significant tensions between these policy goals and the realities of the tourism industry based in global inequalities that provide a large supply of low-paid flexible female workers and entrepreneurs in seasonal, part-time activities such as hospitality, retail, and cleaning (Ferguson 2011: 237). These inequalities in tourism work, globally and regionally, are shaped by the intersectionality of varied axes of power and privilege including race, ethnicity, age, ability, nationality, as well as gender.

This chapter provides an overview of gender in tourism from global agency assessments, and then briefly addresses feminist theory in tourism research. Case studies are referenced to illustrate applications of these theories and various typologies of tourism that frame the analysis of these case studies. Marketing of types of tourism, the poverty/tourism nexus, and the tourist's gaze/motivations intersect to shape a wide range of tourism experiences. A typology of tourism in developing countries includes pro-poor approaches, ethnic performance, homestays, eco-tourism, adventure, mountain and trekking tourism, sex tourism, and volunteer and philanthropic tourism. Types of tourism work include providing food (growing, preparing, and serving), cleaning, making, and selling souvenirs, guiding, and offering entertainment or hospitality management services.

UN and NGO perspectives

The 1999 UN report on *Gender and Tourism* stressed an integrated approach to economic, environmental, and social aspects of sustainable development. The report called on all stakeholders—intergovernmental bodies, national and local government, NGOs, industry, trade unions, and community-based initiatives—to collaborate in tourism that safeguards the natural environment and cultural heritage while increasing social and economic justice. Although very ambitious in scope, the 1999 report's analysis focuses only on formal employment, with no data on the informal sector providing much of the income from tourism for women. For example, the data available for the ILO "restaurant, catering and hotel industry" sector, arguably the largest employers in the tourism industry, were used to represent all tourism work. Few nations disaggregated tourism work data by gender (the USA did not). When these data were missing, generic employment by gender information was used. These partial data show that women's to men's work hours, at 89 percent for wage labor only, were higher in proportion than women's to men's wages (79 percent) in an approximation of tourism work. Women thus work almost as long as men and earn less for their efforts, a wage gap found in most industries. Women comprised on average 46 percent of the global tourism workforce, with the Philippines highest (65 percent), and the lowest in Egypt (13 percent). Occupational segregation was pervasive, and global (male cook vs. female maids for example). A glass ceiling prevailed for administration and management. On average women held 5 percent of these positions and rarely exceeded 20 percent anywhere. Gender stereotypes dominated local descriptions of tourism work: women were seen as having a caring nature, domestic skills, manual dexterity, honesty, concern about physical appearance, willingness to take orders, being accepting of lower wages, and disinclined to supervise others.[3]

In the 1999 study, the statistical data were contrasted with 12 case studies by contributing authors of primarily community tourism development projects from around the globe. One case, located in Quintana Roo, Mexico (Momsen 1999), documents the efforts of an urban Catalan woman philanthropists' NGO to build an indigenous Mayan community tourism project aimed at providing multiple benefits to the village and region. Regrettably the voices of half the population, the women, were not included in the planning or implementation of an elaborate cultural performance for tourist visitors. Proceeds were shared only with the male heads of household. Women, however, were performers, as well as suppliers of souvenirs and food to tourists. While local gender norms upholding male control shaped project design, a subsequent assessment of the project utilizing gender segregated focus groups revealed distinctly different perspectives and priorities of women and men for planned project outcomes.[4]

The 2010 UN *Global Report on Women in Tourism* also provides longitudinal data, with significant limitations. Using ILO figures, the report attempts to understand the informal sector of employment through a data category of "Own Account" labor in the hotel and restaurant sector. Difference between figures given in 1999 and 2010 reflect both change over time and distinct datasets. For examples, an average of the global tourism workforce rose from 46 percent to 49 percent women, while for a specific country the percentage of women in tourism activity could appear to have dropped, as in Egypt going from 13 percent to 3.3 percent. However, different criteria were used for the data, changing from a general assessment of women's employment in 1999 to specific figures kept on the hotel and restaurant industry.

Case studies in the 2010 report feature local and international NGOs that promote women's participation and gender equity in tourism development. One case describing the 3 Sisters Adventure Trekking Company[5] and their affiliated NGO Empowering Women of Nepal (EWN) has been featured in various reports over the years, because of the organization's striking successes. Founded by three Nepali sisters in 1994, they have combined a successful woman-focused trekking business with training programs and paid apprenticeships for rural Nepali women. Stated goals include improving women's work skills and education in health, the environment, and history as well as English. From 1999 to 2012 EWN trained more than 1,800 women from 47 districts of Nepal. They provide regional community development in some of the poorest areas in Nepal through women's employment in a wide range of tourism-related work, anchored by the trekking industry.

In 2013, the International Labour Organization (ILO) published its own report entitled *International Perspectives on Women and Work in Hotels, Catering and Tourism* (HCT). Their definition of tourism in the HCT sector includes specific segments of transport, travel agencies, and tour operators, and notes that hotels, catering, and restaurants are considered by many organizations, governments, and NGOs to be part of tourism industries and may therefore be subsumed under tourism. The ILO narrows its definition to three clustered but separate parts of a sector, while acknowledging such ambiguities impact the quality of statistical data available (Baum 2013: 4–5). Despite these data issues, there is no question that tourism generates significant employment, with estimates by the UNWTO of 235 million jobs worldwide in

2013,[6] with a business volume that equals or even surpasses oil exports, food products, or automobiles, contributing 5 percent of direct global GDP.

In brief segments, the ILO study sketches out the potential of tourism employment for women in HCT cooperatives, a wide range of informal economy activities, and pro-poor development programs. Furthermore, primary data is offered from an online survey of major international hotel and tourism countries and a focus group with high-end transnational hotel management. This research found that all international hotel corporations promoted policies intended to support gender equality in employment, with a commendable record of compliance, particularly in developed countries. The application of such policies, however, may be compromised in meeting the expectations and demands of local owners. A gender-neutral approach to international recruitment, talent management, and succession planning within major hotel companies in practice can actually disadvantage women with child and parental care responsibilities. The study calls for a gender approach and concludes that "change in the utilization and career development of female workers in international HCT companies is an economic and moral imperative" (Baum 2013: 60).

Feminism and gendered approaches to tourism research

Cynthia Enloe's seminal book *Bananas, Beaches and Bases: Making Feminist Sense of International Politics* interrogated legacies of Western colonialism to understand the gender dynamics of travel and tourism. Euro-American "great explorers" and travel writing from the late nineteenth century set the scene for late twentieth-century feminist questions in tourism research. As well documented by other scholars, Western travel writing is full of masculine metaphors—such as "conquering the virgin land" or "penetrating the Dark Continent." Intrepid male adventurers produced travelogues while women were usually cast as staying at home, although a few female adventurers prospered, such as Mary Kingsley in Africa. Others—such as Marianne North in the Malay Archipelago and Anna Leonowens in Siam—had keen eyes for the intersections of local and imperial events in specific locations, and within themselves. While colonized by their own society's gender norms, they were often complicit in the colonization of indigenous peoples. These adventure narratives were supported by systems of human exploitation that marked off citizenship, race, and class—naming people "heathen" or "savage" and numerous other epithets. Such hierarchies are also gendered and sexualized through social and biological reproduction. Colonialism's legacies of impoverishment and political instability affect women, men, and children differently.

Enloe touches on these colonial adventurers to reveal the conflicting roles of women as consumers or as workers. These roles raise complex questions of agency, choice, and empowerment, based in unequal relationships. Gaps between women and men who have leisure time and money to travel and those who must travel to find wage labor or service travel industries, she argues, are all enacted under various gender codes and patriarchy. Enloe (2000: 41) concludes that without enforcement of ideas about masculinity and femininity in departure and destination societies, the tourism industry and its political agenda could not be sustained in its current form. From her perspective, "the very structure of tourism needs patriarchy to survive. Men's capacity to control women's sense of their security and self-worth has been central to the evolution of tourism politics." She suggests that feminist action promoting gender equity in tourism would upset contemporary world power dynamics. Although Enloe first made these arguments in the late 1980s, it was not until her book was reprinted in 2000 that we could say that her predictions were being embraced in Tourism Studies, with a critical turn to transformative scholarship intended to enact and support change for equality.

344

Early studies emphasized the "impacts" of tourism on local people, primarily in developing countries. *Hosts and Guests: The Anthropology of Tourism* (Smith 1977) illustrated the lack of focus on gender issues in Tourism Studies. In 1977, my chapter in this collection was the only one to address women's experiences; by the second edition in 1989 my analysis had developed into "Gender roles in indigenous tourism," while three more of the 15 chapters mention but were not focused on "sex roles" or "women." By the 1980s, terminology in feminist studies had generally shifted in focus from sex roles to gender relations. In some cases women's economic activities in tourism were seen to benefit traditional family structures, while in other locations sharp restrictions were found on women's involvement in tourism development, reflecting current gender relations. Early in the 1990s, gender analysis started to be applied in Tourism Studies to discern power differentials. In addition to gender roles and differential access to employment, research explored the interplay between local gender ideologies, tourists' gendered behavior and expectations, and local, national, and transnational cultural politics

By the late 1990s, Tourism Studies included a clearly focused feminist approach applying ideas such as webs of perspectives, the intersections of sex/gender/sexuality, sites of desire, hegemonic masculinities, and consensual patriarchy. One area of research is on "the body," which links all tourism related activity in one way or another in response to a dominant focus in the field on the tourist gaze, and gazing back. A theoretical task is to embody tourism research rather than continuing to construct conceptual frameworks that objectify individuals as subjects and insist on the detached gaze of the tourism researcher. If we think of "the body" as a cultural artifact, it becomes an interface for the dialectical relationship between biology and culture within the performance of tourism (Swain 2002: 4).

In the early twenty-first century, a re-energized focus on gender emanated from the critical turn shaking up Tourism Studies. Researchers were challenged to think within and outside of our own bodies, be they corporal and/or institutional, about the critical importance of gender equity in our daily world. For the past decade or so, global feminist scholarship has encouraged us to approach our work in terms of "intersectionality." This most cumbersome of words provides a reminder that we are complex beings of many identities, limitations, resources, influences, positions, and perceptions, studying equally complex situations located in multiple truths. The parsing of these truths as we build knowledge in Tourism Studies takes us back time and time again to the diverse facts that shape our motivations to challenge the injustices we find along the way.

One path that I have chosen is to look toward the potential hope of an embodied cosmopolitanism, promoting diversity and equity (Swain 2009). For example, the tourism producer or consumer as an embodied subject forms a complexity of identities, ideologies, life-stage preferences, practices, use and consumption of space and place. Tourism experiences for the tourist may be an escape from the everyday or a space for self-development/ empowerment. It is well documented that women with families rarely have a "holiday," even when tourists, although gendered parenting norms are shifting. Another notable gap is knowledge about the older woman tourist. Generally, older women have not attracted the theoretical attention that younger women have received. While past marginalization of women in the tourism literature could be attributed to sexism, the marginalization of older women in the gender and tourism literature could be considered a function of ageism within feminist analysis.

Among producers, the gendered division of labor in tourism involves differences in quality and types of work; differential access to employment opportunities; and seasonal fluctuation of employment. Thea Sinclair's (1997) edited volume applied feminist theories of gender and work to tourism-related employment for women that could be understood as the stereotypical limits of patriarchal capitalist systems, or as promoting opportunity and autonomy for women that would otherwise not be available. Tourism appears to have provided some women with wider opportunities for paid work and access to higher earning levels, in both formal and informal employment sectors. With income, women have enhanced their status and control over decision-making within the household, enabling a choice of alternative household relations, and providing legitimate access to public space. Such benefits have occurred, however, within the context of: 1) an increased workload for women, 2) a dominant ideology concerning the "normal" heterosexual household structure, 3) an unequal distribution of income within the household, 4) unequal control over expenditure by household members, and 5) an absence of power and control over decision-making at community and higher levels.

The distribution of benefits from tourism for development generally reflects existing inequalities. Scholars are investigating how gender norms and relations shape and are shaped by tourism processes over time, and inform issues of inequality and control in tourism, while relationships between tourism, development, gender, and poverty reduction are much less studied (Tucker and Boonabaana 2012: 437). Ferguson (2011: 239–240) argues that women's economic empowerment through income-generating activities in the tourism industry does not automatically correlate with broader political and social empowerment, in part because of a lack of attention to unpaid social reproduction labor by both the tourism industry and states. The trajectory of gender analysis in tourism is reflected in feminist theories of development over the past decades from a women and development (WID) focus to a gender and development (GAD) approach that reframed analysis in terms of the sociocultural construction of gender and equity in relations, not a universal category of "women." Much development work can be seen as stuck in WID, essentializing "gender" to be synonymous with "women." As Ferguson (2011: 241) warns, the muddling of feminist analysis in development policy and its commitments to gender equality and women's empowerment need to be critiqued: "We should not expect the gender content of tourism development policy to miraculously overcome these constraints, and should bear this in mind when analyzing its potential to contribute to MDG3."

Gradually research has evolved on the gendered appreciation and perception of the natural environment and thus wider issues of environmental sustainability, as well as the acknowledgment of gender dimensions of social and economic sustainability. The pursuit of small rural "self-help" projects by development aid agencies in conjunction with local NGOs raises important questions of social sustainability in developing countries. The promoters of "ecotourism" projects, for example, may emphasize that the lack of local infrastructure offers a pristine, uncommercialized environment that appeals to niche ecotourists. Yet local residents participating in such projects may offer their support on the assumption that tourism will help improve infrastructures and services that will need to be sustained. Ecofeminism encountered ecotourism (Swain and Swain 2004). Scheyvens' (2000) earlier work for example about the impact of ecotourism on local communities parses out the various capitals accumulated as signs of empowerment (economic, psychological, social, and political) as well as disempowerment, but she keeps her analysis at the community level, other than noting the presence of women, youth, and other disadvantaged or special interest groups.

Case studies of gender issues in sustainable tourism development

While there are many useful case studies that could be evoked, I limit my discussion here to Ferguson's (2011) ethnography of the World Bank, and a comparison of their findings by Tucker in Turkey and Boonabaana in Uganda (Tucker and Boonabaana 2012), as they specifically address issues in meeting the MDG3 of women's empowerment through tourism development. I also engage with arguments pioneered by Enloe (2000) about the intersections of global, regional, and local patriarchal cultural systems, politics, and economies, drawing from my several decades of research in southwest China on gender in ethnic tourism development, and the literature on a marriage–sex tourism continuum. Study of the commoditization of bodies in tourism, be it for ethnic display or literal consumption, brings us back to cultural values that often stand in the way of women achieving actual equality through economic aspects alone of tourism development.

By the end of the twentieth century, gender in tourism concerns became located in a variety of alternative, not mass scale, organized types of tourism such as global ecotourism. Research has substantiated that just because a form of tourism has specific alternative goals for the industry and tourist consumers, there are no automatic connections to sustainable improvements in local livelihoods or greater equity in tourism destinations (Ferguson 2011: 236). As Brown and Hall (2008: 841) argue, "sustainable tourism" following the sustainable development movement, has sometimes become a "cynically marketing-based approach to creating further growth in the industry." Thus there is need to distinguish between tourism sold as sustainable in terms of local ecology or culture, and sustainable tourism that incorporates all stakeholders into plans to create alternative conditions ethically framed as bettering the lives of those being toured. Such concerns fed the rather top-down named "pro-poor tourism" (PPT) approach that attempts to align tourism benefits directly towards poverty reduction (Brown and Hall 2008). Various avenues have been analyzed, including linkages of tourism and agriculture through local supplies of food for tourists often grown by women, to achieve pro-poor objectives (Torres and Momsen 2004; Swain 2004). Harrison's (2008) critique of PTT argues that even with these objectives, inequalities are reinforced if power relations in the global political economy are not taken into account and addressed.

Gender and tourism at the World Bank

The World Bank's tourism development policy, in Ferguson's analysis (Ferguson, 2011; Tucker and Boonabaana 2012: 439), constructs women's economic empowerment through tourism microenterprise, to "fix" both gender and poverty problems. The Bank's publications on gender and the MDGs view gender inequality as a barrier to economic development rather than a distinct issue. Such instrumentalist approaches use women's economic empowerment as a means for achieving other development goals, such as poverty reduction or environmental protection. Ferguson (2011) found the World Bank's microenterprise projects to have limited economic empowerment potential for women if basic gender inequalities are not also addressed, in terms of local cultural constructs of gender. Conflicting agendas arise (Ferguson 2011: 243) when the Bank attempts to combine a "gender-friendly" approach with their neoliberal structural adjustment policies.

> Despite an official change in rhetoric from WID to GAD, in practice, gender policy in the Bank has meant getting more women into paid work . . . [A]ttempts to promote women's empowerment through income-generating activities in the tourism industry have not usually led to broader social and political empowerment.
>
> *(Ferguson 2011: 243)*

Ferguson (2011: 244–246) reviews the Bank's "pro-poor" objectives in four recent projects—in Mozambique, Bolivia, Ethiopia, and Honduras—that link tourism employment and gender equality or empowerment. Tourism activity is expected to integrate women into the economy, which in turn allows them to contribute to the family finances and build their personal development. In her assessment of the Honduran project, women trainees found that the program did not take into account broader issues within their lives. Rather, there was little awareness at the Bank about the significant barriers to participation faced by women working in tourism enterprises. Ferguson (2011: 246) concludes that the outcomes of tourism development on gender relations in tourism destinations are complex and uneven, which is not to say that the World Bank's tourism funding is unable to contribute to gender equality and the empowerment of women in tourism destinations. She calls for "more radical starting points and more feminist expertise at all stages of gender and tourism development policies . . . to move beyond narrow understandings of economic empowerment" (Ferguson 2011: 246).

Tourism, gender, and poverty reduction in Turkey and Uganda

In terms of tourism development, poverty reduction, and women's empowerment—MDGs 2 and 3—Tucker and Boonabaana (2012) consider how culturally defined gender relations interact with tourism development in central Turkey and southwest Uganda. They examine how gender relations affect poverty reduction contexts as well as how tourism development induced poverty reduction affects gender relations. In Göreme, Turkey, Tucker found in her longitudinal study that women's earnings do not have a big impact on poverty reduction through microbusiness development, which has generally been more beneficial for men. However, over time women have been able to carve out their own spaces within tourism to achieve personal empowerment "undoing shame." As Ferguson (2011: 238) notes, microenterprise development policy in any country is far from a gender-neutral process, but rather "gender-blind," relying on implicit assumptions about men's and women's work in tourism.

Following Tucker's arguments, Boonabaana's research in Mukono parish, Uganda, on an NGO tourism development project found that women's workload increased, while they spent their earnings on women's usual family concerns of food and children's welfare. Despite their earnings from handicrafts and homestays, women continued to seek permission from their husbands to act. Boonabaana found a gradual negotiation of gender norms in a process of simultaneous power and powerlessness, resistance and participation that must counter the role of gossip in patriarchal control. In Mukono, women's role in poverty reduction from tourism work is becoming valued, with significant implications for gender roles and relations. While in Göreme men are still the predominant income earners and there has been little shift in cultural norms for gender roles and relations. However, women there have a greater possibility than in the past to engage in tourism work and earn income, leading to substantial changes in gender relations in both places. While these findings are hopeful, Ferguson (2011: 794) warns of added burdens to women's social reproduction work, dampening "the limits of a tourism economy that does not provide an infrastructure to support working mothers," which is substantiated by gender in tourism research around the globe.

Commoditized bodies: ethnic, wedding, marriage, and sex tourism

In southwest China, as in many parts of the world, indigenous minority groups have become deeply involved in ethnic cultural heritage tourism development, as a route out of poverty (Swain 2011). Feminization of ethnic tourism providers, who routinely are described in Chinese tourism materials as people who "love to sing and dance" results both in their portrayal as colorful and exotic, and servile, based in centuries of imperial colonization, ethnic stratification, and racial prejudice. The very landscape of indigenous ethnic tourism sites may be sexed female, coded as consumable land and people (Swain 2005). My research on Sani minority women tourism workers in Stone Forest, Yunnan, China documents the vicissitudes of handicraft production, guiding, and performance industries. In 1993 a provincial newspaper story on practicing prostitutes among the women guides at the national Stone Forest Park prompted quick reforms of guide services by regional tourism officials. Free enterprise companies were suppressed, and only government-licensed guides, called "Ashimas" after a Sani cultural hero, were allowed. By 2013, this was still the case, although "Ashima" entertainers serving tourists are also renowned in the area for their sexual enterprise. Local Sani continue to perceive sex workers as outsiders, usually having migrated in.

Gender and sexuality relationships range in a continuum from consensual to contractual to coercive in tourism markets. Scholars have used sexuality, both literal and metaphorical, as a framework for articulating the exploitative aspects of tourism. Early on, Graburn (1983) suggested that the prototypical rich North–poor South tourism dynamic can be thought of through the metaphor of the rich male purchasing or taking the services of the poor female, i.e., of countries that have "nothing to sell but their beauty" (their people, landscapes, nature). From a radical feminist perspective, male sexual power is the root of patriarchal power promoting sex-role stereotyping, heterosexism and compulsory heterosexuality, the institution of marriage, and practices of pornography, prostitution, rape, sexual and other forms of abuse of power including "domestic" violence, sex tourism and trafficking. Postmodern, structuralist academics and Third World feminists have argued for a more localized view of gender hierarchies, while documenting their existence and potential for transformation. An increasingly nuanced literature on the subject, for example Brennan's (2004) study set in the Dominican Republic, suggests that although the overall dynamic is exploitative, the women involved have considerable agency and the men are not uniformly predatory on an interpersonal level. Thinking about this trajectory led me to consider a continuum from destination wedding and honeymoons, to mail-order bride travel, to sex tourism and trafficking. These kinds of gender and sexual relations tourism are often co-located, in beautiful, impoverished locations such as the Dominican Republic, where local exotic cultures are a further draw. Tourism workers provide their services and products to all these tourists, while local people are also commodities in the most exploitative forms. Queer and heterosexual identities are found throughout this continuum, with some marketing aimed at gay and lesbian communities.

From destination weddings to cyber-brides

Destination weddings have become popular globally, not just from the West to developing or exotic locations. These trips include the marital pair, family, and friends who come to witness the wedding and vacation as tourists. Brokers provide "stress-free" package deals of travel, accommodations, and wedding preparations, often at a resort, to destinations including North, Central, and South America, the Caribbean, Australia, New Zealand, Pacific Islands, and Europe. McDonald (2005) documents tourist weddings in Hawaii that are consumed by North Americans looking for an exotic tropical setting and Japanese couples wanting an exotic Christian church location. In Europe, it is no longer unexpected to see Chinese wedding parties at scenic destinations. In the Yunnan, China, mass-ethnic tourism destination of Lijiang, I experienced a destination wedding, where I remarried my husband at a Naxi marriage chapel, in full ethnic regalia, officiated by a genuine Naxi religious practitioner in a very compressed version (three days into three minutes) ceremony. Commodification of Naxi culture is enacted through tourist bodies, perhaps less exploitative of Naxi people, while providing an authentically fake tourist experience.

People drawn to "exotic" locations for potential weddings also include a cohort of primarily white, heterosexual, middle-aged American men, looking for a wife who, in the words of a journalist is both "Madonna and puttana rolled together, an American male desire shaped in equal parts by the Promise Keepers and Internet porn" (Garin 2006). Internet brokers market a pool of potential brides to tourists on standard tours that include roundtrip airfare from the US, hotel accommodations, social events to meet potential "dates," guided tours, interpreters, and promotional materials. Clients access dossiers on hundreds of women in specific Russian, Asian, and Latin American destinations. The potential for cultural miscommunication, isolation, and domestic abuse is high among cyber-brides. Several brutal US murders in the early 2000s prompted the passage of the "International Marriage Broker Regulation" legislation as part of the Violence Against Women Act in 2006. This regulation insures that the prospective "groom" passes a background check when he applies for a fiancée visa, and that this information is given to his fiancée before she immigrates to the USA.

Sex tourism and trafficking

The terrain of sex tourism encompasses children and adults, prostitutes or sex workers, madams and pimps of all genders. Consensual bride or trafficked slave, sex sells and is highly controversial around the world. Boundaries are fluid, images and identities are both globalized and tied to local communities, and the selling of one's body and identity are integral aspects of tourism development. Implications in terms of sustainability and human rights are complex, gendered, and often culture-bound. Former colonies frequently eroticize their colonial heritage through sex tourism as they become increasingly dependent on tourism for foreign earnings at the level of the nation, and for economic survival at the level of the individual (O'Connell Davidson and Sanchez Taylor 2001). Here, it is possible to see both economic and sexual inequality at work (Aitchison 2009).

O'Connell Davidson (2005: 138–139) draws attention to the unequal economic relations of tourism, and posits an inevitability to sex tourism within global capitalist and patriarchal systems. Low wages in the hotel-restaurant industry making employees dependent on tips feed a willingness to turn "a blind eye to the activities of tourists." The social costs of tourism are rarely addressed in the industry. Profits continue to leak back to affluent sending countries and so "will never 'trickle down' to those who pick up tourists' litter, clean their toilets, make their beds, serve their food, and fulfill their sexual fantasies."

Structural links of sexual exploitation of women and children to the tourism industry—as we have seen in the UN Women, WTO, and ILO studies—are often excluded in research and policy intent on the "big picture" of gender issues in tourism. Likewise, Ferguson (2011: 236) acknowledges that looking exclusively at the hyper-exploitation of sex tourism would narrow her analysis of policy, so she excludes this variable. It seems ironic that gender and sexuality, so intertwined in the subjugation of women, would be avoided in analysis. Ferguson draws our attention to strong associations in the popular imagination between gender, tourism, and the sex industry. Why is it in the public imagination? Certainly violence against women must be included as part of the problem. Just as "women" associated with "gender" is inaccurate, the assumption that sex tourism is a separate issue from gender in tourism concerns must be interrogated and power and privilege negotiated.

Conclusions

In conclusion, I return to the proposition that gender concerns must be mainstreamed, addressing the cultural landscapes of gender on global and local scales, to promote sustained tourism development. Articulations of tourism with MDG3 must be considered in light of future directions for environmental, economic, and social sustainability. Organizations such as Equality in Tourism (Ferguson and Moreno-Alacrón 2013) and Gender Responsible Tourism[7] work to integrate research and practice for women's equality in tourism. In their discussion of "why gender matters," Equality in Tourism[8] build from an observation that women are denied a fair future "because men control most resources and decision making processes in tourism . . . [W]ithout a rigorous gender analysis in the thinking, development, practice, and evaluation of tourism, women will continue to be exploited."

Notes

1   http://equalityintourism.org.
2   We have many examples of how regional conflict, government upheaval, or environmental catastrophes can derail tourism development plans, keeping tourists away, destroying potential for economic, social, and environmental benefits. These concerns underlie the fragility of relying on tourism as a development scheme.
3   For a full summary of this report see www.earthsummit2002.org/toolkits/women/current/gendertourismrep.html.
4   See Momsen (2002) for further reporting on this research by Momsen and myself, funded by our UCMEXUS grant.
5   www.3sistersadventuretrek.com.
6   www2.unwto.org/en/content/why-tourism.
7   www.genderresponsibletourism.org.
8   http://equalityintourism.org/why-gender-matters.

References

Aitchison, C. (2009) "Gender and tourism discourses: advancing the gender project in tourism studies," in T. Jamal and M. Robinson (eds.), *The Sage Handbook of Tourism Studies*, London: Sage Publications.

Baum, T. (2013) *International Perspectives on Women and Work in Hotels, Catering and Tourism*, Bureau for Gender Equality Working Paper 1/2013, Sectoral Activities Department Working Paper No. 289, International Labour Organization, Geneva.

Brown, F. and Hall, D. (2008) "Tourism and development in the Global South: the issues," *Third World Quarterly* 29(5): 839–849.

Brennan, D. (2004) *What's Love Got to Do With it? Transnational Desires and Sex Tourism in the Dominican Republic*, Durham, NC: Duke University Press.

Enloe, C. (2000) *Bananas, Beaches and Bases: Making Feminist Sense of International Politics*, second edition, London: Pandora.

Ferguson, L. (2011) "Promoting gender equality and empowering women? Tourism and the third Millennium Development Goal," *Current Issues in Tourism* 14(3): 235–249.

Ferguson, L. and Moreno-Alacrón, D. (2013) *Integrating Gender into Sustainable Tourism Projects*, Equality in Tourism: Creating Change for Women, http://equalityintourism.org/wp-content/uploads/2013/10/Integrating-gender-into-sustainable-tourism-projects_F_Oct_13.pdf.

Garin, K.A. (2006) "A foreign affair: on the great Ukrainian bride hunt," *Harpers*, June.

Graburn, N. (1983) "Tourism and prostitution," *Annals of Tourism Research* 10(3): 437–456.

Harrison, D. (2008) "Pro-poor tourism: a critique," *Third World Quarterly* 29(5): 851–868.

Kinnaird, V. and Hall, D. (eds.) (1994) *Tourism: A Gender Analysis*, Chichester: John Wiley & Sons.

McDonald, M.G. (2005) "Tourist weddings in Hawai'i: consuming the destination," in C. Cartier and A. Lew (eds.), *Seductions of Place: Geographical Perspectives on Globalization and Touristed Landscapes*, London: Routledge, pp. 171–192.

Momsen, J.D. (1999) "A sustainable tourism project in Mexico," in M. Hemmati (ed.), *Gender and Tourism: Women's Employment and Participation in Tourism*, London: UNED-UK, pp. 142–148.

—— (2002) "NGOs, gender and indigenous grassroots development," *Journal of International Development* 14: 859–867.

O'Connell Davidson, J. (2005) *Children and the Global Sex* Trade, Cambridge: Polity.

O'Connell Davidson, J. and Sanchez Taylor, J. (2001) *Children in the Sex Trade in the Caribbean*, Stockholm: Save the Children Sweden.

Scheyvens, R. (2000) "Promoting women's empowerment through involvement in ecotourism: experiences from the Third World," *Journal of Sustainable Tourism* 8(3): 232–249.

Sinclair, T. (1997) *Tourism and Employment*, London: Routledge.

Smith, V. (ed.) (1977) *Host and Guests: The Anthropology of Tourism*, Philadelphia: University of Pennsylvania Press.

Swain, M.B. (1995) "Gender in tourism," *Annals of Tourism Research* 22(2): 247–267.

—— (2002) "Introduction" in M.B. Swain and J.H. Momsen (eds.), *Gender/Tourism/Fun?*, Elmsford, NY: Cognizant Press, pp. 1–14.

—— (2004) "(Dis)embodied experience and power dynamics in tourism research," in J. Phillimore and L. Goodson (eds.), *Qualitative Research in Tourism*, London: Routledge, pp. 102–118.

—— (2005) "Desiring Ashima, sexing landscape in China's Stone Forest," in C. Cartier and A. Lew (eds.), *Seductions of Place: Geographical Perspectives on Globalization and Touristed Landscapes*, London: Routledge, pp. 245–259.

—— (2009) "The cosmopolitan hope of tourism: critical action and worldmaking vistas," *Tourism Geographies* 11(4): 505–525.

—— (2011) "Commoditized ethnicity for tourism development in Yunnan," in T. Forsyth and J. Michaud (eds.), *Moving Mountains: Highland Livelihoods and Ethnicity in China, Vietnam, and Laos*, Vancouver: University of British Colombia Press, pp. 173–192.

Swain, M.B. and Momsen, J.H. (eds.) (2002) *Gender/Tourism/Fun?*, Elmsford, NY: Cognizant Press.

Swain, M.B. and Swain, M.T.B. (2004) "An ecofeminist approach to ecotourism development," *Tourism Recreation Research* 29(3): 1–6.

Torres, R.M. and Momsen, J.H. (2004) "Challenges and potential for linking tourism and agriculture to achieve PPT objectives," *Progress in Development Studies* 4: 294–318.

Tucker, H. and Boonabaana, B. (2012) "A critical analysis of tourism, gender and poverty reduction," *Journal of Sustainable Tourism* 20(3): 437–455.

UN (2010) *Global Report on Women in tourism*, Preliminary Report, www.unwomen.org/2011/03/tourism-avehicle-for-gender-equality-and-womens-empowerment/ Full report: http://ethics.unwto.org/en/content/global-report-women-tourism-2010.

UNED-UK (1999) *Gender and Tourism: Women's Employment and Participation*, Report to the UN Commission on Sustainable Development 7th Session, April 1999, London.

36

IMPACT OF ICTS
ON MUSLIM WOMEN

Salma Abbasi

Introduction – impact of information communication
technologies (ICTs)

This chapter addresses key gaps in the information communication technology for development (ICT4D) debates today that discuss the provision of ICTs for women and the subsequent impact that they have on their lives (Salazar 2009; Hafkin 2002). There is currently a great tension regarding the extent to which ICTs can serve as a "liberator" for women (Pavarala et al. 2006). On the one hand, ICTs are seen as drivers for women's empowerment (Yuen et al. 2010; Hassanin 2009a; Friedman 2006); on the other, some feminists and development practitioners suggest that they widen the gender digital divide and "create new sociological pressures, new cultural paradigms" (Pichappan 2003: 7).

Kyomhendo (2009: 163) argues that "understanding the impact of the use of modern ICTs such as mobile telephones on women's empowerment requires a deep insight into the construction of the sense of empowerment which appears to vary among individuals in similar context". I, too, subscribe to this position. There continues to be very little actual field data to prove or disprove the true impact of ICTs (Unwin 2009, 2005).

However, a 2012 event in Pakistan received tremendous international coverage through news media and social networking sites: the tragic shooting of a 14-year-old girl, named Malala Yousafzai on 9 October in Mingora, Khyber Pakhtunkhwa, because she dared to go to school. Khyber Pakhtunkhwa is one of the most northern states of Pakistan, heavily controlled by the Taliban, with a strict traditional tribal culture of the Pashtoons that strives to keep women and girls restricted in the *char devar* (four walls) of their homes. In fact, there is a continuous stream of violent acts taking place in Pakistan that prevent girls from going to school, that frequently go unnoticed in the media. However, by some twist of fate, this event managed to be broadcast all over the national and international media and became a hot topic for debate, rallying support to oppose this barrier. Thus, in this case, ICTs have greatly helped highlight and propel this terrible event across the world. Furthermore, this has galvanized the demand for substantial actions to be taken by governments around the world to "prevent violence against girls" across multiple fronts.

Additionally, the high visibility of Malala's story and "strength to fight back" against these restrictions has also positively impacted and inspired young girls living in such suppressive societies and communities to fight for their rights for education and access to knowledge.[1] It is also important to note the visible support of Malala's father, Ziauddin Yousafzai, every step of the way; protecting his daughter's right for education and to go to school has also positively simulated great debate on this controversial issue. ICTs have again played a crucial role to enable this situation to be shown around the world.

This chapter discusses the findings of my five-year research on "ICTs and Muslim women", which was primarily conducted in Pakistan across five regions. It examined the "use and subsequent impact of ICTs" experienced by women. The data was gathered through both qualitative and quantitative methods: 99 focus groups, 127 interviews and 768 questionnaire surveys. However, the research was enriched by comparing and contrasting the experiences of 39 elite women, who were interviewed from several Muslim countries: 19 from Pakistan and 20 from seven other Muslim countries. This allowed for reflections and enabled geographical and class implications to be drawn out in the analysis. Furthermore, with the aim of not losing the passion with which comments were shared, I have chosen to integrate the voices of my participants throughout this chapter in their exact words.

This paper discusses the social implications in the context of ICT use and impact on women. However, before delving deeper into this debate, I should like to set the stage by providing an overview of the overall impact of ICT use.

Setting the stage: overall impact of ICTs

This section presents the overall impact of ICTs that the participants identified in field research, in the context of empowerment. The impact was analysed against five categories synthesized from the questionnaire surveys – social; sense of freedom; knowledge; economic; and not empowered. Figure 36.1 indicates the overall impact of ICTs as described by the participants in the questionnaire surveys. Some 69 per cent of the women stated that they feel socially empowered by ICTs, 18 per cent of them felt a sense of freedom due to their use of ICTs, 9 per cent acknowledged that ICTs have supported them in the context of knowledge, but only 1 per cent of the women stated that ICTs had impacted them economically, despite the fact that it has been argued by Stephen (2006) and Sharma (2003) that ICTs will bring economic

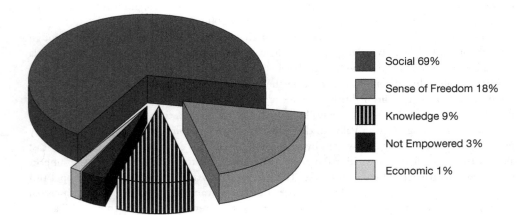

Figure 36.1 How ICTs empowered women
Source: author's survey, October 2009

empowerment to women. Moreover, 3 per cent of the women expressed that they have not been empowered at all in any context. In fact they discussed the risks that they have now been exposed to because of ICTs.

The results indicate that, even though a great number of participants identified that they experience social empowerment from ICT tools, there is evidence that a lower proportion of rural women encountered this experience than those from urban areas. This difference could be explained by Chaudhry and Nosheen's (2009: 224) research in Pakistan, which highlights that women in urban areas enjoy greater freedom because of the increased education of household members, thus creating "less rigid communities and most importantly good knowledge of Islamic teachings and its practice, while this situation is deteriorated in rural and tribal areas". Furthermore, the results also indicate that rural women experienced a far lower sense of freedom than their counterparts in urban areas. This is unaffected by their level of formal or informal education. A possible explanation for this difference is the greater degree of restriction on women's mobility and freedom of choice in rural than in urban areas, as indicated by Chaudhry and Nosheen's (2009) and Saghir et al.'s (2009) research in Pakistan.

My results also show that female heads of households, irrespective of their educational backgrounds, have generally experienced some sense of economic empowerment. This could be due to their necessity to identify creative ways of using ICTs. Nevertheless, it is important to note that there was a small proportion of women within the same group – from both urban and rural areas, and both formally educated (FE) and informally educated (IE) participants – who also reported that they did not feel empowered by ICTs in any way. A possible explanation for this was offered by a formally educated urban participant from Punjab, who stated that: "ICTs may have the key to empowering us by making our lives better, but who will show us how to use them properly and help us find the time."

This reflects the sentiments and frustration of women in Nigeria, as argued by Comfort and Dada (2009). Feminists have long argued that the imbalance of traditionally accepted gender roles places a huge burden on women (Kabeer 2005; Momsen 2004; Rowlands, 1997). My research has confirmed this and further highlighted it, in the case of rural women and those from marginalized communities living in abject poverty, who often have to juggle "quadruple roles" – an example of this is indicated in Figure 36.2. This burden is further exacerbated by the cultural pressures and social structures entrenched in oppressive, patriarchal, male-dominated societies. These factors cumulatively impact women's access to and use of ICTs.

Therefore, in this section I have chosen to highlight and share the reflections that my field research has uncovered, specifically in the context of women in Pakistan, as I feel that if ICT policies, development agencies and sincere well-wishers of women understand the cultural and social challenges and integrate critical threads into ICT policies' implementation, effective positive impact could be greatly advanced.

In the following section, the results from the focus groups and questionnaire surveys have been synthesized in order to understand the type of impact that women have personally experienced in the social context, as this particular aspect appeared to be the most important to understand from the research.

Social implications

It is important to recognize that

> in some cultures women are not permitted to have face-to-face contact with men other than those in their own families, or are expected to stay at home, or indeed

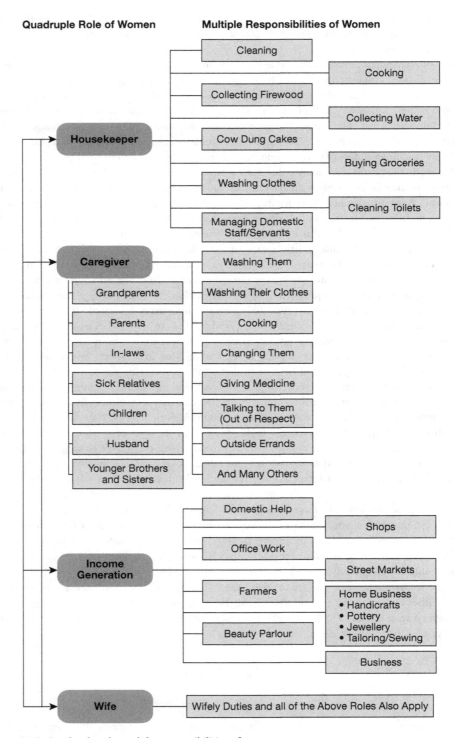

Figure 36.2 Quadruple roles and the responsibilities of women

to be isolated in restricted living facilities. For such cultures, communication technologies may empower women. Telephone, radio, television, and the internet allow women to interact with men without being in the same place, and indeed without any face-to-face contact at all. Especially important in this respect may be distance education and e-commerce.

(Daly 2003: 5)

Thus, it is evident that ICTs could possibly play a pivotal role in bringing the outside world into the home for women (Yusuf 2005). It is in this context that my research has particularly explored the social implications around *if* and *how* ICTs have been used by women living in such environments and, more importantly, has tried to shed light on the possible impact of this on their social context.

As Elnaggar (2007: 4) has argued, "today ICT is the most effective tool in the hands of women enabling them to extend their participation in a variety of productive fields and providing them with an avenue to express the development of their personalities and capacities". Similarly, Primo (2003: 9) asserts that "information and communication technologies could give a major boost to the economic, political and social empowerment of women, and the promotion of gender equality". However, it is difficult to fathom the depth and degree of constraints and restrictions that some women face in Muslim countries. Syed and Ali (2005) have described some of the factors that have shackled women in Muslim societies. They discuss the nature of women's mobility:

Her mobility outside *Chardiwari* is restricted because of the known rationale of modesty (*Haya*), family honour (*Ghairat*), and tribal traditions (*Rivaj*). Modesty is closely related to the concepts of shame (*Sharm*) and humility (*Ijz*). The concept has imposed specific physical and psychological boundaries on the life of a Muslim working woman.

(Syed and Ali 2005: 4)

This fear was also evident in some of the women who participated in each of my focus groups, from all areas and all classes. Furthermore, it was also raised by one of the elite women interviewed who was working for the government of Punjab and who noted: "There is deep fear in the communities that the internet and mobile phones are tools that will spoil their girls by exposing them to boys and Western dangers." Similar concerns were also raised by two elite women from the Gulf region who had noticed that ICTs have begun to have a negative impact on their daughters' and sons' social behaviour. They argued that their children preferred to stay in their rooms and "chat online", rather than joining the family for meals or engaging with relatives when they visited. They further raised concerns that this was destroying Arab traditions and their culture of hospitality. They blamed the increased access to ICTs and globalization for negatively affecting family structures in various ways. Furthermore, recent research by Al Emadi and Ibrahim (2010: 8–9) on Qatari society suggests that "the divorce rate among Qatari families has increased by almost 30% during the last decade, according to statistics released by the Supreme Council for the Family, and family ties have been in constant deterioration lately". They further argue that "the introduction of the new means of communication has also been very harmful for family relationships, especially the internet which has literally confined people to their own rooms for different online activities" (Al Emadi and Ibrahim 2010: 9). The authors' findings appear to reinforce the concerns raised by some of the elite women interviewed in the Gulf.

Thus, keeping this in mind, the next subsections suggest four distinctive ways in which ICTs have been used to provide some form of social empowerment to women from the perspectives of, respectively: access to information; liberty and freedom; connectivity to family and support groups; and access to knowledge of their rights in Islam.

Multi-dimensional access

It was evident from the excitement of some women in the focus group discussions that ICTs are beginning to provide different types of access to a few of them, addressing and helping to overcome some of their social, cultural and traditional constraints. There was a recurring theme across all of the focus groups in the urban areas that expressed gratitude at how ICTs have liberated them somewhat and given them greater freedom to do what they want. An unmarried woman living with her conservative parents in Karachi stated: "I was always interested in poetry but never got the opportunity to show my talent to the world, but then I took part in a poetry competition online and now my poetry is published on their website." Another formally educated unmarried respondent living in Punjab stated: "For me it's everything – a means of gathering information, connecting with relatives. I feel disabled without it. It provides me with knowledge and helps me in handling tasks quickly." Lastly, a young woman living in Punjab praised social media sites, noting that "I use Facebook and Orkut to communicate with my friends and make new friends and this makes me very happy."

These experiences and comments all support Jensen (2006: 11), who argues that ICTs could "have the effect of minimizing and overcoming social divides and injustices, if they are designed, implemented and monitored to do so". Furthermore, Comfort and Dada (2009) have highlighted the positive impact of mobile phones for women in Nigeria and state that "mobile phones have addressed *pardah* by providing an acceptable space for women". Nonetheless, it is very important to point out that these experiences were not shared by the majority of the women participating in the focus groups, despite their class, geographical location or position in the household. In fact, in many cases the respondents giggled as they shared their experiences and others looked on in disbelief, particularly many of the women from rural areas.

The possible causes of these extreme differences could derive from the cultural sensitivity that exists in Pakistani society, which was described by several elite women interviewed as being strongly male-dominated, consistent with its patriarchal social structure (Chaudhry and Nosheen 2009; Kabeer 1999b). An example of this was given by one senior government official from Sindh, who stated:

> One of the causes of this limitation could be fear from their male family members that they would not be able to control or monitor what their women were going to be exposed to . . . both women and men lack the awareness of the benefits of ICTs, and that was also a limiting factor, along with the lack of availability of ICT tools in the poorer parts of both urban and rural areas and the lack of the reliable infrastructure required to make things work.

The impact of this is greatly holding back development in the rural areas.

Liberty and freedom to travel outside the home

Syed and Ali (2005: 4) note that "Chador and Chardiwari are often described as two appropriate domains for women in Pakistani society". *Chador* should be understood as being physically covered up by a large scarf and *Chardiwari* as being restricted to the four walls of the home. Once again, this paints a grim picture for some of the women living in Pakistan. Nevertheless, ICTs appear to have helped create a pathway for women to access the world outside the home. There was considerable discussion, particularly within the focus groups in the urban areas of Punjab, of how the mobile phone has allowed some of them "the freedom to travel outside the home". It should be noted that in some focus groups, many respondents stated that they felt safe whenever they travelled outside their homes when they had a mobile phone with them because they could call for help in case of emergency. An example of this was shown by two unmarried participants from Punjab and Sindh who stated "I take my mobile phone with me and communicate with my family and friends all the time" and "My family also feels comfortable when I have my mobile with me and go outside, as they can contact me at any time in less than a minute".

The value of the mobile phone was also dramatically expressed by a formally educated housewife from Mirpur in AJK, who commented: "Well to my uses, yes – they are actually very useful. For example, once I went outside in my car and it stopped working in the middle of the road. If the cell phone hadn't been invented I would probably have been left crying there."

This phenomenon was also observed in focus groups conducted in rural areas, but to a lesser degree. In fact, a farmer's daughter who was educated in a nearby town in Punjab noted that "before, when I didn't have a mobile, my parents did not allow me to go outside on my own". However, many of the participants in the focus groups in the rural areas stated that their parents had great concerns regarding the use of mobile phones because of the possibility of boys calling and harassing their daughters. The issue of safety and security exists in both the urban and rural areas in Pakistan today. Nevertheless, an elite woman working for an NGO in Punjab commented:

> The threats and dangers to girls and women being kidnapped in the rural areas are higher than in the cities . . . due to the feudal cultures in rural areas, poor men and their families are marginalized and law enforcement agencies are often bought off in the event of any misdeeds. Therefore, it is not so simple to assert that it is only the patriarchal society and male dominance that keep women isolated in their homes, but in some cases the fear of real dangers.

Connectivity with family and support groups

ICTs have transformed the way that people communicate around the world, and the use of the internet for communication was one of the most common uses indicated in the focus groups and surveys that were conducted in urban areas with formally educated women. This is clearly evident from the comments made by two respondents, a professional women and a housewife respectively, who commented that "the internet is very cheap for communicating with family and friends" and "now we can use Skype and other messenger services for calls as they are free of cost and give good quality output". Similar statements were mostly made by women in urban areas, who live in an environment in which broadband is readily available. The one term that was a recurring statement in all of the urban focus groups was "Skype", with the associated joy that women felt in being connected to family and friends at very little cost. The benefits from the connectivity that the internet has brought appear to have two main themes, namely bridging the distance between married girls and their families within and outside Pakistan, and networking with international discussion groups within and outside Pakistan.

This was aptly exemplified by formally educated, married women, from urban and rural areas, in comments such as "I got married and moved to UAE, but ICTs have reduced the distance and now I can talk to my family and friends online through Skype, which is totally free of cost" and "I am living in Lahore and communicate with my family in the UK through the internet".

Nonetheless, this phenomenon was not observed in the focus groups in rural areas, mainly due once again to the limited availability of the internet and lack of awareness regarding the benefits and multiple uses of it. However, this exception was noted in the rural areas of Azad Jammu and Kashmir, primarily due to the fact that every participant had family members living in the United Kingdom who had bought them computers and mobile phones and showed them how to use Skype. This demonstrated some degree of positive influence and moderate views to the subordination and controls on women due to the family member's exposure in the United Kingdom. This level of support was not observed in other rural areas of Pakistan where the external influences are far less. Nevertheless, access to electricity remained a common challenge. ICTs have also been used by some women for connecting with discussion groups, which have allowed transnational discussions and the sharing of experiences (see Ali 2001).

This finding was supported by a women's rights activist living in Lahore, who also shared a possible cause for this difference, noting that:

> The poor women in the rural areas have no idea about ICTs, nor do they know that the internet can be used to make free phone calls on Skype . . . the government needs to do awareness campaigns at multiple levels to ensure that the different types of women and particularly the old generation could benefit from simple things like Skype and SMS, along with ensuring that access for poor women and the elderly is given for free, since they don't have enough money, even to eat; this sort of thinking would transform Pakistan.

Rogerson and Begg (1999) argue that this technology has liberated the silent, voiceless, suppressed community living within Muslim society, namely "women", for whom ICTs have enabled global access and have linked the informed women of the West to the uninformed women of rural areas where ICTs can be accessed. Additionally, access to knowledge through the internet has also reached women who are housebound with little or no mobility. This was demonstrated during a focus group session in the city of Sukkur in Sindh, when a housewife related an experience of her cousin who had discovered a women's NGO in Karachi from an internet discussion forum and had run away to Karachi to a safe house to escape her husband and mother-in-law, who had been burning her and trying to kill her. Similar experiences were recounted during the interviews with women's activists in Lahore stating:

> I have been able to network with international organizations to help protect the abused women in Pakistan and raise awareness of the brutality that the Jirgah system [a tribal assembly of elders that takes decisions by consensus] was enforcing on women in Pakistan.[2]

And continuing with the example of the media blitz that covered Mukhtara Mai's case: "This was the magical power of networking across the globe."[3] The benefits of ICTs can also be seen by the emergence of such organizations as Women Watch,[4] which "provides an online database-driven mechanism which links to websites and web pages" UNDAW (2005: 12). Additionally, the internet is being used as a tool for the monitoring and protection of women's rights through an interesting collaborative project – Women's Human Rights Net.[5] This is an electronic network linking more than 50 international women's human rights organizations for information-sharing, campaigning and capacity-building (Marcelle 2000b: 17). Similarly, Pilch (2006: 114) argues that "in spite of the entrenched patriarchal power structures in many Muslim states, women are slowly becoming more educated, are taking part in worldwide conversations about women's issues through the use of the Internet, are travelling beyond their state boundaries". This view has been effectively utilized by women's activists in Pakistan, but sadly the laws have not changed sufficiently to protect women's rights, as seen most recently in the release of Mukhtaran Mai's rapists from jail, despite it being known that the men were guilty of her cruel gang rape (BBC 2011). The society appears to be willing to accept and tolerate such atrocities against women, despite their supposedly equal rights in Islam.

Women's knowledge of their rights in Islam

This section discusses the results in an area that entails one of the most important uses of ICTs, as it has the potential to transform society by providing women with a knowledge and support network. One respondent noted that:

> The internet provides a vehicle for uncovering and shaming people who are misinterpreting Islam's rights for women and allowing injustices to occur in Pakistani society . . . heartbreakingly, Pakistani women continue to tolerate cruelty and injustices, thinking that it is what Islam says despite the fact that ICTs could provide them with the window to knowing their true rights from an *alim* (religious scholar), not their husband or mothers in law, but they need to know how to use them.

A positive example of the use of ICTs in the context of emails can be seen by the work that the email-based Muslim Women Network (MWN)[6] is doing:

> MWN has been established to provide a forum for engaging in an intellectual discourse on significant contemporary issues that impact on Islam, especially those pertaining to or affecting women. It has also provided a medium for Muslim women to discuss any pertinent issues, as long as the discussion remains within the bounds of Shar'iah.
>
> *(Bastani 2001: 5)*

Despite the fact that "Islamic culture endorses equality of rights for both men and women, Muslim culture, like many traditional African cultures, has in the recent past denied women their basic rights to freedom of choice, education and employment" (Nkealah 2006: 15). Syed and Ali (2005: 1) argue that "in order to improve individuals' overall capabilities, Islam declares the seeking of knowledge (i.e., education) to be a religious duty, which is equally binding on women and men". Nevertheless, my field research demonstrates that women experience multiple types of restrictions, barriers and constraints in their day-to-day lives in all of the above areas. Due to the development of the internet, "a vast wealth of Islamic information is now available through computer facilities [and] supplementary teaching has been revolutionized by using these facilities" (Rogerson and Begg 1999: 2). This is opening up debates across the world between Muslim communities from the North and South and creating a bridge that is bringing an enlightened and balanced perspective to Islam and the position of women within it (Rogerson and Begg 1999).

The focus group discussions revealed that only a few women in some urban areas of Pakistan now use the internet to join discussion groups, "to learn about Islam and what Islam really wants us to do in our homes". However, a participant from Lahore stated: "My sister-in-law sits at home and surfs the internet and chats with other women on blog sites and discussion groups so that she can learn what her rights are, because things in Pakistan are not right." This point catalysed a heated debate in one of the focus groups in Islamabad with formally educated women between the women who use the internet to understand their position in Islam against those who do not. For example, one woman who did not use the internet for any Islamic or spiritual investigation said that "it is all controlled by men and is giving the male interpretation of Islam and therefore cannot be trusted". She particularly referenced a website called Forum Pakistan,[7] which highlights false Islamic information. The elite women interviewed were fully supportive of the use of discussion forums on any topic and in fact, one women's rights activist from Lahore said that:

> These discussion groups need to be voice-activated so that women with poor written English would be able to express themselves . . . It would be wonderful if you could push a button and the computer would automatically translate what has been said in any language.

This was also reported by a few of the other elite women interviewed, who said that they often participated in professional, social and religious discussion forums when they had the time, and one elite woman from the UAE mentioned that "this is an excellent way to stay engaged and participate in multiple areas, despite having a busy lifestyle".

ICTs, and particularly access to the internet, are therefore beginning to provide information to women where they can challenge some of the cultural restrictions within the Muslim society of Pakistan. However, the focus group results suggest that some women are also beginning to question the relevance of content on the internet, as it is not always clear whether the information is from an authoritative source or not. The key finding is that access to information enables a debate within the community and enhances the richness of the knowledge base.

Conclusion

There is no doubt that ICTs have the potential to alleviate some of the social isolation that women face in Pakistan through connecting them to the larger international community and support networks. This can provide opportunities for both social networking and the provision of personal and professional opportunities. In this context, Loh-Ludher et al. (2006) argue that "gender issues interwoven into radio and TV soap operas have a significant impact. However, most programmes still portray women as victims of inequality rather than empowered individuals." Therefore, a prerequisite for women to take advantage of the opportunities offered by ICTs are campaigns to increase awareness of the capabilities that they offer, and a major change in society is needed to give women the rights, equality and respect that they are due.

Furthermore, the social implications of ICTs in the context of providing women access to a deeper understanding of their rights and human rights linked to a support network is beginning to have tremendous impact on the lives of some women. Therefore, it is critical that this aspect be further developed and scaled across oppressive societies, not just Muslim societies, where women are subordinated, marginalized and treated as commodities. We cannot change culture in a day, in fact it has not changed despite decades of debate by many feminists from all over the world (Jensen 2006; Hafkin and Odame 2002; Kabeer 1999a).

Nevertheless, governments need to play their part to ensure that not only ICT policies but other relevant policies that intersect the touch points of cultural social challenges that women deal with and live in need to be carefully integrated and constructed to address the constraints, if ICTs are truly going to be able to bring about a positive change in the lives of women. Furthermore, it is my recommendation that the importance of not just gender sensitization in ICT policies but cultural sensitization is critical if we are to ensure that women truly benefit and are allowed to engage with the digital society.

Notes

1 http://en.wikipedia.org/wiki/Malala_Yousafzai.
2 http://en.wikipedia.org/wiki/Jirga.
3 http://en.wikipedia.org/wiki/Mukhtaran_Bibi.
4 www.un.org/womenwatch/daw/public/w2000–2009.05-ict-e.pdf.
5 www.onlinewomeninpolitics.org/beijing12/mono9-ICT.pdf.
6 http://homes.chass.utoronto.ca/~wellman/publications/muslimwomen/MWN1.PDF.
7 www.forumpakistan.com/false-islamic-websites-t22457.html.

References

Al Emadi, S.A. and Ibrahim, T.J. (2010) *The Social Consequences of Globalization on Qatari Society (Educational Reform)*, www.sec.gov.qa/it/Uploaded/project/PROJ75.docx.

Ali, K. (2001) *Teaching about Women, Gender, and Islamic Law: Resources and Strategies*, paper presented at the American Academy of Religion Annual Meeting, Denver, CO, November.

Bastani, S. (2001) "Muslim women online", *Arab World Geographer* 3(1): 40–59.

BBC (2011) "Pakistan: acquittals in Mukhtar Mai gang rape case", www.bbc.co.uk/news/world-south-asia-13158001.

Chaudhry, I.S. and Nosheen, F. (2009) "The determinants of women empowerment in southern Punjab (Pakistan): an empirical analysis", *European Journal of Social Sciences* 10(2): 216–229.

Comfort, K. and Dada, J. (2009) "Rural women's use of cell phones to meet their communication needs: a study from northern Nigeria", in I. Buskens and A. Webb (eds.), *African Women and ICTs: Investigating Technology, Gender and Empowerment*, London: Zed Books, pp. 44–55.

Daly, J.A. (2003) *ICT, Gender Equality, and Empowering Women, prepared for essay series on Information and Communications Technology Applied to the Millennium Development Goals*, http://topics.development gateway.org/ict/sdm/previewDocument.do~activeDocumentId=840982.

Elnaggar, A. (2007) "The status of Omani women in the ICT sector", *International Journal of Education and Development using Information and Communication Technology* 3(3): 4–15.

Friedman, T.L. (ed.) (2006) *The World is Flat: The Globalized World in the Twenty-First Century*, Harmondsworth: Penguin.

Hassanin, L. (2009a) "An alternative public space for women: the potential of ICT", in I. Buskens and A. Webb (eds.), *African Women and ICTs: Investigating Technology, Gender and Empowerment*, London: Zed Books, pp. 67–76.

—— (2009b) "Egyptian women artisans facing the demands of modern markets: caught between a rock and a hard place", in I. Buskens. and A. Webb (eds.), *African Women and ICTs: Investigating Technology, Gender and Empowerment*, London: Zed Books, pp. 56–63.

Hafkin, N. (2002) *Gender Issues in ICT Policy in Developing Countries: An Overview*, United Nations Division for the Advancement of Women (DAW) Expert Group Meeting, Seoul, Republic of Korea, 11–14 November, www.un.org/womenwatch/daw/egm/ict2002/reports/Paper-NHafkin.PDF.

Hafkin, N. and Odame, H.H. (2002) *Gender, ICTs and Agriculture*, A Situation Analysis for the 5th Consultative Expert Meeting of the CTA's ICT Observatory on Gender and Agriculture in the Information Society, Technical Centre for Agricultural and Rural Co-operation, Wageningen, www.generoyambiente.org/admin/admin_biblioteca/documentos/tic_agricultura.pdf.

Jensen, H. (2006) "Women, media and ICTs in UN politics: progress or backlash?" in A. Gurumurthy, P.J. Singh, A. Mundkur and M. Swamy (eds.), *Gender in the Information Society: Emerging Issues*, UNDP-Asia-Pacific Development Information Programme (APDIP), pp. 3–14.

Kabeer, N. (1999a) "Resources, agency, achievements: reflections on the measurement of women's empowerment", *Development and Change* 30(3): 435–464.

—— (1999b) *The Conditions and Consequences of Choice: Reflections on the Measurement of Women's Empowerment*, Discussion Paper (18), United Nations Research Institute For Social Development (UNRISD), Geneva, Switzerland.

—— (2005) "Gender equality and women's empowerment: a critical analysis of the third Millennium Development Goal", *Gender and Development* 13(1): 13–24.

Kyomhendo, G.B. (2009) "The mobile payphone business: a vehicle for rural women's empowerment in Uganda", in I. Buskens and A. Webb (eds.), *African Women and ICTs: Investigating Technology, Gender and Empowerment*, London: Zed Books, pp. 154–165.

Loh-Ludher, L.L., Sandrakasan, S., Dilling, S., Dilling, J., Kheng, S.M.L., Fong, F.H.F., Abu Hassan, S.A., Yen, A.L.P. and Ravichandran, R. (2006) "Homeworkers and ICTs – Malaysia", *International Development Research Centre (IDRC)* Grant No. 102, pp. 792–801.

Marcelle, G.M. (2000b) "Transforming information and communications technologies for gender equality", *Gender in Development*, Monograph Series # 9, UNDP, New York.

Momsen, J.H. (2004) *Gender and Development*, London: Routledge.

Nkealah, N.E. (2006) *Islamic Culture and the Question of Women's Human Rights in North Africa: A Study of Stories by AssiaDjebar and AlifaRifaat*, Pretoria: University of Pretoria.

Pavarala, V., Malik, K.K. and Cheeli, J.R. (2006) "Community, media and women: transforming silence into speech", in A. Gurumurthy, P.J. Singh, A. Mundkur and M. Swamy (eds.), *Gender in the Information Society: Emerging Issues*, Thailand: UNDP – Asia-Pacific Development Information Programme (APDIP), pp. 96–109.

Pichappan, P. (2003) *Towards Optimizing Mobility in the ICT Sector to Create an International Paradox and Gender Balance*, presented at Global Knowledge Forum, 21 August, 2003, Kuala Lumpur, Malaysia.

Pilch, F. (2006) "The potential role of women in contributing to countering ideological support for terrorism: the cases of Bosnia and Afghanistan", *Connections: The Quarterly Journal* 5(4): 107–129.

Primo, N. (2003) *Gender Issues in the Information Society*, Publications for the World Summit on the Information Society, Paris: UNESCO.

Rogerson, S. and Begg, M.M. (1999) "Islam, ICT and the computer professional", *IMIS Journal* 9(4).

Rowlands, J. (1997) *Questioning Empowerment: Working with Women in Honduras*, Oxford: Oxfam.

Saghir, A., Ashfaq, M. and Noreen, A. (2009) "Gender and information and communication technologies", *The Journal of Animal and Plant Sciences* 19(2): 94–97.

Salazar, J.F. (2009) "Self-determination in practice: the critical making of indigenous media", *Development in Practice* 19(4–5): 504–513.

Sharma, U. (2003) *Women's Empowerment through Information Technology*, New Delhi: Authors Press.

Stephen, A. (2006) *Communication Technologies and Women's Empowerment*, New Delhi: Rajat Publications.

Syed, J. and Ali, F. (2005) *A Historical Perspective of the Islamic Concept of Modesty and its Implications for Pakistani Women at Work*, www.historians.ie/women/syed.PDF.

UNDAW (2005) *Women 2000 and Beyond, Gender Equality and Empowerment of Women Through ICT*, United Nations Division for the Advancement of Women, Department of Economic and Social Affairs, New York, www.un.org/womenwatch/daw/public/w2000–2009.05-ict-e.pdf.

Unwin, T. (2005) *Partnership in Development Practice: Evidence from Multi-Sector ICT4D Partnership Practice in Africa*, Paris: United Nations Educational, Scientific and Cultural Organization.

—— (2009) "ICT4D Implementation: polices and partnerships", in T. Unwin (ed.), *ICT4D, Information and Communication Technology for Development*, Cambridge: Cambridge University Press, pp. 125–176.

Yuen, S.C.-Y., Duan, X. and Yuen, P.K. (2010) "M-Learning: a new wave of learning", *International Journal of Intercultural Information Management*, 2(1): 24–39.

Yusuf, D. (2005) *Islam and Muslims in Cyberspace, From (Re) Presenting to (Re) Understanding*, www.islamonline.net/servlet/Satellite?c=Article_C&pagename = Zone-English-ArtCulture/ACE Layout&cid = 1158658426110.

37

GENDERED COSTS TO THE "LEFT BEHIND"

A challenge to the migration and development nexus

Rebecca Maria Torres

Introduction

More women than ever before are migrating and, in many cases, providing the primary source of financial support for their families at home through remittances. As governments and international organizations continue to tout transnational migration as a strategy for economic development and poverty alleviation, women are increasingly gaining attention as agents of development. While migration and development discourse focuses heavily on migrants themselves, the women "left behind"[1] play a central role in the reproduction of migrant labor. They care for families in sending communities and contribute to economic development by managing remittances and overseeing local projects funded by money transfers. While migration is viewed optimistically for its potential to generate household income and to spur development, the unseen costs to the left behind are rarely explored in-depth by state and international agencies that tend to concentrate primarily on perceived benefits. Yet, it is often the "immobile" who are most affected by migration, with the heaviest burdens falling on the women, children, elderly, sick and disabled who remain behind in origin communities.

While some scholars suggest the so-called "feminization of migration" is nothing new, the fact that women make up approximately half of all international migrants has garnered considerable recent attention. The gendered nature of migratory flows varies greatly across space. Countries such as Mexico demonstrate a more masculine migration, particularly among those who migrate independently or are undocumented. Just as the flow of migrants is gendered, so too is the population of those who remain to shoulder the productive and reproductive work in sending communities. Indeed, in most parts of the world women are overrepresented among the left behind. This imposes great economic, social and bodily costs on them and their children.

The costs to the left behind must be well understood and factored into the equation before migration can be considered seriously as a tool for poverty alleviation, improved well-being and development. Contemporary policy discussions on the potential of migration and development often take a celebratory tone, failing to adequately address fundamental questions such as: What is development, and for whom? Who are the winners and losers? In the plethora of migration and development policy reports and discourse, major development brokers such as the World Bank, the IMF, the UN, the Inter-American Development Bank, the OECD states, DFID and USAID focus heavily on migrants, remittances, and sending/origin nation-states—with little attention paid to those family members who remain.[2] For example, the International Organization for Migration's (IOM) *World Migration Report 2013: Migrant Well-Being and Development*, which moves beyond simple, gross measures such as remittances to take a more enlightened migrant centered approach, devotes only one page to "Well-being of migrant families back home."

In many cases, the negative impacts of migration on families remaining are mentioned in passing or given cursory treatment under the category of "human" or "social" costs (Kunz 2006)—a peripheral footnote on unintended consequences. Recently scholars have argued that dominant discourses on migration and development silences the left behind by concentrating on the macro scale such as nation-states; emphasizing the experience of only the migrants themselves; and privileging economic and financial issues (Kunz 2006). This has been countered with a call to "bring the left behind back into view" and moving beyond the heavy focus on broad conceptions of migration as a conduit for development and poverty alleviation through remittances (Toyota et al. 2007).

This chapter challenges dominant migration and development discourse by arguing that transnational migration can exact a considerable toll, not only on the migrants themselves, but on the left behind (particularly women, children and the elderly), which must be given greater attention in current dialogue, planning, policy and research. Following a brief review of migration and development, as well as literature on the left behind, an example of the uneven gendered impacts of migration on the left behind will be presented. This case study draws on research from rural Veracruz, Mexico, which employed narrative interviews, participatory appraisal workshops and a Mexican Migration Project (MMP)[3] survey sample to examine the intersection between neoliberal rural transformation and migration to the US. Findings challenge assumptions of remittances leading to poverty alleviation and development, and illustrate the various social, economic, emotional and embodied costs borne disproportionately by women, elderly and children. This critical account of the gendered consequences of migration seeks to complicate the migration and development discourse, and open the space to bring in the left behind. The chapter concludes with a call for policy and research aimed at inserting the left behind into a more central position within the migration and development policy dialogue and agenda.

The migration and development nexus

Over the past decade and a half a new development mantra has emerged among governments, international development agencies, civil society and certain academic circles, optimistically touting migration as a tool for development[4] (Castles and Delgado Wise 2008; Faist 2008). With United Nations General Assembly High-level Dialogues (HLD) on International Migration and Development in 2006 and 2013 as well as the creation of the Global Forum on Migration and Development (GFMD) in 2007, migration has become a key component of the global poverty reduction agenda focused on achieving the Millennium Development Goals (IOM 2013). The notion of harnessing migration for development is predicated on cash transfers to households; collective remittances generated through migrant organizations, diasporic initiatives and hometown associations to fund public projects in origin nations; and circular or temporary labor migration flows where migrants eventually return and/or invest in sending communities. Migrants, in this discourse, serve as agents of development (Faist 2008; Raghuram 2009; Geiger and Pécoud 2013). They become, in effect, ideal neoliberal subjects—hard-working, self-reliant, individually responsible subjects who subsidize the state by providing: 1) a social safety net for families remaining (particularly medical care and educational expenses); and 2) investment in community infrastructure through remittance matching schemes such as the Mexican "tres por uno" program.

Scholars critical of the celebratory migration and development discourse attribute the recent surge of interest to the failure of mainstream development policy, as Raghuram (2009: 104) so aptly observes: "These development paradigms have come and gone, but inequality and poverty remain a haunting presence in a globalizing world." Castles and Delgado Wise (2008: 9) blame neoliberal policies such as structural adjustment programs promoted by the World Bank and IMF, suggesting that remittances are looked to as "a way of alleviating the ravages caused by globalization: growing inequality, impoverishment and marginalization of large sections of the population." Also associated with neoliberal globalization are the marked disparities in wealth both within and between nations which helps fuel South–North migration in particular. The growth of South–North migration has led to heightened security concerns and a desire to control and restrict migration by wealthier more powerful nations, which is yet another reason behind the intensified interest in migration and development (Castles and Delgado Wise 2008; Faist 2008; Raghuram 2009; Geiger and Pécoud 2013).

Despite the enthusiasm surrounding migration and development at the policy level it is recognized that the relationship between the two remains unclear (Castles and Delgado Wise 2008; Faist 2008; Raghuram 2009); and even the IOM (2013), an avid proponent, acknowledges that migration has not been adequately integrated into development frameworks now being applied at national and local scales. A number of scholars have not only challenged uncritical notions of migration and development but have argued that migration can be detrimental to migrants, their families and sending communities. At the most fundamental level, critics argue that this discourse rarely engages with contemporary critical development studies debates, and serves primarily to reinforce hegemonic Northern-biased Western conceptions of development that have become naturalized—thus leaving no space for alternative visions (Dannecker 2009; Raghuram 2009; Geiger and Pécoud 2013). Development is conceived in primarily liberal economic macro-scale terms with little consideration of sustainability, dependency, place-specific context, hidden social and bodily costs; differential impacts and the role of migration in exacerbating existing inequities; and/or generating new asymmetries within regions, communities, households and families.

Given the centrality of remittances to the migration and development nexus, many critiques focus on the weaknesses inherent in equating money transfers with sustainable economic and social development. One common concern is that the bulk of remittances are consumed for basic needs such as food, housing, health and education rather than serving as investments in "productive" entrepreneurial activities that might generate revenues and jobs that may diminish the need for future migration. Aside from the problematic nature of the "productive/ unproductive" binary that situates reproductive household expenditures as unproductive, this does suggest that remittances form the basis of a social safety net back home, which raises troubling questions. Prominent among these is, should the price of education, health care, infrastructure and other public projects (via matching programs) be placed on the backs of migrants, thus effectively relieving the state of responsibility and transferring the cost to private individuals? Also, even if remittances were sufficient to extend beyond domestic subsistence needs to finance "productive" endeavors, there is an inherent shortage of such profitable investment opportunities within rural origin communities that have few existing, viable businesses, and are suffering from the debilitating effects of population decline and moribund agriculture.

Among other concerns related to remittance-driven development is that it promotes dependency that can lead to and perpetuate a vicious cycle of migration. This is exacerbated by migration-related distortions such as labor shortages, inflation, abandonment of economic activities, depopulation, land concentration and remittance-related house-to-house inequities that affect local sending economies (Castles and Delgado Wise 2008; Dannecker 2009). The latter, in particular, can lead to patterns of uneven development that result in communal dissonance and even conflict. Additionally, it is not unusual for remittances to decrease or dry up altogether over time as migrants establish new lives and families in destination nations. This can leave women, children and elderly parents who depend on remittances from distant or estranged family members in a precarious situation. Remittance-dependent development is also vulnerable to global political economic shifts. For example, in 2009 Mexico experienced a sharp decrease in remittances (Maldonado and Hayem 2013) related to declining migration, the US economic downturn, border militarization, drug-related violence and increasing numbers of returnees. Remittance based development begs the question whether an approach that is so susceptible to global political economic shifts can be considered to be sustainable. Is it not just another form of dependency?

Bringing in the "the left behind"

After many decades of scholarship that virtually ignored gender issues, scholars have increasingly come to recognize the highly gendered nature of migration and its multiple outcomes (Pessar 2005). Indeed, during the past couple of decades, research on women's migration has flourished. Despite this, the experiences of non-migrant women (as well as children and other family) remain understudied (Pessar 2005; Kunz 2006; Toyota et al. 2007). With its heavy focus on migrants, "the migration literature can be said to have thus far 'left behind' the 'left behind'" (Toyota et al. 2007: 158). Kunz (2006: 7) argues, "the people who stay in the communities of origin are not considered relevant actors in this debate and their voices are ignored." While existing literature has focused on the welfare of migrant families and their material conditions, scholars argue that research has been superficial in its treatment of non-migrant family members, particularly women, as passive recipients of remittances (Kunz 2006; Toyota et al. 2007). Toyota et al. (2007) argue that the left behind must be integrated into the center of migration research, and call for a focus on the "migration–left behind nexus," which disentangles the complex relationships between migration and those who remain. Non-migrant families in origin communities, particularly women, often play a key role in facilitating the success of migrants through their reproductive and care work, outside employment and income earning activities, as well as managing remittance (Wilkerson et al. 2009). Moran-Taylor (2008: 91) support this with the contention that "Caretakers are central to the maintenance of transnational migration. These unsung heroes and heroines help migrant parents to work for longer periods in the United States by caring for their offspring and occasionally even sharing the financial costs of social reproduction."

Most of the research on the left behind focuses, in some manner, on the positive or negative impacts of remittances on household income, health, education, children's welfare and women's empowerment (Toyota et al. 2007; de la Garza 2010; IOM 2013). Results have been ambiguous and vary considerably according to the specificities of place. For example, there are studies where migration and remittances appear to have helped children remain in school, while in other cases it serves as an incentive for youth to leave school (Cortes 2008). While children may experience immediate material gains, studies have reported serious negative impacts, including increased juvenile delinquency, vulnerability to abuse, drugs, risky behavior, teenage pregnancy, emotional problems, susceptibility to trafficking, poor academic performance, educational deficiencies, withdrawal from classes, discrimination targeting, increased child labor and increased migration (Cortes 2008). With respect to gender roles and empowerment, results are often mixed depending on existing gender ideologies and patriarchal structures. In some instances women experience new autonomy in the absence of partners, while in others they are subjected to greater surveillance, dependency and restriction (Pessar 2005; IOM 2013). Among other costs noted in the literature are family separation and disintegration; increased inequality; declining/disappearing remittances; community instability; growing numbers of female-headed households; high stress and poor mental health; burdens placed on elderly caregivers; labor shortages; and a self-perpetuating "culture of migration" (Moran-Taylor 2008; Wilkerson et al. 2009).

The risk of conventional analyses that focus primarily on remittances at the exclusion of the diverse embodied experiences of being left behind is that they "over-estimate the gains resulting from emigration and under-value the costs emigration imposes on the overall well-being of families left behind, and on sending communities in general" (de la Garza 2010: i). According to de la Garza (2010), examination of the relationship between migration and development must be approached from a multifaceted perspective that encompasses the multilayered impacts of migration. In order to achieve this, there is a need for multi-scalar, place-specific and intersectional understandings of the consequences of neoliberal globalization and migration; not only for nation-states, communities and households but for different individuals distinguished by traits such as gender, (im)mobility, legal status, age, race, ethnicity and disability, among others. Understanding these differences, across multiple scales in diverse places, is essential to gauging the uneven impacts of migration. A feminist frame of analysis focusing on the different ways in which household members experience, negotiate and respond to neoliberalism and migration through everyday practices, exposes the unequal costs borne by the left behind. Life-history narratives of both immigrants and their families who remain allows us to uncover and understand these unseen costs—notably the harmful and unfair gendered impacts. These "externalities" are typically absorbed by the families of international migrants left behind and are experienced differently by individual members of the household.

Gendered costs of migration in rural Veracruz, Mexico

The following is a brief example of the gendered costs of migration to family members left behind in one community in rural Mexico. While not intended to be generalizable to all of rural Mexico, the migration story in this community is not unusual, and it serves to challenge many commonly held assumptions regarding migration and development. Neoliberal restructuring and development has served to deepen socioeconomic inequality within Mexico, and to usher in a "new geography" of migration to non-traditional destinations in the US, such as the rural south, from new sending communities in rural Mexico, such as southern indigenous communities. This has resulted in marked regional asymmetries, as well as highly differentiated local responses and reconfigurations of social, familial and economic relations involving both internal and transnational migration. Currently, at over $22.4 billion (Maldonado and Hayem 2013), remittances are critical to the Mexican economy, constituting a significant portion of the nation's foreign exchange earnings, as well as an important stopgap measure for a significant number of households. The proposition that migration should serve as a vehicle for development is consistent with Mexico's transition from a statist to a neoliberal paradigm of governance and development. For these reasons, and the likelihood that immigrants to the US are more likely to be male and married with children, Mexico thus provides an ideal context for examining the gendered costs of migration to families remaining.

Migration through a gender lens of analysis

Gender analysis is critical to the study of migration, not only because of the gendered nature of mobility and labor, but also because it is the key social construct upon which we organize our lives and society. Scholars increasingly recognize the need to apply gender analysis to understand migration processes that are mediated by socially constructed gender relations, and that in turn reshape gender dynamics at individual, family, household and community scales. Men and women experience neoliberal practices and migration in deeply different ways, even within the same family and community. Understanding these differences is critical to assessing the feasibility of the increasingly popular proposal that migration should serve as a tool for rural development.

Feminist migration scholarship challenges researchers to avoid generalized essentialist notions of the actors, institutions and structures, recognizing that they are unique and fraught with internal differences, power asymmetries, conflicts and negotiations between individuals. A feminist geopolitical approach contrasts significantly with conventional geopolitical analysis of migration and development, which tends to focus primarily on issues of borders, labor markets, economic effects and broad social impacts. Such macro-scale approaches can mask the localized, asymmetrical and embodied ways migrant families bear the costs, and subsidize the US/Mexico (inter)national integration project through labor, social reproduction, mobility, consumption and extended family networks. A feminist approach reinforces the need to consider the specificities and differences across space, place, gender and race/ethnicity in development policy, planning and implementation.

Mixed methods approach to gender and migration

This case study is based upon three months of intensive fieldwork in the Totonacapán region of northern Veracruz (Mexico). This newer non-traditional sending region is characterized by the decline of the small farming sector resulting from neoliberal agricultural reforms. One consequence has been emigration to the US and domestic urban centers that have become pronounced over the past 15 years. Departing from a gender frame of analysis we employed a mixed methods approach including 66 in-depth interviews with women, elderly, youth, local leaders and officials in 16 different communities (indigenous, mestizo and mixed ethnicity); a Mexican Migration Project (MMP) survey sample of 150 households in one community; and a series of participatory appraisal workshops with women, men, youth, children and local leaders.[5]

To understand how mobility transformed the lives of both migrants and their non-migration family members we interrogated not only the material geographies of migration but also less tangible dimensions such as subjectivities, hopes, fears and desires. For a more complex and nuanced analysis we have relied heavily on narratives in which people provide their own interpretations of place and experience. Respondents' stories can uncover the deeper meanings of less tangible dimensions of transnationalism such as identity, power relations, exclusion and belonging. Evidence of the hidden costs of international migration brought forth in (non-) migrants' stories produce a significant challenge to the notion of migration and development.

These narratives demonstrate the gendered consequences of migration and how they are confronted, negotiated and resisted in people's everyday lives.

Neoliberal restructuring, migration and gender in Veracruz

Results of the Veracruz study reveal how the combined effects of a reduction in state support for small farmers; eliminating agrarian land reform; competition from cheap US commodity imports produced by heavily subsidized American growers; and intensified domestic competition from the well-capitalized Mexican NAFTA "winners" (the larger commercial farms) has dealt a powerful blow to Mexico's small farmers. This resulted in accelerated outmigration to urban centers (i.e., Mexico City), the border maquila zone (i.e., Reynosa and Ciudad Juarez) and the US from new sending regions. Tierra Azul, the MMP sample community, has become a regional center for emigration from the Totonacapan region—housing well-known coyotes (local migration facilitators) and serving as a focal point for the export of labor. Given that Veracruz is a non-traditional sending destination with most migration being post-IRCA, the majority are undocumented, experiencing more restricted mobility and thus more permanent and less circular temporary migration.

Migration is a highly gendered process in the region with about three times as many males as females migrating according to our survey of Tierra Azul. Male migration to the US started in the late 1980s to destinations in North Carolina, Tennessee, Alabama, California and Washington, among other places. Women started emigrating in the late 1990s often accompanying partners or family members. We found that approximately 20 percent of our households were headed by women (for details see Torres, under review). Migration is clearly implicated in the creation of these female-headed households—with more than three-quarters having at least one or more members with migration experience in contrast to only 50 percent in male-headed households.

Gendered costs of migration to the left behind in Tierra Azul

Study results suggest that female-headed households in Tierra Azul[6] (the majority of which are a product of emigration) are economically more vulnerable than male-headed households, and they depend more on remittances for survival. One possible explanation for this result might be that poorer households are those with migrants, but that was not the case. More than half of female-headed households received remittances in contrast to a quarter of male-headed households. Women indicated that these remittances were more important to their livelihoods than did their male counterparts; however, at an average of 355 pesos a month, they received less than men (445 a month) (see Torres, under review). Several female-headed households with migrants, often partners, in the US receive no remittances. Our interviews revealed that this is not unusual either because of poor job conditions in the US or partners who, over time, may lower or stop remittances altogether. In this sense they are also more subject to the ebbs and flows of the US economy and shifting labor markets that are often driven by global economic restructuring.

Much of the research on migration remittances indicates it is common for money transfers to decline over time—notably in the instance of adult children who marry or start families in the US. Findings from our qualitative work indicate a growing problem of estrangement and abandonment of women and children left behind by their partners, often resulting in a reduction or elimination of remittances. Several study participants relayed stories of male partners disappearing or starting second families with "other wives" in the US. In these cases when remittances are diminished or not forthcoming, women must find a way to raise children and maintain households as single parents, while also seeking to sustain themselves through outside employment.

We also found that the female-headed households (including those where spouses are remitting from the US) scored significantly lower than male-headed households on most development and poverty indicators (see Torres, under review). For example, one of the principal arguments supporting migration as an engine for development is that remittances enable children to obtain a better quality and higher degree of education. In the case of Tierra Azul, the survey revealed disturbing disparities in education levels of youth between female- and male-headed households with the former achieving significantly lower educational levels than their counterparts in male-headed households (Torres, under review).

According to our interviews in the community, as well as observations during our youth-oriented participatory appraisal activities, a large number of teenagers drop out of school to migrate or leave immediately upon graduation. In part this is a reflection of a "culture of migration" that is developing in Tierra Azul, but it is also a function of households' inability to support further studies, or the need to seek outside employment to help with household expenses or the education of younger children. This becomes a vicious cycle because, as remittances dry up, youth step in to assist. In many cases families receive remittances when youth emigrate to the US—but those dry up once adult children establish their own family in the "otro lado." This leads to household crises, which in turn perpetuates the cycle with younger siblings.

We found that migration takes a toll on migrant bodies with a significant proportion reporting poorer health after migration. With respect to the left behind, female household heads reported significantly poorer health than male heads. They indicated significantly higher incidence of stress, anxiety, heart problems and hypertension. Indeed, more than half of female household heads reported that they suffered from emotional, nervous or psychiatric problems —a considerably higher proportion than either males or females in male-headed households. These quantitative results were highly significant, and supported much of what we had learned through qualitative interviews and life stories. Women described extreme physical, psychological and emotional stress, which they ascribed to their requirement to play multiple roles: income earners, parents, managers, farmers and caregivers. They not only cared for children and elderly but there were several cases of migrants returning ill or injured when they were unable to access health care in the US. In these instances migrants have become "disposable bodies," byproducts of a system that extracts value for profit without granting them the rights of "legal" workers. It is often the women who pick up the pieces and provide the care once such injured or otherwise damaged migrants return. The lower socioeconomic status of female-headed households is also a likely factor in that group's significantly lower health indicators.

Through interviews and participatory workshops we learned that some of the women's greatest concern with migration was the impact on youth being raised by one parent, grandparents or other family members. Similar to other studies on children left behind they described abusive or inadequate caregivers, drug addiction, alcoholism, teenage pregnancy, low self-esteem and feelings of abandonment. Even young children relayed stories of family separation and abandonment, growing up without fathers who started "other families" on "the other side."

These stories[7] illustrate how women and children disproportionately bear the burden of the "externalities" of transnational migration that has partially been fueled by neoliberalism. In Tierra Azul, many families were found to be in a state of crisis, with women and children enduring the sacrifices of migration while receiving few of its benefits. Indeed, for many women, migration served to place them in a more vulnerable condition, exacerbating their poverty and lack of autonomy. This counters other research conducted in communities where migration has been interpreted by researchers as a source of empowerment for the women remaining, stimulating a more active role in financial management and decision-making, while allowing development of a strong sense of independence. These consequences bring into question generalized notions of migration for development—suggesting, to the contrary, that it can have grave repercussions for non-migrant family members who remain. This also begs the question of sustainability that is so rarely addressed in conversations that uncritically equate remittances with development.

Conclusions

Given their central role in the production and reproduction of migration and the wellbeing of origin community households, non-migrant family members should figure prominently in the migration and development nexus. Migration places added demands on women who remain behind as *de facto* household heads. They are required to shoulder the principal responsibility for the social reproduction and maintenance of labor, as well as outside wage labor and income earning activities, farming/gardening, community work, managing remittances and caring for children, elderly, sick and injured (often return migrants) household members. There is growing recognition of the social and economic value of the largely unpaid investment of "care work" by women that enables the migration and the sale of household labor to generate cash; and that forms a foundation for the global neoliberal economy—an economy that is contingent upon a stable supply of flexible, mobile and cheap labor. The increasing number of single mothers and female-headed households reveals the precarious position of the left behind, often women and children who are highly dependent upon remittances for survival. They face dire economic and social consequences when long-term family separation (a function of constrained mobility given current border/immigration policy) leads to weakened ties, and a reduction or total loss of remittance income over time. Greater risk of poverty, exploitation and marginalization result as a consequence.

Gross macro-scale analyses of exclusively quantitative variables such as remittance earnings or development indicators fail to address the hidden costs of migration, especially to non-migrating family members. These "externalities" extend beyond conventional, capitalistic economic welfare variables that often fail to ascribe value to the reproductive responsibilities and unpaid work of women. Feminist perspectives sensitive to gendered differences can help move analysis and resulting policy away from overgeneralized and reductionist accounts of migration that gloss over key differences and inequities within families, households, communities, regions and nations. In order to more accurately weigh the potential of migration to promote sustainable development and improve wellbeing, it is necessary to understand the experience not only of migrants themselves, but also those family members who remain. Myopic policies and programs that prescribe tapping remittances to fuel development can encourage migration without giving adequate consideration to the place-specific, uneven nature of the costs and benefits of migration at multiples scales to different stakeholders.

Notes

1   The term "left behind" is problematic because, while it suggests that those who remain in origin communities have little agency or choice, that is often not the case. The term also positions non-family members as peripheral to the migrants themselves, a proposition that runs counter to the central argument made in this chapter. Nevertheless, because the term is widely employed in policy and academic circles I use it for convenience and clarity, while acknowledging its shortcomings.

2 UNICEF is a notable exception with several policy documents on the impacts of migration on women and children in sending countries such as Cortes (2008) cited in this chapter.
3 For details on the Mexican Migration Project (MMP) methodology, see http://mmp.opr.princeton.edu/home-en.aspx.
4 Geiger and Pécoud (2013) point out that migration and development has been an object of policy for several decades despite contemporary debates that present the issue as "new."
5 For more detail on the study methodology and the analytical definition of female-headed household employed in this chapter see Torres (under review).
6 This is a pseudonym used to protect the anonymity of the study community.
7 Given space constraints I was not able to present specific women's narratives. To read some selected stories documenting the experience of the left behind see Torres (under review).

References

Castles, S. and Delgado Wise, R. (2008) *Migration and Development: Perspectives from the South*, Geneva: International Organization for Migration.

Cortes, R. (2008) *Children and Women Left Behind in Labour Sending Countries: An Appraisal of Social Risks*, Division of Policy and Practice working paper, New York: United Nations Children's Fund.

Dannecker, P. (2009) "Migrant visions of development: a gendered approach," *Population, Space and Place* 15(2): 119–132.

Faist, T. (2008) "Migrants as transnational development agents: an inquiry into the newest round of the migration–development nexus," *Population, Space and Place* 14: 21–42.

de la Garza, R. (2010) *Migration, Development and Children Left Behind: A Multidimensional Perspective*, Social and Economic Policy working paper, New York: United Nations Children's Fund.

Geiger, M. and Pécoud, A. (2013) "Migration, development and the 'migration and development nexus'," *Population, Space and Place* 374: 369–374.

IOM (2013) *World Migration Report 2013: Migrant Well-Being and Development*, Geneva: International Organization for Migration.

Kunz, R. (2006) "The 'social cost' of migration and remittances: recovering the silenced voices of the global remittance trend," *Second International Colloquium, 'Migracion y Desarollo*, pp. 1–27.

Maldonado, R. and Hayem, M. (2013) *Remittances to Latin America and the Caribbean in 2012: Differing Behavior Across Subregions*, Washington, DC: Multilateral Investment Fund.

Moran-Taylor, M.J. (2008) "When mothers and fathers migrate north: caretakers, children, and child rearing in Guatemala," *Latin American Perspectives* 35(4): 79–95.

Pessar, P. (2005) "Women, gender, and international migration across and beyond the Americas: inequalities and limited empowerment," *Expert Group Meeting on International Migration and Development in Latin America and the Caribbean* 30: 1–26.

Raghuram, P. (2009) "Which migration, what development? Unsettling the edifice of migration and development," *Population, Space and Place* 117: 103–117.

Torres, R.M. (under review) "Migration and development? The gendered costs of migration on Mexico's rural 'left behind'."

Toyota, M., Yeoh, B. and Nguyen, L. (2007) "Editorial introduction: bringing the 'left behind' back into view in Asia: a framework for understanding the 'migration–left behind nexus'," *Population, Space and Place* 13: 157–161.

Wilkerson, J.A, Yamawaki, N. and Downs, S.D. (2009) "Effects of husbands' migration on mental health and gender role ideology of rural Mexican women," *Health Care for Women International* 30(7): 614–628.

38

WOMEN AND PUBLIC SPACES IN RURAL CHINA

Li Sun

Introduction

Rural women are easily ignored by the public and studies on them are few and far between, although the estimated number of rural women in China was around 320 million in 2010 (NBS 2011). This research studies Chinese rural women by exploring their participation in public spaces in rural China. I consider public spaces as places that are open and freely accessible to people (including venues for special events) with face-to-face interaction as an essential element.

In China, different communities have different kinds of public spaces depending on cultural, social, religious, and other factors. I categorize public spaces into four main types in this research based on their different functions: first, "recreational public spaces," which are places of leisure, such as parks and beaches; second, "daily life public spaces," which are closely related to people's daily life, such as markets and malls; third, "religious public spaces," which are locations for religious worship and are organized by believers in order to practice their religious beliefs; fourth, "event-based public spaces" where the main characteristic is that there is no fixed physical place—they are made up of the venues where events might be held, such as sport games and public exhibitions.

The fieldwork site for this research, H village, is located in a mountainous area of southern China with a total number of 667 households and 2,510 people, of whom 1,343 are male and 1,167 are female. The total labor force of the village is 1,292, of whom 690 are male and 602 are female. Of these, there are 650 villagers, mainly male, who have migrated to cities for employment. Therefore, on average, one member of each household has migrated. This phenomenon exists widely in China: nationally, 242 million out of the 671 million rural inhabitants have migrated to urban areas for employment. In 2006, I collected data in H village using qualitative methods, including participant observation, focus group discussions, and semi-structured interviews. I did participant observation in public spaces such as temples and laundry locations. Five focus group discussions were organized in public spaces with female members and 30 interviews were conducted in total.

In the beginning of this chapter, I give a brief introduction to women's participation in public spaces such as the motivation for their visiting, forms of interaction, etc. Second, I point out the various impacts of public spaces on rural women's life and work, including as emotional substitute, providing access to information, transmitting traditional culture, and enhancing social networks. I discuss the differentiated function of public spaces between China and democratic countries and make a few policy recommendations in order to promote public spaces in rural China.

Women's participation in public spaces

Located in a mountainous area with little infrastructure, H village is one of the many closed and isolated communities in rural China. Since there is no road suitable for motorized transportation, it takes six hours for a peasant living on the top of the mountain to walk to the market in the nearest town. Due to this lack of infrastructure, some peasants, especially women, have never even visited the town during their whole life. This restricted environment means that rural women regard public spaces as an integral part of their daily life, because public spaces offer them a chance to meet people and have social interaction.

What public spaces are there in H village?

Through empirical research, I found there were eight kinds of public spaces in H village: the village square, the teahouse, several laundry locations, several grocery stores, a temple, a church, and various venues for weddings and sports meetings. These public spaces were categorized into the four types as identified above. First, the washing locations and grocery stores are daily life public spaces because the main function of washing locations is to fulfill residents' daily need for doing the laundry, and the grocery store is the only place where local residents can buy daily necessities, as there is no market in H village. Second, the square and the teahouse are recreational public spaces, since these places meet residents' needs for leisure and entertainment where people can chat and relax. Third, due to the main functions of the temple and the church, they are classified as religious public spaces. It is important to note that due to financial constraints, there is no formal church building in H village. The local Christian church is situated in one of the members' living rooms, which is regarded as a house church. Finally, whenever there is a wedding or sports meeting, these venues are regarded as event-based public spaces, because the appearance and disappearance of these public spaces depends highly on the event. Although in rural China a wedding is usually held at a villager's house, everyone, even strangers, can attend because rural residents are simply interested in taking a look at the wedding ceremony. Therefore, although the wedding venue is a private place, people regard it as a public space due to both the way they want to use it and the way they think about it (Staeheli and Mitchell 2006).

Who visits public spaces?

Because of men's emigration, women, children, and the elderly are left behind in villages, which is called the "386199" group (March 8 is Women's Day, June 1 is Children's Day, and the ninth day of the ninth lunar month is "Respect the Elderly" Day in China) (Bai and Li 2008: 97). In rural China, the number of women whose husbands migrate to cities for employment reached 47 million in 2010, resulting in what is referred to as "left-behind women" in China. Compared with children and the elderly, women are more likely to visit public spaces, making up the largest group in public spaces.

Generally speaking, poor accessibility due to the local geography excludes elderly people's participation in public spaces. On the one hand, a large majority of older people with poor

Figure 38.1 One of the washing spots in H village

mobility do not have a wheelchair or even crutches because they cannot afford these. On the other hand, the poor infrastructure renders such items useless since one cannot get down a narrow mountain path in a wheelchair and, even with road access, most roads are quite bumpy and muddy. In H village, it is rare to find children appearing in public spaces alone because adults worry about young children's vulnerability to what they described as "stranger-dangers in public space" (Katz 2006). For example, in 2003 one three-year-old girl in a nearby village was kidnapped by human traffickers. Since then, adults in H village appear to be especially concerned about their children's safety in public spaces and guardians do not allow children to go to public spaces alone.

It is worth noting that gender difference is distinctive in some public spaces. Men have difficulties entering a "gendered" public space based on the traditional labor division in households. For example, in washing locations (see Figure 38.1), it is common to find that only women do laundry there, because laundry, cooking, and other household chores are regarded as women's responsibilities in rural China. If a man were to appear in a washing spot to do his family's laundry, other people might think that the husband of that family is "useless" and even the wife of that family might get a bad reputation, as it might appear that she could not take care of her domestic work. Therefore, despite efforts to promote gender equality, nowadays people in rural communities such as H village still hold on to traditional views on labor division according to gender, although this is not so common anymore in cities.

Why do women visit public spaces?

Women have several motivations for going to these public spaces due to public spaces' different functions, which are, first, the pursuit of leisure opportunities: as compared to urban women, rural women have less access to diverse methods of leisure such as malls or mass media. Therefore, they go to public spaces for recreation. Second, to worship one's religion, Christians or Buddhists visit the local church or temple to practice their beliefs (Jaschok and Shui 2011). Third, the need to do domestic chores: for example, some women visit nearby washing locations to do laundry for their family. Fourth, to attend special events, such as weddings, which are quite common occurrences in rural China (Wu 2002). It is interesting to explore women's choices to attend some forms of public spaces. For example, quite a few women went to church due to either their own poor health, or that of a family member. Because they did not have enough money to go to the hospital, they chose to pray in the church instead and had the strong faith that god could help them get healthy again. In another case, it was surprising to find that women participated in sports or games for diversity or variation from the otherwise mundane activities they engaged in.

The primary social networks, such as kinship and the neighborly relationships, play a positive role on the dissemination of information of public spaces in H village, and exclusively oral communication is the only channel through which participants come to know about the public spaces. Most women visit public spaces on their own initiative, while some women are invited to attend, especially if they are sick or unhealthy: these women were invited to attend the activities in temple and church by senior members, who believed that the patient would recover through religious practice.

What are the duration and frequency of their visiting?

The duration and frequency of women's visits are varied. For example, one Christian had already attended church activities for more than 45 years, which is the longest duration of involvement in public space in H village. On the other hand, the shortest duration of involvement was less than a year, which was in the annual Peasants' Games. In general, the elderly got involved in public spaces much earlier than the young people, so the duration for them is longer; local women also have longer duration than migrants from outside. Meanwhile, different functions of the different public spaces result in the different frequency of visiting. For example, women logically go to wash locations or grocery stores almost every day, which means such places have the highest frequency among public spaces. However, they only have the opportunity to attend Peasants' Games once each year. Religious public spaces, on the other hand, have a stable schedule and hold activities regularly, so believers have a fixed frequency of visiting these places. For example, Christians go to church to pray every Sunday, so the frequency is once per week, while Buddhists go to temple on the first and fifteenth day of each month according to the lunar calendar, so the frequency is twice per month.

When asked how long they stayed in a particular public space during each visit, the answers varied by space. The longest was weddings lasting several days; the shortest one was a visit of only a few minutes to the grocery store for buying food and goods. In general, time spent during a visit is quite flexible for people who attend recreational and daily life public space, ranging from several hours to only a few minutes, depending on the individual preference of participants. However, in religious public spaces, peasants usually stayed one whole day, for example, every Sunday, Christians stay in church from morning until late afternoon, including a lunch break. The length of time peasants are involved in event-based public spaces is not flexible either, being decided by the specific arrangements of those responsible for organizing the event.

How do they interact with each other?

Chatting and activities are the two most common forms of members' interaction in public spaces. In general, chatting is the main form of interaction in recreational public space and daily life public space. The topics of conversation vary from one public space to another. For instance, in the washing locations where there are no men around, the topics are private and exclusively among women, with a focus on gossip and domestic affairs. In washing locations, women feel very happy and relaxed, free to talk and laugh naturally since, according to the social standard for a good woman in traditional China, women are supposed to control their laughing and speaking voices in public spaces. However, the interactive atmosphere in the grocery store is different from that in the washing locations. Women are usually much quieter when men are present and the chatting topics become more generalized.

Besides chatting, shared activity is the other form of interaction in event-based public spaces as well as religious public spaces, such as prayer and worship in the latter. Taking the activities in the church as an example, throughout the day Christians participate together in different activities: they start their worship at 8am by singing for half an hour, and follow it with an hour of praying. Then, after a short break, the pastor tutors the participants on Christianity for an hour. Everyone (except those who are illiterate) has a copy of the Bible as well as a book with poetry. After Bible study, they sing again for half an hour until 11:30am. Then they cook and have lunch together in the house, after which they repeat that morning's activities, singing, and Bible study, until 4:30pm.

The impacts of public spaces on rural women

The most obvious function of public spaces is that women can satisfy their initial motivations, for example, domestic chores, leisure, or religious activities. However, what rural women get out of public space might actually go beyond their expectations. I find that public spaces have a significant impact on rural women's daily life and work, which can be explained through four aspects.

First, public spaces have a positive influence on women's mental health. As mentioned, due to men's emigration, there are a large number of left behind women in rural areas, who take on various responsibilities: continuing to perform all domestic tasks as before as well as taking over tasks that men did before emigration, such as agricultural production. This new form of labor division changes rural women's status in the household from an invisible contributor to an active laborer in agriculture, which is called agriculturalization of females (Gao 1994). Besides the heavy physical burden, left behind women suffer emotionally due to their husbands' long-term absence. It is common that emigrated men visit the village only once a year, during the period of the Chinese New Year, during which they stay at home for two weeks. Without husbands' substantial emotional support, left behind women feel socially isolated and are more likely to have mental illness such as depression. In this context, public spaces offer them a place to meet people and find emotional support to reduce their loneliness, which is beneficial to their mental health.

Second, as they have little access to news or information through mass media, rural women always rely on word-of-mouth communication. Therefore, public spaces are essential for women's access to information. The public space as a physical site has the function of disseminating information to participants in diverse ways. On the one hand, through chatting with other members in public spaces, women can share or get information and news of farm activities or regional news, etc. On the other hand, it is common to see commercials or notices

Figure 38.2 Yangko dancing in the square

hanging on the walls of the grocery store or other venue. In H village, the teahouse is a place for public information campaigns; for instance, the village committee organized the open air show to spread healthcare information on severe acute respiratory syndrome (SARS) and the swine flu pandemic.

Third, public spaces provide venues for women to transmit traditional culture. Public spaces and culture are interrelated: on the one hand, the public space is one part of culture. For example, Wang (1998) did a study about public space in Late-Qing Chengdu, pointing out that popular culture absorbed elements such as public theaters, public parks, new teahouses, and public exhibitions. On the other hand, public spaces have a positive impact on cultural transmission (Abaza 2001; Amin 2008; Low 1997). This research finds that public spaces in H village, such as the square and the teahouse, are important sites for rural women to share traditional Chinese culture. For instance, in the evening, a group of middle-aged women get together in the square and practice "Yangko" (see Figure 38.2), a traditional Chinese dance. One member commented, "This square was built in 2005; before that, there was no place for us to get together for dancing. Thanks to the square, now, we can practice Yangko regularly."

Fourth, women' social networks are enhanced through participating in public spaces because they can meet up with old friends or build new friendships there. Since kinship or local relationships with families, relatives, and neighbors make up most of rural women's social networks, public spaces in H village provide a platform and a chance for them to enhance their social networks through intensive and close interaction. It is important to note that the number of participants in a public space is not equal to the number of people women get to know, because no single person in public spaces can have direct interaction with all participants.

For example, the largest number of participants in the temple at any one time was about 500, but each participant may have had the chance to get to know only a few other members. However, in the church, participants get to know each other and they have a close relationship with each other because, first, the average number of participants is around 20, a manageable number for all participants to have one-on-one interaction with each other. Second, as the church is situated in one of the members' living rooms, the limited size forces participants into close interaction. Third, during prayer, each member speaks out about their true feelings and ideas in turn. Members gain a deep understanding of each other through this practice.

Public spaces promote the emergence of rural women's new social networks as well as the maintenance of their existing social networks. Enhanced social relationships can provide resources that could facilitate rural women's life and work both physically and emotionally. Both observation and interviews made it clear that finding friends in public spaces resulted in mutual assistance during critical events such as illness, house construction, or agricultural harvest. This enhanced social network from public spaces is essential, especially for left behind women, in that it can even sometimes substitute for the function of kinship. For example, friends gained in public spaces help left behind women with agricultural activities during the agricultural high seasons, substituting for the emigrating men who would have normally performed this function (Sun 2012).

From the above, we can see that public spaces play an important role in rural women's everyday life. However, there are a small number of women who are not willing to go to public spaces because of various reasons. On the one hand, some women feel embarrassed when they are in public spaces because of poverty or family problems. For example, a 57-year-old woman who lives only 500 meters away from a washing location hasn't been there for more than four years. The main reason is that her three sons died in accidents within two years, leaving her childless. Since her sons died, she tries to avoid going to any public space, anticipating strong nostalgia for her sons:

> At washing locations other women always talk about their sons or daughters, and I feel very sad and miss my sons when I hear these topics. In order to control myself, not to miss or think about my sons, I like to carry water back to my home and wash clothes at home.

On the other hand, some women prefer to stay at home rather than go to public space because they have an introverted personality. For instance, one 41-year-old woman seldom goes to public spaces, although she lives near the main road, which is very convenient to go to the grocery store, the square, or teahouse. If she needs to buy something from the grocery store, she goes there only to shop without staying to chat, simply because she does not like chatting with people in groups.

Conclusion

Through empirical research in H village, this chapter has sought to reveal women's participation in public spaces in rural China, and explore how public spaces affect rural women's everyday life. There are eight kinds of public spaces in H village, which are identified into four types based on their main functions: daily life public spaces (washing spots and grocery stores), recreational public spaces (the square and teahouse), religious public spaces (the temple and the church), and event-based public spaces (sites for weddings and sporting events). Due to the large number of men migrants, the majority of the participants in public spaces in rural communities consist of village females. I find public spaces play an essential role in rural women's everyday life: first, public spaces have a positive influence on women's mental health; second, public spaces are vital for women's access to information; third, public spaces provide venues for women to transmit traditional culture; fourth, women' social networks are enhanced through participating in public spaces, and these enhanced social networks significantly facilitate rural women's life and work both physically and emotionally. As public spaces play an important role in rural women's everyday life, it is necessary for the government to promote the development of public spaces in rural communities. One measure of the "New Countryside Construction" policy, carried out by the Chinese central government since 2005, is to improve rural infrastructure and villagers' everyday lives, and it is clear that public space is one kind of rural infrastructure. Therefore, public funds from this policy could be allocated to these projects for public spaces. There are a few specific suggestions with regard to public spaces in rural China. Second, when constructing public spaces, the demands of different groups of villagers should be taken into consideration. Because of the various characteristics of different villagers (such as gender and age), their demands for public spaces vary. For example, women would like some special places where they can talk freely and relax outside of men's presence. Third, diversified collective activities should be organized in public spaces more frequently in H village, such as the Peasants' Games and open air movies. This would make rural women's daily lives more colorful, significantly improving their quality of life. Last but not least, rural women's opinions should be taken into account when designing and building public spaces in rural communities. Due to men's migration *en masse*, women make up the majority of residents in rural China, so their voices are vital in order to maximize the potential functions and use of public spaces.

References

Abaza, M. (2001) "Shopping malls, consumer culture and the reshaping of public space in Egypt," *Theory Culture Society* 18(5): 97–122.

Amin, A. (2008) "Collective culture and urban public space," *City* 12(1): 5–24.

Bai, N. and Li, J. (2008) "Migrant workers in China: a general survey," *Social Science in China* 29(3): 85–103.

Gao, X. (1994) "China's modernization and changes in the social status of rural women," in C.K. Gilmartin, G. Hershatter, L. Rofel, and T. White (eds.), *Engendering China*, Cambridge, MA: Harvard University Press, pp. 80–100.

Jaschok, M. and Shui, J. (2011) *Women, Religion, and Space in China: Islamic Mosques and Daoist Temples, Catholic Convents and Chinese Virgins*, New York: Routledge.

Katz, C. (2006) "Power, space, and terror: social reproduction and the public environment," in S. Low and N. Smith (eds.), *The Politics of Public Space*, New York: Routledge, pp. 105–122.

Low, S.M. (1997) "Urban public spaces as representations of culture: the Plaza in Costa Rica," *Environment and Behavior* 29(1): 3–33.

NBS (2011) *China Statistical Yearbook2010*, Beijing: China Statistics Press.

Staeheli, L.A. and Mitchell, D. (2006) "USA's destiny? Regulating space and creating community in American shopping malls," *Urban Studies* 43(5/6): 977–992.

Sun, L. (2012) "Women, public space, and mutual aid in rural China," *Asian Women* 28(3): 75–102.

Wang, D. (1998) "Street culture: public space and urban commoners in Late-Qing Chengdu," *Modern China* 24(1): 34–72.

Wu, Y. (2002) "Public space," *Zhejiang Academic Journal* 2: 93–94.

39

THE INFLUENCE OF GENDER AND ETHNICITY IN THE CREATION OF SOCIAL SPACE AMONGST WOMEN IN RURAL SRI LANKA

F. Munira Ismail

This chapter deals with an analysis of the results of an empirical study that documented and analyzed the construction, negotiation and maintenance of social space amongst rural women in Sri Lanka. The study compares the spatial movement of women belonging to the three major ethnic communities in Sri Lanka and tries to problematize the concept of gender and space by analyzing the influence of ethnicity, religion, age, marital status and family structure on the creation of a gendered and (a)gendered social space. The study highlights the fractured, fragmented and nested social spaces that are representative of the lived realities of many non-Western women (Lefebvre 1991).

A growing body of literature examines the mechanisms by which spatial boundaries and the intersection between male and female spaces are crossed and constructed in the 'third world'. This literature draws attention to the fact that although the term 'gendered space' provides a useful analytical tool for looking at the 'gendering of space and the spacing of gender' (Huang and Yeoh 1996), this and other dualisms such as the public/private dichotomy are based on Western experiences of the creation of social space. Therefore while the term gendered spaces with its inherently bifurcated focus may provide a useful analytical category to study the lives of Western (urban) women, arguably due to the historic manifestation in the wake of Western industrialization and capitalist modes of production that segregated wage earning occupations by gender, it fails to uncover the gendered social and spatial lived realities of the non-Western women whose social and economic histories do not always lend themselves to an analysis that is bounded by primarily dichotomous restraints. Many times there exist, as we shall see, spaces of social reality that are created independent of the need of a contrasting space to define it. As articulated by Rose (1993), there is a need to displace the exhaustiveness of gendered perspectives by asserting the importance of other axes of identity such as ethnicity, age and religion. The flexibility of (a)gendered social space and its ability to morph into social spaces that use as their axes other identities such as ethnicity, religion and language is thus the focus of this paper.

Research for this paper was conducted in a village near Kandy in the Central Province of Sri Lanka (Ismail 1999). It was a small village with a total of 243 households of whom 68 were interviewed. The village was chosen because it had the three major ethnic communities of Sinhalese, Muslims and Tamils, albeit 'Estate Tamils' of Sri Lanka, in rough proportion to national ratios. Sample households were selected with the help of the 'village headman' (also known as the Gramasevaka) using a sorting process based on family last name. For every household chosen a female was also chosen for interviewing. The research techniques used ranged from focus group discussions, interviewing, participant observation, digital tracking of 20 women using a GPS recorder, daily diaries and observation maps that were used to verify GPS data. The sample of households was stratified by ethnicity, age, marital status, number of children, age and level of education. The sample of women was made up of women between the ages of 12–55 years with age playing an important role in defining the extent of spatial mobility. As a general rule it followed that the older the women the greater the range of spatial mobility. However, issues of ethnicity, physical safety, sexuality, religion, language, dexterity and dress had a profound impact on this simplistic generalization. Women between the onset of puberty and menopause were found to be the most spatially restricted in their movements. Prepubescent and postmenopausal women and children were significantly freer to create their own spatial boundaries. More pragmatic considerations of safety rather than socially imposed gendered spatial constraints such as the need for an appropriate escort or particular forms of dress, as was the case with Muslim women who felt the need to cover their hair, were used by women to access unsafe areas or gender neutralize some of the perceived restrictions of spaces around the village.

The household size of the sampled population, particularly in terms of the number of younger children, was found to also have a profound impact on the creation and expression of social space in the village. Families with younger children felt forced by the needs of the children to adjust their daily paths around issues of schooling, play and late afternoon lunches. Mothers tried to pack in all of their chores in the morning in order to get back home to cook lunch and await the arrival of their children from school. Families with older non-school age children were less motivated by the pressures of their children and thus developed a more village-based spatial agenda that revolved around issues of paid work, which in the case of all the women interviewed meant either resizing cement bags into smaller bags and pasting them, or rolling cow dung and perfume into 'jossticks', aka incense sticks. Families with no children were much more eclectic in their spatial and social interactions. With no family induced spatial dictates their interactions were more a reflection of personal preference. In these cases ethnic affiliations were more pronounced (De Silva 1986).

This was a re-settlement area established by the government in the 1970s on an old tea estate for landless workers as part of a land reform program. The village was new and only one-quarter mile from a pre-existing town which happened to be predominantly Muslim and hosted many facilities and amenities such as a post office, businesses, restaurants and clothing stores etc. that were not available in the study village. This should have been the obvious town of choice for all the women of the village to conduct their shopping and other business. However, the Sinhalese women of our sample rejected this space and preferred to access an alternate space by either walking or taking the bus over a hilly two miles to the neighboring town which was predominantly ethnic Sinhalese and where Sinhalese was commonly spoken. In this case ethnicity, and more importantly language, was more of an incentive than distance or time. The closest town to the village was predominantly Muslim and used Tamil as the lingua franca and this intimidated the majority Sinhalese who did not, for the most part, speak Tamil. The Estate Tamils and some of the Muslim women in the sample did all of their shopping and trade in the nearest town, saving themselves time and money. This ethnic segregation of social space set up alternate spaces of mobility and social interactions that etched an interesting spatial dynamic amongst the women. The Sinhalese created an alternative geography of resistance by banding together and making the rather arduous walk or bus ride to the more distant town. Many times they had the men carry their jossticks on the backs of their bicycles whilst the women all walked to town together. The trek back to the village, rather than being a chore, was looked forward to as a women only social group where they compared purchases, traded stories and bonded. The Muslim women who for the most part were bi-lingual would straddle these alternate gendered spaces by speaking in Sinhalese or Tamil and using language as a passageway into the ethnically sensitized social space. Unfortunately for the Estate Tamils this was not an option because very few spoke Sinhalese and if they did it was so heavily accented that it carried with it the mark of an outsider. Given the wide ethnic distrust between Sinhalese and Tamils on a national scale there was not much interaction between the Estate Tamil and Sinhalese women. The Muslim women however used language as a social lubricant to enter the gendered spaces of the Tamil women by speaking in Tamil and shopping in the Muslim town. When it suited them they entered the ethnic enclaves of the Sinhalese women and men by speaking in non-accented Sinhalese and or using modest dress to desexualize social spaces when in the presence of some men. Although the Muslim women might have donned a head scarf to venture out of the village thus marking themselves as ethnically different, it was not seen as a deterrent to socializing with other ethnic groups at the time of this research in 1999 (Ismail 1999). This may not be the case fifteen years later .

The household space

The lived spaces of the homes of women belonging to all three ethnic communities followed a typical floor plan. In most cases, the front door opened into a long room that spanned the breadth of the house. It often had two doorways leading off of it: one to the bedroom occupied by the male members of the family and the other to a short corridor that led to a second bedroom and kitchen. The house was surrounded by a compound that was usually a Kandyan forest garden with vegetables, fruit and spice trees grown for subsistence and at times for sale in the village store. An outhouse or bathroom was located within the compound. A clothes line was a common feature as well. Inside the house an extra bedroom or possibly a dining cum family room added variation to an otherwise remarkably similar floor plan used by all communities. The social spaces created within these structures however varied considerably by ethnicity.

The Muslim household

In the case of the Muslim household one of the most notable features observed was the 'sexualization' of social space especially in the presence of unrelated members of the opposite sex and the consequent gender orientation of that space. In most rural Muslim households interactions between the sexes are highly controlled and regulated. The conceived spaces of the Sri Lankan Muslim household mirror these conservative ideologies by creating flexible gendered spaces that are evoked in the presence of a 'trigger' such as a visit by a male or older female relative. For example, based on our interviews, most of the women identified the verandah, which is the room into which the front door opens, as gender neutral. However under certain circumstances, such as in the presence of non-filial male visitors who were entertained in this room, this space became sexualized and consequently gendered albeit only for a select subset of women of the household who were of child bearing age. Most interesting is that even though these spaces are sensitive to gender identities, these identities are flexible and can be transformed by mitigating factors. In the case of Muslim women, appropriate dress with a head scarf or the presence of an older relative might be the conduit through which gendered spaces are neutralized. This flexibility is also seen in the lived realities of these women outside of their homes.

The Tamil household

Amongst the Estate Tamils, the creation and maintenance of household space is very different to that of the Muslims or Sinhalese. Hindu inspired patriarchal caste sanctioned family traditions bring with them gendered family practices that dominate the household life of Tamil women. These women are influenced in their everyday lives by the internal dynamics and cultural expectations of married life that is taught to them by their mothers and instilled after marriage, somewhat harshly, by their mothers-in-law. Internalized gender socialization through various cultural practices such as religion, caste, wealth, status and ethnic identity create privileged and separate social spaces exclusively for the husband whose culturally superior status over his wife acts as a catalyst in fragmenting the social spaces of the home into a complex quilt of gendered space. According to the daily diaries kept by the women interviewed, the Tamil wife wakes up at dawn to make her husband's cup of tea and breakfast. He is served his breakfast before anyone else in the household. It is only after he is taken care of that the children and wife take care of themselves. If the sons are older, then they might join their father at the table. Mothers and daughters eat later. The social spaces created within the structure of the home of the Indian Tamil household is thus nested within issues of cultural and social identities that manifest themselves in the form of marriage. Within this structure the wife is subordinate to her husband and his mother. The oldest male child enjoys privileged sleeping arrangements, food and clothing and is often allowed into discussions amongst elders that are closed to younger members of the family and sometimes even to older females. Factors such as age, gender, order of birth and even physical beauty combine to provide a complex variety of social arrangements that act as catalysts to the creation and maintenance of social space.

The Sinhalese household

This was the least spatially segregated household. In fact when the women were asked to identify the spaces of the home, by gender, they classed all of them as being 'neutral'. In the absence of strong gender inspired social spaces the two most active identities responsible for the fragmenting of social space were age and filial ties. Sinhalese women found it amusing to hear us insinuate inappropriateness in talking to and accommodating a non-filial man in the kitchen or veranda. The usual response to our question of appropriateness was 'oh no, he is like my brother' or 'oh that is so silly, he is like a father to me'. In general Sinhalese women were for the most part unrestrained either physically or mentally from accessing any of the spaces of the home. Individual family issues were of course the exception as they were in any other house as well. Because the Muslims were so eloquent in Sinhalese, in many instances the Muslim men, both young and old, were entertained in the kitchen as well. We never saw an Estate Tamil male being socially accepted as a friend within the home of a Sinhalese.

Missing men

The majority of the men in the households sampled were not permanent residents of the village. Singhalese made up four-fifths of the households sampled and in these families some

70 percent worked as laborers in Colombo or Kandy returning home either at the weekend or less frequently. Those who did live in the village worked as shop assistants in other villages or in the tea factory. Twelve percent of the sample were Muslim households and of these half the Muslim men worked in the Middle East or in Kandy and Colombo, returning home occasionally. The remaining households were Tamil and these men were all residents of the village working in the tea factory. Overall, few men were permanently present in the village.

The village

The women frequented the village store for their daily supplies. Longer trips were very much motivated by language and ethnicity as explained above. The well for water and bathing, the public stand-pipe for drinking water and neighbors' houses were all part of the daily spatial patterns traversed by these women and rarely exceeded a total of 200 meters. Both males and females were cautious about leaving home after dark, citing fear of snake bites and other animals and a general apprehension. These socially created spaces and times of spatial restriction were somewhat less intimidating for the older women past menopause. They had no fear of drunks, having acquired a pseudo-male status with age, and often a much more feisty spirit to that of the older men in the village. Age and gender was an ideal spatial lubricant that could traverse and negate some of the other variants of spatial constraint

Conclusion

This study problematizes concepts of gender and space by analyzing the factors of gendered identity that have a real impact on the way women spatially organize and live their lives in poor rural communities. By studying the creation and maintenance of social space in a Sri Lankan village it became obvious that although gender was important in the creation of social space it was not the dominant criterion. Other social variables such as ethnicity, religion, age, household size, sex, filial rank and status were sometimes equally or more influential in the creation and maintenance of social space amongst women belonging to the three ethnic communities studied.

This chapter would be incomplete without a few words regarding the lived realities of women in post-war Sri Lanka. With the end of the civil war in 2009, Sri Lanka saw a large rise in the number of female-headed households in both the Sinhalese and Tamil ethnic groups. Whilst it is true the end of the war led to a freedom to travel widely throughout the island not felt in over 30 years, for the many widowed women left behind to look after young children, this meant they had to venture into a public domain they might not have otherwise entered. To many, especially Tamil households, conservative spatial restrictions, as were observed in the village, had to be ignored as women had to go out to work. However, it also left many of these female single heads of families feeling vulnerable to sexual assault by the jubilantly victorious Sinhalese soldiers, many of whom were discharged from the army where they had been indoctrinated with decades of ethnic hatred.

Also to be mentioned are the brewing ethnic tensions between the Muslims and Sinhalese. Many pro-Sinhala Buddhist nationalists who would ideally like a Sri Lanka void of any minorities are now finding cause to focus on the Muslims. In the past the Muslims of Sri Lanka had maintained a very low profile within Sri Lankan society and shared the public space. However, recently a conspicuous affiliation to fundamentalist 'Wahabbi'[1] Islamic interpretations of the religion promoted by Saudi Arabia, together with the physical exposure to this brand of Islam by many Sri Lankans who travel to Saudi Arabia for work, have made the Muslims very conspicuous within Sri Lankan society. Many Sri Lankan Muslim women have adopted the Saudi Arabian dress called the 'Abaya' as an ethnic identity. This has created distinct gendered social spaces that exclude other ethnic groups. In a country that has just crawled through a long and costly ethnic war that ended with the minority being violently subdued by the State, this might lead to another unfortunate bout of ethnic violence (Tyler and Cohen 2010).

Note

1   A form of fundamental interpretation of Islam made popular by the Saudi Arabian clergy that received royal support.

References

De Silva, Kingsley M. (1986) *Managing Ethnic Tensions in Multi-Ethnic Societies: Sri Lanka 1880–1985*, New York: University Press of America.

Huang, S. and Yeoh, B.S.A. (1996) "Gender and urban space in the tropical world," *Singapore Journal of Tropical Geography* 17(2): 105–112.

Ismail, F. Munira (1999) *A Multi-Ethnic Analysis of Gendered Space amongst Rural Women in Sri Lanka*, unpublished doctoral dissertation, Department of Geography, University of California, Davis.

Lefebvre, Henry (1991) *The Production of Space*, translated by Donald Nicholson Smith, Cambridge, MA: Blackwells.

Rose, G. (1993) *Feminism and Geography: The limits of geographical knowledge*, Oxford: Polity Press.

Tyler, M. and Cohen, L. (2010) "Spaces that matter: gender performity and organizational space," *Organizational Studies*: 193–198.

PART VI

Conflict and post-conflict: victims or victors?

40

INTRODUCTION
TO PART VI

In April 2003, just after the "end" of the Iraq conflict, the Kosovan Women's Network sent an open email message to the women of Iraq. They described how initially they had welcomed the UN peacekeeping force because the UN had passed important resolutions on women's rights and promoted women's participation in all levels of decision–making. But they were soon disillusioned; most of the agencies "did not recognize that we existed." Instead of being able to focus on helping fellow women rebuild their lives, their energies went on fighting to be heard and for their perspectives and contributions to a peaceful future to be taken into account. They urged the women of Iraq to be forceful and proactive from the start (quoted in el Jack 2003).[1]

The key UN document that sets out women's role in conflict situations is Security Council Resolution 1325 of 2000 on "Women, peace and security." It is a remarkably comprehensive document. It recognizes that civilians, especially women and girls, are the vast majority of those affected by armed conflict and that increasingly they are targeted by combatants. It calls on all parties involved in armed conflict to take special measures to protect women and girls from violence, notably gender-based violence and especially rape. Importantly, it also recognizes the significant role that women play in the prevention and resolution of conflicts, calling on the institutions involved to include women as equal partners in peace-building at all levels. Sadly, as the women of Kosovo learnt, there is a gap between such resolutions and the reality on the ground. The gender-sensitive training that the institutions need, and the support that may be required to give women's organizations an effective place at the table, are frequently lacking and their voice in peace-building and planning for their societies' future goes unheard. (For the similar exclusion of women from high level decision-making, this time after disasters, see also Vu et al. in Part VIII). Arguably 1325 has had more effect in relation to protecting women than in involving them in peace-building.

In recent decades many conflicts have occurred in places characterized by long, ongoing low-intensity violence, which then flared up into armed conflict. When this episode is over, without good peacekeeping and the rebuilding of societies, violence continues in the aftermath. There is no clear distinction between civilians and combatants; modern weapons and communications penetrate even "safe zones," causing widespread damage and displacement of people, especially in densely populated urban areas. While armed combat may spread beyond state boundaries and while external players may encourage the conflict, much occurs internally as civil war. (Old-fashioned stereotypes of international wars, with confronting armies and well demarcated front lines seem inappropriate in such circumstances.) The present actors include, as well as armies, rebel groups, informal militias, drug traffickers, and the like (Green and Sweetman 2013).

As Resolution 1325 recognized, those most affected at all stages are likely to be women. This was brought to public attention, especially after the atrocities in Rwanda and the Balkans in the 1990s. Afterwards the relevant tribunals recognized rape as a weapon of war. Sexual violence of all types typically increases dramatically in conflict situations. Women and girls are especially vulnerable because their male kin are fighting elsewhere. Combatants and, later, even peacekeepers, have been able to seize opportunities because normal societal constraints have broken down. Raping women can be a way of triumphing over the opposing side by dishonoring women and degrading their men's masculinity. Rape that leads to pregnancy is particularly devastating. It can cause women and their offspring to be abandoned and render girls unmarriageable. However, not all sexual violence is directed against women. Male-on-male rape may also be used to humiliate men (see chapter by Nibbe).

Conflict reaches women in their homes and not only because of sexual violence. Their workload increases enormously. So do their responsibilities. They have to hold the family together, both as providers and protectors. Growing and sourcing food and other necessities are more difficult. Health and education services may be disrupted but children (both boys and girls) may be needed anyway to help with these tasks or earn additional income. Displacement is disorienting and disrupts normal social networks and support systems. Leaving home may be especially difficult if it involves many dependents, the elderly, infants and toddlers. Some women may resort to selling sex to maintain themselves and their families. In such circumstances camp life may become preferable (see Nibbe) even though sexual abuse continues.

However, women are not always helpless victims of conflict. In some cases women have virtually sustained an area's economy during hostilities, using innovative strategies to maintain trade and communications. They may also be active participants, indirectly or directly, as propagandists, as supporters of male fighters at individual and national level, as undercover informers or as combatants. The proportion of women in armed forces is 10–30 per cent (Green and Sweetman 2013). Many are girl soldiers, sometimes enlisted by force, sometimes escaping family troubles. While a very few are temporarily empowered by the bearing of arms, most are degraded and brutalized, even while brutalizing others. Many become virtual domestic slaves, sexually abused and taken as army "wives." It may be particularly difficult for them to return to "normal life" after the conflict. Families may be reluctant to take them back. Compared with rehabilitation and reintegration schemes for boy soldiers, there are few for girls. In particular girl soldiers, but also sexually abused women and girls, need comprehensive reproductive health services that are usually lacking.

Women contribute to peace-building in many ways. They provide humanitarian relief and contribute to social welfare, making use of their traditional roles as women. They help create space for negotiation through advocacy at the national and local levels, for example in Northern Ireland (see also chapter by Parra-Fox). They have often spearheaded reconciliation activities. Women are necessary to the peace process because they have experienced the conflict differently from men and have a different perspective on what peace means and what needs to be changed to achieve it. They are often the "voice of civil society," bringing a strategy of greater inclusiveness and consensus building to the political arena, but also working outside the formal negotiating table. In this they can make use both of their traditional roles and their specific perspective on how society might be improved to achieve a sustainable peace. They make an important contribution to social justice. As Gardener and El Bushra (2013: 11) write, "Women's goals in peace-building have been both to improve society in general and to improve women's participation within that society."

There are, however, formidable hurdles to be overcome if women are to be successfully involved in formal peace negotiations. Continuing unsettled violent conditions make it hard or dangerous to travel to meet or take part (Moosa et al. 2013). Generally women lack the political experience necessary and the understanding of the processes involved and may lack the self-confidence to engage with the elitist male environment. They lack the resources, capacity-building, money, and time for the necessary sustained engagement. Networking may be needed to present a common view as priorities may differ between groups, for example, between elite women and grassroots organizations. The report by Accord on *Women Building Peace* (Accord Conciliation Resources 2013: 16) builds on the insights and initiatives from nine case studies from 1998 to the present to provide recommendations for policymakers to enable the peace process to involve women as participants. These include providing resources to women's movements to deal with the hurdles facing them, maximizing the benefit of their contribution by timely support. Women should be recognized as valid political actors. They should be involved from the start since it is much harder to add them later. Given women's experience of the violence, much of it sexual, immediate security issues certainly require their input. They should "support women's own varied and broad-based initiatives" and their ability to network across the divide. Such an approach can, the report suggests, succeed only if all the parties adopt a gender mainstreaming inclusive approach.

Two chapters in this section, those by Parra-Fox and Harris, report on very different approaches to promoting peace, illustrating the range of innovative and practical strategies that women may employ. Parra-Fox gives an unusually detailed description of the emergence of a women's peace movement. As one of its leaders, she provides a vivid account of La Ruta, a peace-building alliance of women's groups formed in Colombia, where the women's movement had grown since the 1980s, and had been, briefly, politically influential. La Ruta itself was founded in the mid-1990s as a response to the upsurge of violence that had afflicted the country for several decades, so that Columbia "had the dubious distinction" of being the country with the most displaced people in the world. Parra-Fox's story of La Ruta has the enthusiasm and emotion of a personal account and conveys the excitement of an inclusive grassroots movement, but valuable reflections set it in the context of political theory. Its strategy for addressing violence, displacement and rape is innovative. Confronting the establishment, La Ruta highlights women's own perspectives on war. Its means have included symbolic, artistic, and theatrical events and rallies. The violence has continued,[2] but La Ruta has increased the self-confidence and power of its participants, and has raised the consciousness of the plight of Columbian women both internally and internationally, especially within Latin America.

Harris describes her own key role in three initiatives to reduce continuing violence between community members in the aftermath of conflict, using a participatory approach rather similar to that of Welbourn's Stepping Stones (2013). Both involved using separate groups of young men, young women, older men, and older women exploring the causes of the violence, particularly sexual violence. Harris worked in Tajikistan after the peace in 1997, in Kaduna city, Nigeria during a lull in the tension in 2011, and in Acholiland, Uganda, between 2007 and 2011. In all three areas, poverty and uncertainty in a post-conflict situation increased intra-familial and intergenerational tensions. These were exacerbated by pressures on grown men's masculinity, as socioeconomic disruption and poverty challenged their patriarchal status as family heads. To help communities deal with this, Harris set up the education programs, shaping the approach and training local facilitators. Participatory gender analysis played a significant role. The effect has been to reduce domestic and community-level violence, showing the crucial importance of underlying gender norms and demonstrating that, with encouragement, communities can decide to mitigate and change these.

Nibbe's study also took place in Acholiland towards the end of the conflict. Nibbe, like Harris, touches on the onslaught on male masculinity that occurred in Uganda's camps. In these, men were not only humiliated by their inability to provide for their families but by the violence and frequency of male-on-male rape. This accounted for why men, presumably even non-combatants, avoided camps at all costs. Nibbe's chapter provides a nuanced and contrasting view of life in these camps with a case study that spans the end of the Opit Internally Displaced Persons (IDP) camp and the return home of the detainees, 70–80 percent of whom were women and children. While recognizing that violence against women occurs in camps, she nevertheless shows how camps can empower women and disempower men, suggesting that women may be better equipped to negotiate the camp experience. Given that the camp boundaries were too close to allow much farming, economic activity focused on women's entrepreneurship, petty trading, and especially brewing. After release, women missed the neighbors and female support systems experienced over the previous decade or so. The traditional dispersed settlement pattern and patriarchal culture were disempowering and rendered them vulnerable to violence, especially because of the unsettled conditions in the aftermath of hostilities. An estimated 30 percent of camp dwellers stayed, mostly older women with no family support and child ex-combatants outlawed by their families.

Nibbe's chapter shows women exerting agency in the difficult circumstances of displacement and camp life, a view that partially contradicts the stereotype of women as "helpless victims." Examining this commonly accepted perception, she shows how women and children in camps are portrayed, not only as helpless but as innocent, compared with men who are seen as perpetrators of conflict—a view necessary to sustain the flow of humanitarian aid and one that also serves to justify the caring role of external relief workers.

Daley's chapter takes a different approach. She critiques the collection of data on sexual violence in the Congo, which has been described as "the rape capital of the world." The country is effectively a "warscape," with many no-go areas. She also analyzes the stereotype of women as helpless victims of conflict. She shows how both the media, with its penchant for sensationalism, and humanitarian organizations, keen to attract funds for relief purposes, have motives for exaggerating the numbers affected. Data collection is inevitably difficult, but it is needed because decisions concerning external interventions and humanitarian aid depend on it. Policymakers require quantitative data in a situation where little exists. Daley reviews the methods used, the shortcuts taken, the shortcomings of these, the ethical considerations raised, and the power imbalance between researcher and researched. She shows how, even in these horrific circumstances, victimized women can still show agency, confounding interviewers by resorting to strategies to avoid providing information they wish to withhold. Importantly Daley queries not only the motives but also the perspectives of the researchers. Historically, Western perceptions of African sexuality were "full of caricatures and imbued with racism," something still likely to be reflected in researchers' imaginative geographies and thus in their attitudes to their work.

Acknowledgement

Anne Coles acknowledges lively discussions with Lilian Volcan (IGS) on this topic.

Notes

1   "A cautionary tale from Kosovan's women to women in post-war Iraq" quoted by El Jack (2003).
2   As this goes to press, guerrilla activity is much reduced.

Bibliography

Accord Conciliation Resources (2013) *Women Building Peace*, London: Accord.
el Jack (2003) "Overview report," *Gender and Armed Conflict*, Bridge Cutting Edge Packs, Sussex: IDS.
Gardener, J. and El Bushra, J. (2013) "From the Forefront of Peace and Reconciliation: testimonies from women building peace," in *Women Building Peace*, London: Accord Conciliation Resources.
Gender and Development (2013) Special Issue "Violence and conflict", *Gender and Development* 21(3).
Green, C. and Sweetman, C. (2013) "Introduction to violence and conflict," *Gender and Development* 21(3): 423–431.
Moosa, Z., Rahmani, M., and Webster, L. (2013) "From the private to the public sphere: new research on women's participation in peace-building," *Gender and Development* 21(3): 453–472.
Welbourn, A. (1995) *A Training Package in HIV/AIDS, Communication and Relationship Skills*, London: Strategies for Hope.
World Vision International (1996) *The Effects of Armed Conflict on Girls*, discussion paper for the UN study on the impact of armed conflict on children, California and Geneva: World Vision International.

41

LA RUTA, THE PACIFIC WAY

Women for a Negotiated Solution to the Armed Conflict

Adriana Parra-Fox

Introduction

Colombia has experienced a low-intensity armed civil conflict for the past 50 years. A significant escalation in this conflict during the 1990s, marked by a wave of gruesome massacres, led to massive internal displacements affecting millions of people, disproportionately women and children belonging to Afro-Colombian and indigenous ethnic groups (Kirk 2003; IDMC 2011). Attuned to the plight of these groups, an alliance of women's organizations, The Pacific Way: Women for a Negotiated Solution to the Armed Conflict (La Ruta Pacífica de las Mujeres por la Solución Negociada al Conflicto Armado), arose quickly to accompany them in a series of activities, often alone, frequently opposed by the established power structure, and generally ignored by the media. They came from all the regions and levels of society: academics, peasants, wealthy, poor, black, white, indigenous, and mestizo, but all with the firm conviction of the need to work for peace and to build a dialogue from women's perspectives. La Ruta, as it came to be known, not only confronted the established power base, but offered a new strategy for dealing with the phenomena of violence, displacement, and rape affecting women. From the beginning La Ruta has demonstrated an unconventional approach to peace activism by spotlighting a woman's perspective on war. Working diligently at the grassroots while maintaining a global vision, it raised awareness among Colombians as well as in the international community (Ruiz 2003). This brief account draws on a short history published by La Ruta (Ruiz 2003), interviews with some of the early participants, as well as the firsthand observations of the author.

The formation of La Ruta

The idea of doing something for the women of the Urabá region of Colombia first came up in a meeting in early 1995 with the Women's House (Casa de la Mujer) and two union organizations. The Women's House, an innovative, some would say hardline, feminist organization

396

working out of Bogotá since 1982, nurtured La Ruta through its early years and maintains a strong relationship with it today. The idea was simply to travel to the sorely afflicted region of Urabá and "embrace each of the suffering women" (Ruiz 2003: 13).

In early 1996 the idea reemerged during a regional Security Council meeting in the state of Antioquia in which a nun representing a community organization in the state of Antioqua in NW Columbia reported that "in one of its villages 95% of the women had been raped." A number of women's organizations joined together to assume the challenge of mobilizing 1,000 women from all over Colombia to converge on Mutatá, a major center for refugees in Urabá, with the simple design of embracing there those women "who suffered in silence the shame of war" (Ruiz 2003: 13–14). The date they selected for the event, November 25, held a special significance for Colombian women.

The Women's House, which had formed in the wake of the First Feminist Encounter of Latin America and the Caribbean convened in Bogotá in 1981, found itself well-positioned to make the connection between domestic violence and the horrific stories of women suffering at the hands of armed groups in Urabá. In the intervening years it had worked hard for grassroots women's education and against domestic violence. However, it lacked the reach to organize the mobilization of women from all across the country on its own. A number of women's networks had emerged starting in the 1980s including the National Women's Network (Red Nacional de Mujeres), the Colombian Network for Sexual and Reproductive Rights (Red Colombiana de Mujeres por los Derechos Sexuales y Reproductivos), the Women's Network of the Southwest (Red de Mujeres del Suroccidente del País), the District Health Network of Women from the Southwest (Red Distrital de Salud de las Mujeres del Suroccidente del País), and the District Health Network of Women from Poor Neighborhoods (Red Distrital de Salud de las Mujeres de Sectores Populares). Many of these, in turn, had made contact with networks operating at the international level, principally in Latin America (Sánchez Gómez 1995: 388). The alliance of these networks and other organizations with the Women's House made it possible to bring together women from many places outside of the principal urban centers.

When the organizers of La Ruta's first activity set the goal of mobilizing 1,000 women to descend *en masse* on Mutatá, considered the most dangerous place in all of Colombia, they wanted to bring together women not only willing but conscious and joyful. They wanted women who understood the individual and collective significance of their action, who could talk about it and explain it to others, friends and strangers. They wanted women who knew they might have to look down the barrel of an automatic rifle.

To ensure a more equitable geographic representation of women the ad hoc group of women leaders later known as "La Ruta" established a presence in eight regions, including several frequently excluded from such national groups, such as Putumayo and Chocó. The group recruited a leader from each of these regions and met monthly to develop and consolidate the philosophy of La Ruta, strategize, and make specific plans. Those leaders connected with bigger and better-financed organizations subsidized the travel expenses of those from less well-financed organizations. They met in the centrally located cities of Pereira or Medellín in order to reduce travel expenses and facilitate attendance. Several of those who participated in these meetings remember them as a time of exceptional creativity, energy, and learning.

Beyond the development of a strong philosophy uniting feminism and pacifism and planning for the event itself, this group had to face two important practical matters: carrying out the multiple preparatory workshops in each of the regions to recruit, train, and select the women who would participate in the mobilization; and the procurement of the necessary funds. To both tasks the group found elegant and serendipitous solutions. The execution of the preparatory workshops was left in the hands of each region; the coordinating group restricted itself to elaborating a thematic outline for the workshops and establishing performance goals. The participants should have the capacity to express a clear understanding of violence in their own lives and its relationship to the situation in Mutatá, be able to discourse on peace and specific measures to achieve it, and exhibit a respectful and cautious awareness of the dangers of traveling to a zone of conflict.

Funding for the mobilization, in large portion, came indirectly from the national government through the newly created National Directory for Women's Equity (Dirección Nacional para la Equidad de la Mujer). The newly elected president, Ernesto Samper, created this agency within the Office of the Presidency in August 1995, shortly after taking office, to fulfill a promise he made to the women's organization and out of gratitude for the solid and effective support that they had given his candidacy. He named Olga Amparo Sánchez Gómez as its first director. In the words of La Ruta, as director she "perceived the potential of this proposal to mobilize [women], and supported it in every way possible" (Ruiz 2003: 14). Although La Ruta lacked legal status and could not receive government funds directly to cover the costs of the mobilization, the associated, legally established women's organizations could, and did so. These organizations also supported the mobilization with their own discretionary funds and logistics.

The National Women's Network had been established in 1992 in part as a result of the enormous effort a group of women made to produce a new Political Constitution for Colombia in 1991. This constitution confirmed an ongoing process of decentralization and opened many doors for underrepresented and marginalized groups such as women and blacks. The ratification and dissemination of the new constitution raised an exceptional level of hope and expectation throughout the country. As members of other organizations, many of these feminist leaders had actively worked on the project of a new Constitution since 1988. That same year they produced a specific proposal, many parts of which were subsequently incorporated into the new Constitution. They organized the National Women's Encounter in October 1990, an event preceded from 1988 through 1990 by numerous preparatory encounters at the regional level.

The National Women's Network extended rapidly to different regions of the country through the formation of departmental subgroups including, in September 1992, the Departmental Network of the Women of Chocó. Four years later this provided the framework for carrying out the preparatory workshops in Chocó for the first mobilization. A delegate from the National Women's Network reached the town of Quibdó in 1992, having heard rumors about some interesting work with women going on there. In her meetings this delegate proposed the formation of an organization in Chocó that would bring together the various existing groups working independently with women, create a space for sharing, and establish a common agenda. On a second visit two representatives of the National Network joined a three-day event with 105 women from Chocó at the end of which ten organizations committed themselves as founders of the Departmental Women's Network of Chocó. It thereby became a node in the National Women's Network, connected with a wide spectrum of women's initiatives in the rest of the country.

The Departmental Women's Network of Choco elected five women to serve as a coordinating team. Many of them had opportunity to travel within the department as part of their regular jobs. They took advantage of these trips to publicize the Departmental Network and to promote the formation of new women's organizations. For its first four years the Network sustained itself principally from dues paid by its member organizations. By the time it achieved legal status in 1996 it counted on 39 affiliates. The National Directory for Women's Equity financed its first project to develop a comprehensive and integrated approach to reaching the women of Chocó. This project, in turn, increased the visibility of the Departmental Network at the national level.

The fluid process of conformation, consolidation, and search for a coherent philosophy taking place at the regional level of Chocó mirrored the process taking place at the national level. The invitation for one member of the coordinating team to participate in the planning meetings of the group which would become La Ruta fit harmoniously with the evolving vision of the Departmental Women's Network. The fact that both the National Network and the Departmental Network emphasized women's education for the exercise of civil rights and participation in the political process and that neither placed peace activism high on their agendas in no way deterred them from facilitating the participation of women from the Chocó in this new initiative.

Preparations for travelling to Mutatá

Around the beginning of October 1996, the Departmental Women's Network sent out word, through its formal and informal channels, of the planned mass mobilization of women to Mutatá. All those who wished to participate had to attend two half-day preparatory workshops held at the small office near the center of town which the Departmental Women's Network had established and maintained for nearly four years. The Network offered to reimburse the women for bus fares and to provide a simple snack for the afternoon session. However, many women scheduled their errands in the town center to coincide with the day of the workshop and some brought food to share so as to minimize costs. The first workshop started with a clear exposition of the mobilization to Mutatá, its purpose and objectives. This required giving some background information concerning the situation of violence and conflict at the national level and, specifically, in the Urabá region and Mutatá. Many of the women had family and friends who had fled the Middle Atrato region for Urabá instead of Quibdó. Almost all had heard blood curdling stories of violence against the rural population.

The discussion of violence and its causes in other places led the women naturally to speak of violence in their own lives and how it affected them, their families, and their communities. An examination of fear as a motivating force followed. The workshop acknowledged that people have many ways to deal with fear. It focused on talking with each other about fear as a very effective mechanism for dominating fear. The first workshop ended with the discussion of two important points. Those considering traveling to Mutatá must understand that they make the decision individually, autonomously, and fully conscious of its implications, and that La Ruta unlike other mass movements, does not tolerate a mob mentality.

The second workshop emphasized specific actions that the women could take to avoid conflict and promote peace. Together they imagined conflictive scenarios and play-acted ways to deal with them. Each of these situations offered an opportunity to teach through actions that peace can overcome war and love can defeat hatred. Since the symptoms of fear can resemble signs of threat, they talked about controlling fear and preparing their minds to speak calmly but firmly. A nervous young man with an automatic rifle and grenades might be reminded that he has a mother. They also demonstrated and practiced techniques of self-defense in case of physical attack. In general, the session emphasized teaching the oppressor by practically putting oneself in his shoes.

The previous subject led logically to the use of symbolic gestures and art. Aided by Vamos Mujer, a woman's organization in Medellín, part of La Ruta, with a rich experience in the field, the coordinating team had selected a variety of symbols to incorporate into the meeting with the women in Mutatá. These were to be: aromatherapy to strengthen intuition and overcome fear; masks and face paint, not to hide but to assume the identity of the other and to escape for a moment in a carnival-like atmosphere; and colorful placards and banners with rallying cries such as: "no fruit of our womb will feed the war." After explaining these symbols, the women prepared and practiced them with each other. Finally, they discussed and agreed upon the logistic details of the trip.

The journey to Mutatá

All those who participated in the workshops found them useful and uplifting, even those unable to travel. In the end, 30 women left from Quibdó early in the morning of November 23, 1996, on a chartered bus bound for Medellín. After more than 12 hours of travel they arrived at La Plaza de Banderas where they met up with more than 1,000 other women who had come from all over Colombia and arrived the same day. The women of Medellín had prepared a warm reception and an evening display of artistic creativity. The travelers had the opportunity to make new friends and share their stories. One group from Caicedo, Putumayo, described how most of them had left their hometown for the first time in their lives and traveled 36 hours straight in an act of rebellion, against their families' wishes. The parish priest had supported the idea of the mobilization from the moment he first heard about it and had helped the women overcome the opposition of their husbands and children (Ruiz 2003: 16).

After less than six hours in Medellín the women from Chocó left for Mutatá, this time in a caravan of 40 buses with more than 1,500 women. Despite the dangerously winding road, stops at two police checkpoints, and one landslide, the 12-hour trip seemed more like a party with storytelling, songs, and laughter. As they approached Mutatá in the heat of early afternoon they could not believe their eyes. A crowd of more than 400 colorfully dressed people lined the street waiting for them. In a flash all the tiredness of the hours of traveling passed away, and the two groups merged in one great embrace. The encounter lasted only six hours, barely time to share a fraction of the stories, the lessons, the legends and traditions, the symbols, the rituals, and the caresses that each woman had prepared and stored up. Time seemed to stand still, yet also raced by, and all seemed especially aware that something special had happened. At eight o'clock the women of La Ruta started their return to Medellín in an atmosphere of peaceful jubilation (Ruiz 2003: 19).

Reflections

As the journey drew to a close, the task of evaluating and understanding what had just happened began. The French historian, philosopher, and social scientist, Michel de Certeau argues that change, what might be called a new story, requires a "founding event" that serves to establish a field of action (de Certeau 1984: 125). This act is not juridical; it precedes the establishment of a new law. He compares it to a ritual carried out by the Romans before any important undertaking involving a foreign nation. "The ritual was a procession with three centrifugal stages, the first within Roman territory but near the frontier, the second on the frontier, the third in foreign territory . . . it is designed to create the field necessary for political or military activities"(de Certeau 1984: 124). He goes on to point out several distinctions between the story and the Roman ritual of founding concluding that

> A narrative activity, even if it is multiform and no longer unitary, thus continues to develop where frontiers and relations with space abroad are concerned . . . it is continually concerned with marking out boundaries. What it puts in action is once more the *fãs* [founding] that "authorizes" enterprises and precedes them.
>
> *(de Certeau 1984: 125)*

The story of La Ruta's first activity, the journey of 1,500 women to embrace the battered women in Mutatá, becomes its founding event and corroborates the ideas of de Certeau.

Although Chocó had a relatively small numerical participation in this first and founding activity of La Ruta, the telling and retelling of the stories of those who participated became a source of pride, identity, and inspiration for those who heard it, young and old, men and women, feminists and union workers, blacks, whites, and Indians. There exists a danger of viewing La Ruta as a mere series of events carried out by a group of daring, eccentric women. The annual observation of the International Day for the Elimination of Violence against Women on November 25 may help foster this erroneous view. The original founding event of La Ruta took place in the road to Mutatá or, more broadly, to Urabá. In a sense, La Ruta has never left Mutatá physically, psychologically, or spiritually. Each succeeding activity of La Ruta, whether one month later or ten years later, built on the creation of a theater of action that took place on November 25, 1996. Each one is both "a renewal and a repetition," "a recitation and a citation" (de Certeau 1984: 124).

La Ruta summarized its learning from this first event in the following words:

> La Ruta to Urabá left several lessons. Perhaps the first is the organic manner in which pedagogy became incorporated with a political mobilization. The preparatory workshops were just as important as the march itself. The participatory methodology, the planning, the evaluation, and the broad convocation gave as a result a process of consciousness raising which was as much personal as collective and integral because it offered not only an approach of rational discourse, but also of symbolic, non-verbal, and artistic languages which enriched the polyphony of voices that one hears in the villages and urban neighborhoods, the unions and the universities.
>
> *(Ruiz 2003: 19, author's translation)*

After Mutatá

The fact that the following year La Ruta carried out not one but two major events testifies to the strength and flexibility of this young organization. On November 17 women from the department of Antioquia carried out a mass mobilization, similar to the one nearly a year before, to one of its municipalities named Andes. This remote mountainous region on the eastern slopes of the Andes Mountains bordering with Chocó, once a prosperous coffee-growing area, had become an epicenter of conflict between guerillas and paramilitary groups. Entire villages had been abandoned out of the fear of imminent attack. The Embera Indians had declared their reservation, located within the municipal boundaries, as a zone of active neutrality, the term used at that time for the effort communities made to prohibit the presence of all armed groups, including the army and the police, from their territories. On this occasion La Ruta selected tenderness for the "other" as its theme, and left the town square filled with dozens of bigger-than-life dolls, made of papier-mâché and cloth, as symbols of solidarity, sisterhood, and peace (Ruiz 2003: 24).

During the course of the year since Mutatá, the members of La Ruta had continued discussing the concept of active neutrality and the different forms of pacifism. The consensus of the group had evolved to embrace a position of radical pacifism in the style of Martin Luther King, Jr. and Gandhi. It felt a strong need to invite women from other countries, especially those that had suffered through civil wars, negotiations, and peace processes, to talk together about this new position of feminist pacifism and to share with them their experiences. So it convened for November 24–25, 1997, what it called the International Town Council of Women for

Peace. As the site La Ruta selected the most important public library (Luis Angel Arango) in the historic center of Bogotá. Women, some of them ex-combatants, from Nicaragua, El Salvador, Guatemala, and Mexico were invited. The Colombian secretary of the interior, representatives from the Colombian offices of the United Nations, and the Colombian High Commissioner of Peace also attended. The size of the hall limited the number of attendees. Nonetheless, more than 600 women from across the country attended, each of whom, prior to traveling, had participated in a six-hour workshop in her region reflecting on the daily life of the civil population confronted by armed conflict and the role played by women (Ruiz 2003: 21).

As if two major events in less than a month were not enough, La Ruta responded to the needs of 49 communities, mostly from the Chocó side of the Urabá region, who had recently been forced to abandon their homes, lands, and crops, and had taken refuge in Parvarandó, by sending a delegation, with the support of CINEP (Centro de Investigación y Educación Popular), to accompany the women during Christmas time. They had declared themselves a "peace community." During the visit the delegation held nine workshops with 100 women (Ruiz 2003: 28).

The Truth Tribunal

One of the striking features of the violence against women in Colombia is the complete impunity with which it is carried out. No one ever goes to jail or pays retribution. No one ever goes to trial. No one is ever charged with a crime. Extremely few are even investigated. In 1998 La Ruta decided to organize one of its most creative and controversial events: a symbolic trial, a political and ethical trial, an opportunity for women to stand in a safe and decorous setting and tell the stories of the crimes committed against them and their families that they had kept silent about for so long before a jury of their peers; intelligent, experienced, and compassionate women ready and willing to listen to every last word. La Ruta called this event the "Truth Tribunal Opposing Crimes Committed against Women: Denouncing Impunity and Recovering Memory."

La Ruta deliberately chose Cartagena as the site for this symbolic trial. For its leadership and resistance during the time of independence from Spain, Cartagena had earned the designation of "the heroic city." On the darker side, Cartagena served for many years as the principal slave port for the region. Hundreds of thousands of human beings were bought and sold on the auction blocks of Cartagena. The founders of the famous stockade city of Palenque San Basilio, which resisted the Spanish government so ferociously that it was finally granted independence, were Cimarrons, or runaway slaves from Cartagena. But at the time of the Truth Tribunal, Cartagena had become the principal arrival city for the forcibly displaced from an extensive rural area devastated by violence.

La Ruta, as in the past, prepared exhaustively for the event. Eight preparatory workshops were held in the different regions (Ruiz 2003: 33). These gave women the chance to tell their stories in front of other women and the opportunity to select those cases to present in Cartagena. The author attended the Truth Tribunal as part of the delegation from Chocó. Alongside the senator from Antioquia, Piedad Córdoba, she had the privilege of serving as one of the jurists. The trial, held in the Cartagena Theatre, lasted two days, November 24–25. More than 1,200 attended the first day and nearly 2,000 the second day.

On the morning of the first day, tension hung over the hall like a thick cloud despite the best efforts of the organizers to provide a safe atmosphere. Who would speak first? Would women have the courage to tell stories that they had kept secret for years before such a large crowd? Would their voices be swallowed up by shame in the large theater? The judge politely called the first witness. A nondescript middle-aged woman named Fabiola Lalinde climbed the stage and started to tell her story clearly and concisely. On October 4, 1984, her son, a known communist youth activist, disappeared. For 12 years she fought, peacefully, respectfully, and persistently to discover what had happened and to recover her son's body. When she exhausted all resources at the national level she turned to international human rights organizations. The forced disappearance of Luis Fernando Lalinde, her son, became the first case reported to the newly formed Inter-American Commission for Human Rights of the Organization of American States, and the first case in which Colombia received a sanction for the violation of international human rights, specifically forced disappearance, torture, and inhumane and degrading treatment by the Colombia Army (Mejia 1999).

The persistence of Fabiola came with a high price for her and her family. They were vilified, harassed, and forced into exile. Shortly after the sanction, the police broke into her home and "discovered" a significant quantity of cocaine in a closet. She was arrested and sent to prison where she would have stayed for 20 years if not for the respect and admiration she had earned from the international community. She showed her good humor by calling her effort "Operation Cirirí," the cirri being a small bird that doggedly and fearlessly pursues the much larger hawk until it drops its prey. Finally, on November 19, 1996, Fabiola was able to bury the remains of her son (Mejia 1999).

At the end of her presentation the theater burst into applause. The atmosphere had magically shifted from one of tension and uncertainty to one of sisterhood, solidarity, and trust. The confidence of the women who followed Fabiola could be seen in the way they walked. One after another, they shared their secret stories of horror, not to terrify, but as humble gifts, the offering of which lightened their loads and strengthened the unity of the group. One was raped by the army and then made to wash clothes and cook for them. Another watched as the guerilla killed her husband and burned her home before they made her flee, and on and on. Freed, if only momentarily, from fear and silence, even some not scheduled to testify arose to speak. The moment seemed to say, "I am not alone. We all have similar stories. When we join together we have a powerful voice and we can confront the perpetrator." Everyone wanted to embrace the women who had shared their stories.

Symbolic gestures and acts were woven into the event. Many women wore white headbands to symbolize justice. They wove a maypole with ribbons of different colors: yellow for truth, white for justice, and blue for reparations. On the last day the women paraded to the central plaza of Cartagena dressed in white, carrying their placards and banners, chanting their slogans, and formed a human chain to embrace the plaza. The jury passed down its judgment condemning equally the patriarchal culture, the Colombian state, the guerillas, and the paramilitaries.

Reflections

At this point I want to introduce my personal observations of this event.

By the end the initial tension had evaporated and much of the hopelessness had disappeared. As a group we developed an environment of familiarity, trust, and even victory. There was a magical feeling in the evening after the session. I could hear and see small groups of women telling each other the stories of their suffering. We were staying in an educational center where we slept in big rooms that made it easy for groups of women to continue sharing their stories late into the night. This supports the hypothesis that narrative and collective memory encourage growth and make people more resistant to the trials of war. I suggest that these places became spaces of healing that made the women believe in their own powers again: the power of unity, the power of determination, the power of words, the power of accompaniment, and the power of resistance.

These women came from different places, had different ethnic backgrounds, different economic statuses, and various levels of education, but all were like one soul supporting each other. Their stories differed in details, but they all had two things in common: fear and the name of the perpetrator. In silence these two things separated and isolated women; but brought out into the open, fear and the name of the perpetrators lost their power and joined the rest of the details. The shared stream of narrative broke down the barriers between the women.

The tribunal helped to create a safe environment in which the women could concentrate their energies. They brought this energy to the place where they spent the night. In one group I was surprised and delighted to hear how a woman could pick up the thread of a story and continue telling it as if it were her own. Or perhaps it was her own. The stories merged and blurred until there seemed to be just one story. It did not seem to matter what story was told or who told it. The essence of the story was the same with the same perpetrators and the same victims. La Ruta had created a safe environment for the women to share their stories not once but twice, and surely they would continue telling their stories when they returned home. They understood the power of being united, and with the support of La Ruta they were ready to form a new organization or strengthen the one they had before. They now formed part of a national movement called La Ruta Pacífica de las Mujeres. They had found a new identity as women.

The following quote comes from the preamble of the principal talk delivered at the Tribunal:

> We will not renounce memory of the dead. We will not renounce solidarity. We will not renounce the right to name injustice and iniquity. We will not renounce our history. We will not renounce looking life in the face. Because the future is impossible without memory and there will be no peace without memory. Without memory forgotten crimes threaten to be reedited. Because the memory of sufferings belongs to the cultural heritage of every people. Because of memory our longings and aspirations to persist in the defense of life will allow us to construct and recuperate the collective ties that guarantee the peace for which we all yearn.
>
> *(Ruiz 2003: 34)*

La Ruta today

The very success achieved by La Ruta over the first three years stretched to the limits the loose alliance formed among the participating organizations and obliged La Ruta to take on a formal administrative structure and a legal status in order to continue its campaign against war and violence. Although it continued to use mass mobilizations of women into zones of conflict as one of its principal instruments, reaching places abandoned to violence such as Barrancabermeja, Quibdó, and Puerto Caicedo, it diversified its strategies to good effect. It established permanent offices in nine regions: Antioquia, Bogotá, Bolívar, Cauca, Chocó, Putumayo, Risaralda, Santander, and Valle del Cauca; and now represents more than 300 women's organizations nationwide. It held a massive demonstration of women for peace in the capital city, thought to be the largest of its kind at the time. It researched, edited, and published books that set out clear and convincing arguments for peace and women's rights in the current crisis. It developed and applied a series of training manuals for political education for peace and democracy. It entered into partnerships with other organizations with similar interests such as the Working-

Women's Organization (Organización Feminina Popular, OFP) in Barrancabermeja, and facilitated the presence of international observers in areas previously hidden from the eyes of the world. In 2001 representative of La Ruta traveled to New York to receive the Millennial Peace Prize from the United Nations Fund for the Development of Women. Finally, it leveraged its international exposure to assure stable funding for its activities from organizations such as the Suisse Program for the Promotion of Peace in Colombia (SUIPPCOL).

From mid-July to mid-November of 2012, La Ruta engaged in at least three well-publicized campaigns: in support of the peace negotiations between the national government and the FARC guerilla, in defense of the Native American population held hostage by Colombian army operations in the northern part of the state of Cauca and deprived of food and medicine, and presenting a brief to the Constitutional Court of Colombia arguing in favor of the right of the civil society to contact non-state military groups without prior governmental permission in order to mitigate the impact of armed conflict. In each of these diverse cases La Ruta strongly and clearly emphasizes the key role played by women in achieving peace.

Conclusions

Extreme, widespread, irrational violence produces the most challenging conditions for any kind of positive development, especially for women and children. At the same time, it can present a kind of stark contrast in which courage, determination, and sacrifice may emerge, spread, and eventually lead to lasting change. By any objective, material standard La Ruta has failed to achieve its goal. Indeed, massive forced displacement, like a stubborn cancer, continues in Colombia, destroying the lives of tens of thousands of women each year, and giving Colombia the dubious distinction of competing with Sudan as the country with most internally displaced people in the world. The Internal Displacement Monitoring Centre (IDMC 2011) reported that 89,000 people were displaced during the first six months of 2011.

However, the author wishes to examine the experience of La Ruta using other standards. In her opinion, La Ruta embraces a greater geographical, ethnic, and economic diversity of Colombian society than any other feminist organization. At the same time it has effectively demonstrated its capacity to speak truth to power at all levels, including on the international stage. Its early and consistent strategy of valiant, resilient democracy makes it extremely difficult for the existing Colombian power structure to corrupt, coopt, or crush La Ruta the way it has with other threats. The current strategy of the establishment seems to be to actively ignore it. Curiously, the English-speaking academic research community has taken little note. The few who mention it perceive little distinction between La Ruta and more homogeneous and opportunistic women's organizations (Cockburn 2007).

The coincidence of the theoretical work of de Certeau with the practice of La Ruta in the employment of narrative and symbols was mentioned above. The controversial field of collective memory combined with the recent, even more controversial work of Loftus may provide another key to appreciating the experience of La Ruta (Hutton 1987; Loftus 1999). If memory is not set in stone but plastic, then it offers a fertile area of human creativity and recreation of both past and future, most productively carried out in affinity groups. Parallels to the constant participatory mode of learning demonstrated by La Ruta are best found in the works of Freire and Fals-Borda as well as in the sensitive and painful narrative of Behar (Freire 2009; Fals-Borda 1980; Behar 1993).

While La Ruta has made significant practical contributions to the understanding of feminist activism and peace-building, it is not one to rest on its achievements.

References

Behar, R. (1993) *Translated Woman: Crossing the Border with Esperanza's Story*, Boston: Beacon Press.

Cockburn, C. (2007) *From Where We Stand: War, Women's Activism and Feminist Analysis*, London: Zed Books.

de Certeau, M. (1984) *The Practice of Everyday Life*, Berkeley: University of California Press.

Fals-Borda, O. (1980) *Mompox y Loba: Historia Doble de la Costa, Tomo 1*, Bogotá: Carlos Valencia Editores.

Freire, P. (2009) *Pedagogía del Oprimido*, Bogotá: Siglo XXI.

Hutton, P. (1987) "The art of memory reconceived: from rhetoric to psychoanalysis," *Journal of the History of Ideas* 48(3): 371–393.

IDMC (2011) *Colombia: Improved government response yet to have impact for IDPs*, Internal Displacement Monitoring Centre, www.internal-displacement.org/8025708F004BE3B1/(httpInfoFiles)/4C85108 1FBE3FB10C1257975005E685E/$file/colombia-overview-Dect2011.pdf.

Kirk, R. (2003) *More Terrible Than Death: Massacres, Drugs, and America's War in Colombia*, New York: Public Affairs.

Loftus, E. (1999) "Lost in the mall: misrepresentation and misunderstanding," *Ethics and Behavior* 9(1): 51–60.

Mejia, P. (1999) *Online Video*, http://lockerz.com/u/20884287/decalz/9329164/la_vigencia_de_la_ operaci%C3%B3n_ciriri_fab.

Ruiz, M. (ed.) (2003) *La Ruta Pacífica de las Mujeres: No Parimos Hijos ni Hijas para la Guerra*, Bogotá: Servigraphic.

Sánchez Gómez, O.A. (1995) "El Movimiento Social de Mujeres: La Construcción de Nuevos Sujetos Sociales," in M. Velásquez Toro (ed.), *Las Mujeres en a Historia de Colombia , Tomo 1 : Mujeres, Historia y Política*, Bogotá: Consejeria para la PolíTica Social.

42

GENDER AND POST-CONFLICT REHABILITATION

Colette Harris

Introduction

The aftermath of armed conflict and the breathing spaces between violent episodes such as riots leave affected communities in a state of confusion and disorder. It can be very difficult to know how to move on or even in which direction to go, especially for those recently returned from displacement, who are often struggling both materially and psychologically to survive. In such situations the habit of violence is not easily relinquished; once it has been integrated into everyday life it becomes part of the repertoire of how to achieve certain aims. At this point it is only too easy for violence to become a way of life in the post/intra-conflict setting too, at both community and domestic levels. The question then for development professionals is whether and how to intervene to support communities to address this.

This chapter investigates three community development projects I established in (post-)conflict settings – southern Tajikistan, northern Uganda and northern Nigeria – between 1997 and 2011. After summarising each of these contexts, I describe the approach used, looking mainly at how it supported communities to improve the overall wellbeing of all their members – young and old, male and female – especially in regard to violence reduction. I explain how community-based gender analysis allowed participants to identify the relevant characteristics of masculinity and femininity for their own sociocultural groups, leading them to consider the roles each trait played in underpinning their social system. This enabled them to choose whether or not it made sense to adjust each trait to fit with their current (post-)conflict situation, as part of their strategy for tackling social ills, including domestic and community-level violence. The chapter concludes by analysing the impact of these projects and drawing lessons for improving the international community's approach to working in (post-)conflict settings.

The setting

In some ways the title of this chapter is a misnomer since it implies that a post-conflict exists, that there are times when conflict and violence are not a part of the experience of the socioeconomically deprived, and that it is thus possible neatly to differentiate peaceful pre- and post-conflict environments from a clearly defined conflict situation (Pankhurst 2003). Nevertheless, in all three settings that are the focus of the chapter, the experiences undergone by the populations during the formally recognised conflict period left them particularly vulnerable and unsure about the future and thus in recognisably different circumstances from before.

This chapter does not discuss the conflicts themselves, neither attempting to explain their causes nor the reasons for individual or community participation. Rather it explores the situations those concerned found themselves in at the end of the formal conflicts and the interventions I helped establish to support them to improve their own circumstances, concentrating particularly on interpersonal relationships and the reduction of violence. In all three cases the most significant barriers to doing this, identified through ethnographic research carried out before the interventions were established, turned out to be attempts to live up to gender norms, most especially those pertaining to masculinity (Harris 2004, 2012c). This suggested the importance of making gender central to the interventions and I argue in this chapter that this approach contributed enormously to the success of the interventions.

The three settings dealt with in this chapter are – southern Tajikistan at the end of the civil war (1992–1997), Kaduna city in northern Nigeria after a series of episodes of sectarian violence (2000–2011) and Acholiland in northern Uganda in the return from internal displacement (IDP) camps after two decades of civil war (1986–2006). It discusses the interventions I was responsible for putting into place; in each setting my role was to shape the projects and train local facilitators who carried out most of the interactions, although I spent far more time *in situ* in Tajikistan than elsewhere. In the first and last cases we worked in a largely rural environment, in Nigeria in an urban one. Nevertheless, in all three cases a similar approach was taken and this particularly applies to the gender aspects. The chapter explains the contexts, lays out the principles and practices of the interventions, describes what occurred and evaluates the impact, drawing out lessons for the future.

The contexts

Tajikistan

Tajikistan was the poorest and least developed of Soviet Union republics (Harris 2006). While nominally the state claimed to be developing its Muslim Asian peoples, among the Gharmi peoples of Khatlon Province, the group that most suffered in the civil war, relatively little social development took place. Gharmi women were virtually secluded, despite working in the cotton fields on collective farms, with the men employed as farm labourers or in other semi-skilled or unskilled jobs (Harris 1998). After independence in 1991, the economy of Tajikistan collapsed and political turmoil followed, in 1992 turning into a civil war fought largely in Khatlon. Some 50,000 were killed and more than 250,000 fled, many Gharmis ending up as refugees in Afghanistan (Heathershaw 2009: 21). By late 1994 the main fighting was over and most Gharmis had returned and were rebuilding their homes that had been destroyed during the war along with most of their possessions. They came back to a new situation, one dominated by externally imposed proto-capitalist economic relations and to educated adults with illiterate teenage children as a result of the destruction of the schools and the loss of teachers, greatly exacerbated by the post-Soviet economic collapse (Harris 1998).

Although the adult population was mostly literate, even most women having undergone eight to ten years of schooling, this seemed to have taught them few practical skills that could have helped them cope with this new world. The authoritarian nature of the Soviet state together with its provision of basic amenities in the shape of a plot of land or public housing, subsidised basic commodities, free education and health care had left them unused to self-reliance.

By 1997 the Gharmis had rebuilt their homes but were still feeling lost and bewildered. They were waiting for the Soviet state to return so their lives could revert to the former levels of security or failing that were hoping the UN would step in and establish new structures to make peace and provide for their basic needs (Harris 1998).

This was the point at which our project started in April 1997, implemented by a small group of local facilitators, trained and supported by myself, and funded by Christian Aid of London. Later, with increased funding, we expanded the project and formally registered as an NGO named Ghamkhori. In 1999 we expanded further with the support of a two-year grant from the EU's TACIS LIEN fund allowing us to work in many more villages, mainly among the Gharmi population.

Nigeria

Nigeria was formed in 1914 as a British colony produced by cobbling together what under their rule were turned into a largely Muslim north and a mainly Christian south. The resultant religio-ethnic split was politicised, leaving Nigeria at independence in 1960 a heritage of divisiveness. The state's positioning in the global political economy as a rentier oil-producing state made it a valuable property for elite capture, while structural adjustment forced on it by the international community in the mid-1980s produced serious economic difficulties including high unemployment. In Kaduna city, former capital of northern Nigeria and current seat of the government of Kaduna state, with a population divided fairly equally between Muslims and Christians, hostilities between ethnic groups as an indirect result of British policies turned into sectarian conflicts, exacerbated by competition between the religions over access to the state. The result has been that sectarian violence has now become a defining feature of the landscape (Falola and Heaton 2008).

Since the late 1980s Kaduna has been seriously affected by such violence. The worst episode occurred in February 2000, as a result of clashes over the imposition of *sharia* law in the state's criminal code. It left well over 2,000 dead and did billions of Naira worth of damage. Three further major episodes have ensued – in May 2000, November 2002 and April 2011. Besides the physical destruction, each successive episode further injured intergroup relationships, resulting in significant tensions between members of the two religions to the point that youths from each side frequently clash in the streets and it is dangerous for them to enter each other's neighbourhoods. Meanwhile, the perpetrators of violence have allegedly been some of the city's poorest young men (HRW 2003).

At the time the first project in Kaduna was established, in June 2007,[1] the last episode of serious violence had occurred long enough in the past that it no longer occupied everyone's minds but suspicion of the 'other' was nevertheless palpable. Thus, ameliorating relations between the two groups was one of the main goals of the project. For this reason it concentrated particularly on male youths, although women were also incorporated. Most of the youths had completed secondary education and a few were even embarking on further/higher education. The Christian women were semi-literate, most Muslim women, however, were completely illiterate. The interventions were facilitated by local staff, again with my training and support.

Uganda

Acholiland was incorporated into the British Protectorate of Uganda at the start of the twentieth century and its men were generally treated by the colonial power as mainly suitable for working as soldiers and policemen, although they also formed a significant proportion of civil service employees. After independence in 1962, the Acholi became prominent in national politics too. In the 1970s, Acholiland was attacked by Idi Amin, who perpetrated extreme violence on the region and the population has lived with violence for most of the ensuing decades. From the mid-1980s, this took the form of a civil war between the government and the Lord's Resistance Army (LRA), with atrocities against the Acholi people committed by both sides. In the mid-1990s, the government forced the rural population of Acholiland into camps, where they were subject to considerable levels of abuse until, after a provisional peace accord in 2006, they were permitted to return to their villages (Harris 2012c).

During their time in the camps, Acholi men felt emasculated by their treatment and determined to reverse this situation as soon as possible in the return; in so doing they tried to repress both wives and children, including adult sons, leading to considerable friction. This was exacerbated by the fact that they no longer controlled the same level of resources as in the past, forcing their wives and children to provide much greater input into farm labour than before while receiving little personal benefit from this as the men continued unilaterally to make all household decisions including in relation to crops farmed largely by these other family members (Harris 2012d). To make things worse, some of the population had returned from abduction during which they had been forced to fight with the LRA. They were strongly resented by their fellows who blamed them for their suffering. One aim of the intervention, therefore, was the reintegration of these former fighters. A group of local facilitators and I started work in July 2009 after a year of ethnographic research in two villages in Gulu district, followed by evaluations at the end of the project in 2010 and in 2011 (Harris 2012c). An informal external evaluation was carried out in early 2013 by International Alert.[2]

Violence

In all these settings it was clear that multilevel violence had long been endemic in their communities (Falola and Heaton 2008; Girling 1960; WHO 2000). This included structural violence that had disadvantaged these communities, thus contributing significantly to direct forms of violence at community and family levels, the latter consisting not only of spousal abuse but also maltreatment of children and youths by parents of both sexes (see, for instance, Harris 2004). The economic instability and deprivation resulting from the influence of neoliberal ideology applied via globalising capitalism had destroyed the ability of male family heads to provide for their families in the traditional manner through control over land in subsistence farming communities without replacing this with appropriate alternatives, such as formal jobs. This was especially problematic since the very notion of the masculine breadwinner was a result of capitalist colonialism, while the effects of implementing neoliberal capitalist ideology have been to prevent most men from ever achieving it (Harris 2012b). I suggest this contradiction, which has deeply affected men psychologically as well as economically, has been at the heart of much of the violence of recent decades. The contradictions between traditional notions of appropriate masculinity and contemporary economic reality described above for postwar northern Uganda exist in similar forms in many other parts of the Global South and even today perhaps in the West as well.

In the three target settings, thus, contact with colonialist forces had done much to wrench traditional lifestyles out of shape, whether through attempts to 'civilise the natives' as in the African settings or through deliberate projects aimed at gender re-engineering as in Soviet Tajikistan (Harris 2004). Traditional notions of household organisation and the roles and obligations of the different classes of family members no longer cohered with the practices needed for survival in an environment ontologically distinct from anything previously imaginable

– that is, coming from a fundamentally different philosophy of life. This has made it very difficult for the populations to come up with new childrearing practices and notions of how to construct family relationships appropriate for the altered conditions, especially given the rapid rate of change and in the absence of external support. Moreover, for many men, conforming to traditional styles of masculinity may be the only status symbol now available to them (Harris 2012b, 2012c).

As it is, clashes and misunderstandings between parents and young people, particularly but not solely the young men, have intensified as the generation gap widens with the increasing pace of change and each setting becomes increasingly involved in the global political economy. In the African settings, fathers needing help from their sons with farming or other tasks may prioritise this, while youths might be desperate to complete their schooling to a level where their chances of future employment would be significantly improved. In Khatlon in the late 1990s, the clashes tended to be more around the issue of who had the right to make decisions affecting the future of the youths; themselves or their parents.

It seemed likely that during the formal conflicts, in all three settings, experiences of violence within the family had coalesced with socioeconomic marginalisation to facilitate young men's participation in the fighting. The post-conflict interventions then needed to tackle these different sites of violence. Merely to consider the superficial level of the public sphere, we believed, would be insufficient to create the fundamental change we hoped to achieve.

The interventions

The principles

The interventions took the form of non-formal, community-based education projects, using discovery-based learning[3] (combined with participatory gender analysis[4] (Harris 2014). This entailed an approach aimed at helping participants through a process of self-empowerment, based on a combination of mutual support and exercises/focused discussions, pitched at group level, including group decision-making around potential solutions. It was seen as crucial for achieving change for the groups concerned to consist of a significant proportion of each village's population.[5]

In addition to the basic pedagogic principles, the application of a gender analytical framework proved crucial to the success of the interventions as a whole. The effectiveness of such an approach has been demonstrated through a rigorous evaluation of the Stepping Stones method that uses very similar principles of working with the majority of rural community members over a period of months, carrying out gender analysis and discussing interpersonal violence (Jewkes et al. 2007).

By gender I do not mean a proxy for women's nor men's rights but rather the sociopolitical construction of sets of norms males and females in a particular social grouping are expected to conform to (Harris 2004: 14). Since the societies concerned were gerontocratic, the norms differed to a significant extent between generations, in particular between youths and their parents, with older women holding significant levels of power over their sons, for instance (Harris 2006, 2012d). In such societies, gender norms differ between same-sex members of different generations, such as fathers and sons, mothers(-in-law) and daughters(-in-law) (Harris 2004). The aim of the pedagogy then was to bring the populations to carry out their own gender analyses in order to understand the implications of the norms for actual behaviour patterns and the acceptance within their own communities of practices they themselves considered deleterious (compare also Jewkes et al. 2007).

The practice

The participants were from deprived and poverty-stricken groups who already before the conflicts were largely marginalised from any real benefits accruing from the state. The difficulties they faced in dealing with their post-conflict situations arose in part through exposure to globalising capitalism, the effects of which they were ill-equipped to deal with. This was especially so for the Tajik and Acholi populations whose pre-conflict lives had been embedded in very different political and economic contexts.

Interestingly, their different educational levels left these three sets of participants in a similar position regarding family and community relationships, their understanding of their sociopolitical environment, and grasp of health and other immediate issues they were faced with. In other words, the difference in coping abilities and survival skills between illiterate Hausa Muslims or Acholi on the one hand and Tajiks or Christian Nigerians with eight or more years of schooling, even university education, on the other was considerably slighter than might be expected. This suggests a very serious need to revisit the entire basis of formal education as necessarily being an empowering experience and to rework the elements it should contain, as well as to interrogate the power of literacy in itself to produce social change (Robinson-Pant 2004).

The interventions took the shape of weekly meetings lasting 90–120 minutes over the course of from six months to a year. In the rural cases, the entire village population was invited to attend, after which it was divided into groups by sex and age – producing older and younger men's and women's groups. In Nigeria our participants came to us via local leaders and each group met on a different day. In Tajikistan, while in each village all groups met simultaneously, social constraints kept them from interacting. However, in Uganda sessions ended with all groups reporting to one another the salient points of their discussions and jointly deciding how to put into practice what they had learned that week (Harris 2012c).

The introductory session for each group always began with participants devising lists of the problems that preoccupied them as a group, after which all would consider which could usefully be tackled through the programme. Thus, the topics to be dealt with would emanate from the members themselves. In all cases, the need to tackle violence was seen as crucial.

As the earlier section on violence suggested, the single most important element for the overall improvement of community life turned out to be gender, in particular supporting participants to deconstruct the meaning of masculinity and femininity for their own communities and using the characteristics identified as the basis for gender analyses. This was aimed at supporting participants to grasp the role the traits played in exacerbating their problems, including poverty and violence (Harris 2012c, 2014; see also Welbourn 1995).

Targeting gendered change for violence reduction

The practice

Since it appeared that tensions between local sets of gender ideals and the material transformations that had occurred over the past few decades were responsible for many of the problems experienced, community-based gender analysis was placed at the centre of the education projects.

The starting point for the analysis was participants identifying the most salient facets of gender norms for their own communities and, if literate, writing them down. They then served as the basis of discussion. In all three settings stereotypical notions emerging from this exercise showed men as leaders, decision-makers and heads of household, women as bearers of children and obedient, submissive and domesticated spouses. Men were encouraged to be polygynous, whether formally or informally, women expected to be monogamous and, in Tajikistan at least, virginity was essential at first marriage. Divorce was relatively easy, except perhaps for Nigerian Christians. Tajik and Hausa women had limited mobility, requiring specific permission from their husband to leave the house but Christian women too were constrained in mobility. While women were seen as weak and emotional, men were supposed to show themselves strong, brave and protectors of their own.

Discussing these notions during small group discussions brought participants to focus on their consequences and thus on the issue of their desirability. It was a small step from there to questioning how far these were set in stone, whether changes could or should be made and what they might entail. In order to accomplish this, various pedagogic techniques were used.

Role plays proved to be extremely fruitful for supporting discussions around current practices and potential changes in social relations, especially among the less educated women who claimed this greatly facilitated their conceptualisation and comprehension and allowed them to work out how to put fruitful changes into practice.

In Kaduna, women blamed their husbands for their inability to provide adequately for their families and said they frequently accused them in front of their children and others of failing to fulfil their masculine responsibilities. After all, the women complied with their domestic labour and childrearing obligations, why could the men not do likewise? A group discussion around the difficulties of finding employment in today's circumstances, however, led the women to consider their husbands' perspectives for the first time. They started to realise that, while women's work was time-consuming and tiring, it was usually within their capacity to achieve. Their husbands, on the other hand, had to manage to be hired at a sufficiently high rate so they could provide for their family's needs before they were even able to start their labour and if this were not a permanent job they would have to do this over and over. This was far easier said than done, especially since most of their husbands were low-skilled and lacked connections through whom they could find decent employment. There was also the question of status. If men accepted work seen as inappropriate for their status they would be subject to considerable social pressures to abandon it.[6] The upshot was they were often very constrained in their abilities to provide, which already made them feel failures as men. When in addition their wives publicly upbraided them and showed no understanding of their struggles, this exacerbated the problem, frequently leading to domestic violence (Harris 2012b). Role playing helped the women realise they needed to rethink the ways in which they discussed these issues with their husbands. As a result they said they had dropped their aggressive behaviour and started holding private discussions and showing understanding of the men's predicament, which produced a significant improvement in their relationships.

Women were also involved in violence against their children, for instance hitting babies who bit them while breastfeeding and older children who did not instantly obey them. Especially in Khatlon, where patrilocal marriage was the norm, daughters-in-law were frequently subjected to abuse too (Harris 2004, 2006). Here role plays were useful in helping women hark back to the harsh treatment they had experienced when young and thereby learn to empathise with the young people they had power over. Role plays further allowed the women to consider the consequences of their children learning violent behaviour in the home and subsequently repeating it in their future lives.

Young people were usually also amenable to using role plays in this way. Something else that proved exceptionally powerful in Uganda was young men listening to the testimonies of young women from their own villages on the issue of sexual abuse. This brought them to rethink their notions of masculine privilege in relation to sexuality and for the first time ever to consider that girls might have a similar subjectivity to that of their own (Harris 2012c). As in the project directed by Kandirikirira (2002) in post-apartheid Namibia, in Uganda too this led to a significant reduction in sexual violence against girls and the rethinking of how to attain manhood without premarital sexual activity, particularly considering the high HIV rates in the region.

Adult men, however, found it beneath their dignity to do anything other than hold discussions. This was a particularly complicated group to work with since it was a matter of pedagogy for the privileged and this is far from easy to deal with, requiring extremely sensitive handling (Curry-Stevens 2007), since blaming adult men for gender-based inequalities and demanding rights for women and young people, approaches often used in development projects influenced by international actors, often lead to a stubborn refusal to make change (Wendoh and Wallace 2006). For these men, therefore, we needed very carefully to devise productive methods to encourage them to focus on relevant issues without feeling they were being blamed but in such a way as to bring them to reflect seriously on their own roles in producing violence, damaging family relationships and harming their overall wellbeing and that of their families, hoping thereby to produce 'cognitive restructuring' (Curry-Stevens 2007: 44).

In Tajikistan, we used clippings from a particular newspaper that published stories supposedly contributed by local individuals reflecting on their own situations as starting points for discussions with men. In one session on domestic violence, for instance, we asked someone to read aloud a story of an old man reflecting on the violence he had perpetrated on his family after returning from the front during World War II. The result was his wife left him and his children refused to have anything to do with him. The article had him lamenting this behaviour that had led to his present lonely and difficult old age. One listener, about 35 years old, seemed particularly moved by this. He told the group he had almost ended up in the same situation. He had believed what others had told him about the importance of showing one's wife who was boss, even to the point of using physical violence at times. This produced chilly relations between them and later with his children too that had made him deeply unhappy. Eventually he realised this was the result of his treatment and the only way to improve things was for him to stop privileging his masculine entitlement to authority. His changed behaviour slowly brought his family to relate to him differently; they now got on very well and he was happier than ever before.

The fellow villager reacting to the newspaper story produced a heated discussion that focused on the crucial issue of adult male authority irrespective of consequences or the creation of warm human relationships with family members (see also Harris 2012d). While there were divergent opinions, the discussion raised vital issues about masculinity and led to further discussions among the men, which eventually produced significant change in a goodly proportion of village families, reported by wives and children as well as husbands/fathers.

Similarly, in one of the two Ugandan villages, the result of several hours of discussion on whether adult men should continue with their project of reinstating their supposed pre-war superior power positions despite economic conditions that forced them into much greater dependence on the labour of wives and children than in the past, was that they really could not continue on this path. It was unfair, unproductive and already fracturing their families. This too resulted in a significant improvement in domestic relationships (Harris 2012c).

Impact

Clearly these sessions mainly affected those who had participated in them. However, they frequently led to further informal discussions in which others would join, thus affecting a far higher population. In the rural projects, after sensitisation to the issue of domestic violence, village committees consisting of both female and male members were established in the majority of places we worked, to police situations in their own communities.

Because in Kaduna the conflict is continuing, the aim of working with the young men was somewhat different from in the other two settings where the wars were over and the issues were mainly how to improve the capacity of our participants to cope with the post-conflict situation. In the Nigerian setting it was felt to be crucial to reduce the propensity of the youths to join in future episodes of sectarian violence. We wished to arm them against pressures to do so by helping them reconsider whether the masculine attributes of bravery and protection of their own could best be met by fighting or by taking a Gandhian non-violent stance. The youths suggested the importance of incorporating a far larger number of young men than only our direct participants if we were to make a meaningful impact.

One way of doing this was through street dramas using theatre-for-development approaches such as audience discussion (Harris 2012a). In mid-2009, Christian and Muslim youths combined to put on plays on violence-related issues in their two neighbourhoods. A definite reduction in violence could be seen in relation to the specific issues raised, but the most significant change, according to the local chiefs, was that for the first time in more than a decade, youths were able freely to enter each other's neighbourhoods. In April 2011 this again became problematic after Kaduna experienced a further episode of sectarian violence in which hundreds were killed as a consequence of disagreements over the outcome of the presidential election. However, it seems at least that the direct project participants and their family members did not participate, and some even managed to prevent others from doing so. I have no information on whether youths were discouraged from joining in by the dramas (Harris 2012a).

The impact of these programmes then was significant. Although it is impossible to assess the proportion of participants who changed, it was high enough to make a difference to the rural communities. It is obviously far harder to make a meaningful difference to urban communities such as Kaduna, especially as we were unable to work with the adult men, but even there changes could be seen.

Conclusion

Due to the crucial role played by pressures to conform to gender norms in maintaining social order and in regard to interpersonal relationships, focusing on violence through a gender lens proved a particularly efficacious approach to sensitising communities to the issues concerned. This was especially true for those elements that privileged the masculine over the feminine and so gave men power over women and in gerontocratic societies such as those with which this chapter deals, older people over youths.

In all three settings our analyses showed that much of the violence that was such a major part of everyday experience could be directly attributed to the pressures to conform to gender norms. Here, increasing levels of economic instability have contributed to the violence in multiple ways but especially problematic has been the coupling of masculinity and breadwinning in an environment that prevents young men in particular from gaining access to employment considered appropriate for their social status and reduces the ability of older men to maintain their traditional authority.

The populations concerned lived in environments in which former lifestyles already bent out of shape by colonialist forces were being further buffeted by exposure to neo-colonial capitalism, without any substitute being offered that could support populations to develop positive intrafamilial relations. This situation was exacerbated by gendered power imbalances that led to domestic violence becoming normalised and extended into communities.

The interventions discussed in this chapter suggest the importance of understanding the causes of violence at these sociocultural and psychological levels in order to support communities to tackle it. They also indicate that helping participants discover the influence of gender traits for themselves and realise that they are malleable and can thus be purposefully changed is crucial for making a real difference.

Notes

1 With funding from DFID in connection with the Citizenship Development Research Centre run by the Institute of Development Studies, Sussex and subsequently a further project was established funded by the AHRC/ESRC's Religion and Society Programme run by the University of Lancaster (see Harris 2012a, 2012b).
2 www.international-alert.org/news/gender-and-peacebuilding-research-northern-uganda.
3 Whereby participants are helped to arrive at their own ideas and solutions via a process of guided learning.
4 This has much in common with the approach to gender used in the Stepping Stones programme (Welbourn 1995).
5 The process has been described in some detail for the northern Ugandan setting in Harris (2012c).
6 This happens in the West as well.

References

Curry-Stevens, A. (2007) 'New forms of transformative education: pedagogy for the privileged', *Journal of Transformative Education* 5(1): 33–58.
Falola, T. and Heaton, M. (2008) *A History of Nigeria*, Cambridge: Cambridge University Press.
Girling, F.K. (1960) *The Acholi of Uganda*, London: Her Majesty's Stationery Office.
Harris, C. (1998) 'Coping with daily life in post-soviet Tajikistan: the Gharmi villages of Khatlon Province', *Central Asian Survey* 17(4): 655–671.
—— (2004) *Control and Subversion: Gender Relations in Tajikistan*, London: Pluto Press.
—— (2006) *Muslim Youth: Tensions and Transitions in Tajikistan*, Boulder, CO: Westview Press.
—— (2012a) 'Community-based pedagogies, religion and conflict resolution in Kaduna, Nigeria', in L. Marsden (ed.), *The Ashgate Research Companion to Religion and Conflict Resolution*, Farnham: Ashgate, pp. 501–530.
—— (2012b) 'Masculinities and religion in Kaduna, Nigeria: a struggle for continuity at a time of change', *Journal of Religion and Gender* 2(2): 207–230.
—— (2012c) *The Importance of Post-Conflict Socio-Cultural Community Education Programmes: A Case Study from Northern Uganda*, MICROCON Research Working Paper 64, www.microconflict.eu/publications/RWP64_CH.pdf.
—— (2012d) 'Gender-age systems and social change: a Haugaardian power analysis based on research from Northern Uganda', *Journal of Political Power* 5(3): 465–493.
—— (2014) 'The use of participatory gender analysis for violence reduction in (post-)conflict settings: a study of a community education project in northern Uganda', *Gendered Perspectives on Conflict and Violence, Advances in Gender Research*, 18B: 145–170.
Heathershaw, J. (2009) *Post-Conflict Tajikistan: The Politics of Peacebuilding and the Emergence of Legitimate Order*, London: Routledge.
HRW (2003) 'The "Miss World riots": continued impunity for killings in Kaduna', *Human Rights Watch*, www.hrw.org/reports/2003/07/22/miss-world-riots.
Jewkes, R., Nduna, M., Levin, J., Jama, N., Dunkle, K., Wood, K., Koss, M., Puren, A. and Duvvury, N. (2007) *Evaluation of Stepping Stones: A Gender Transformative HIV Prevention Intervention*, Pretoria: Medical Research Council, www.mrc.ac.za/policybriefs/steppingstones.pdf.
Kandirikirira, N. (2002) 'Deconstructing domination: gender disempowerment and the legacy of colonialism and apartheid in Omaheke, Namibia', in F. Cleaver (ed.), *Masculinities Matter!*, London: Zed Books, pp. 112–137.
Pankhurst, D. (2003) 'The "sex war" and other wars: towards a feminist approach to peace building', *Development in Practice* 13(2/3): 154–177.

Robinson-Pant, A. (ed.) (2004) *Women, Literacy and Development: Alternative Perspectives*, Abingdon: Routledge.

Welbourn, A. (1995) *Stepping Stones: A Training Manual on HIV/AIDS, Communication and Relationship Skills*, Oxford: Strategies for Hope.

Wendoh, S. and Wallace, T. (2006) *Living Gender in African Organisations and Communities: Stories from The Gambia, Rwanda, Uganda and Zambia*, Transform Africa, www.transformafrica.org/docs/gender-research-report.pdf.

WHO (2000) *Violence Against Women: Report on the 1999 WHO Pilot Survey in Tajikistan*, Copenhagen: World Health Organization, http://who.int/violence_injury_prevention/media/en/150.pdf.

43

WOMEN, CAMPS, AND "BARE LIFE"

Ayesha Anne Nibbe

Introduction

The problem of forcible displacement due to war, natural disaster, or other crisis is a rising problem around the world. The humanitarian aid community, spearheaded by the United Nations High Commission on Refugees (UNHCR), offers protection and services to millions of displaced people in camps and elsewhere in more than 125 countries.[1] In 2012, 45.2 million persons in the world were forcibly displaced and uprooted by humanitarian crisis—a 7.6 million person increase from 2011.[2] A subsection of this group of both refugees and IDPs lives in camps that are located all about the world, and up to four-fifths of those displaced people in camps are women and children. This chapter is an investigation into the gendered implications of the interplay between camps, humanitarian aid organizations, and internally displaced persons in the context of conflict in northern Uganda.[3] First, this chapter will define the terms "refugee" and "IDP," and draw a picture of the diversity of camps around the world. Next, I will outline the general discourse about camps and women in the academic and practitioner literature. The literature suggests that women are seen as a specialized subgroup of displaced persons that are "forgotten" and therefore particularly vulnerable and in need of special attention in camps. In reality, women are in no way a specialized or forgotten subgroup within camps. While roughly half of refugees around the world are female, women and children make up an overwhelming majority of the population of camps worldwide. Why are camps so full of women? The case of northern Uganda suggests that social, political, and economic dynamics in camps has the effect of "leveling the playing field" between men and women, and often raises the social status of women in comparison to men in camps. This chapter does not make the claim that camps are beneficial for women's empowerment, but rather that camps create strange and unique sociospatial conditions that women seem to be better equipped to navigate and exploit to their relative advantage vis-à-vis men.

Refugees, IDPs, and camps

Refugees and IDPs are both groups that are displaced either due to war, famine, or natural disaster. However, the key difference between the two categories is that a refugee is a person displaced *outside* of the border of his or her country of citizenship, whereas an IDP is displaced *within* national borders. Of the total number of currently displaced persons, 10.5 million are classified as refugees, and 17.7 million are considered internally displaced persons (IDP).[4] One in four refugees hail from Afghanistan with nearly three million Afghanis dispersed around the world, 1.6 million of whom live in Pakistan alone.[5] The country with the largest numbers of internally displaced persons is Colombia, with four million IDPs (see chapter by Parra-Fox in this volume). Since refugees are viewed as "stateless" people, under the UN Convention on Refugees, the United Nations (via UNHCR) has full jurisdiction over refugees—and therefore refugees are subjects of the international community. On the other hand, IDPs are *citizens* of the country who are displaced within state borders so IDPs are primarily the responsibility of the state.[6] The UN may *assist* in IDP humanitarian concerns, but only at the request of (and in coordination with) the host government. In general, a subset of refugees and IDPs live in camps. According to UNHCR, 81 percent of refugees are in "developing countries," and may be absorbed into urban areas. Worldwide, UNHCR reckons in their 2012 report that the "total population of concern" was more than 35 million, that is people seeking refuge living in or outside camps.[7]

Every camp is different in terms of shape, size, and internal dynamics, but in general camps tend to be a collection of tents or other temporary structures. Once officially recognized by the United Nations, camps are essentially supported through foreign aid. As a result, camps operate under a strange peri-urban political economy that is based on infusions of aid-based capital—including food aid, seeds, tools, cash payments, and other non-food items. It is common for camps to be run as if they are "short-term" living spaces, but the reality is that displacement camps tend to exist much longer than anyone ever expects—for decades, in many cases—and have populations similar to towns or small cities. For example, the world's biggest refugee camp is now 22 years old, located in Dabaab in northeastern Kenya. This camp was designed to hold 90,000 people but today holds more than 300,000 people;[8] on average, camp populations are about 11,400 people.[9] The oldest camps in the world host Palestinian refugees who were driven out in 1948 after the Arab–Israeli war and fled to Jordan, Lebanon, Syria, the West Bank, and the Gaza Strip, and this coalition of camps currently houses about four million displaced persons in total.[10] At first those Palestinian camps were essentially collections of tents, but after 55 years of existence they have morphed into what appear to be ramshackle urban slums.[11] Many camps in Africa also exist for decades, and after camps have been in place for several years, the housing tends to evolve into makeshift mud-walled, thatched huts. Essentially whatever supplies are available within the camp setting (rocks, sticks, mud, plastic, tarpaulins, or tents) may be used as building materials for houses. Refugees in camps are generally barred from citizenship, most employment opportunities, landownership, and state services. IDPs, like those in northern Uganda prior to 2007, are citizens so they have rights to employment and may own land, but while living in the camps they oftentimes are restricted from full freedom of movement. As a result, the right and ability of an individual to practice full economic, political, and social agency is seriously infringed upon in a displacement scenario, if not essentially eliminated altogether.

Methodology

This analysis is based on more than two years of research in both the towns and camps of northern Uganda between 2005 and 2008. Working in displacement areas and camps required a multipronged approach that included research with aid workers and displaced people in Gulu town and in the camps. To gain an understanding of aid operations and the role of donor agencies, I worked with two aid agencies: Action Contre La Faim, a prominent humanitarian aid non-governmental organization (NGO), and the United States Agency for International Development (USAID), the largest donor agency in northern Uganda. During my first six months in Uganda I conducted informal interviews and developed relationships with all the critical players in the conflict zone including UN representatives, donor agents, NGO workers in key agencies, critical players in the Acholi Religious Leaders group, Human Rights Watch, the International Criminal Court (ICC), and local and regional government officials. During this period in Gulu and Kampala, I took advantage of archival resources available at Gulu and Makerere Universities, on human rights (old humanitarian reports, newspapers, and other documents) to reveal more about the aid enterprise, especially from an historical perspective. Through participant observation, informal and formal interviews, meetings with expatriates and Acholi aid workers, and meetings with Acholi in the camps and in Gulu, I learned how the emergency started, how it developed, and how it affected political, economic, and social networks in the area. While the conflict was ongoing and security would not allow me to live in the camps, I trained three displaced people to record ethnographic data about life in the camps, peace talks, decongestion, resettlement camps, and the resettlement process. In my final phase of research, I focused almost entirely on the Acholi community in the camps, and when security conditions finally allowed me to experience life in the camps I stayed in Opit IDP camp for five months. There was an effort made to balance the data garnered in as many ways as possible by social class, spatial considerations such as region, camp size, organization, and the different temporal contexts of war and peace, including gender.

Camps and women

Humanitarian aid organizations exist to fill gaps in basic needs created by disruptions such as war and natural disaster. Camps are convenient service points to operationalize emergencies as a "means to determine what one should monitor, count and take into account" (Ophir 2010: 72) when providing materials such as food aid, tents, jerry cans, seeds, tools, etc., but camps are much more than mere functional humanitarian spaces. In a situation of diminishing nation-state power and the decentering spatial effects of the globalized political economy, humanitarian aid and camps are tools for maintaining control over resources and people, with a particular focus on Foucaultian *biopower*. Giorgio Agamben describes the camp as the ultimate tool in this bodily form of governance, which reduces human existence in a camp to *bare life*, that is, a being that solely exists in a bare physical state. Agamben would call the camp system in northern Uganda a *zone of exception*, a space where the rule of law does not hold and where people exist in a situation where "life ceases to be politically relevant . . . and can as such be eliminated without punishment" (Agamben 1998: 139). So within any humanitarian aid-dominated zone, camp dwellers are essentially bound in an unspoken contract with foreign agents to play the role of "bare life." Ideally people in camps should not display social, economic, or political preferences, opinions, biases, or agency. If people in camps do anything but play the role of victim, they threaten the symbolic base upon which their physical wellbeing rests.[12] In taking on the "refugee" or "IDP" label, displaced persons lose their agency and become helpless, dependent victims, a morass of humanity clamoring for basic physical survival. The "refugee" is essentially rendered a "speechless emissary" (Malkki 1996).

In a disaster zone with refugee camps, particularly in Africa, humanitarian actions are based on a strange cocktail of Western donor demands, national politics, and local negotiations. But in order to garner funds and support to intervene in a messy situation like this, a clear story must be constructed to sell the case for action to the international community that avoids political tangles and instead focuses on vulnerable populations caught in the crossfire (Roberts 2001). And as the most vulnerable of the vulnerable, images of women and children are fundamental marketing tools to authenticate victimhood and sell a humanitarian cause to the international community (see Rosenblatt 1996). With a conceptualization of the woman-and-child as blameless and helpless, there is a clear moral stance, a clear victim, and therefore a clear point of action. While there is a substantial amount of literature that critiques the common use of victim imagery in humanitarian discourse (de Waal 1997),[13] the use of woman-and-child images persists in practice because of the digestibility and marketability of this particular narrative.

The humanitarian world operates under a heteronormative worldview, that is, families are supposed to be male-headed, and children are normatively bonded to their parents in camps. Therefore any household structure that deviates from this norm is named and measured—non-normative categories in the aid world include: CHH (child-headed household) and FHH (female-headed household). This provides evidence that the refugee or IDP is normatively gendered as a male, so female displaced persons are singled out and discussed as a specialized category in camps that tends to be "forgotten" and more vulnerable to violence (and therefore women in camps require special needs and protection).[14] Humanitarian aid organizations are accused of and constantly attempting to rectify their failure "to adopt a gender-aware approach" (Carpenter 2005: 323). Still, with all the self-flagellating, sensitizing, discussing, and redirecting of aid programs to incorporate female refugees and IDPs, women are not "forgotten" and in fact, they are a central focus of aid organizations for two reasons. On a practical level, mothers are perceived as the primary caretakers of children, and child images are the most important fundraising tools for humanitarian aid agencies. As a result, agencies that focus on and market their work with children (such as UNICEF) in reality primarily work with *women* in order to target their desired demographic of children. And on a symbolic level, mothers and their children are critical to the sustenance and funding of the humanitarian project, because they fall more neatly into the blameless-and-helpless category, as opposed to men who are framed as the perpetrators, not victims, of violence. In a sense, women and children are bundled into a solitary humanitarian object—"womenandchildren" (Enloe, 1990)—that is critical to the survival of the humanitarian project.

Interestingly, a simple perusal of refugee and IDP statistics makes it difficult to understand how women could possibly be "forgotten" or seen as a special subgroup of refugees and IDPs because, in fact, they make up the majority of camp populations around the world—approximately 70–80 percent of camp-dwellers worldwide are women and children. While it is clear that women and children are symbolically important to humanitarian aid organizations in terms of articulating need and justifying projects for children, this still does not fully explain how and why the symbolics translate into physical reality. In other words, if one assumes that in general the gender balance of populations are roughly equal around the world, why are refugee and IDP camps teeming with women (and children) and not men? If one says that camps are set up to provide safety, do the men not need safety too? Is the presence of men not also critical for the wellbeing of the women and children?[15] And if one makes the argument that men need to relocate to other areas for work to support the family, could this argument not also apply to women who presumably also need to make money to support their families, especially if it's a female-headed household? Even if there is a situation of a forced displacement where there might be less choice in terms of mobility, as was the case in northern Uganda, why is it that only the men flee the camps and not the women? To attempt to answer these questions, let us consider the case of the camp system in northern Uganda.

The camps of northern Uganda[16]

War began in northern Uganda in 1986. For most of the time since the start of conflict, the main adversaries have been the northern-based Joseph Kony's Lord's Resistance Army (LRA) and the southern-based government of Uganda. Most analysts see the conflict as an historical power struggle between ethnic Bantu groups in the south and Nilotic groups in the north. In a nutshell, this multiphased counterinsurgency was launched in 1986 when the current president, Museveni, triumphed in his rebellion against his predecessor. As fears of an anticipated retribution petered out over time, the conflict started to wane and might have died out in the mid-1990s. But assistance from Sudan at that time transformed the northern Uganda conflict from a local skirmish to a regional conflict in which the LRA essentially became a pawn in the larger Sudanese quagmire.[17]

After ten years of conflict, the Ugandan government employed a far-reaching military tactic to defeat the rebels. The government forcibly removed the Acholi from their homesteads into camps that were called "protected villages." There was a security perimeter enforced by the Ugandan military, with an average radius of two kilometers, around each camp. Within this boundary the people in the camp were allowed to move and farm. Of greater priority than *protecting* civilians, the government had several strategic military aims. For one, the government wanted to clearly identify the rebels by removing civilians from the countryside. Acholi were first given a warning by the military to evacuate, but then there were subsequent harassments, beatings, and sometimes killings if Acholi did not move off their homesteads and into the camps. After the government forced people off their lands and into camps, they informed them that if they were caught outside the boundary of the camps, they would be considered to be rebel collaborators and would be dealt with accordingly. By 2002, virtually all Acholi in northern Uganda (and even parts of Teso and Lango groups in northeastern Uganda) were interned in camps. At the height of the displacement, the total number of northern people in camps was estimated to be 1.6–2 million persons.

The second part of the military strategy was to cut off the rebels from their source of supplies. These supplies included anything from durable goods to child soldiers, but the most essential supply that the rebels sought was food. By moving people from their homesteads, the government aimed to cut off the LRA from their food supply in the countryside. Eventually all food sources in Acholiland were depleted, and the government effectively stopped agricultural production in the north for years in order to starve the rebels and force their surrender. Within six months, the United Nations World Food Programme (WFP) stepped in to provide food aid to the Acholi in the camps. Despite their good intentions, the WFP essentially became a complicit player in this forced illegal displacement of 1.6 million people because it provided the material means needed to maintain the camp system for decades (Branch 2011).

Once in the camps, there were other spatial dimensions that the Acholi had to contend with. The most significant of these was a two-kilometer security perimeter that was set around the camps. Humanitarian organizations cited the "two-kilometer security perimeter" in almost every official document, but in reality this figure varied from camp to camp and the perimeter varied even within any given individual camp. So this two kilometer figure was more of an average. No one was allowed to move outside of this boundary for any purpose or they would suffer reprisals from the military. The donors asserted that if the perimeter was set at 5km, 80 percent of people could access farmlands, as well as food, water, health care, and schools.[18] Instead people were crowded into a small area with few resources. The lack of food, water, and medicine caused a huge spike in disease and death. It was estimated in 2005 that 1,000 people per week died in northern Uganda, not due to the violence of the conflict itself, but rather due to the effects of living in the camps.[19] So it was actually this narrow security perimeter around the camps that caused most of the deaths in northern Uganda, not the conflict itself.

This forced displacement was undertaken in the name of "protection" of civilians, but the Acholi people constantly argued they were more at risk in the camps than outside. For one thing, many Acholi felt more vulnerable in the camps since it was much easier for rebels to loot and abduct when people were conveniently assembled into underprotected centralized locations. Poor military protection stemmed in part from the placement of the military barracks relative to the camps. In some camps, the garrison was set at a distance away from the camp with only a road connecting the two areas. In other camps the barracks were often set in the dead center of the camps so that it almost appeared that the displaced people were acting as a buffer between the rebels and military. "Is the military protecting us or are *we* protecting the military?" was a common question that many Acholi raised. The inconsistencies in military protection strategies and the high mortality rates in the camps also made people wonder if the displacement was really about security, or was it about something else? Instead of being protected, people in the camps felt there were other more sinister plans. Was the displacement really a plot to grab land from the Acholi to redistribute to investors or government officials? Or was it a genocidal project on the part of the president of Uganda to effectively eliminate his main political opponents, the Acholi?

For all these reasons, when peace talks started and resettlement followed in 2007, I expected a rapid mass migration of people back to their homesteads to escape this situation. It was at this time, as the conflict waned and security improved, that I lived in Opit IDP camp to capture the last moments of the camps as people returned to their lands. But instead of hearing a uniform denouncement, many people almost seemed positive about the camps. One woman told me, "If the war starts up again, I'll be the first person back there!" Another person even said "In the end, the camps were a good thing." I noted that some men expressed positive sentiments about the camps, but it was mainly women who made these affirmative comments about the camp experience.

The positive strain in people's memories of the camps severely challenged Agamben's bare life thesis of the camp as a zone of exception (Agamben 1998: 139) where "life ceases to be politically relevant . . . and can as such be eliminated without punishment." Were the camps truly a "zone of exception," or were outside observers simply preconditioned to see "bare life" there? And of the people who spoke in favor of the camps, why did women make up the overwhelming majority?

I found in my conversations about life in the camps that there was always some level of political, economic, and social agency that individuals could exercise in the camps. For example, on the economic front, people set up businesses in the camp and in Opit IDP camp there was a long line of shops running down the middle of the camp. These establishments ranged from bars (with digital satellite television, where they would show videos and football matches), to stores with sundries such as soap, matches, candies, and slippers. There were several small restaurants in the camps, a motor parts/gasoline/car oil stand, and even a dance club. Several food grinding businesses operated in the camp, because of the influx of food aid grains that were difficult to grind by hand. There were women selling charcoal, fired-clay pots, and home-made brew. And in the center of the camp was a fairly large market, full of fresh vegetables, dried fish, beans, rock salt, and meat. Many of the larger businesses in the camp were started with funds that were covertly given to families by their abducted children who were fighting with the rebels. Ironically, for some, the loss of their child to the rebels in some cases ended up being a boon for the family, especially if the child soldier crept home every now and then with some looted food or money to give to their parents. By discussing the presence of businesses I do not mean to insinuate that there was a booming economy in the camps. This was not the case, and people were indeed struggling. However, there was enough money floating around in the camps to sustain many businesses. And the survival of all those businesses was made possible because of the change in spatial relationships; with 29,000–50,000 people in a 1.4km² area, there was enough concentrated demand to allow this kind of economic agency. Many of these businesses were run by women, most notably the alcohol-brewing establishments, which were exclusively run by women and almost exclusively patronized by men. While grain has always been used to brew alcohol in northern Uganda, it was never done as unilaterally as a cash-earning strategy as it was during the displacement, so much so that it superseded agriculture as the traditional main source of Acholi household income. Even devout, born-again Christian women, who abstained from drinking alcohol, found themselves distilling and selling local brew because they had few other money-earning options. Alcohol brewing from food-aid grains became a main source of the small incomes earned in the camps. One-third of Acholi women in camps brewed as a primary source of income in northern Uganda versus 30 percent who worked in agriculture. Alcohol consumption increased, becoming a chronic and debilitating activity for idle and depressed men. This was a factor in the rise of the level of alcoholism of men in particular in Acholiland, as well as alcohol-induced sexual abuse and violence in the camps. So the camps and the food aid that kept the camps running provided a situation where women became the main entrepreneurs and almost all the men were rendered into a state of chronic drunkenness.

It is difficult to assess what transpired with regard to grassroots political organization at the height of the displacement because movement and access for researchers was limited during this period. But interviews carried out during the ceasefire reveal an interesting picture of political life within the camps. In a blunt sense, political agency *was* thwarted very directly by the creation of these camps. Elected leadership essentially became irrelevant during the displacement. At first the elected official within the camp was put in charge of the camp, but this official would refer disciplinary or other political matters to the elected representative from the area of origin for the person in question. But locally elected officials were also displaced and moved into the camps. Meanwhile a camp leader was installed by the United Nations to manage food aid in the camps. But over time, the practice of local officials referring cases to one another diminished, and the UN camp leader became the *de facto* head of the camp for all issues, not only those related to food aid. The camp leader managed, allocated, and coordinated the main resources in the camps, namely humanitarian food aid. Because food aid is essentially the anchor for all other aid operations, over time the camp leader ended up becoming the main point of contact for all aid operations over the 12-year period of the displacement. On one level, this appears to be completely in line with Agamben's thesis of the camp as a place where political agency and accountability does not exist. But a story told to me suggested that, despite the lack of democratic processes, people still were able to have their voices heard, notably including women. The first camp leader appointed by the UN in Opit IDP camp was suspected of misallocating resources and other forms of corruption. A group of people in Opit took a vote and chose another man to take the role of camp leader. With this vote result in hand, a group of men and women from the camp went directly to the United Nations camp leader and demanded that another person be put in charge of the camp. The presence of women as part of this delegation is in part a product of funding requirements in humanitarian aid organizations as oftentimes aid organizations require participation of women in petitions or proposals as a prerequisite for funding.[20] The UN complied and a new camp leader was installed in Opit. This is not to suggest that political agency was fully manifested for women (or anyone else) in the camps, but there still was *some* level of political agency that existed. Furthermore, the political structure of the camps as a humanitarian aid zone where the UN was the main provider of goods and services definitely created a space for greater political engagement for women than existed outside the camp context.

When I spoke to returnees about their memories of the camps I found many people missed their friends and neighbors from the camp. In Acholi, living spaces reflect spheres of relatedness; closer kin live in your inner ring, and as one moves out of the living space one also moves into the extended family, the clan, and so forth into the larger identity of being Acholi. In the camps, these living spaces were altered, and while the norms of family living were mirrored in the camps as much as possible, available space did not allow people to set up their huts in the same configuration as in their homesteads. This meant that while immediate and close kin *tended* to live together in the camps, extended families were often spread out across the camp (e.g., extended families were often in different sections of the camp depending on where space was available to set up a hut at the time of their displacement). As a result, people ended up living next-door to people who had no kin relationship with them. After living next to one another for 6–12 years, many people created kin-like relationships with their neighbors, helping each other with house chores and raising children. When the camps were disbanded, these alternative kin groups were split apart. Back in their homesteads away from each other they missed the support and company of their former neighbors in the camps. And again, most of the people who commented on these lost relationships were women.

Surprisingly, while many people immediately left the camps some people stayed, even after the camps were officially decommissioned by the government. It was estimated by the United Nations and by local political leaders that up to 30 percent of people would stay in the camps indefinitely, perhaps permanently. To some humanitarian and government actors, the urbanization of northern Uganda was a surprising effect of the camps. Some displaced people stayed in the camps because they were designated as "vulnerable groups" by the aid agencies and were therefore still eligible for aid. But some people who stayed in the camps did so to maintain those non-familial social ties in order to survive. These non-familial kin included former child soldiers who returned from the bush but were not welcomed back into their communities (because perhaps they had killed a family member or a neighbor). Or this could include outcasts of society, such as elderly women without sons to care for them in old age. I met a group of four elderly ladies who were not related to one another and lived in separate huts, but they ate together every day. I asked why they ate together. The four women explained to me that they all produced only daughters who ended up marrying into families with "no good, drunkard" husbands (as they put it). As a result, these old women were abandoned and fell through the cracks of the Acholi social system. Sadly, they also fell through the humanitarian aid social net as only one of the four women was on the rolls for food aid. That elderly woman on the rolls shared her food rations with the others as they banded together to survive in the camp. So for vulnerable people, the camps provided a space of sustenance when family and clan support networks failed them.

The proximity of neighbors in the camps was not only an important survival strategy for obtaining material goods, but it also provided a more secure environment in some respects. If someone tried to rob, rape, or molest another person in the camps, all the victim had to do was scream and many people around could hear, and there were many witnesses who could corroborate their story. But soon as people started leaving the camps, there was an upswing in banditry in the countryside. In Opit, the residents heard about child molestations, rapes, lootings, a mysterious beheading of a man, and a murder of a prominent tobacco grower and his wife. All of this non-conflict-related violence was blamed on *boo kec* (Acholi for "bitter spinach"), which roughly translates to "bad elements of society." *Boo kec* is one part of Acholi life that was almost completely annihilated during the camp time, for while people could become victims of violence or harassment by the rebels or the military, *boo kec* was not as active. Now back at home, people were very fearful about the *boo kec* and, for this reason, some people missed the relative security in the camp due to the spatial setup and the compressed living conditions. Ironically, being in the camps provided security not because of the presence of the military, but rather because they had neighbors in close spatial proximity. When I spoke to the police chief after the camps were disbanded, he told me that his main job was to move around the area "sensitizing" people about how to identify a crime when they encountered one. According to the police officer, a majority of the crimes he confronted in the field were child molestation, rape, and domestic violence—all crimes in which the victims were overwhelmingly women and children. So violence among the general population, but in particular towards women (and children), increased after people left the camps.

Based on this discussion about life in Opit IDP camp, one should not come to the conclusion that women enjoyed favorable conditions in the camps in northern Uganda. There was an extraordinarily high level of violence waged within the camp system for everyone who lived there. There was the structural violence of the camp itself as the camp, in just keeping people barely alive, forced them into a situation of illness, undernourishment, and disempowerment for more than a decade, and in some cases two decades. Since women (and children) were the main dwellers in the camps, by default this meant that they were the main victims of these forms of structural violence. In addition, there also was a considerable amount of physical abuse and sexual violence. However, it is important to remember that in the context of the camps in northern Uganda, sex became an act and a site of socioeconomic agency. For example, women regularly became "wives" (or sexual partners) of soldiers posted in the camps in order to secure special protection and material wellbeing. Many Acholi men in the camps felt deeply threatened by the loss of their social standing and the discomfort from this loss often spilled over into camp violence. I was told on numerous occasions about violent clashes between soldiers and male IDPs over women. This was often the underlying reason for some of the worst atrocities recorded during the entire period of the conflict—a violent mix of alcohol and sexual jealousy.

That said, all sexual related incidences in the camps were not consensual or jealousy-fueled. Much of the gender-based violence involved outright, forcible rape. This is one of the reasons cited by civilians as to why northern Uganda had a higher prevalence of HIV/AIDS than in other parts of the country. When discussions about rape come up in the context of war (or perhaps in any context), the general assumption is that this violation is a male-on-female violation. However, Chris Dolan, a prominent scholar of northern Uganda, points out that in reality, much of the rape that occurred in northern Ugandan camps involved male-on-male sexual violence.[21] Sverker Finnström, another scholar of northern Uganda, documents evidence of widespread rape of Acholi men by Ugandan soldiers.[22] This act of male-on-male rape was a way by which Acholi men were essentially stripped of their masculinity and shaped into docile and agentless subjects. For these and other reasons related to the emasculating structural forms of violence that existed in northern Ugandan camps, men often refused to stay, even though the displacement was forced and the camps were heavily guarded. Many men avoided the camps at all costs to secure their gender positionality. And this might explain, at least partially, why so many men were absent from the camps.

Concluding thoughts

People in camps are not just helpless victims. Acholi, and displaced people around the world, find fractures in the "bare life" construct and are able to express economic, social, and political agency even in the most trying of situations. But the case in northern Uganda suggests that it is mainly *women* who are adept at using the social tools to locate and exploit those fractures, thus the camps became a space that in a sense "leveled the field" between men and women. One reason for this might be that it is much easier for women than men to play the "bare life" role necessary to live in a camp. Women are socialized to comply and play certain feminized roles and performances of helplessness, dependency, and subjugation in patriarchal society, so the "bare life" performance and role of a refugee is less of a departure for women than it is for men. Women have historically developed skills to locate fractures in patriarchy in order to exploit alternative modes of power and influence, and these are the very same skills, I argue, that make women better-equipped to negotiate life in a camp.

But beyond a generalized system of patriarchy, there is another notable point in this case: the role that humanitarian aid organizations played in raising the profile of women and diminishing the relative power of men. In the case of northern Uganda this manifested itself in three ways: 1) aid organizations forced gender equality in political action through their policies; 2) women's physical security, economic agency, and social networks expanded as a result of the population density in the camps (and humanitarian food aid was the mechanism that kept the camps in place for long enough to solidify those networks); and 3) food aid surpluses spawned a huge alcohol-brewing industry in the camps that was an economic boon for women and rendered the men drunk. To put it in blunt terms, men were completely emasculated through the camp experience in northern Uganda and women were oddly empowered vis-à-vis their male counterparts. I hypothesize that this unintended relative empowerment is the reason women spoke wistfully about camps, while their male family members shook their heads in a defeated manner when women made those positive comments.

Humanitarianism particularly focuses on technologies of the body to wield new decentered forms of global economic, social, and political power. Thus, embodied social constructions such as gender, race, and class are not only interesting footnotes in a discussion about humanitarianism, but instead these categories are actually central component parts of this modality of power. Any analysis about camps, refugees, or humanitarian aid must be tightly integrated using a feminist theoretical lens. The camp, instead of being a space of "exception," is instead a microcosm of a larger system of power that exists far beyond war zones and is fundamentally intertwined with race, class, gender, and other systems of social inequality and subjugation.

Notes

1 "Where are the 50 most populous refugee camps?" *Smithsonian Magazine*, www.smithsonianmag.com/ideas-innovations/Where_are_the_50_Most_Populous_Refugee_Camps.html
2 UNCHR Global Trends 2012, *Displacement: The New 21st Century Challenge*. This figure only reflects refugee counts; according to www.internal-displacement.org/statistics there are also more than five million IDPs within Syria, in addition to approximately two million refugees who have fled from the Syrian civil war. I do not include this in the main text because it is not clear whether these five million people are accounted for in UNHCR's total global displacement count.
3 I use gendered categories as they are understood in northern Ugandan society, i.e., the biological sex (male/female) corresponding with socially constructed, patriarchal male/female roles in society.
4 UNCHR Global Trends 2012, *Displacement: The New 21st Century Challenge*.
5 Ibid.
6 This, of course, presented a big challenge in northern Uganda where the state government actually displaced the IDPs in the first place.

7 For the purposes of this paper we shall conflate these two terms "population of concern" and "camp dwellers," since almost all of the UNHCR "population of concern" was residing in camps in northern Uganda.

8 www.theatlantic.com/infocus/2011/04/the-worlds-largest-refugee-camp-turns-20/100046.

9 UNCHR Global Trends 2012, *Displacement: The New 21st Century Challenge*.

10 www.internal-displacement.org/statistics.

11 www.economist.com/news/middle-east-and-africa/21587846-some-palestinians-want-their-people-abandon-refugee-camps-without-demanding.

12 An example: the Western media was shocked to see Haitian earthquake victims summarily rejecting a truck of "food aid." The truck was actually full of biscuits, but talking-heads expressed retaliatory sentiments like: "If they're so hungry shouldn't they just eat whatever they can get?"

13 Also see work on this topic by Peter Redmond, David Rieff, and Graham Hancock.

14 An example, of many, is www.zakat.org/blog/refugee-women . . . invisible-and-often-forgotten. This rhetoric about women in camps as "invisible" and "forgotten" is commonly used as a fundraising strategy, but it is evident in the practitioner literature of the United Nations as well as in academia.

15 Several scholars highlight that civilian children and women become more vulnerable in wartime when their men go missing (e.g., Shoemaker 2001: 19).

16 Parts of this section are either adapted or excerpted from "Camps and bare life" in Nibbe (2011).

17 The LRA essentially became guns-for-hire for the Sudanese government, and since then the reasons for the conflict have transformed and the battleground has shifted to the Congo and Central African Republic.

18 Personal communication with Jeff Drumtra, USAID, April 2006.

19 www.oxfam.org/sites/www.oxfam.org/files/uganda.pdf.

20 Community groups looking for funding sometimes struggle to find an interested woman to join projects, because without at least one woman they cannot proceed.

21 Dolan's work on male gender issues in the context of the northern Uganda conflict is extensive, including his 2011 book *Social Torture: The Case of Northern Uganda, 1986–2006*.

22 Finnström's work references rape as a weapon in northern Uganda in several places, most prominently in his article "Gendered war and rumors of Saddam Hussein in Uganda" (Finnström 2009).

References

Agamben, G. (1998) *Homo Sacer: Sovereign Power and Bare Life*, Palo Alto: Stanford University Press.

Branch, A. (2011) *Displacing Human Rights: War and Intervention in Northern Uganda*, New York: Oxford University Press.

Carpenter, R.C. (2005) "'Women, children and other vulnerable groups': gender, strategic frames and the protection of civilians as a transnational issue," *International Studies Quarterly* 49(2): 295–334.

de Waal, A. (1997) *Famine Crimes: Politics and the Disaster Relief Industry in Africa*, Oxford: James Currey.

Dolan, C. (2001) *Social Torture: The Case of Northern Uganda. 1996–2006*, New York and Oxford: Berghahn Books.

Enloe, C. (1990) "Womenandchildren: making feminist sense of the Persian Gulf crisis," *Village Voice*, September 25.

Finnström, S. (2009) "Gendered war and rumors of Saddam Hussein in Uganda," *Anthropology and Humanism* 34(1): 61–70.

Malkki, L. (1996) "Speechless emissaries: refugees, humanitarianism, and dehistoricization," *Cultural Anthropology* 11(3): 377–404.

Nibbe, A.A. (2011) *The Effects of A Narrative: Humanitarian Aid and Action in the Northern Uganda Conflict*, dissertation, Abstract International, 72(8), UMI No. AAT 3456852.

Ophir, A. (2010) "The politics of catastrophization: emergency and exception," in M. Pandolfi and D. Fassin (eds.), *Contemporary States of Emergency: The Politics of Military and Humanitarian Interventions*, New York: Zone Books.

Roberts, A. (2001) "Humanitarian issues and agencies as triggers for international military action," in S. Chesterman (ed.), *Civilians at War*, Boulder, CO: Lynne Reinner, pp. 177–196.

Rosenblatt, L. (1996) "The media and the refugee," in R. Rotberg and T. Weiss (eds.), *From Massacres to Genocide*, Washington, DC: Brookings Institution Press, pp. 136–148.

Shoemaker, J. (2001) "Women and wars within states: internal conflict, women's rights and international security," *Civil Wars* 4(3): 1–34.

44

RESEARCHING SEXUAL VIOLENCE IN THE EASTERN DEMOCRATIC REPUBLIC OF CONGO

Methodologies, ethics, and the production of knowledge in an African warscape

Patricia Daley

Introduction

'DR Congo: 48 rapes every hour, US study finds' – thus stated a BBC News headline in May 2011 (BBC 2011). Such headlines have become commonplace in media representation of the conflict in eastern Democratic Republic of Congo (DRC). Shock statistics abound in the Western media, and, as they are derived from articles published in reputable science journals and by human rights organizations, their authority is rarely questioned. Having done research in a war-afflicted country, I am acutely aware of the difficulty of data collection, poor databases, and wide variations in estimated data.

Let me recount an incident that has provoked my thoughts on researching sexual and gender-based violence (SGBV) in Africa. In June 2006, I was conducting research in Burundi and visited a refuge to obtain admittance data on various forms of violence, including SGBV. The medical staff were respectful and were keen to show me the gravity of the problem and the importance of their work. They took me to a room where there were young girls who had been raped and asked one girl to recount what had happened to her. The girl was visibly distraught. I wanted to curtail her testimony, but others in the room thought her story was important for me to hear – despite the trauma it was obviously invoking. The refuge was doing its best to heal the physical wounds of the women but had no psychological care. I felt uncomfortable, ethically, but also emotionally. Realizing that visits like mine from NGOs, donor governments, researchers, and journalists were a frequent occurrence, I wanted to find out whether other researchers were confronted with the ethical challenges that such encounters posed.

Since 1996, the eastern provinces of the DRC have experienced widespread warfare carried out by numerous rebel groups. An estimated four million people have been killed and thousands

429

of survivors have been subjected to horrific forms of violence, including sexual violence that has mainly affected women. Since 2002, there has been a proliferation of publications on SGBV, especially rape, sponsored by a range of non-governmental organizations (NGOs) – human rights, peace and development NGOs, Western governments' aid agencies, UN peacekeeping missions and the American military establishment in the form of the Africa Command (AFRICOM). SGBV in Africa has also become a popular dissertation topic among masters and doctoral students in the West (see Henry 2013). However, the scholarly work has struggled to keep pace with the policy documents.

Warscape, a term coined by anthropologists to reflect the complexity of lived experiences in conflict-affected areas, is a useful concept to apply to the eastern DRC where local, regional and global factors intersect to produce multiple manifestations of violence that includes SGBV (Nordstrom and Robben 1995; Hoffman and Lubkemann 2005; Hoffman 2003). Geographers Korf et al. (2010: 385) define warscapes as 'landscapes characterised by brutal violence, political volatility, physical insecurity and the disruptions and instabilities that exist in many civil war zones that different social actors navigate through'. In the case of the eastern DRC, such actors are gendered and class differentiated, local, regional and international, as well as civil society, military and private.

War zones are spaces where direct combat between armed opponents is deemed legitimate and sanctioned by international norms. Violence committed outside the established rules of warfare are seen as violations, labelled uncivilized and barbaric, and can constitute war crimes. Violence against civilians, especially in African wars in the post-Cold War era, has received much attention in recent years, such that research has addressed the rationale for the targeting of civilians and questioned the association of the term itself with women and children (Sjoberg and Peet 2011). Gender-based violence in warscapes has only recently received international and scholarly attention, despite decades of campaigning by feminists. Attention to sexual violence in African warscapes radically increased with the Rwandan genocide and the abuse carried out by Hutu *génocidiaires* of Tutsi women who symbolized the Tutsi nation (Baines 2003; Kombo 2009).

Following a spate of international legal instruments aiming to protect women in wartime and case law making rape a war crime, humanitarian agencies are expected to bring women and SGBV within their remit.[1] This focus on SGBV has led to the protection of women being used as a justification for external intervention in civil wars and postwar reconstruction.

The rigour of the research process on which these SGBV reports are based requires critical scrutiny for two main reasons. First, warscapes are affected by insecurity, displacement and rumours that can distort information – making the reliability of data sources questionable. This combines with personal safety issues, especially for Westerners coming from increasingly risk-averse institutions. Second, propaganda is central to the prosecution and interpretation of wars and is used both by all warring factions and their supporters. Acts of atrocities can give international recognition to a ragtag rebel group and can become bargaining tools in peace negotiations.

The questions posed by this chapter range from who is sponsoring and conducting SGBV research projects in the DRC; what are their methodologies; how are ethical considerations dealt with; and, finally, what ontological foundation informs them, what do these articles/reports say about African women's bodies as the objects of research? The chapter is not aimed as a critique of any specific researcher, although some are used as exemplars. Instead, the paper is a discussion of the trends that seemingly neutral research can take in the context of differential power and geographical otherness.

The chapter is divided into three sections. The first examines the type of research that is being conducted; identifies the researchers, and the purpose of their research. In the second section, bearing in mind my experience in the field, I consider the specificities of the methodologies, how they relate to those used in other contexts, and how far they fit with contemporary guidelines on researching sexual violence, especially pertaining to research ethics. In the final section, I consider the ontological foundations of the research, especially how the intersection of race, gender, power, and advocacy combine to shape research on the DRC. Evidence comes from research articles published in refereed journals, newspaper articles, and the grey literature produced by policy elites.

Type of research and researchers

Research on sexual and gender based violence (SGBV) in the DRC has been conducted by three primary groups of researchers and/or consultants. Firstly, some of the earliest research was carried out by human rights advocacy groups, such as Human Rights Watch (2002, 2005) and Amnesty International (2004). Some reports are collaborative, involving local NGOs. For example, the London-based International Alert has produced reports with the Congolese women's networks, Réseau des Femmes pour un Développement Associatif and Réseau des Femmes pour la Défense des Droits et la Paix (Ohambe et al. 2005), and with the Kampala-based Refugee Law Project (Dolan 2010). Numerous parliamentary groups (e.g., Swedish Foundation for Human Rights 2008), Western governments' aid agencies (e.g., Pratt et al. 2004), United Nations agencies and peace-keeping missions (e.g., MUNUSCO 2009), and international and local humanitarian NGOs have commissioned consultancy reports on SGBV in the Congo.

Secondly, research is carried out by medical professionals, particularly North American-based or South-African-based (Johnston et al. 2010; Bartels et al. 2010; Peterman et al. 2011). The project by Johnston et al. (2011) was supervised by a team of medical doctors associated with McGill University Department of Health, Harvard Humanitarian Institute and sponsored by US Department of Defense Africa Command (AFRICOM). Bartels et al.'s (2010) research was sponsored by Oxfam America and carried out by the Harvard Humanitarian Initiative. Such research often involves collaboration with Congolese medical professionals who often conduct the actual data collection (Kalisya et al. 2011). Findings from medical research tend to receive wide publicity in the international media. Finally, research by academic social scientists is relatively recent and constitutes a minority of the published work. Sponsorship often comes from development agencies seeking information to clarify the need for humanitarian interventions (Eriksson Baaz and Stern 2010). Independent research is rare due to the logistical difficulties of working in a warscape.

Purpose of the research

Research in the context of warscape and low intensity violence is high-risk and often difficult to conduct independently of local political actors and humanitarian agencies. Fact-finding for policymaking is often the primary purpose of such research. Johnston et al. (2010) suggest that their aim is to assess the prevalence of sexual violence and human rights violations in specific territories in eastern DRC. Similarly, Peterman et al. (2011: 1060) sought 'to provide data-based estimates of sexual violence in the DRC and describe risk factors for such violence'. Undoubtedly, the magnitude of sexual violence is beyond the expected norm in any society. Until recently, there was no systematic collection or collation of data on sexual violence in the DRC. Assessing the scale of death, displacement, and SGBV is challenging for a number of reasons: recordkeeping by state bureaucracy is extremely limited geographically and poor in quality, and episodes of violence are not consistently reported, due to the failure of disclosure by women and households, fearing stigmatization and potential social exclusion from family and community.

Policy-makers tend to respond to quantitative data; justified by one researcher, Lynn Lawry (2011) as 'because numbers talk . . . by being able to extrapolate it to a number, you can say you need x number of dollars to be able to effect change'. Estimates of the scale of sexual violence become critical for pressurizing donors to provide adequate financial resources for humanitarian relief, to take preventative action, and to sustain peacekeeping missions in the very situations where accurate wide-scale quantitative data are particularly hard to come by.

Published data on sexual violence in the DRC originate from two main sources: the Panzi hospital in Bukavu, South Kivu, and the Heal Africa hospital in Goma. Such data are derived from the medical records of these treatment centres and questionnaire-based interviews with rape survivors who present themselves at the clinics. In the context of war, institutionally based sources such as these become the primary databases used by researchers. For example, Bartels et al. (2010) carried out research at the Panzi hospital's Victims of Sexual Violence Programme, and Kalisya et al.'s (2011) study of paediatric rape used hospital referral data from Heal Africa's hospital in Goma. Recognizing that only a fraction of rape survivors are likely to seek medical attention, researchers seeking to gain a more accurate representation consider wider population-based studies that are more likely to be conducted in areas adjacent to the warscape. Therefore, numerical extrapolations are often accompanied by spatial generalizations.

Methodologies and research ethics

The literature on sexual violence pertaining to methodologies is quite extensive, especially concerning research conducted in Western societies.[2] Methodologies used in stable Western countries, as well as in post-conflict countries such as Bosnia, have been applied in war-affected African countries. Most SGBV researchers do not interrogate openly the contextual relevance of the methods deployed in Africa, with the exception of South Africa where there is extensive scholarly interest in sexual and interpersonal violence and where intersecting hierarchies of race and class have propelled ethical and methodological issues to the fore. Jewkes et al. (2000), based on their experience of researching sexual violence in Southern Africa, provide useful recommendations on how the researcher and the researched can be protected. They identify issues, such as obtaining privacy in a village setting, having an alternative set of questions if interrupted by family members, and embedding SGBV questions among other more general ones. Beyond these exemplary cases, 'African exceptionalism' and humanitarian imperatives often lead to some of the dangers for the survivors and researchers being downplayed.

Differences in the methodologies used by social and medical science researchers are apparent, though not always contradictory in their agenda. Harrington (2006) links the developments in trauma science to the international focus on SGBV as a security issue after the ending of the Cold War. Trauma science, she argues, has extended its application to victims of war via medical understanding of post-traumatic stress disorder (PTSD) theory, which 'has provided scientific authority to the liberal problematization of authoritarian forms of government by classifying experiences that render people helpless as a cause of mental illness' (Harrington 2006: 358). Harrington claims that PTSD enabled feminists campaigning against sexual violence to interpret it as a form of integrity violation requiring 'rehabilitative international intervention', and thus to rationalize it as an international security problem. Arguably, languages such as 'sexual terrorism' and 'rape as a weapon of war' pander to international securitization in the post 9/11 era.

In warscapes, a significant proportion of SGBV research has shifted to producing the statistical information required to detect PTSD and within it evidence of sexual trauma. The project of Johnston et al. (2010) used 'randomly-selected sampled population-based survey' in which households are surveyed using the PTSD Symptom Scale Interview (PSS_1), and the Patient Health Questionnaire Interview-9 to detect symptoms of major depressive disorder (MDD). The questionnaire includes questions on sexual violence among a range of health-related ones. The authors, extrapolating from one such database and using log linear regression modelling, were able to estimate that about 400,000 Eastern Congolese women were raped per annum between 2004 and 2007. When this information was published in 2010, it was represented as new findings and implied that the scale of violence was ongoing.

Techniques

While feminist research methods have informed much research on sexual violence in the West, in Africa, they are underutilized. Sexual violence researchers in the West, adopting post-structuralist feminist methods, have moved away from quantitative surveys with predetermined cultural definitions of violence. They advocate the use of qualitative techniques, unstructured and semi-structured interviews, and focus groups.

In warscapes, ethnographic research that depends on the interpersonal skills of the researcher is seen as valuable in recording the 'events' and narratives of people's everyday experiences (Hoffman and Lubkemann 2005). Boonazier and van Schalkwyk (2011: 281), investigating intimate partner violence, advocate the narrative approach as a way of giving agency to women. Narratives, they argue, can be political, as the women in their study fluctuated 'between investing in identities that are consistent with broader social discourses of "abused women" and also identities that challenged these scripts'. Narratives can unveil the complexities of violence and serve as a mechanism for addressing wider structural causes of SGBV. For Tolia-Kelly (2010: 361) narratives, even when sensitive to race [gender], tend to be 'disconnected from material geographies'. Although most human rights reports of sexual violence tend to rely on specific testimonies of survivors, their relatives, or medical personnel, such narratives are presented as apolitical. Such advocacy materials, aiming to draw public attention to the atrocities and to provoke action from national and international organizations, tend to de-emphasize the political factors behind the violence as illogical and irrational.

Increasingly researchers recognize that survivors of violence might be less inclined to reveal all, even though, because of the power hierarchies in the research process, the interview might not be curtailed by the interviewee. Fujii (2010: 232) notes the importance of 'meta-data – spoken and unspoken thoughts' among her respondents in post-genocide Rwanda. She argues that lack of attention to the discursive strategies, such as rumours, inventions, denial, and evasions, can affect the interpretations of interviews. People living in unstable political situations may possess a range of 'communicative toolkit' that includes misinformation, dissimulation, and silence. However, attending to the metadata, researchers can make 'sense of the ambiguities and complexities such strategies generate'.

An analysis of the SGBV literature on the DRC reveals limited problematization as to the most appropriate methods for researching SGBV or of the epistemologies that shape the research and knowledges produced. Methods that generate quantitative data are privileged, with questionnaire surveys as the primary technique for data collection. Where interviews of rape survivors have been conducted, as in the treatment centres, questionnaires are often completed by medical personnel during interviews with women presenting themselves for sexual violence-related medical treatment. Lawry justifies using population-based survey and a quantitative approach by critiquing treatment centre-based research. She states:

> In the case of the DRC, what we're seeing is one hospital that has a huge amount of press, stars going to it, has computers and lighting and everything in the periphery; doesn't even have any lighting or tables to actually examine patients. So there is a negative effect and you have to be very careful to understand that qualitative is not extrapolated beyond.
>
> *(Lawry 2011)*

In Lawry's project, the sampled population is obtained from randomly-selected village-based households, using election registers. This method assumes a certain level of competence of state bureaucracy and the continued reliability of such databases in a warscape of extended period. The project selects interviewees from households sampled by the following method.

> Interviewers began in the geographical center of the village. The interviewer chose which direction to go by tossing a pen into the air. The number of houses to pass to reach the first sampling unit (and the sampling interval) was chosen by selecting a number from 1 through 10 from a hat. A Congolese interviewer interviewed 1 adult (18 years) per household in the sample. At each house, the interviewer requested to speak with a male or female adult household member, randomly chosen by coin toss before entering the household. If that person was unavailable then the next adult in the household was approached. If only 1 adult or only 1 male or female was present at the time a household was visited, that person was interviewed regardless of sex. Records were kept of refusals, ineligible households, and lack of availability after 2 attempts. One-on-one interviews were conducted anonymously in a setting that offered privacy and confidentiality, typically inside the housing unit.
>
> *(Johnston et al. 2010: 555)*

Such sampling of households and women seems to be wholly inappropriate for an African village setting, unless conducted after repeated visits, or with more direct measures taken to ensure privacy. The validity of data acquired through this method is questionable, as well as the ethics of fieldwork implementation.

An alternative to direct sampling of households for sexual violence is to use pre-existing surveys, such as the domestic violence module completed by women participating in the Demographic and Health Survey (DHS). Peterman et al. (2011) used the 2007 DHS for the DRC and population estimates to assess levels of sexual violence across the country; concluding that the provinces of North and South Kivu have a higher rate of women reporting sexual violence.

Social science researchers, using a range of ethnographic methods, produce a more complex interpretation of SGBV – shifting the discourse away from rape being a weapon of war. Eriksson Baaz and Stern (2010) conducted interviews with Congolese soldiers and officers and were able to produce a more nuanced understanding of the causes and consequences of SGBV, contextualizing it within the wider framework of militarism and poverty. Dolan (2010) also preferred the use of key informants' interviews and focus groups in Congolese communities to explore factors that contribute to rape in wartime and the ongoing postwar sexual violence.

Ethical considerations

With respect to ethics, scholarly research points to the need to exercise extreme care and sensitivity in the conduct of research among victims who may be severely traumatized. With no national guidelines on interviewing the vulnerable, researchers in the DRC tend to use international guidelines; the most prominent being the World Health Organization ethical guidelines for the conduct of research on sexual violence (WHO 2007). Some of its key recommendations include the need to ensure that 'information gathering and documentation . . . be done in a manner that presents the least risk to respondents' and the necessity to ensure that there is care and support for survivors. From the published reports, it is unclear the extent to which these recommendations are followed in the DRC's context. Johnston et al. (2010: 555) claim 'referrals to appropriate local psychological services were given if requested by the participant or if the participant was perceived as a threat to himself or herself assessed by answers of yes to any red flag questions in the survey'. However, the probability of appropriate referral services being available to interviewees is extremely low, especially in remote areas, away from internationally sponsored treatment centres.

Most research projects use Congolese interviewers who have been trained specifically. Only limited information is provided on their selection and training, and there is no discussion of sexual violence advocates in Africa and their vulnerability to rape myths. Research by Maier (2012) on sexual violence advocates in the USA found that victim-questioning attitudes were prevalent in a significant proportion of those sampled (24 per cent). Maier (2012: 1428) concludes that 'systematic and societal questioning of rape victims is so pervasive that it even invades the thoughts and attitudes of some individuals dedicated to assisting and advocating for rape victims. Most who exhibited victim questioning attitudes during interviews were completely unaware that they were doing so.' Hypothetically, the stigma of rape in Africa and the prevalence of rape myths suggest that local interviewers may not be immune.

Research into sexual violence can be implicated in increasing dominant ideologies about ideal femininity. In SGBV research more generally, it is argued that non-hierarchical relationships between participants and researchers is most valid in the interview process. Hierarchical relationships persist in some places in rural Africa, where race, gender, and class affect women's experience of treatment centres and of bureaucracy – even when the researcher appears with a clipboard and a questionnaire. Of the Congolese fieldworkers, in some cases only their gender and areas of origin are mentioned. Furthermore, virtually all the research teams are headed by white women and men – whose sympathy is not in doubt. Yet it is Congolese research assistants, whose lives are presumed dispensable, who enter the danger zone to conduct fieldwork, leaving the analysis and interpretation to researchers protected by white privilege and whose sojourn is brief. This may explain the prevalence of caricatures and stereotypes about how Congolese people survive in the warscape.

Informed consent and confidentiality

The WHO guidelines recommend that confidentiality and informed consent is protected. As in the extract from Johnston et al. (2010), where interviews have been used, it is doubtful if more than cursory attention has been paid to the willingness of household members to participate and to confidentiality. In the DRC, formal permission tends to be obtained from a hierarchy of bodies: ministries of health, provincial authorities, and international humanitarian agencies, before the affected population. This can be an arduous process and is done to give recognition to a bureaucracy that is remote from the affected population. The survivors' permission may be considered the least problematic, being at the bottom of the power hierarchy.

More importantly, the context in which survivors' permission is obtained can be problematic, especially in a hospital or after tossing a coin outside a door as part of a random sampling exercise. They may reveal information in hospital but, especially since local authority structures and social hierarchies make it difficult for them to openly refuse participation, they may see themselves as not having much choice as to whom they talk to. And, while interview material may have been shredded, reports that make victims visible through photographs and film immediately compromise the notion of confidentiality.

Lawry drew on prevailing donor narratives to deny written consent to women. She claimed:

> We use verbal consent, the reason being that any time you ask for somebody's signature,
> a thumbprint or anything, there is automatic assumption that you'll get something
> for that, no matter what you say, because that's how aid is distributed.
>
> *(Lawry 2011)*

It is ironic that financial incentives were ruled out for desperately poor Congolese women, when the researcher had the financial resources to carry out such an extensive population-based survey. Giving poor women a small sum of money would not have unbalanced the research or worsened corruption, especially since, in some research in the USA, small financial compensations are made to women. This is often an incentive to victims, as the researchers recognize that the majority of victims tend to be from poor backgrounds (Campbell and Adams 2008). In addition, researchers of rape in Rwanda were able to obtain written consent and did not see this as problematic in the context of humanitarian assistance (Mukamana and Brysiewicz 2008).

It is the case that some survivors of sexual violence often want to speak of their experiences in order to help others. In the DRC, reports sometimes highlight rape survivors who come forward to tell their stories. There are multiple reasons why survivors might speak openly. Campbell and Adams (2008) contend that it is psychologically beneficial for victims to share their stories with engaged empathic listeners. Survivors might participate to raise awareness, to help themselves, and to help other survivors. In the DRC specifically, the vast international interest, including the presence of the peacekeeping troops, might give the impression to survivors that external actors are in a position to stop the violence.

Race, gender, and power in knowledge production in warscapes

Humanitarian interventions have spawned a form of scholarship that claims to be scientifically objective, informative, and promotes rights-based outcomes. Such forms of research allow little room for independent thought and tend to reproduce donor ideology. In the world of policy, attention to local differences and sensitivities is often dismissed as time-consuming amid the urgency of dealing with crises. Consequently, humanitarian researchers have been reluctant to think reflexively about the power dynamics in research, especially in a global space of white privilege and power hierarchies. Therefore, issues pertaining to race privilege, gendered hierarchies, and the power dynamics in respect to the consumers of such research are silenced. Lati Mani (1990), in presenting her research on domestic violence in India, found herself challenged by preconceived assumptions from her audience both in India and North America.

She notes: 'there have always been multiple investments and diverse audiences' (1990: 26), and warns of the 'politics of intellectual work in the neo/post-colonial contexts, and the difficulties of achieving an international feminism sensitive to the complex articulations of the local and the global' (1990: 25).

The plethora of international reports on SGBV in the DRC suggests a thriving market for information. Such reports provide a double-edged sword; while they may be beneficial to a few women, they can produce perverse outcomes. Eriksson Baaz and Stern (2011: 42) write, 'one problem of the current reporting of sexual violence in the DRC is that it reproduces and strengthens existing gender power inequalities and stereotype' in the minds of the international community. Emphasis is placed on women as vulnerable, needing protection, and as victims of men's brutality.[3]

Yet, there is no evidence that rape was tolerated in pre-war Congolese society. Eriksson Baaz and Stern (2011: 42) note that 'rape was considered a serious crime before the war and was punishable in different ways, as it was seen as an assault against the individual, woman or girl, the family and the community'. However, they admit that such communal-legal structures may have broken down as a result of the war.

Those presenting African societies as fixed in time ignore transformations in African culture under colonialism, especially the attempts to domesticate women under the guidance of Christian missionaries and the reinforcement that gave to African patriarchal structures. Missionaries were keen to stop women's entrepreneurship – a mainstay in the lives of poor women (Hunt 1990). Not surprisingly Pavlish (2007: 30), in her study of Congolese refugee women in a Rwandan camp, writes that 'women talked proudly of "doing business" in Congo; they bartered, sold and bought food and other items they needed. Apparently, "doing business" was a significant aspect of their good lives in the Congo'. How such essential livelihood activities have been affected by the war is rarely discussed in contemporary SGBV literature, neither are the inevitable changes to gender relations in warscapes.

Congolese women are also presented as an undifferentiated mass. Yet we know class matters. A surgeon providing reconstructive surgery to rape survivors remarked that 'rape is a problem for poor women' (Amnesty International 2004: 28). Congolese observers have drawn attention to the class position and international linkages to corporate capital of one of eastern DRC's female advocates (Snow 2007). Before the proliferation of SGBV literature on the DRC, Puechguirbal (2003: 1271), reviewing the situation of women, writes that 'the war provided Congolese women with opportunities as well as burdens. They took over leadership positions and revived local networks. They were not mere victims as they fought for their survival.' She continued: 'while we can get a lot of documentation about women as victims, there is little documentation of what many observers believe was a great increase in women's independence and self-confidence'.

Historically, Western research on the sexed body in Africa was full of caricatures and imbued with racism. Sylvia Tamale (2011: 15), writing about researching sexuality in Africa, notes that 'not only was African sexuality depicted as primitive, exotic and bordering on nymphomania, but also it was perceived as immoral, bestial and lascivious'. For colonials, controlling African sexuality was a battle between primitivity and modernity. In more recent times, Tamale notes how HIV/AIDs research led to 'a resurgence of the colonial mode of study of sexuality in Africa . . . racist, moralistic and paternalistic' (2011: 21).

What then of sexual violence research? Here, I would argue that one needs to consider the recommendations and the silences in the discourse as possible indicators of the underlying epistemologies guiding SGBV research. Interventions to stop rape are threefold: first, reforming legislation and upholding the law through prosecutions and convictions for rape; second, improving the status of women in Congolese society, such as giving them inheritance rights, and third, educating soldiers.

The reporting back of findings reproduces particular tropes of the DRC. Describing the sexual violence in the DRC to members of the American military (AFRICOM), Lawry (2011) states: 'there is rape. There is gang rape . . . And rape, I mean any orifice, and it can be either instrumentation or penetration.' She continues: 'what are we talking about? Women raping men? They're using instrumentation, cutting off genitals, requiring them to have bush wives – I mean bush husbands – as combatant rebel groups. So it does exist.' Even if these dehumanized and violent representations have some factual basis, they are atheoretical and are not contextualized as the outcome of people ravaged by warfare and forced to coexist in a particular warscape.

Conclusion

This chapter seeks to investigate the origins and nature of the research on SGBV that has been conducted in the warscape of eastern DRC. It questions the methods, techniques, and approaches that have been deployed, especially the scientific objectivism of SGBV research in a humanitarian warscape. It seeks to reaffirm that the production of humanitarian knowledge is social, ideological, and also of economic value.

It reminds us that there are different modes of knowing that are influenced by our imaginative geographies of place, as certain spaces are deemed to be exceptional, and require less rigour and sensitivity to local subjectivities. Researchers may speak to a particular audience – at some distance from the locus of the field. Much of the SGBV research has been done for an international audience, who, through the media, have preconceived assumptions about the practice and causes of sexual violence in the DRC. SGBV reports are repetitive and draw on an ontological foundation that reproduces a particular geographical imagination of Africa that finds easy acceptance in the West. Stereotypes abound – the people of the DRC are represented as being trapped in traditionalism and barbarism, and resistant to the liberal values and norms of Western modernity.

Finally, the issue of positionality is rarely engaged with in SGBV research in the DRC. How far is this because the researchers embody race privilege, status, wealth, and power? For some influential women from the Global North, sisterhood and solidarity become almost as commodities to be marketed on the global stage. Solidarity, when there is a consciousness of differential and hierarchical power relations, can be empowering to both groups.

Notes

1   UNSC Resolution 1325 (2000) was adopted by the United Nations Security Council at its 4213th meeting on 31 October 2000 and addresses the impact of war and women and the role of women in peace-building. Four other related resolutions followed: 1820 (2008), 1888 (2009), 1889 (2009), and 1860 (2010), specifically addressing sexual violence in wartime. Regional conventions and protocols have been enacted and states are encouraged and funded to replicate these legislations in domestic law to protect women.
2   Some useful methodological literature on researching in sexual violence include: Hlavka et al. (2007), Campbell and Adams (2008), Campbell et al. (2009), Thoresen and Overlien (2009).
3   For a discussion of SGBV amongst other narratives on the DRC see Autesserre (2012).

References

Amnesty International (2004) *Democratic Republic of Congo: Mass Rape – Time for Remedies*, www.amnesty.org/en/library/asset/AFR62/018/2004/en/618e1ff2-d57f-11dd-bb24-1fb85fe8fa05/afr620182004en.pdf.

Autesserre, S. (2012) 'Dangerous tales: dominant narratives on the Congo and their dangerous consequences', *African Affairs* 111(443): 201–222.

Baines, E. (2003) 'Body politics and the Rwandan crisis', *Third World Quarterly* 24(3): 479–493.

Bartels, S., Scott, J., Mukwege, D., Lipton, R., Van Rooyen, M.J., and Leaning, J. (2010) 'Patterns of sexual violence in eastern Democratic Republic of Congo: reports from survivors presenting to Panzi Hospital in 2006', *Conflict and Health* 4(9): 1–10.

BBC (2011) 'DR Congo: 48 rapes every hour, US study finds', BBC News, 12 May, www.bbc.co.uk/news/world-africa-13367277.

Boonazier, F.A. and Samantha van Schalkwyk (2011) 'Narrative possibilities: poor women of color and the complexities of intimate partner violence', *Violence against Women* 17(2): 267–286.

Campbell, R. and Adams, A.E. (2008) 'Why do rape survivors volunteer for face-to face interviews? A meta-study of victim's reasons for and concerns about research participation', *Journal of Interpersonal Violence* 24(3): 395–405.

Campbell, R., Adams, A.E., Wasco, S.M., Aherns, C.E., and Sefl, T. (2009) 'Training interviewers for research on sexual violence: a qualitative study of rape survivors' recommendations for interview practice', *Violence Against Women* 15(5): 595–617.

Dolan, C. (2010) *War is Not Yet Over: Community Perceptions of Sexual Violence and its Underpinnings in Eastern DRC*, London: International Alert.

Eriksson Baaz, M. and Stern, M. (2011) *The Complexity of Violence: A Critical Analysis of Sexual Violence in the Democratic Republic of Congo*, Uppsala, Sweden: Nordiska Afrikainstituet/SIDA.

Fujii, L.A. (2010) 'Shades of truth and lies: interpreting testimonies of war and violence', *Journal of Peace Research* 47(2): 231–241.

Harrington, C. (2006) 'Governing peace-keeping: the role of authority and expertise in the case of sexual violence', *Economy and Society* 35(3): 346–380.

Henry, M. (2013) 'Ten reasons not to write your Master's dissertation on sexual violence in war', https://the disorder of things.com/2013/06/04/ten-reasons-not-to-write-your-masters-dissertation-on-sexual-violence-in-war.

Hlavka, H.R., Kruttschnitt, C., and Carbone-Lopez, K.C. (2007) 'Revictimizing the victims? Interviewing women about interpersonal violence', *Journal of Interpersonal Violence* 22(7): 894–920.

Hoffman, D. (2003) 'Frontline anthropology: research in a time of war', *Anthropology Today* 19(3): 9–12.

Hoffman, D. and Lubkemann, S.C. (2005) 'Warscape ethnography in West Africa and the anthropology of "events"', *Anthropological Quarterly* 78(2): 315–327.

Human Rights Watch (2002) *The War Within the War: Sexual Violence Against Women and Girls in Eastern Congo*, www.hrw.org/reports/2002/drc/Congo0602.pdf.

—— (2005) *Seeking Justice: The Prosecution of Sexual Violence in the Congo War*, www.hrw.org/sites/default/files/reports/drc0305.pdf.

Hunt, N. (1990) 'Domesticity and colonialism in Belgian Africa: Usumbura's *Foyer Social*, 1946–60', *Signs: Journal of Women in Culture and Society* 15(31): 447–474.

Jewkes, R., Watts, C., Abrahams, N., Penn-Kekana, L., and Garcia-Morena, C. (2000) 'Ethical and methodological issues in conducting research on gender-based violence in Southern Africa', *Reproductive Health Matters* 8(16): 55–65.

Johnston, K., Scott, J., Rughita, B., Kisielewski, M., Asher, J., Ong, R., and Lawry, L. (2010) 'Association of sexual violence and human rights violations with physical and mental health in territories of the eastern Democratic Republic of Congo', *Journal of the American Medical Association* 304(5): 553–562.

Kalisya, L.M., Justin, P.L., Kimona, C., Nyavandu, K., Eugenie, K.M., Jonathan, K.M.L., Claude, K.M., and Hawkes, M. (2011) 'Sexual violence toward children and youth in war-torn eastern Democratic Republic of Congo', *Science* 6(1): 1–5.

Kombo, E.M. (2009) 'Their words, and meaning: a researcher's reflection on Rwandan women's experience of genocide', *Qualitative Inquiry* 15(2): 308–323.

Korf, B., Engeler, M., and Hagmann, T. (2010) 'The geography of warscape', *Third World Quarterly* 31(3): 385–399.

Lawry, L. (2011) *Research Study on Sex and Gender Based Violence*, www.africom.mil/Newsroom/Transcript/7957/transcript-research.

Maier, S.L. (2012) 'The complexity of victim-questioning attitudes by rape victim advocates: exploring some gray areas', *Violence Against Women* 18(2): 1413–1434.

Mani, L. (1990) 'Multiple mediations: feminist scholarship in the age of multinational reception', *Feminist Review* 35: 24–41.

Mukamana, D. and Brysiewicz, P. (2008) 'The lived experience of genocide rape survivors in Rwanda', *Journal of Nursing Scholarship* 40(4): 379–384.

MUNUSCO (2009) *Comprehensive Strategy for Combatting Sexual Violence in DRC, Executive Summary*, http://monusco.unmissions.org/Portals/MONUC/ACTIVITIES/Sexual%20Violence/KeyDocuments /Comprehensive%20Strategy%20Executive%20Summary.pdf.

Nordstrom, C. and Robben, A.G.C.M. (1995) *Fieldwork Under Fire: Contemporary Studies of Violence and Survival*, Berkeley: University of California Press.

Ohambe, M.C.O, Muhigwa, J.B.B., and Wa Mamba, B.M. (2005) *Women's Bodies as a Battleground: Sexual Violence Against Women and Girls During the War in the Democratic Republic of Congo, South Kivu (1996–2003)*, Réseau des Femmes pour un Développement Associatif (RFDA), Réseau des Femmes pour la Défense des Droits et la Paix and International Alert.

Pavlish, C. (2007) 'Narrative inquiry into life experiences of refugee women and men', *International Nursing Review* 54: 28–34.

Peterman, A., Palermo, T., and Bredenkamp, C. (2011) 'Estimates and determinants of sexual violence against women in the Democratic Republic of Congo', *American Journal of Public Health* 101(6): 1060–1067.

Pratt, M., Werchick, L., Bewa, A., Eagleton, M.-L., Lumumba, C., Nichols, K., and Piripiri, L. (2004) *Sexual Terrorism: Rape as a Weapon of War in Eastern Democratic Republic of Congo: An Assessment of Programmatic Responses to Sexual Violence in North Kivu, South Kivu, Maniema, and Orientale Provinces*, USAID/DCHA, http://pdf.usaid.gov/pdf_docs/PNADK346.pdf.

Puechguirbal, N. (2003) 'Women and war in the Democratic Republic of Congo', *Signs: Journal of Women in Culture and Society* 28(4): 1271–1281.

Sjoberg, L. and Peet, J. (2011) 'A(nother) dark side of the protection racket', *International Feminist Journal of Politics* 13(2): 163–182.

Snow, K.H. (2007) 'Three cheers for Eve Ensler? Propaganda, white collar crime and sexual atrocities in Eastern Congo', http://allthingspass.com/uploads/html-230THREE%20CHEERS%20for%20Eve %20ENSLER[8].htm.

Swedish Foundation for Human Rights (2008) *Justice, Impunity, and Sexual Violence in Eastern Democratic Republic of Congo*, www.humanrights.se/upload/files/2/Rapporter%20och%20seminariedok/DRC%20 SGBV%20Mission%20Report%20FINAL.pdf.

Tamale, S. (2011) 'Researching and theorising sexualities in Africa', in S. Tamale (ed.), *African Sexualities*, Oxford: Pambazuka Press, pp.11–36.

Thoresen, S. and Overlien, C. (2009) 'Trauma victim: yes or no? why it may be difficult to answer questions regarding violence, sexual abuse and other traumatic events', *Violence Against Women* 15(6): 699–719.

Tolia-Kelly, D.P. (2010) 'The geographies of cultural geography 1: identities, bodies and race', *Progress in Human Geography* 24(3): 358–367.

WHO (2007) *Ethical and Safety Recommendations for Researching, Documenting, and Monitoring Sexual Violence in Emergencies*, www.who.int/gender/documents/OMS_Ethics&Safety10Aug07.pdf.

PART VII

Economies: empowerment and enrichment

45

INTRODUCTION
TO PART VII

Economic restructuring as a result of globalization tends to reinforce and exacerbate existing gender inequalities (Marchand and Runyan 2000). The impacts of globalization for women and men have been multiple and contradictory, and there have been conflicting interactions between local and global economies, cultures, and faiths (Afshar and Barrientos 1999). It has been argued that globalization is gendered into two different worlds: one is a structurally integrated world of international finance and postmodern individuality associated with Western capitalist masculinity and high salaries; the other is explicitly sexualized and racialized and based on low-waged, low-skilled jobs often done by female migrants for the rich cosmopolitans of the first globalized world (Momsen 2010).

The process of globalization involves a search for cheap labor and is reinforced by national and international trade agreements and multinational corporations. Such changes undermine the patriarchal gender contract under which families are supported by a male breadwinner as financial stress forces more women into the labor market in the face of increasing poverty. Transnational manufacturing companies have created a new market for female labor. Industrialization has become as much female-led as export-led. Miraftab points out that the footloose industries that relocated to the Global South in search of the cheapest production costs were attracted by the supposed "nimble fingers and docile attitudes" of women. In many cases, in recent years, education has enabled daughters to move out of such jobs to "pink collar" positions as secretaries and in call centers, while mothers remain in factories. Miraftab links this global restructuring to the international care chains provided by women and discussed in Part IV and so this chapter makes a good transition between mobilities and economies. She further stresses the multi-directionality of female resource flows as outlined also by Torres in Part V and the emotional importance of "home" to which one can return that makes working for low wages overseas bearable.

On the whole, Miraftab's chapter takes a negative view of women's economic activity seeing it as externally imposed by global processes. The main theme of the other papers in this section is that of economic empowerment for women through entrepreneurship. As Cornwall and Edwards (2014: 1) point out: "From its origins as a feminist strategy for social transformation, women's empowerment has come to be championed by corporate CEOs, international NGOs, powerful Western governments and the financial institutions they preside over – and, or so it would seem, the entire global development apparatus." Corporate social responsibility such as

Nike's girl power project and Primark's financial assistance to the female victims of the collapse of the clothing factory in Dhaka, provide examples of such activities. Power is multidimensional and influenced by context and environment. The process of providing assets and resources to individuals does not in itself empower women. Empowerment lies in the way these assets and institutions come together and enable individuals to have the capacity to change their social relations and to see new possibilities in their lives. Aladuwaka feels that microcredit programs for women can meet both their practical interests and help them achieve their strategic interests so empowering them.

A second underlying theme is that of the financial resources provided to enable entrepreneurial activities. The reliability of these resources, the ability to repay loans, and the influence of the source of the finance on the entrepreneur all influence the success and empowerment brought about by entrepreneurship. In Romania the source of finance is usually family or friends, or money earned during a period of overseas migration or a secondary occupation. As Lelea shows, this makes such small businesses very unstable and so they are usually only undertaken when paid employment cannot be found because of living in an isolated area or feeling too old to find other jobs. In Africa, as Spring describes, financing for women's businesses usually comes from family, personal savings, or rotating savings and credit associations (ROSCAs), utilized as a way of accumulating savings over a longer period. Donors also sometimes provide loans and in a few cases financing is obtained from banks. Smith et al. point out that women working in Fair Trade groups benefit from advice and training and low cost loans from the producer organizations and the establishment of ROSCAs. For women using microcredit to start an income-earning activity repayment can be a problem as explained by Aladuwaka although the Samurdhi Bank in Sri Lanka does insist that women taking out microcredit loans set up savings accounts into which they deposit money on a regular basis. Huq-Hussain indicates that the Grameen Bank has brought great benefits to poor women in Bangladesh but interest on some loans is still quite high and basically the bank is transferring funds from the rural poor to the wealthy in Dhaka. Recently interest rates for all microfinance institutions (MFIs) have gone up from 30 percent in 2004 to 35 percent in 2011 globally but this is still much lower than local moneylenders ask. Even as small loans have become more expensive they have also become more common worldwide growing by 30 percent per annum from 2004 to 2011 while the average loan size has fallen, suggesting that MFIs are lending to a wider range of clients including the very poor (*Economist* 2014a).The recent transfer of control of the Grameen Bank from Professor Yunus to the government is seen negatively by many poor women clients of the Bank. On April 6, 2014, the Bangladesh government took the power to appoint its own board members away from Grameen and gave it to the central bank (*Economist* 2014b). In Bangladesh there are more than 500 microcredit providers and almost a third of rural households are members of more than one (*Economist* 2014b). A recent study by the World Bank shows that these loans do benefit women more than men, increasing the female labor supply and raising school enrolment by eight percentage points (Khandker and Samad 2014).

A third theme is shortage of time. Many poor women see time as their most precious resource. In broad terms, time poverty can be understood as the burden of competing claims on a person's time that constrains their ability to choose how individual resources are allocated. In most time-use studies, women spend more hours than men working in the household and in economic activities with generally less sleep and leisure than men (Momsen 2010: 161–164). Several of the chapters mention that even if individual women become successful entrepreneurs they rarely get help with household chores from other family members. Lelea, in one of her case studies, was told that the woman found running a shop difficult when her children were small as she had no help with childcare. One of the individuals described by Huq-Hussain in

Bangladesh started a shop on her own when her husband died but when she was successful her sons began to help out so giving her some time off. One of the advantages of running their own businesses, noted by several people interviewed by Aladuwaka, was that they could run a shop or keep a cow close to home so reducing the time needed to travel between home and workplace, and thus giving themselves flexibility in the time demands in their lives.

For those women involved in Fair Trade groups and microcredit groups, working with other women was seen as generally supportive and providing a social network. Husbands allowed women to become more mobile since they had to attend regular meetings of the groups if they were to obtain loans (Aladuwaka) or continue with Fair Trade work and benefit from the Fair Trade Premium (Smith et al.). However, the high repayment levels of these microcredit loans was achieved through social pressure from other members of the group since all had to repay loans before others could get additional loans. Some found this pressure too much to cope with. In addition, the benefits of self-help groups and other participatory development projects can be limited, as the poorest and other marginalized women such as migrants and those with very young children can sometimes be excluded from groups as being thought unable to contribute, as we found in Kerala (see chapter by Kunze and Momsen in Part II).

Success for individual entrepreneurs is varied and tends to be greater among those with the most education. Spring identified a business network of a New Generation of African Entrepreneurs set up between 1993 and 2004. These women were well-educated and often had professional qualifications that they used as the basis for building their companies. Many formed links with companies in other countries for trade. In Romania, Lelea interviewed one woman entrepreneur who was successful because her parents had been leaders in the village under Ceauşescu and so were able to help her financially. She was well-educated but wanted to stay in the village because she felt she owed it to her neighbors to provide a service through her shop. The village was far from the nearest town and most people could not afford to travel there to buy goods so she helped by bringing items to the village. However, the women in Sri Lanka and Bangladesh who set up businesses with microcredit loans rarely made much money but they did gain confidence and respect within the family. They often were able to make decisions for the family and tended to make sure their children were educated and healthy (Huq-Hussain). The women working on Fair Trade projects also appreciated being able to send their children to school, buy furniture for their homes, and replace the thatched roofs of their houses with corrugated iron. They felt their ability to earn had also improved their status with their husbands according to Smith et al.

In general, women have begun to benefit from global restructuring if only through increased confidence, new opportunities for their children, and improved status in the family.

References

Afshar, H. and Barrientos, S. (eds.) (1999) *Women, Globalization and Fragmentation in the Developing World*, London and Basingstoke: Macmillan; New York: St Martin's Press.

Cornwall, A. and Edwards, J. (eds.) (2014) *Feminisms, Empowerment and Development*, London: Zed Books.

Economist (2014a) "Poor service," *The Economist*, February 1, p. 67.

—— (2014b) "Rehabilitation and attack," *The Economist*, April 19, p. 72.

Khandker, S. and Samad, H. (2014) *Dynamic Effects of Microcredit in Bangladesh*, Policy Research Working Paper 6821, Washington, DC: World Bank.

Marchand, M.H. and Runyan, A.S. (eds.) (2000) *Gender and Global Restructuring: Sightings, Sites and Resistances*, London and New York: Routledge.

Momsen, J.H. (2010) *Women and Development*, London and New York: Routledge.

46

CRISIS OF CAPITAL ACCUMULATION AND GLOBAL RESTRUCTURING OF SOCIAL REPRODUCTION

A conceptual note

Faranak Miraftab

Feminist scholarship has helped us understand the intricacies and intimately interconnected nature of production and social reproduction for the accumulation of capital. With the crisis of capitalism, production and social reproduction processes are restructured to facilitate accumulation. In this chapter I outline the processes I call the global restructuring of social reproduction. Global restructuring of social reproduction refers to processes that socially, temporally and spatially reorganize workers' biophysical, social and cultural reproduction responsibilities, and create new sources of expectation and obligation for the provision of collective social reproduction. I articulate how, in this current crisis of capitalism, social reproduction is yet again restructured, now in its global scope. Parts of the social reproduction activities are outsourced to women, families and communities across the world to be performed at low social, economic and political cost to employers. In these processes of restructuring, not only production but also aspects of social reproduction are fragmented and outsourced. Women are at the center of this transnationally performed social reproduction work. This new articulation of their gendered roles within the capitalist accumulation processes requires closer examination.

The literature on gender and development in relation to processes of globalization has chiefly focused on the restructuring of production processes and how this taps into women's cheap labor to advance the accumulation of capital—a process of exploitation concealed by terms such as 'economic development.' This literature brought to light how the global restructuring of production, which fragments the work to be performed in different parts of the world, has brought women in large numbers to the industrial labor force—a process labeled as feminization of labor. In the 1970s and 1980s women, many of whom had never participated in the labor market, became targeted for manufacturing jobs that were now performed in a global assembly line. Feminization of the manufacturing labor force in particular presented an intense trend in the Global South. The footloose industries that had relocated to the Global South in search of the cheapest production costs were in particular attracted to the recruitment of women for they were believed to have nimble fingers and docile attitudes. Female workers were more likely to accept lower wages than their male counterparts, based on the ideological conviction within patriarchal societies that women's labor is worth less, and they were less likely than male workers to organize or protest their working conditions. Feminist political economic analysis of globalization highlighted the massive and hierarchically positioned integration of women into the global labor market, a process that was lubricating capital's abilities for accumulation and lucrative surplus creation. These are processes that Mies and colleagues eloquently saw as global capital exploring and exploiting its last colony: the female labor force (see Mies et al. 1991).

In the context of globally restructured production processes, there is another aspect to this 'last colony' that needs further attention. That is the realm of social reproduction. Marxist critics have long analyzed and discussed the important role the capitalist state and patriarchal family play in the social reproduction of the labor force and in sustaining capitalism and its ability to accumulate. For capitalism to sustain processes of accumulation, the laborers' class needs to be biophysically and ideologically reproduced. The former concerns laborers and their families' cost of living, housing, food, shelter—the resources and processes needed to biophysically regenerate the labor force (Engels 1972 [1884]). The latter concerns the role played by the education system through schools and curriculum to ideologically socialize laborers to social relations that sustain or perpetuate capitalist production (Bowles and Gintis 1977; Willis 1982). Beyond items provisioned within the family, there are also items key to the social reproduction of the working class that need to be provisioned for collective use by the state—Castells (1983) calls these collective consumption items. These are basic services and resources such as roads, water and sewage that are consumed collectively by the working class in the city. The failure of the state to provide these items intensifies the class struggle in the city and catalyzes the grassroots movements around access to neighborhood and urban services (Castells 1983). Feminist scholarship contributes to this debate by articulating how specifically the patriarchal gender ideologies facilitate the work of social reproduction within the family and at large in the city and neighborhoods.

Since the 1980s, this order of relationships between the state, capital and social reproduction has undergone significant stress and reconfiguration, what feminist scholars recognize as a crisis of social reproduction most heavily weighing on women. They credit the crisis to two related processes: structural adjustment policies and neo-liberal reforms (Lawson and Klak 1990; Smith 2002; Katz 2001). They argue that the state withdrawal or redefinition of its role in the provisioning of social care, and city and state support for social reproduction, has diminished and precipitated a crisis of social reproduction (see contributions in Benería and Feldman 1992; Miraftab 2010; Chant 2010). Capitalism, feminists argue, seeks to resolve the crisis by reprivatizing social reproduction into the domestic realm of unpaid women's activities (Bakker and Gill 2003; Katz 2001; Kunz 2010). It is the free labor of care women provide not only to their families in the domestic realm but also to their un-serviced neighborhoods and towns in the public realm (referred to as municipal housekeeping) that makes social reproduction of low-income populations possible (Miraftab 2004; Mitchell et al. 2004). For a working class that is healthy and able to return to work each day, an army of women invest their free labor not only in domestic chores to care for their family but also in collective chores for the sake of the municipality and to care for their neighborhoods.

Feminist sociologists further articulate the transnational dimensions of such reorganization of social reproduction. They highlight the contemporary version of an old and dirty system of care that was performed by enslaved and domesticized women and wet nurses who, deprived of their own offspring and families, cared for and raised the children of colonizers and slave masters (Hontagneu-Sotelo and Avila 1997; Hontagneu-Sotelo 2001; Arat-Koc 2006). In its contemporary provisioning, care is structured hierarchically and displaced along a global chain, from service provided to less affluent families to those who work for the more wealthy (Ehrenreich and Hochschild 2003; Parreñas 2001; Benería 2008). This chain of care could take place within national boundaries—for example, rural–urban trans-local care— or across national boundaries through transnational care.

The literature that explores the transnational dimensions of social reproduction in capitalism, however, has paid less attention to how the global provisioning of care integrates with the global restructuring of production. I argue that the restructuring of production processes has involved certain restructuring of social reproduction that is precisely connected to the ability of global capital to deal with its crisis of accumulation. To accumulate, capitalism has had to rely on restructuring social reproduction processes and practices of households, some within and others across national borders. This process, I argue, involves fragmenting the lifecycle and outsourcing segments of social reproduction work to the communities of immigrants' origin— to communities elsewhere, where the cost of social reproduction is cheaper and can be performed at a lower cost or for free by families and women. I conceptualize this as a closely interlinked global restructuring of production and social reproduction. In this analytic framework, we notice that not only production but also aspects of social reproduction are fragmented and outsourced. Let me explain.

The globalization literature has comprehensively documented instances where firms outsource part or all of their production offshore in order to reduce costs. For some industries such as food and agriculture, or jobs like commercial or domestic services and care work, outsourcing or relocation offshore is not a viable option. For these place-based jobs, it is the laborers who relocate and temporally and spatially reorganize their biophysical, social and cultural reproduction. I argue that industries and services that did not cross the border to tap into the cheap labor of female and male workers and their families in the Global South, by hiring migrant laborers they are still tapping into the unpaid or underpaid labor of women and families back in migrant workers' communities of origin. Migrant workers' families, with women at their center, contribute to this accumulation process through strategies that fragment the worker's lifecycle and allow certain segments of their social reproduction to be performed in distinct geographic locations. For that, they temporally and spatially reorganize social reproduction work.

For instance, let's take meat processing, an industry known for its reliance on a migrant and minority labor force. These jobs are at the same time more hazardous and lower paid than other manufacturing jobs in the US—conditions that explain the industry's high labor turnover rates and high incentive to recruit among immigrants. My ethnographic study of a meat packing town in Illinois, for example, reveals that, while production is performed in the meat processing plant in the Midwest of the US, the social reproduction for segments of the migrant laborers' life cycle is performed in remote communities by their trans-local or transnational families. Namely for many of these workers, the care at the beginning and the end of their lifecycle takes place in their communities of origin. For many of the Mexican migrant workers, for example, it is the Mexican state, as dysfunctional, corrupt and autocratic as it might be, that takes care of childbirth, plus the limited health care or education before the Mexican children reach an age to journey away to sell their labor to US employers at the most productive moments of their lifecycle. Since many of these workers have no access to social security or health care in the US, they return to Mexico when no longer able to work. In other words, the beginning and the end of the lifecycle rely on practices and processes that take place abroad.

An army of people, with women at their center, are involved in and contribute to the trans-local social reproduction processes of the immigrant worker (Parreñas 2001; Hochschild

2000). These range from family members nursing children and caring for family and elderly immigrants, to neighbors caring for the property immigrants left behind. Women involved in these processes include not only female spouses of male immigrants but aunts, grandmothers, sisters and daughters of female and male immigrants that act as the protagonists of immigrants' transnational and trans-local families. For migrant workers to remain in these high-risk, low-paid jobs, parts of processes we know as social reproduction are outsourced to extended families, to malfunctioning or disfunctioning schools, governments, churches, NGOs and a whole industry of so-called 'development programs' back home. These smooth the accumulation of capital and insure the supply of workers 'willing' and able to sell their labor power to do hazardous work at low cost.

Biophysical reproduction of immigrant workers and the free work that their transnational families invest in the care of their children or their injured, old or tired bodies is only one part of the cheapening of the labor force's social reproduction. The promise of a place in their home country to which they will return with their savings and be secure for life is an important force in this story. The imagination of an 'elsewhere' where a person would 'be set for life' has a material power and exchange value that needs to be taken into account. This imagination can make a wage that is unviable for one worker viable for another. Imagination and/or reproduction of an alternative place, a place for retreat or, ultimately, retirement, is an important aspect of this process. 'Home' here as a 'physical and social infrastructure' to go back to hence becomes an important asset for the immigrant worker. Imagined or real, home community as an alternative place that workers create or dream of creating becomes an asset that distinguishes the viability of wages across workers' groups.

In this process, family members who take care of the migrant workers' children or elderly back home need to be recognized as subsidizers of the industry in places of production.[1] The trans-locally and transnationally restructured families of migrant workers engage in complex practices, processes and imaginations that compensate for the low-wage and hazardous work the industry offers. Like outsourcing of production, I argue, social reproduction also is fragmented and outsourced, to be performed by families, neighbors and institutions abroad.

The important point here is to recognize that the global restructuring for social reproduction meshes intimately with the restructuring taking place in the realm of production. In other words, dispossession and displacement are two processes that work together in the new global order of labor. Take the case example above. Policies such as NAFTA and free trade produce a migrant labor force by devastating their prior forms of livelihood. Mexican farmers who no longer could compete in the 'free' market for sale of their products, be it milk, corn or beans, face neoliberal policies of *ejido*[2] privatization that promote the sale of their previously communally-owned land. Through this transaction, the former *ejidatorio* earns a small amount of cash upon sale of the *ejido* share but before long joins the army of surplus labor willing to take jobs anywhere, be it a footloose industry within Mexico or across the border in the meat processing plants of the Midwestern US. The dispossession of this migrant labor force is a precondition of their displacement as is the forming of transnational or trans-local families that restructure their social reproduction work. The two processes that restructure production and social reproduction are interconnected. They join forces to address the crisis of accumulation that the meat industry began facing three decades ago.

In closing, I would like to stress the significance of conceptualizing interconnected restructurings of production and reproduction. This is important in many ways for gender and development scholarship. First, it brings to light that as women are increasingly incorporated into the labor force it is still other women within their familial and social network who take over the work of social reproduction for them, whether trans-locally in villages or other towns of origin or transnationally across the border. Second, it brings to light the multidirectional flow of resources between communities of origin and destination as places where the activities associated with social reproduction and production are performed respectively. It allows us to recognize that, unlike the narratives of globalization where workforce migration facilitates unidirectional resource flows from north to south as in remittances, the contributed resources flow from south to north. Existing literature, however, predominantly explores the social reproduction–immigration nexus in terms of the role immigration plays in the development of immigrants' communities of origin, and not the other way around.[3] Gillian Hart's (2006) notion of understanding the world relationally and Harvey's (2005) articulation of accumulation by dispossession are helpful here in understanding the multi-directionality of resource flows— not only remittances immigrants send home but also subsidies their trans-local and transnational families provide to migrant wage earners in the Global North.

As the crisis of capitalism deepens, we can expect the restructuring of social reproduction to become more complex over time. We need a more sophisticated analytic optic to see the connections (in this case between production and social reproduction) and multidirectionalities (in this case in respect to the flow of resources across communities of origin and destination). This chapter is a humble effort towards that goal.

Notes

1   In classic rural urban studies De Janvry (1981) wrote about rural families subsidizing urban workers by visits they made home to their families, and food and care items they benefited from in the rural areas.
2   *Ejido* is an Aztec system of communal landownership reintroduced and institutionalized as a component of the Mexican land reform programs of the revolutionary governments, 1911–1934. *Ejidos* were by and large dismantled by the neoliberal privatization policies of President Salinas in the 1990s that amended Article 27 of the Constitution in ways to allow privatization of communally owned *ejidos*.
3   Exceptions include the work of Klooster (2005) on how Mexican families subsidize the cheap reproduction of laborers in cities and commercial agriculture in both Mexico and the US.

References

Arat-Koc, S. (2006) 'Whose social reproduction? Transnational motherhood and challenges to feminist political economy,' in M. Luxton and K. Bezanson (eds.), *Social Reproduction: Feminist Political Economy Challenges Neo-Liberalism*, McGill: Queens University Press.

Bakker, I. and Gill, S. (2003) 'Global political economy and social reproduction,' in I. Bakker and S. Gill (eds.), *Power, Production and Social Reproduction*, London and New York: Palgrave Macmillan.

Benería, L. (2008) 'The crisis of care, international migration, and public policy,' *Feminist Economics* 14(3): 1–21.

Benería, L. and Feldman, S. (eds.) (1992) *Unequal Burden: Economic Crises, Persistent Poverty and Women's Work*, Boulder, CO: Westview Press.

Bowles, S. and Gintis, H. (1977) *Schooling Capitalist America: Education Reform and the Contradictions of Economic Life*, New York: Basic Books.

Chant, S. (2010) 'Gendered poverty across space and time: introduction and overview,' in S. Chant (ed.), *The International Handbook on Gender and Poverty: Concepts, Research and Policy*, Northampton, MA: Edward Elgar, pp. 1–28.

Castells, M. (1983) *The City and the Grassroots*, Berkeley: University of California Press.

De Janvry, A. (1981) *The Agrarian Question and Reformism in Latin America*, Baltimore: Johns Hopkins University Press.

Ehrenreich, B. and Hochschild, A. (2003) 'Introduction,' in B. Ehrenreich and A. Hochschild (eds.) *Global Woman: Nannies, Maids, and Sex Workers*, New York: Henry Holt and Company.

Engels, F. (1972 [1884]) *The Origin of the Family, Private Property and the State*, New York: Pathfinder Press.

Hart, G. (2006) 'Denaturalizing dispossession: critical ethnography in the age of resurgent imperialism,' *Antipode* 38(5): 977–1004.

Harvey, D. (2005) *New Imperialism*, Oxford: Oxford University Press.

Hochschild, A.R. (2000) 'Global care chains and emotional surplus value,' in W. Hutton and A. Giddens (eds.), *On the Edge: Living with Global Capitalism*, London: Jonathon Cape.

Hontagneu-Sotelo, P. (2001) *Domestica: Immigrant Workers Cleaning and Caring in the Shadow of Affluence*, Berkeley: University of California Press.

Hontagneu-Sotelo, P. and Avila, E. (1997) '"I'm here, but I'm there": the meanings of Latina transnational motherhood,' *Gender and Society* 11: 548–571.

Katz, C. (2001) 'Vagabond capitalism and the necessity of social reproduction,' *Antipode* 33: 709–728.

Klooster, D.J. (2005) 'Producing social nature in the Mexican countryside,' *Cultural Geographies* 12(3): 321–344.

Kunz, R. (2010) 'The crisis of social reproduction in rural Mexico: challenging the re-privatization of social reproduction,' *Review of International Political Economy* 17(5): 913–945.

Lawson, V. and Klak, T. (1990) 'Conceptual linkages in the study of production and reproduction in Latin America,' *Economic Geography* 66: 310–327.

Mies, M., Bennholdt-Thomsen, V. and Von Werlhof Mies, C. (1991) *Women: The Last Colony*, London: Zed Books.

Mitchell, K., Marston, S.A. and Katz, C. (2004) 'Life's work: an introduction, review and critique,' in K. Mitchell, S.A. Marston and C. Katz (eds.), *Life Works: Geographies of Social Reproduction*, Malden, MA: Blackwell, pp. 1–26.

Miraftab, F. (2004) 'Neoliberalism and casualization of public sector services: the case of waste collection services in Cape Town, South Africa,' *International Journal of Urban and Regional Research* 28(4): 874–892.

—— (2010) 'Contradictions in the gender-poverty nexus: reflections on the privatisation of social reproduction and urban informality in South African townships,' in S. Chant (ed.), *The International Handbook on Gender and Poverty: Concepts, Research and Policy*, Northampton, MA: Edward Elgar.

Parreñas, R.S. (2001) *Servants of Globalization: Women, Migration, and Domestic Work*, Palo Alto: Stanford University Press.

Smith, N. (2002) 'New globalism, new urbanism: gentrification as global urban strategy,' *Antipode* 34(3): 427–450.

Willis, P. (1982) *Learning to Labor: How Working Class Kids Get Working Class Jobs*, New York: Columbia University Press.

47

WOMEN PRODUCERS, COLLECTIVE ENTERPRISE AND FAIR TRADE

Sally Smith, Elaine Jones and Carol Wills

This chapter considers the benefits of women's collective enterprise and fair trade as a route to overcoming gender-based constraints to engagement in markets. It gives a brief overview of the context and existing literature, then presents the findings of action research with women producers in Asia, Africa and Latin America.

Introduction

Vast numbers of women in developing countries are involved in small-scale production of agricultural and consumer goods for sale on local, national or international markets. For many this engagement in trade is the source of considerable material and social gain, sustaining households and giving women status within their families and communities. However, all too often the benefits women derive from trade are curtailed by a range of gender-based constraints. Women typically have more limited access than men to productive resources (land, labour, capital) as well as markets. They may be neglected by business development and agricultural extension services, and excluded from producer cooperatives and associations, which have historically orientated their activities with the stereotypical male-headed household (and male producer) in mind. Women's bargaining position with buyers is frequently weakened by a lack of information, knowledge and organization, particularly in the context of informal economies. While globalization and market liberalization have created new opportunities to participate in global production networks, the dynamics of power in these trading relationships often lead to costs and risks being pushed down to producers and women are often found clustered in low-return, high-risk activities (Carr and Chen 2001). Gender norms and conventions can also limit women's freedom or ability to trade beyond their communities, including through effects on the division of reproductive labour within households and resulting time poverty. These and other gender inequalities, which are perpetuated through intra-household gender relations and the institutions of society (education, law, media, etc.), reflect the gendered nature of economies and present significant challenges for women producers (Elson 1999).

Responding to the challenges: collective enterprise and fair trade

Some women producers have responded by organizing into collective enterprises, through which they jointly produce and market their goods, seeking to benefit from economies of scale and to gain access to resources, skills, knowledge and institutional support. These enterprises take various forms, including artisan associations, producer cooperatives, networks of home-based workers and informal community-based groups. In many cases they are focused on selling into local or regional markets, but in others they are linked into broader networks of trade that offer expanded opportunities.

One such network is the fair trade movement, which aims to provide better trading opportunities to disadvantaged producers while also advocating for broader change in global trade policy and practice, in line with the following definition:[1]

> Fair Trade is a trading partnership, based on dialogue, transparency and respect, that seeks greater equity in international trade. It contributes to sustainable development by offering better trading conditions to, and securing the rights of, marginalized producers and workers – especially in the South. Fair Trade Organizations, backed by consumers, are engaged actively in supporting producers, awareness raising and in campaigning for changes in the rules and practices of conventional international trade.

Fair trade has its origins in 'solidarity trading' and 'trade not aid' initiatives of the 1950s and 1960s, with a range of Northern organizations historically involved including social enterprises, non-governmental organizations, religious bodies and charitable foundations. Early examples of fair trade involved close relationships between these organizations in the North and poor and marginalized producer groups in the Global South, with the focus on providing a direct market for their goods as a way to strengthen their livelihoods. Trade was based on relationships of trust and support, rather than commercial imperatives, and most sales took place in non-profit 'World Shops' and charity shops, as well as within local communities and churches.

Since the 1990s there has been a formalization and expansion of fair trade through the establishment of two international bodies – Fairtrade International (FLO) and the World Fair Trade Organization (WFTO) – and the adoption of formal rules and principles for the production and sale of fair trade goods. FLO is the owner of the 'Fairtrade Mark', a product label that signals to consumers that goods have been produced and traded according to an agreed set of standards. The standards for traders include requirements to provide upfront guarantees of purchases to producers and to make advance payments to help cover their costs, and to pay at least the Fairtrade minimum price (calculated to cover the cost of sustainable production) plus an additional premium for collective investment. Producer organizations are also required to meet certain standards such as democratic, transparent, participatory organization, environmentally-friendly production and respect for labour rights. The introduction of the Fairtrade Mark and associated third-party certification of Fairtrade supply chains meant that a broader range of actors could get involved in trading and selling Fairtrade goods, including conventional businesses that include Fairtrade as part of their overall product portfolio. This led to rapid growth in Fairtrade markets, with global sales of Fairtrade-labelled goods reaching €4.9 billion in 2011, and sales in more than 120 countries.[2]

WFTO is a membership-based organization uniting more than 450 fair trade organizations (FTOs) – producers, marketing organizations, wholesalers and retailers – that trade exclusively in fair trade goods and commit to abide by WFTO's ten principles of fair trade.[3] These principles include many of the same concepts as the Fairtrade standards, but go further in terms of the nature of trading relationships expected and the types of support buyers should offer to producers. Members are monitored for compliance with the principles and can use the WFTO logo to signpost their commitments, but, unlike FLO, WFTO does not currently use product labels.[4] Although there is some overlap between Fairtrade-labelled and WFTO member goods, FLO principally deals in agricultural commodities such as coffee, cocoa, sugar and cotton, while many WFTO members are focused on artisan goods. As well as facilitating trade and knowledge sharing between members, WFTO aims to be a global network and advocate for fair trade, providing a voice for small farmers and artisans on the international stage.

Together these two entities involve more than 1,000 producer organizations across Asia, Africa and Latin America.[5] Many of these are collective enterprises involving women, especially within the WFTO network given the high percentages of women in artisan production (with notable exceptions by country). Often women producers are specifically targeted by FTOs as part of their core mission to support the most disadvantaged in global trade. Under the FLO system there is less focus on women, but all certified producers are required to be organized democratically and subscribe to principles of non-discrimination.

Assessing the benefits of collective enterprise and fair trade for women producers

The value of organizing for access to resources, services and markets has long been recognized, with agricultural cooperatives being just one example of an extensive history of collective enterprises in industrialized and developing countries alike. This has been reinforced with the emergence of modern, globalized agrifood systems and their associated needs for highly coordinated supply chains, leading to significant resources being invested in producer organizations by both public and private institutions.[6] However, relatively little is known about the differential benefits for women and men. Studies across various countries have found that women and men participate in different types of groups and networks, with women more likely to be involved in informal groups, such as savings and credit associations, self-help groups and local civic associations, while men have a greater tendency to be part of formal organizations with an economic focus, including collective marketing groups. It has been argued that these differences reflect gender inequalities that determine levels of inclusion and exclusion in networks and groups of different types, with women often excluded from those that bring economic advantage and resorting to (rather than preferring) community-level collective action as a 'coping strategy' to secure basic needs (Molyneux 2002). This is not to deny that women make choices about the types of organization they devote time to, based on perceived and actual costs and benefits.

One of the few examples of in-depth research in this area is an extensive study led by Oxfam International on women's collective action in agricultural markets in Ethiopia, Tanzania and Mali (Baden 2013). This research found significant differences between countries and subsectors[7] in the types of organizations women belong to, the functions they perform and the benefits associated with them. There were also differences depending on a woman's age, marital status and levels of wealth, with older, married, more wealthy women in general more likely to be part of collective enterprises as they have greater access to resources and fewer demands on their time. Common economic benefits include increases in the volume and quality of production and greater access to markets and credit, leading to higher income. However, in most cases, time poverty, limited mobility and social norms remain key barriers to women's engagement in markets. Informal and formal groups are associated with different types of benefits: informal groups help women develop leadership skills and build savings, while formal groups build access to inputs, services and markets. As such there are complementarities between them, with the empowerment impact often greater from the combined effect of membership in different groups. In addition, membership of women-only groups allows women to develop the confidence and skills to participate effectively in mixed-sex groups. At the same time, the research highlighted the importance of inclusive, supportive leadership for positive outcomes, no matter what form an organization takes and whether leaders are women or men.

Turning now to the Fairtrade system and knowledge about impacts for women producers and their collective enterprises, the existing literature suggests mixed effects: in some situations and contexts Fairtrade has supported improvements in women's income, wellbeing and status, strengthening their position within the household and organizations, while in others it has exacerbated pre-existing gender inequalities (Smith 2013). Factors that have been shown to be important determinants of Fairtrade gender impacts include: the local context of gender relations and pre-existing levels and types of inequality; the type of Fairtrade product and production system; the outlook and commitment of certified producer organizations in relation to gender, and number of women members; and the extent and type of support provided by the Fairtrade system and other actors such as trading partners and development agencies. For example, in the West African context, where women cultivate their own fields as well as working on crops controlled by male relatives, Fairtrade cotton and associated awareness-raising by trading partners and NGOs around women's right to be paid directly for their own production have enabled many women to earn, and retain control over, more income from cotton. However, in India, where women typically farm alongside male relatives, these positive effects are not found and instead women are marginalized from any direct benefits (Nelson and Smith 2011). Some studies have found that the high quality required for Fairtrade markets, in addition to restrictions on the use of agrochemicals, can increase women's burden of work as they are often responsible for related tasks such as weeding, harvesting and post-harvest processing. At the same time, premiums earned on Fairtrade sales are frequently invested in processing equipment that reduces their work. For women in male-headed households, the effect depends on the extent to which their contribution to Fairtrade production is recognized and rewarded; again, there are mixed findings on this depending on the context of gender relations (Lyon et al. 2010; Ruben et al. 2008).

Requirements to comply with Fairtrade standards on democratic organization and non-discrimination have been associated with increased numbers of women participating in meetings and trainings and a greater percentage of elected positions being held by women. However, when this is done without genuine commitment to, or understanding of, gender equality, women's role in the organization may remain marginal. For example, they may continue to be excluded from more senior positions or lack the confidence, skills or education to have a real influence over decision-making.[8] The Nicaraguan coffee cooperative SOPPEXCCA demonstrates the difference made when commitment and understanding among the leadership is in place. Its holistic approach includes assisting women to acquire land, sensitizing men and women on issues such as gender-based violence and reproductive rights, and supporting women to take on leadership roles. The role of Fairtrade in this context is to provide the economic foundations for this work (Dilley 2011).

The Women Organizing for Fair Trade project

The remainder of this chapter focuses on women's collective enterprises within WFTO fair trade networks. It draws on the findings of an action research project with women producers in Kenya, Tanzania, Uganda, India, Nepal, Nicaragua and Mexico between 2009 and 2011. The 'Women Organizing for Fair Trade' project was coordinated by Women in Informal Employment: Globalizing and Organizing (WIEGO), a global action-research-policy network that seeks to improve the status of the working poor, especially women, in the informal economy.[9] The project involved documenting the experiences of women producers alongside sharing and learning events at enterprise, national and international levels, with the dual aims of adding to the knowledge base and strengthening the fair trade movement through providing opportunities to reflect and exchange.[10]

Sixteen collective enterprises linked to Fair Trade took part in the research, plus five national Fair Trade and women's networks which provided coordination at the country level (see Table 47.1).[11] All but four of the enterprises are all-women groupings, with a further two involving a majority of women producers. Some are small community-based groups or collectives of small groups which are largely dependent on FTOs for linking to markets, others are themselves FTOs involving self-governing groups of artisans. Two are multi-purpose cooperatives with direct links to Fair Trade markets overseas. Although all are formal organizations in that they are legally registered with the relevant authorities, many involve semi-formal groups at the community level (i.e. they have a structure and regular meetings, but no legal status).

Across all seven countries women reported many positive outcomes from their experiences with collective enterprise, although also numerous challenges. Some benefits are derived from specific activities and services undertaken by the enterprise and partner organizations, while others relate to more general effects of being part of a group. The latter meant that, even in cases where the degree of commercial success was relatively limited, women gained social value from collective action. Below we discuss first the economic benefits and then the social effects, then look at the role of fair trade in bringing about change.

Economic benefits associated with being part of collective enterprises

The main commercial activities and services women access through their collective enterprises are: input supply; product-related services (training and information on production, processing, design and packaging); financial services; marketing of goods; and organizational development. Depending on the size and degree of sophistication of the enterprise, these are more or less extensive and professionalized. In the case of the small community-based groups, financial services may simply be the establishment of rotating savings and credit schemes, product-related services relates to peer-to-peer learning, and marketing may focus on local markets only. However, these groups are also able to access additional services through umbrella organizations (such as secondary level cooperatives) or FTOs, to a greater or lesser extent depending on how well resourced these are. The larger enterprises (mostly FTOs and cooperatives in Asia and Latin America) have evolved into multipurpose organizations linking between 35 and 60 small groups, with a range of professional services and activities to meet members' needs.

Women producers reported that working together to buy raw materials, solve production problems, develop skills and increase volumes has made them more attractive to buyers, while also enhancing their bargaining power. They value the opportunity to share ideas and learn from each other, especially in relation to production skills and techniques. They also have a better understanding of markets, which helps them to make better decisions about who to sell to, at what price and when. For example, the Artisans Association in India has given training on costing and pricing that has enabled its members to set their own prices based on real costs. This and other types of training, including on production and business management, is widely appreciated, as described by Nezuma Simai Juma, member of Tusife Moyo in Tanzania:

Table 47.1 Overview of national networks and collective enterprises that participated in the research

Country	Lead organization	Participating fair trade organizations (FTOs)
Kenya	Kenya Federation of Alternative Trade (KEFAT): network of 90 FTOs	– Baraka Women's Group – 38 members engaged in organic agricultural production and marketing through Undugu Fair Trade. – Turkana Women's Group – 180 women palm leaf basket weavers, marketing through Undugu Fair Trade. – Mathima Women's group – about 50 members weaving sisal kiondo baskets and marketing through the Machakos Cooperative Union, which has 78 primary cooperative societies and more than 60,000 individual members.
Tanzania	The Tanzania Fair Trade Network (TANFAT): reformed in 2010 with four founder FTOs	– Tusife Moyo Women's Cooperative, Kidoti, Zanzibar – 25 active members making soap, marketing through Kwanza Collection. – Wawata Njombe – 300 active members organized in five village-based groups who weave reed baskets, marketing through Kwanza Collection.
Uganda	National Association of Women Organisations of Uganda (NAWOU): network of 70 national and 1,500 community-based women's organizations	– Kazinga Basket Makers – 60 members weaving baskets, marketing through NAWOU. – Ngalo and Kanyanya textile handicraft groups – 58 members, marketing through NAWOU. – Patience Pays Initiative, Kayunga – about 100 members; and Kangulumira fruit dryers – about 30 members; both sell solar-dried fruit to the Fairtrade-certified export company, Fruits of the Nile.
India	Fair Trade Forum – India (FTF-I): network of 80 Fair Trade organizations	– Artisans Association, Kolkata – linking 30 self-organizing rural and urban groups of artisans (mostly women) and women-headed enterprises, marketing through Sasha (study involved six subgroups). – SABALA, Bijapur, Karnataka – 1,000 women in 60 village community groups. – Sadhna, Udaipur, Rajasthan, – 700 women members in 49 self-help groups (study involved subgroups). – All these groups work with textiles.
Nepal	Fair Trade Group Nepal (FTGN): network of 17 Fair Trade organizations	– Association for Craft Producers (ACP), Kathmandu – more than 1,000 women producers (study focused on Kirtipur Weavers Group – approximately 50 members). – Women's Skills Development Organization (WSDO), Pokhara – 400 home-based weavers and hand-stitchers in self-organizing village groups (study involved three subgroups).
Nicaragua	PRODECOOP: cooperative union representing 39 primary cooperatives	– PRODECOOP – a Fairtrade-certified agricultural cooperative union with around 2,400 members (study involved eight primary cooperatives).
Mexico	Ya Munts'i B'ehña: primary cooperative	– Ya Munts'i B'ehña, Valle del Mezquital, Hidalgo – a primary cooperative with about 250 members from five communities, markets body scrubs made from maguey cactus fibre.

> Tusife Moyo has enabled me to network and meet others for learning and exchange of ideas which eventually makes life easier as I understand more and I am not alone. Just being a member I now know the use of a bank where I am saving money for my business and can access loans when necessary. I got a bit of basic business training, which now helps me when doing business as an individual and for the groups . . . Business engagement for a woman frees women from the dominance of men.
>
> *(cited in Jones et al. 2011: 28)*

Some women talked about the saving of money and time by acquiring inputs through their groups, either through collective harvesting and processing of raw materials or through joint (bulk) purchasing from independent suppliers. Access to lower cost loans and participation in savings schemes were other important economic benefits, as women producer's access to financial services is often very limited. For example, WSDO in Nepal allows members to take out loans at nominal interest rates or channel a portion of the payments they receive into a savings account, while PRODECOOP in Nicaragua provides members loans at the beginning of the production season that can be paid back upon harvesting coffee.

Together these services and activities have enabled many women to strengthen their livelihoods, with increased skills and knowledge, greater ability to diversify products and markets, and a stronger position with buyers. The resulting impact on household income has in many cases led to marked improvements in wellbeing and reduced vulnerability to poverty.

> Through the weaving of baskets we have been able to get money and thus managed to send children to school, and we are able to buy furniture. Through the project women have also benefited by changing the houses from grass thatch to corrugated iron sheets. Thus the basket weaving has really become our own source of income. This is supplementing our farming efforts. This activity has also contributed to the change in our husbands' attitude as they see now the benefit. This has made them to be supportive of the weaving activities as women contribute to the wellbeing of the family.
>
> *Yesekina Joseph Mwinami, Wawata Njombe, Tanzania*
> *(cited in Jones et al. 2011: 12)*

Sometimes proceeds from the initial enterprise have been used to invest in other, more profitable income-generating activities, as with the Turkana Women's Group in Kenya, which used money saved from selling baskets to invest in a fishing boat that has now become their main source of income. Traditionally fishing is a man's activity in Turkana society, but the experience of working collectively to produce and market baskets has given women the confidence to challenge this norm. A common theme from the experiences of women producers in all countries is that they have often been enabled through their collective enterprises to earn or retain their own income for the first time. This has been particularly important for unmarried, divorced or widowed women, especially those in South Asia who often lack support from their families and are stigmatized in their societies. Being able to provide for themselves and their children has not only often protected them from extreme poverty, it has improved their confidence and self-esteem.

> Surya is separated from her husband and has been a sole provider for her two children and herself for the last 13 years. Surya has a sense of contentment when she looks back and recalls that working with [the Women's Skills Development Organization]

has enabled her to raise her daughters, both of them now married. She has a savings account in a local bank and is also extremely proud that the gold necklace and earrings she wears are made from her own income.

From the personal testimony of Surya Pandit, Banjhapatan
subgroup, WSDO, Nepal (Jones et al. 2011: 23)

It is important to acknowledge that not all enterprises in the study are providing women with sustainable livelihoods, with many women expressing concerns about irregular sales and increased costs of living and production that are not adequately reflected in the prices they receive for their goods. However, even in these cases women derived social benefits from their engagement in collective enterprises, as discussed below.

Social benefits associated with collective enterprises

Some social benefits are the direct result of non-commercial services provided by their organizations or FTO partners, such as literacy classes, health education, life skills training and community development projects. Others relate to the effects of being part of a group, such as feelings of belonging and solidarity, being able to voice opinions and share concerns, and access to support in times of difficulty.

I was married and divorced once and left with four children. My ex-husband left with two children and I take care of the other two. I face many challenges, among them is being economically independent. I knew being independent requires enduring so much in our culture but . . . Tusife Moyo Kidoti has encouraged and facilitated me to face all . . . My children are living comfortably and I can afford to pay the school fees. There was a time I got ill for a long time but was able to pay my hospital bills through the business though I was not working. Even if I am stuck financially, my group supports me to access loan . . . Tusife Moyo has enabled me to network and meet others for learning and exchange of ideas which eventually makes life easier as I understand more and I am not alone.

Nezuma Simai Juma, Tusife Moyo, Tanzania
(cited in Jones et al. 2011: 30)

Improvements in confidence and self-esteem were also widely reported, derived not only from economic achievements but also from the experience of participating in group meetings and taking on leadership roles, travelling to distant markets and engaging with input suppliers and buyers, and attending workshops and trade shows, sometimes overseas.

Before [us] women didn't have value or power to decide things, because of the fear that one had. Women weren't worth anything, they should just stay at home, cleaning or looking after children. I think that it's something really important that in the group one can learn . . . how to value [ourselves as] women, because before we didn't talk or go out with other women, other artisans.

Josefina Oliva, Ya Munts'i B'ehña, Mexico
(cited in Jones et al. 2011: 30)

Some women told how their contributions to household income and their success in fulfilling new roles has redefined them in the eyes of others, including among their extended families and communities, earning them respect and enhancing their status and influence. In Mexico, for instance, members of the cooperative Ya Munts i B'ehña faced a culture of 'machismo', with men disparaging of women's abilities and women themselves doubting their worth. The success of their cooperative in selling sisal scrubs to an international retailer has changed attitudes towards them, as well as their sense of self-worth.

> Before the men criticized, 'How are women going to do it . . . how can you think they are going to be able to use a computer, how can you think that they will make progress?' But now they see us and they ask us, 'In little time you have learned, how have you done it? Teach me.' And our children also say, 'Mama, teach me or we will teach you so that you learn more.' So we have advanced . . . because we all can, we externalise it and we have the power . . . In meetings we voice our opinions and sometimes we don't shut up!
>
> *Concepción Flores, Ya Munts'i B'ehña, Mexico*
> *(cited in Jones et al. 2011: 27)*

> My work wasn't recognized much because it is always the routine for women to work in the house . . . But now I feel happy because women's work is noticed and it's valued by men. And I'm very happy because before my family didn't receive any income from my work, I worked and worked without salary . . . and now it's noticed that with my work comes income. It's helped a lot with my family because with this money I have studied a bit and others have studied. Without the organization, none of this would have been possible.
>
> *Lucía Acuña, Cooperative Nuevo Amanecer, PRODECOOP,*
> *Nicaragua (cited in Jones et al. 2011: 53)*

Although a change in attitudes was fairly widely reported, the extent of change in gender relations was often quite limited. Most women still bear the brunt of responsibility for household work, in addition to their paid work, although in a few cases (particularly in Nicaragua and Nepal) married women did report that their husbands are now helping more with domestic chores. As a result, many women appreciated being able to earn money close to their homes, allowing them to combine productive and reproductive work more easily, and reducing the need to migrate to find seasonal or casual work, or engage in petty trade. For single mothers this is particularly important.

The role of fair trade in supporting women's collective enterprises

Many of the benefits discussed above are linked to women's participation in fair trade markets and networks. There are four main areas in which linkages can be identified: opportunities for marginalised women; supportive trading relationships and fair pricing; enterprise development; building networks and alliances.

The social mission of fair trade organizations means that they specifically set out to support poor and vulnerable people. This often involves women producers, including those who are marginalized within their societies and within global trade. FTOs support women to organize and develop their skills, providing ongoing capacity-building and market information, and working jointly with them to resolve problems. This sets them apart from most conventional businesses, which seek to keep costs to a minimum and are therefore biased towards producers who already have the necessary skills and capacity to deliver, as well as often working through networks of intermediaries rather than forming direct relationships. Having said that, mainstream companies are increasingly engaging with fair trade as well, as part of their corporate responsibility agenda, and in some cases they target their support towards women. This is the case for Ya Munts'i B'ehña in Mexico, which since its inception has received wide-ranging assistance from The Body Shop International, a UK company now owned by the global cosmetics giant L'Oreal. Importantly, however, this support is aligned with the ten principles of fair trade, including a commitment to direct, long-term trading relationships and adaptation of supply chain practices to meet the needs of producers.

The development of supportive relationships in supply chains and fair pricing are central to fair trade and have a critical impact on the commercial (and organizational) performance of collective enterprises. Fair trade markets recognize the intrinsic costs associated with enabling poor producers to trade, and as such provide higher or more stable prices than are typically available in conventional markets. Under the Fairtrade-labelling system this is formalized through the minimum price guarantee and premium, while under WFTO prices are calculated jointly by producers and buyers in accordance with fair trade principles. Being able to access these 'fair' markets has been important for the enterprises involved in the research, providing the economic foundations for business development as well as for investments in social protection and development. Among the case study enterprises, those that have evolved into solid, multipurpose organizations all have a history of strong linkages with fair trade markets, while many of the smaller, less well-resourced enterprises have fairly weak links in comparison.

However, some producers and FTOs complained that fair trade market prices do not always keep up with changing costs of production and living, suggesting that fair pricing systems are not working as well as intended. There are also risks of over-dependency on fair trade markets, with some enterprises clearly suffering as a result of fluctuating levels of demand as economic conditions and tastes in consumer countries change. The more successful enterprises and partner FTOs have developed diversified marketing strategies, including developing local retail and online sales and/or linkages with alternative high-value markets, to offset this risk. Diversification into other products and activities, such as ecotourism, has also been pursued by some. However, for many enterprises this diversification has come on the back of success in fair trade, in terms of support during the early stages of enterprise development and allowing the accumulation of investment capital. As such, fair trade can play an important role as an 'incubator' for enterprise development, even if it cannot by itself deliver long-term commercial success.

Finally, being part of fair trade means women producers have both commercial and sociopolitical networks that they can tap into. They have opportunities to meet and share experiences, receive advice and support to solve problems, and get information about potential buyers and sources of assistance. Some country-level fair trade networks are active in fundraising and building alliances to provide more intensive capacity-building support to members, as well as providing opportunities to participate in workshops, trade shows and conferences. Fair trade networks also provide a platform for advocacy in favour of small-scale producers, women and trade justice, at both national and international levels. Being part of a wider group, with the sense of belonging and solidarity that this entails, is something that many women producers valued.

> With Fair Trade I have seen that we have advanced a lot, we have achieved a lot of things. We can deliver the work and receive the money, a good price, they don't take from us or steal from us . . . It's also really nice to get to know more organizations, because we were self-absorbed, closed, we didn't open the door and put time in to get to know women from other organizations . . . This way, perhaps we can help

another organization by showing them how to grow, or vice versa. So this is the important thing and the great thing – when we share with other organizations we can grow more.

Luciana Bautista, Ya Munts'i B'ehña, Mexico
(cited in Jones et al. 2011: 34)

Learning what makes women's collective enterprises successful

As part of the action research project, women producers and partner organizations reflected on factors that were important for the success of their enterprises (which were not always achieved). Two categories emerged: factors related to commercial performance and factors related to group functioning and dynamics. The main success factors related to commercial performance were strong and diversified links to markets, adding value to products through processing, packaging or brand development and good quality control. In terms of group functioning and dynamics, the importance of strong leadership was emphasized, with many organizations being led by highly committed and visionary leaders, but also recognition of the challenges of leadership renewal. A clear vision with social as well as economic goals was identified as critical to outcomes, with a strong commitment among the leadership to gender equity particularly important for mixed groups. This means that as well as designing commercial activities and services around women's realities (e.g., time poverty, limited mobility, limited education), non–commercial activities that address broader socioeconomic and political needs are undertaken (e.g., literacy classes, self-esteem workshops, gender sensitization with men). Good governance based on democratic principles was seen as important for building trust and accountability within organizations, with the other side of the coin being commitment and active participation of members. At least a certain degree of formality is desirable, in the sense of holding regular meetings, electing officials and keeping records, with the benefit of more formal registration with relevant authorities being potential access to support and greater recognition from governments.

Most of these success factors are relatively well-known but they do indicate how participation in fair trade markets may make a difference to the success of women's collective enterprises. This especially relates to the focus on social as well as economic objectives as with fair trade, unlike most conventional markets, there is a willingness to contribute to the costs of working with women who require tailored and sustained support to overcome the barriers to their engagement in trade. While this support may also be provided through development agencies, the aim of fair trade is to, as far as possible, internalize the costs within markets as a more sustainable solution. There is also considerable synergy with key success factors and fair trade principles, such as the requirement for buyers to engage in long-term relationships with producers, and for producer organizations to be democratic, transparent and participatory, suggesting that these principles can provide a useful reference framework beyond fair trade.

However, it is also clear that fair trade alone is not the solution to all the challenges women producers face. Furthermore, participants in the research identified multiple challenges for the fair trade movement, including a lack of awareness of the concept and practice of fair trade at the grassroots level, poor marketing and branding by some FTOs leading to weak market growth, and a need to develop awareness and support among consumers and governments in the South. There were also concerns about fair trade pricing systems and that the requirements for participation in Fair Trade (particularly the Fairtrade system, but increasingly also WFTO fair trade) may be too stringent and too costly for many producer organizations.

Conclusions

Being involved in collective enterprises can bring a range of benefits to women in developing countries, although the extent and nature of change varies considerably from one organization, and one context of gender relations, to the next. Economic benefits may include increased access to resources, services, information and markets, improved production and quality, and enhanced bargaining power, which helps women overcome gender-based constraints to effective participation in markets. In addition, there are important social benefits associated with being part of a collective enterprise, such as increased confidence and self-esteem, enhanced status and influence within households and communities, and a sense of solidarity and access to support in times of need. These benefits hold particular importance for single mothers and those with little access to family support, although such women often also face severe challenges in combining productive with reproductive work and participating in group activities. Adapting the services and activities of collective enterprises and their trading partners to the needs and realities of women (in their context) is critical, but this has cost implications which are not readily absorbed in conventional markets. Linking to fair trade markets and networks is one part of the solution, as they place value on social as well as economic outcomes and provide access to sources of financial, technical and organizational support. However, additional strategies will always be needed, especially engagement with governments to create a more conducive policy environment for women producers. This includes ensuring they can access productive resources and be members of formal producer organizations, both in principle and in practice, and targeted policies and resources to support both men and women in undertaking reproductive work.

Notes

1 This definition was agreed in 2001 by a consortium of four major fair trade networks: Fairtrade International (www.fairtrade.net), World Fair Trade Organization (www.wfto.com), Network of European World Shops and European Fair Trade Association (http://www.european-fair-trade-association.org/efta).
2 Figures cited in Fairtrade International Annual Report 2011–2012, *For Producers, With Producers*, www.fairtrade.net/fileadmin/user_upload/content/2009/resources/2011–12_AnnualReport_web_version_small_FairtradeInternational.pdf
3 These principles can be accessed on WFTO's website, www.wfto.com
4 The difficulties of standardizing artisan products have previously prevented the development of product labels within the WFTO system. However, without product labels there is a limit to growth in markets for WFTO member products, as mainstream markets typically want third-party guarantees of product integrity. As such, WFTO is currently developing a third-party certification system for fair trade organizations that will then be able to use the WFTO logo on all the goods they produce.
5 Fairtrade International's Annual Report for 2011–12 (cited above) reports 991 certified producer organizations, while membership numbers stated on the websites of the regional chapters of WFTO in Africa (www.cofta.org), Asia (www.wfto-asia.com) and Latin America (www.wfto-la.org) combine to a total of 215 members, the majority of which are producer organizations (rather than marketing organizations or other types of members). Some producer organizations are associated with both Fairtrade and WFTO, but the precise number is not available.
6 For related discussion, see Berdegué et al. (2008).
7 Subsector is used in relation to different agricultural outputs, such as honey, vegetables and shea butter.
8 Details of studies documenting increased participation of women as well as limits to their participation are given in Smith (2013).
9 For more information on WIEGO and its Global Trade Programme, see www.wiego.org.
10 Full details of the project and its findings were published by WIEGO as *Trading Our Way Up: Women Organizing for Fair Trade* (Jones et al. 2011). The report is available on WIEGO's website,

www.wiego.org. Additional outputs from the project include documentary films that are available online at http://fairtradeforwomenproducers.wordpress.com.

11 In Mexico and Nicaragua this role was played by the collective enterprises themselves.

References

Baden, S. (2013) *Is Women's Collective Action in African Agricultural Markets the Missing Link for Empowerment?* Oxford: Oxfam International.

Berdegué, J.A., Biénabe, E. and Peppelenbos, L. (2008) *Innovative Practice in Connecting Small-Scale Producers with Dynamic Markets*, IIED: London.

Carr, M. and Chen, M. (2001) *Globalization and the Informal Economy: How Global Trade and Investment Impact on the Working Poor*, Cambridge, MA: WIEGO.

Dilley, C. (2011) 'Women and Fair Trade coffee production in Nicaragua', *Malaysia Journal of Society and Space* 7(1): 74–84.

Elson, D. (1999) 'Labor markets as gendered institutions: equality, efficiency and empowerment issues', *World Development* 27(3): 611–627.

Jones, E., Smith, S. and Wills, C. (2011) *Trading Our Way Up: Women Organizing for Fair Trade*, Cambridge, MA: WIEGO.

Lyon, S., Aranda Bezaury, J. and Mutersbaugh, T. (2010) 'Gender equity in fair-trade-organic coffee producer organisations: cases from Mesoamerica', *Geoforum* 41: 93–103.

Molyneux, M. (2002) 'Gender and the silences of social capital: lessons from Latin America', *Development and Change* 33(2): 167–188.

Nelson, V. and Smith, S. (2011) *Fairtrade Cotton: Assessing Impact in Mali, Senegal, Cameroon and India*, Chatham: Natural Resources Institute, University of Greenwich.

Ruben, R., Fort, R. and Zuniga, G. (2008) *Impact Assessment of Fair Trade Programs for Coffee and Bananas in Peru, Costa Rica and Ghana*, Fair Trade Programme Evaluation: Final Report, CIDIN, the Netherlands

Smith, S. (2013) 'Assessing the gender impacts of Fairtrade', *Social Enterprise Journal* 9(1): 102–122.

48

THE ENTREPRENEURIAL LANDSCAPE FOR AFRICAN WOMEN

Sectors and characteristics from microenterprises to large businesses

Anita Spring

Women's entrepreneurial landscape in Africa: informal and formal sectors

This chapter discusses the ability of women's entrepreneurial and business landscape in sub-Saharan Africa (SSA) to create employment, provide services/products, and increase wealth in both the formal and informal sectors. Gender-differential access to education, business skills, and enterprise resources influence the entrepreneurial landscape in terms of business arrangements and financial scale. To analyze women in this landscape, the informal sector (unregistered, unregulated, and untaxed businesses and service enterprises, production activities, and sales) and the formal sector (taxed, registered and regulated businesses, usually recognized by governments and officials) are compared (Spring and McDade 1998; Spring 2009). The range of characteristics of traditional women entrepreneurs and emerging globally oriented business-women is a particular focus, and a series of models and diagrams graphically display the characteristics that limit or enhance the outcomes.

Generally, the literature and conceptual notions of entrepreneurial prowess present a picture of upward mobility ("rags to riches") based on hard work. However, large numbers of micro-entrepreneurs are concentrated at the bottom, and top-of-the-scale medium/large African-owned firms are few. One question is "Why has upward mobility been more difficult in Africa?" The limitations on enough capital and education, and the lack of business networks for product sources and sales, especially for women, are shown here to be factors. A second question is "Where is the missing-middle in Africa?" The data shows that it is not missing, but the sector is smaller, and the playing field is not gender-neutral in Africa.

Women micro- and small-scale entrepreneurs in the informal sector fill the local markets in African countries and have been studied extensively (e.g., Clark 1994; Coquery-Vidrovitch 1997; Horn 1994; House-Midamba and Ekechi 1995; Spring 2000, 2009). Many microfinance programs for women support these types of entrepreneurs. Currently, governments in countries such as Kenya and Ghana have also begun to assist the informal sector by providing "sites and services" to tax and monitor small enterprises. And due to economic downturns from structural adjustment programs in most African countries in the 1980s and 1990s, as well as the effects of the global recession after 2008, many salaried workers have entered the informal sector, increasing competition for the pre-existing enterprises. Another smaller group is that of medium- and large-scale traders (Coquery-Vidrovitch 1997; Spring 2009), whose sales and net worth range from thousands to millions of dollars, but are difficult to locate. They are not listed and are untaxed, but sell agricultural products, textiles, household goods, appliances, and electronics throughout the continent.

In the formal sector, a few women own small- or medium-size businesses. They have been studied even less (e.g., Buskens and Webb 2009; Rutashobya and Olomi 1999; Snyder 2000; Spring 2000, 2009). Little research has also been conducted on women owners of large-scale formal-sector businesses (who own large factories or service companies) that are globally and/or locally based (e.g., Fielden and Davidson 2010; Fick 2006; Hallward-Driemeier 2013; McDade and Spring 2005).

Consequently, there is little clarity about these different types of African women entrepreneurs in terms of their business scale of work, enterprise sector, and financial needs. In the author's research in Uganda, some women owners of formal-sector manufacturing, retail, and professional services companies said that donors and some scholars were confused about levels and sectors and wanted to link them with micro-entrepreneurs in the informal sector. One business owner declined a donor agency's invitation to join a weekly forum with traditional female micro-entrepreneurs. She said the advice and assistance to take her business to the "next level" would not be found in such a group. Other women owners of medium and large companies said they sympathized with micro- and small-scale entrepreneurs in the informal sector, but suggested there were many organizations that focused on them. They noted that donor agencies provided small loans (e.g., US\$50–100) to micro-entrepreneurs and banks loaned large amounts to large corporations, but the medium-sized loans (US\$50,000–1,000,000) needed to expand their businesses were hard to find. Hallward-Driemeier (2013: 129) writes that the study of women entrepreneurs must "look at gender gaps across these three dimensions—size, formality, and industry—and . . . examine why these gender-differentiated patterns of entrepreneurial activity exist. At the same time, different ways of defining and measuring women's ownership of enterprises affect the results and have different policy implications."

What facilitates or constrains upward mobility within and between sectors? The paradigms below heuristically conceptualize the formal-informal contrast and the types of movement within and between categories/sectors. Brief data are given on women micro-entrepreneurs, large-scale traders, retail shops, small industries, medium-sized companies, large companies, and service-sector providers. Formal-sector businesswomen frequently state that they feel obstructed by the "gender divide" that prevails in virtually all African countries, due to "traditional African attitudes" that inhibit women being outside the "old boys" networks and social clubs where businessmen congregate. Fick's (2006) thumbnail cases of well-established and well-known individual African women in a "continent of economic opportunity," and McDade and Spring's (2005) research on women provide brief case studies of women from West Africa (Ghana, Nigeria, and Senegal), East Africa (Ethiopia, Kenya, Tanzania, and Uganda), and Southern Africa (South Africa, Botswana, and Mozambique), although there are examples written from most of the 47 countries in SSA. These entrepreneurs are owners of formal-sector farms and factories for food processing, clothing, textiles, timber, and cement, as well as service-sector businesses in printing, accounting, computer, tourism, transportation, and construction.

There is also an emerging sector of globally oriented entrepreneurial women, termed by themselves and donors, the New Generation of African Entrepreneurs (NGAEs). These younger, formal-sector business globalists of both sexes of which 23 percent were women, and my research sample included 33 percent women, were supported by the World Bank, USAID, and the Club de Sahel in the 1990s and 2000s. Linked into 31 country networks, three regional networks, and a continent-wide governing body (see Table 48.1), they provide an "endpoint" of the formal sector in their scale of operation and revenues. Current networks such as the Africa Businesswomen's Network constituted by Vital Voices Global Partnership, ExxonMobil Foundation, and local African business women's organizations are helping to create the larger Africa Businesswomen's Network to construct local networks, raise profiles of women in business, advocate for policy changes, and expand economic opportunities for women (Hallward-Driemeier 2013).

Tables 48.2–48.7 below show the range of women's enterprises and businesses. The size ranges from micro- to large-scale levels in the informal sector, and from small- to large-scale in the formal sector, with the NGAEs shown as the end point of the latter. A number of variables are analyzed including:

- Demographic status (education, marital status).
- Types of enterprises and firms.
- Sources of start-up capital.
- Product sources and markets.
- Memberships in networks and associations.
- Economic and physical movement within and between the sectors.

In addition, Table 48.8 shows usage of information and communications technologies including cell phones, internet, email, mobile banking, and social media by entrepreneurs at all levels.

The next section briefly describes the characteristics of the informal- and formal-sector entrepreneurial landscape. This is followed by a section that proposes an analytical framework to consider the factors that enhance or restrict women's upward mobility in terms of scale and remuneration.

Brief sector characteristics of women entrepreneurs

Snyder (2000) interviewed women entrepreneurs in Uganda who established and operated successful businesses that span the entire spectrum of the entrepreneurial landscape. They have used entrepreneurial activities to empower themselves, and include market women (who sell everything from water to curtains), manufacturers of custom leather shoes, owners of private clinics and supermarkets, as well as major hoteliers and tourism industry operatives. Snyder concluded that despite women's activism and its positive impacts, the situation for the majority of African women showed gender disparities across all socioeconomic and political indicators.

Micro- to small-scale enterprises and large-scale traders in the informal sector

A large literature exists on the vast and diverse group of women-owned micro- and small-scale enterprises in the informal sector (Clark 1994; Horn 1994; Spring and McDade 1998). They are often single-owners with limited products (agricultural products, foodstuffs, crafts, manufactured goods for resale), who are self-employed with family or low paid helpers.

Exceptions include the medium and large-scale traders in agricultural products (Clark 1994); cloth and textiles (Coquery-Vidrovitch 1997); and appliances and electronics (author's field research 2000 and 2006). This sector is discussed in greater detail under types of enterprises.

Small- to medium-scale formal sector

The myth of the "missing middle" is contradicted by studies pointing to viable small- and medium-scale formal-sector enterprises (Fielden and Davidson 2010; Fick, 2006; Rutashobya and Olomi 1999). Snyder's (2000) recognized growing segments of women owners of middle-level businesses in Uganda who formed business associations and availed themselves of government programs for private-sector development.

Spring and Rutashobya (2009) describe significant progress in women's ability to have diverse business networks. Women's business networks initially have more family and friends as members, but then gradually include other businesspeople, while men have higher proportions of business partners and colleagues from the start. Women's businesses are in all sectors, not just "pink-ghetto" enterprises. Women tend to cherish the flexibility that self-employment allows compared with salaried jobs. Their businesses are smaller than men's and they have fewer employees. Men have more education than women, but this is not the case for the women and men globalists of the NGAEs. The Africa Businesswomen's Network is enabling women's networks to overcome obstacles such as lack of business education and training, shortage of bank financing, lack of visibility and credibility of women leaders, and gender inequity in the workplace (Hallward-Driemeier 2013).

Large-scale formal sectors

African large-scale businesses generally employ hundreds of workers and earn annual revenues in the millions of dollars. Most owners are men, but some are owned by women. The medium to large formal sector enterprises owned by African women are a critical part of the entrepreneurial landscape because they point in the direction of economic growth and gender equity, and they provide inspiring role models. These businesses conform to regulations, exhibit high levels of human and financial capital, and are integrated into the structures of the formal economy. Fick (2006) provides examples of such women owners for most African countries. In Ghana, women own factories for food and beverage processing, for manufacturing sanitary products, furniture, textiles, and clothing, as well as large retail stores and transport companies. In Ethiopia, a woman owns the largest supermarket chain, and leather factory. In South Africa women are involved in mining, construction, transportation, and professional services; in Uganda a husband and wife own the largest cell phone company. Women in the NGAE category in Kenya and Zambia own large export horticultural operations (roses, cut flowers, and vegetables).

Business networks: The New Generation of African Entrepreneurs (NGAEs)

The NGAE networks were formally constituted between 1993 and 2004 to advocate for private-sector policy changes in the entrepreneurial landscape. They are a subset of contemporary business owners who use global business practices similar to Western businesses. Interviews were carried out by the author between 2000 and 2009 in Botswana, Ethiopia, Ghana, Kenya, Mozambique, Senegal, South Africa, Tanzania, and Uganda on a network of formal sector business globalists; other formal-sector entrepreneurs were interviewed for comparisons. After

Table 48.1 NGAE women interviewed by business positions and acquisition

Country	Members interviewed N = 57	National network member-ship	Business position		Business acquisition		
			Owner	Manager	Self	Family	SOE*
Botswana	6	14	6		6		
Ethiopia	3	7	2	1	1		1
Ghana	13	44	13		9	3	1
Kenya	11	11	11		10	1	
Mali	2	17	1	1	1		
Senegal	5	18	5		5		
South Africa	6	9	5	1	5		
Uganda	8	10	8		6	2	
Zambia	1	10	1		1		
Zimbabwe	2	8	2		2		
Total	57		54	3	46	6	2
Percent of category			95%	5%	85%	11%	4%

* State Owned Enterprises

termination of donor funding for their formal networks in 2004, these formal-sector businesses and network contacts were maintained. NGAEs emphasized financial transparency, being apolitical, using merit rather than kinship- and ethnic-based hires, adhering to ethical business practices, advocating for the private sector with governments, and having global perspectives. Members often succeeded in influencing business conditions by implementing cross-national joint business ventures; creating a professional sub-network; gaining official observer status at the regional economic organizations (ECOWAS, COMESA, and SADC); signing memoranda of understanding with multilateral agencies (World Bank and Ecobank); forming venture capital funds; and changing government regulations. Table 48.1 details the countries, members interviewed, business position, and how businesses were acquired (self, through family or by taking over state-owned enterprises, SOEs). The data on this group is discussed further in McDade and Spring (2005) and Spring (2009).

Paradigms to consider the range of the entrepreneurial landscape

Demographic profiles

Table 48.2 diagrams demographic characteristics such as education, formal-sector work experience, marital status, and husband's financial contribution for women's formal and informal sector enterprises. Women micro-entrepreneurs may have little or no education, while large-scale informal entrepreneurs may have primary to secondary to some college education, and some are now assisted by their educated children in doing e-commerce (author's interviews with Ethiopian and Uganda traders; Coquery-Vidrovitch 1997; Snyder 2000). By contrast, small and medium business owners, (e.g., most Uganda women entrepreneurs and the Ghanaian Association of Women Entrepreneurs (GAWE)) have completed secondary school; some attended college and have degrees (author's data). Conventional large-scale businesses owned by African women are mostly in the manufacturing and agricultural sectors (Spring and McDade 1998; McDade and Spring 2005, Spring 2009). Some women owners in Uganda (Snyder 2000), Kenya (author's fieldnotes 2001), and Ethiopia (author's fieldnotes 2000, 2006), have college

Table 48.2 Demographic characteristics: informal and formal sectors

	Informal sector*		Formal sector			
	Micro ⟶	Large	Small ⟶ Medium ⟶	Large	NGAEs and globalists
Education	None or little	Primary to secondary to some college	Secondary Some primary, some college	Secondary and college	BAs, MAs, PhDs
Formal work experience	None	None to retrenched workers	None, retrenched workers, some salaried in the formal	Salaried, formal	Salaried, formal, CEOs, COOs
Marital status	Married and female head of household				Married, some divorced legally
Husband's contribution	None to some	Income may reduce husband's contribution (varies)		Share financial responsibilities (varies)	May be business partners

Source: Spring (2009)

*Informal sector entrepreneurs may have primary education and higher. Some now use the internet for sales, often assisted by their educated children.

degrees. By contrast, almost all NGAEs (both women and men) have bachelor's degrees, some have masters, and a few have PhDs (McDade and Spring 2005).

Coquery-Vidrovitch (1997) noted that some West African women passed down their businesses to their offspring for hundreds of years, but now their children are well-educated and prefer formal-sector professional work. Owners of small to large businesses have several pathways: salaried workers who then start their own companies or, more commonly, takeovers of family businesses. By contrast, 85 percent of the NGAEs started their businesses, while 11 percent took over SOEs (Table 48.1; McDade and Spring 2005). Most had formal-sector positions (including in the US, UK, and Europe) before starting their companies; a few began companies after college graduation. Some were CEOs and chief operating officers (COOs) in their own countries. Many NGAEs' husbands and wives had their own businesses; in some cases, both spouses were network members.

In some places in Africa, village and peri-urban women expect to receive money from their husbands for their enterprises and marketing. This was true of Yoruba and Igbo women in Nigeria (House-Midamba and Ekechi 1995) and some Ghanaian women (Clark 1994; Spring and McDade 1998), while in other places, this contribution by husbands was not automatic (House-Midamba and Ekechi 1995; Horn 1994; Spring and McDade 1998). Often women were supposed to turn over profits to husbands but in fact, many hid money to use for household expenses (Spring 2000). As well, in some households, when wives earned income their husbands reduced household allowances. By contrast, salaried women workers range from holding separate purses and checking accounts, to having separate responsibilities for household, children's expenses, and large purchases, to having joint-bank-accounts (author's data; Spring 2000, 2009). However, for those in the formal sector, this reduction in allowances does not occur. Medium- and large-scale company owners are wealthier, often coming from affluent families. Women easily spend money to support the household. Some globalists and NGAEs are business partners with spouses and family members (e.g., daughter and father, mother and son, and siblings).

Types of enterprises and firms

Table 48.3 provides examples of enterprise types. Small-scale traders and marketers sell agricultural produce, cooked food, beer, crafts, and cloth. In West Africa, some women specialize in selling palm and kola nuts, salt, fish, and shea butter, while in Central/East Africa they sell grains, tubers, vegetables, and fruits. In certain areas (e.g., Lagos) women once specialized in selling herbals and household utensils, but these products and lucrative sales were taken over by men (Coquery-Vidrovitch 1997: 99). Traders of any scale buying from rural areas to sell in peri-urban and urban areas often provide forward and backward linkages in terms of moving agricultural and craft products in one direction (towards the city) and manufactured items (agricultural tools and inputs, cloth, and household utensils) in the other (towards the countryside) (Clark 1994; Spring and McDade 1998; Spring 2000).

In most places in SSA, there is an economic gender division of labor. Women trade produce, food, cloth, crafts, and pottery. Metal product-making is controlled by men. Commodities that require capital and direct connections to international markets are usually controlled by men; items to do with daily subsistence and local markets are usually female-controlled because of women's limited cash resources (Clark 1994; Horn 1994). Still, there are exceptions for large-scale traders, such as women in the Democratic Republic of Congo who sell gold, diamonds, and electrical appliances (Spring and McDade 1998).

Some women involved in large-scale trading have extensive networks; some operate internationally dealing both in wholesale and "cash-and-carry" in salt, fish, palm and kola nuts, shea butter, cloth, and gold (Coquery-Vidrovitch 1997) in West Africa. In the Horn of Africa, such women move manufactured products from Dubai, and in East Africa, women sell crafts made regionally in Kenya, Uganda, and Tanzania. Generally these large-scale vendors handle cloth/clothing and household goods, but are currently being undermined by Chinese vendors. Industrial factory production of foodstuffs, as well as bakeries and grocery stores provide

Table 48.3 Types of enterprises and firms

Informal sector		Formal sector			
Micro $\longrightarrow$	Large Enterprises	Small $\longrightarrow$	Medium $\longrightarrow$	Large Firms/ companies	NGAEs and globalists firms/ companies
• Agricultural produce • Prepared foods • Crafts • Herbals • Household goods • Hairdressing	• Textiles • Food products • Transport • Furniture • Household goods	• Manufacturing of clothing • Furniture • Bakeries • Retail shops	• Manufacturing of textiles, clothing, agro-processing • Retail shops • Transport firms	• Manufacturing of textiles, clothing, agro-processing, industrial products • Supermarkets • Transport firms • Retail shops chains	• High-value exports • Tourism • Computers • Real estate • Public relations • Manufacturing • Information technology • Manufactured products • TV/radio

Source: Spring (2009)

examples of small–medium women-owned businesses. Women own large-scale agro-processing firms (fruits, juices, and water), and may link with women owners of grocery stores and supermarkets. Some women have expanded from successful retail operations to multiple stores and chains. NGAE women and men are also heavily involved in the service sector (tourism, ICT, public relations, and consultancy companies), as well as in growing high-value agricultural exports (e.g., vanilla in Uganda and roses in Kenya and Uganda). Others have transport companies (growing from several trucks and busses to fleets), and own factories for manufacturing construction materials, household products, and clothing. One Ghanaian owner of a small dress factory needed transport to deliver goods to her customers, so she bought a minibus. Subsequently, and after gaining increased capital, she purchased six more buses and started a taxi service as a second business. South African Dora Ndaba, described below, also moved into this sector.

Startup capital

Table 48.4 models the sources of capital, probably the single largest factor. While women at all levels use their own funds and family money, some inherit family businesses. The model shows that NGAE women usually start their own firms. Donor loans and projects are usually directed to micro-entrepreneurs, although donors (e.g., British, Chinese, and American) have also targeted small formal-sector industries, and USAID, World Bank, and Club du Sahel targeted the NGAEs. Rotating savings and credit associations (ROSCAs) are utilized by micro- and small-scale businesswomen as a means of accumulating funds above weekly and monthly revenues. But formal-sector and medium- to large-scale businesses require significant capital. NGAEs have overcome the problems of limited capital markets by often using previous salaries and bank credits. Small–medium women-owned firms have greater difficulty: they may have to resort to bank overdrafts and supplier credit to finesse payments to suppliers. In Uganda and elsewhere, women were not as good as men at negotiating loans and interest rates, as noted below.

Table 48.4 Sources of startup capital

	Informal sector		Formal sector			NGAEs and globalists
	Micro ⟶ Large		Small → Medium → Large			
Own funds	x	x	x	x	x	x
Family money	x	x	x	x	x	
ROSCAs, informal institutions	x		x			
Inherited businesses	(x)	x	x	x	x	
Donor loans	x		x			
Retirement funds		x	x			
Previous salaries		x	x	x	x	x
Diversified investments		x	x	x	x	
Bank credit		(x)		(x)	x	x
Bank overdrafts		x	x	x	x	

Sourcing and markets

Table 48.5 models the differences and similarities between entrepreneurial sectors and levels in terms of where they find produce sources and markets. Micro-entrepreneurs use local resources mostly and sell locally, while globalists can source and sell regionally and internationally (and can use any level of resources and markets). Entrepreneurs noted that the most successful international formal-sector companies also have successful national bases. Large-scale informal-sector traders are global in their purchases of (inexpensive) manufactured goods for local, informal-sector distribution. Women traders in Ethiopia, Kenya, and Mozambique go to Dubai and Asia to obtain electronics. Women in Senegal and Ghana bring back appliances and clothes from France and the US. They perform a valuable service of bringing products to local people, but do not contribute to GDP in terms of taxes, creating industries, and purchasing local materials and products.

Table 48.5 Sourcing and markets

	Informal sector		Formal sector			NGAEs and globalists
	Micro —→ Large		Small → Medium → Large			
Local sources	x	(x)	(x)			
Local markets	x	x				
National sources			x	x	x	
National markets		x	x	x	x	x
Regional sources		x		x	x	
Regional markets		x		x	x	x
International sources		x			x	x
International markets					x	x

Networks and associations

Table 48.6 models the kinds of networks and associations to which women in business belong in order to disentangle the "confusing" range of women entrepreneurs. One type of network or organization cannot hope to accommodate informal and formal-sector businesswomen. For example, memberships in national and local chambers of commerce in Africa tend to include a large range of business people and sectors. But large-scale informal business women who do not pay taxes or contribute much to GDP might be omitted. Similarly, the ultra-modern NGAEs with their new management techniques, Western business practices, and fiscal transparency might not want to participate. In fact, many noted that they were not members of these "too inclusive" chambers of commerce, meaning that companies with corrupt practices might also be chamber members. They much preferred memberships in employers' federations and professional organizations.

Women's entrepreneurial groups, large-scale market women associations, and business and professional women networks have been prominent features for decades in Africa. Small-scale traders often are members of local organizations of market women, and large-scale wholesalers may be members of some types of national women's organizations (e.g., GAWE and the African Federation of Women Entrepreneurs (AFWE) have small businesses and large-scale traders

Table 48.6 Membership in networks and associations

	Informal sector		Formal sector			NGAEs and globalists Firms
	Micro ⟶ Large Enterprises		Small → Medium → Large Firms			
Local organizations	x	x	x			
Chambers of commerce			x	x	x	
Manufacturer's associations				x	x	
Employers' federations			(x)★	x		(x)★
Professional associations				x	x	x
Regional and international associations				(x)★		x

★ Only some women are in these networks

among their members). GAWE and AFWE members are a mixture of large-scale traders and small- to large-scale formal-sector businesswomen who were initially concerned with women's social issues but then went on to focus on wider economic issues. Founder Lucia Quachey (personal communication, 2000) noted that "women were not being invited to anything important or considered an economic force . . . Women just came in handy to be present or to speak on women's issues." Subsequently, GAWE and AFWE, based in Accra, Ghana, organized proposals and received funding from donors. AFWE sponsored three international trade fairs and investment forums (in Ghana, Ethiopia, and the US), and received funding from United Nations organizations such as Norway/NORAD and UNIFEM, that targeted both informal and private-sector entrepreneurs. Quachey noted considerable changes a decade later (personal communication, 2012, 2013) due to World Bank conferences and donor funding.

The NGAEs were already a network at pan-African, regional, and country levels. Members traveled to attend national, regional, and international meetings, as well as to visit other countries to conduct business. They focused on trade and investment while advocating business climate improvements in their own countries, thereby building South–South capacity. Some of their achievements included cross-national joint ventures and investments in regional and national funds. Marie Ba created a professional West African network of accountants across the English- and French-speaking countries. Women members used the networks to "meet like-minded women" with whom they could freely discuss business problems. Many mentioned the social aspects of being able to interact with women at their same educational and wealth levels.

Upward movement and changing sectors

Tables 48.2–48.6 show the range of women in entrepreneurial and business activities. But how do women entrepreneurs move within or between sectors? Data from interviews and business histories show some fluidity within and between sectors. However, there are barriers to upward movement within and between the categories as shown in Table 48.7. These include: 1) the inability to expand due to lack of capital, management skills, formal education, and formal-sector business experience; 2) the failure to source necessary materials and find profitable markets due to limited networks; 3) the inability to extend business endeavors outside their local area due to a lack of business connections and supplier channels; and 4) a political landscape that may limit private enterprise, present extensive government regulations, and may require bribes. In addition, the gender gap that limits women in general prevails. Many new in-roads are a result of newly configured networks and greater empowerment for women due to ICT usage, international travel, and increased access to funding sources.

However, recent analyses of small firms show that Ugandan women are limited in their negotiation skills compared to men resulting in dependence on smaller and higher-interest loans to take their businesses to the next level (Kibanja and Munene 2009). Boohene's (2009) analysis of Ghanaian women entrepreneurs in informal and small businesses showed they lacked the management skills, access to resources, and confidence in strategic planning abilities needed to get to the next level. By contrast, the NGAE women with their college education and global business skills did not have these barriers (McDade and Spring 2005; author's fieldnotes, 2000–2008).

Table 48.7 tracks the movement within and between sectors. In terms of movement between small- and large-scale informal trading businesses, there are women who have made such moves, but they are rare. Esther Occlu from Ghana, whose grandmother gave her some "small change" with which she bought oranges and made marmalade and then went on to own agro-processing and textile factories (first informally and then in the formal sector), is a well-known example (personal communication, July 2000). Dora Ndaba from South Africa qualified as a nurse, but obtained a taxi permit, and with great perseverance was subsequently propelled into what was called "the Black Taxi Industry" (Bagshawe 1995). Subsequently, she became vice-president of the National Association of Business Owners. By contrast, Marie Ba from Senegal, who earned degrees in accounting locally and in France, owned the fifth largest accounting company and the six-story building that housed it (Spring, fieldwork 2003). Higher education and formal business sector experience are usually requirements for entry into formal sector businesses (Hallward-Driemeier 2013; Hallward-Driemeier and Hasan 2012).

Lisa Daniels' case study from Zimbabwe (Spring and McDade 1998) argued that there are both high- and low-profit small- and medium-scale industries with men and women having different strategies and capital requirements: few entrepreneurs (and fewer women than men)

Table 48.7 Changes within and between sectors by size of business enterprise

Informal sector	Formal sector	NGAE medium and large firms
Micro ⟶ Large enterprises/ traders*	Small → Medium → Large firms/companies	
Movement possible to formal sector but rare ⟶	Movement upward and downward, dependent on capital, education, networks, etc. ⟶ Retrenched salaried workers may move to informal sector ⟵	Movement upwards and downward, dependent on capital ⟶ ⟵
Movement possible but rare. Lack education, networks, desire to pay taxes, be registered ⟶	Stay in sector	
	Movement possible but need to be financially transparent, highly educated, and apolitical ⟶	
	Movement possible but most don't want to go back to informal sector or lower levels of business ⟶	

Source: Spring (2009)

* Agricultural and artisanal producers to traders moving products

rise from low- to high-profit industries. High-profit small industries include retailing garments, auto works, carpentry, and electrical repair; low-profit industries include tailoring, vending foods and farm products, knitting, and woodcarving. They represent 19 percent and 81 percent, respectively, of the micro- and small-enterprises, and most women tend to be in the latter. Women's particular micro-enterprises tended to be limited to trade in foodstuffs, food processing, and household goods, and to certain occupations (e.g., domestic service, beer-brewing, hairdressing, sewing, knitting, and crocheting, and pottery and craft-making). They faced competition from men, as well as from women, who lost formal-sector jobs and who subsequently entered the informal sector, albeit usually in the high-profit category.

From cell phones to internet to social media to ICT careers

Table 48.8 tracks the rise and expansion of technology usage among entrepreneurs in all sectors and levels (see chapter by Abbasi in this volume). Buskens and Webb (2009) address various ICT usages that affect women's economic gain, independence, empowerment, and network formation; they also distinguish marginal versus regular usage in public places (telecenters) and private spaces (institutions, businesses, and residences). The literature tends to demonstrate that cell (mobile) phones are the largest contributor to enhancing women's abilities to be entrepreneurs and they add to women's empowerment as well. Cell phones assist women's strategies to deal with product sourcing and market sales; time and labor management; family needs and crises; and financial aspects of home and business. Internet usage depends on education and literacy, as well as wealth

Table 48.8 Cell phones, computers, internet access, and ICT networks and careers

	Informal sector		Formal sector			NGAEs and globalists
	Micro ⟶	⟶ Large	Small ⟶	Medium ⟶	⟶ Large	
Cell phone access only	x		x			
Cell phone ownership	some	x	x	x	x	x
Cell phone banking	?	x	x			
Computer access: telecenters multimedia centers, institutions		x	x			
Computer ownership		x		x	x	x
Computer mentoring: family friends		x		x	x	
Computer training		tele-centers	?		x	x
Internet access	x	x		x	x	x
Internet from own computer	x			x	x	x
Internet email	x	x		x	x	x
Internet website		Craft sales			x	x
Internet social media*				x	x	x
ICT networks					x	x
ICT career						ICT service firms

★ Data on internet and social media are unknown, so the checked boxes are estimates only levels to buy computers or use commercial centers. Women's business and social organizational networks currently link members in many countries such as Ghana, Senegal, South Africa, and Zambia. Company websites link up suppliers and clients in cases from Kenya, Uganda, Cameroon, and Ghana (author's fieldnotes; Hallward-Driemeier 2013). Women are also moving into careers in computer science in industry and private businesses.

Conclusion

The informal–formal distinction and delineation of sectors have been used heuristically here to disentangle the complex landscape of women entrepreneurs in sub-Saharan Africa. The data and case study materials show discreteness and stability of sectors and types rather than substantial movement of women entrepreneurs from the informal to the formal sector, other than reclassification due to registering or collecting taxes for sites and services. This is because the "entrance requirements" of capital, education, and network affiliations are real barriers that are difficult to overcome. Within the informal sector, the movement from microenterprises to large enterprises is rare and depends on capital inflows and networks, as well as on innovation and new product development. However, there are exceptions, often dramatic, of women's entrepreneurial spirit, intensive hard work, good fortune, and business acumen that allow women to move upward.

Women traders and importers and exporters selling large quantities of products they did not manufacture may have operations that are substantial. But their methods of conducting business separate them from those in formal industry and service sectors, even though they may share the same revenues and business scale. Within the traditional formal sector (from small to large businesses), there are some leaps to the next levels, but most are limited by access to capital, networks, market intelligence, product innovation, and specialized niches. Business people such as the NGAEs have moved from the conventional formal sector to a new global level within the continent and around the world through their types of businesses, networks, goals and methods, tertiary education, entrepreneurial drive, and access to capital. African Business Women networks and their partners are supporting business organizations to promote formal-sector expansion and the use of improved business practices to promote new paths and the possibilities for better outcomes for women entrepreneurs and business owners.

Scholars and donors must not confuse small-scale micro-entrepreneurs and formal sector small industries, or medium to large businesses and new generation businesspeople in terms of their needs for business training, credit and loans, and associations. For informal sector women micro-entrepreneurs in the 1990s, Horn (1994) listed ten guiding principles: 1) entrepreneurship is a gendered activity; 2) market women take risks; 3) they diversify income-earning activities; 4) they create microenterprise niches; 5) they lack access to formal capital opportunities but rely on spouses, kin, money-lenders, and savings; 6) entrepreneurship requires market intelligence and reliable wholesalers, as well as knowledge of clientele and their preferences; 7) women apprentice themselves to experienced traders to learn entrepreneurial skills; 8) they adapt trading techniques to available locales and spaces; 9) women strategize in terms of ways to make a profit and devote much time to their business; and 10) some women find freedom from domestic chores by engaging in entrepreneurial activities.

Through time, there have been changes to these principles for women in larger-scale informal and formal-sector businesses. The ten guiding principles that can be gleaned from women formal-sector entrepreneurs are: 1) women enter non-traditional, formal-sector enterprises based on their education and training; 2) women members linked in enterprise networks are considered as businesspeople in mainstream arenas, and not as a separate category of "women" entrepreneurs;

3) women informal-sector traders do not become new generation businesswomen; 4) many women still use their own and family money, but some obtain loans and funding, often with more difficulty than men; 5) women obtain market intelligence through global methods (e.g., cell phones, internet, personal networks, international travel); 6) entry into business is based on formal business training, college education, and formal-sector employment backgrounds; 7) women belong to national and international associations/networks, that may be gender- and profession-specific; 8) new entrepreneurs separate businesses and family financially and in time and space; 9) business women adapt global standards of accounting and financial transparency; and 10) these women differ from salaried personnel in that they have taken risks to create businesses. It is worth mentioning that women who are successful in the formal sectors and those who adopt more global methods and strategies do not want to return to the informal sector or to the more conventional formal sector. Many factors constrain movement upward but they also prevent downward spiraling.

References

Bagshawe, P. (1995) *Viva South African Entrepreneurs*, Musgrave, South Africa: Lifespan Publications.

Boohene, R. (2009) "The relationship among gender, strategic capabilities, performance and small retail firms in Ghana," *Journal of African Business* 10(1): 121–138.

Buskens, I. and Webb, A. (2009) *African Women and ICTs: Investigating Technology, Gender and Empowerment*, London and New York: Zed Books.

Clark, G. (1994) *Onions Are My Husband: Survival and Accumulation by West African Market Women*, Chicago: University of Chicago Press.

Coquery-Vidrovitch, C. (1997) *African Women: A Modern History*, Boulder, CO: Westview Press.

Fick, D. (2006) *Africa: Continent of Economic Opportunities*, Johannesburg: STE Publications.

Fielden, S. and Davidson, M. (2010) *International Research Handbook on Successful Women Entrepreneurs*, Cheltenham and Northampton, MA: Edward Elgar.

Hallward-Driemeier, M. (2013) *Enterprising Women: Expanding Economic Opportunities in Africa*, Washington, DC: World Bank.

Hallward-Driemeier, M. and Hasan, T. (2012) *Empowering Women: Legal Rights and Economic Opportunities in Africa*, Washington, DC: World Bank and Agence Française de Développement.

Horn, N. (1994) *Cultivating Customers: Market Women in Harare, Zimbabwe*, Boulder, CO: Lynne Rienner Publishers.

House-Midamba, B. and Ekechi, F. (1995) *African Market Women and Economic Power*, Westport, CT: Greenwood Press.

Kibanja, G. and Munene, J. (2009) "A gender analysis of bank loan negotiations," *Journal of African Business* 10(1): 105–119.

McDade, B. and Spring, A. (2005) "The new generation of African entrepreneurs: changing the environment for business development and economic growth," *Entrepreneurship and Regional Development* 17: 17–42.

Rutashobya, L.K. and Olomi, D.R. (eds.) (1999) *African Entrepreneurship and Small Business Development*, Dar es Salaam: DUP Ltd.

Snyder, M. (2000) *Women in African Economies: from Burning Sun to Boardroom*, Kampala: Fountain Publishers.

Spring, A. (ed.) (2000) *Women Farmers and Commercial Ventures: Increasing Food Security in Developing Countries*, Boulder, CO: Lynne Rienner Publishers.

—— (2009) "Empowering women in the African entrepreneurial landscape," in M. Ndulo (ed.), *Power, Gender, and Social Change in Africa and the African Diaspora*, Oxford: James Currey Press.

Spring, A. and McDade, B. (eds.) (1998) *African Entrepreneurship: Theory and Reality*, Gainesville, FL: University Press of Florida.

Spring, A. and Rutashobya, L.K. (2009) *Journal of African Business*, special issue: Gender and Entrepreneurship in Africa, 10(1).

49

GENDERING ENTREPRENEURSHIP IN ROMANIA

Survival in a post-communist borderland

Margareta Amy Lelea

Introduction

Women, as a group, are more likely to live in poverty than men, less likely to own land than men, more likely to be paid less for the same work and more likely to do unpaid care work (Seager 2009). Adding to this list of economic disadvantages, women are less likely to become entrepreneurs (Xavier et al. 2013).

In this chapter, I demonstrate how becoming an entrepreneur is a gendered process situated in place. The first section introduces the concept of entrepreneurship and situates it in the context of feminist scholarship. The second section briefly reviews global statistics available on women's entrepreneurship. The third and fourth sections situate this analysis in a case study that represents a spatial and temporal crossroads, a Romanian border region a decade after the fall of communism and prior to European Union accession. In the fifth section, the gendering process of entrepreneurship is discussed using specific examples from interviews in the Romanian case study. In conclusion, the last section discusses the implications of gendering and situating entrepreneurship in the face of growing global inequalities.

What is entrepreneurship?

The concept of entrepreneurship is central to capitalism. The classic typology of an entrepreneur is masculine – the "economic man" – a risk-taker, innovator and hero that manipulates the free market to his advantage following neoclassical standards described by Adam Smith. A list of various terms associated with entrepreneurship are "'small business owners', 'entrepreneurs' or 'micro-entrepreneurs', 'independent contractors', 'free agents', and 'self-employed'" although these "have conflated often widely diverging work situations" (Hughes 2005: 36). The definition of an entrepreneur varies across cultures. Even within the same country, variations may exist between legal and tax codes. In more restrictive definitions of entrepreneurship, it is more frequently businesses run by women that are excluded because many women's businesses are smaller and have no employees (Hughes 2005: 37–40).

For this reason, feminists have often called for a broader recognition of what is considered entrepreneurial activity. The implications of this recognition can be linked to access to credit and inclusion in training programs. For example with a case study from Peru, Maureen Hays-Mitchell (1999: 112) argues that "the widespread exclusion of women informal workers from conventional programs for micro-enterprise development is the result of gender biases deeply embedded in the neo-liberal model of development and upheld by patriarchal interests, ideologies and institutions". She further links this to the issue of human rights via the 1986 United Nations Declaration of the Right to Development.

Other feminists specifically promote the concept of entrepreneurship for women as a means for challenging gender stereotypes, economic independence and survival. Dafna Kariv (2013: 9) points out that

> for women, many of whom have experienced underemployment due to environmental constraints, such as the need to sacrifice professional or job-related aspirations to fulfil traditional social and domestic roles, as well as barriers in the labour market, entrepreneurship opens a door to a venture that allows women entrepreneurs to fulfil their needs and employ their job-related and professional experiences, knowledge and capabilities. More specifically, the independence of entrepreneurship allows women to decide on and control their time schedule, tasks, work pace and workload relative to their family and personal time.

Kariv's teaching text extensively connects life narratives to the motivation and preparation for entrepreneurial activities with case studies contributed by scholars based in Brazil, Canada, China, France, India, Laos, Russia, Sri Lanka, Sweden and the USA among others. While explicitly including international comparisons, the spatial language was restricted to terms such as ecology, environment and context.

Geographer Susan Hanson (2009: 245) extends this transformative argument to specifically include place in a reciprocal relationship. She also encourages studies of entrepreneurship to include gender. She says that "women are using entrepreneurship to change their lives and those of others, and, in the process are changing the places where they live". She uses four illustrations – from her own research in the USA and from Botswana, India and Peru – to show how the social spaces of inclusion and exclusion, personal networks, gendered norms and expectations can shift through the process of entrepreneurship. This shift may further entrench patriarchal gender norms or it may challenge gender norms.

Returning again to the debates over what is included in the concept of entrepreneurship (i.e., is informal market trading entrepreneurship? Does an individual need to demonstrate innovation to be an entrepreneur? Is the business activity undertaken "needs-based" or "opportunity-driven"? etc.) and how more restrictive definitions tend to exclude women, the argument that Alice Hovorka and Dawn Dietrich make, resonating with that of Susan Hanson, cuts through by clearly asserting, "entrepreneurship is a gendered process whereby the specific context of such activities, not simply the entrepreneurs and enterprises themselves, must be studied and articulated" (Hovorka and Dietrich 2011: 55). This claim urges analyses to go beyond individualistic assessments and to more deeply contextualize the gendered structures that are formed through social, cultural, political and physical spaces and places in which different entrepreneurial endeavours occur. Hovorka and Dietrich use field research from Botswana that shows how programmes enacted there need to be more responsive to how women have less access to credit, have smaller firms and the relationship that they have as married in or out "of the community", among other issues.

Momsen et al.'s earlier study (2005) demonstrates both the importance of place and the gendered process of entrepreneurship by comparing entrepreneurial activity in localities along the eastern and western borders in Hungary. The presence of the borders, whether there was a border crossing or not, proximity to urban areas and relative isolation of rural areas revealed different patterns of uneven development and associated inequalities all of which influence the gendered relationship to place of entrepreneurial activities. They found that rural areas with more poverty had a higher ratio of women's entrepreneurship as compared to men's (Momsen et al. 2005). Thus the rural border regions of eastern Hungary had more women entrepreneurs in the population than in rural western Hungary because the east offered fewer alternative sources of employment. They also found that, "entrepreneurship in villages has not changed the traditional division of labour [but rather, because] . . . it involves women working from their homes and with more flexible hours than when they had worked as employees under socialism, it may have allowed men to feel that there was less need for their help in the house" (Momsen et al. 2005: 113) and in this way reinforced binaries. Revealing gendered differences in the rationalizations for entrepreneurship, men in their study tended to describe themselves as profit-driven and women that they interviewed either

> chose to become self-employed or were forced into it because of economic necessity. For this latter group women saw entrepreneurship as a means of family survival and as supporting their husbands while those choosing to become self-employed saw their business in terms of service to the community and as a means of self-fulfilment. Thus, entrepreneurship either reinforces the dominant patriarchal culture or creates a counter-cultural discourse on women's role in civil society.
>
> *(Momsen et al. 2005: 118)*

Very broadly, the gendering process of entrepreneurship varies over time and place.

Women in the Global South are more likely to become entrepreneurs

To record some aspects of these variations, the Global Entrepreneurship Monitor (GEM) has created a regularized system for measurement and analysis including research teams in 69 countries, or "economies" (Xavier et al. 2013: 6). Their statistics capture both formal and informal entrepreneurial activity because it relies on a survey of more than 2,000 people administered by a national team for each economy rather than official lists of registered businesses. From the survey responses, they then differentiate attitudes towards entrepreneurship generally and gauge businesses reported temporally. Their key indicator to represent "dynamic entrepreneurial activity" is a composite of "nascent" (up to three months of activity) and "new" (up to 3.5 years of activity) firms (Xavier et al. 2013: 14). Gender differentiated statistics available of this key indicator have been mapped in Figure 49.1, where the female nascent and early stage entrepreneurs as a percentage of the total adult female population have been subtracted from the male nascent and early stage entrepreneurs as a percentage of the total adult male population. Figure 49.1 highlights countries where there are more women early entrepreneurs (Thailand, Ghana, Nigeria, Panama and Ecuador) and where there is gender parity among early entrepreneurs (Mexico and Uganda).

In Thailand, 21 per cent of adult women in the population are estimated to have started a business in the past three years, whereas 17 per cent of adult men have done so. In Ghana, 38 per cent of adult women and 35 per cent of adult men are early entrepreneurs. In Nigeria, 36 per cent of adult women and 34 per cent of adult men are early entrepreneurs.

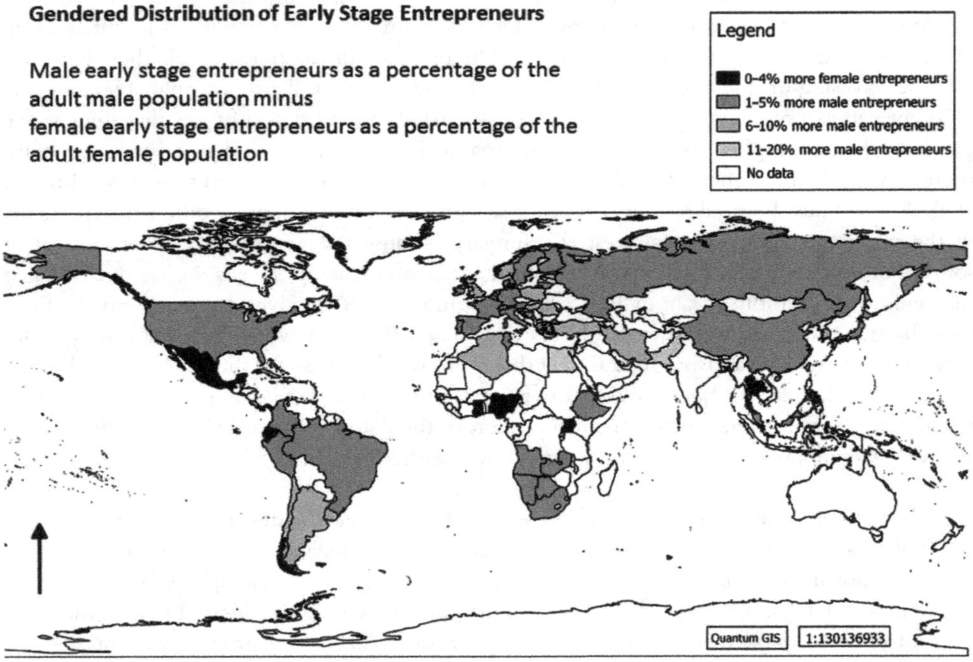

Gendered Distribution of Early Stage Entrepreneurs

Male early stage entrepreneurs as a percentage of the
adult male population minus
female early stage entrepreneurs as a percentage of the
adult female population

Legend

■ 0–4% more female entrepreneurs
■ 1–5% more male entrepreneurs
▨ 6–10% more male entrepreneurs
▨ 11–20% more male entrepreneurs
□ No data

Quantum GIS   1:130136933

Figure 49.1 Gendered distribution of early stage entrepreneurs, 2012 (Xavier et al. 2013: 60–61)

In Panama, 10 per cent of adult women are early entrepreneurs and 8 per cent of adult men. In Ecuador, 27 per cent of adult women are early entrepreneurs and 26 per cent of adult men. At parity, Mexico is at 12 per cent and Uganda is at 36 per cent (Xavier et al. 2013: 60–61).

On the opposite end of the spectrum, in Pakistan, 21 per cent of adult men are early entrepreneurs and only 1 per cent of women. Following are Latvia, where 19 per cent of adult men are early entrepreneurs and 8 per cent of adult women, and Egypt, where 13 per cent of adult men are early entrepreneurs and 2 per cent of women (Xavier et al. 2013: 60–61).

As reflected in the statistics listed above with the relatively high rates for Nigeria, Ghana and Uganda all over 30 per cent of the adult population engaged in early entrepreneurship, the GEM survey shows that geographically, there is the highest percentage of entrepreneurs in sub-Saharan Africa. This region also reports the highest proportion of people interviewed who felt that they had all of the skills necessary to start a business in the next six months. The GEM survey uses the three designations from the World Economic Forum of factor-driven economies, efficiency-driven economies and innovation-driven economies to differentiate levels of development with the conclusion that factor-driven economies (which have the highest levels of "subsistence-based agriculture, extraction industries and unskilled labour") have the highest rates of entrepreneurship (Xavier et al. 2013: 14). Particularly in sub-Saharan Africa, women of the Global South are more likely to become entrepreneurs than women in the Global North. Maps such as the one in Figure 49.1 make entrepreneurialism seem ubiquitous around the world, but this is a more recent phenomenon.

Gender and entrepreneurship from communism to capitalism in Romania

From 1947 to 1989, it was not possible to be an entrepreneur in Romania nor, give or take a couple of years, in any other country in the communist bloc. Entrepreneurs who had businesses were forced to stop activity and their assets were confiscated and nationalized during the communist period. Rather than a market-based demand economy, the economy of the Second World[1] was a planned supply economy controlled by a central political party, the communist party. In Romania, this power was concentrated in Bucharest where orders were given for rapid industrialization, urbanization and collectivization – particularly collectivization of agricultural lands. Citizenship was linked to being a worker and if someone refused to work, they faced imprisonment. One of the many extremist goals of totalitarian dictator Nicolae Ceaușescu (1965–1989) was to bulldoze all individual homes in favour of large-scale apartment blocks, a process that was started even in rural areas but fortunately not completed. Any trading of autonomous food production or any other goods was done illicitly in the underground economy.

Particularly by the end of the 1980s, an austere quota system was enforced whereby the limited quantities of sugar, oil, flour, meat, cheese and other food items were rationed. If possible, people would stockpile as much as they could, and some would then trade these items in personal networks. The way that some individuals navigated between these layers of economy can be described as a kind of entrepreneurialism. Goods such as particular types of blue jeans and higher-end televisions were also brought in from Western Europe. For example, Denise Roman (2007: 62) describes how a VCR purchased on the black market cost 30,000 Lei while monthly salaries ranged from 1,000 to 4,000 Lei. This VCR would then be used to make more illegal copies of movies and music to be further distributed in the unofficial economy. Surveillance of citizens by the *Securitate* meant that there was risk involved in such transactions and some were punished and some were not.

The official rhetoric of the Communist Party recognized women as equal to men, but reality was mixed with benefits and oppressions. Benefits of the system included access to education, quotas to increase political participation and state-provided childcare. Oppressions experienced specifically by women during the communist period included surveillance of their bodies through forced pregnancy examinations for the purpose of policing against abortion with the goal of increasing the future workforce (Kligman 1995). Rather than create equality, communism produced a "patriarchy without fathers" where the state held power over women and feminized men, while men were maintained as symbolic heads of household with the expectation that women would do domestic duties after their full day of work (Miroiu 1998).

During the post-communist period, there was a backlash where some women actively chose to retreat to the private sphere when they could and there was a decrease in political participation. This chaotic period unveiled rising inequality as many found themselves out of work when the old factories were closed and sold for scrap metal and inflation skyrocketed. Many people left the country to join temporary migration circuits to work in Western Europe (often on three-month tourist visas), Israel, the US and Australia, among other places. Women's labour was in demand for care work abroad and many women participated in cross-border trading networks. Remittances sent home have shown that "women have not only remained economically active outside the household, but become increasingly 'burdened' by the responsibility for sustaining multi-generational households through their earnings" (Cassidy 2012: 104).

Romania's European Union (EU) accession in 2007 raised hopes for economic stability and increased standards of living. Institutions and legal frameworks were harmonized with other EU countries. Borders became porous for the free movement of capital within EU territory, while agreements regarding the free movement of Romanian (and Bulgarian) labour within the EU have been delayed into the future and negotiated on a per country basis. Improvements include increased opportunity for educated youth. However, the cost of living increased more quickly than wages. The optimism of the period declined with the global economic crisis that started in 2008 and with austerity that accompanied a loan from the International Monetary Fund in 2010.

Returning to the statistics on new entrepreneurs from the Global Entrepreneurship Monitor which is available for Romania starting in 2007, Figure 49.2 illustrates that there have been more male early entrepreneurs for each year measured. The sharp decrease in male early entrepreneurs coincides with the 2010 austerity. In 2010 the gender gap in early entrepreneurship narrowed with 5.1 per cent of the adult men identifying as early entrepreneurs and 3.2 per cent of adult women. In 2010, the gender gap among early entrepreneurs was 1.9 per cent, in 2011 it was 5.2 per cent and in 2012 it had widened to 7.9 per cent. Figure 49.1 represents this gender gap as relative to other countries in the world and Figure 49.2 represents this gender gap temporally for Romania since its EU accession.

Relative to other EU countries, Eurobarometer surveys on entrepreneurship show that there are higher perceptions of entrepreneurialism in Romania (Eurobarometer 2012). However, as shown in the section above, this is consistent with the findings from the Global Entrepreneurship Monitor that use World Economic Forum typologies to designate Romania as a middle-level development country in the "efficiency economy" group, and designate the majority of other EU countries as "innovation economies" (Xavier et al. 2013).

Neither Eurobarometer surveys on Entrepreneurship nor Global Entrepreneurship Monitor data are available for Romania prior to EU accession in 2007. Prior to accession, national level data on entrepreneurship is not gender de-aggregated. The gendered experiences of entrepreneurship during the transition/transformation from communism to capitalism can only be estimated from partial information. For example, Ioan Ianoş and Alexandru Gavril's study (2012) on women business owners in Bucharest, comparing 1992 and 2002, analyses data for

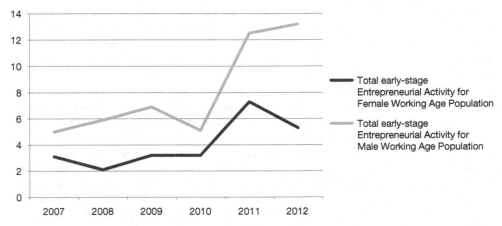

Figure 49.2 Gender gap among early entrepreneurs in Romania, 2007–2012 (Global Entrepreneurship Monitor 2013)

urban self-employed women registered using the PF (*Persoana Fizica*) tax code. In my qualitative research (Lelea 2009) in the mostly rural western border of Romania in 1999 and 2003–2005, I included interviews with entrepreneurs in the self-employed category as well as family business category and limited liability business category (the latter two categories do not include gendered information that can be retroactively tracked). Studies such as these offer some insight into the gendered dynamics of entrepreneurship during the crossroads of post-communism between the 1989 revolution and the 2007 European Union accession.

Researching post-communism along Romania's western border

The study of gender and entrepreneurship presented in this chapter is part of a larger dissertation research project on livelihood strategies along Romania's western border (Lelea 2009). Economic change in communities adjacent to this western border is highlighted because of the stark shift from being a region of deliberate disinvestment during the communist era to being a magnet for investment of factories seeking to utilize relatively cheap labour in Romania with minimized transport costs to consumers in Western Europe. The border influences inclusion and exclusion, barriers and flows of formal and informal trade and movement and restriction of ideas, goods and people.

Along Romania's 443km western border shared with Hungary, a selection of localities were chosen to reflect a range of accessibility and isolation. All of the localities were rural with the exception of one small city that had recently earned this designation by crossing the threshold of 10,000 residents. From north to south, the localities included are Bercu, Pelişor, Petea, Dara, Borş, Varşand, Sânmartin, Curtici, Dorobanţ, Iratoşu, Variaşu Mare, Variaşu Mic and Beba Veche. Four of these had either rail or automobile border crossings.

Ethnographic field research included in-depth interviews. In 1999,[2] I interviewed 81 entrepreneurs. Of these, 62 were non-agricultural entrepreneurs (13 female entrepreneurs, 29 male entrepreneurs, and 20 businesses with owners of both genders). From 2003 to 2005,[3] I conducted 68 interviews with predominantly micro-entrepreneurs, but also included factory workers, farmers and pensioners. Of these, 36 were women, 20 were men, and 12 were focus groups with a mix of men and women. In one county, I started with entrepreneurs that had open storefronts, and in the other three counties, I started with a list of registered businesses from the Chamber of Commerce. I then used a snowball technique to find other people to interview. I used a questionnaire with each interview and took handwritten notes. With additional permission, I was able to audio record one-quarter of the interviews. Interviews were conducted in the Romanian language with respondents of both Romanian and Hungarian ethnicities. Select quotes were later translated into English during the coding and analysis process.

Gendering the process of entrepreneurship: surviving by reconciling and challenging roles

The patterns of people's life experiences shape whether and how they will engage in entrepreneurialism as emphasized by Kariv. Because of this, interviews for this project included open-ended questions to encourage entrepreneurs to discuss their life course and particularly their work experience starting in the late 1980s to describe their experiences of change from communism to capitalism.

Some entrepreneurs were emphatic that regardless of the system that was in place, they would be resilient and adaptable. One woman who described herself as "among the best in my section" in her factory in the communist era later started two different thriving businesses along with her husband. They actually chose to move from one of the larger western cities to a busy locality with a border crossing to take advantage of the expansion and opened a specialized business. She embodied the classic "opportunity entrepreneur" when she exclaimed: "Wherever I was and for any amount of time, I could get by, because I was the type of person who if they threw me out the door, I would return through the window. I don't ever give up." The same entrepreneur also described how the good relationship she had with her partner was part of her business success. She said: "I can't really say that I had many obstacles because we, as husband and wife, get along very well, and we have worked together at his work and in the home – and starting a business – and everything, everything, everything, with my husband."

The end of communism meant that the nuclear family could again act as an economic unit. The revolution in 1989 signalled the end of "patriarchy without fathers" as Miroiu had described it. The type of familial relationships that one had, whether they were egalitarian or oppressive, was no longer just a part of the private sphere of the home, but also the public sphere when the newly created workplace was shared. Again, a woman entrepreneur who emphasized her positive relationship stated that, "I started this business in the city together with my husband. We no longer worked in different places, but in our own business together . . . We opened this small kiosk selling snack foods, drinks and candies."

Although some women described their businesses as flourishing because of their supportive partnerships, other women described how their entrepreneurship increased their negotiating power within their families. A woman who started a small grocery store in a relatively isolated village described how her husband and son consulted her regarding starting a new entertainment business venture. She said, "I am the one with the official [business] paperwork and so they must ask me. I am normally the one who decides in the family: they always ask me what to do as I am the one with the money and who takes care of the family budget."

In other cases, women entrepreneurs supported their husbands through their businesses. In an example in a relatively isolated village, a woman was able to open a bar through a loan from the male entrepreneur for whom she had worked for nine years. With the proceeds from the alcohol sales, she would repay the loan to her former employer, support her disabled husband and their multiple children. She was a classic 'necessity entrepreneur' and expressed dismay about the type of business that she ran. She was frustrated by the loud men and drunken behaviour and wished that she had enough money to open a small bakery and coffee shop. However, she described how, even if she had the money to invest, her village was too poor. The women of the village did not have enough disposable income to go out for coffee and cakes so she was stuck with the men in the bar.

After the dissolution of an intolerable marriage, a mother with small children turned to her parents who had continued to have prominent positions in the village during and after communism for investment capital to start a small grocery and convenience store. She was one of the few women entrepreneurs that I interviewed who described the importance of a network of supportive women mentors.

> It is hard for a woman to succeed in a man's world. However, men do treat me as an equal. They do not regard me as a housewife. There are few women who succeed in doing it all. The majority of women stay home with the children in the old tradition and the men resolve everything. Housewives – I don't want this! There is another woman entrepreneur in a different town who has been a mentor to me. She taught me to toughen up. She didn't allow me to stop – she encouraged me to develop more; to continue growing.

This same woman entrepreneur also described driving to the nearest city to take computer classes to improve her skills. When asked if she had ever considered international migration to earn income, she was decidedly against it. She not only wanted to work on her own self-improvement, but she wanted to be a part of improving her village and her community. As Hanson (2009) described it, the relationships and social networks in a place help to form an entrepreneur and, in turn, the activities of an entrepreneur also contribute to shaping a place.

Inevitably, the flows of people and goods over the border are part of shaping the place. In 1999, 13 per cent of the entrepreneurs that I interviewed described getting investment capital from unofficial cross-border trade during the embargo with the former Yugoslavia. From 2003 to 2005, there were a few entrepreneurs who described getting funds from international migration to start their businesses. The only one with children described how the older children had not assimilated to life in Spain and, with the birth of her third, it was too difficult to be a migrant. Rather than sink the money they had earned abroad on consumer items, she was determined to create something that would help them survive in a village that was relatively isolated and had particularly high unemployment.

> We worked in Spain, and with two kids, it was really hard. With three kids, I could no longer work at all, I needed to return to Romania and that is when I decided to start this shop. It is going well. My husband still needs to go to Spain. We started this business with money from working over there . . . My husband went first, and afterwards, the whole family. But I didn't stay after the baby was born. We had stayed for two and a half years over there. And the kids were in school for one year. It was hard for them to get along. They had to repeat a year of school. The children didn't want to stay any longer. After I returned to Romania, I wanted to do something with the money that would be long-lasting. We needed to buy this house, and since then, slowly, slowly . . . I opened this store while my husband was gone working in Spain. It was really hard for the first one and a half months with a six-month-old baby. I took her with me and also to my mother-in-law. I didn't know what else to do. I needed to put something aside for the children so that we can move forward and create something for them.

Framing motivation in terms of motherhood was common among entrepreneurs whether they had small children or grown children. Another woman described wanting to support her daughter's aspirations to become a medical doctor. She said: "We are not really satisfied with this business because there aren't really enough clients and I want to earn more to pay for my children's university education. But what can I do? Unfortunately, my daughter didn't qualify for a scholarship . . . so we are paying for her tuition. She wants to be a medical doctor."

Like in the Momsen et al. (2005) study, where fewer job opportunities in the eastern border region of Hungary resulted in more women entrepreneurs, women in isolated parts of Romania's western border region were creating jobs for themselves. Even without profit, some would say it was worthwhile just to have work years to count towards retirement. One woman described age as a barrier in the job market. She said, "I am making a modest living so I can survive. There is a lot of discrimination against age – at over 40, who will hire me? When I opened the store, I sold over 80 loaves of bread a day, and now only 15 with the same competition. This shows that people are only buying what is strictly necessary."

The survival of an entrepreneur with a retail business is connected to the prosperity of their neighbours. A particularly exasperated woman entrepreneur, who, in 1999 had been full of hope and optimism building her new business with funds from her family working abroad, was full of despair when I interviewed her again in 2003. She exclaimed:

> It is going from bad, to worse. The bureaucratic obligations are expanding while the cost of living and the salaries are cemented where they are so people can't allow themselves to buy too much. When you let them have things on credit and they don't repay you, what can you do? There is little recourse, and nobody to follow up with ... People would buy, but with what?! The lower class does not have money ... And meanwhile, the foreign investors in the free zone don't pay any taxes.

Although the free zone had been part of the border factory boom, it had not expanded as quickly as government officials had hoped. Workers from the free zone complaining to her about not getting their paycheques added to her ire about the discrepancy between her struggle to pay taxes and their tax exempt status. Unless they have a more specialized store, location in a higher traffic section of the border can serve to offer more competition creating greater barriers to entry and continuation. A woman who started her business after experiencing age discrimination, along with another female family member in a house that they inherited along a trucking route, thought that they had something positive to work towards with building their business. Rather, at the time of the interview in 2005, they were contemplating closure. She lamented: "From the year 2000 until now, it has been pretty slow just like this, pretty weak. The other gas station ... opened up ... [and] they have all kinds of automatic coffee machines, and it is cheaper over there than in a bar. I have lost a lot from that competition as well ... I don't even have a chance anymore."

As Hovorka and Dietrich (2011) explained, gendering the process of entrepreneurship – analysing the broader gender structures specific to a place counters the prevailing tendencies to focus only on the individualism of the entrepreneur.

Conclusion: implications for inequality?

This case study from a post-communist borderland offers a window of understanding to the social, cultural, economic and political processes that shape ideas about entrepreneurship and serve as a reminder of how economic processes are about relationships between people and places. With the increasing gender gap among early entrepreneurs in Romania, perhaps this qualitative research from an earlier period of dramatic changes can shed light on how to create programmes, policies and projects that are more responsive to the needs of women and situated in specific places.

By reviewing the concept of entrepreneurship, mapping global trends whereby some regions have high rates of women starting new businesses and others have few, and then discussing a specific case study from a place that is a cross-roads between two political and economic systems, the gendered process of entrepreneurship is situated.

Notes

1   This terminology was used during the Cold War to designate capitalist countries as the First World, communist countries as the Second World and those non-aligned and developing as the Third World.
2   Appreciation to the National Science Foundation for a Supplementary Research Grant for Undergraduate Research conducted under Janet Henshall Momsen, PhD.
3   Research made possible through a Fulbright Junior Fellowship to Romania, 2003 with extension to 2005.

References

Cassidy, K. (2012) 'Gender relations and cross-border small trading in the Romanian-Ukrainian borderlands', *European Urban and Regional Studies* 20(1): 91–108.

Eurobarometer (2012) *Flash Eurobarometer 354: Entrepreneurship – Country Report: Romania*, European Commission, http://ec.europa.eu/enterprise/policies/sme/facts-figuresanalysis/eurobarometer/index_en.htm.

Global Entrepreneurship Monitor (2013) *Key Indicators*, www.gemconsortium.org/key-indicators.

Hanson, S. (2009) 'Changing places through women's entrepreneurship', *Economic Geography* 85(3): 245–267.

Hays-Mitchell, M. (1999) 'Gender, informal employment and the right to productive resources: the human rights implications of micro-enterprise development in Peru', in T. Fenster (ed.), *Gender, Planning, and Human Rights*, New York: Routledge, pp. 111–124.

Hovorka, A. and Dietrich, D. (2011) 'Entrepreneurship as a gendered process', *Entrepreneurship and Innovation* 12(1): 55–65.

Hughes, K.D. (2005) *Female Enterprise in the New Economy*, Toronto: University of Toronto Press.

Ianoş, I. and Gavril, A. (2012) *Intra-Urban Spatial Changes of the Business Women Population During the Transition Period in Bucharest, Romania*, conference presentation, www.idm.at/index.php?download=2022.pdf.

Kariv, D. (2013) *Female Entrepreneurship and the New Venture Creation: An International Overview*, New York: Routledge.

Kligman, G. (1995) 'Political demography: the banning of abortion in Ceausescu's Romania', in F. Ginsburg and R. Rapp (eds.), *Conceiving the New World Order: The Global Politics of Reproduction*, Berkeley: University of California Press.

Lelea, M. (2009) *On the Margins of the European Union: A Feminist Geography of Changing Livelihood Strategies in Romania's Western Borderlands, 1999–2005*, dissertation, UMI Publishing.

Miroiu, M. (1998) 'Feminismul ca politica a modernizarii', in A. Mingiu-Pippidi (ed.), *Doctrine Politice*, Iasi: Editura Polirom.

Momsen, J., Szörényi, I.K. and Timár, J. (2005) *Gender at the Border: Entrepreneurship in Rural Post-Socialist Hungary*, Hampshire: Ashgate.

Roman, D. (2007) *Fragmented Identities: Popular Culture, Sex, and Everyday Life in Postcommunist Romania*, Lanham, MD: Lexington Books.

Seager, J. (2009) *The Penguin Atlas of Women in the World*, 4th edition, New York: Penguin.

Xavier, S.R., Kelley, D., Kew, J., Herrington, M. and Vorderwulbecke, A. (2013) *2012 Global Report: Global Entrepreneurship Monitor*, www.gemconsortium.org/docs/download/2645.

50

GENDER EMPOWERMENT AND MICROCREDIT IN BANGLADESH

Shahnaz Huq-Hussain

Introduction

Bangladesh has been a pioneer in the microfinance movement since the 1980s. This was triggered by the Nobel Laureate Professor Dr. Mohammed Yunus with his Grameen Bank model. He became worried about the state of the poor in Bangladesh during the famine of 1974 and so set up the Grameen program in January 1977 (Yunus 1997). He identified the vital importance of access to capital as a necessary element to facilitate women's empowerment. By extending banking facilities to the poor women of the villages, the Grameen Bank removed their dependency on moneylenders who charged high interest rates and exploited the poor. The Bank aimed at making those from the poorest households operate all necessary credit transactions by themselves. Despite initial worries it soon became clear that this scenario posed negligible credit risk to financial institutions in general. Bangladesh was the first country to encourage women's empowerment through microcredit. The Grameen Bank (village bank) known as the bank for poor women, along with other major Bangladeshi non-governmental organizations (NGOs) such as the Bangladesh Rural Advancement Committee (BRAC), have shown the world how microfinance can change the lives of poor women. Gender empowerment is the process by which women take control of their lives by increasing their choices and thereby expanding their horizons in different fields. Empowerment is a multidimensional concept that can be viewed from different angles (Friedmann 1992; Kabeer 1998). Social mobilization and economic security are vital to gender empowerment. The first microcredit summit held in Washington, DC, in February 1997 launched a nine-year campaign to reach the world's poorest 100 million families, particularly the women in the family, and provide them with credit for self-employment or opportunities for starting small businesses. Reed and Maes (2012) in the *State of the Microcredit Summit Campaign* report for 2012 indicated that throughout the world the number of poor women accessing microcredit went from 10.3 million in 1997 to 113.1 million in 2010.

Within the traditional societal structure of Bangladesh, women have always been suppressed. They had a very inferior position in family decision-making, access to properties/assets, credits or other instruments with which to change their lives. It was through the introduction of the microcredit system that Bangladesh tried to change the situation. Microcredit is an important development paradigm, which has spread an entrepreneurial culture to millions of Bangladeshi women. As we shall see it is used as an effective tool for poverty alleviation. It has played a valuable role in reducing the vulnerability of the poor through employment generation, asset creation, emergency needs protection and income and consumption enhancement. It has empowered women by giving them a voice, resulting in an increase in their self-esteem. Successful women entrepreneurs have a positive effect on the power structure of the society. This in turn results in improvements in women's economic wellbeing, which strongly reflects on the wellbeing of their family members, particularly of children. Use of children as income-earners for the household has been reduced significantly and the average literacy rate has also notably increased in recent years as a result of the microcredit efforts made by the government and the NGOs. Family sizes have also come down considerably. Available studies point out that the most notable dimension of women's empowerment is indicated by improvements in their literacy, health, nutrition, labor force participation, contraceptive use, mobility and ownership of assets. Women who were once oppressed in their villages have not only become financially independent through microcredit loans and entrepreneurial practices, but have gained a certain level of voice to fight oppression through the support of their microcredit group members.

Many governmental organizations—such as the Palli Karma Sahayak Foundation (PKSF) and the Bangladesh Rural Development Board (BRDB)—plus NGO agencies such as BRAC, the Grameen Bank, Proshika and ASA and many more small- and medium-sized organizations are working for women's empowerment through their different programs. However, this chapter will only focus on the role of the Grameen Bank on women's empowerment in Bangladesh on the basis of secondary materials and some personal interviews with Grameen Bank loan recipients. The first section of the paper discusses the history of the microcredit movement and the Grameen Bank in Bangladesh, the second section analyses the dimensions of empower-ment of women through microfinance in Bangladesh, the third section deals with the debates on the microcredit model, the fourth section discusses the recent Bangladesh government interventions and the Grameen Bank dispute, followed by the conclusion.

The microcredit movement and the Grameen Bank in Bangladesh

Professor Yunus with his Grameen Bank model identified the vital importance of access to capital as a tool to facilitate women's empowerment. The traditional banking system of Bangladesh left poor women totally outside of its credit system as they were illiterate and possessed no assets to be mortgaged or used as collateral. Poor women were undermined by their small deposits and were considered by commercial banks as risky clients who might not be able to return the loan. Moreover, such small loan accounts were seen to be costly to maintain by the commercial banks. These constraints made poor Bangladeshi village women totally dependent for small loans on local moneylenders who charged them very high interest.

In 1976, Professor Yunus, then a professor of economics at the University of Chittagong in Bangladesh, launched an action research project to examine the possibility of designing a credit delivery system to provide banking services targeted at poor rural women in the villages around the university (Yunus 1997). The success of the project inspired him to extend it to some other villages in Tangail district near the capital city Dhaka. His concept of helping village women using small credits for income generation proved that women could improve their earning capacity and return the loan in installments within the stipulated time. These successful results inspired Professor Yunus to form an independent bank for the poor called the Grameen Bank in October 1983 under government legislation. As per the Bank's regulations, the borrowers owned 97 percent of the Bank's shares while the remaining 3 percent was owned by the government. The Bank was entirely based on principles of trust and solidarity and it aimed at making the poor operate all necessary credit transactions by themselves. The system of this bank runs on non-kin based small homogenous groups consisting of at least five members and the group members have to adopt 16 rules laid down by the Bank that are related to family wellbeing, empowerment and social change. The Grameen Bank's high rate of repayment of loans has also led to the rejection of the fallacy that poor women pose credit risks to financial institutions.

Dimensions of empowerment of women through microfinance in Bangladesh

The impacts of microcredit on women's empowerment have been examined by several researchers (Mahmud 2003; Nazneen et al. 2011). Pitt et al.'s (2003) study found several positive impacts of credit programs on enhancing physical and human capital. They also noted that the participation of women in microcredit programs helps to increase women's empowerment by encouraging them to have a greater role in household decision-making, by providing greater access to financial and economic resources, developing better social networks, building greater bargaining power and having greater freedom of mobility. Female credit also tended to increase spousal communication in general about family planning and parenting concerns. Hashmi et al. (1996), Kabeer (1998) and Reza (2002) noted that women who are involved with microcredit facilities experienced less spousal violence. This was due to women's group support and their fallback position of links to the Grameen Bank. There are many case studies of the effects of microcredit and most of the studies have indicated that women's empowerment has been reflected in the areas of family wellbeing, poverty alleviation, fertility reduction, increased children's education, better access to services such as health, education and legal services, and bringing about social change through participation in group meetings. Women involved with microcredit are more likely to vote in local and national elections, have a higher than average level of entrepreneurial activity and to be more active in family decision-making. They are more likely to deal independently with financial institutions, speak out on key public decisions and have greater bargaining capacity and overall confidence. It was noted by Mizan (1994) that access to credit has allowed poor women to make monetary savings as well as empowered them with knowledge enhancement, negotiating power, maintaining social networks, as well as exerting their say in family decisions. Todd (1996) indicated that the Grameen Bank clients were taking family decisions jointly with their spouses, which gives evidence of a change in the traditional system when men were the main decision-makers in the family. With respect to family wellbeing, Khandker et al. (1995) stated that women's income contributes more to their families' needs than does that of their male counterparts, especially for children's education and the health care needs of their family.

Some studies felt that interest rates on loans were high and they have recently increased worldwide from an average of 30 percent in 2004 to 35 percent in 2011, but village moneylenders often charged 100 percent (*Economist* 2014a). Grameen Bank loans were also criticized because they represented a transfer of funds from poor rural women to a prosperous urban institution dominated by men and by targeting women it may just transfer the burden of household debt and subsistence onto women (Momsen 2010).

However, women's access to credit has allowed them to start several home-based activities or strengthen existing ones such as poultry and cattle raising. They have also been able to build some assets by utilizing loans for different activities. For instance, Ayasha Begum, aged 33 of

Dasra village, belonged to a well-off family that had a considerable amount of landed property. But she became helpless after her husband's death. Her land was grabbed by her brother-in-law. Her two sons were studying at the local school. She received no help from her family. After struggling for some time Ayesha heard about the Grameen Bank. She contacted the Grameen Bank and applied for a loan of 20,000 taka to be used for poultry and fisheries. She said:

> I purchased two cows from the loan money and from them I have been getting seven liters of milk daily that I sell in the local market. I pay my loan installments every month by selling milk. Beside this I have also bought some hens and ducks. Now I have five cows. I have also made a biogas plant from the cow dung that meets the energy demands of my family. I am sure soon I will be able to pay back all the Grameen Bank loan. Then I am planning to purchase some cultivable land in my village. My success has inspired many people to do something by their own effort.

Regular participation of women in group meetings has empowered them to make contacts to sell or market their products by themselves. Their access to market allowed them to gain negotiating, bargaining and economic power and become confident. Such exposure also gives them an opportunity to acquire new knowledge, and become less dependent on the male family members for marketing. They are able to start new businesses of their own that have often accelerated the economic progress of families. The following narrative illustrates this.

Monowara Khatun, aged 48, of Dasra village in Naogaon district, has three sons and two daughters. Suddenly her life became difficult and uncertain when her husband died in a road accident leaving only the *kutcha* house for them. She was just a homemaker at that time. Her sons were married and living separately and did not help her in such a situation. But Monowara did not give up her hope. She contacted the local Grameen Bank group in May 2006 and applied for a loan of 15,000 taka to start a small grocery shop at the corner of Sarawastipur Bazaar in her local area. Her application was favorably considered despite her lack of collateral. When she started the shop it came as a big surprise to the villagers to see a housewife running a shop. They made fun of her but she was determined to lead her life independently and to educate her daughters. Initially her sons ignored her but seeing the success of her shop they came forward to help her and it is now run on rotation so that Monowara gets some free time. She has already paid back the third installment of her loan and thinks she can pay back the entire amount within a year. She is also supporting her daughters' education. She is now respected by people of her community for her strength and determination.

Although there are debates on the interest rates of the Grameen Bank, which will be discussed in the following section, some of its schemes, particularly the education loan, have helped many girls and women to get qualifications and take up jobs and so mitigate family poverty. For example, Taslima Khatun, 32, is a primary school teacher. She narrated that after passing her Higher Secondary Examination (HSC) her family could not support the expenses of her university education because of poverty. She tried to get support in various ways but without success. Then one of her neighbors told her about the Grameen Bank education loan. On her advice she went to the Grameen Bank office and was able to receive an education loan from the Bank. "I was getting 4,000 taka on a quarterly basis. I gained admission to a local college on a BA degree course. I have been getting the loan for five years. While in the college I managed to get a job in a primary school, which helped me to repay the loan within a very short period of time. Now I am self-reliant. I can also give financial support to my family and reduce their poverty."

Women's consciousness about their political rights is another dimension of their empowerment. For example, there was a very high female voter turnout in the recent national elections (about 76 percent). This enormous turnout is even more than that in many developed countries. Is was found that almost all the Grameen Bank women participated in the voting process, some traveling long distances and standing for hours in the queue for their turn to vote, irrespective of their age.

The introduction of village Grameen cell phones is another example of women's empowerment (see chapter by Abbasi in this volume). Lack of information is one of the main causes of inequality and poverty. Telephone services are influential in the production, marketing and other economic decisions of rural households. Ownership of this technology gave village women the power to earn a livelihood. Individuals who are unable to buy a phone have access to phone services through the Grameen phone ladies and can be connected globally to their families. These women have proven that, given the opportunity, they can make good use of technology to help improve their personal status. The impact of Grameen phones on women's empowerment and their mobility has been noted by Hultberg (2008). These phones tend to be concentrated in northwest Bangladesh between Dhaka and West Bengal, where trade and migration require easy communications (Momsen 2010). But the expansion of other cell phone service providers has come as a challenge to the Grameen phone ladies. The village phone business of the women is threatened and they have almost been wiped out by cheap deals from different companies. But mobile phones have enabled them to receive and send money from home and overseas through a new technology for quick money transfers.

Debates on the microcredit model

Debates surround the microcredit model in Bangladesh as some economists favor microcredit as a development tool against those who think that it is not effective enough for development. Bateman (2009: 3) argued that, while microfinance favors a small number of people by giving them a quick opportunity for income generation, the medium- and long-term benefits are not clear. He said that the positive effects of such a model may have been exaggerated and concludes by saying "that microfinance in practice is actually more associated with adverse development trajectories, negative knock-on effects and major opportunity costs." However, a new study of more than 3,000 households in 87 villages carried out over 20 years did not find these negative effects but did find that borrowing does benefit women more than men (*Economist* 2014b). Controversy also surrounds the use of women's Grameen Bank loans by male members of their family. Karim (2008) noted from her research that, in the majority of cases, husbands and male kin of the women used the loans. The women acted as the recipient of the loan while the actual user was the man and the control and management of the loan lay with men. In one field interview I also noted a similar pattern: Santona Biswas, 26, of Chackrajapur village of Nawabgonj district, is a member of Grameen Bank. She belongs to a landless family. Her husband is a day laborer and the income he earned was not enough to provide three meals a day. Santona said "two years ago I took a 5,000 taka loan from the Gameen Bank payable in installments over four years. I had purchased a van for my husband, which he runs. Now his daily income is about 120–150 taka, which enables me to arrange three meals a day for my children. I have also paid back the entire amount of the loan." When asked who is now the owner of the van and who handles the money she did not answer. She did, however, mention that family decisions are taken jointly. There are several such mixed results of women being empowered by microcredit in available studies. Critics of Grameen Bank in Bangladesh think that, although the Bank has done fairly well in Bangladesh and many other countries, the interest rates of between 24 and 36 percent, including service charges, are rather high. Grameen Bank sources say that they have differential rates for different schemes. According to them there are four interest rates; 20 percent for income-generating loans, 8 percent for housing loans, 5 percent for student loans that are payable after completion of education and 0 percent (interest-free) loans for "struggling members" (beggars). All interests are simple interest, calculated on the declining balance. This means, if a borrower takes an income-generating loan of 1,000 taka and pays back the entire amount within a year in weekly installments, she will pay a total amount of 1,100 taka, i.e., 1,000 taka as principal, plus 100 taka as interest for the year, equivalent to 10 percent flat rate (Grameen Bank Website). The chapter by Aladuwaka in this volume discusses the problems women have in repaying such microcredit loans.

Bangladesh government interventions and the Grameen Bank dispute

The Grameen Bank and Professor Yunus in particular have come under attack recently by the present government. The government removed Professor Yunus from the post of the managing director (MD) in March 2011, stating that the retirement age was 60 years, which he had passed. Although the government retains the power to monitor and regulate the Bank, the entire responsibility of managing the Bank was vested in the board of directors as per the ordinance of 1983. The government held only three of the 12 board seats. But the present government amended the ordinance of 1983 and passed a new ordinance in 2013 to have more control of the Bank. This ordinance has given more power to the chairman, who is one of the three government nominees on the board of directors. The board chairman, in consultation with the board of directors, will now form the committee of not less than three and not more than five members for the purpose of selecting a candidate for appointment as managing director, and preference shall be given to persons having knowledge and experience in rural economy and finance or in the field of microfinance, as per the amended ordinance. The enactment of this ordinance created furious resentment among the Grameen Bank members, women activists and intellectuals belonging to civil society. A news conference was arranged by a Grameen board member at the National Press Club on August 29, 2012 to protest against this amendment. The borrowers thought the powers of the governing body to appoint a managing director had been snatched away by the present government in trying to appoint a MD of its choice and that this would not bring any benefit to the Bank and would deprive the Grameen Bank members of their legitimate rights.

In Bangladesh, where the government is headed by a woman, many feel this amendment has seized power from the poor women board members as the criteria set in the amended ordinance may not allow a Grameen woman to become MD of the governing board. On April 6, 2014, the government took a further step in its assault on the bank by taking the power to appoint its own board members away from Grameen and giving it to the Central Bank of Bangladesh (*Economist* 2014b: 72).

Conclusion

Microfinance programs for women's empowerment grew dramatically from the second half of the 1990s in Bangladesh and are now found in more than 50 countries throughout the world. The most notable dimension of women's empowerment through microcredit is indicated by improvements in their literacy, health, nutrition, labor force participation, contraceptive use, mobility and ownership of assets. Women who were once oppressed in their villages have not only become financially independent through microcredit loans and entrepreneurial practices, but have gained a certain level of empowerment and can now fight oppression through the support of their group members. Despite all the criticism, microfinance has had a positive effect on the progress of women's empowerment. However, this progress can only be fruitful if the socioeconomic conditions of women's households are sustained and their entrepreneurial skills are protected. The fact is that women's workload has sometimes increased as a result of their economic work. Now they have to bear the double burden of domestic tasks as well as income-generating activities. This has, in various cases, promoted family tension as some men of the family tried to control them, and became jealous of the economic freedom and success of women.

The empowerment process of poor village women in Bangladesh through the Grameen Bank model also needs careful watching as it is sometimes difficult to measure women's decision-making position in a family that still remains within the sphere of male control. It has been noted in this chapter that some researchers think that microcredit cannot be the only tool for women's development as well as for their empowerment and wellbeing. Some have indicated that agencies such as the Grameen Bank should not only target women's survival needs but their actual empowerment through enhancing their ability to control. My thinking is that in a society such as Bangladesh, where a large proportion of adult village women are illiterate, the meaning and understanding of empowerment is still in its infancy. Although the Grameen Bank and other NGOs are working in this field, newspaper evidence suggests that women are still struggling with power relations, poverty and personal security. But one must understand that with all its weaknesses and criticisms the Grameen Bank model, and Professor Yunus in particular, must be appreciated for bringing women out of their traditional domestic sphere, organizing their group power and bringing greater women's empowerment by using microcredit that, in the near future, might lead to gender equality through better self-confidence, resource and asset possessions, enhanced coping capacities, more freedom of choice and better power-relations. Finally it should be mentioned that the achievements of women's microcredit have also impacted on many male members of their families. For example husbands' attitudes are changing as they are allowing their women to attend group meetings, and to go to public places, taking family decisions jointly and listening to their voices and valuing the worth of women. Recently Professor Yunus has advocated a model of social business. This may add a new dimension to make people entrepreneurs, more self-reliant and may act as a further catalyst to reduce poverty.

References

Bateman, M. (2009) "Locked in: microfinance needs radical change before it's too late," *Microfinance Insights* 11: 1–3.

Economist (2014a) "Microfinance: poor service," *The Economist*, February 1, p. 67.

—— (2014b) "Microfinance in Bangladesh: rehabilitation and attack," April 19, p. 72.

Friedmann, J. (1992) *Empowerment: The Politics of Alternative Development*, Oxford: Blackwell.

Grameen Bank Website (no date) *Grameen Bank Website*, www.grameen-info.org/index.php?option=com_content.

Hashmi, S.M., Schuler, S.R. and Riley, A.P. (1996) "Rural credit programmes and women's empowerment in Bangladesh," *World Development* 24(4): 635–653.

Hultberg, L. (2008) *Women Empowerment in Bangladesh-A Study of the Village Pay Phone Program*, C-thesis 15 hp Media and Communication Studies, School of Education and Communication (HLK) Jönköping University.

Kabeer, N. (1998) *Money Can't Buy Me Love? Re-evaluating Credit and Empowerment in Rural Bangladesh*, IDS Discussion paper 363, IDS, Brighton.

Karim, L. (2008) "Demystifying micro-credit: the Grameen Bank, NGOs and neoliberalism in Bangladesh," *Cultural Dynamics* 20(1): 5–29.

Khandker, S., Khalily, B. and Khan, Z. (1995) *Grameen Bank: Performance and Sustainability*, World Bank Discussion Paper 306, Washington, DC: World Bank.

Mahmud, S. (2003) "Actually how empowering is micro credit?" *Development and Change*, 34(4): 577–605.

Mizan, N.A. (1994) *In Quest of Empowerment: The Grameen Bank Impact on Women's Power and Status*, Dhaka: University Press.

Momsen, J. (2010) *Gender and Development*, London: Routledge.

Nazneen, S., Hossain, N. and Sultan, M. (2011) *National Discourses on Women's Empowerment in Bangladesh: Continuities and Change*, IDS Working Paper No 368, Institute of Development Studies.

Pitt, M., Shahidur, M., Khandker, R. and Cartwright, J. (2003) *Does Micro-Credit Empower Women? Evidence from Bangladesh*, World Bank Policy Research Working Paper No 2998.

Reed, L. and Maes, J. (2012) *State of the Microcredit Summit Campaign*, 2012 Report, Washington, DC: Microcredit Summit Campaign.

Reza, M.H. (2002) *'Community' and the Empowerment of Women: The Role of Microfinance in the Changing Status of Gender Relations in Bangladesh*, research funded by Special Initiative International Development Research Centre, December, https://idl-bnc-idrc.ca/dspace/bitstream/123456789/33419/1/Hasan Reza.pdf.

Todd, H. (1996) *Woman at the Center: Grameen Bank after One Decade*, Dhaka: University Press Limited.

Yunus, M. (1999) *Banker to the Poor: Micro-lending and the Battle Against World Poverty*, New York: Public Affairs.

51

WOMEN, MICROCREDIT PROGRAMS AND REPAYMENT CHALLENGES

The Sri Lankan experience

Seela Aladuwaka

Introduction

Microfinance programs have gained significant recognition from both developed and developing countries in the past few decades. Microcredit for microenterprise development is promoted as a key strategy to alleviate poverty and empower women (Mayoux 1997). In particular, microcredit programs have been shown to empower women by providing them access to material resources and increasing their participation in household decision-making (Amin et al. 1998). The success of such programs has also been measured by the repayment rate. Yet, the ability of microcredit to empower women cannot be measured through client lists or timely repayments (Schurmann and Johnston 2009). Women may repay their loans but they may end up in more serious debt as they are forced to seek additional income earning ventures to repay their loans (Mayoux 1997).

Most studies on microcredit and empowerment focus on the final outcomes of micro-enterprises rather than the process through which they are achieved (Fernando 1997). One of the critical aspects of this is the process of repaying loans. The process of repayments and related problems has been neglected and needs more detailed understanding because it has direct impact on women's economic independence and empowerment. Better perception of the repayment process could have direct implications for the design of microfinance programs and repayment problems could be addressed. Using a case study of the Samurdhi credit program in Sri Lanka, this paper discusses some of the experiences women borrowers face in repaying their loans.

The chapter is organized into four sections. After the introduction, the second section discusses literature related to issues of repayments in microcredit programs. The third section provides a brief introduction to the case study of the Samurdhi Bank and to the research methodology used in the study. Findings related to repayment issues are presented in the fourth section followed by a conclusion.

Theories of microcredit and the related loan repayments

The theoretical background for the study draws from the empowerment approach. The empowerment perspective is based in the subordination of women and offers alternative approaches to gender-sensitive development. It also encourages context specific initiatives that eliminate women's subordination and improve their lives. The empowerment approach to development and planning provides a useful framework to understanding issues relating to women's microenterprise credit programs. This approach utilizes the experiences of women at the grassroots level and critiques development that ignores women. Women's credit and microenterprises are a part of many women's everyday life and their daily struggle to deal with economic difficulties. Through economic independence, women can become more involved in decision-making in their families, thus changing the traditional role of women.

The empowerment approach also recognizes the distinction between women's practical and strategic interests. In addressing poor women's practical interests, strategic gender interests can be met by microenterprise finance programs. Women's microenterprise credit programs can meet both the practical needs and strategic interests of women by allowing them to become economically independent and self-confident. Experiences have shown that women's financial dependency is one of the sources of subordination of women. The empowerment approach is a strategy to help women to understand and get control over social and economic forces that help to improve their status in society. Access to credit is one way women can improve their economic status and gain more control over their lives. In rural areas poor women have less access to productive resources, such as land and credit, than men and their economic power is limited without access to such resources.

In relation to the impact of microcredit on women's empowerment, research suggests that the provision of microcredit to women entrepreneurs has led to improvements in women's income and status (Amin et al. 1998). Some researchers, however, claim that microcredit has not helped credit borrowers as much as expected and question its impact on women's empowerment. These critics claim that male family members tend to control the loans women receive from credit programs (Goetz and Gupta 1996) and such loans can put pressure on women, when they cannot repay the loans so becoming more indebted (Rahman 1999). In understanding the impact of credit for women's empowerment, it is imperative to address issues relating to repayments because the process of repayments itself could hamper women's ability to accomplish economic and social gain and even limit their ability to improve their social and economic condition. Thus identifying some of the problems women face in the process of repayment would be highly valuable for interventions and improvements in lending and repayment policies and practices.

Poor women became a target for a majority of credit programs because they are seen as both creditworthy and trustworthy clients and efficient agents for the welfare of the household and low-risk borrowers (Schurmann and Johnston 2009). Various studies about Grameen Bank and similar programs in Asia and other parts of the world demonstrate that women are better at repayments (Armendariz and Morduch 2005). Thus targeting women is being increasingly advocated because of the evidence of their high repayment rates (Mayoux 1997).

The Grameen Bank in Bangladesh became a model for many micro-credit programs worldwide (see chapter by Huq-Hussain in this volume). Group based credit provision is used mainly because the poor lack collateral and such systems are used as peer pressure to ensure repayment and help build solidarity among group members. Thus the group is used as a joint liability (referred to when a member of a group of borrowers is liable for the loans of the others) and it is the most celebrated feature of Grameen Bank contracts (Armendariz and Morduch

2005). Even though loans are provided to individual members, all members are responsible and must support each other when difficulties arise (Armendariz and Morduch 2005).

Besley and Coate (1995) studied the impacts of repayment rates by lending to groups that are made jointly liable for repayments. According to their findings, positive impacts include that the successful members may have an incentive to repay loans for group members' projects that have unsuccessful returns. The negative impact would be when whole groups default, whereas some members would have repaid under individual lending.

As Schurmann and Johnston (2009) assert, the success of microfinance depends on peer pressure (social collateral) as a substitute for material collateral and insurance against late or non-payment and peer pressure is used both as a disciplinary and support mechanism within microcredit. The balance between these two functions is a fine one. The group based system is very important for borrowers and the repayment of their loans is critical for other members' ability to borrow money. Espallier et al. (2011) analyzed women and repayments in microfinance in 350 microfinance institutions in 70 countries in order to study the common belief that women are generally better credit risks for microfinance than men. Their findings suggest that female clients are associated with lower portfolio risk.

Research also suggests that female-based groups tend to have better repayment rates. Even though high repayment rates have been one of the measures used in identifying the success of microfinance programs, Mayoux (1997) argues that the impact on empowerment cannot be inferred from repayment levels because women may repay through taking loans from elsewhere and getting into serious debt. Research also reveals that in some cases men control women's loans and women are used to get loans from credit programs eventually creating greater dependency on male family members. Conflicts may also occur between women and men to fulfill repayment targets (Goetz and Gupta 1996). Fernando (1997) raises concerns over the feasibility of high payment rates and the long-term sustainability of microfinance programs. He points out that, of the majority of microenterprises, there is at least a three-month gestation period between the disbursement of credit and the earning of income. Yet women are required to begin loan repayments only a week after obtaining the loan even if it is not sure that their investment would yield an income immediately after they borrow money. Thus Fernando emphasizes the importance of examining the methods borrowers use to meet their repayment requirements during the gestation period and their consequences for the larger goals of empowerment.

The Grameen Bank credit program has not improved poor women's lives as much as expected according to Rahman (1999). Rahman asserts that to ensure timely repayment in the loan centers, bank workers put pressure on women clients and to maintain their regular repayment schedules, women became more indebted at the individual level with increased tension and frustration among household members, producing new forms of domination over women and increased violence in society. He pointed out that in some cases the husband used the wife's loans and investment and when the husband migrated abroad for work the wife continued to pay her weekly dues from her husband's remittances.

In their qualitative research on women borrowers of Grameen Bank loans, Goetz and Gupta (1996) found that significant proportions of women's loans were controlled by male relatives. They conclude that the contribution of Grameen Bank credit to changing gender relations may be minimal insofar as loans are kept relatively small and loan activities remain highly traditional (see chapter by Huq-Hussain in this volume). Their study showed that women borrowers systematically lost control over their loans and stated that many women bear the risk and burden of loans without directly benefiting from them. Furthermore, research on Bangladeshi women borrowers has shown problems associated with repaying loans. Findings indicated that, even though men benefit from loans by controlling the money women get, women are responsible for the weekly repayments creating stress and dependency (Goetz and Gupta 1996; Rahman 1999). However, recent studies have shown that borrowing does increase personal expenditure, household assets, labor supply and children's education and that loans benefit women more than men (Khandker and Samad 2014).

In assessing the success of microcredit programs on poverty alleviation, various measures have been used. A high repayment rate is used to indicate the positive impacts of credit programs. High repayment levels by women do not mean that women borrowers are successful in their economic activities. In some cases, loans are repaid from male earnings. Evidence reveals that, even though women's repayment records are better and group liability helps them in their repayments, their loan repayments sometimes come from sources other than income from microenterprises and when women cannot repay they are forced into a suppliant relationship with their husbands (Fernando 1997). This is a common problem when earning activities fail to produce income to repay the loans. A thorough understanding of the repayment process is critical in analyzing empowerment as well as for policy changes. The following section sheds light on this issue by using a case study from rural Sri Lanka.

Case study and methodology

The Samurdhi Bank in Sri Lanka adopted the Grameen Bank model employing a group-based credit system for loans to poor families. As a grassroots organization it recognizes the financial needs of the poor and promotes economic empowerment through a process of consciousness-raising and small group organization. Started in 1996, more than 68 percent of Samurdhi credit borrowers were poor women. The Samurdhi Bank continues to increase its membership. For example, the Samurdhi Bank had a membership of more than 2.5 million in 2009 and had the highest number of borrowers with nearly 700,000 in 2008 when compared to other credit programs available in the country.

The Samurdhi program consists of three related areas. First, the income transfer program provides consumption support to Samurdhi beneficiaries. Poor families identified on the basis of their monthly income receive a set of stamps every six months and these stamps are validated to use at government cooperative stores at the beginning of the month. The savings and credit program is the second component that aims to both reduce the vulnerability of the poor and to provide them with the capital necessary to participate in economic development. The final component of the program is improvements to rural infrastructure. The goal of all three programs is get poor families out of poverty. The consumption grant provides financial assistance to low income families to maintain a better standard of living, while the second and the third components focus on long-term poverty reduction strategies.

The Samurdhi Bank system is part of the saving and credit programs that were established to provide credit for the poor in Sri Lanka. By integrating these main components, the Samurdhi Bank system has aimed to improve both economic and social empowerment of poor communities. The organizational structure of the Samurdhi Bank involves family, group, village and zonal levels (a zone generally includes about 15 villages). Four subsystems are the small groups, Samurdhi society, the Samurdhi General Union and Bank Union. Samurdhi society membership is a requirement for joining the Bank, and small group membership and savings experience is a prerequisite for loan application. Small groups meet weekly to discuss each other's problems and build solidarity. Some of the main activities in the small groups (the group being five women) are to maintain weekly reports, build group funds, develop consumer services, exchange labor, guarantee bank loans and follow-up on loans. They also learn to maintain records of their work.

Small groups within the Samurdhi credit program also have a group saving component and intra-group credit system. Group-based saving and credit programs aim to promote savings habits among poor households and improve their financial management skills. There is no fixed amount that they should save but the group can decide how much is saved. All members of the group must contribute to these savings. The majority of Samurdhi Bank assets is made up of share capital. All Samurdhi group members have to purchase 500 rupees worth of shares in order to be a member of the Samurdhi Bank society. Field work for this research was conducted in The Pathahewahata Assistant Government Agent Division (AGA) in Kandy District. Three Gramasevaka Niladhiari (GN) divisions (GN divisions are the lowest administration divisions in the country) located in Pathahewahata were the study locations: the villages of Haputale-Egodagama, Etulagma-East and Pothgoda represent three different Samurdhi zones and Samurdhi Banks in Pathahewahata AGA Division. The research locations were chosen in Vehgaldeniya zone (16 GN divisions); Talatuoya zone (16 GN divisions) and Milapitiya zone (15 GN divisions). These villages represent different Samurdhi Banks in the area and the most important reason to select these particular villages was that they have supportive GN officers as well as Samurdhi Bank personnel that helped the researcher as key informants, making access to these communities easy.

The methodologies utilized in this research are guided by feminist perspectives, specifically, participatory research that promotes the use of a variety of methods and encourages involvement of research participants. Several qualitative methods such as semi-structured interviews, focus group discussion, participant observation and gender activity profiles were employed in this study. The time spent conducting the fieldwork was four and half months, from January 2002 to mid-April 2002. Thirty women credit borrowers who had received credit under the Samurdhi Bank program were selected using a snowball-sampling method. This method is considered a non-probability sampling technique. The chosen women were interviewed using semi-structured questionnaires. A socioeconomic survey was conducted among the respondents (credit borrowers). The mean age of the borrowers was 41 years. The vast majority of borrowers, 27 out of 30, were married. The majority of them had received at least a fifth grade education, although five of them had education up to the twelfth grade. In terms of employment, most borrowers were self-employed.

Women credit borrowers utilized their loans in diverse economic activities such as agricultural work, self-employment (sewing clothes, subcontracting work, establishing small stores/boutiques and raising livestock such as cattle or chickens). Many women, 13 of them, invested their money in agriculture, mostly in vegetable farming. These women, in many cases, were farming jointly with their husbands. Except for a few cases, their husbands owned the land, therefore women did not have to rent land to grow vegetables. Second, five women invested their money to start small stores in villages. Investing in a store is attractive because that gives women the chance to combine their household work and income-generating activities. Many women run the store with the help of the husband, whose contribution is mainly to transport stock from town. Cattle farming, poultry and subcontracting work such as making *suwandakuru* (incense sticks) are some other microenterprises in which women chose to invest their loans.

Discussion: process of repayment and challenges

Repayment is an important aspect of the credit program that reflects the success of credit programs. Many microcredit programs claim a high rate of repayments and many of these programs are considered successful. However, borrowers face various challenges in repaying their loans and based on a qualitative study this section explains women's repayment experiences.

Table 51.1 Status of repayment

Repayments	Number of borrowers
Paid on time	10
Could not pay on time, but repaid later with additional interest	9
Still paying back with additional interest	7
Could not pay and the bank has taken their monthly compulsory savings to recover their loans	4

It is crucial to understand the challenges borrowers encounter in the loan repayment process when analyzing the success of microcredit programs and also their impact on poverty alleviation and women's empowerment.

The study data indicates that repayment of credit is high among borrowers (see Table 51.1). Nineteen borrowers repaid their loans, of which ten borrowers managed to repay their loans on time while nine borrowers paid their loans late. Out of the ten women who paid their loan on time, six of them were able to get income from their investment, and so paid back their loans using the income from their investment, without any difficulties.

About 15 borrowers had taken loans from the Samurdhi Bank more than once. About six of those women had taken loans twice. Four of the women had borrowed three times while three women had borrowed four times. The fact that they have access to more loans shows that they have repaid earlier loans. Those who get credit more than once continue to make progress in their income-earning work. Once they establish their work and repay their loans, they are eligible to get more loans.

Borrowers who paid their loans on time were successful in their income generation. One such borrower from Pothgoda obtained a total of five loans. Each time she repaid her loan, she was able to apply for a higher amount (Figure 51.1). For example, her first loan was 2,000 rupees (equal to $20) and the second one was 5,000 rupees. Both loans were used for vegetable farming and she was able to make a good income. Because she was able to repay her loans from the income from vegetable farming, she applied for a third loan of 10,000 rupees and she bought a cow with that money. She was able to sell the milk and was very happy about her investment choices. She mentioned that she had no difficulties in repaying her loan as she got money from milk (she sells milk daily to the village milk van). Her fourth loan was 15,000 rupees and she invested again in vegetable farming and managed to expand by investing in a motor to mechanize irrigation on her vegetable farm. With this new investment, she managed to earn a higher income from the vegetable farm by being able to water the farm more efficiently than before. Her fifth loan was for 20,000 rupees and this she invested in improving the vegetable farm and cattle holding.

Her husband owned the land where she had the vegetable farm and she could take care of both the vegetable farm and the cattle as both are located very close to their home. She said "I paid all my loans on time; I got income from my work and was able to pay on time. That is why I was able to get more loans and invest in my work" (interviews, 2002). Borrowers who pay their loans on time are eligible to get more loans. Every time they pay their loans on time, they can apply for a larger amount of money. Her story is a success because she was able to repay her loan through the money she generated from her work. She has been so successful that she has given up her food ration that she receives from Samurdhi because she felt she no longer needed it and another poor family can benefit from it. Several other borrowers have had similar experiences and they continue to increase their income because they repaid

Figure 51.1 Successful entrepreneur with dairy cows supplying milk for sale

Figure 51.2 Mother and daughter in village shop started with microcredit loan
Source: Field survey (2002)

their loans from the income they received from their investments. In Haputhale Pallegama village a mother invested in operating a small shop and she managed to repay the loan from the income she got from the store and her daughter now has a job after successfully completing her twelfth grade education. Mother is happy because she was able to get loans and provide an employment opportunity for her daughter in the shop she started using her loans (Figure 51.2). In Pothgoda, a borrower invested her loan in sewing handbags, which she sells in town. She managed to repay her loan from the income she gets without any problem.

Women who are successful in earning income and repaying loans increase their financial independence and empowerment. The borrowers who get income from their microenterprises have better access to financial resources in the household. This can be linked to their bargaining power in the household. It is believed that women's bargaining power increases with their access to material resources.

The following comment from a respondent who earns money from farming demonstrates the borrower's stronger position within the family after they receive credit and income: "Getting our own income is a great help to us as women. It is better to be financially independent, so we do not have to ask for money from our husbands all the time. That way we can avoid dependency."

These women borrowers also have more choices in personal decision-making. They could decide if they want to travel alone, especially to visit family or friends or go outside the village. As women become involved in the credit program, women's mobility has increased because of the requirement that they attend regular meetings. For example, a borrower explains, "this credit program gave us an opportunity to come out from home. Otherwise, we are trapped inside the house, doing all the work. We never get to go out."

Their increased mobility leads to their greater visibility in the community and a gain in their confidence. This was evident from some women's comments and also indicates the changing traditional/cultural perceptions of women such as "women should be in the kitchen." Improvements in their mobility have given them confidence and the ability to participate in community work that has provided the opportunity to establish a strong social network. This is an important aspect of empowerment as women move away from isolation in their own household, they are able to meet other women and build up organizations among them.

Overall, access to credit from the program gave many women the opportunities to not only improve their access to economic resources, but also their position within their family, and to build self-confidence. Women's participation in the program also provided them with an active role in their communities. Access to financial resources such as credit, especially for poor women, can be seen as a starting point to challenge women's disadvantageous positions both economically and socially. The social gains are a long-term investment in their families and communities.

However, there are some drawbacks with the credit program. Problems associated with repayments are one of them. Not all borrowers are able to repay their loans on time from their investments. Two borrowers repaid their loans on time although they did not get an income from their investment. They borrowed money from moneylenders or relatives (in most cases from the husband) to repay their loans on time because they feared having to pay higher interest or losing their food stamps (a provision of the consumption grant given to eligible families provided by the Samurdhi program).

According to a Pothgoda borrower, "because we could not get a good income from our farm, we had difficulty in repaying our money; many times I used my husband's money." When they borrow money from moneylenders, they have to pay high interest, and this increases their debt. Borrowers express dissatisfaction with this type of debt cycle.

Two borrowers who paid on time did not get sufficient income from their investment, and so had to find additional income-earning work, in most cases a low-paying job such as vegetable harvesting, to earn money to repay their loan on time or even in some cases sell their belongings (one borrower had to sell her saris). She expressed her disappointment with regard to repayments as her chicken farm did not produce any income at all. As she put it:

> Some animals came into my chicken farm and ate most of the chickens in one night. I could not afford to buy more chickens. I could not get any money from the farm and I did not have any other income to pay my loan in time. I was afraid to lose our food stamps [food rations], because I have four children to feed. I once sold my saris to get some money to repay my loan on time. I have to make sure that I pay my loans on schedule because I cannot afford to lose my food ration. It is very stressful for us who are so poor. I somehow repaid my loan, but it was a struggle, it took me one year, I have to pay another 500 rupees as additional interest. I am very unhappy about this whole thing.

It is apparent that this type of situation makes borrowers discouraged from participating in such programs.

Another nine borrowers could not pay on time, but repaid their loan later with additional interest. These borrowers were disappointed that they had to pay additional interest because they failed to pay their loans on time. They could not pay their loans on time because they did not get the expected income from their investments. Many borrowers invested their loan in farming (mostly vegetable farming). Various reasons such as weather conditions (drought) or the low prices in the market make it hard for them to earn a decent income. As stated by one borrower from Ethulagama East, "I used my loan to grow vegetables. We did not have a field, so we rented a plot. Due to bad weather, we could not get any income. It was very hot and most of the plants died before we could harvest." A borrower from Pothgoda mentioned that even though her tomato farm was a success, she could not get a good income because of the very low price she got. She still has to pay her loan on time, and she has taken an additional income-earning job in the village to repay her loan. They explained that it is a struggle because the additional work does not pay much at all. In some cases for vegetable harvesting payment was as little as 200 rupees (less than $2 a day). The related problem is that when they cannot pay their loan on time, the interest rate goes up and as a result, they have to pay more money in additional interest later.

Another borrower from Vehegaldeniya borrowed money to buy a cow and she was able to sell milk and earn an income. Yet, she explained she ended up paying additional interest because some days her milk got rejected by the milk-collecting people in the village because the quality of the milk was not good. She is frustrated because if that happens for a few days, she gets behind repaying her loan. According to the borrower:

> I paid my loan already; I got income from selling milk and was able to pay it, yet I paid 400 rupees additional interest because I could not pay on time. For me, the main problem with this type of loan is that even if you do not get an income, you still have to pay your loan on time. If you cannot pay by the required time limit, you end up paying more money than you got and that is an additional burden for poor people like us.

Some borrowers could not earn income from their work, but had to repay their loans, so they had to borrow money from their husbands. According to a borrower in Ethulagama East:

> I paid my credit, it was hard as I did not get a good income from my work, and I also had to get some money from husband. The main problem is repayments on time. They expect us to repay the loan from the month after we take the money – we cannot get income that fast, but if we did not pay on time, we will have to depend on our husband's income.

As her comments suggest, women experience considerable pressure due to repayments related issues. Repayments and debt collection lead to significant pressure on women because they are less likely to control household resources (Goetz and Gupta 1996).

Borrowers' ability to get more loans depends on their payments. The Bank expects to have better records of repayments for borrowers. This is problematic for many of the borrowers as shown in this study sample, as twenty borrowers had various difficulties in repaying their loans on time. A borrower from Pothgoda states:

> I am still paying it, I had to repay my loan by one year's time, but I could not pay it. This is my second year. I still could not finish paying my loan. So far, I have paid an additional 1,000 rupees. The main problem I think is if we cannot pay, we will lose our food stamps. Also, everybody in the group has to pay their loans before we apply for another one, but it does not work like that. We need money and we have to wait. Sometimes people cannot pay even if they like to pay; simply because they do not get a good income.

These comments from credit borrowers indicate the diverse difficulties they face in the process of repayments and show that this process is not an easy task for poor people.

As is evident from this study, some borrowers had problems because they did not make any income from their investment in agricultural work. Their decision to get a loan and improve their income became a burden for them. Their frustration shows in some of comments they made: "I am very unhappy about this program because at the end I gain nothing but disappointments with more trouble. I would not do this type of risk taking again in my life. These programs are supposed to help poor women like us, but they get us into more trouble."

Some borrowers, about four in the sample, could not repay their loans because they did not invest their loans in income-generating work or did not get any income from their investment at all. Those who did not invest their loans in income-earning work have used the loans to buy material to build houses or gave their loan to a family member. In such a situation, the Bank has used their compulsory savings. These borrowers go through a hard time as they cannot repay their loans without any income, but the peer pressure they get from group members disturbs them. Group pressure for repayment could lead to an uneasy situation among members. "If one of your group members did not pay her loan, other members of that group cannot get loans. Thus, group members use pressure to pay our loans." Even though research has shown group liability encourages repayments, it can also lead to tension among group members and impact negatively on their social network and support system.

Samurdhi Bank data from the research demonstrates that there are high repayment rates among credit borrowers. Sample data also suggest that there is a high repayment rate, even though some respondents could not pay on time. However, more detailed information and qualitative analysis from the sample show that respondents had a difficult time repaying their loans. Some issues raised by borrowers are also critical. Women borrowers often talk about difficulties they face because they are required to repay their installment within a few weeks of getting the loan. They find it extremely difficult because most of the income-generating activities do not produce income immediately. As shown by the study, even though some borrowers repay their loans on time with income received from the investment, others have various difficulties.

Conclusion

This paper analyses issues related to repayments in micro-credit programs using a case study from rural Sri Lanka. The information gathered from the qualitative study shows that even if the repayment rate is relatively high, borrowers encountered various problems when repaying their loans. They had to find low paying additional income work or borrow money from their husband or family members in order to repay their loans. This research shows that it is important to look beyond the high repayment rates to evaluate a successful project and assess the impacts of microcredit on poverty alleviation and women's empowerment.

Microcredit programs can play a positive role in the alleviation of poverty (Yunus 1999). However, in order to achieve the full potential of these programs, they must be designed to guide women and provide flexibilities in the repayment process. As pointed out by Schurmann and Johnston (2009), the ability of microcredit to empower women cannot be measured through client lists or timely repayments. Failing to recognize critical issues with regard to repayments could lead to indebtedness and powerlessness instead of supporting women. An in-depth understanding over high repayment rates and clients' experiences in the process is vital for the sustainability of such programs.

References

Amin, R., Becker, S. and Bayes, A. (1998) "NGO-promoted micro credit programs and women's empowerment in Bangladesh: quantitative and qualitative evidence," *Journal of Developing Areas* 32(2): 221–236.

Armendariz, B. and Morduch, J. (2005) *The Economics of Microfinance*, Cambridge, MA: MIT Press.

Besley,T. and Coate, S. (1995) "Group lending, repayment incentives and social collateral," *Journal of Development Economics* 46: 1–18.

Espallier, B.D., Guerin, I. and Mersland, R. (2011) "Women and repayment in microfinance: a global analysis," *World Development* 39(5): 758–772.

Fernando, J.L. (1997) "Nongovernmental organizations, micro-credit, and empowerment of women: a new orthodoxy in development," *Annals of the American Academy of Political and Social Science* 554: 150–177.

Goetz, A.M. and Gupta, R.S. (1996) "Who takes credit? Gender, power, and control over loan use in rural credit programs in Bangladesh," *World Development* 24(1): 43–63.

Khandkar, S. and Samad, H. (2014) *Dynamic Effects of Microcredit in Bangladesh*, Policy Research Working Paper 6821, Washington, DC: World Bank.

Mayoux, L. (1997) *The Magic Ingredient? Microfinance and Women's Empowerment*, briefing paper prepared for the Micro Credit Summit, Washington, DC.

Schurmann, A.T. and Johnston, H. (2009) "The group-lending model and social closure: microcredit, exclusion, and health in Bangladesh," *Health Population and Nutrition* 27(4): 518–527.

Rahman, A. (1999) "Micro-credit initiatives for equitable and sustainable development: who pays," *World Development* 27(1): 67–82.

Yunus, M. (1999) *Banker to the Poor: Micro-Lending and the Battle against World Poverty*, New York: Public Affairs.

PART VIII

Development organizations: people and institutions

52

INTRODUCTION
TO PART VIII

General books on gender and development seldom include the organizations whose specific purpose is to encourage and fund the development process. These include local and international NGOs, women's networks, grassroots movements, bilateral and multilateral agencies, development banks, philanthropic foundations, and faith-based organizations. These players differ in their motives, purposes, and procedures. To them must be added consultants (both individuals and firms) that make their living from managing, monitoring, and contributing to the preparation of, and sometimes evaluating, the programs and projects involved. How each deals with or ignores the goal of equality and social justice for women is vital to the success of the development process. In this section we provide a glimpse, although only a glimpse, of that development world. We look at how these organizations have incorporated understandings of gender into their development work, at the priority given to gender equality in policies, in developing programs and projects, and at the ways in which gender issues are addressed in the field. Our interest is also in the experiences of women staff, and particularly those of gender specialists (overwhelmingly women) who strive to take forward the goal of gender equality. Study of the people who work in the development sector and how they affect the development process is attracting increasing attention. Indeed, during the writing of this introduction, three new edited collections have been published and they are discussed below.

As Holden points out, understanding how donor development organizations work, their structures and their processes (both formal and informal) requires a steep learning curve for the newcomer. Depending on the organization's ethos and internal culture, these systems can be highly formal or more flexible (see also Smyth 2013). The development of programs or projects in a large organization typically builds on its previous experience of the country or the sector concerned. The process involves nurturing a project or program from its origins as a "twinkle in the eye" through the development and negotiation stages to official agreement between the relevant parties concerned. Then comes implementation, requiring different management skills and regular monitoring with, only occasionally, an external evaluation on completion (Coles 2007). In fact a frequent worry among practitioners has been an organization's limited use of lessons learnt in the field as a result of poor feedback mechanisms. The gender specialist's effectiveness lies in how she, or occasionally he, learn to manipulate and take advantage of these processes and the internal cultures that may initially be seen as discouraging (chapter by Holden in this volume; Eyben and Turquet 2013).

Teams involved in development planning seldom have the opportunity to engage in long-term fieldwork, even though the complete program planning process may easily take a couple of years to reach the stage of implementation. The social analysis that provides the basis for many projects and programs is likely to be required near the start of the process. Thus there is a need to make best use of existing material, including large scale censuses, surveys, and government records, as well as academic research, including micro field studies. Importantly, officials generally respond most readily to quantitative data, which are, helpfully, usually disaggregated by sex (small qualitative studies are apt to lack credibility). A further constraint is that projects require indicators against which success can be measured so that the preliminary information provided needs to include some suitable for this purpose.

For social development and, more specifically, gender professionals, the overall objective may be to provide a sound understanding of men's and women's roles, responsibilities, relationships, and powers so that the development program can achieve maximum participation and benefits for both the women and men concerned. Nevertheless, additional work, particularly fieldwork, is often needed to provide an adequate gender analysis for the population concerned. Sometimes this is done by hiring women academics, local or foreign, sometimes by engaging consultants. Timeliness is usually important, placing constraint on traditional anthropological methods. In these circumstances, focus groups provide a useful complement. While quantitative data provide answers on the scale and distribution of the problem to be tackled, focus groups can answer questions about why and how, attitudes and opinions. They are comparatively quick to carry out since they concentrate on aspects relevant to the proposed program, and, by carrying them out in different parts of the country or with different social groups, they can provide representativeness. Since groups are typically divided by sex and generation, differences in women's and men's knowledge and opinions are revealed and some assessment of trends and change can be made.

Stakeholder analyses can be used to identify the interests of the various groups that may be affected both positively and negatively by the program. These range from those of the government and department officials concerned, to the intended beneficiaries. Measures can be taken to maximize the gains and number of beneficiaries, and to reduce the disadvantage or hostility of those experiencing "disbenefits." With beneficiaries normally analyzed by sex, the opportunity can be taken to promote greater equality. Stakeholder analyses can be used not only at the initial stage of project development but throughout its life. For managers, they provide a useful complement to more participatory methods.

Eyben, a feminist development professional with broad bilateral and international experience, starts this section. She provides a balanced overview of the ups and downs of aid organizations promoting gender equality since the first UN Conference on Women in 1975, with particular emphasis on the past 20 years. Two chapters in Part I might well be read in conjunction with Eyben's. Sweetman interrogates the reasons why international and local development agencies have met with limited success in carrying out the gender mainstreaming strategy that evolved from the Beijing Conference on Women of 1995. Lund uses autobiography as a means of tracing both milestones in the women's development movement and how she both learnt from it and, as an academic and researcher, contributed to it.

At the end of her chapter, Eyben reflects on the present challenges facing women's rights: the dominant framing of international aid as "technical"; the conservative backlash, itself partly a response to the success of women's movements; and the overconcentration on economic growth She is not alone. Porter et al. (2013), with long experience of the NGO community, are also deeply concerned. They too emphasize the risks inherent in instrumentalism, the narrow focus on technocratic solutions and overemphasis on "value for money," which lead to more onerous procedures. These trends, despite the current interest in girls and their future, seem to be silencing concerns for broader approaches to women's rights. Porter et al. call for less short-termism and for realistic understanding of the time taken to achieve the necessary transformational social changes, particularly for women.

Holden writes of her time as an insider within the UN system, describing not only what it is like to be a woman advocating gender equality but analyzing how this can be done within a remarkably complex system. Given that the UN is *the* international body promoting human rights, including women's rights, and that its agencies play a major role in international development, understanding how it operates is vital. So, too, is negotiating successfully within it. While outlining the structures addressing women's concerns, her account is personal and ethnographic. A recent book by Eyben and Turquet (2013) provides more perspectives from feminists working within large bureaucracies. Their focus is on internal change, the ways in which these women seek to alter their organizations' norms, values, and informal practices, so as to contribute to the ultimate aim of achieving real world improvements in gender equality.

This book and Holden's chapter can usefully be read by younger colleagues. Often operating in a patriarchal, bureaucratic, stressful environment and not infrequently considered "difficult" by their colleagues, the most successful of these "femocrats" adopt a highly political approach. In the book, Eyben offers good advice, including building internal and external alliances, leveraging outside pressures, looking for win–win solutions, and preparing for and seizing opportunities, while coping with bureaucratic resistance. To these, one editor (Coles) would add retaining a sense of humor in a situation where male colleagues have described women gender specialists as "threatening," "embarrassing," "terrier-like," "dangerously radical," "emotional," "so stressful to talk to," and as "both extreme feminists and sort of fluffy!"

Far from the arena of headquarters and macro-policy, Vu et al. move to field level with a practical case study. They explain how Oxfam first engendered a widely used analytical framework, popular with practitioners, to assess the impact of natural disasters, and then piloted it in a vulnerable area of Vietnam. This chapter is concerned with a project to prevent a slow-onset disaster. It is forward-looking in its approach, thus contrasting with the short-termism of emergency aid in a post-crisis situation. Together with partners, they were able to identify practical initiatives to both improve the representation and widen the livelihood opportunities of local women and girls. Oxfam's long engagement with gender issues and experience with the methods of participatory development may have contributed to their success in engaging with local government and other bodies—providing a contrast to the discouraging experience reported by Sultana (in Part IV) She, in a different situation, describes how this participatory approach, long favored by donors, failed to include women's views in new projects, because of cultural differences and mutual misunderstanding. For other examples of NGOs working on the ground see Porter et al. (2013).

Fechter is an anthropologist who, for some years, has been studying "development people," a largely neglected group. Through applying a gender lens to these practitioners, both local and national, the chapter examines the ways in which women and men, respectively, influence and are influenced by the development paradigm. She considers how gender matters for development workers, including how power and gender interrelate at work, and how the personal values of feminists, as individuals, shape aid policy and practice (see also Eyben and Turquet 2013). The chapter asks to what extent gender equality is implemented within organizations, and how this is affected by the constant mobility of personnel. Fechter valuably reviews the hitherto rather scattered literature on how aid work affects the personal lives of those involved, including the seminal earlier work by Goetz. She raises particular concerns over sexual abuse and gendered security risks, and recommends further research on those working in aid.

In her edited book (Fechter 2013) she has now done this, throwing further light on the personal lives and values of a range of humanitarian aid and development practitioners. She suggests that, until recently, the personal tended to be marginalized as recipients are foregrounded. The implicit narrative of aid work as altruistic may make it inappropriate to call attention to the provider. Nevertheless, as the chapters illustrate, the relationships and behavior of these workers influence how recipients perceive the assistance given and, importantly can improve its effectiveness. The chapters are certainly relevant to women aid professionals, although there is only one specifically on feminists.

The final chapter by Fiddiyan-Qasmiyeh is widest in scope—both in relation to the past and, importantly, for the future. Writing as an academic, she provides an historical analysis of the faith-gender-development nexus since colonial times. She studies the gendered nature and implications of faith-based organizations (FBOs) working in humanitarian aid and development. As she points out, hierarchies and oppressive structures in both FBOs and secular environments frequently, although not always, serve to reduce women's agency (see chapter by Jashok in Part I). Largely forgotten during a period of aid secularism, FBOs are now receiving more resources from donors. The author recommends caution against their premature idealization. Feminists argue that further analysis is needed to understand the complexity of FBOs (whose purposes may range from charity, through social welfare and development, to proselytizing, and, sadly, even terrorism). And the use of religious rhetoric may sometimes mask deeper more complex cultural or economic motivations, or indeed vice versa.

The authors' varied positions on the insider-outsider continuum and the perspectives from which they write have enriched this section. So, too, have the range of organizations about which they write.

Bibliography

Coles, A. (2007) "Portrait of an aid donor: a profile of DFID" in S. Ardener and F. Moore (eds.), *Professional Identities: Policy and Practice in Business and Bureaucracy*, New York and Oxford: Berghahn Books, pp. 125–142.

—— (2001) "Men, women and organisational culture: perspectives from donors' in C. Sweetman (ed.) *Men's Involvement in Gender and Development Policy and Practice*, Oxford: Oxfam Working Papers.

Eyben, R. and Turquet, L. (eds.) (2013) *Feminists in Development Organizations: Change from the Margins*, Rugby: Practical Action Publishing.

Fechter, A. (ed.) (2013) *The Personal and the Professional in Aid Work*, London: Third Worlds, Routledge.

Goetz, A. (2001) *Women Development Workers: Implementing Rural Credit Programmes in Bangladesh*, London: Sage.

Porter, F, Wallace, T., and Ralph-Bowman, M. (eds.) (2013) *Aid, NGOs and the Realities of Women's Lives*, Rugby: Practical Action Publishing.

Smyth, I. (2013) "Values and systems: gender equality work in different organizational settings" in F. Porter, T. Wallace, and M. Ralph-Bowman (eds.), *Aid, NGOs and the Realities of Women's Lives*, Rugby: Practical Action Publishing, pp. 127–144.

53

PROMOTING GENDER EQUALITY IN THE CHANGING GLOBAL LANDSCAPE OF INTERNATIONAL DEVELOPMENT COOPERATION

Rosalind Eyben

Introduction

This chapter is about the achievements and challenges relating to the half-century of feminist engagement with the global actors and institutions of international development cooperation – otherwise known as 'aid'. Some feminists have refused to have anything to do with what they perceive to be an instrument of patriarchal global capitalism. Yet others, including the present author, have struggled to realize aid's transformative potential. Although success has been variable and compromises frequent, international aid has nevertheless provided manifold opportunities for individual and collective feminist agency, including in the global spaces and institutions that an unavoidably partial (in both meanings of the word) account explores in this chapter.[1]

This introduction provides a brief history of aid. The rest of the chapter is in two parts. The first covers a 30-year period up to 2000. The four world conferences on women are used as a framing device to pace events, successes and setbacks in promoting gender equality. Part two brings us up to the present and includes the impact of the aid effectiveness agenda and the shift to a multi-polar world. The conclusion identifies three current challenges with respect to international aid's potential for promoting gender equality, already discernible at the fourth world conference in 1995.

A brief history of aid and development approaches

International aid is often defined as the flow of concessionary resources from richer to poorer countries. Government agencies in rich countries finance governments and non-governmental organizations (NGOs) in poorer ones. NGOs in rich countries also raise money through voluntary contributions and government grants that they pass on to their counterparts in poorer countries. Governments also give aid via multilateral organizations including international finance institutions such as the World Bank and regional banks like the Asian Development

Bank, the European Commission and United Nations agencies such as the United Nations Development Programme and UN Women. Multilaterals in turn finance recipient country governments and NGOs.

Aid, however, should not just be thought of in terms of the movement of money. Aid is also about flows of ideas, values and practices and it is these latter that are the focus of the present chapter. Moreover, phases in the history of international aid can be variously determined, depending on the subject matter.[2] For present purposes, it is useful to understand the evolution of aid within the wider history of international economic and political relations since the Second World War:

- the achievement of independence by most colonial countries and the establishment of most of today's international development institutional arrangements;
- the impact of the 1970s oil crisis and the resultant global economic recession that led to the indebtedness of many low-income countries in the 1980s, resulting in rigorous treatment by IMF and World Bank stabilization and structural adjustment measures, including cuts to public services, privatization of state-owned enterprises and reduction of tariff barriers;
- the end of the Cold War and a decline in aid flows; and
- in the past decade, the emergence of a multi-polar world in which erstwhile aid recipients such as China, India and Brazil emerged as major global powers.

Equally important is to look at the evolution of aid as it relates to the history of policy ideas – themselves influenced by changes in international relations – which have shaped development cooperation discourse and practice:

- modernization in the 1960s;
- the realization of basic needs in the 1970s;
- the Washington Consensus of the 1980s, involving a swing from state-led to market-oriented policies;
- the people-centred and poverty-reducing focus of the 1990s; and
- the return since then to market-oriented policies, along with a framing of development as a security matter.

Since the disappearance in the early 1980s of Southern-led Third World dependency theory – to which the Southern remedy was a New International Economic Order[3] – policy ideas in development aid have largely been generated in the North, as a consequence of changes in international relations and/or domestic policy concerns. These ideas – for example, neoliberalism that was a response by politicians such as Reagan and Thatcher to economic stagnation in their own countries – have tended to be applied with greater purity and ideological vigour to aid-recipient countries than at home.

Likewise, the history of promoting gender equality in development cooperation could serve to illustrate how Northern intellectuals have overlooked ideas originating in the South. The 'othering' of Third World women into a homogeneous group limited the possibilities of trans-national coalitions between women of different classes and colour (Mohanty 1988). Although most Northern-based feminist scholar-activists were sensitive to this critique, gender advisers in development agencies were less so, driven by the need to simplify complexity in making the case to their agencies to support gender equality. Moreover, rather than involve Southern expertise, agencies usually relied on Northern development consultants' analyses of the problems of women in aid recipient countries. The only African women permitted to have a voice were those at the grassroots as Everjoice Win (2007), an African development professional the agencies side-lined, has observed.

Nevertheless, constructive relations between activists, scholars and feminists in aid agencies (Eyben and Turquet 2013) have generated ideas and practices about gender equality that have significantly influenced approaches to its promotion in Northern as well as Southern countries. For example, in the run-up to the 1995 Beijing Women's Conference, gender mainstreaming, promoted by feminists working in multilateral development agencies became a shared agenda for Southern feminists and their Northern counterparts, moving into the domestic policy agendas of aid-giving countries and then out again into development cooperation practice.

Women, gender and development cooperation 1975–1995

The early history of promoting gender equality in development cooperation can be recounted from many perspectives. I have chosen to do so through examining the discourses and effects of the four United Nations conferences and parallel NGO forums between 1975 and 1995 as perhaps the most useful. Women from all over the world met in large numbers to discuss solutions to their gendered subordination.

Mexico 1975

The first women's conference, in Mexico (1975), launched the International Decade on Women. Organized by the UN Commission on the Status of Women, all member governments of the United Nations were invited to participate. It was the culmination of years of patient work by a small dedicated group of women within the UN (Jain 2005), supported by a larger group on the outside, leading to the UN General Assembly's adoption in 1967 of the Declaration of the Elimination of all Forms of Discrimination Against Women, followed 12 years later by the Convention (CEDAW). Despite Sweden having been quietly supportive for many years, the overall non-reflexive intellectual dominance of Northern agencies and activist feminists gave the impression that the Mexico conference was *their* idea, while ideas and actions of Southern networks got overlooked. In the early 1970s North American feminists had coined the phrase 'women in development' (WID), and the Percy Amendment to the 1973 US Foreign Assistance Act became the model for bilateral and UN agencies for integrating women into development (Fraser 2004). In 1975 the Development Assistance Committee (DAC) – which brings together countries in the Organisation for Economic Co-operation and Development (OECD) to discuss aid-giving countries' development cooperation policies and practices – held its first meeting of WID specialists working in government aid agencies.

The Mexico conference established the principle that governments needed an institutional machinery to integrate women's issues into policy and programming; within ten years the majority of UN member states had established some kind of bureaucratic arrangement. The 1975 creation of the UN Women's Fund (UNIFEM) was mirrored in the WID units set up in other United Nations organizations. 'A good working definition of WID is simply the taking of women into account, improving their status, and increasing their participation in the economic, social and political development of communities, nations and the world' (Fraser and Tinker 2004: ix). WID, commonly referred to as the 'equity' approach (Jain 2005), posits that women should receive the same rights and privileges as men. Although this approach enhanced the understanding of women's development needs and led to a recognition (at least in principle) that women should be included in development projects, Jain argues that it led to women in developing countries being wrongly seen as the equivalent of housewives in developed countries – an irony, bearing in mind that the housewife image was subject to feminist challenge in the North. However, most WID advisers in development agencies were oblivious to such criticism and even more so to Okello Pala's that 'integrating women into development was inextricably linked to the maintenance of the economic dependency of Third World and especially African countries on the industrialized countries' (cited in Rathgeber 1990: 492).

Copenhagen 1980

The purpose of the second UN women's conference on women in Copenhagen was to review mid-term progress in the international women's decade. The tensions at Mexico between First and Third World feminists reappeared. Third World caucuses framed their feminism as a liberation struggle from neo-colonialism related to the demands for a New International Economic Order. A participant commented 'I found North American feminists surprised to discover that not everyone shared their view that patriarchy was the major cause of women's oppression and that Third World women held views closer to Marx than Friedman' (Jain 2005: 86). Civil society did not have any officially recognized organized voice at the conference and in protest activists gate-crashed the formal proceedings.

The feeling that women from developed countries were using development assistance to define what Third World women wanted was unintentionally strengthened by the growing institutionalization of WID among bilateral donor agencies. Following the Copenhagen conference, WID advisers from the agencies set to work to produce the DAC's Guiding Principles for Supporting the Role of Women in Development, published in 1983, and in that same year a formal WID Expert Group in the DAC was created.[4]

The consequent relative invisibility of Southern feminism in development cooperation thinking reinforced the arguments of conservative aid officials in Britain and elsewhere that feminism was an idea created and promoted by the North, which should not be imposed upon aid-recipient countries. In practice, there was no common understanding of WID among development agencies. Whereas USAID focused on women's economic potential, others such as the British aid administration did not see women as economic development actors until the late 1980s. Before then, WID was about women's special needs for such basics as water and firewood (men apparently never being thirsty or hungry); women were a category of the population deserving 'consideration' as a 'vulnerable' group.

Nairobi 1985

The conference in Nairobi marked the end of the international women's decade. Before it met, the Third World caucuses that had been active in Mexico and Copenhagen agreed on a common framework – Development Alternatives for Women in a New Era (DAWN) – that emphasized the centrality of women living in poverty as development's objective and the need to reform macro-economic structures (Jain 2005). The WID Group in the DAC responded to and was sympathetic to the DAWN agenda. However, whether the WID advisers in the group were able to influence their agencies' development policies depended largely on their government's ideological slant and on the level of domestic concern about the issue in parliament, the women's lobby and non-governmental organizations.

In addition to structural adjustment debates, Nairobi is remembered for the maturity of international NGO activism in seeking to shape development policy agendas. The NGOs, now formally recognized as key players, organized in parallel to the official conference their own forum as a transnational space for consciousness-raising, networking and strategy formulation.

The majority of forum participants were from developing countries, their participation largely funded by international NGOs and bilateral agencies. Although the sharp divisions at Copenhagen had not entirely disappeared, there was much greater preparedness to accept different feminisms along with a shared concern to move beyond arguments to agreement that development was not working for women (Çağatay et al. 1986). Nairobi paved the way for Beijing in 1995, the apex of 20 years of sustained endeavour.

Beijing: the watershed

Following the end of the Cold War, the United Nations Conference on Human Rights in Vienna (1993) made a breakthrough in recognizing women's rights as human rights. As the preparations for Beijing developed thereafter, a vision of global social transformation grew. The coalition of grassroots activists, politicians and bureaucrats meeting in Beijing was emboldened by a favourable international climate of a return of parliamentary democracy in many countries and an increased emphasis on civil society voice. Although the macroeconomics of the Washington Consensus and the associated structural adjustment policies of the 1980s did not disappear, they ceased to be a central preoccupation. There appeared to be much that governments *could* agree about.

The preparations for Beijing were a significant moment for feminists in aid agencies. They appreciated its importance for strengthening the global women's movement – of which they felt part – and for influencing their own agencies to put resources into gender equality. They used the DAC to coordinate their funding of Southern civil society participation in the forum and influenced their regional and country programme offices to finance preparatory activities in aid-recipient countries. They also funded their own development NGOs to prepare position papers, to support civil society in the South and to be members of the official delegations. The DAC-WID Group provided a representative to sit on the informal committee meeting in New York to advise the UN Division on the Advancement of Women on the drafting of the Platform for Action.[5] The group also commissioned a major study of DAC member states' effectiveness in promoting women in development and, at their annual meeting in 1995, ministers of development cooperation endorsed gender equality as a vital goal for development and development assistance efforts. Although the statement from the meeting repeated some of the by now well-rehearsed efficiency arguments that investing in women is good for development, for the first time reference was made to the 'transformation of the development agenda'. Meanwhile, in the multilateral agencies, UNDP had made gender equality the subject of its 1995 Human Development Report and the World Bank produced a new policy statement that announced a switch that most development agencies were now making from WID to gender equality.

No longer just the radical fringe but the mainstream also was arguing that systemic improvement to the status of women could only be achieved by transforming gender relations. Historically derived structures that sustained these relations, including state bureaucracies, were failing to deliver gender-equitable policies. Gender mainstreaming, as defined at Beijing, was both a strategy for infusing mainstream policy agendas with a gender perspective and for transforming the institutions associated with these agendas. Changes in values, attitudes, practices and priorities were required at all levels.[6]

Yet, Beijing also witnessed debates over the meanings of gender. Feminist radicals, particularly from the South, were subsequently vindicated when they warned that should aid agencies adopt 'gender' it would be de-politicized, 'stripping away consideration of the relational aspects of gender, of power and ideology and of how patterns of subordination are reproduced' (Baden and Goetz 1997: 7). The radical promise of gender mainstreaming in development agencies rapidly dimmed as it became increasingly bureaucratized and as predicted gender advisers in aid agencies downplayed the political implications of gender mainstreaming to avoid scaring senior management.

Another challenge was a growing conservative backlash objecting to the very notion of gender. Conservative civil society groups had been active in the Beijing Forum, while at the official conference, the Vatican and its supporters strengthened an alliance with conservative Islamic states (forged the previous year at the Cairo Population Conference) to reject the idea of gender equality and women's reproductive rights. The increasing influence of these forces prevented the holding of a fifth international conference in 2005 for fear of a concerted push back against Beijing's agenda. The next section considers the implications of these and other political and economic trends that since Beijing have shaped the promotion of gender equality in the context of the changes taking place in the international aid system.

Gender in the international aid system 1996–2012

International aid's promotion of gender equality since Beijing is a complex story of progress and reversals – the outcome of broad global social, political and economic trends, and of changes in the international aid system, responding in part to these broader trends. I start by considering the impact on gender equality work of the evolution of aid practices and discourses from the seminal DAC position paper *Shaping the 21st Century*[7] to the implementation of the Paris Declaration on Effective Aid. As these still play out in donors' country programming and policymaking, the past decade has been marked by a more complex and emergent global landscape of development cooperation, with growing influence of large middle-income countries that used to be aid recipients.

The evolution of the OECD's approach to aid

In 1996, when donor agencies were busy preparing their strategies in support of the implementation of what had been agreed at Beijing and the DAC-WID Group was starting work on revising its Guiding Principles, development ministers approved *Shaping the 21st Century*. This document was both a product of the optimism of First World countries following the end of the Cold War and a response to what donors now recognized as the negative impact of structural adjustment policies on low-income countries. Moreover, enthusiasm for 'good governance' and promoting democracy abroad was accompanied by a belief that aid-recipient countries should own their development policies rather than have these imposed upon them. Participatory approaches to development (including women's participation) had become mainstream and poverty reduction a central objective. With greater consensus than ever before among donor agencies, and public opinion in the North supportive of aid, *Shaping the 21st Century* set 'international development targets' (the precursors of the UN Millennium Development Goals). Gender advisers in the aid agencies were confident that gender equality would be among these targets but the Japanese (unlike their North American, Antipodean and most European counterparts), had refused to make the policy switch from WID to gender equality. In high-level negotiations they successfully insisted that if gender equality were to be included at all, it should be restricted to parity in education. Just one year after Beijing, the DAC had rejected the broad-based challenge of the Platform for Action.

Three years later, international aid's growing focus on poverty reduction led to agreement to provide debt relief to highly indebted poor countries on condition that those countries approved national poverty reduction strategies that were to guide World Bank IDA credit and IMF support. Many bilateral donors committed themselves to realign their own programmes accordingly. National women's movements and their feminist allies in donor agencies saw these poverty reduction strategies as the opportunity for implementing the Beijing agreement but in most countries women's voices were often muted and gender issues marginalized. In Bolivia, for example,[8] the broad-based National Dialogue had not favoured equally the voices of women and men, and last-minute efforts by the vice-ministry for gender (90 per cent dependent on aid from the Dutch and Swedish governments) to incorporate gender issues into the strategy failed. This pattern repeated itself in other low-income countries where gender equality ministries were largely dependent on donor funding. The often good relationship between such a ministry and the donor community could be to the detriment of stronger links with the rest of the government or with its own civil society constituency.

Aid flows in decline since the end of the Cold War began to pick up again with the commitment to poverty reduction and the Millennium Development Goals. Accompanying the increase was a drive to make aid expenditure more efficient through strengthening recipient country ownership, integrating donor-funded programmes into national budgets, harmonizing donor procedures, placing greater emphasis on measurable results and establishing mechanisms for mutual accountability. These were the aims of the Paris Declaration on Effective Aid (2005) that coincided with a decline in donor interest in supporting gender equality.

After Beijing, many aid agencies had established a 'twin-track' strategy of integrating gender equality goals into the large public sector programmes they financed, while also financing small initiatives by civil society groups or policy advocacy units in government. Now funds available for the second track shrunk, negatively affecting transformative change work on women's rights.[9] Moreover, emphasis on recipient-country policy ownership allowed governments indifferent to gender equality to drop it from their agendas. In this environment, gender advisers struggled to maintain supporting women's rights activism as integral to their agencies' gender and development policies. Donors' increasing emphasis on tangible, measurable short-term results also made it harder for women's organizations to work for long-term political and social change (Mukhopdahay and Eyben 2011).

Accordingly, 'Beijing plus Ten' provoked a moment of reflection. Evaluations confirmed that aid agencies were not meeting their Beijing commitments. In response to evidence that funding allocated to 'mainstreaming' gender into general budget and sector support was not working, some donors (e.g., the Netherlands, Norway and Ireland) re-established earmarked budgets in support of the second track. The DAC GenderNet secured a commitment to gender equality in the final text of the 2008 agreement from the conference in Accra that reviewed progress in the implementation of the Paris Declaration. By 2010 the mood had shifted again. Gender equality and women's empowerment had re-established themselves in international development agencies as important goals to which senior management appeared to be paying serious attention. After a dip in expenditure in the first half of the decade, the proportion of aid money committed to gender equality increased. The General Assembly's agreement to establish UN Women was an impressive result.

A change in discourse accompanied this improvement. There was less emphasis on a rights and more on an efficiency approach to gender equality. The World Bank set the tone in its 2006 action plan – 'Gender equality is smart economics' – financed by bilateral agencies. Some gender advisers believed that progress in incorporating gender equality into mainstream official development assistance depended on representing women as 'economic agents', a return to the WID equity approach. Some agencies turned back the clock even further. In its 2011–2015 business plan, the British development ministry reverted to language last used in the early 1980s

– 'empower and educate girls, recognize the role of women in development and help to ensure that healthy mothers can raise strong children'.[10]

A global multi-stakeholder partnership

So far this chapter has focused on bilateral and multilateral agencies, but the institutional web of development cooperation involves many more actors. The current economic crisis and the austerity measures in many OECD countries have accelerated the transition from the post-World War Two global political settlement (which included the establishment of the OECD, the United Nations and the international finance institutions) towards an emergent future in which formerly aid-recipient countries became powerful economic and political players. The DAC's response has been to invite these and other actors into the erstwhile trilateral relationship of OECD government donors, recipient-country governments and multilaterals. I look briefly at how three of these institutional actors have approached gender equality in the context of the outcome of the Busan high-level conference on aid.

INGOs

International non-governmental organizations (INGOs), usually based in OECD countries, are major funders of development programmes in their own right, as well as beneficiaries of government grants. Their gender equality units and specialist staff seek to ensure that gender is 'mainstreamed' in their organizations' programmes and policy advocacy on development issues. At the start of the century some INGOs, such as bilateral agencies, manifested a declining interest in gender. As with the bilaterals, there has since been a revival. INGOs with a longstanding commitment to women's rights, such as Oxfam and Action Aid, hired more staff, while others such as CARE and Plan began to develop and implement more comprehensive approaches. At the same time, more INGOs have opened offices in aid-recipient countries, leading to accusations of them crowding out national women's organizations in the competition for donor funding.

INGOs often collaborate to advocate for policy change by governments and multilaterals, including on gender equality matters. In the UK, for example, the Gender and Development Network lobbies the British development ministry and also participates in European and international advocacy networks such as AWID, the international women's rights alliance, as well as broader alliances such as Better Aid, a global civil society consortium. Directly or indirectly, the gender advocacy of these networks is supported by governments, such as the Swiss, Swedish, Dutch and Norwegian, which believe that strong civil society coalitions are vital for promoting gender equality in international development cooperation. However, global civil society networks working on other issues, such as economic justice, are still far from identifying gender as a priority concern.

Private-sector foundations

'Philanthrocapitalism', typified by the large amounts of money spent by the Gates and other new foundations, sees development as a technical challenge best met through employing business management approaches for achieving improvements in health, agricultural production, etc. The Gates Foundation, in response to criticisms about its failure to monitor impact, recruited in April 2010 a gender specialist as a senior advisor. That same year UN ECOSOC hosted a forum to discuss 'the role, progress and difficulties of engaging philanthropic organizations to promote gender equality and women's empowerment'. Some smaller foundations have chosen women's empowerment and/or rights as a central theme, including the Nike Foundation, which has entered into partnership with the UK development ministry and influenced its business approach to gender equality. Multilateral development banks have also established partnerships with private foundations and encouraged corporate social responsibility to include gender issues. In 2009 the Inter-American Development Bank set up a gender and diversity fund, which brings together the various gender equality trust funds it receives from bilateral donors but also invites private-sector contributions.

Rising powers

Middle-income countries that were former aid recipients are now practicing South–South cooperation. The focus on supporting economic development is welcomed by low-income country governments that felt that the Millennium Development Goals had ignored growth as a central development objective. Only a small proportion of South–South cooperation is devoted to social development including gender equality. It appears that even in countries such as Brazil, where domestically considerable progress has been made, gender equality has not yet been integrated into development cooperation approaches. So far, South–South cooperation avoids posing human rights conditions on recipient governments.

The global policy environment in 2013

In December 2011 the DAC organized a conference, hosted by South Korea in Busan, which aimed to bring the rising powers into a new Global Development Partnership. The outcome reflected a consensus (only civil-society participants objected) that development cooperation should prioritize economic growth. Gender equality secured two paragraphs in the outcome document – seen as a triumph by gender advisers in aid agencies – and US secretary of state Hillary Clinton spoke at a special session where the USA and Korea circulated a draft Busan Joint Action Plan on Gender. Better Aid, the civil society consortium, refused to endorse the plan because of the absence of a rights focus, and arguing that women's empowerment requires fundamental shifts in social, political and economic structures.

Nowadays many bilateral and multilateral staff committed to rights-based approaches have difficulty influencing their organizations' policy agendas, and it is largely global civil society networks such as Better Aid who maintain a rights perspective in multi-stakeholder dialogues. But the future is unclear; much of the financing for such global networks originates from the efforts of these same bilateral and multilateral staff who may find it harder to access the resources for funding global advocacy. Meanwhile civil society actors from rising power countries are distancing themselves from INGOs that have made a major contribution to financing Southern civil society participation in global development spaces. And those from the low-income South who continue to accept Northern support for global advocacy run the risk of being judged in their own countries as foreign agents promoting Western views on development.

Another section of global civil society seeks to undermine international agreements on gender equality. The neo-conservative administration in the United States from 2000 to 2008 and its alliance with fundamental religious movements led to feminists struggling to maintain hard-won gains (Sen 2005). Islamist governments in the Middle East are also contributing to a global policy environment pushing back against women's rights. Feminists were appalled about the absence of any reference to such rights in the Rio Plus 20 outcome document in 2012. That same year the meeting of the UN Commission on the Status of Women ended in disarray because of disagreement over reproductive rights, and at the 2013 meeting Iran, Russia and the Vatican led a group of countries pushing back on Beijing commitments to tackle violence against women and girls. An earlier suggestion from UN Women to organize a Beijing Plus 20 conference has been shelved for fear that agreements made in 1995 would be rolled back at any major inter-governmental gathering on gender equality.

Conclusion: three challenges and an opportunity

Contradictions have marked the half-century of feminist engagement with the global actors and institutions of international development cooperation. Its chequered history of promoting gender equality reflects the broader challenges facing the global women's movement described as a 'perplexing equation of progress and backlash' (Sandler and Rao 2012). Findings from the Pathways of Women's Empowerment programme indicate that women's organizing – at the grassroots, nationally and globally – has been central to achieving greater gender equality. Similarly, it was women's movements in aid-giving countries that demanded gender equality be an objective of their official aid programmes and influenced the policies and programmes of development NGOs. In turn, donor success in securing gender equality objectives relied heavily on the organizing capacity of recipient-country women's organizations to work at societal change. Effective gender advisers in aid agency country offices developed long-term trust relationships with the local women's movement, supporting processes of locally-led change rather than imposing their own world view or strategic objectives (Mukhopdahay and Eyben 2011).

However, Goetz and Hassim (2003) argue that the nature of civil society and the status, capacity and approach of gender equality advocacy within it are only one of three factors for securing policy change. The political system and political parties, on the one hand, and the nature and power of the state, including its bureaucratic machinery, are also key factors in the ability of gender advocates and donor agencies in aid-recipient countries to achieve change. There is yet another factor, highlighted in this chapter – ideas and the policy practices associated with them. With respect to ideas, three challenges and one opportunity present themselves with respect to the continuation of development cooperation support for the socially transformative Beijing agenda.

The first challenge is internal to the aid sector and concerns the current dominant framing of international aid as 'technical' best-practice interventions associated with the growing emphasis on delivering measurable results (Eyben 2013). This approach exacerbates aid's tendency to see people as subjects requiring treatment rather than as citizens with political voice. It forecloses analysis and debate about the structural causes and consequences of inequity and how these should be tackled. Arguably in response to taxpayers' lack of enthusiasm for aid programmes during a time of domestic financial austerity, approaches securing quick, easily measurable results are favoured over long-term support to locally-generated complex processes of social change. Widespread reductions in aid agencies' staff numbers also make it harder to develop the long-term, facilitative relationships required for supporting such processes.

The second challenge – the conservative backlash – is a direct consequence of the success of the global women's movement. Opponents of gender equality have made common cause across religious and other divides to push back against the achievements of the past 50 years (Sen 2005). Women's rights activists surveyed by AWID have identified this push back as absolutist and intolerant: patriarchal, anti-women, anti-human rights, often violent, opposing democratic politics and the vision of an egalitarian society.

The third challenge concerns the current consensus between old and new development actors about the centrality of economic growth. Whereas traditional aid-giving countries (other than the USA during the Bush administration) have continued to support reproductive and sexual rights and the struggle to reduce gender-based violence, as Sen (2005) argues their progressive position on these issues has been contradicted by their support to neoliberal policies that sustain and reproduce gender inequalities. In today's multi-polar world, the growth consensus – and the associated 'gender equality is smart economics' – among new and old development-cooperation actors has made the challenge even bigger for feminists advocating a transformative approach to gender equality, one that tackles structural issues such as care, which remain largely invisible in economic policy.

Nevertheless, the changing global landscape of development cooperation also provides an opportunity to be seized. Until now, women's rights movements in countries such as India, Brazil and South Africa have focused on domestic inequities, but as these countries become more important global players we may anticipate a growing interest among civil society to influence their countries' cooperation policies in support of the realization of rights worldwide.

The global landscape has changed radically since the international women's decade that started in 1975; international aid is less significant for many countries that previously depended heavily upon it. At its best, when feminists from South and North – activists and bureaucrats – learnt to work together, development cooperation facilitated real and lasting social change for greater gender equality.

Notes

1   The material for this chapter derives from a review of the secondary literature (particularly for the early years), from the author's own experience in the UK aid ministry and the DAC network (1987–2000) and from research undertaken as part of the Pathways of Women's Empowerment Research Program (2006–2011).
2   See De Haan (2009), Chapter 4 provides a good overall introduction to the history of aid.
3   The Declaration for the Establishment of a New International Economic Order, adopted by the United Nations General Assembly in 1974, covered a wide range of trade, financial, commodity and debt-related issues between industrial and developing countries, focusing on the restructuring of the world's economy to permit greater participation by and benefits to developing countries.
4   Re-named GenderNet in 2003. See www.oecd.org/dac/gender-development.
5   The author.
6   www.un.org/womenwatch/daw/beijing/platform/plat1.htm.
7   www.oecd.org/dac/2508761.pdf.
8   Where I headed the DFID office between 2000 and 2002.
9   www.awid.org.
10 www.publications.parliament.uk/pa/ld201012/ldselect/ldeconaf/278/27809.htm.

References

Baden, S. and Goetz, A.M. (1997) 'Who needs [sex] when you can have [gender]? Conflicting discourses on gender at Beijing', *Feminist Review* 56: 3–25.

Çağatayy, N., Grown, C. and Santiago, A. (1986) 'The Nairobi Women's Conference: toward a global feminism?', *Feminist Studies* 12(2): 401–412.

Cornwall, A., Harrison, E. and Whitehead, A. (eds.) (2007) *Feminisms in Development: Contradictions, Contestations and Challenges*, London: Zed Books.

De Haan, A. (2009) *How the Aid Industry Works: An Introduction to International Development*, West Hartford, CT: Kumarian Press.

Eyben, R. (2013) 'Uncovering the politics of evidence and results', www.bigpushforward.net.

Eyben, R. and Turquet, L (eds.) (2013) *Feminists in Development Organisations: Change from the Margins*, Rugby: Practical Action Publishing.

Fraser, A. (2004) 'Seizing opportunities: USAID, WID and CEDAW', in A. Fraser and I. Tinker (eds.), *Developing Power: How Women Transformed International Development*, New York: Feminist Press, pp. 164–175.

Fraser, A. and Tinker, I. (eds.) (2004) *Developing Power: How Women Transformed International Development*, New York: Feminist Press.

Goetz, A.M. and Hassim, S. (eds.) (2003) *No Shortcuts to Power: African Women in Politics and Policy Making*, London: Zed Books.

Jain, D. (2005) *Women, Development, and the UN: A Sixty-Year Quest for Equality and Justice*, Indiana: Indiana University Press.

Mohanty, C.T. (1988) 'Under Western eyes: feminist scholarship and colonial discourses', *Feminist Review* 30: 61–88.

Mukhopdahay, M. and Eyben, R. (2011) *Rights and Resources. The Effects of External. Financing on Organising for Women's Rights*, Brighton: Institute of Development Studies.

Rathgeber, E.M. (1990) 'WID, WAD, GAD: trends in research and practice', *The Journal of Developing Areas* 24: 489–502.

Sandler, J. and Rao, A. (2012) 'The elephant in the room and the dragons at the gate: strategising for gender equality in the 21st century', *Gender & Development* 20(3): 547–562.

Sen, G. (2005) *Neolibs, Neocons and Gender Justice: Lessons from Global Negotiations*, Geneva: United Nations Research Institute for Social Development.

Win, E. (2007) 'Not very poor, powerless or pregnant: the African woman forgotten by development', in A. Cornwall, E. Harrison and A. Whitehead (eds.), *Feminisms in Development: Contradictions, Contestations and Challenges*, London: Zed Books, pp. 79–85.

54

GENDER EQUALITY, WOMEN'S EMPOWERMENT AND THE UN

What is it all about?

Patricia Holden

Introduction

This chapter attempts to provide readers with an insider's view of some of the social and political complexities that both promote and impede work on gender equality and women's empowerment in the United Nations (UN) system. It also shows how some of these factors were involved in the setting up in 2011 of UN Women, the new UN gender organization.

Methodology

The information used in this chapter draws upon my experience of working with different parts of the UN system over a period of some 30 years including 20 years as a senior social development adviser with the UK Department for International Development (DFID). For part of this time I was based in the UK Mission to the UN (UKMIS) in New York performing the roles of both diplomat and international development specialist as well as supporting the work of various UN development organizations such as UNIFEM, UNICEF, UNDP, UNFPA and the social policy section of UNDESA (UN Department of Economic and Social Affairs). From 2001 to 2003 I was seconded by DFID to work as an adviser to the ILO in Geneva. In 2009 I undertook a consultancy for DFID to look at ways in which a new UN entity for women could further contribute towards improving the lives and livelihoods of women in developing countries (Holden and Earle 2009). I have, therefore, gained some understanding and insight into the work of the UN in a variety of different contexts. As I happen to be an anthropologist by background, this chapter also attempts to consider the UN as an ethnographic field, which has its own esoteric language, symbolic systems and collective identity.

Brief overview and recent developments

The UN has a long history of working on women's rights and gender equality going back to the founding of the Commission on the Status of Women (CSW) in 1946. Recognition of the rights of women was enshrined in the UN Charter of 1945 when it referred to 'the equal rights of men and women'. Work on 'women's rights' and later on 'gender equality' has been ongoing in the UN since its inception. The terminology 'women's rights' has now been largely replaced by that of gender equality and women's empowerment (GEWE), which also includes women's human rights. I shall use the acronym GEWE throughout this chapter.

July 2010 was a significant moment for GEWE as the UN General Assembly created a new UN body called UN Women. This new body merged four previously distinct parts of UN system: the Division for the Advancement of Women (DAW); the United Nations Development Fund (UNIFEM); the Office of the Special Adviser on Gender Issues and the Advancement of Women (OSAGI); and the International Research and Training Institute for the Advancement of Women (INSTRAW). The CSW remains outside this entity and continues to carry out its normative function.

Merging these various entities was particularly challenging because they had all been founded at different times with differing mandates and governance structures. It was also significant because the setting up of any new UN body is usually strongly opposed by member states. As Williams (1995) notes in a guide to the UN:

> For fifty years the UN agencies and programs have been growing in a haphazard way, with little coordination and lots of overlap. There are around three dozen separate agencies and programs – some of which exist just because their staff would hate to be jobless or their host country likes the dollars and prestige of having a UN office . . . It is well-known that is it is almost impossible or unheard of to close a UN body once established.

Box 54.1 The gender entities merged into UN Women

THE DIVISION FOR THE ADVANCEMENT OF WOMEN (DAW) was created in 1946 to provide administrative support to the Commission on the Status of Women (CSW). It has played an important role in organizing global conferences, and undertaking research and policy advocacy including the *World Survey of Women and Development* report produced every five years.

UNIFEM was founded in 1986 after the Mexico Conference as a voluntary fund. It was set up to fund projects for poor women in the poorest countries. In addition it was always intended that it should have a gender mainstreaming role across the UN system. It was primarily an operational body carrying out programmes and projects and had a strong constituency amongst women's grassroots organizations in developing countries.

INSTRAW was established in 1976 and is based in the Dominican Republic. It has a mandate to carry out research and training but has generally been poorly resourced. It is one of the few UN agencies based in a Southern country.

OSAGI was created in 1997 to 'promote, strengthen and implement' the Beijing Declaration and PFA. It was also was established to provide leadership and an 'ombudsperson' type role for gender in the UN system as a whole. However, this post was never at a sufficiently high level to exercise real influence – and was not located in the secretary-general's office as had been originally requested.

At the time of writing it is still difficult to tell how effective this new body is proving to be in bringing about real and significant change globally to the lives and livelihoods of women. Nevertheless, the creation of UN Women is seen as a significant achievement in the history of the UN and was the result of concerted lobbying efforts by the global women's movement.

The UN as a political and social space

All efforts to bring about changes in GEWE in the UN are inevitably deeply rooted in the ways in which the UN functions. Its ways of working are embedded in politics and in its elitist, hierarchical and patriarchal structures.

In this section I attempt to show that many of the ways in which the UN operates are 'hidden' and those seeking to promote and implement GEWE have had to become very adept at ' understanding and working' the UN system.

It is impossible to adequately describe how the UN system functions in a few brief paragraphs. It has headquarters offices in New York, Geneva, Vienna and Nairobi. There are UN organizations based in Rome and Paris. It has country offices in numerous countries; and subregional and regional offices in different parts of the world.

The UN has a multiplicity of functions. A key distinction is made in UN terminology between its normative and operational work.

> Normative work in the United Nations is the support to the development of norms and standards in conventions, declarations, regulatory frameworks, agreements, guidelines, codes of practice and other standard setting instruments at global, regional and national level . . . Normative work also includes support to the implementation of these instruments at the policy level.
>
> *(United Nations Evaluation Group)*

Operational work refers to the international development assistance, humanitarian and peacekeeping work that the UN routinely provides.

It is also important to understand the distinction that is made between the funds and programmes (UNDP, UNICEF, UNFPA and UNIFEM) and the specialized agencies (ILO, WHO, FAO, etc.). The funds and programmes were set up originally as temporary arrangements in the expectation that a time would come when they would no longer be needed. They all deal solely with countries of the South. The specialized agencies are global in coverage and play a key role in monitoring global standards as well as implementing projects and programmes.

In this paper I shall discuss GEWE in both the normative and operational work of the UN.

Normative work: the UN headquarters building in New York

I begin by describing the work that goes on in the main UN HQ building in New York (also a major New York tourist attraction), situated on First Avenue by the East River. This is the main location for the UN's normative work and it is the home of the intergovernmental bodies that come under the auspices of the General Assembly (GA). It is here that the UN acts as a kind of global parliament, with all the decisions that are made subject to agreement or otherwise by the multiplicity of countries that make up the UN.

When I first began attending UN meetings, in common with the majority of people who are unfamiliar with its workings, I often found them long and boring. I admired the patience of those who attended them daily and were often obliged to sit there until the early hours of the morning. There seemed little point to processes that involved haggling endlessly over small details of text or that locked countries in power struggles and intransigent positions. (Although I had been a theology student at one time and I should have realized that the length of time that the UN can take to come to an agreement was nothing compared to the length the early Church took to negotiate, for example, the Nicene Creed!) However, once I had begun to understand what was going on and to engage in the 'game' that was being played, I soon realized 'that beyond the event horizon in that gray hole on the East River all sorts of things were happening ' (Williams 1995).

My job in New York was concerned with influencing people. I quickly learnt that if I wanted to bump into the people I needed to meet I would have to leave my desk and ensure that I went regularly to the crossing at First Avenue where, sooner or later, most people of significance in the UN system would need to cross, either going back to their missions or homes; or to their desks in the UNDP, UNICEF, UNIFEM, UNFPA, DESA buildings, all situated on the other side of the road.

Box 54.2 The six committees of the UN

The First Committee (the security council) deals with disarmament.

The Second Committee deals with economic and financial matters.

The Third Committee deals with social, humanitarian and cultural matters.

The Fourth Committee deals with decolonization.

The Fifth Committee deals with administrative and budgetary affairs.

The Sixth Committee deals with legal matters.

(Williams 1995)

The committees are mainly normative bodies although they also oversee or mandate some of the operational work of the UN. The main participants in these committees are the accredited country delegates. They are usually diplomats with the titles, in order of seniority, of ambassadors, deputy ambassadors, counsellors and first, second and third secretaries. Officials from country governments also routinely attend. Observers from other UN bodies and from civil society may also have the right to be present. However, there are strict rules about who can make interventions in meetings. At the top of the ranks is the Security Council, the most prestigious committee, which is attended primarily by ambassadors.

The UN diplomats are still predominantly male especially at the senior levels. However, it is important to note that the Third Committee, which is the main committee for dealing with GEWE and rights, is frequently serviced by third secretaries. These are often young women diplomats on their first postings who have less experience of the more complex workings of the UN and hence less ability to influence their own colleagues who service other committees; and generally carry less weight all round within the system.

What is really happening?

The UN is primarily a place where information is collected and traded both inside and outside the main committee rooms. It is an arena where deals are constantly done. It is also a place of 'secrecy' and reputedly of espionage. Official records exist of UN committee agreements and conclusions but there are no official records of the endless discussions that take place both formally and informally behind closed doors. While some delegates sit behind their country nameplates and debate the wording of texts, others can be seen working the room and going into huddles with 'friends' and 'foes' in order to achieve the desired result. The cafeteria area of the UN building and the cafés in the surrounding areas outside are also crucial informal spaces for striking the most difficult deals. (I assume, at the time of writing, that electronic social networks may have now changed the way information flows to the outside world.)

The UN is also a 'theatre'. It is a place where the diplomats and delegates take on different roles. They are required to cajole, deceive and disguise what they are really thinking and to assume postures of pleasure or annoyance depending on the situation. The script is quite often pre-written and negotiations often seem to simply act out what has long been understood by all parties. Delegates do most of the work and prepare the scene for the lead speeches of the 'stars' – ambassadors, ministers, prime ministers and presidents.

The diplomatic residencies and restaurants of New York provide other stages for performance and exchange of information. It always surprised me that some delegates would negotiate brutally and viciously (to the extent of shedding tears) and then speak amicably and warmly to each other at the constant round of drinks parties, dinners and lunches. It is hardly surprising that many families of diplomats feel constant frustration over the boundaries between official business and the domestic space. It can also have its humorous side. It was well-known that negotiations in one of the committees, no matter how acrimonious, would always reach agreement by 11pm so that the chair could catch the last train to the New York suburb where he lived.

Outsiders can find it puzzling that the delegates and diplomats do not necessarily act as the mouthpiece of policies that emanate from their capital cities and that there is often a disconnect between different ministries within countries with regard to agreed policy. Many diplomats appear to have been left to their own devices. And some from smaller, less developed or resource-poor countries may take no instructions whatsoever from their capitals. For example, people have been surprised to have visited a country and to have understood that a particular policy may be espoused by a Ministry of Women's Affairs in that country and yet a completely opposite policy may be being put forward in the UN by the diplomat representing that country.

There is a shared language of code words for the purposes of negotiating texts, which can mean nothing to the newcomer to the UN and it can cause surprise that an apparently innocuous word can cause many hours of negotiation.

The business of the UN committees largely revolves around the main political divides of the North and the South; between the G77 and China; and the EU and JUSCANZ countries (Japan, United States, Canada, Australia and New Zealand).There are, in addition, many regional sub-blocks operating within those groups. Sometimes sub-blocks will form unexpected alliances in order to achieve certain outcomes.

Prior to the negotiations held in the main committees, each political block carries out its own closed sessions. So finding out what the other block or blocks are likely to come up with becomes a major preoccupation of all groups, with information being leaked or prized out of the 'opposition'.

Negotiations start with all parties in the committee rooms in what are called 'plenary' sessions. After this increasingly smaller groups referred to as 'formals' and 'informals' are set up to tease out the tricky issues in the text. There are clearly power relationships between Northern and Southern blocks. For example, poorer countries may be staffed by only a handful of people and they are thus unable to cover all of the smaller formal and informals, hence reducing their influencing power. Countries that are well-staffed with people and expertise can produce what are called 'non-papers', which are think pieces usually distributed within a particular block that set out various arguments for discussion. These are not for negotiation but can be quite influential in the process of agreeing a position. Getting hold of these can be hot currency among NGOs. The eagerness, with which pieces of paper – often totally incomprehensible to anyone outside the UN system – are coveted, is probably unparalleled anywhere else in the world.

Inside the UN building there is a spatial divide between the diplomats and other officials who occupy country seats. Representatives from the funds and programmes, specialized agencies and other UN bodies are allocated designated seats, usually at the back of the committee rooms. However, they generally have limited rights to intervene in negotiations, or to attend the closed sessions of the UN.

NGOs have access on a controlled basis and generally have no right to speak except in specially agreed circumstances. They have to obtain special accreditation in order to attend meetings. Most NGOs do their lobbying and networking in the UN cafeterias.

Many NGO representatives have become 'old hands' at attending UN meetings but even they can feel a constant sense of frustration at the lack of transparency around UN business and the difficulties they experience in trying to influence proceedings and outcome statements and resolutions. For many countries (even those with a strong record of openness) this can be a deliberate tactic for managing the NGO community. The NGO community can, of course, be very varied in its objectives, ways of working and ideologies. This has been very apparent over the past few years with the constant battles acted out in the UN to maintain the Beijing language on 'reproductive health'. However, the alliance between the NGOs who regularly attend GEWE meetings and governments can also be very productive.

There is also a professional divide; between the diplomats and other specialists. For example, many people employed by UN agencies are international development specialists and they may have very different agendas from those of the politically driven country delegates.

The convention is that countries should not speak outside the agreed position of their 'block'. However, there are always 'rogue' countries that regularly raise their flag to make an intervention and for those NGOs who want to influence proceedings these countries can become important allies in shifting what might appear to be intransigent positions.

Normative work on GEWE: the Commission on the Status of Women (CSW)

The UN's normative work on GEWE is mainly dealt with through the work of the CSW. It was founded as a mechanism to promote, report on and monitor issues relating to the political, economic, civil, social and educational rights of women. It meets annually to discuss previously agreed agenda items and produces agreed conclusions. It has not been incorporated into UN Women but continues to have a monitoring and oversight role. Since 1995, the work of CSW has focused largely on follow up and implementation of the Beijing Platform for Action.

In contrast with many other UN bodies, the delegates to the CSW are predominantly female and my own memories of the CSW are those of seeing the UN space filled with women – both official delegates and civil society – dressed in brightly coloured clothing representing many different parts of the world, laughing and talking, networking, and forming very long queues for the comparatively small number of female toilets. All this in contrast with the sombre, grey-suited male atmosphere that generally characterizes the UN.

The CSW helped convene the four world conferences on women held between 1975 and 1995. Mexico City in 1975 marked the beginning of a new global dialogue on discrimination against women. Copenhagen in 1980 reviewed progress and set out new goals. Nairobi in

1985 was significant because of the large attendance of NGOs (some 15,000 at a parallel conference) and it marked a new coalition between governments and civil society. Beijing in 1995 also saw a large attendance of 17,000 NGOS and was significant because it moved the discourse away from a focus on women to the importance of changing the structure of society and relationships between men and women if real change was to be achieved. It also affirmed that women's rights were human rights and that gender equality mattered to everyone. The Beijing Platform for Action set out 12 critical areas of concern for implementation by governments. However, the political context in which the UN operates has meant that attempts to 'roll back' the gains achieved at Beijing systematically continue.

An earlier major achievement for CSW was the adoption by the UN General Assembly in 1979 of the Convention on the Elimination of All Forms of Discrimination Against Women (CEDAW). This requires countries who are signatories to report regularly to the UN on the implementation of the Convention. The optional protocol to CEDAW allows individual women to take complaints to the CEDAW committee. This is extremely important for women in countries with poor human rights records and mechanisms.

There have also been important achievements for women that have been produced in committees outside those UN bodies specifically designated to deal with gender equality and women's empowerment. Most notably, the Security Council passed resolutions 1325 in 2000 and 1820 in 2008. These are important and comprehensive commitments to means for eliminating violence against women in conflict and recognizing their role in peace-building. Both resolutions resulted from concerted lobbying by countries and by representatives from civil society. Again implementation of these resolutions by member states remains weak.

Why was there a need for a new UN body for GEWE?

Many reasons were put forward over the years as to why the UN gender entities needed 'reforming'. These included overlapping mandates, competition for scarce financial resources and lack of real leadership. The latter was generally attributed to the fact that the posts held by directors of these entities were not at a sufficiently high level to carry weight at the most senior levels of the UN.

The creation of UN Women was rooted in the reform process in the UN that began in the late 1990s. The reforms focused primarily on the UN's operational work in countries. The UN has a wide range of agencies based mainly in developing countries. Any one country may house the offices of UNDP, UNICEF, UNFPA, UN Women, FAO, WHO, ILO, WFP, UNHCR etc. These agencies are generally divided into those dealing with routine development work, and those dealing with relief and humanitarian assistance, under the overall leadership of resident and humanitarian coordinators. The reforms were based on the idea that the UN family in country should have 'one programme, one budgetary framework, one leader and one office'.

A new instrument called UNDAF (United Nations Development Assistance Framework) was introduced as a means for better coordination of the various agencies working at the country level. The overall objective was to cut down competition for resources and to reduce duplication and overlap. The introduction of the UNDAF process did assist in bringing greater focus to GEWE in UN country programmes as gender analysis is considered to be a very important part of the country analyses that inform the preparation of the UNDAF.

UNIFEM

A key issue in the setting up of UN Women was the issue of what role UNIFEM would have (see Sandler 2013 and Rao 2013 for excellent accounts of the politics involved in this). It was the only entity that had been set up originally to have an operational role and to implement projects and programmes in developing countries. It was also meant to have a 'catalytic and mainstreaming role' exerting influence across the whole UN system. In effect, its position and funding base were never conducive to doing this.

UNIFEM's funding base was always extremely small compared with other funds and programmes. Its role was also subject to UNDP's rules and regulations as it was technically part of UNDP. By 2009 it had evolved from a structure that was largely based on regional offices to one in which there was a presence in many UN country offices. However, the posts of director at UNIFEM HQ and at country level were not senior enough to provide them with access to the key decision-making bodies in the UN system.

In many countries, the annual UNIFEM country allocation was often spent almost entirely on maintaining the office and paying core staff. This resulted in a situation in which UNIFEM often had to look to additional voluntary funding from donors who were happy to see it being drawn into implementing projects, while at the same time complaining that it was failing to exert its gender mainstreaming and influencing role across the UN system.

In spite of these difficulties, UNIFEM did manage to take the lead on gender and established itself with governments in country as the lead UN voice on gender. However, in some countries, other agencies such as UNFPA and UNICEF have increasingly taken a lead role on gender. (It should be noted that many UN agencies have a strong focus on gender and have developed their own internal gender strategies.)

There was nothing intrinsically wrong with different agencies leading on gender but it pointed to the persisting need for an overarching strategy on GEWE understood and implemented by all agencies and backed up by leadership at a recognized senior level.

UN Women and the renewed efforts at reform

In the event the creation of UN Women resulted from the renewed UN reform initiative in 2006 when the UN secretary-general (USG) established a high-level panel (HLP) on system-wide coherence to explore ways in which efforts could be renewed to make the UN more effective, efficient, coherent and coordinated. The areas chosen for this were development, humanitarian assistance and the environment.

In the end these three areas proved problematic and a subsequent report produced by the HLP recommended 'the establishment of one dynamic UN entity focused on gender equality and women's empowerment'. This was widely welcomed by many as an important opportunity to promote gender equality and women's empowerment more effectively through coordination and collaboration of UN agencies backed up by increased resources.

Cynically, it could be argued that 'gender' was chosen because it was seen as a 'soft' area where political agreement was more likely to be achieved, rather than as a genuine commitment to GEWE in the UN. Nevertheless the idea quickly received widespread backing and civil society rallied to mobilize support.

Ultimately the process of setting up UN Women was more protracted and difficult than had been expected. There were endless debates about its governance structure and whether CSW should be an integral part of it. Countries used the process as a bargaining chip for other unrelated issues such as membership of the Security Council. The Western donor countries were very anxious to see the setting up of the new body, as evidence of a more effective and efficient UN that would improve the delivery of programmes and policies for women. Some countries of the South saw it as yet another donor driven cost-cutting exercise that would reduce funds being given to recipient countries. Many countries wanted UNIFEM to remain as it was. There were also tensions over types of funding being proposed. A key achievement was the establishment of the post of executive director at under-secretary general level.

Gender mainstreaming

It would be unfair not to point out that there has been a long history of efforts at 'gender mainstreaming' throughout the UN system

So, for example, delegates attending preparatory meetings for global conferences and routine meetings on a wide range of topics often come armed with text on GEWE that they try to get inserted into UN outcome documents. Agreement by the meeting to the insertion of such text can give a great sense of euphoria. Realistically the implementation of such 'text' will always depend on the energy, commitment and resources of member countries; or of those in civil society who make it their job to hold countries to account.

Over the years, UN organizations have tried to 'mainstream' gender through various methods. Most UN bodies have appointed gender advisers who may or may not be in a position to exert influence over programmes and resources. The UN has also mandated that all UN bodies should have gender focal points. This usually means that an official (female or male) is nominated to take on the role of monitoring and advising on gender mainstreaming in their particular agency or organization. Although there are some success stories, this system has not generally been seen as successful. The person nominated to this role is often a woman at a junior level, who may or may not have the required skills and who receives no additional remuneration. Hence the quality of their contribution is very variable and dependent on their skill-level, the time they can allocate to the role, their level of seniority and support they receive from the head of agency.

UN partners for GEWE at the country level

Governments

Developing country governments feel strongly that they have a stake in the UN and like to feel that they are the primary clients for the UN's services in their respective countries. In some countries the UN system is known to be very close to government and goes out of its way not to upset it. This can make it difficult for the UN to exert its monitoring role on adherence to international agreements and conventions on issues related to women's rights.

Women's machineries

Almost all countries will claim to have 'women's machineries' that is departments of government or ministries that are devoted to promoting women's issues and gender equality. Many of these machineries deal with a wide portfolio of responsibilities that go beyond gender (e.g., youth, sports, media, etc.). Women's machineries both in the North and the South have been consistently shown to lack core resources and technical capacity. Representatives from women's machineries usually attend the annual CSW meetings. They will also be responsible for the preparation of periodic reports to CEDAW.

Women's machineries are generally primary partners for the UN in country. They can have a particularly important role in ensuring that governments adhere to the various women's rights and gender conventions and agreements such as the implementation of the Beijing Platform for Action. In many cases they tend to be used as implementing agents by donors and this can distract them from their core tasks and mandates.

Civil society

Another client for the UN's services in country is civil society. However, this has to be handled carefully, as in many countries civil society is resented as an instrument of opposition to governments. Civil society often looks to the UN to lead on the implementation of agreements and conventions such as CEDAW and the Beijing Platform for Action. The UN generally makes an effort to involve civil society in UN-led consultation processes.

Donors

Bilateral donors play an important role in providing funding to the UN in country. However, there are many examples of donors advocating for reform and coordination, while at the same time distorting this by providing funding for 'pet' projects and this can sometimes work against effective coordination of work on GEWE. There are often tensions between donors, the government and the UN when donors try to exert influence over the UN's operations through providing finance for human rights and women's empowerment programmes.

Some concluding thoughts

Can UN Women bring about change to the lives and livelihoods of women globally?

As noted it is too early to make an informed judgement on this. However, the setting up of UN Women has laid the foundation for a number of things. The appointment of its leaders at more senior levels should provide the scope for better advocacy with partners and within the UN system as a whole. It should have improved capacity to provide strong technical advice which the whole UN system in country can call upon.

It can take a stronger lead role in promoting the implementation of international agreements and conventions on GEWE and it can provide greater support to women's machineries to strategize, plan and integrate gender into national plans and to access greater levels of resources. It can work more strategically with civil society instead of simply providing ad hoc funding. However, it seems that after all the negotiations, tears and joy that accompanied the establishment of UN women, many donors have not yet met their promised financial commitments.

Conclusion

There is a mass of recorded history and documentation on the work of the UN in addressing gender equality and women's empowerment both through its normative work and through support to projects and programmes throughout the world. There have been many success stories; and the UN has consistently sought in varying degrees to promote women's rights.

I have tried to show in this chapter that the UN is a sometimes messy and complicated institution with its workings often impenetrable to outsiders. However, in spite of this, advocates for GEWE in the UN have produced many significant achievements and, in spite of difficulties, women have over a long period 'colonized' UN space and made enormous progress in establishing a globally agreed consensus on women's human rights and gender equality (Timothy 2005). The UN has effectively set global standards that have enabled women to advocate for change in countries in the North and South. Many women in the South continue to look to the UN to affirm their rights and to raise awareness of their situation.

But the UN remains a patriarchal and hierarchical, rule-bound institution. Most senior positions in the UN are held by men and collectively it holds deeply entrenched views about the role of gender in its everyday business. The process of negotiations I have described engender a 'macho' style of behaviour that women may not feel entirely comfortable with. This often merely replicates the style of government of the diplomats' home countries. It is arguable that, whatever progress is made with GEWE, little will really change until these institutional structures are changed.

Nevertheless, women as individuals and as part of groups have been struggling since the UN's inception to establish the importance of women's human rights and gender equality as an integral part of its business.

I conclude with a question I have often been asked: is it worth the effort for women (and men) to regularly attend the CSW? To find a cheap ticket to travel to New York every year in March at a time when the weather is often unreliable; to stand in long queues to get UN passes and often to discover that they cannot meet in the main HQ building but have to crowd into smaller meeting rooms in the buildings across the road?

My answer would be yes. Because it is here that networks and friendships are formed, where ideas are exchanged and developed, where women especially can get renewed energy for taking on the intransigence of their own governments. The presence of women themselves in those often dark and gloomy rooms is a constant symbol and reminder that they are integral to international debates, whether they are in the context of the Security Council, the environment or humanitarian assistance.

To the outsider it may all look as if it is just about 'language', a talk shop that has no relation to the real lives of women and the challenges they face. But it is difficult to describe the pain, the euphoria, the tears and the applause that this language can evoke. There are no other fora where the gains in women's human rights can be discussed in the same way by women whose backgrounds are so far removed from each other. It is a place where women have discovered and continue to rediscover their common ground.

References

Holden, P. and Earle, L. (2009) *The UN Delivering for Women at Country Level: Case Studies of UN Practice in Ecuador, Pakistan, Tajikistan, Tanzania and Uganda*, unpublished report for the UK Department for International Development.

Rao, A. (2013) 'Feminist activism in development bureaucracies: shifting strategies and unpredictable results', in R. Eyben and L. Turquet (eds.), *Feminists in Development Organisations: Change from the Margins*, Rugby: Practical Action Publishing, pp. 177–192.

Sandler, J. (2013) 'Re-gendering the United Nations: old challenges and new opportunities', in R. Eyben and L. Turquet (eds.), *Feminists in Development Organisations: Change from the Margins*, Rugby: Practical Action Publishing, pp. 145–163

Timothy, K. (2005) 'Defending diversity, sustaining consensus: NGOs at the Beijing World Conference on Women and beyond',. *Journal of Women, Politics & Policy* 27(1–2): 189–195.

Williams, I. (1995) *The United Nations for Beginners*, New York: Writers and Readers Publishing Inc.

55

BUILDING GENDER INTO VULNERABILITY ANALYSIS

An example using the "Crunch Model"

Vu Minh Hai, Ines Smyth and Anne Coles

Introduction

Disasters are news—visually exciting, extreme events. Images show the physical force of the hazard, the damage done and the victims. Disasters are a crisis for the communities concerned; the result is disruption and damage that go beyond their ability to cope. And, rightly, at the time of the crisis, the focus is on the immediate need for practical assistance. These realities mask a much more complex scenario, the role that people as well as physical events play in contributing to disasters, although this has become much more widely appreciated with understandings of climate change.

In this chapter disaster is defined as "a natural hazard which has consequences in terms of damage, livelihoods/economic disruption, and/or casualties that are too great for the affected area and people to deal with properly on their own" (Wisner et al. 2012), thus omitting conflict situations. We begin with a brief overview of how perceptions of disasters have changed and how, gradually, the importance of adopting a gender perspective in understanding the impact of disasters is being recognized. This section also explains how and why disasters affect men and women differently and analyses women's contribution in the aftermath. In the second part we stress the usefulness of available tools and frameworks to understand vulnerabilities to disasters, and we introduce a "gendered" version of the "Crunch Model" developed by Oxfam. Oxfam suggests detailed revisions to the contents of the original model to make it fully and explicitly gender sensitive, and then illustrates the importance of these changes using the example of a slow onset disaster in Vietnam. The methodology is important for understanding disasters both for policymakers and for practitioners in the field.

The understandings of disasters has changed over the past 30 or so years with a greater emphasis placed on the human element. There has been recognition of the historical, political and socioeconomic causes that have shaped how communities and individuals are affected. This has several implications. Extreme events need to be "factored in," not treated as isolated "acts of God" when planning for development. A more people-centered approach has led to closer involvement of local communities in disaster risk reduction (DRR) and to the recognition that women experience disasters differently from men but are frequently underrepresented in the planning process, despite playing a major role in the aftermath.

Application of the Crunch Model provides an opportunity to reflect on the use of the gendered version to involve women in a practical way in planning DRR in a marginal area liable to marine hazard. It looks particularly at the local community's vulnerabilities, at the difficulties facing men and women in a situation where, in a low lying coastal area, both unsustainable practices and a deteriorating natural environment make this marginal area increasingly liable to disaster. It also provides an example of participatory planning, recognizing the role of local authorities in harnessing the contribution that women can play in reducing community poverty, in making families more resilient in time of hazard, and specifically in giving them a voice to contribute to disaster risk reduction.

Context and understandings

The literature on disasters is huge and embraces different disciplinary approaches. Several recent books are important but there are also valuable collections of material accessible online. Since 2009 there has also been the very useful Gender and Disasters Network. *The Routledge Handbook of Hazards and Disaster Risk* Reduction is particularly helpful for those unfamiliar with the subject as a whole (Wisner et al. 2012) and *Gender Development and Disasters* provides a gender perspective that spans both subjects (Bradshaw 2013). This section touches briefly on the general nature of disasters. It then concentrates specifically on gender and disasters, in particular, on how women are affected, discussing both their vulnerability and their coping abilities.

Disasters have long been classified as rapid or slow in onset; thus volcanic eruptions, earthquakes, coastal storms and tsunami are rapid, whereas drought might be the cumulative result of several seasons of inadequate rainfall, and soil erosion or salinity developing over many years before reaching crisis severity. But disasters are not simply physical events. Some 40 years ago White wrote, "By definition, no natural hazard exists apart from human adjustment to it. It always involves human initiative and choice" (White 1974: 3).

Recognition of the role that people may play in contributing to disasters is vital. While inappropriate cultivation of semi-arid land may be seen as the immediate cause of soil erosion, the underlying cause is the social, political and economic structures or population pressures that may have led farmers to settle in these risky, marginal areas. Livelihood strategies are thus intimately bound up with perceptions of opportunities, with limited opportunities generally being associated with high risk. Appreciation of the underlying historical and socioeconomic factors has come to be seen as fundamental to our understanding of why disasters happen. In the 1990s and subsequently, authors drawing on fresh understandings of poverty have emphasized concepts such as vulnerability and risk and their converse, capacity and resilience. While the people affected by disasters had tended to be treated as an undifferentiated group, now it was recognized that both individuals and communities vary in these respects (Blaikie et al. 1994).

There have been several separate strands in the writing about disasters, reflecting very different organizational cultures. Humanitarian aid practitioners have tended to focus on the emergency itself. Particularly in the past, and motivated by the need to provide speedy relief, Western workers have too often ignored the coping strategies of local communities despite the fact that it is those on the spot who inevitably deal with the immediate consequences of crises. And, unsurprisingly, in the aftermath, women's specific capacities and needs have often gone unrecognized (Enarson 2000; Fordham 2012).

Meanwhile, development practitioners have tended to see disasters as a "problem" or a "setback" to the development process. Despite work on sustainable livelihoods, environmental degradation and vulnerability, the likelihood of disasters occurring is often poorly incorporated into their planning strategies. Those involved in disaster relief and those involved in development, particularly in the developing world, need to work more closely together to reduce the risk of disasters. Disaster risk reduction (DRR)—a comparatively new term covering strategies, policies and practices to reduce risk—is seen as "the conceptual framework of elements considered with the possibilities to minimize vulnerabilities and disaster risks throughout a society, to avoid (prevention) or to limit (mitigation) the adverse impact of hazards, within the broad context of sustainable development" (Bradshaw 2013: 157).

Consideration of gender has been slow to enter the disaster literature and draws on much from gender and development theory. The accounts of the Bangladesh floods of the 1970s were an exception.[1] The higher mortality of women was very evident and widely reiterated (one Western NGO even recommending that affected men should be given aid in cash to acquire new wives). When, from the late 1970s, occasional women-specific articles began to be published, they had a medical perspective. Women's biological vulnerability was repeated in the emergency management manuals of the time. The concept of women as "victims" has persisted; for fundraising purposes humanitarian agencies emphasize the special vulnerability of a frequently undifferentiated group of "womenandchildren" (Fordham 2012).

There is general agreement among practitioners and policymakers that disasters are likely to have a disproportionate impact on women (DARA 2011; Fordham 2012), a consensus that owes much to the persistence of women practitioners, researchers and rights lobbyists. While much of the evidence for this was originally anecdotal and scattered, a statistical study (Neumayer and Pluemper 2007) of events in 141 countries from 1981 to 2002 found that, on average, natural disasters (and their subsequent impact) killed[2] more women than men both during the disaster and in its aftermath. The stronger the disaster, and the lower women's socioeconomic status, the greater was the gender gap in mortality.

The study built on earlier published work to suggest that it is women's socially constructed gender specific vulnerability, built into socioeconomic everyday patterns, that is the main cause of the relatively higher female disaster mortality rates compared to men, although men, as combatants, are likely to suffer more in times of conflict. Biological and physiological differences inevitably affect women's and men's survival chances during the disaster itself but arguably these are less important than is often construed. Depending on the hazard, men's greater physical strength, size and speed may give them advantage but women may survive famine better, except when pregnant or lactating. Rather it is social norms and behavior that more seriously influence outcomes and these reflect women's "normal" inequality in many societies.

The importance of these factors was illustrated by contemporary and later studies of floods in Bangladesh (Cannon 2002). Women drowned because as girls they had not learnt to swim or climb trees and perhaps because their modest clothing hindered flight. More relevant were their responsibilities as carers of children, the elderly and the home itself that hampered their departure. Seclusion (purdah) often made it difficult for them to go to communal shelters or to leave the house without explicit permission from a husband who might be temporarily absent, illustrating these women's lack of autonomy. In contrast there are situations where men, brought up with cultural expectations of masculinity and often seen as greater risk-takers, may engage in dangerous search and rescue activities. Differences in occupation are relevant. In the 2004 tsunami, Oxfam reported that, in some coastal areas of Indonesia, men fishing from boats were able to ride out the storm at sea, while women, at home or waiting by the shore to handle the catch, were swept away. In one Indian earthquake occurring at night, men asleep outside in the heat survived better than women sleeping in the privacy of the house, as poorly constructed buildings collapsed. Thus the type and characteristics of the hazard as well as its place and timing influence who is most affected (Neumayer and Pluemper 2007).

In the aftermath of a disaster, where physical environments are shattered, social services destroyed and families uprooted, women's conventional responsibilities to provide for the immediate needs of dependents inevitably multiply and become more arduous. They bear a heavier burden than men. The female networks through which women may bypass or compensate for discriminatory norms and practices in normal life may be disrupted, while men may be psychologically devastated by being unable to continue their role as providers for their families, which can be reflected in an increase in alcoholism and aggressive behavior. There may be what has been called a "second disaster" for women. Emergency shelters seldom make specific provision for the needs and security of women, and in such a crisis, where community constraints have broken down, violence—including sexual violence against women and girls—can become commonplace. In such circumstances, longer-term consequences for women are reported to include abandonment, an increase in transactional sex and a lowering of the age of marriage (Fordham 2012; Bradshaw 2013).

There are other reasons why women, particularly women on their own, may find recovery particularly difficult. They tend to have fewer resources to fall back on than men. Their access to and control over resources, whether tangible (such as land) or otherwise (such as information and, recognition), are more limited than those of men, earnings are usually lower and the burden of caring is mostly on their shoulders (Enarson 2000). Their exclusion from much decision-making and positions of authority may adversely affect their access to public bodies and benefits, including food, and therefore their ability to recover after a crisis.

While acknowledging women's specific vulnerabilities is essential, so too is recognizing the skill of poor women, like poor men, to adapt to changing environmental hazards and disasters and to have the ability to contribute in the aftermath. Women and men often perceive disasters rather differently and, given their different roles, have different insights and priorities on what may be needed to deal with the hazard. For effective disaster risk reduction, the involvement of the affected communities is critical and incorporating the capabilities and understandings of both women and men is important. Two recent publications highlight women's contributions. These have included risk mapping in Honduras, improving health services in post-tsunami Tamil Nadu and education services in areas of Turkey affected by earthquake, establishing revolving funds and other livelihood boosting activities in Cameroon and in Manila, as well as contributing to food security and safe housing in several places (Groots International and UNDP 2011; ISDR et al. 2007). Sometimes women's initiatives have been replicated in other disaster situations. Women or their NGOs are increasingly represented on local disaster-preparedness committees, although much less frequently at higher organizational levels.

While this evidence is emerging, it is probable that much of women's resilience and knowledge still remains unnoticed and undocumented. As Fordham (2012: 395) writes: "Recent studies of disasters have shown that the suffering of women and girls is often more severe and yet, paradoxically, less visible in reports, policies and scholarship. On the other hand their capacities and capabilities are even less in evidence." Moreover, while disasters may offer opportunities for strategic or transformational changes in gender relations,[3] after a disaster many, both men and women, understandably long for things to "get back to normal."[4] Thus even women who have taken initiatives and leadership roles during and after a disaster may find themselves expected, and even prepared, to resume a subordinate position in home and community. Ironically, Bradshaw (2013) suggests that, even when women's contribution is recognized, this can be counterproductive. For example, an agency may devolve extra responsibility onto women, and the additional tasks are typically just extensions of their caring roles: practicality takes precedence over "gendered rights."

Making analysis of vulnerability more gender-sensitive: the Crunch Model

Oxfam, with its long tradition as an international NGO engaging in both humanitarian and development aid and of adopting a gender perspective, has for some years been working to mitigate the risk of disasters. The emphasis is on strengthening the capacity of communities to cope. Recognizing both women's socioeconomic vulnerability and their key roles in the face of hazards, they have worked to strengthen women's resilience and to make audible their "voice," for example in Nepal (Dhungel and Nath Ojha 2012). Oxfam was therefore well qualified to "engender" the Crunch Model, a framework that sets out to explain the causal progression of vulnerability. It was first developed by Blaikie et al. (1994).

Various tools and methodologies exist to analyze the different vulnerabilities and capacities of given communities. Different prominence may be given to the various causal factors, the physical, the social, the economic and the environmental (Birkman 2006). Models such as these have both theoretical value and practical utility in enabling people from different backgrounds to gain an understanding of the complex processes involved. The Crunch Model is one of the best known conceptual frameworks worldwide. The diagram deals with causal factors, putting "heavy emphasis on the national and global levels" (Birkman 2006: 31). It has been widely used since the 1990s, particularly by practitioners, to understand the less visible factors that contribute to vulnerability and so to identify effective strategies to prepare and reduce the risks of disasters (for which purpose a reverse model is also provided). It serves as a counterweight to the earlier emphasis on the nature of the hazard and seeks to clarify the range of human factors and the changing processes, both underlying and recent, which have led to the vulnerability. The root causes and dynamic pressures leading to increasingly unsafe conditions then interact with the hazard, resulting in disaster risk (see Figure 55.1) Curiously, although Blaikie et al. (1994) had numerous references to women's vulnerability, the model is undifferentiated with respect to gender. The model forms the backbone of Oxfam's participatory capacity and vulnerability analysis (PVCA) guidelines (Turvill and de Dios 2009) and other organizations also rely on it for their approach.

Many development and humanitarian organizations now realize that it is important to incorporate gender awareness in their work of responding to or preparing for disasters, both as a means of being more effective and, for the more progressive, to contribute towards gender equality and women's rights. Yet at least some analysts conclude that, despite increased awareness and progress in policy development, no tangible or sustainable progress has resulted, with the exception of some ad hoc activities (ISDR et al. 2007). Furthermore, there has not been much substantial progress in mobilizing resources for mainstreaming gender perspectives into disaster risk reduction processes. Finally, interventions often struggle to identify useful methodologies to understand the relationship between gender inequality and vulnerability and plan accordingly. This poor understanding of gender in DRR linkages at the policy and practitioner levels is what made rendering the Crunch Model more gender-sensitive a worthwhile undertaking. Practitioners using the Crunch Model to analyze information gathered from local level assessments realized that it would be more effective if gender aspects were also incorporated. The Humanitarian Program staff in the East Asia Region (now Asia) of Oxfam took the initiative and collated their ideas into a gender sensitive Crunch Model, as shown in Figure 55.1.

Figure 55.1 takes the three elements that according to the original model cause people's vulnerability (root causes, dynamic pressures and unsafe conditions) and adds to them the details that explain how men and women are differently vulnerable. For example, considering ideologies that regulate political and economic systems as root causes is important, but introducing

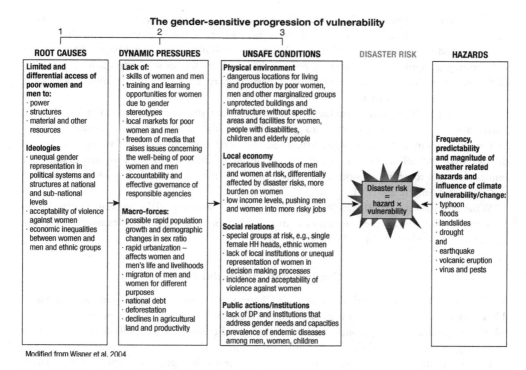

Figure 55.1 Application of the gender-sensitive Crunch Model for participatory capacity and vulnerability analysis in Vietnam

"acceptability of violence against women" among those helps highlight a very specific cause of women's vulnerability that may have remained invisible. This promotes a better understanding of the central fact that men and women experience different and unequal levels and types of vulnerability to disasters and, more importantly, helps plan accordingly.

This section provides a practical example of how the gender-sensitive tool has been used in practice. The gender-sensitive Crunch Model was used in 2012 to collect information on and analyze the vulnerabilities and capacities of men and women in Dai Hoa Loc and Dinh Trung communes, in Binh Dai district, Ben Tre province, which are two of the 15 communes covered by Oxfam's project, "Building resilience of women and men to disaster and climate risks from 2012 to 2017," funded by New Zealand Aid. This project has three objectives: to improve the skills of both poor communities and local authorities to identify climate and disaster risks; to increase the productivity and incomes of farmers through various adaptation options; and to increase access to clean water of poor people, and especially female headed households.

In the project design, only the main strategies on how to achieve each expected result were spelled out, with the specific activities being identified on the basis of the results of a participatory capacity and vulnerability analysis (PCVA) to be carried out in each commune. Oxfam and partners' staff from the provincial, district and commune levels were trained as PCVA facilitators on how to use the tools and methodologies required to collect and analyze information utilizing the gender-sensitive Crunch Model. On the basis of this modified model, questions for the PCVA were revised to ensure that sex and gender disaggregated information was collected and gender differences in vulnerability and capacity discussed and taken into account in planning.

The five-day training program[5] included government officials from the commune, district and provincial levels as well as project partners' staff from the Vietnam Women's Union, Vietnam Red Cross Society (VNRC)—a Vietnamese NGO—and from Oxfam. Most of the staff from government bodies were more familiar with a "top-down" approach to planning and decision-making; therefore utilizing a participatory approach incorporating a gender perspective was new to them.

After the training event, participants undertook a four-day field exercise to "identify how a hazard affects men, women, boys and girls differently; the specific aspects of gender relations and inequalities in their vulnerability; their specific needs, concerns and priorities in reducing disaster risks; as well as to examine the implications of these issues for program design and implementation to promote more effectively gender equality, address poverty and lessen suffering" (Vu and Smyth 2012: 9).

What follows are summaries of the main hazards—and especially saltwater intrusion—identified in the area in question, and the main socioeconomic vulnerabilities found using the gender-sensitive Crunch Model (see Figure 55.1).

Hazards and vulnerability

The two communes in question are prone to a number of natural and human-made hazards. These include hydro-meteorological hazards (e.g., annual tropical storms, typhoons and whirlwinds), saltwater intrusion, river bank erosion and environmental pollution, e.g., hazards caused by excessive use of chemicals in shrimp farming.

In 2006 typhoon Durian caused significant damage to the communes and climate change scenarios indicate that the frequency of such events is likely to increase. However, saltwater intrusion is a more prevalent and damaging hazard with far-reaching consequences already observed in the whole province (Vietfish International 2011). For the purpose of this example, and for the sake of brevity, the gendered analysis that follows focuses on this saltwater intrusion in Dai Hoa Loc commune only, rather than on the full range of hazards and the whole geographical area.

Locally, most people rely on shrimp farming and rice cultivation for their livelihood. Many poor people, however, have no land and often have to seek work as daily laborers. These livelihoods are dependent on natural resources such as land and water and thus are increasingly affected by saltwater intrusion, partly by sea-level rise but also by a dam built in Ba Lai River that is about 7km upstream from the commune. The commune in the past had approximately six months of saltwater and six months of fresh water but now, in one out of the four villages in the commune, the water is saline throughout the year. This has affected both shrimp and rice farming badly.

Due to the potential for high income from shrimp farming, many people had previously converted their rice land into shrimp ponds, using bank loans. However, few farmers were able to make a regular profit and many lost their investments repeatedly, getting into serious debt. In addition, such shrimp farming had received no formal approval from the local authorities.

In recent years, the loss in productivity of shrimp and rice farms has led pond owners to borrow even more heavily from the bank and from high-interest moneylenders (almost all are men), the latter often charging an interest rate of 10 percent per month. This situation has pushed many people further into poverty. According to the commune officials, more than 60 percent of the shrimp farmers are in debt.

Mr. Luong from Binh Hoa 2 village has been raising shrimp for the past ten years. He could earn good money only once when he started but after that he lost his investments more often than he gained. He did not say how much money he owed the bank but said it was a lot. He has had to sell some of his land to repay his debt. To gain further income, his wife went out daily to catch small fish in the irrigation channels and in the rice fields. When he managed to earn VND 40million (nearly US$2,000) from seven months' work for other shrimp pond owners in other areas, he came back and invested in his pond and lost again. When asked if he would change to another livelihood, he said no as he still hoped that his pond would give him a fortune with the next investment to help him repay his debts.

The gendered impacts

Men are involved mainly in shrimp farming activities such as preparing ponds, feeding, guarding and observing the shrimps, switching the oxygenating machines on and off at the right time. Women help also in these activities but are more involved with classifying and selling shrimps. This is in addition to their daily household tasks. Often, both poor men and women looked for opportunities to work for shrimp pond owners for daily wages, but with the increasing prevalence of shrimp farm failures, their chances of finding such work have become reduced.

For rice farmers, the majority of whom are women, this area only allows one crop in the six-month period of fresh water supply, and even this is being affected by saltwater intrusion. This means that of the rice produced, little if any is available for sale; harvests are only partly sufficient even for household consumption and households have to use their income to purchase more rice to satisfy their dietary needs.

Given this situation, the main coping strategy adopted by households is for men to migrate to urban areas or to other shrimp farms out of the commune to look for work as manual laborers. Women, especially after getting married, generally are not able or do not want to go away from their homes, and so stay behind with very limited opportunity for a good income. Many women work for an agent on a daily basis to shell cashew nuts, which is arduous and provides the extremely low income of VND 4,000 (less than US$0.2) per kilogram. On average, a woman can shell a maximum of 3kg of cashew nuts per day. Often women combine this with household work, perhaps spending up to six hours a day undertaking this monotonous and unhealthy work. The nut is surrounded by an acid in the shell, which burns the skin so that gloves need to be worn but usually women cannot afford this protection.

Loan, also from Binh Hoa 2 village, is a 21-year-old woman who could not pass her Grade 12 graduation exam. Her family has 4,000m² of shrimp pond and no land for rice. Due to the unstable income from the shrimp farm, her father has to go fishing daily in the nearby rivers. She has to stay at home to help take care of the household work and together with her mother peels cashew nuts for extra income. Together they can peel 4–5kg of cashew nuts per day, which gives them about VND 20,000 (less than US$1/day for both). Her dream was to go to a nursing school in the city but the fees for two years would cost VND 18million (about US$850), excluding her food and accommodation. With the current economic situation of her family, her dream is not likely to be realized unless there would be some program that support young people like her to find jobs. Loan said that there were many young women like her, who would get married and have a similar life to that of their parents and see their aspirations unrealized for lack of opportunities.

Table 55.1 Initial utilization of the gender-based Crunch Model for the Dai Hoa Loc commune, Binh Dai district in Ben Tre province

Root causes	Dynamic pressure	Unsafe conditions	Hazards
Power • Poor women and men lack access to decision-making bodies/and decision making, e.g., upstream dam construction, upstream pollution. • Limited financial resources are available (e.g., for shrimp farming) and do not allow diversification. • Access to potable water is limited, especially for poor people and those far from pipelines. **Ideologies** • There is unequal gender representation on the Commune, District and Provincial Committees for Flood and Storm Control as well as in the Peoples' Committees at these administrative levels. • Men mostly control assets, are able to access bank loans. • Gender stereotypes are still common: men make the final decisions on important matters (e.g., investment in shrimp farming). Alcohol consumption and domestic violence socially acceptable.	**Lack of** • Skills and training opportunities for both men and women; vocational training for men (only) exists in water pump repair; women can learn catering. • Limited opportunity for both men and women to voice their concerns. • Access to media for men and women is limited, although the plight of shrimp farmers is well documented in the media. **Macro forces** • Men are able to migrate to seek labouring work in the city, leaving women behind to look after rice cultivation, and look after the children/parents. • Decline in agricultural productivity due to saltwater intrusion: shrimp farms particularly badly affected and thus the earnings of men and women as well as their dependence on money lenders. • Saltwater intrusion leads to limited available of water and increase of women's domestic burden.	**Physical environment** • Coastal communes are vulnerable to annual storms, sea surge and saltwater intrusion. **Local economy** • Dependency on rice cultivation: women (who take the burden of responsibility) are more often impacted by poor rice crops due to saline water intrusion. • Women take on additional and arduous tasks, e.g., shelling cashew nuts, damaging their health. • Women's work in rice fields and catching fish in rivers requires them to spend many hours in polluted water, leading to skin and gynaecological problems. **Social relations** • The absence of women in local decision-making bodies at commune and district levels means that women's issues are not considered. • Widespread incidents of domestic and gender-based violence. **Public actions and institutions** • Local Commune Committees for Flood and Storm Control have limited resources (financial, human) to deal with the hazards. • Many women and younger children and men show signs of under nourishment and have limited choice of diet.	• Saltwater intrusion is a major problem; annually brackish water is creeping 5km farther up the local rivers, exacerbated by decreased flow down the Mekong from large upstream dams and severe dry seasons. Salinity of water has risen to twice the proper salinity for shrimp, affecting benefits to farmers despite increases in market price. • River bank erosion. • Tropical storms. • Typhoons. • Sea surges. • Whirlwinds. • Drought. • Rice pests and diseases. • Shrimp diseases.

Saltwater intrusion has affected fresh water for drinking and household consumption. Although there is piped water in the communes, it is only accessible to those who are better-off and living along the main roads. For six months of the rainy season, local people can collect rainwater to use but the remainder of the year they have to buy water from unidentified sources that vendors transport in trucks to sell.

Much of this work is women's responsibility. In addition, the research showed that regular failure of the shrimp farms is leading to increases in domestic and gender-based violence, with men often resorting to drinking readily and cheaply available rice wine, and abusing verbally and physically their wives and other family members.

Gendered vulnerability analysis

Table 55.1 summarizes the findings of the application of the gendered Crunch Model to the local Vietnamese reality.

The contents of Table 55.1, as well some of the descriptions given earlier, provide a picture of the gendered division of labor among local farming households, as well as some of the relevant cultural characteristics. For example, migration is an option open to men but not to women, especially those who are already married. Men and women are actively engaged in production, although with different roles and responsibilities, with women also responsible for reproductive tasks. While both men and women are affected by the three layers of social processes (root causes, dynamic pressures and unsafe conditions) that determine their vulnerability, gender relations appear to make the latter more acute for women, especially through their care responsibilities, limited access to assets and absence from decision-making.

In this project, the use of the gendered Crunch Model allowed Oxfam staff and partners to tease out and analyze the information gathered through the PVCA at community level, with positive results for the project and for the longer-term efforts of communities and officials to fend off the many hazards threatening the area.

The training and the participatory nature of the activities associated with the collection of data, and the analysis of the PVCA results and of the gender-sensitive Crunch Model meant that local people and staff of the local authorities were able to identify several gender-specific disaster risk reduction and climate change adaptation measures that could be further explored. For example, they proposed that local authorities should restrict production to one shrimp and one rice crop per year to reduce overuse of chemicals. They also suggested that better awareness of the danger of pesticide use should be promoted, especially to men involved in rice farming, and that vocational training opportunities should be created for women to overcome the current lack of opportunities, alongside job creation schemes

For the longer term, discussions were undertaken with the Commune People's Committee on the use of the PCVA results. It was agreed that they will be used for the development of commune Committees for Flood and Storm Control (CFSCs) and gender-sensitive social economic development plans. Oxfam and its partners intend to follow this process closely.

Finally, the entire process led the existing planning procedures of the local authorities to be more participatory and thus meet specific needs and capacities of local men and women, and of vulnerable groups, and help ensure the sustainability of the project results.

Conclusions

Climate change and the recent focus on extreme weather related events have emphasized the importance of incorporating the risk of natural hazards into development planning. Disasters are increasingly seen "not as outside development processes, but as linked to them, or even as products of them" (Bradshaw 2013: 21). A concern for and an awareness of the different ways in which women are involved in and experience the consequences of such transformation are also becoming more established, although with a too strong emphasis on women's vulnerability. Given the need for a holistic approach that includes the socioeconomic, it is now even more important that women should be seen not merely as a vulnerable group but as active participants, with the capacity to make a distinct contribution towards the solution. The development and use of frameworks such as the adapted Crunch Model, with gender considerations at their core, will allow for progress in this field, thus contributing to both more effective prevention of and response to disaster and the realization of women's rights.

Notes

1   The only specific reference to women in White's book (1974) was in Anne White's chapter on floods.
2   The measure used for mortality was expectation of life.
3   This is despite the fact that international agencies working with women after disasters may provide leadership training. It seems that this is seldom accompanied by gender sensitization for men.
4   Unfortunately "getting back to normal" has often been the aim of humanitarian aid organizations because failure to adopt a more developmental perspective has sometimes masked the need for changes to reduce future risk.
5   The main facilitator was an experienced trainer from the VNRC, who worked closely with the Oxfam senior technical advisor for building resilience and a project officer from Oxfam, to ensure that the organizational ambition in gender equality was taken well into the process.

References

Birkman, J. (ed.) (2006) *Measuring Vulnerability to Natural Hazards: Towards Disaster Resilient Societies*, Tokyo, New York and Paris: UN University.
Blaikie, P., Cannon, T., and Davis, I. and Wisner, B. (1994) *At Risk: Natural Hazards, People's Vulnerability and Disasters*, London: Routledge.
Bradshaw, S. (2013) *Gender Development and Disasters*, Cheltenham: Edward Elgar.
Cannon, T. (2002) "Gender and climate hazards in Bangladesh," *Gender and Development* 10(2): 45–50.
DARA (2011) *The Humanitarian Response Index: Addressing the Gender Challenge*, Madrid: DARA.
Dhungel, R. and Nath Ojha, R. (2012) "Women's empowerment for disaster risk reduction and emergency response in Nepal," *Gender and Development* 20(2): 309–321.
Enarson, E. (2000) *Gender and Natural Disasters*, Geneva: Recovery and Reconstruction Department, ILO.
Fordham, M. (2012) "Gender, sexuality and disaster," in B. Wisner, J.C. Gaillard and I. Kelman (eds.), *The Routledge Handbook of Hazards and Disaster Risk Reduction*, London and New York: Routledge.
Groots International and UNDP (2011) *Leading Resilient Development: Grassroots Women's Priorities, Practices and Innovations*, New York: UNDP.
ISDR, UNDP and IUCN (2007) *Making Disaster Risk Reduction Gender Sensitive*, Geneva, United Nations.
Neumayer, E. and Pluemper, T. (2007) "The gendered nature of natural disasters: the impact of catastrophic events on the gender gap in life expectancy, 1981–2002," *Annals of the Association of American Geographers* 97: 551–566.
Turvill, E. and de Dios, H. (2009) *Participatory Capacity and Vulnerability Analysis Training Pack*, Oxford: Oxfam GB.
Twigg, J. (2004) *Disaster Risk Reduction: Mitigation and Preparedness in Development and Emergency Programming*, London: Overseas Development Institute.
Vietfish International (2011), *Southern Provinces: Saltwater Intrusion and Pollution Hurting Shrimp Farmers*, http://vietfish.org/20110420014956825p49c70t80/southern-provinces-saltwater-instrusion-and-pollution-hurting-shrimp-farmers.htm.
Vu, M.H. and Smyth, I. (2012) *Crunch Model: Guidelines for a Gendered Approach*, Oxford: Oxfam.

White, G. (ed.) (1974) *Natural Hazards: Local, National, Global,* London and Toronto: Oxford University Press.

Wisner, B., Gaillard, J.C. and Kelman, I. (eds.) (2012) *The Routledge Handbook of Hazards and Disaster Risk Reduction,* London and New York: Routledge.

56

DEVELOPMENT PEOPLE

How does gender matter?

Anne-Meike Fechter

The bulk of research and policy on 'gender and development' has been concerned with improving the situation of women across the Global South, aiming to achieve equity through gender mainstreaming. The purpose of this chapter is to broaden this paradigm by bringing 'development people' into view. Recognizing that the development process encompasses institutions, policies and aid recipients, but also aid workers as important actors, it is only consequent to consider their lives and working practices too, through a gendered lens.

Offering a complement to more tightly framed understandings of how gender matters in development, my aim here is to explore some avenues of existing as well as future research. This approach takes as its starting point national and international workers, and thus turns the attention to the 'developers' themselves. Such a perspective is part of what has been described as 'ethnography of aid'. Among other issues, this aims to provide ethnographically based accounts of the experiences and practices of individual aid workers, and asks how these shape development processes.

The question how gender matters for development people is thus situated at the intersection of the fields of 'gender and development' and 'ethnographies of aid'. While the former is long-established and extensive, the latter has emerged more recently; furthermore, there is no comprehensive body of research on how gender matters for development workers. While bringing development workers into the analysis has broad scope for future research, here I am addressing three particular themes. The first concerns how gender matters for aid workers as individuals, including the question how femininity and masculinity have been theorized in the context of aid work. The second explores how gender matters in development organizations, specifically with regard to their staff members and internal policies, rather than programming and aid recipients. Finally, I draw attention to a comparatively neglected field, that is the interrelations between gender, sexuality and development people.

Given that efforts at gender mainstreaming have endeavoured to influence development policies and programmes based on ideals of gender equity, the question arises how these efforts relate to development organizations and aid workers themselves. A crucial issue here is consistency: if one of the Millennium Development Goals was to promote gender equity and empower women in developing countries, then this needs to be reflected in the policies and practices of development as a sector, and thus constitutes a legitimate, and perhaps neglected, concern for research and policy.

Gender among development people: individual dimensions

Based on an initial survey of the literature, it appears that gender is not particularly prominent in much of the 'ethnographies of aid'. It is perhaps symptomatic that several recent accounts of development workers, while employing a life-history approach, are hardly attentive to gender dimensions at all. Yarrow's (2011) otherwise informative study of the social and professional lives of Ghanaian NGO workers is a case in point, as is McKinnon's (2011) monograph on international development workers in Thailand. This is also the case for early works, such as Frank (2002), Stirrat (2008) and others, which cast categories such as race, ethnicity and professional status as more relevant than gender for aid workers.

Given this relative lack of interest, it is perhaps unsurprising that the few existing studies focusing on gender to some extent mirror the early preoccupations of the 'gender and development' paradigm, in the sense that being about 'gender' is interpreted as being about 'women'. The most pertinent studies here are therefore concerned with female development workers. A seminal case study of national women development workers is that of Anne-Marie Goetz (2001). This is an empirically rich work, which documents Bangladeshi women employed by a state agency and a national NGO respectively, which implement rural microcredit programmes. It is complemented by Mokbul Morshed Ahmad's study (2002), which discusses both male and female fieldworkers in Bangladesh, and confirms many of Goetz's findings with regard to gender issues. Goetz argues that implications arising from their gender are highly relevant for women workers, in ways that are challenging as well as potentially empowering. She found that the main reasons for taking up this work include economic necessity; the desire to make use of their education; and the perceived status and independence that they gain as a result of their work. This is despite the fact, as both Goetz and Ahmad describe, that these women face specific hindrances and hardships in their daily work as a function of their gender. Many of these obstacles are rooted in traditional views, especially common in rural areas, of the proper place for and comportment of women. For example, the unsupervised mobility especially of unmarried women using bikes or motorcycles in the countryside, in order to visit clients in villages, is often seen as inappropriate and potentially damaging to the women and their families. Similarly, being accommodated near the villages where they work, rather than with their native kin or spouse, means that women are often not able to fulfil what are considered their domestic duties, including childcare. These women therefore often face hostility not only from the villagers who are the clients of their organization, but also from their own families. This includes their husbands, who may partly rely on their income, but are often critical of their work. Due to practical difficulties, such as lack of appropriate eating places or latrines for women, the demands made by their work are described as exhausting and physically dangerous.

In spite of these hardships, Goetz points out that, although many women did not necessarily enter this sector with a particular interest in gender equality, changes in their attitudes would often occur during their employment. For example, over time women espoused an increased commitment to gender-related development aims. That means that despite, or perhaps because of their continued involvement in work that positioned them outside of traditional norms, and cast them as deviant from conventional gendered behaviour, this brought about changes in their own gendered identities (Goetz 2001: 139). As Goetz demonstrates, even though commitment to gender-related values was not initially a driver for becoming an NGO worker, these women's professional aims and personal values eventually intersected, resulting in their increased critical awareness of gender equality issues.

It is useful to relate these insights with those emerging from research on international development workers, as exemplified by the work of Hero (2007) and Cook (2007). Here, the development activities of white, Western women are analysed through a framework of moral commitment. These studies examine the relevance of a helping imperative, and questions of female solidarity of affluent women with women in the Global South. This seems to speak of an assumption that the moral responsibilities and dilemmas related to development work are most appropriately discussed in relation to white female, rather than male development workers. One might argue that this is justified insofar as poor, non-white women can be seen as doubly disadvantaged, on the basis of race and nationality as well as gender; while privileged white women are seen to bear a particular, gender-based responsibility, which has been critically rendered as 'the white woman's burden' (Syed and Ali 2011). The tendency to consider white women's development work in the framework of solidarity and charity, rather than as a professional career, although the two are obviously not mutually exclusive, is also evidenced in de Jong (2010). Her analysis of female employees of a Northern NGO similarly foregrounds issues of values and commitment, a focus which remains implicit rather than being reflected on. A notable exception is the work of Eyben (2010), who examines how feminist-inspired values matter for female aid bureaucrats or 'femocrats', how they motivate them, and how these women are finding ways to translate them into policy and practice. I will return to this issue below, in the context of organizational cultures.

To complement these approaches, I argue that what is needed is a gender-focused analysis of development as a professional sector, such as exists for domestic third sectors and for masculine-oriented professions such as finance; and one that is comparable to other overseas-based occupations such as expatriate management. One initial example in this vein is Roth (2013), who examines how gender matters for the everyday experiences of aid workers. Some of her findings echo insights gained from the corporate sector, such as the need for women to sometimes 'act like a man' in order to achieve their professional aims; others are specific to the aid sector, such as the difficulty of establishing a family in conditions of frequent mobility.

Further, I suggest that future research needs to heed the call for making visible the role of masculinity in development. Even though this call is in the first instance aimed at programmes and recipients, it is similarly valid for aid workers. So far, there appear to be no ethnographies that specifically examine how masculinity and aid interrelate. This is a curious lacuna, reinforcing the status of development work as an area where maleness and masculinity remain pervasive, yet unmarked characteristics. Overall, I argue that future research will need to undertake a more wide-ranging analysis of how gender matters among aid workers. This would include problematizing how femininity and masculinity are conceptualized, and questioning lingering tendencies to link white female aid workers with issues of morality and solidarity. Insofar as such broader topics have been addressed, they have not necessarily been positioned within the 'gender and development' paradigm. One of the prominent themes, however, is how gender matters in development organizations; and less visibly so, the relevance of the body and sexuality among development people. In the following, I discuss these in turn.

Gender in development organizations

One feature that makes the issue of gender equality more urgent in the development sector than in others is the alignment between the values held by organizations regarding working with clients on the one hand, and their own internal practice on the other. As Caroline Sweetman (1997: 2) states, 'taking on a "gender agenda" has implications for the internal running of development organizations as well as for the development interventions they undertake'. She explains that 'working on gender issues obliges organizations to set their own houses in order, and change aspects of the organizational culture which discriminate against women staff, and women "beneficiaries"' (1997: 2). This is especially pertinent for organizations that have a strong feminist or activist base among their founders and staff, and the overall values they hold as an organization.

Noticeably, the most coherent body of work on gender and development people has emerged with regard to gender in aid organizations, rather than to aid workers as individuals, although the two are clearly connected. While the former is undeniably important, it also conforms to a more general tendency in development studies to consider individual staff through the lens of the organizations in which they work. How they are affected by and implicated in gender issues has thus often been articulated as part of the discussion on gender mainstreaming in development agencies. While this comprises a broader agenda, it also relates to their policies and practices with regard to their staff. The challenges of implementing gender mainstreaming are multi-fold and characterized by intersectionality. As Coles (2001:5) reminds us, 'gender is, of course, only one aspect of a person's identity. Other aspects such as ethnicity and class are also important, and gender cuts across them, often in complex ways.' Consequently, in the context of gender mainstreaming in organizations, the issues arising may vary depending on whether the organizations are Southern or Northern-based, whether the staff are male or female, white or non-white, affluent or not. In the following, I indicate some features of this debate.

Sweetman (1997) suggests that the theme of 'gender and organizations' comprises three key dimensions. The first, substantial one is a shared vision of gender equality; the second is structural, relating to the procedures and regulations to ensure the transformation of that vision into practice; and the third is the dimension of internal organizational culture, including staff's beliefs and attitudes, which are crucial for the implementation and sustainability of the transformation. While the contributors to Sweetman's collection consider gender mostly, although not exclusively, through female staff based in Southern NGOs, a separate strand of research has considered the workings of gender in global or Northern governmental agencies, such as Miller and Razavi (1998).

Within this field, different strands of literature demonstrate how the three dimensions as outlined by Sweetman matter in a range of different contexts. Relating to the first dimension, the 'substantial vision', Northern NGOs and governmental organizations have been examined in relation to how their female staff – especially those inspired by feminist ideals – are able to implement such visions and personally held values within NGOs and, crucially, in large governmental aid agencies. The attitudes and strategies of the latter have been documented by Eyben (2010) and Eyben and Turquet (2013). Eyben aims to highlight the relevance of the individual, which, she suggests, is often overlooked in development policy. There, change is understood as driven by the system, not by individuals. In contrast, Eyben (2010: 7) claims that 'paying closer attention to agency brings into focus the changes that can occur through bureaucratic activism'. She describes how policy activists use situations of discursive as well as strategic ambiguity in order to further their aims. Other key tactics involve networking and alliance-building, as well as insider activism. Overall, Eyben (2010: 5) suggests that, in order to implement their substantial visions, 'the effective policy activist identifies the opportunities for introducing discursive shifts within the dominant rules of the game'. On this theme, but on a more pessimistic note, Porter (2012: 311) questions the agency of individual volunteers within VSO with regard to how personal gender values can inform organizational practice. She concludes that despite token gestures, the organization offered no 'formal recognition of the idea that the behaviour of volunteers is a crucial part of the "meaning" of gender equality in volunteering'.

While a shared vision is clearly important, Sweetman reminds us that both the structural dimension – the formal policies held by the organization – as well as the informal organizational culture are crucial for how successfully this vision can be implemented. Beginning with the latter, there are several insightful accounts of the institutional cultures of bilateral and multilateral agencies. Coles (2007), for example, discusses the roles that men can play for gender mainstreaming in large governmental agencies, in this case the Department for International Development (DFID) in the UK. As Coles (2001: 6) points out, with regard to the organizational culture at DFID and elsewhere, 'bilateral development organizations are largely male, middle-class and white – or rather the institutional culture has tended to display these characteristics', which has implications for the possibilities of gender mainstreaming in these agencies. Men's roles can consist, for example, in influencing other men; contributing male or masculine-based views to a produce a balanced gender analysis; and identifying elements of personal and institutional resistance to a 'gender agenda' that may be less visible to women. Overall, however, Coles (2001: 10) reiterates the centrality of organizational culture: 'the culture of the development agency in which they work is critical: its mandate, which determines how gender policies will "fit", its norms and values, its formal structures, and (especially important) its informal working behaviours, which may be very powerful.'

The organizational culture is also the focus of a comprehensive report by Wigley (2005) on the situation at UNHCR. Wigley (2006: 26) describes the shortcomings of UNHCR in achieving gender equity within the organization, stating that 'the dilemma continues for an organization endeavouring to mainstream awareness of age, gender and diversity across all operations and functions, when it so clearly cannot redress gender inequity within its own ranks'. Even though the lack of women at senior levels is often explained with reference to women being unwilling to go to non-family duty stations, she argues that apart from lack of concrete data on this issue, a more likely reason is that 'UNHCR is like everywhere else in that it establishes systems, often operating outside of awareness, that continue to make it more difficult for women to advance' (Wigley 2006: 27). As an antidote, she recommends a number of initiatives, such as adopting a gender equity policy, integrating a gender equality perspective into all the Global Strategic Objectives and several others that match the gender mainstreaming strategy the organization has adopted with regard to its outside operations.

Furthermore, there are important insights to be gained from a related sector, namely humanitarian aid and relief work. Even though comprehensive data on this is not available, a report by the human resource agency People in Aid suggests that humanitarian organizations are often perceived as fostering a culture of masculinity with regard to the staff they recruit, and with regard to their ways of working. The psychologist Alessandra Pigni, who deals with humanitarian workers at risk of burnout, suggests that 'the majority of aid workers are tired of the "macho culture" that still permeates many of humanitarian organisations' (interview, www.peopleinaid.org/news/241.aspx). A consultant to People in Aid, Sara Davidson, also finds that 'action on bullying and harassment is vital to tackle the "macho management" found in the humanitarian sector' (www.peopleinaid.org/pool/files/newsletter/2006-jan-en.pdf). Davidson explains that such macho culture is partly explained as a result of 'psychological self-defence by stressed field staff', but other reasons, such as lack of managerial competence, also play a role, as indicated by an expatriate staff survey. As a result of such macho culture, it emerged that women who were as qualified as men left the organization sooner; they felt less valued with regard to their professional experience; and an 'old boys network' was cited as a source of frustration, contributing to a lack of women in senior management positions. These insights gained from disciplines such as organizational psychology are a reminder to utilize their existing scholarship on gender mainstreaming in aid organizations. Further evidence of this need comes from non-traditional development actors, such as military and security forces. I will return to this below in the context of sexual harassment among aid workers.

Finally, an aspect that· has attracted relatively little scholarly attention is the 'structural dimension', namely the role of organizational policies in relation to how gender matters for development workers. Key issues here include staff recruitment, retention and promotion as well as the effects of mobility and relocation. A recent UNDP (2008) report found, for example, that female staff were significantly underrepresented in senior management positions within the organization. Although there is insufficient research on women's careers in international organizations, preliminary evidence suggests that more policy efforts are needed in order to attract and retain women workers in multilateral organizations which require a high degree of mobility. Papan (2006), for example, in a small-scale study of single women working for the UN, argues that the organization's relocation policies could have a detrimental effect particularly on women, as they not only find it difficult to establish long-term relationships and families while they are pursuing a career within the UN, but that they may be more likely than men to prematurely leave a career with the organization because of such considerations. She suggests that many women had given little thought to these potential consequences of career choice when they embarked on it. Among the most problematic features that she identifies is an unsatisfactory work–life balance that does not easily accommodate people's aspirations to a fulfilled family life, and a lack of targeted mentoring and career-planning, especially for younger female staff. In order to counter this, she suggests that key policy changes need to include reduced mobility demands at strategic stages in the life course, as well as creating so-called 'off-ramp' and 'on-ramp' strategies, that is, mechanisms to allow staff to suspend their careers temporarily to devote time to raising a family, and to return at a later stage. While Papan's explorations are insightful, she highlights the need for more systematic research, since it is not clear to what extent these relocation policies may disadvantage women disproportionately in comparison to men. In addition, it would also be pertinent to investigate to what extent the educational needs of children, or caring for ageing parents of those employees, might have particular gendered effects with regard to their careers.

At the same time, the high mobility that is a characteristic of much aid work does not necessarily mitigate against female employment in this sector. Based on ethnographic fieldwork in Cambodia, I have argued that the development sector, with its flexible working practices, also offers opportunities for increased gender equality through enabling dual-career households, accompanying male spouses, and partners taking turns in being lead migrants (Fechter 2013). Nevertheless, aid workers' mobility cannot be understood as simply overcoming gender inequalities. As constant mobility places strains on intimate relationships, preliminary evidence from my informants suggests this disproportionately affects women, as they are more likely to suspend a career that reduces their opportunities for maintaining a relationship or having a family. Further, any gender equalities gained are partly based on local female domestic workers, who support these international households. This is not unique to the development sector; but it is worth noting that such global and racial divisions of labour may be at odds with the aims and sensitivities of many development initiatives.

Gender and sexuality

As the lives of development workers have been comparatively underresearched, it is perhaps not surprising that one aspect of their lives has been near-invisible: the complex and multivalent intersections between aid work, gender and sexuality. There has been a strong call for (re)considering the role of the body in development (Cornwall et al. 2008). This includes a critical discussion of how sexuality has been framed in development policy and practice (see also Adams and Pigg 2005; Lind 2010; Jolly 2010). However, this reframing of the body and sexuality has first and foremost been conceptualized with regard to the bodies of aid recipients. As argued above in relation to gender and development more generally, however, this reframing needs to be equally applied in relation to aid workers themselves. In the following, I highlight a few of these emerging aspects.

One, deeply troubling, issue concerns child sexual abuse by aid workers and peacekeepers, as evidenced in a report by Save the Children (2008). Based on data from Haiti, Côte d'Ivoire and Southern Sudan, the author states that perpetrators of sexual abuse 'exist in every type of humanitarian, peace and security organization, at every grade of staff, and among locally recruited and international staff'. These findings are augmented by a case study from Afghanistan, documenting sexual misconduct by international aid workers (Fluri 2012). Even though their vulnerabilities and circumstances can differ, it is important to note that the victims of sexual abuse or harassment are not only members of local communities, but also include international, mostly female aid workers. Harris and Goldsmith (2010) and Appleby (2010), for example, found that Australian women who were working with an international policing operation and as English language teachers respectively in post-conflict Timor-Leste, experience sexual harassment from their male Australian colleagues as well as from Timorese men. Appleby notes, however, that the women responded to these incidents very differently. While they condemned the behaviour of members of the international community as an expression of 'hegemonic masculinity', their views of Timorese men were more ambiguous, as 'discourses of gender equality and sexual freedom came into conflict with discourses of tolerance towards perceived cultural difference' (Appleby 2010: 1).

It is significant that sexual harassment experienced by aid workers has been highlighted by researchers who contribute to development studies from a cross-disciplinary perspective. Both Appleby's and Harris and Goldsmith's work is concerned with non-traditional development actors, such as international policing units and English language teachers. Their comparative outsider perspective might have enabled them to discuss aspects of aid work that tend to be downplayed or ignored in mainstream development literature. This is possibly facilitated by being able to draw on established analyses of sexual harassment among a workforce, such as policing, and import these insights into the aid sector. The value of viewing development people not as unique, but as facing similar problems as other, well-studied sectors, is therefore considerable.

More broadly speaking, the issues outlined above signal entanglements of power, aid and sexuality which have not been extensively discussed so far. While there is a broad critique of power imbalances in aid work (e.g., Baaz 2005), these critiques do not usually address sexual politics. Apart from sexual abuse and harassment, there are a range of other intimate involvements, which gain particular significance as they take place under the auspices of international aid work. That the nature of these relations can be ambiguous, and boundaries between consensual and coerced practices can be contested, is demonstrated for example in Laurie Charlés' autobiographical account of her time as a Peace Corps volunteer in Togo, *Intimate Colonialism* (2007). Here, she describes how her advocacy for adolescent girls does not always sit easily with Togolese understandings of appropriate gendered behaviour, and her own intimate encounters with Togolese men.

At the same time, anecdotal evidence from humanitarian and relief work suggests that sexuality – understood as a life-affirming force – becomes especially important in situations where aid workers are routinely faced with risk of violence and death. The popular aid memoir *Emergency Sex and Other Desperate Measures* (Cain et al. 2006) describes sexual intimacy as a vital outlet for those working in high-pressurized emergency relief operations. This involves consensual relations between international aid workers as well as with members of local communities.

A similar sentiment is voiced in a recent documentary film on Médecins sans Frontières, *Living in Emergency* (2010). There, the Italian doctor Sara, working in sub-Saharan Africa, explains the attraction of the erotic in the field by stating that 'sex is life' – which matters in a situation where doctors constantly battle to save, and often lose, the lives of others.

Less borne out of pressure, and perhaps more aptly described as 'transnational jouissance' (Hacker 2007), Ara Wilson (2010) describes NGOs as 'erotic sites'. Based on her fieldwork with female-oriented anti-trafficking NGOs in Bangkok, she observes the multiple erotic engagements between women, especially in intense working situations in preparation of major conferences or events. According to Wilson, 'there was an erotic component to the unprecedented scale of female homosociality within the 1990s to 2000s boom in NGOs' (Wilson 2010: 87), offering a further dimension of how in this case, NGO work, gender and sexuality interrelate and affect women's personal lives as well as the work in their organizations. While this may not be remarkable in the sense that many workplaces are sites of erotic attraction, it is worth acknowledging that aid agencies are not exempt from this, and tracing how personal values, sexual intimacy and advocacy-related goals may intermingle.

Finally, the body also matters in relation to dress. For instance, Western female aid workers in some predominantly Muslim countries have faced the question to what extent they need to adopt local dress codes in order to effectively carry out their work. As Cook describes in her study of female VSO workers in Pakistan, these women recognize 'the professional benefits of winning respect by wearing Pakistani clothing', but also find this personally useful, as 'wearing *shalwar kameez* signals their modesty and respectability to Gilgitis' (Cook 2005: 363). By extension, appearing 'respectable' is also considered as a safeguard against harassment in male-dominated spaces such as the bazaar. Extending the concern about women as victims of abuse, the question has been raised how gender matters in the context of aid workers' security. This has become more urgent since attacks on aid workers have increased over the last few years (Roth 2011). Even though data is scarce, in a preliminary report Wille and Fast (2011) found that both women and men were affected by all types of security events. It emerged, however, that 'a higher proportion of men face specific vulnerabilities that proportionally fewer women experience and vice versa' (Wille and Fast 2011: 6). In particular, this meant that women seemed more vulnerable to small-scale events such as crime and threats; whereas men were subject to more 'serious security' events, such as those involving weapons and those taking place on roads. As a result, men were more likely to suffer serious injury or even death than women. Wille and Fast conclude that there is a need for systematic reporting and analysing security events with a particular view on gender implications. An assessment of risks as they pertain to women and men therefore needs to become part of future training, they suggest.

Conclusion

This exploration of how gender matters for development people has attempted to outline some existing and prospective avenues of enquiry. First, I have noted the relative gender-blindness of many 'ethnographies of aid'. It is possible that foregrounding aid workers' professional status – as expert consultant, aid bureaucrat or fieldworker – as well as their race, ethnicity and national status has rather obscured the relevance of gender. Insofar as relevant studies exist, they are often framed with female aid workers at their centre, thus narrowing a focus on gender to women in particular. In addition, there appears to be an assumption that ethical responsibility and the 'desire to help' are most appropriately discussed in relation to women – thus establishing a close connection of femininity and morality that needs to be questioned, rather than taken for granted. Furthermore, the imperative to make the role of masculinity in development more visible also holds for aid people: the interrelations between masculinity and aid work are rather under-researched and require proper scrutiny.

A second theme concerns the role of gender among aid workers not only in organizations – although that is important too – but viewing them in their capacity as individuals in addition to, and outside of being part of institutions. The question how feminist-inspired values can be implemented within aid organizations remains pertinent. There are also more operational concerns, such as how personnel policies and practices of aid organizations may enable or obstruct gender equality. The constant mobility required, for example, by UN organizations is likely to mitigate against retention of female staff. At the same time, the mobile nature of aid work is not necessarily detrimental to achieving gender equality, but its flexibility can foster dual-career households among aid workers, albeit often with the support of local domestic workers. Such intersections between gender, aid work and the life course need to be more systematically investigated.

A third, rather less visible theme is the role of the body and sexuality in aid. The issue of sexual abuse committed by international aid workers, for example, is almost certainly underreported. At the same time, it emerges that sexual harassment, both by aid workers and by men in local communities can constitute a problem for female aid workers. Further, the significance of the body and of intimate relations in aid work is not well understood. It emerges from popular accounts that development workers' bodies and desires are an important part not just of their personal lives, but can have professional implications. Overall, this provisional survey hopefully demonstrates that a gendered perspective on aid workers contributes an important, albeit rather overlooked, dimension to the gender and development paradigm.

References

Ahmad, M.M. (2002) *NGO Field Workers in Bangladesh*, Aldershot: Ashgate.

Appleby, R. (2010) '"A bit of a grope": gender, sex and racial boundaries in transitional East Timor', *PORTAL: Journal of Multidisciplinary International Studies* 7(2).

Cain, K., Thomson, A. and Postlethwaite, H. (2006) *Emergency Sex and Other Desperate Measures: True Stories from a War Zone*, London: Ebury Press.

Charlés, L. (2007) *Intimate Colonialism: Head, Heart, and Body in West African Development Work*, Walnut Creek, CA: Left Coast Press.

Coles, A. (2001) 'Men, women and organisational culture: perspectives from donors', in C. Sweetman (ed.), *Men's Involvement in Gender and Development Policy and Practice*, Oxford: Oxfam Working Papers.

—— (2007) 'Profile of DfID and other aid agencies', in S. Ardener and F. Moore (eds.), *Professional Identities: Policy and Practice in Business and Bureaucracy*, Oxford: Berghahn.

Cook, N. (2005) 'What to wear, what to wear? Western women and imperialism in Gilgit, Pakistan', *Qualitative Sociology* 28(4): 349–367.

—— (2007) *Gender, Identity and Imperialism: Women Development Workers in Pakistan*, New York: Palgrave Macmillan.

Cornwall, A., Jolly, S. and Correa, S. (2008) *Development with a Body: Sexualities, Development and Human Rights*, London: Zed Books.

Eyben, R. (2007) 'Becoming a feminist in Aidland', in A. Coles and A.-M. Fechter (eds.), *Gender and Family Among Transnational Professionals*, London: Routledge.

—— (2010) 'Subversively accommodating: feminist bureaucrats and gender mainstreaming', *IDS Bulletin* 41(2).

Eyben, R. and Turquet, L. (2013) *Feminists in Development Organisations: Change Comes From the Margins*, London: Practical Action Publishing.

Fechter, A.-M. (2013) 'Mobility as enabling gender equality? The case of international aid workers', in T. Bastia (ed.), *Migration and Inequality*, London: Routledge.

Fluri, J. (2012) *Sexual Misconduct and International Aid Workers: An Afghanistan Case Study*, http://journals. cortland.edu/wordpress/wagadu/files/2014/02/Fluri.pdf.

Goetz, A.-M. (2001) *Women Development Workers: Implementing Rural Credit Programmes in Bangladesh*, London: Sage.

Hacker, H. (2007). 'Developmental desire and/or transitional jouissance: re-formulating sexual subjectivities in transcultural contact zones', in K. Browne, J. Lim and G. Brown (eds.), *Geographies of Sexualities*, Aldershot: Ashgate.

Harris, V. and Goldsmith, A. (2010) 'Gendering transnational policing: experiences of Australian women in international policing operations', *International Peacekeeping* 17(2): 292–306.

Heron, B. (2008) *Desire for Development: Whiteness, Gender, and the Helping Imperative*, Toronto: Wilfrid Laurier Press.

Lind, A. (2010) *Development, Sexual Rights and Global Governance*, London: Routledge.

McKinnon, K. (2011) *Development Professionals in Northern Thailand: Hope, Politics and Power*, Singapore: Singapore University Press.

Miller, C. and Razavi, S. (1998) *Missionaries and Mandarins: Feminist Engagement with Development Institutions*, London: ITDG Publications for UNRISD.

Papan, A.S. (2006) *'I Didn't See This Coming . . .': The Impact of an International Civil Servant Career on Prospects for Partnership, Marriage and Family*, paper presented at workshop on 'Development People', University of Oxford, UK, April.

Porter, F. (2012) 'Negotiating gender equality in development organizations: the role of agency in the institutionalization of new norms and practices', *Progress in Development Studies* 12: 301.

Roth, S. (2011) 'Dealing with danger: risk and security in the everyday lives of aid workers', in A.-M. Fechter and H. Hindman (eds.), *Inside the Everyday Lives of Development Workers*, Sterling: Kumarian.

—— (2013) *'Sometimes I Need to be a Man': Doing Gender in Aid Organisations*, paper presented at the Annual Meeting of the British Sociological Society, April.

Save the Children (2008) *No One to Turn to: The Under-reporting of Child Sexual Exploitation and Abuse by Aid Workers and Peacekeepers*, www.savethechildren.org.uk/resources/online-library/no-one-to-turn-to-the-under-reporting-of-child-sexual-exploitation-and-abuse-by-aid-workers-and-peacekeepers.

Sweetman, C. (ed.) (1997) *Editorial: Gender in Development Organisations*, Oxford: Oxfam Focus on Gender Series:.

Syed, J. and Ali, F. (2011) 'The white woman's burden: from colonial civilisation to Third World development', *Third World Quarterly* 32(2): 349–365.

UNDP (2008) *Gender Parity in UNDP*, www.peacewomen.org/assets/file/PWandUN/UNImplementation/ ProgrammesAndFunds/UNDP/undp_genderparityreport_17mar08.pdf

Wigley, B. (2005) *The State of UNHCR's Organization Culture*, report, www.unhcr.org/428db1d62.pdf.

—— (2006) *The State of UNHCR's Organizational Culture: What Now?* report, www.unhcr.org/ 43eb6a862.pdf.

Wille, C. and Fast, F. (2011) *Aid, Gender and Security*, InSecurity Insight, www.insecurityinsight.org/ files/Security%20Facts%202%20Gender.pdf.

Wilson, A. (2010) 'NGOs as erotic sites', in A. Lind, *Development, Sexual Rights and Global Governance*, London: Routledge.

Yarrow, T. (2011) *Development Beyond Politics: Aid, Activism and NGOs in Ghana*, London: Palgrave.

57

ENGENDERING UNDERSTANDINGS OF FAITH-BASED ORGANIZATIONS

Intersections between religion and gender in development and humanitarian interventions

Elena Fiddian-Qasmiyeh

Introduction

This chapter engages with debates about the gendered nature and implications of faith-based non-governmental organizations (FBOs) working in the fields of development and humanitarianism. It starts by tracing the role of faith-inspired interventions designed to "improve" the lives of Others around the world. It then introduces feminist critiques of Christian missionary societies' support for Western colonial projects ostensibly designed to "protect" women and children from what were labelled "traditional" and "barbaric" religious and cultural structures. This preliminary discussion provides the foundations for the remainder of the chapter, which examines the faith-gender-development nexus via a series of ruptures and continuities in discourse and practice. Hence, although Christian discourses, doctrines and actors were pervasive in colonial-cum-development programmes, after World War II the newly institutionalized development industry prioritized secularism as the strongest means to secure socioeconomic development and good governance, including gender equality. In turn, academic and policy interest in faith-based development increased dramatically throughout the 2000s, and it is now broadly recognized that faith continues to motivate and inspire responses to poverty, crisis and human rights violations across the Global South and Global North alike.

With increasing funds and resources being allocated by states and international agencies to faith-based development actors, however, many feminist analysts have warned against the premature idealization of faith-based development actors. Inter alia, they note that FBOs may exclude women from decision-making processes and refuse to engage with individuals and social groups who do not comply with norms regarding gender and sexuality. While such concerns are valid on many levels, the chapter notes that there is an overall lack of comparative analyses examining the intersections between gender, faith and development, and more evidence is therefore urgently required to evaluate the gendered motivations, nature and implications of initiatives developed by FBOs. Equally, however, the chapter argues that more evidence is also necessary in order to assess the *assumptions* that continue to be held by many secular and faith-based actors that faith-based development initiatives will necessarily be more conservative than secular programmes with regards to gender relations and gender equality. Indeed, despite the apparent shifts in the official space granted to religion and secularism in development discourse, policy and practice, continuities with European colonial assumptions regarding religious barriers to women's rights remain. In contrast, the chapter argues that neither FBOs nor secular organizations are a priori "conservative" or "liberal" with regards to gender roles and relations, and that critical analysis is therefore necessary in order to overcome the diverse hierarchies and structures of oppression which underpin the development industry as a whole, whether these hierarchies exist within faith-based *or* secular organizations' operations and programmes for Others.

Definitions and typologies

A "faith-based organization" can be defined as "any organization that derives inspiration from and guidance for its activities from the teachings and principles of faith or from a particular interpretation or school of thought within a faith" (Clarke and Jennings, 2008: 6). Just as "secular" organizations are highly diverse, so too are FBOs involved in development and humanitarian activities, ranging from small-scale local-level religious congregations, to national inter-denominational coalitions and networks, to international faith-based humanitarian agencies with multibillion dollar budgets; in turn, organizations may combine the provision of assistance and protection with proselytization and/or faith-centred delivery strategies, or reject these processes and strategies in respect of the international humanitarian principles that prohibit this. In line with this heterogeneity, FBOs have different histories, underlying motivations, fundraising mechanisms and modes of operation.

Clarke (2006) identifies five "functions" guiding FBOs' activities around the world, leading to the following typology: faith-based representative organizations; faith-based charitable or development organizations; faith-based sociopolitical organizations; faith-based missionary organizations; and faith-based radical, illegal or terrorist organizations. Clarke's typology is helpful because it recognizes the diverse aims and modes of operation of organizations broadly motivated by "faith", highlighting the potential role of FBOs in tackling poverty and social exclusion via charitable or development initiatives. Nonetheless, such classificatory systems need to be critically examined, including for the following reasons.

First, the label 'faith-based organization' may not be used by members of a given organization or network, since faith principles are often conceptualized as a foundational part of "a community's heritage, culture and broader way of life", rather than as a "religious" framework per se (interview with Yossi Ives, Tag Development, cited in Fiddian-Qasmiyeh and Ager 2013). Indeed, UNFPA notes that the term "community organization" is often used rather than "faith-based organization" (interview with Henia Dakkak, UNFPA, cited in Fiddian-Qasmiyeh and Ager, 2013). This is because "faith" may not be explicitly identified by community members themselves as the core motivating or organizational principle, and "faith" may be indistinguishable from the community's broader social, cultural and political life.

Second, an organization's aims and objectives, whether it is denominated "faith-based", "community-led", or indeed "secular", may be difficult to identify and delimit. As such, organizations and networks can simultaneously fall under the categories of faith-based charitable organizations, which aim to implement development and humanitarian programmes, and faith-based missionary organizations, which combine the provision of development support with spreading "key faith messages beyond the faithful, by actively promoting the faith and seeking converts to it" (Fiddian-Qasmiyeh and Ager 2013). Furthermore, as I have argued elsewhere with reference to American Evangelical humanitarian organizations active in the Sahrawi and Palestinian protracted refugee situations in the Middle East and North Africa (Fiddian-Qasmiyeh 2012), FBO charitable missionary organizations may simultaneously be identified as "faith-based socio-political organizations, which organize and mobilize social groups on the basis of faith identities but in pursuit of broader political objectives" (Clarke 2006). As such, FBOs often have overlapping motivations, including "charitable", "missionary" and "sociopolitical" objectives that might be difficult to separate in theory or practice. Furthermore, classifying FBOs can also be challenging due to the "difficulty in sometimes determining the nature of an FBO's gender agenda, because often a single organization takes different standpoints on various gender issues" (Tadros, 2010b: 1). Equally, however, critical analyses of contemporary secular NGOs and agencies often highlight their overlapping motivations and aims: charitable objectives may exist alongside the promotion of sociopolitical and ideological priorities in the name of what Kandiyoti (2004: 134) refers to as "the trinity of democratization, good governance and women's rights".

Faith-based organizations, gender and development: a brief introduction

Faith principles have long inspired individual and communal responses to the socioeconomic and spiritual needs and rights of members of their own and other communities. Indeed, extensive studies document the ways in which diverse faiths have motivated responses to human needs throughout history and around the world. For instance, Islam provides an obligation for Muslims to provide financial or material assistance to care for widows and orphans, and to offer protection and sanctuary to both Muslims and non-Muslims who are fleeing conflict and persecution. In turn, followers of Buddhism known as *bodhisattva* have purposefully delayed or relinquished their personal quest for enlightenment in order to alleviate the suffering of others, and philanthropy has also historically played a central role in Confucianism and other religions. As the connections between faith, gender and development are many, two bodies of literature exploring these intersections are particularly relevant for the purposes of this chapter.

Protection narratives and civilizing missions: the "colonialism-as-development" nexus

Numerous studies have examined the gendered dynamics surrounding Christian missionary societies' interventions in support of Western colonial and imperial projects and their "civilizing mission".[1] Such faith-based interventions included missionaries' roles in implementing paternalistic and Orientalist colonial policies that ultimately aimed to "save brown women from brown men" (Spivak 1993: 93). Officially in the name of promoting women's rights, these initiatives included missionary groups' support for anti-*sati* ("widow immolation") campaigns in India, strategies to "liberate" Muslim women across the Middle East and North Africa by "unveiling" them, "morality" campaigns to promote Victorian models of marriage, sexuality and reproduction in the colonies, and to "save" illegitimate or "miscegenated" children through forced adoptions and/or internment programmes.

These programmes and campaigns formed part of the foundational discourses of what we currently refer to as "development", and they were infused with religious discourses and imagery, often supported and implemented by actors explicitly motivated by faith, and were intrinsically gendered in nature. For example, specific notions of femininity and womanhood were promoted through these programmes, and campaigns were developed to protect women from social and religious systems and practices that were labelled as "barbaric" in nature. Simultaneously, colonial systems developed strategies to redress the perceived characteristics of colonized men, ranging from feminized and infantile males requiring paternal guidance, to "deviant" or inherently violent men needing to be civilized and controlled.

As a whole, these feminist studies reveal that colonial "development" discourses and policies have historically been based upon a "discursive strategy that constructs gender subordination as integral only to certain [non-Western] cultures," solidifying a separation and hierarchy between "us" (liberal, equal) and "them" (illiberal, barbaric and oppressive of women), for highly political purposes (Volpp 2001). By opposing race and religion with gender in such debates, the "positional superiority" of Western culture has historically been reinforced over Other cultures (Nader 1989). Western actors have thus established violence against Other women as a central concern, proposing the need to "save" these women from "their" "religion" and "culture", and perceiving the West as being responsible for liberating and empowering women across a range of geographies through development programmes and foreign policy frameworks alike.

The *official* position of religious belief, practice and imagery within contemporary mainstream development discourse, policy and practice has shifted significantly since the colonial period. Nonetheless, the broader discursive frames that have historically constituted certain societies, social groups and individuals as being in need of external interventions to "develop", save and protect them, have continued to date.

From secular development paradigms to the "rediscovery" of faith in/and development

A second, increasingly extensive body of literature regarding the role of faith-based and faith-inspired actors in development activities has emerged since the 2000s. However, the intersections with *gender* have tended to be less prominent than in the above-mentioned analyses of the colonial era.

Christian discourses, doctrines and actors were pervasive in colonial-cum-development programmes, and yet from the birth of the professionalized aid industry in the post-World War II era to the early 2000s, conceptualizations of development prioritized secular approaches as the strongest means to secure democratic political structures, good governance and women's rights. Related debates at the time also argued that "multiculturalism was bad for women" in Western liberal democracies since allowing "minority" women's religious and cultural frameworks to exist in "our" liberal democracies would perpetuate "their" abuse by "their" men. Academics and policymakers subsequently argued that it was "our" responsibility to "protect" Other women from practices defined by Western observers as "abusive", "illiberal" and "violent" (including "forced marriage", "child marriage" or "female genital surgeries") (Okin, 1998; Cohen et al. 1999; Fiddian-Qasmiyeh 2013). Indeed, with reference to the latter, religion "is often perceived to be conservative, steeped in tradition, and invariably resisting change. For example, while modern secular values are invariably presented as espousing gender equality, religion is assumed to confine women to traditional roles" (Ferris 2011: 623). The official promotion of secularism prior to the 2000s was thus effectively justified through a continuation of colonial assumptions that "traditional" religious and cultural frameworks were barriers to sociocultural "change" (read "modernization") in general, but also to women's empowerment and women's rights more specifically.

However, broader debates within social theory throughout the 1990s and 2000s questioned longstanding assumptions that modernization and modernity would be characterized by the entrenchment of rationality and secularization. Furthermore, academics increasingly argue that we live in a "post-secular" age in which religious belief and practice are becoming more, if differently, important for individuals and communities around the world. In the field of international development, there has been a notable increase since the 2000s in academic and policy attention to the role of religion and spirituality on the one hand, and faith-based organizations on the other. This interest can be perceived in the policy and practice of mainstream development organizations such as the World Bank, UNDP, UNFPA and UNAIDS, and by states such as the UK and the US. These organizations and states have officially continued to promote what I refer to as "secularism at home" through the separation of religion from the public sphere, while increasingly acknowledging the potential benefits of funding development initiatives and programmes implemented by FBOs domestically and in the Global South, and supporting initiatives designed to strengthen Southern civil societies, including faith-based communities and networks.

Such attention is understandable on many fronts, including for pragmatic reasons: for instance, in 2006 the World Health Organization reported that one in every five responses to HIV/AIDS is related to faith. Furthermore, numerous studies have noted that the renewed academic and policy interest in faith and development is partly related to the belief that faith-based initiatives have the potential to promote a holistic model of human development including through notions of spiritual development and spiritual capital. It has also been argued that faith-based development may be more relevant to beneficiaries for whom religious identity, belief and practice are pivotal elements on individual and collective levels. Aid delivered by organizations – and individuals – inspired by faith may be trusted more than assistance offered by secular institutions, and faith leaders may themselves be well-positioned to provide information and assistance to potential beneficiaries. For instance, the Islamic Foundation of Bangladesh offers training to Imams that includes topics on reproductive health, gender empowerment and HIV/AIDS, with 40,000 Imams having been trained to promote HIV prevention among their local communities (Berkley Center for Religion, Peace and World Affairs 2010). Indeed, it has been argued that local faith leaders and local faith communities are often well-positioned to engage with issues that are considered too sensitive, taboo or stigmatized to openly share with external actors (Fiddian-Qasmiyeh and Ager 2013). For instance, Parsitau's (2011) study of female internally-displaced Kikuyu victims of sexual and gender-based violence in Kenya highlights that faith communities were the only actors able to provide trauma counselling in that context. Equally, Roy demonstrates the special access that Muslim female medics working with the Islamic FBO al-Wafa (an FBO affiliated with Hamas) had to address the sexual problems experienced by severely disabled Palestinian men in the Gaza Strip (Parsitau 2011: 158).

However, this increased attention to the roles of FBOs, especially those inspired by Islam, has often been associated with a securitization framework that questions the motivations and aims of such actors, and denominates their activities as political and ideological rather than motivated by humanitarian principles. In effect, the renewed interest in supporting faith-based development actors has often been "tied" to a range of non-economic conditionalities; these conditionalities include an official commitment to secular principles institutionalized within the postcolonial development and humanitarian industry, such as a commitment to non-proselytization; the universal, neutral and impartial delivery of aid; and an explicit commitment to principles of gender equality and female empowerment.

Indeed, gender-based conditionalities associated with international funding for faith-based and secular development actors alike are widespread: in essence, grants will only be awarded if gender has been mainstreamed throughout development programmes. And yet very few comparative or theoretical studies explore the *gendered* impacts of faith-based development actors. Even fewer critically evaluate the connections between faith-based organizations, gender and humanitarian situations. The latter is a key area requiring further analysis by academics, policymakers and practitioners alike.

Engendering our understandings of faith-based organizations in development and humanitarian contexts

The sex workers reportedly appreciate the non-judgmental approach adopted by the nuns [of the Antonio Center in the Philippines], who do not aim to persuade them to leave the sex industry, but rather aim to protect them during their time as sex workers, and to support them to find alternative livelihoods if and when they choose to leave sex work.

(Kaybryn and Nidadavolu 2012)

It has been argued by Tadros that the majority of the recent literature on religion and development frames FBOs "as positive agents for the advancement of gender equality . . . highlighting the positive role faith and faith-based initiatives can play in eliciting social change" (Tadros 2010b: 1). She subsequently draws on a wide range of case studies, primarily of Christian and Muslim FBOs, to warn against the premature idealization of faith-based development actors. For instance, she maintains that "A critical dimension of women's agency and power has to do with the conditions and terms of [women's] participation in FBOs" (Tadros 2010a). This concern resonates with Islamic Relief Worldwide's reflection that "women's participation in planning stages is considered low in comparison to implementation stages" in its operations in Sudan (Survey Response, Islamic Relief Worldwide, cited in Fiddian-Qasmiyeh and Ager 2013). Islamic Relief Worldwide continued by noting that certain Muslim religious leaders in Pakistan and Afghanistan have hindered women's involvement in recovery and reconstruction following natural disasters and conflict situations; in such instances, faith leaders have argued that it is culturally inappropriate for women to work in this area. Equally, a Christian organization admitted that church hierarchies in certain contexts across sub-Saharan Africa foster the exclusion of women and other social (and sexual) minorities in decision-making (Fiddian-Qasmiyeh and Ager 2013).

Importantly, however, when critiquing women's unequal participation in FBO decision-making processes, Tadros notes that although "women are often the majority of paid workers in third sector organizations" around the world, "in most cases they do not occupy leading positions" in *either* secular *or* faith-based organizations" (Tadros 2010b:14). This qualification is significant because "*little evidence* is available about the gender-related implications of current development policies and practical initiatives that actively engage with religion" (Tomalin 2013: 193; emphasis added). Nonetheless, Tomalin (2011: 6) asserts that "there is a danger that the uncritical adoption of dominant (usually male) perspectives and voices within religious traditions may result in the marginalization of alternative voices and positions, for example feminist or gender-equal interpretations within religious traditions". She continues by warning that "in prioritizing religion, other identities and alternative approaches may be ignored". Such a warning is in line with the argument that hegemonic religious attitudes to sexuality and marriage may endorse "gender inequality in relationships", with such attitudes reinforcing "practices that increase women's vulnerability to domestic and sexual violence, and their inability to access appropriate health and legal support" including sexual and reproductive health services (Smith and Kaybryn 2012).

While these and other concerns are valid, it is clear that more evidence is needed to assess whether *assumptions* held by secular actors about local faith communities and national and international FBOs can or cannot be maintained, and to what extent. These include *beliefs* that FBOs are *automatically* more "conservative" and "patriarchal" than their secular counterparts; that LFCs and faith leaders will *necessarily* hinder the participation of women and girls as decision-makers, as aid and service providers and as beneficiaries alike; and that FBOs will *undoubtedly* refuse to engage with individuals and social groups who do not comply with norms regarding gender and sexuality. In addition to the example from the Philippines cited above, this last presumption has also recently been challenged by a survey of attitudes toward lesbian, gay, bisexual, transgender and intersex (LGBTI) asylum seekers,[2] which concluded that FBOs' views on providing services to LGBTI people are *no better or worse* than the attitudes held by secular institutions (Survey Response, UNHCR-Geneva, cited in Fiddian-Qasmiyeh and Ager 2013). Examples of FBOs engaging with gender non-conforming individuals and social groups in development and humanitarian contexts abound, including the Ojus Medical Institute in Mumbai, which provides services to people living with HIV, including "men who have sex with men, injecting drug users and transgender people" (Herwadkar, cited in Smith and Kaybryn 2012: 51).

While such examples are increasingly being recognized, can it be asserted, as Tadros (2010b) does, that FBOs have been prematurely idealized by the development industry? I would argue that neither the academic literature nor mainstream policy discourse and practice have taken it for granted that FBO involvement is necessarily positive regarding gender roles and relations. While this may now increasingly be the case in *official* declarations by international (secular) *development* actors and agencies, Clarke's research with DFID officials clearly reveals "significant concerns about the erosion of DFID's traditional secularism . . . They fear donor entanglement in sectarian or divisive agendas" (Jennings and Clarke 2008: 262). Furthermore, official engagement with FBOs is a very new recent phenomenon within *humanitarian* contexts. Research with many of the largest secular and faith-based organizations (including UNHCR, Oxfam, UNFPA, Islamic Relief, Christian Aid, CAFOD, Tag International and Anglican Alliance) confirms that development and humanitarian operations in the field often continue to be characterized by tension and mistrust (Fiddian-Qasmiyeh and Ager 2013). Importantly, this tension exists between secular and faith-based organizations on the one hand, but also between different FBOs, including in particular those that variously denounce or enact proselytization in assistance and service delivery contexts, on the other (Fiddian-Qasmiyeh and Ager 2013).

Indeed, while Tomalin (2011: 6) posits that "One of the perceived problems facing secular development organizations is wariness about being openly critical of religious organizations for their attitudes towards gender, or indeed probing very far at all into their values and policies on gender equality", recent research with mainstream secular and faith-based humanitarian organizations reveals high degrees of self-reflection and self-critical approaches by both secular and religious organizations (Fiddian-Qasmiyeh and Ager 2013). This includes the recognition of the challenges and opportunities arising in their own and other faith-based organization's approaches to gendered divisions of labour within FBO structures, and the gendered nature of FBO's aims, objectives and outcomes vis-à-vis gender relations and gender equality.

FBOs can therefore be critical not only of secular organizations' work but also of their own and other faith-based organizations' approaches to gender. In many ways, this transcends

Marshall and Taylor's (2011) positive interpretation of Tearfund's critical position towards evangelical churches' conservative approach to sexual activity and behaviour in the context of HIV/AIDS programming. Marshall (herself employed by Tearfund) and Taylor argue that Tearfund was able to be critical towards its evangelical partners due to their identity as "insiders". Equally, Tomalin (2011: 7) suggests that such critical projects "have the potential to be replicated in contexts where secular organizations find it difficult to gather information about the gender attitudes in particular religions, or to critique them when they are found to be problematic". Beyond these conclusions, Fiddian-Qasmiyeh and Ager's (2013) study offers an example of multidirectional critiques and the possibility for open debate and mutual learning within and across secular and faith-based organizations working in humanitarian contexts.

Recognizing this possibility highlights the importance of academics, policymakers, practitioners, as well as beneficiaries themselves, being critical of both faith-based *and* secular responses to development and humanitarian settings, especially when the protection of women and female empowerment continue to be invoked as key motivating factors for interventions in contexts of peace and conflict, thereby perpetuating the Western colonial legacy discursive and policy frameworks above.

Concluding remarks

Given the long history of faith-based interventions in the name of charity and development, a range of continuities and shifts can be identified in the ways faith-based actors have designed and implemented development and humanitarian responses around the world. Despite conceptual and programmatic shifts within social theory and the development industry throughout the twentieth and twenty-first centuries, faith principles have continued to motivate individual and collective responses to others in need, whether these others are co-religionists, members of other faiths, or of none. Equally, faith-based discourses are still regularly invoked to justify diverse forms of intervention to address poverty and social deprivation, women's oppression and conflict-induced displacement. Continuity with the diverse activities undertaken by missionaries during the colonial era to "save" Other women is perhaps particularly notable with reference to the geographical areas and the thematic issues that FBOs often become active in and, of course, through the continuation of missionaries' presence and activities in peace, conflict and post-conflict situations around the world. The latter has been most visible over the past decades across the Middle East and North Africa, where international commitment to support the protection of women and promote women's rights has mobilized gendered religious symbols such as the burqa. Indeed, with the image of forcibly veiled Afghan women having been invoked by Western politicians as not only justifying but even demanding military intervention, exploring the role of faith-based responses to development and humanitarian situations is particularly pertinent in cases where intersecting discourses regarding gender and faith have been amongst the factors *causing* conflict-induced displacement. With the relatively late emergence of interest in the FBO-gender-humanitarianism nexus, future research into this area will be particularly important for academia, policy and practice over the coming years.

Distinctions and distinctiveness?

It may indeed be the case that "religion is [often] used to legitimize patriarchal hierarchies" in FBOs (Tadros 2010b: 14), and yet patriarchal hierarchies and Orientalist priorities are also prevalent throughout "secular" organizations and the overarching development industry. As such, analyses of the gendered nature and impacts of FBO programmes must be paralleled by ongoing investigations into secular organizations, and of the development and humanitarian industries more broadly. This is especially urgent since both secular and faith-based organizations arguably embody problematic continuities with the faith- and gender-based dynamics under-pinning the colonial era's "civilizing mission".

Indeed, it could be argued that FBO interventions draw particular attention to key dilemmas about what is defined as "development" or "empowerment", and what position gender and religious identity and practice can or should play when attempting to achieve these "goals".

Despite the increased interest in FBOs' potential to promote human development since the 2000s, in this chapter I have argued that faith-based actors are often perceived as being likely to maintain or reinforce the gendered status quo by reproducing patriarchal structures. In this denomination, FBOs' relationship with patriarchy is implicitly (and often explicitly) contrasted with a firmly held *assumption* that secular organizations have overcome these oppressive frameworks, practices and dynamics and are therefore ideally positioned to promote the empowerment of women. This is visible, for instance, when Greany points to the key question explored in a number of articles in her 2006 Special Issue of *Gender and Development*: "how faiths and institutions which have a history of repression of, and discrimination against women, and which continue to be dominated by patriarchy in many areas of belief and practice, can act as catalysts for and supporters of positive social change for women" (Greany 2006: 346). Through framing the key question in this manner, the reader is led to believe that *secular* organizations, unlike FBOs, *are* well positioned to "act as catalysts for and supporters of positive social change for women" since they do *not* have a "history of repression" or "discrimination", and are *not* "dominated by patriarchy" (Greany 2006). Such an assumption is highly problematic on numerous levels, and is clearly contradicted by numerous examples of patriarchal dynamics pervading secular organizations and agencies: indeed, it is precisely *because* of the prevalence of gender bias across the development industry that proponents of gender and development have developed sophisticated critiques of the androcentric foundations and implications of main-stream development theory, policy and practice. It is also highly problematic given that secular development programmes designed to "empower" women often have paradoxical impacts, ultimately reproducing systems of oppression (i.e., Fiddian-Qasmiyeh 2014b).

If secular organizations often fail to promote gender equality and female empowerment through their programmes, it is equally the case that certain FBOs have officially promoted the empowerment of women and the transformation of female subjectivities. For instance, Parsitau (2011) argues that female-led Pentecostalist and charismatic churches in Kenya aim to transform women's expectations of their potential, in addition to advancing the spiritual and material empowerment of both single and married women. However, this in turn raises the question of whether spiritual empowerment is to be considered to be a form of "development", and if so (or if not), by whom.

Another pertinent example derives from UNFPA's awareness that the very definition of what may be considered to be a "basic need" in a humanitarian situation is both highly gendered and intimately related to the faith-identity and belief system of affected communities. This conclusion is supported by UNFPA's account of its response to the Indian Ocean tsunami of 2004: local conceptualizations of "basic needs" transcended secular organizations' perceptions, since many Muslim women affected by the tsunami held that headscarves were essential to maintain their dignity and were a prerequisite to be able to access other services in public fora (Fiddian-Qasmiyeh and Ager 2013). That some individuals, communities and organizations might prioritize the provision of veils, or indeed the reconstruction of a mosque or temple as a "basic need" to be prioritized over the delivery of food or medicine, may be perceived by external analysts as promoting the continuation of the status quo, rather than maximizing the opportunity to promote "women's rights" at a time when traditional social, political and religious structures have been disrupted by the processes preceding and characterizing humanitarian crises – the latter is a prevalent view in conflict and displacement studies, and is even codified as an international obligation for UN agencies such as the United Nations High Commissioner for Refugees (Fiddian-Qasmiyeh 2014b). Indeed, although many individuals and communities would prioritize "saving a way of life" over "saving a life" (Allen and Turton 1996), much of the development industry considers that certain "ways of life" are effectively at the root of discrimination and abuse, and that these ways of life are precisely why international interventions are necessary and not only morally justifiable, but effectively morally obligatory.

Key questions emerging in this regard include whose perspectives are to be prioritized in development and humanitarian interventions, and whether beneficiaries' beliefs (including those pertaining to religion and gender) and priorities (including on spiritual and material, personal and collective levels) are accepted by international actors, or are rejected on the assumption that beneficiaries are so deeply embedded in "their" patriarchal, oppressive structures that "they" are suffering from false consciousness that only "we" can overcome. It is this overarching hierarchy that underpins the development industry that needs to be critically analysed and overcome, whether the hierarchy exists within faith-based *or* secular organizations' operations and programs for "Others".

Notes

1   In other instances, including during the Spanish "discovery"/occupation of the Americas, missionaries drew upon Christian doctrines to herald abolitionist movements and to argue in favour of the common humanity of colonized persons, often advocating for political and legislative change within colonies and the metropolis alike.
2   The survey was conducted by the Organization for Refugee Asylum and Migration.

References

Allen, T. and Turton, D. (1996) "Introduction: in search of cool ground", in T. Allen (ed.), *In Search of Cool Ground: War, Flight and Homecoming in Northeast Africa*, Geneva: UNRISD in association with James Currey, pp. 1–22.

Berkley Center for Religion, Peace and World Affairs (2010) *Faith-Inspired Organizations and Global Development Policy: A Background Review "Mapping" Social and Economic Development Work in South and Central Asia*, Washington, DC: Berkley Center for Religion, Peace, and World Affairs.

Clarke, G. (2006) "Faith matters: faith-based organizations, civil society and international development", *Journal of International Development*, 18: 835–848.

Clarke, G. and Jennings, M. (2008) "Introduction", in G. Clarke and M. Jennings (eds.), *Development, Civil Society and Faith-based Organizations: Bridging the Sacred and the Secular*, Basingstoke: Palgrave Macmillan, pp. 1–16.

Cohen, J., Nussbaum, M. and Howard, M. (eds.) (1999) *Is Multiculturalism Bad for Women? Susan Moller Okin with Respondents*, Princeton: Princeton University Press.

Ferris, E. (2011) "Faith and humanitarianism: it's complicated", *Journal of Refugee Studies* 24(3): 606–625.

Fiddian-Qasmiyeh, E. (2012) *Conflicting Missions? The Politics of Evangelical Humanitarianism in the Sahrawi and Palestinian Protracted Refugee Situations*, Max Planck Institute for the Study of Religious and Ethnic Diversity Working Paper, April 2012.

—— (2013) "Transnational abductions and transnational jurisdictions? The politics of 'protecting' female Muslim refugees in Spain", *Gender, Place and Culture*, iFirst article 2013.

—— (2014a) "Gender and forced migration," in E. Fiddian-Qasmiyeh, G. Loescher, K. Long and N. Sigona (eds.), *The Oxford Handbook of Forced Migration and Refugee Studies*, Oxford: Oxford University Press.

—— (2014b) *The Ideal Refugees: Gender, Islam and the Sahrawi Politics of Survival*, Syracuse: Syracuse University Press.

Fiddian-Qasmiyeh, E. and Ager, A. (eds.) (2013) *Local Faith Communities and the Promotion of Resilience in Humanitarian Situations*, RSC/JLI Working Paper 90, Oxford: Refugee Studies Centre, February 2013.

Greany, K (2006) "Editorial", *Gender & Development* 14(3): 341–350.

Jennings, M. and Clarke, G. (2008) "Conclusion: faith and development – of ethno-separatism, multi-culturalism and religious partitioning?", in G. Clarke and M. Jennings (eds.), *Development, Civil Society and Faith-based Organizations: Bridging the Sacred and the Secular*, Basingstoke: Palgrave Macmillan, pp. 260–274.

Kandiyoti, D. (2004) "Political fiction meets gender myth: post-conflict reconstruction, 'democratisation' and women's rights", *IDS Bulletin* 35(4): 134.

Kaybryn, J. and Nidadavolu, V. (2012) *A Mapping of Faith-Based Responses to Violence Against Women and Girls in the Asia-Pacific Region*, Asia-Pacific Women, Faith and Development Alliance and the United Nations Population Fund.

Marshall, M. and Taylor, N. (2011) "Tackling HIV and AIDS with faith-based communities: learning from attitudes on gender relations and sexual rights within local evangelical churches in Burkina Faso, Zimbabwe, and South Africa", in E. Tomalin (ed.), *Gender, Faith and Development*, Oxford: Oxfam, pp. 25–36.

Nader, L. (1989) "Orientalism, Occidentalism and the control of women", *Cultural Dynamics* 2(3): 323–355.

Okin, S.M. (1998) "Feminism and multiculturalism: some tensions", *Ethics* 108: 661–684.

Parsitau, D. (2011) "The role of faith and faith-based organizations among internally displaced persons in Kenya", *Journal of Refugee Studies* 24(3): 473–492.

—— (2012) "Agents of gendered change: empowerment, salvation and gendered transformation in urban Kenya", in D. Freeman (ed.), *Pentecostalism and Development: Churches, NGOs and Social Change in Africa*, Basingstoke: Palgrave Macmillan.

Pearson, R., and Tomalin, E. (2008) "Intelligent design? A gender sensitive interrogation of religion and development", in G. Clarke and M. Jennings (eds.), *Development, Civil Society and Faith-Based Organizations: Bridging the Sacred and the Secular*, Basingstoke: Palgrave, pp. 46–71.

Roy, S. (2011) *Hamas and Civil Society in Gaza: Engaging the Islamist Social Sector*, Princeton: Princeton University Press.

Smith, A. and Kaybryn, J. (2012) "HIV and maternal health: faith groups' activities, contributions and impact", Joint Learning Initiative on Faith and Local Communities, http://jliflc.com/resources/hiv-maternal-health-faith-groups-activities-contributions-impact.

Spivak, G.C. (1993) "Can the subaltern speak?" in P. Williams and L. Chrisman (eds.), *Colonial Discourse and Post-Colonial Theory*, New York: Harvester Wheatsheaf, pp. 66–111.

Tadros, M. (2010a) "Faith in service: what has gender got to do with it?", *Open Democracy*, 16 November, www.opendemocracy.net/5050/mariz-tadros/faith-in-service-what-has-gender-got-to-do-with-it.

—— (2010b) *Faith-Based Organizations and Service Delivery: Some Gender Conundrums*, Geneva: UNRISD.

Tomalin, E. (2011) "Introduction", in E. Tomalin (ed.), *Gender, Faith and Development*, Oxford: Oxfam, pp. 1–12.

— (2013) "Gender, religion and development", in M. Clarke (ed.), *Handbook of Research on Development and Religion*, Northampton, MA: Edward Elgar, pp. 183–200.

Volpp, L. (2001) "Feminism versus multiculturalism", *Columbia Law Review* 101(5): 1181–1218.

INDEX

Printed in the United States
by Baker & Taylor Publisher Services